Principles of Geochemistry

Principles of Geochemistry

GIULIO OTTONELLO

Columbia University Press / *New York*

Columbia University Press
Publishers Since 1893
New York   Chichester, West Sussex

Library of Congress Cataloging-in-Publication Data

Ottonello, Giulio.
    [Principi di geochimica. English]
    Principles of geochemistry / Giulio Ottonello.
        p.   cm.
    Includes bibliographical references (p. 817) and index.
    ISBN 978-0-231-09985-1 (acid-free paper)
        1. Geochemistry.   I. Title.
    QE515.08813   1997
    551.9—dc20   96–23987
                                                               CIP

CONTENTS

CHAPTER SEVEN
Introduction to Petrogenetic Systems   **449**

PART THREE
Geochemistry of Fluids

CHAPTER EIGHT
Geochemistry of Aqueous Phases   **479**

PREFACE

A quick look at the reference list of this textbook (initially conceived for Italian university students) is sufficient to appreciate the leading role played by North American scientists in the development of geochemistry, and it is a great honor for me to present my appraisal of this fascinating discipline to readers whose mother tongue is English.

The fascination of geochemistry rests primarily on its intermediate position between exact sciences (chemistry, physics, mathematics) and natural sciences. The molding of the quantitative approach taken in physical chemistry, thermodynamics, mathematics, and analytical chemistry to natural observation offers enormous advantages. These are counterbalanced, however, by the inevitable drawbacks that have to be faced when writing a textbook on geochemistry: 1) the need to summarize and apply, very often in a superficial and incomplete fashion, concepts that would require an entire volume if they were to be described with sufficient accuracy and completeness; 2) the difficulty of overcoming the diffidence of nature-oriented scientists who consider the application of exact sciences to natural observations no more than models (in the worst sense of that term).

To overcome these difficulties I gradually introduce the various concepts, first presenting them in a preliminary way, and then, whenever possible, repeating them more rigorously and in greater detail.

The organization of this book follows the various states of aggregation of the earth's materials, in an order that reflects their relative importance in geology. Five chapters deal with the crystalline state. The first chapter is preparatory, the second and third are operative. The fourth summarizes some concepts of defect chemistry, the role of which in geochemistry is becoming more and more important as studies on kinetics and trace element applications advance. The fifth chapter is a (necessarily concise) state-of-the-art appraisal of the major silicate minerals.

Two chapters are devoted to silicate melts. One is an introduction to petrogenetic diagrams, extensively treated in petrology. Aqueous solutions are covered in a single chapter that basically deals with electrolyte solution theory and its applications, since any further subdivision seemed unnecessary. A single chapter was deemed sufficient to describe the up-to-date information about gases. The decision not to treat chemistry and equilibria in the earth's atmosphere was dictated by the consideration that such treatment would involve a preliminary evaluation of organic chemistry concepts that could not be adequately covered owing to lack of space. The last two chapters concern trace element geochemistry and isotope geochemistry.

As regards trace elements, whose widespread use is due to the simplicity of operation, I have attempted to show that their simplicity is often only apparent and that deductions on a planetary scale must be made with caution. Isotope geochemists may find the last chapter rather concise, but again, space restrictions did not allow more extended treatment, and I tried to favor the basic concepts, which should always be borne in mind when dealing with global systematics.

This book is dedicated to my wife Elisabeth and my son Giulio Enrico.

Principles of Geochemistry

PART ONE

Geochemistry of Crystal Phases

CHAPTER ONE

Elemental Properties
and Crystal Chemistry

1.1 General Information on Atomic Structure and the Nature of Electrons

1.1.1 Atom Building

An atom is composed of a nucleus of protons and neutrons surrounded by an electron cloud. Theoretically, electrons may be found at any distance from the nucleus, although they preferentially rotate around low-energy orbits or *levels*. Within a single level, various *sublevels* can be distinguished. [The term "level" corresponds to "electron shell" in the Bohr model. The terminological analogy is: shell K = level I ($n = 1$); shell L = level II ($n = 2$); shell M = level III ($n = 3$); shell N = level IV ($n = 4$); and so on.] Electron levels are established according to four quantum numbers:

1. The **principal quantum number (n)** is related to the energy of the electron levels. An increase in quantum number n corresponds to an increase in the distance of the electron from the nucleus.
2. The **orbital or "azimuthal" quantum number (l)** defines "form" (i.e., "eccentricity of elliptical orbit"; cf. Pauling, 1948) and indicates which sublevel is occupied by the electron. It assumes integer values between 0 and $n - 1$.
3. The **magnetic quantum number (m_l)** indicates which orbital of a given sublevel is occupied by the electron. For a given orbital, quantum number l assumes integer values between -1 and $+1$. An electron with a magnetic quantum number m_l subjected to a magnetic field **B** increases its energy by a term $E = \mu_b m_l \mathbf{B}$, where μ_b is Bohr's magneton (Zeeman effect).
4. The **spin quantum number (m_s)** gives the projection of the intrinsic angular momentum of the electron, with values $+\frac{1}{2}$ and $-\frac{1}{2}$. Two oppositely oriented spins on the same orbital are graphically represented by arrows in opposite directions ($\uparrow\downarrow$).

The development of quantum mechanics has led to the formulation of various principles governing the structure of atoms:

1. The **Aufbau or "build-up" principle** states that electrons occupy orbitals of progressively increasing energies.
2. The **Pauli exclusion principle** states that no more than two electrons may occupy the same orbital, and they must have opposite spins. Based on this principle, $2n$ is the maximum number of electrons compatible with a given level.
3. **Hund's rule** states that, before two opposite-spin electrons can occupy the same orbital of a given sublevel, all the other orbitals of the same sublevel must have been occupied by single electrons of equal spin.

Table 1.1 shows the quantum numbers for the first four levels. Table 1.2 lists the complete electron configurations of all the elements.

1.1.2 Wavelike Behavior of Electrons

Like any particle, an electron has a definite mass, corresponding to $\frac{1}{1850}$ of the mass of the hydrogen atom. Nevertheless, experiments with electron beams accelerated by known voltage differences show that a wavelength λ can be associated with the mass of the electron, so that

$$\lambda = \frac{h}{mv},\tag{1.1}$$

where h is Planck's constant and v is velocity. Equation 1.1, which is one of the fundamental results of quantum mechanics, establishes the dual nature of electrons as particles and wavelike elements.

The propagation of a wave in one dimension is obtained by the partial derivative of the wave function

$$\psi = \psi_0 \sin 2\pi\left(\frac{x}{\lambda} - vt\right),\tag{1.2}$$

where v is frequency ($v = v/\lambda$) and t is time. Function 1.2 is periodic in t when x is constant (each point vibrates) and periodic in x when t is constant (instantaneous picture of a wave). Partial differentiation of function 1.2 on x results in

$$\frac{\partial^2 \psi}{\partial x^2} = -\frac{4\pi}{\lambda^2}\psi.\tag{1.3}$$

Table 1.1 Quantum numbers.

n	l	m_l	m_s
1 (level K)	0 (sublevel s)	0	$+\frac{1}{2} -\frac{1}{2}$
2 (level L)	0 (sublevel s)	0	$+\frac{1}{2} -\frac{1}{2}$
	1 (sublevel p)	$+1$	$+\frac{1}{2} -\frac{1}{2}$
		0	$+\frac{1}{2} -\frac{1}{2}$
		-1	$+\frac{1}{2} -\frac{1}{2}$
3 (level M)	0 (sublevel s)	0	$+\frac{1}{2} -\frac{1}{2}$
	1 (sublevel p)	$+1$	$+\frac{1}{2} -\frac{1}{2}$
		0	$+\frac{1}{2} -\frac{1}{2}$
		-1	$+\frac{1}{2} -\frac{1}{2}$
	2 (sublevel d)	$+2$	$+\frac{1}{2} -\frac{1}{2}$
		$+1$	$+\frac{1}{2} -\frac{1}{2}$
		0	$+\frac{1}{2} -\frac{1}{2}$
		-1	$+\frac{1}{2} -\frac{1}{2}$
		-2	$+\frac{1}{2} -\frac{1}{2}$
4 (level N)	0 (sublevel s)	0	$+\frac{1}{2} -\frac{1}{2}$
	1 (sublevel p)	$+1$	$+\frac{1}{2} -\frac{1}{2}$
		0	$+\frac{1}{2} -\frac{1}{2}$
		-1	$+\frac{1}{2} -\frac{1}{2}$
	2 (sublevel d)	$+2$	$+\frac{1}{2} -\frac{1}{2}$
		$+1$	$+\frac{1}{2} -\frac{1}{2}$
		0	$+\frac{1}{2} -\frac{1}{2}$
		-1	$+\frac{1}{2} -\frac{1}{2}$
		-2	$+\frac{1}{2} -\frac{1}{2}$
	3 (sublevel f)	$+3$	$+\frac{1}{2} -\frac{1}{2}$
		$+2$	$+\frac{1}{2} -\frac{1}{2}$
		$+1$	$+\frac{1}{2} -\frac{1}{2}$
		0	$+\frac{1}{2} -\frac{1}{2}$
		-1	$+\frac{1}{2} -\frac{1}{2}$
		-2	$+\frac{1}{2} -\frac{1}{2}$
		-3	$+\frac{1}{2} -\frac{1}{2}$

The three-dimensional analog of function 1.2 is

$$\psi = \psi_0 \sin 2\pi\left[\frac{(x^2 + y^2 + z^2)^{1/2}}{\lambda} - vt\right], \tag{1.4}$$

and it can be shown that partial derivation of equation 1.4 over the three dimensions results in

$$\frac{\partial^2 \psi}{\partial x^2} + \frac{\partial^2 \psi}{\partial y^2} + \frac{\partial^2 \psi}{\partial z^2} = -\frac{4\pi}{\lambda^2}\psi. \tag{1.5}$$

Let us now consider a particle with mass m, total energy E, and potential energy Φ. The kinetic energy of particle E_k will be

Table 1.2 Electron configuration of elements.

		1	2		3			4				5				6				7
Z	Element	s	s	p	s	p	d	s	p	d	f	s	p	d	f	s	p	d	f	s
1	H	1																		
2	He	2																		
3	Li	2	1																	
4	Be	2	2																	
5	B	2	2	1																
6	C	2	2	2																
7	N	2	2	3																
8	O	2	2	4																
9	F	2	2	5																
10	Ne	2	2	6																
11	Na	2	2	6	1															
12	Mg	2	2	6	2															
13	Al	2	2	6	2	1														
14	Si	2	2	6	2	2														
15	P	2	2	6	2	3														
16	S	2	2	6	2	4														
17	Cl	2	2	6	2	5														
18	Ar	2	2	6	2	6														
19	K	2	2	6	2	6		1												
20	Ca	2	2	6	2	6		2												
21	Sc	2	2	6	2	6	1	2												
22	Ti	2	2	6	2	6	2	2												
23	V	2	2	6	2	6	3	2												
24	Cr	2	2	6	2	6	5	1												
25	Mn	2	2	6	2	6	5	2												
26	Fe	2	2	6	2	6	6	2												
27	Co	2	2	6	2	6	7	2												
28	Ni	2	2	6	2	6	8	2												
29	Cu	2	2	6	2	6	10	1												
30	Zn	2	2	6	2	6	10	2												
31	Ga	2	2	6	2	6	10	2	1											
32	Ge	2	2	6	2	6	10	2	2											
33	As	2	2	6	2	6	10	2	3											
34	Se	2	2	6	2	6	10	2	4											
35	Br	2	2	6	2	6	10	2	5											
36	Kr	2	2	6	2	6	10	2	6											
37	Rb	2	2	6	2	6	10	2	6			1								
38	Sr	2	2	6	2	6	10	2	6			2								
39	Y	2	2	6	2	6	10	2	6	1		2								
40	Zr	2	2	6	2	6	10	2	6	2		2								
41	Nb	2	2	6	2	6	10	2	6	4		1								
42	Mo	2	2	6	2	6	10	2	6	5		1								
43	Tc	2	2	6	2	6	10	2	6	6		1								
44	Ru	2	2	6	2	6	10	2	6	7		1								
45	Rh	2	2	6	2	6	10	2	6	8		1								
46	Pd	2	2	6	2	6	10	2	6	10										

Table 1.2   *(continued)*

Z	Element	1	2		3			4				5				6				7
		s	s	p	s	p	d	s	p	d	f	s	p	d	f	s	p	d	f	s
47	Ag	2	2	6	2	6	10	2	6	10		1								
48	Cd	2	2	6	2	6	10	2	6	10		2								
49	In	2	2	6	2	6	10	2	6	10		2	1							
50	Sn	2	2	6	2	6	10	2	6	10		2	2							
51	Sb	2	2	6	2	6	10	2	6	10		2	3							
52	Te	2	2	6	2	6	10	2	6	10		2	4							
53	I	2	2	6	2	6	10	2	6	10		2	5							
54	Xe	2	2	6	2	6	10	2	6	10		2	6							
55	Cs	2	2	6	2	6	10	2	6	10		2	6			1				
56	Ba	2	2	6	2	6	10	2	6	10		2	6			2				
57	La	2	2	6	2	6	10	2	6	10		2	6	1		2				
58	Ce	2	2	6	2	6	10	2	6	10	2	2	6			2				
59	Pr	2	2	6	2	6	10	2	6	10	3	2	6			2				
60	Nd	2	2	6	2	6	10	2	6	10	4	2	6			2				
61	Pm	2	2	6	2	6	10	2	6	10	5	2	6			2				
62	Sm	2	2	6	2	6	10	2	6	10	6	2	6			2				
63	Eu	2	2	6	2	6	10	2	6	10	7	2	6			2				
64	Gd	2	2	6	2	6	10	2	6	10	7	2	6	1		2				
65	Tb	2	2	6	2	6	10	2	6	10	9	2	6			2				
66	Dy	2	2	6	2	6	10	2	6	10	10	2	6			2				
67	Ho	2	2	6	2	6	10	2	6	10	11	2	6			2				
68	Er	2	2	6	2	6	10	2	6	10	12	2	6			2				
69	Tm	2	2	6	2	6	10	2	6	10	13	2	6			2				
70	Yb	2	2	6	2	6	10	2	6	10	14	2	6			2				
71	Lu	2	2	6	2	6	10	2	6	10	14	2	6	1		2				
72	Hf	2	2	6	2	6	10	2	6	10	14	2	6	2		2				
73	Ta	2	2	6	2	6	10	2	6	10	14	2	6	3		2				
74	W	2	2	6	2	6	10	2	6	10	14	2	6	4		2				
75	Re	2	2	6	2	6	10	2	6	10	14	2	6	5		2				
76	Os	2	2	6	2	6	10	2	6	10	14	2	6	6		2				
77	Ir	2	2	6	2	6	10	2	6	10	14	2	6	7		2				
78	Pt	2	2	6	2	6	10	2	6	10	14	2	6	9		1				
79	Au	2	2	6	2	6	10	2	6	10	14	2	6	10		1				
80	Hg	2	2	6	2	6	10	2	6	10	14	2	6	10		2				
81	Tl	2	2	6	2	6	10	2	6	10	14	2	6	10		2	1			
82	Pb	2	2	6	2	6	10	2	6	10	14	2	6	10		2	2			
83	Bi	2	2	6	2	6	10	2	6	10	14	2	6	10		2	3			
84	Po	2	2	6	2	6	10	2	6	10	14	2	6	10		2	4			
85	At	2	2	6	2	6	10	2	6	10	14	2	6	10		2	5			
86	Rn	2	2	6	2	6	10	2	6	10	14	2	6	10		2	6			
87	Fr	2	2	6	2	6	10	2	6	10	14	2	6	10		2	6			1
88	Ra	2	2	6	2	6	10	2	6	10	14	2	6	10		2	6			2
89	Ac	2	2	6	2	6	10	2	6	10	14	2	6	10		2	6	1		2
90	Th	2	2	6	2	6	10	2	6	10	14	2	6	10		2	6	2		2
91	Pa	2	2	6	2	6	10	2	6	10	14	2	6	10	2	2	6	1		2
92	U	2	2	6	2	6	10	2	6	10	14	2	6	10	3	2	6	1		2

continued

Table 1.2 *(continued)*

Z	Element	1	2		3			4				5				6				7
		s	*s*	*p*	*s*	*p*	*d*	*s*	*p*	*d*	*f*	*s*	*p*	*d*	*f*	*s*	*p*	*d*	*f*	*s*
93	Np	2	2	6	2	6	10	2	6	10	14	2	6	10	4	2	6	1		2
94	Pu	2	2	6	2	6	10	2	6	10	14	2	6	10	6	2	6			2
95	Am	2	2	6	2	6	10	2	6	10	14	2	6	10	7	2	6			2
96	Cm	2	2	6	2	6	10	2	6	10	14	2	6	10	7	2	6	1		2
97	Bk	2	2	6	2	6	10	2	6	10	14	2	6	10	8	2	6	1		2
98	Cf	2	2	6	2	6	10	2	6	10	14	2	6	10	10	2	6			2
99	Es	2	2	6	2	6	10	2	6	10	14	2	6	10	11	2	6			2
100	Fm	2	2	6	2	6	10	2	6	10	14	2	6	10	12	2	6			2
101	Md	2	2	6	2	6	10	2	6	10	14	2	6	10	13	2	6			2
102	No	2	2	6	2	6	10	2	6	10	14	2	6	10	14	2	6			2
103	Lw	2	2	6	2	6	10	2	6	10	14	2	6	10	14	2	6	1		2

$$E_K = \frac{1}{2} mv^2 = E - \Phi. \tag{1.6}$$

From equation 1.6 we obtain

$$m^2 v^2 = 2m(E - \Phi), \tag{1.7}$$

which, when combined with equation 1.1, gives

$$\frac{1}{\lambda^2} = \frac{2m(E - \Phi)}{h^2} \tag{1.8}$$

Substituting the right side of equation 1.8 into wave equations 1.3 and 1.5 leads, respectively, to

$$\frac{d^2\psi}{dx^2} + \frac{8\pi^2 m}{h^2}(E - \Phi)\psi = 0 \tag{1.9}$$

and

$$\nabla^2\psi + \frac{8\pi^2 m}{h^2}(E - \Phi)\psi = 0, \tag{1.10}$$

where ∇^2 is the Laplacian operator

$$\nabla^2 = \frac{\partial^2}{\partial x^2} + \frac{\partial^2}{\partial y^2} + \frac{\partial^2}{\partial z^2} \tag{1.11}$$

(see appendix 2). Equation 1.9 is the *Schrödinger equation for one-dimensional stationary states,* and equation 1.10 is the tridimensional form. Differential equations 1.9 and 1.10 possess finite single-valued solutions, within the domain of squared integrable functions, only for definite values of energy (*eigenvalues*). For vibrational systems, such values are given by

$$E_{(n)} = \left(n + \frac{1}{2}\right)hv,$$ (1.12)

where *n* is the *vibrational quantum number.*

Applying to equation 1.9 the potential of the *monodimensional harmonic oscillator* (i.e., a particle constrained to a certain equilibrium position by a force that is proportional to its displacement from the origin), we obtain

$$\frac{d^2\psi_{(x)}}{dx^2} + \frac{8\pi^2 m}{h^2}\left(E - \frac{K_F x^2}{2}\right)\psi_{(x)} = 0.$$ (1.13)

If position *x* of the particle corresponds to its displacement from the origin, the restoring force acting on the particle by virtue of the harmonic potential is given by *Hooke's law:*

$$F = -K_F x,$$ (1.14)

where K_F is the force constant.

Because, in a conservative field, the force is linked to the gradient of the potential by

$$F = -\frac{d\Phi_{(x)}}{dx},$$ (1.15)

by integration of equation 1.15 it follows that

$$\Phi_{(x)} = \frac{K_F x^2}{2},$$ (1.16)

which, substituted into equation 1.9, leads to equation 1.13.

The eigenvalues of the energy that satisfy equation 1.13 are given by

$$E_{(n)} = \left(\frac{K_F h^2}{4\pi^2 m}\right)^{1/2}\left(n + \frac{1}{2}\right).$$ (1.17)

We will later adopt function 1.17 when treating the vibrational energy of crystalline substances and its relationship to entropy and heat capacity.

A more detailed account of the Schrödinger equation can be found in physical chemistry textbooks such as those of Hinshelwood (1951) and Atkins (1978), or in more specialized texts such as that of Hirschfelder et al. (1954). An excellent review of the applications of quantum mechanics to geochemistry has recently been proposed by Tossell and Vaughan (1992).

1.2 Periodic Properties of the Elements

The periodic law, discovered by Dmitri Mendeleev, establishes "an harmonic periodicity of properties of chemical individuals, dependent on their masses" (Mendeleev, 1905). To highlight the existence of periodic properties, the various elements are arranged, according to their increasing masses, in groups (8), subgroups (8), and periods (7) (table 1.3). A given group comprises elements with similar properties. The periodicity of elemental properties is easily understood if we recall the laws governing the occupancy of atomic orbitals. The periodic table shows the electron structures of the various elements. For the sake of brevity, the structure of the inner shells is indicated by referring to that of the preceding inert gas. An inert gas displays no chemical reactivity (i.e., no tendency to enter into other states of combination). This "stable" configuration corresponds to completion of all sublevels of the principal quantum number n. As we can see, a period begins with the occupancy of a new level; elements in this position are ready to lose one electron, with the formation of a univalent positive ion. At the end of the period, immediately preceding the inert gas, the atoms lack a single electron to complete the stable configuration. Such atoms tend to be transformed into univalent negative ions by electron capture. The number of electrons that complete the period is readily calculated from the following summation:

$$\sum_{l=0}^{n-1}\left(4l + 2\right) = 2n^2 \tag{1.18}$$

—i.e., two electrons in the first period ($n = 1$), eight electrons in the second ($n = 2$), 18 electrons in the third ($n = 3$), and so on.

The first period is completed by only two elements (H, He), because the first level has a single sublevel that can be occupied by two electrons with opposite spin (because $n = 0$, l is also zero; m_l may vary from $-l$ to $+l$ and is also zero; $m_s = +\frac{1}{2}$ and $-\frac{1}{2}$). The second period begins with an electron occupying the lowest energy orbital of sublevel $2s$ (Li) and terminates with the addition of the sixth electron of sublevel $2p$ (Ne). The pile-up procedure is analogous for the third period, but in the fourth period this regularity is perturbed by the formation of transition elements (Sc-Zn) and by progressive pile-up of electrons in the $3d$ orbitals, whereas the fourth shell is temporarily occupied by one or two electrons. Pile-up in the fourth level starts again with Ga, and the period terminates with the completion of the stable octet (Kr). The same sequence is observed in the fifth period (Y-Cd transition elements). In the sixth period, the transition is observed over sublevel f of the fourth shell ($4f$), generating a group of elements (rare earths, La-Lu) of particular importance in geochemistry for the regularity of progression of reactive properties within the group (the most external shells are practically identical for all elements in the series; the progressive pile-up in the inner fourth shell induces shrinkage of the atomic radius, known as "lanthanidic contraction"). In the seventh period, the transition begins with actinium (Ac) and thorium (Th), whose external electrons occupy sublevel $6d$, and continues with the subsidiary series of protactinium (Pa-Lw), with electrons progressively occupying sublevel f of the fifth shell ($5f$).

1.2.1 Periodic Properties, Ionization Potentials, and Energy of External Orbitals

In terms of energy, the existence of periodicity is easily understood by analyzing the expressions that relate the energy of external orbitals to quantum number n.

Under the *hydrogen ion approximation* (ions of nuclear charge Ze with a single electron progressively occupying the various orbital levels), the potential acting on the electron by virtue of the nuclear charge is given by

$$\Phi_C = -\frac{Ze^2}{r},$$
(1.19)

where Z is the atomic number and r is the distance from the nucleus expressed in Cartesian coordinates:

$$r = \left(x^2 + y^2 + z^2\right)^{1/2}.$$
(1.20)

Substitution of equation 1.19 into equation 1.9 leads to

$$\nabla^2 \psi + \frac{8\pi^2 m}{h^2}\left(E + \frac{Ze^2}{r}\right)\psi = 0.$$
(1.21)

The eigenvalues of equation 1.21 are

$$E_{(n)} = -\frac{2\pi^2 m e^4 Z^2}{n^2 h^2},$$
(1.22)

where $E_{(n)}$ is the energy of the various orbital levels in the hydrogen ion approximation.

Table 1.3   Mendeleev's periodic chart and electron configuration of elements.

TRANSITION ELEMENTS

Group	I a	II a	III b	IV b	V b	VI b	VII b	VIII		
First Period	1 H $1s^1$									
Second Period He+	3 Li $2s^1$	4 Be $2s^2$								
Third Period Ne+	11 Na $3s^1$	12 Mg $3s^2$								
Fourth Period Ar+	19 K $4s^1$	20 Ca $4s^2$	21 Sc $4s^23d^1$	22 Ti $4s^23d^2$	23 V $4s^23d^3$	24 Cr $4s^13d^5$	25 Mn $4s^23d^5$	26 Fe $4s^23d^6$	27 Co $4s^23d^7$	28 Ni $4s^23d^8$
Fifth Period Kr+	37 Rb $5s^1$	38 Sr $5s^2$	39 Y $5s^24d^1$	40 Zr $5s^24d^2$	41 Nb $5s^14d^4$	42 Mo $5s^14d^5$	43 (Tc) $5s^24d^5$	44 Ru $5s^14d^7$	45 Rh $5s^14d^8$	46 Pd $4d^{10}$
Sixth Period Xe+	55 Cs $6s^1$	56 Ba $6s^2$	57* La $6s^25d^1$	72 Hf $6s^24f^{14}5d^2$	73 Ta $6s^24f^{14}5d^3$	74 W $6s^24f^{14}5d^4$	75 Re $6s^24f^{14}5d^5$	76 Os $6s^24f^{14}5d^6$	77 Ir $6s^24f^{14}5d^7$	78 Pt $6s^14f^{14}5d^9$
Seventh Period Rn+	87 (Fr) $7s^1$	88 Ra $7s^2$	89** Ac $7s^26d^1$							

	III b	IV b	V b	VI b	VII b	VIII	VIII
* Lanthanides Xe+	58 Ce $6s^25d^14f^1$	59 Pr $6s^25d^14f^2$	60 Nd $6s^25d^14f^3$	61 (Pm) $6s^25d^14f^4$	62 Sm $6s^25d^14f^5$	63 Eu $6s^25d^14f^6$	64 Gd $6s^25d^14f^8$
** Actinides Rn+	90 Th $7s^26d^2$	91 Pa $7s^26d^15f^2$	92 U $7s^26d^15f^3$	93 (Np) $7s^26d^15f^4$	94 (Pu) $7s^25f^6$	95 (Am) $7s^25f^7$	96 (Cm) $7s^26d^15f^7$

I b	II b	III a	IV a	V a	VI a	VII a	Inert gases
							2 He $1s^2$
		5 B $2s^2 2p^1$	6 C $2s^2 2p^2$	7 N $2s^2 2p^3$	8 O $2s^2 2p^4$	9 F $2s^2 2p^5$	10 Ne $2s^2 2p^6$
		13 Al $3s^2 3p^1$	14 Si $3s^2 3p^2$	15 P $2s^2 3p^3$	16 S $3s^2 3p^4$	17 Cl $2s^2 3p^5$	18 Ar $3s^2 3p^6$
29 Cu $4s^1 3d^{10}$	30 Zn $4s^2 3d^{10}$	31 Ga $4s^2 3d^{10} 4p^1$	32 Ge $4s^2 3d^{10} 4p^2$	33 As $4s^2 3d^{10} 4p^3$	34 Se $4s^2 3d^{10} 4p^4$	35 Br $4s^2 3d^{10} 4p^5$	36 Kr $4s^2 3d^{10} 4p^6$
47 Ag $5s^1 4d^{10}$	48 Cd $5s^2 4d^{10}$	49 In $5s^2 4d^{10} 5p^1$	50 Sn $5s^2 4d^{10} 5p^2$	51 Sb $5s^2 4d^{10} 5p^3$	52 Te $5s^2 4d^{10} 5p^4$	53 I $5s^2 4d^{10} 5p^5$	54 Xe $5s^2 4d^{10} 5p^6$
79 Au $6s^1 4f^{14} 5d^{10}$	80 Hg $6s^2 4f^{14} 5d^{10}$	81 Tl $6s^2 4f^{14} 5d^{10} 6p^1$	82 Pb $6s^2 4f^{14} 5d^{10} 6p^2$	83 Bi $6s^2 4f^{14} 5d^{10} 6p^3$	84 Po $6s^2 4f^{14} 5d^{10} 6p^4$	85 (At) $6s^2 4f^{14} 5d^{10} 6p^5$	86 Rn $6s^2 4f^{14} 5d^{10} 6p^6$

65 Tb $6s^2 5d^1 4f^8$	66 Dy $6s^2 5d^1 4f^9$				
97 (Bk) $7s^2 6d^1 5f^8$	98 (Cf) $7s^2 5f^{10}$				

III a	IV a	V a	VI a	VII a
67 Ho $6s^2 5d^1 4f^{10}$	68 Er $6s^2 4d^1 4f^{11}$	69 Tm $6s^2 4d^1 4f^{12}$	70 Yb $6s^2 4d^1 4f^{13}$	71 Lu $6s^2 4d^1 4f^{14}$
99 (Es) $7s^2 5f^{11}$	100 (Fm) $7s^2 5f^{12}$	101 (Md) $7s^2 5d f^{13}$	102 (No) $7s^2 5f^{14}$	103 (Lw) $7s^2 6d^1 5f^{14}$

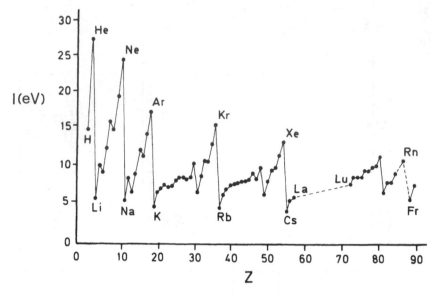

Figure 1.1 First ionization potentials as functions of atomic number. Adapted from Harvey and Porter (1976).

The energy of external orbitals in real atoms (which is numerically equivalent to the first ionization potential expressed in eV) is reproduced by an empirical expression of the form:

$$E = \frac{(Z - S_e)^2 C_e}{n^2} = \frac{Z_{eff}^2 C_e}{n^2}$$

(1.23)

where S_e is a screen constant, depending on the number of inner electrons, and C_e is a conversion factor. Equation 1.23 establishes that, with an increase in the quantum number (table 1.2), we should observe a jump in the energy values of the external orbitals. We recall for this purpose that the external or "valence" electrons are those involved in the formation of atomic bonds (because the absolute values of their orbital energies are much lower than those pertaining to the inner electrons). As figure 1.1 shows, the first ionization potentials* plotted as functions of atomic number Z exhibit marked discontinuities, corresponding to the variations in the principal quantum number, and hence corresponding to the change of period in Mendeleev's periodic chart. Within the same period, we ob-

---

* The "ionization potential" is the amount of energy that is necessary to subtract an electron from an atom and to bring it at infinite distance at rest. The "first ionization energy" is defined as the amount of energy necessary to subtract a valence electron from a neutral atom. The "second ionization energy" is the amount of energy required to subtract a second electron from an ion from which a first electron has already been subtracted, and so on. The "total electronic energy" is the sum of all ionization potentials for a given atom. The "electron affinity" is the amount of energy required to add an electron to a neutral atom or to an anion. Ionization energies for all atoms are listed in table 1.6.

serve a progressive increase in the first ionization potential, with minor discontinuities for transition elements (due to the progressive filling of inner shells, which
modifies the value of screen constant S_e in eq. 1.23). Recalling the definition of
"valence electron" and considering that bonding energies are related to external
orbitals, the periodicity of the reactive properties of the elements is readily understood in light of equations 1.22 and 1.23.

1.3 Generalities on the Concept of Chemical Bond in Solids

"There is a chemical bond between two atoms or groups of atoms in the case that
the forces acting between them are such as to lead to the formation of an aggregate with sufficient stability to make it convenient for the chemist to consider it
as an independent molecular species" (Pauling, 1948).

We will consider four distinct types of bonds:

1. ionic bonds
2. covalent bonds
3. van der Waals bonds
4. metallic bonds

Although in nature it is rare to see solid compounds whose bonds are completely
identifiable with one of the types listed above, it is customary to classify solids
according to the dominant type of bond—i.e., *ionic, covalent, van der Waals,* and
metallic solids.

In ionic solids, the reticular positions are occupied by ions (or anionic
groups) of opposite charge. In NaCl, for instance (figure 1.2A), the Cl^- ions are
surrounded by six Na^+ ions. The alternation of net charges gives rise to strong
coulombic interactions (see section 1.11.1).

In covalent solids, two bonded atoms share a pair of electrons. Covalent
bonds are particularly stable, because the sharing of two electrons obeys the
double duty of completing a stable electronic configuration for the two neighboring atoms. In diamond, for instance (figure 1.2B), each carbon atom shares
an electron pair with four surrounding carbon atoms, generating a stable giant
molecule that has the dimension of the crystal.

In van der Waals solids (e.g., solid Ne; figure 1.2C), the reticular positions
are occupied by inert atoms held together by van der Waals forces. The van der
Waals bond is rather weak, and the bond does not survive a high thermal energy
(*Fm3m* solid Ne, for instance, is stable below 20 K). The van der Waals bond also
exists, however, in molecular compounds (e.g., solid CO_2). The reticular positions
in this case are occupied by neutral molecules (CO_2). Within each molecule, the
bond is mainly covalent (and interatomic distances are considerably shorter), but

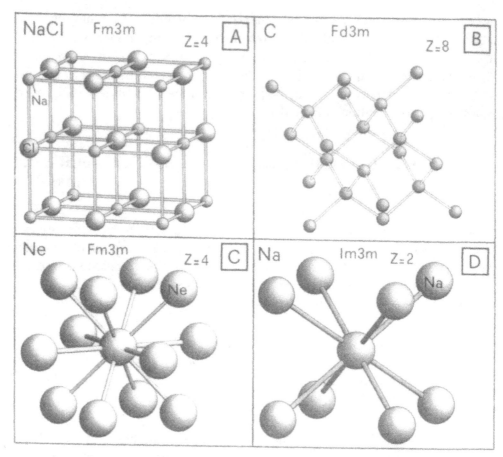

Figure 1.2 Various types of solids. For each solid, a sketch of the bond is given.

the various molecules are held together by van der Waals forces whose nature will be discussed later.

In metallic solids, the reticular positions are occupied by cations immersed in a cloud of delocalized valence electrons. In solid Na, for instance (figure 1.2D), the electron cloud or "electron gas" is composed of electrons of the s sublevel of the third shell (cf. table 1.2). Note that the type of bond does not limit the crystal structure of the solid within particular systems. For example, all solids shown in figure 1.2 belong to the cubic system, and two of them (NaCl, Ne) belong to the same structural class (*Fm3m*).

1.4 Ionic Bond and Ionic Radius

The concept of "ionic radius" was derived from the observation of crystal structures understood as "regular arrays of ions of spherical undeformable symmetry." The structure of ionic crystals obeys pure geometrical rules and corresponds to

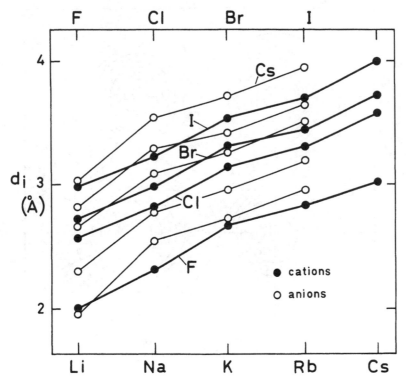

Figure 1.3 Interatomic distances in alkali halides, arranged in cationic and anionic series.

the energetically optimal condition of an ion surrounded by the maximum number of ions of opposite charge. The assumption of "undeformability" makes it possible to consider interionic distance as a constant, at parity of ionic charge. The concept of ionic radius is thus related to that of "interionic distance" (d_i) by simple addition, according to

$$d_i = r_+ + r_-,$$
(1.24)

where r_+ and r_- are, respectively, the ionic radii of a cation and an anion at the interionic distance observed in a purely ionic crystal.

The concept that the ionic radius is relatively independent of the structure of the solid arose intuitively from experimental observations carried out on alkali halides, which are ionic solids par excellence. Figure 1.3 shows the evolution of interatomic distances in alkali halides as a function of the types of anion and cation, respectively. Significant parallelism within each of the two families of curves may be noted. This parallelism intuitively generates the concept of constancy of the ionic radius.

The first attempt to derive empirical ionic radii by geometrical rules applied to ionic compounds (and metals) was performed by Bragg (1920), who proposed a set of ionic radii with a mean internal precision of 0.006 Å. Further tabulations

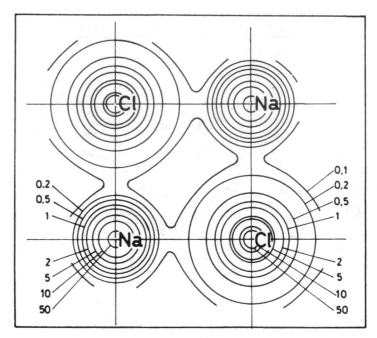

Figure 1.4 Electron density (in electrons per Å³) in the (100) plane of NaCl. Reprinted from Witte and Wölfel (1955), with kind permission of R. Oldenbourg Verlag GmbH, München, Germany.

were proposed by Landé (1920), Wasastjerna (1923), and Goldschmidt et al. (1926).

However, we have already seen that the Schrödinger equation denies the existence of definite limits in the geometry of atomic orbitals, and that the presence of an electron in a particular portion of the space surrounding the nucleus is a probabilistic concept. These two facts apparently conflict with the concept of an ion as a spherical undeformable entity. But this conflict is only apparent, as may be deduced from the "electron density maps" obtained for crystalline compounds by X-ray diffraction techniques. Figure 1.4 illustrates the electron density distribution for the (100) plane in NaCl, as determined by Witte and Wölfel (1955). Concentric lines delineate zones of equal electron density (electrons per Å³), increasing toward the nuclei (to a maximum of 70 electrons per Å³ for Na^+ and 130 electrons per Å³ for Cl^-). The absolute minimum electron density between the two nuclei defines the limits (hence the "dimensions") of the two ions. By integration of the electron densities, it can be calculated that 17.70 electrons are present in the region surrounding the nucleus of Cl^- and 10.05 electrons are present in the region surrounding the nucleus of Na^+. These values, compared with the theoretical limits of 18 and 10 electrons, respectively, indicate that, within experimental approximation, there has been the complete transfer of one electron—and, hence, that a completely "ionic bond" is present in the cubic structure of NaCl. We also observe that the electron density map defines almost perfect sphericity for both ions. The "ionic radii" of Na^+ and Cl^- in the crystalline cubic compound

NaCl thus correspond to the radii (radial extensions) of the two elements in the condition of "ionic bond" when there is complete transfer of one electron from Na to Cl.

1.5 Pauling's Univalent Ionic Radii

The first attempt to derive ionic radii independently of the geometrical concept of simple additivity was made by Pauling (1927a). His method was based on the assumption that the dimension of an ion is determined by the distribution of its outer electrons. As we have already seen (eq. 1.23), the "effective nuclear charge" of an ion (Z_{eff}) is lower than its atomic number, due to the shielding effect of inner electrons, expressed by screening constant S_e. Values of S_e may be obtained theoretically (Pauling, 1927b), and partially by results of X-ray diffraction studies (Pauling and Sherman, 1932). Based on such values, the univalent ionic radius (r_{+-}) can be expressed as

$$r_{+-} = \frac{C_n}{Z - S_e} = \frac{C_n}{Z_{\text{eff}}},$$
(1.25)

where C_n is a constant determined by the principal quantum number of the outermost electrons in the inert gas configuration. Applying equation 1.25 to couples of isoelectronic ions such as NaF, KCl, RbBr, and CsI, which have, respectively, the same C_n, Pauling determined the dimension ratios of the isoelectronic ions in the corresponding solids and derived the corresponding ionic radii by combining obtained with observed values of interionic distance. For ions of the neon structure, for example, S_e has a value of 4.52, and the values of Z_{eff} for Na^+ and F^- are thus $6.48e$ and $4.48e$, respectively. Because C_n is identical for the two ions, knowledge of the interionic $Na^+ - F^-$ distance (2.31 Å) gives the univalent radii solving the system

$$\frac{r_{F^-}}{r_{Na^+}} = \frac{Z_{\text{eff},Na^+}}{Z_{\text{eff},F^-}} = \frac{6.48}{4.48},$$
(1.26)

$$r_{Na^+} + r_{F^-} = 2.31$$
(1.27)

which gives 1.36 Å for F^- and 0.95 Å for Na^+. If we now continue to apply equation 1.25 to multivalent ions, assuming that C_n remains unchanged at the value of the intermediate inert gas, we obtain values for all ions with He, Ne, Ar, Kr, and Xe structures. These radii, known as "univalent ionic radii in VI-fold co-ordination," "do not have absolute values such that their sums are equal to equilibrium inter ionic distances" (Pauling, 1960) but rather "are the radii the

Table 1.4 Ionic radii in VI-fold coordination, based on univalent ionic radii of Pauling (Huheey, 1975).

Z	Ion	r_Z (Å)	Z	Ion	r_Z (Å)	Z	Ion	r_Z (Å)
3	Li^-	0.60	24	Cr^{6+}	0.52	48	Cd^{2+}	0.97
4	Be^{2+}	0.31	25	Mn^{2+}	0.80	49	In^{3+}	0.81
5	B^{3+}	0.23	25	Mn^{3+}	0.66	50	Sn^{2+}	0.93
6	C^{4+}	0.15	25	Mn^{4+}	0.60	50	Sn^{4+}	0.71
7	N^{5+}	0.13	26	Fe^{2+}	0.74	51	Sb^{5+}	0.62
11	Na^+	0.95	26	Fe^{3+}	0.66	55	Cs^+	1.69
12	Mg^{2+}	0.65	27	Co^{2+}	0.72	56	Ba^{2+}	1.35
13	Al^{3+}	0.50	28	Ni^{2+}	0.69	57	La^{3+}	1.14
14	Si^{4+}	0.42	29	Cu^+	0.96	58	Ce^{3+}	1.07
15	P^{5+}	0.35	29	Cu^{2+}	0.72	58	Ce^{4+}	0.94
16	S^{6+}	0.30	30	Zn^{2+}	0.74	72	Hf^{4+}	0.78
17	Cl^{7+}	0.27	31	Ga^{3+}	0.62	74	W^{5+}	0.62
19	K^+	1.33	32	Ge^{4+}	0.53	80	Hg^{2+}	1.10
20	Ca^{2+}	0.99	33	As^{5+}	0.46	81	Tl^+	1.47
21	Sc^{3+}	0.81	34	Se^{6+}	0.42	81	Tl^{3+}	0.95
22	Ti^{3+}	0.76	37	Rb^+	1.47	82	Pb^{2+}	1.20
22	Ti^{4+}	0.68	38	Sr^{2+}	1.13	82	Pb^{4+}	0.84
23	V^{2+}	0.88	39	Y^{3+}	0.92	90	Th^{4+}	1.02
23	V^{3+}	0.74	40	Zr^{4+}	0.79	92	U^{4+}	0.97
23	V^{4+}	0.63	41	Nb^{5+}	0.69	92	U^{6+}	0.80
23	V^{5+}	0.59	42	Mo^{6+}	0.62			
24	Cr^{3+}	0.63	47	Ag^+	1.26			

multivalent ions would possess if they were to retain their electron distributions but enter into Coulomb interaction as if they were univalent" (Pauling, 1960). Univalent radii may be opportunely modified to obtain effective ionic radii in real structures (Pauling, 1960; Huheey, 1975):

$$r_Z = r_{+-} z^{-2/(n-1)},$$

(1.28)

where n is the Born exponent (cf. section 1.11.2) and z is the ionic charge. Ionic radii in VI-fold coordination derived by Huheey (1975) are listed in table 1.4.

1.6 Covalent Bond and Covalent Radius

According to Lewis's (1916) definition, a "covalent bond" exists whenever two atoms are linked in a stable molecule by sharing two outer electrons. We distinguish between "homopolar" and "heteropolar" covalent bonds on the basis of whether or not the bonded atoms are of the same type. Pauling (1960) has shown that in covalent compounds the distance between two elements A and B, independent of the nature of the two atoms, is the same in all compounds containing

Table 1.5 Nonpolar covalent radii of Sanderson (1960).

Z	Ion	r_Z (Å)	Z	Ion	r_Z (Å)	Z	Ion	r_Z (Å)
1	H	0.32	23	V	1.18	49	In	1.44
2	He	0.93	24	Cr	1.17	50	Sn	1.40
3	Li	1.23	25	Mn	1.17	51	Sb	1.40
4	Be	0.89	26	Fe	1.17	52	Te	1.36
5	B	0.82	27	Co	1.16	53	I	1.33
6	C	0.77	28	Ni	1.15	54	Xe	2.09
7	N	0.75	29	Cu	1.17	55	Cs	2.35
8	O	0.73	30	Zn	1.25	56	Ba	1.98
9	F	0.72	31	Ga	1.26	57	La	1.69
10	Ne	1.31	32	Ge	1.22	58	Ce	1.65
11	Na	1.54	33	As	1.20	59	Pr	1.64
12	Mg	1.36	34	Se	1.17	72	Hf	1.44
13	Al	1.18	35	Br	1.14	73	Ta	1.34
14	Si	1.11	36	Kr	1.89	74	W	1.30
15	P	1.06	37	Rb	2.16	79	Au	1.34
16	S	1.02	38	Sr	1.91	80	Hg	1.49
17	Cl	0.99	39	Y	1.62	81	Tl	1.48
18	Ar	1.99	40	Zr	1.45	82	Pb	1.47
19	K	2.03	41	Nb	1.34	83	Bi	1.46
20	Ca	1.74	42	Mo	1.30	86	Rn	2.14
21	Sc	1.44	47	Ag	1.34	90	Th	1.65
22	Ti	1.22	48	Cd	1.48	92	U	1.42

both elements, regardless of the state of aggregation of the compound (i.e., solid, liquid, or gaseous). Covalent radii are easily calculated by dividing by 2 the interatomic distances observed in homopolar covalent compounds (e.g., P_4, S_8, etc.). An internally consistent tabulation was derived by Sanderson (1960) from interatomic distances in heteropolar covalent gaseous molecules (e.g., $AlCl_3$, CH_4, SiF_4, etc.). The covalent radii of Sanderson (table 1.5) are defined as "nonpolar," because they are obtained from compounds that do not have charge polarity.

1.7 Electronegativity and Fractional Ionic Character

The concept of "electronegativity" is widely used in geochemistry to define the combinational properties of the various elements in a qualitative manner. However, it is important to stress that this concept is largely empirical, because it varies from author to author and cannot be univocally quantified.

According to Pauling's (1932) definition, electronegativity is "the capacity of an atom of a molecule to attract the electrons of another atom." The electronegativity scale of Pauling (1960) is based on the differences in the binding energies in diatomic molecules AB with respect to homeopolar compounds AA and BB, according to

$$\chi_A - \chi_B = \gamma^{1/2}\left[E_{AB} - \frac{1}{2}\left(E_{AA} + E_{BB}\right)\right],$$ (1.29)

where χ_A and χ_B are the electronegativities of atoms A and B, E_{AB}, E_{AA}, and E_{BB} are the binding energies of compounds AB, AA, and BB, and operative constant γ is a term that constrains the adimensional χ_A and χ_B to change by 0.5 with each unit valence change in the first row of the periodic table ($\gamma^{1/2} = 0.208$).

According to Mulliken (1934, 1935), electronegativity is the algebraic mean of the first ionisation potential and of electron affinity.

According to Gordy (1946), electronegativity is represented by the value of the potential resulting from the effect of the nuclear charge of an unshielded atom on a valence electron located at a distance corresponding to the covalent radius of the atom.

According to Sanderson (1960), electronegativity represents the mean electron density per unit volume of an inert element:

$$D_e = \frac{3Z}{4\pi r_c^3},$$ (1.30)

where Z is the atomic number and r_c is the covalent nonpolar radius.

Iczkowski and Margrave (1961) consider electronegativity as the local value of a potential defined by the partial derivative of the energy of the atom W with respect to ionic charge z:

$$\chi_{(z)} = \frac{\partial W}{\partial z}.$$ (1.31)

The most immediate use of the concept of electronegativity is in the definition of "fractional ionic character" (f_i). The more or less ionic character of a bond is determined by the distribution of charges within the diatomic molecule. For a purely ionic bond, there is complete charge transfer between the two neighboring atoms, whereas in a purely covalent bond there is no charge transfer. The relationship between electronegativity and percentage of ionicity in diatomic molecules has been proposed by several authors in different ways. According to Pauling (1932, 1960), the more or less ionic character of a bond is expressed by parameter f_i, derived as follows:

$$f_i = 1 - \exp\left[-\frac{1}{4}\left(\chi_A - \chi_B\right)^2\right].$$ (1.32)

According to Gordy (1950), the same parameter can be obtained by applying

$$f_i = \frac{1}{2}\left(\chi_A - \chi_B\right). \tag{1.33}$$

Hinze et al. (1963) suggest that charge transfer in diatomic molecules obeys the principle of equality of energies involved in the process. If we adopt equation 1.31 as a definition of electronegativity and imagine that the amount of energy involved in the transfer of a fraction of charge dz from atom A to atom B is quantifiable as

$$dE_A = \left(\frac{\partial W_A}{\partial z}\right)_{z_A} dz, \tag{1.34}$$

then, by application of the equality principle, we have

$$dE_A = \left(\frac{\partial W_A}{\partial z}\right)_{z_A} dz = \left(\frac{\partial W_B}{\partial z}\right)_{z_B} dz = dE_B. \tag{1.35}$$

A similar concept was developed by Sanderson (1960).

Electronegativity values according to Pauling and Sanderson are listed in table 1.6, together with the first four ionization potentials and the electron affinities of the various elements.

At present, the most widely used definition of fractional ionic character in solid state chemistry is that of Phillips (1970), based on a spectroscopic approach. Phillips defined fractional ionic character as

$$f_i = \frac{E_i^2}{E_i^2 + E_c^2}, \tag{1.36}$$

where E_i ("ionic energy gap") is obtained from the "total energy gap" (E_g) between bonding and antibonding orbitals (see section 1.17.1):

$$E_g = \frac{\hbar \omega_P}{\left(\varepsilon_\infty - 1\right)^{1/2}} \tag{1.37}$$

$$E_i = \left(E_g^2 - E_c^2\right)^{1/2}, \tag{1.38}$$

where $\hbar \omega_P$ is the plasma frequency for valence electrons, ε_∞ is the optic dielectric constant, and E_c corresponds to E_g for the nonpolar system in the same row of the periodic table, with a correction for interatomic spacing.

Table 1.6 Elemental properties: first four ionization potentials (I1, I2, I3, I4, from Samsonov, 1968; values expressed in eV; values in parentheses are of doubtful reliability); electron affinity (e.a., from Samsonov, 1968; eV); Pauling's electronegativity (P, from Samsonov, 1968; eV); Sanderson's electronegativity (S, from Viellard, 1982; adimensional).

| Z | Atom | Ionization Potential | | | | e.a. | Electronegativity | |
		I1	I2	I3	I4		P	S
1	H	13.595	–	–	–	0.747	2.15	3.55
2	He	24.58	54.40	–	–	−0.53	–	–
3	Li	5.39	75.62	122.42	–	0.82	1.0	0.74
4	Be	9.32	18.21	153.85	217.66	−0.19	1.5	2.39
5	B	8.296	25.15	37.92	259.30	0.33	2.0	2.84
6	C	11.264	24.376	47.86	64.48	1.24	2.5	3.79
7	N	14.54	29.60	47.426	77.45	0.05	3.0	4.49
8	O	13.614	35.15	54.93	77.39	1.465	3.5	5.21
9	F	17.418	34.98	62.65	87.23	3.58	4.0	5.75
10	Ne	21.559	41.07	63.5	97.16	−0.57	–	–
11	Na	5.138	47.29	71.8	98.88	0.84	0.9	0.70
12	Mg	7.644	15.03	78.2	109.3	−0.32	1.2	1.99
13	Al	5.984	18.82	28.44	119.96	0.52	1.5	2.25
14	Si	8.149	16.34	33.46	45.13	1.46	1.8	2.62
15	P	10.55	19.65	30.16	51.35	0.77	2.1	3.34
16	S	10.357	23.4	34.8	47.29	2.07	2.5	4.11
17	Cl	13.01	23.80	39.9	53.3	3.76	3.0	4.93
18	Ar	15.755	27.6	40.90	59.79	−1.0	–	–
19	K	4.339	31.81	45.9	61.1	0.82	0.8	0.41
20	Ca	6.111	11.87	51.21	67.3	–	1.0	1.22
21	Sc	6.56	12.89	24.75	73.9	–	1.3	1.30
22	Ti	6.83	13.57	28.14	43.24	–	1.5	1.40
23	V	6.74	14.2	29.7	48.0	–	1.6	1.60
24	Cr	6.764	16.49	31	(51)	–	1.6	1.88
25	Mn	7.432	15.64	33.69	(53)	–	1.5	2.07
26	Fe	7.90	16.18	30.64	(56)	–	1.8	2.10
27	Co	7.86	17.05	33.49	(53)	–	1.7	2.10
28	Ni	7.633	18.15	36.16	(56)	–	1.8	2.10
29	Cu	7.724	20.29	36.83	(59)	0.29	1.9	2.60
30	Zn	9.391	17.96	39.70	(62)	–	1.6	2.84
31	Ga	6.00	20.51	30.70	64.2	–	1.6	3.23
32	Ge	7.88	15.93	34.21	45.7	–	2.0	3.59
33	As	9.81	18.7	28.3	50.1	–	2.0	3.91
34	Se	9.75	21.5	32.0	42.9	3.7	2.4	4.25
35	Br	11.84	21.6	35.9	47.3	3.54	2.9	4.53
36	Kr	13.996	24.56	36.9	52.5	–	–	–
37	Rb	4.176	27.56	40	52.6	–	0.8	0.33
38	Sr	5.692	11.026	43.6	57.1	–	1.0	1.00
39	Y	6.38	12.23	20.5	61.8	–	1.2	1.05
40	Zr	6.835	12.92	24.8	33.97	–	1.4	1.10
41	Nb	6.88	13.90	28.1	38.3	–	1.6	–
42	Mo	7.131	15.72	29.6	46.4	–	1.8	–
43	Tc	7.23	14.87	31.9	(43)	–	1.9	–
44	Ru	7.36	16.60	30.3	(47)	–	2.2	–

Table 1.6 (continued)

Z	Atom	Ionization Potential				e.a.	Electronegativity	
		I1	I2	I3	I4		P	S
45	Rh	7.46	15.92	32.8	(46)	–	2.2	–
46	Pd	8.33	19.42	(33)	(49)	–	2.2	–
47	Ag	7.574	21.48	36.10	(52)	2.5	1.9	2.57
48	Cd	8.991	16.904	44.5	(55)	–	1.7	2.59
49	In	5.785	18.86	28.0	68	–	1.7	2.86
50	Sn	7.332	14.6	30.7	46.4	–	1.8	3.10
51	Sb	8.64	16.7	24.8	44.1	>2.0	1.9	3.37
52	Te	9.01	18.8	31.0	38	3.6	2.1	3.62
53	I	10.44	19.0	33	(42)	3.29	2.5	3.84
54	Xe	12.127	21.2	32.1	(45)	–	–	–
55	Cs	3.893	25.1	34.6	(46)	–	0.7	0.29
56	Ba	5.810	10.00	37	(49)	–	0.9	0.78
57	La	5.614	11.433	19.166	(52)	–	1.1	0.88
58	Ce	6.54	12.31	19.870	36.7	–	1.1	–
59	Pr	(5.76)	(11.54)	(20.96)	–	–	1.1	–
60	Nd	(6.31)	(12.09)	(20.51)	–	–	1.1	–
62	Sm	5.56	11.4	(24.0)	–	–	1.1	–
63	Eu	5.61	11.24	(24.56)	–	–	1.1	–
64	Gd	5.98	(12)	(23)	–	–	1.1	–
65	Tb	(6.74)	(12.52)	(22.04)	–	–	1.2	–
66	Dy	5.80	(12.60)	(21.83)	–	–	1.2	–
67	Ho	5.85	(12.7)	(22.1)	–	–	1.2	–
68	Er	6.11	(12.5)	(23.0)	–	–	1.2	–
69	Tm	5.85	(12.4)	(24.1)	–	–	1.2	–
70	Yb	5.90	12.10	(25.61)	–	–	1.2	–
71	Lu	6.15	14.7	(21.83)	–	–	1.2	–
72	Hf	5.5	14.9	(21)	(31)	–	1.3	1.05
73	Ta	7.7	16.2	(22)	(33)	–	1.5	–
74	W	7.98	17.7	(24)	(35)	–	1.7	1.39
75	Re	7.87	16.6	(26)	(38)	–	1.9	–
76	Os	8.7	17	(25)	(40)	–	2.2	–
77	Ir	9.2	17.0	(27)	(39)	–	2.2	–
78	Pt	8.96	18.54	(29)	(41)	–	2.2	–
79	Au	9.223	20.5	(30)	(44)	2.1	2.4	2.57
80	Hg	10.434	18.751	34.2	(46)	1.54	1.9	2.93
81	Tl	6.106	20.42	29.8	50	2.1	1.8	3.02
82	Pb	7.415	15.03	31.93	39.0	–	1.8	3.08
83	Bi	7.277	19.3	25.6	45.3	>0.7	1.9	3.16
84	Po	8.2	19.4	27.3	(38)	–	2.0	–
85	At	9.2	20.1	29.3	(41)	–	2.2	–
86	Rn	10.745	21.4	29.4	(44)	–	–	–
87	Fr	3.98	22.5	33.5	(43)	–	0.7	–
88	Ra	5.277	10.144	(34)	(46)	–	0.9	–
89	Ac	6.89	11.5	–	(49)	–	1.1	–
90	Th	6.95	11.5	20.0	28.7	–	1.3	–
91	Pa	–	–	–	–	–	1.5	–
92	U	5.65	14.63	25.13	–	–	1.7	–

Catlow and Stoneham (1983) have shown that ionic term E_i corresponds to the difference of the diagonal matrix elements of the Hamiltonian of an AB molecule in a simple LCAO approximation ($H_{AA} - H_{BB}$; cf. section 1.18.1), whereas the covalent energy gap corresponds to the double of the off-diagonal term H_{AB}—i.e.,

$$E_g = \sqrt{\left(H_{AA} - H_{BB}\right)^2 + 4H_{AB}}.$$

(1.39)

1.8 Ionic Polarizability and van der Waals Bond

The electron cloud of an ion subjected to an electric field undergoes deformations that may be translated into displacement "d" of the baricenters of negative charges from the positions held in the absence of external perturbation, which are normally coincident with the centers of nuclear charges (positive). The non-coincidence of the two centers causes a "dipole moment," determined by the product of the displaced charge (Z_e) and the displacement d. The displacement is also proportional to the intensity of the electrical field (**F**). The proportionality factor (α) is known as "ionic polarizability":

$$\mu_d = Z_e d = \mathbf{F}\alpha.$$

(1.40)

Let us consider two neighboring atoms A and B (figure 1.5) and imagine that the electronic charge distribution is, for an infinitesimal period of time, asymmetrical. In this period of time, atom A becomes a dipole and induces a distortion in the electron cloud of atom B by attracting electrons from its positive edge: atom B is then transformed, in turn, into a dipole. All the neighboring atoms

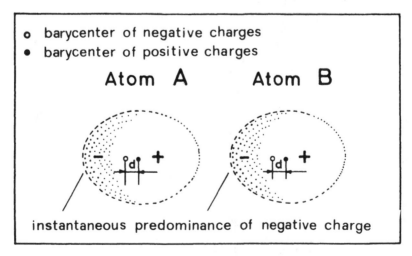

Figure 1.5 Polarization of charges in neighboring atoms.

undergo the same process, generating attractive forces among themselves. Due to electron orbital motion, the induced forces fluctuate with a periodicity dictated by the rotational velocity of electrons. The *van der Waals bond* is the result of the induced attractions between fluctuating dipoles. The attractions diminish with increased molecular distance and, when distances are equal, increase with the number of electrons within the molecule.

Free-ion polarizability calculations have been proposed by various authors, based on the values of molecular refraction of incident light (R_m), according to

$$\alpha = \frac{3R_m}{4\pi N_0}, \tag{1.41}$$

where N_0 is Avogadro's number.

Free-ion polarizabilities, arranged in isoelectronic series, are listed in table 1.7. A semilogarithmic plot of the listed values (figure 1.6) reveals a functional dependence on atomic number. This dependence, quite marked, allows us to estimate the free-ion polarizability for ions for which there are no precise experimental data (values in parentheses in table 1.7; estimates according to Viellard, 1982).

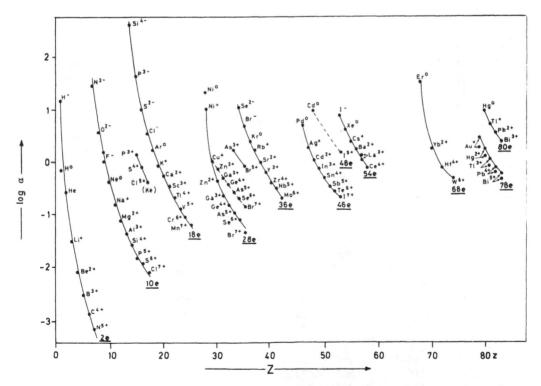

Figure 1.6 Free-ion polarizability as a function of atomic number. Curves are drawn for isoelectronic series. Reprinted from Viellard (1982), Sciences Geologiques, Memoir n°69, Université Louis Pasteur, with kind permission of the Director of Publication.

Table 1.7 Free–Ion polarizability (α_f) arranged in isoelectronic series. Data in Å^3. N = number of electrons (adapted from Viellard, 1982).

Ion	N	α_f	Ion	N	α_f	Ion	N	α
H^-	2	16.86	Sc^0	21	26.8	Se^{2-}	36	10.63
He^0	2	0.221	Ti^+	21	1.67	Br^-	36	4.82
Li^+	2	0.030	V^{2+}	21	(0.55)	Kr^0	36	2.475
Be^{2+}	2	0.008	Cr^{3+}	21	(0.33)	Rb^+	36	1.415–1.704
B^{3+}	2	0.003	Mn^{4+}	21	(0.24)	Sr^{2+}	36	0.978
C^{4+}	2	0.0014				Y^{3+}	36	0.550
N^{5+}	2	0.007				Zr^{4+}	36	0.3704
			Ti^0	22	13.603–22.4	Nb^{5+}	36	0.293
N^{3-}	10	28.80	V^+	22	1.41	Mo^{6+}	36	0.19
O^{2-}	10	2.75–3.92	Cr^{2+}	22	(0.50)			
F^-	10	0.978	Mn^{3+}	22	(0.32)			
Ne^0	10	0.4162				Pd^0	46	4.928–6.91
Na^+	10	0.148–0.54	V^0	23	11.395–19.10	Ag^+	46	1.856–2.682
Mg^{2+}	10	0.069–0.075	Cr^+	23	1.141–1.16	Cd^{2+}	46	0.926–1.012
Al^{3+}	10	0.0419	Mn^{2+}	23	0.499–0.7172	In^{3+}	46	0.702
Si^{4+}	10	0.0255	Fe^{3+}	23	0.234–0.2475	Sn^{4+}	46	0.453
P^{5+}	10	0.014				Sb^{5+}	46	0.326
S^{6+}	10	0.0112	Cr^0	24	6.8–16.6	Te^{6+}	46	0.259
Cl^{7+}	10	0.008	Mn^+	24	1.13	I^{7+}	46	0.206
			Fe^{2+}	24	0.6875			
P^{3+}	12	1.35	Co^{3+}	24	0.548	Cd^0	48	6.001–9.187
S^{4+}	12	0.792				In^+	48	–
Cl^{5+}	12	0.502	Mn^0	25	8.61–14.8	Sn^{2+}	48	(3.98)
Fe^+	25	1.03	Co^{2+}	25	(0.66)	Sb^{3+}	48	(2.88)
Si^{4-}	18	376.9				Te^{4+}	48	(1.995)
P^{3-}	18	41.66				I^{5+}	48	1.449
S^{2-}	18	10.32	Fe^0	26	12.60			
Cl^-	18	3.556	Co^+	26	0.8446–0.896	I^-	54	7.41
Ar^0	18	1.674	Ni^{2+}	26	0.771	Xe^0	54	4.03
K^+	18	0.811				Cs^+	54	2.415
Ca^{2+}	18	0.511	Co^0	27	6.80–11.113	Ba^{2+}	54	1.689
Sc^{3+}	18	0.320	Ni^+	27	1.329	La^{3+}	54	1.05
Ti^{4+}	18	0.200	Cu^{2+}	27	0.971–1.289	Ce^{4+}	54	0.74
V^{5+}	18	0.123						
Cr^{6+}	18	0.084	Ni^0	28	6.51–10.00	Er^0	68	31.9
Mn^{7+}	18	0.0756	Cu^+	28	0.982–1.115	Tm^+	68	–
			Zn^{2+}	28	0.4186–0.726	Yb^{2+}	68	1.718
K^0	19	37.30–42.97	Ga^{3+}	28	0.213–0.477	Lu^{3+}	68	–
Ca^+	19	8.34	Ge^{4+}	28	0.1269–0.347	Hf^{4+}	68	0.750
Sc^{2+}	19	–	As^{5+}	28	0.083–0.243	Ta^{5+}	68	–
Ti^{3+}	19	(0.46)	Se^{6+}	28	0.190	W^{6+}	68	0.4603
V^{4+}	19	(0.32)	Br^{7+}	28	0.043–0.132			
						Au^+	78	2.792
Ca^0	20	25.04–33.80	As^{3+}	30	1.572	Hg^{2+}	78	1.615
Sc^+	20	2.06	Se^{4+}	30		Tl^{3+}	78	1.074
Ti^{2+}	20	(0.72)	Br^{5+}	30	0.8054	Pb^{4+}	78	0.775
V^{3+}	20	(0.40)				Bi^{5+}	78	0.590

Table 1.7   *(continued)*

Ion	N	α_f	Ion	N	α_f	Ion	N	α
Cr^{4+}	20	–						
						Hg^0	80	9.129
						Ti^+	80	5.187
						Pb^{2+}	80	3.439
						Bi^{3+}	80	2.47
						Th^{4+}	86	1.75
						U^{6+}	86	1.141

1.9   Crystal Radius and Effective Distribution Radius (EDR)

The first satisfactory definition of crystal radius was given by Tosi (1964): "In an ideal ionic crystal where every valence electron is supposed to remain localised on its parent ion, to each ion it can be associated a limit at which the wave function vanishes. The radial extension of the ion along the connection with its first neighbour can be considered as a measure of its dimension in the crystal (crystal radius)." This concept is clearly displayed in figure 1.7A, in which the radial electron density distribution curves are shown for Na^+ and Cl^- ions in NaCl. The nucleus of Cl^- is located at the origin on the abscissa axis and the nucleus of Na^+ is positioned at the interionic distance experimentally observed for neighboring ions in NaCl. The superimposed radial density functions define an electron density minimum that limits the dimensions or "crystal radii" of the two ions. We also note that the radial distribution functions for the two ions in the crystal (continuous lines) are not identical to the radial distribution functions for the free ions (dashed lines).

The "crystal radius" thus has local validity in reference to a given crystal structure. This fact gives rise to a certain amount of confusion in current nomenclature, and what it is commonly referred to as "crystal radius" in the various tabulations is in fact a mean value, independent of the type of structure (see section 1.11.1). The "crystal radius" in the sense of Tosi (1964) is commonly defined as "effective distribution radius" (EDR). The example given in figure 1.7B shows radial electron density distribution curves for Mg, Ni, Co, Fe, and Mn on the M1 site in olivine (orthorhombic orthosilicate) and the corresponding EDR radii located by Fujino et al. (1981) on the electron density minima.

Table 1.8 shows the differences among EDR radii, ionic radii, and crystal radii for selected elements.

It also must be stressed that EDR may assume different values along various crystallographic directions. This is clearly evident if we compare electron density maps obtained

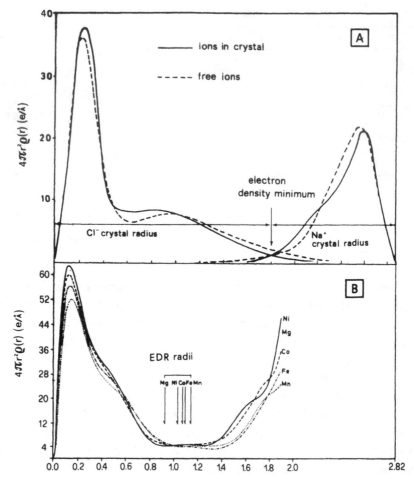

Figure 1.7 (A) Radial electron density distribution curves for Na⁺ and Cl⁻ in NaCl. (B) Radial electron density distributions for Mg, Ni, Co, Fe, and Mn on M1 site in olivine. Arrows indicate electron density minima defining EDR radii. Reprinted from Tosi (1964), with kind permission of Academic Press Inc., Orlando, Florida (A); and from Fujino et al. (1981), with kind permission of The International Union of Crystallography (B).

by calculation of electron density functions of neutral, spherically averaged atoms (known as Independent Atom Model or IAM) with observed total electron densities. Figure 1.8, for instance, makes this comparison for BeO: subtraction of IAM densities (B) from total electron densities (A) (pseudoatom approximation; see Downs, 1991 for details) results in a deformation density chart (C), which shows how the electron density of the IAM reference model has migrated due to bonding effects. (Solid contours in the figure are positive accumulations of charges; short dashes are negative; long dashes mark unperturbed zones.)

Table 1.8 Comparison of EDR radii (Fujino et al., 1981) with ionic radii (IR) and crystal radii (CR) (Shannon, 1976) in various structures (adapted from Fujino et al., 1981; values in Å).

Cation	Phase	Site	EDR	IR	CR	Coordination
Li	α-spodumene	M2	0.83	0.92	1.06	VIII
Na	Giadeite	M2	1.30	1.18	1.32	VIII
Mg	Periclase		0.92	0.72	0.86	VI
Mg	Forsterite	M1	0.93	0.72	0.86	VI
Mg	Forsterite	M2	0.94	0.72	0.86	VI
Mg	Diopside	M1	0.91	0.72	0.86	VI
Mg	Enstatite	M1	0.89	0.72	0.86	VI
Mg	Enstatite	M2	0.96	0.72	0.86	VI
Ca	Diopside	M1	1.15	0.83	0.97	VI
Mn	Tephroite	M2	1.16	0.83	0.97	VI
Fe	Fayalite	M1	1.10	0.78	0.92	VI
Fe	Fayalite	M2	1.12	0.78	0.92	VI
Co	CoO		1.00	0.745	0.883	VI
Co	Co-olivine	M1	1.03	0.745	0.883	VI
Co	Co-olivine	M2	1.07	0.745	0.883	VI
Ni	Bunsenite		1.08	0.69	0.83	VI
Ni	Ni-olivine	M1	1.04	0.69	0.83	VI
Ni	Ni-olivine	M2	1.05	0.69	0.83	VI
Si	Forsterite		0.96	0.26	0.40	IV
Si	α-spodumene		0.96	0.26	0.40	IV

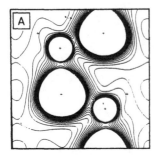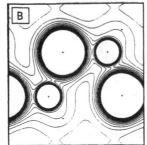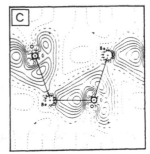

Figure 1.8 Electron density plots along (110) plane of BeO: (A) effective total electron density (pseudoatom approximation); (B) total electron density of IAM; (C) deformation density (pseudoatom-IAM). From Downs (1992). Reprinted with permission of Springer-Verlag, New York.

1.10 Effective Ionic Radii

For crystal chemical purposes, it is useful to have at hand tabulations of "effective ionic radii" to allow the reproduction of interionic distances in real crystals, independently of precise knowledge of the structure of interest. The construction of such tables is a stepwise process that needs to be outlined to some extent.

1.10.1 Ionic Radii and Coordination

Goldschmidt (1923) was the first to note that ionic radii vary as a function of coordination number—i.e., with the number of first neighbors surrounding the ion. The observed variation is fundamentally identical for all cations and, in the case of cations coordinated by anionic ligands, results in a progressive increase in ionic radius with increasing coordination number. As an example, figure 1.9 shows the increases in ionic radius for the various rare earth elements (and In^{3+}, Y^{3+}, and Sc^{3+}) as a function of coordination number with O^{2-}.

1.10.2 Ionic Radii and Ionization Potentials

Ahrens (1952) proposed the first extended tabulation of ionic radii, partially modifying the univalent radii in the VI-fold coordination of Pauling (1927a) on the basis of the observed correlation between ionic radius (r) and ionization potential (I), which can be expressed in the forms

$$I \propto \frac{1}{\sqrt{r}} \tag{1.42}$$

and

$$r \propto \frac{1}{I^2} \tag{1.43}$$

for ions of homologous charge, and in the form

$$I^n \propto Z_c, \tag{1.44}$$

where Z_c is valence and $n \leq 2$, for isoelectronic series.

By interpolation techniques based on equations 1.42 to 1.44, Ahrens determined several ionic radii not known at that time (Ag^{2+}, At^{7+}, Au^{3+}, Br^{5+}, Cl^{5+}, Cu^{2+}, Fr^+, Ge^{2+}, I^{5+}, N^{3+}, Np^{7+}, P^{3+}, Pa^{5+}, Pd^{2+}, Po^{6+}, Pt^{2+}, Re^{7+}, S^{4+}, Se^{4+}, Sn^{2+}, Tc^{7+}, U^{6+}, V^{2+}); compare these with the preceding tabulation of Pauling

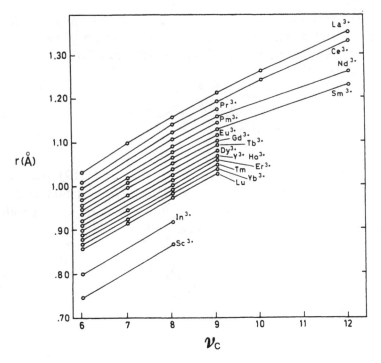

Figure 1.9 Ionic radii of rare earth ions (and In^{3+}, Y^{3+}, and Sc^{3+}) as a function of coordination number with oxygen v_c. Reprinted from Shannon (1976), with kind permission from the International Union of Crystallography.

(table 1.4). Ahrens's (1952) tabulation was a common tool for geochemists until the 1970s, when the extended tabulations of Shannon and Prewitt (1969), Shannon (1976), and Whittaker and Muntus (1970) appeared.

1.10.3 Ionic Radii, Bond Strength, and Degree of Covalence

Pauling (1929) introduced the concept of "mean strength of the bond" $(\overline{S}_l)$ as a ratio of the valence with coordination number v_c. He also formulated the "principle of electrostatic valence," which establishes that the sum of the mean bond strengths for a given coordination polyhedron (P) coincides with the valence state—i.e.,

$$\sum_P \overline{S}_l = Z_c .$$

(1.45)

This last principle is not strictly observed in natural compounds, and, particularly in silicates (Baur, 1970), deviations of up to 40% over theoretical values have been observed. These deviations must be attributed to partial covalence of the bonds.

Pauling (1947) has also shown that, in purely covalent bonds, modification of bond distance within a given coordination polyhedron may be associated with the "bond number" (n_l; number of shared electrons per bond) through arbitrary constant K_l—i.e.,

$$\Delta R_l = R_l - R_{1,l} = -2K_l \log n_l, \tag{1.46}$$

where $R_{1,l}$ is characteristic distance.

Combination of the above principles led Donnay (1969) and Donnay and Allmann (1970) to propose a generalized scheme for variation of bond length with bond strength in mixed bonds (not purely ionic or purely covalent). The equation proposed by the above authors is

$$S_l = S_{0,l} \left(\frac{\overline{R}_l}{R_l} \right)^{N_l}, \tag{1.47}$$

where S_l is the strength of a bond with length R_l, $S_{0,l}$ is the ideal bond strength of a bond of length $\overline{R}_l$, coinciding with the mean of bond distances within the coordination polyhedron, and N_l is a constant that has a single value for each cation-anion pair and, in some instances, for each single coordination number within a fixed cation-anion pair.

A family of empirical covariance curves between bond strength and bond length was then derived by Brown and Shannon (1973) on the basis of equation 1.47 coupled with analysis of a large number of crystal structures. In solving the system by means of statistical interpolation procedures, Brown and Shannon kept the bond strength sum for a given polyhedron as near as possible to the theoretical value postulated by Pauling's (1929) "electrostatic valence principle" (cf. eq. 1.45). They also showed that the degree of covalence for a given bond and its relative bond strength are related by an empirical equation in the form

$$f_c^l = \alpha_l S_l^{M_l}, \tag{1.48}$$

where α_l and M_l depend on the number of electrons of the cation and are constant for isoelectronic series. The α_l and M_l constants for cation-oxygen bonds are listed in table 1.9.

Figure 1.10 shows the relationship between bond strength and degree of covalence in cation-oxygen bonds for cations with 18, 36, and 54 electrons (note that in table 1.9 the parameters of equation 1.48 are identical for the three isoelectronic series). By combining equation 1.48 with a modified form of Donnay's (1969) equation:

Table 1.9 Parameters relating covalence
to bond strength and bond length in
equations 1.48 and 1.50 (adapted from
Brown and Shannon, 1973).

e	α_l	M_l	$M_l N_l$
0	0.67	1.8–2	4
2	0.60	1.73	7.04
10	0.54	1.64	7.04
18	0.49	1.57	7.04
28	0.60	1.50	9.08
36	0.49	1.57	
46	0.67	1.43	
54	0.49	1.57	

Table 1.10 General parameters for
relationship between bond strength and
bond length in cation-oxygen bonds
(adapted from Brown and Shannon, 1973).

Cation	e	$R_{1,l}$	N_l
H	0	0.86	2.17
Li, Be, B	2	1.378	4.065
Na, Mg, Al, Si, P, S	10	1.622	4.290
K, Ca, Sc, Ti, V, Cr	18	1.799	4.483
Mn, Fe	23	1.760	5.117
Zn, Ga, Ge, As	28	1.746	6.050

$$S_{0,l} = \left(\frac{R_l}{R_{1,l}} \right)^{-N_l} , \tag{1.49}$$

where $R_{1,l}$ and N_l are general parameters valid for isoelectronic series (with the
same anion; see table 1.10), Brown and Shannon obtained an empirical equation
relating bond length and degree of covalence:

$$f_c^l = \alpha_l \left(\frac{R_l}{R_{1,l}} \right)^{-M_l N_l} . \tag{1.50}$$

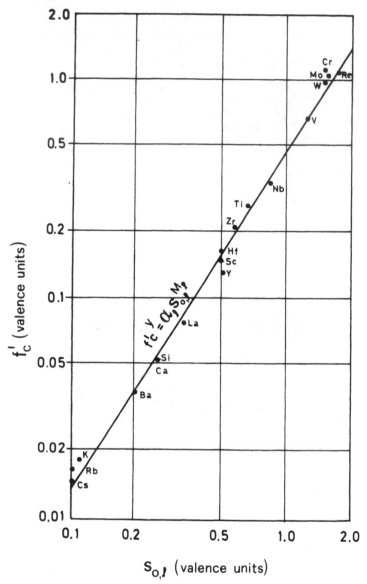

Figure 1.10 Degree of covalence (f_c^i) vs bond strength ($S_{o,l}$) in M-O bonds for cations with 18, 36, and 54 electrons. Values are in valence units and scales are logarithmic. Reprinted from Brown and Shannon (1973), with kind permission from the International Union of Crystallography.

1.10.4 Present-Day Tabulations: Effective Ionic Radii and Crystal Radii

Table 1.11 lists the *effective ionic radii* (IRs) and *crystal radii* (CRs) obtained by Shannon (1976), updating the preceding tabulation of Shannon and Prewitt (1969). In deriving ionic radii based on the values of interionic distances for more

Table 1.11 Effective crystal radii (CR) and ionic radii (IR) of Shannon (1976). CN = coordination number; SP = spin; sp = square planar; py = pyramidal; HS = high spin; LS = low spin; data in Å.

Ion	CN	SP	CR	IR	Ion	CN	SP	CR	IR
Ac^{3+}	VI		1.26	1.12	Bi^{5+}	VI		0.90	0.76
Ag^{1+}	II		0.81	0.67	Bk^{3+}	VI		1.10	0.96
	IV		1.14	1.00	Bk^{4+}	VI		0.97	0.83
	IVsp		1.16	1.02		VIII		1.07	0.93
	V		1.23	1.09	Br^{1-}	VI		1.82	1.96
	VI		1.29	1.15	Br^{3+}	IVsp		0.73	0.59
	VII		1.36	1.22	Br^{5+}	IIIpy		0.45	0.31
	VIII		1.42	1.28	Br^{7+}	IV		0.39	0.25
Ag^{2+}	IVsp		0.93	0.79		VI		0.53	0.39
	VI		1.08	0.94	C^{4+}	III		0.06	−0.08
Ag^{3+}	IVsp		0.81	0.67		IV		0.29	0.15
	VI		0.89	0.75		VI		0.30	0.16
Al^{3+}	IV		0.53	0.39	Ca^{2+}	VI		1.14	1.00
	V		0.62	0.48		VII		1.20	1.06
	VI		0.675	0.535		VIII		1.26	1.12
Am^{2+}	VII		1.35	1.21		IX		1.32	1.18
	VIII		1.40	1.26		X		1.37	1.23
	IX		1.45	1.31		XII		1.48	1.34
Am^{3+}	VI		1.115	0.975	Cd^{2+}	IV		0.92	0.78
	VIII		1.23	1.09		V		1.01	0.87
Am^{4+}	VI		0.99	0.85		VI		1.09	0.95
	VIII		1.09	0.95		VII		1.17	1.03
As^{3+}	VI		0.72	0.58		VIII		1.24	1.10
As^{5+}	IV		0.475	0.335		XII		1.45	1.31
	VI		0.60	0.46	Ce^{3+}	VI		1.15	1.01
At^{7+}	VI		0.76	0.62		VII		1.21	1.07
Au^{1+}	VI		1.51	1.37		VIII		1.283	1.143
Au^{3+}	IVsp		0.82	0.68		IX		1.336	1.196
	VI		0.99	0.85		X		1.39	1.25
Au^{5+}	VI		0.71	0.57		XII		1.48	1.34
B^{3+}	III		0.15	0.01	Ce^{4+}	VI		1.01	0.87
	IV		0.25	0.11		VIII		1.11	0.97
	VI		0.41	0.27		X		1.21	1.07
Ba^{2+}	VI		1.49	1.35		XII		1.28	1.14
	VII		1.52	1.38	Cf^{3+}	VI		1.09	0.95
	VIII		1.56	1.42	Cf^{4+}	VI		0.961	0.821
	IX		1.61	1.47		VIII		1.06	0.92
	X		1.66	1.52	Cl^{1-}	VI		1.67	1.81
	XI		1.71	1.57	Cl^{5+}	IIIpy		0.26	0.12
	XII		1.75	1.61	Cl^{7+}	IV		0.22	0.08
Be^{2+}	III		0.30	0.16		VI		0.41	0.27
	IV		0.41	0.27	Cm^{3+}	VI		1.11	0.97
	VI		0.59	0.45	Cm^{4+}	VI		0.99	0.85
Bi^{3+}	V		1.10	0.96		VIII		1.09	0.95
	VI		1.17	1.03	Co^{2+}	IV	HS	0.72	0.58
	VIII		1.31	1.17		V		0.81	0.67

continued

Table 1.11   *(continued)*

Ion	CN	SP	CR	IR	Ion	CN	SP	CR	IR
	VI	LS	0.79	0.65	Eu^{3+}	VI		1.087	0.947
	VI	HS	0.885	0.745		VII		1.15	1.01
	VIII		1.04	0.90		VIII		1.206	1.066
Co^{3+}	VI	LS	0.685	0.545		IX		1.260	1.120
		HS	0.75	0.61	F^{1-}	II		1.145	1.285
Co^{4+}	IV		0.54	0.40		III		1.16	1.30
Cr^{2+}	VI	LS	0.87	0.73		IV		1.17	1.31
	VI	HS	0.94	0.80		VI		1.19	1.33
Cr^{3+}	VI		0.755	0.615	F^{7+}	VI		0.22	0.08
Cr^{4+}	IV		0.55	0.41	Fe^{2+}	IV	HS	0.77	0.63
	V		0.72	0.58		IVsp	HS	0.78	0.64
	VI	LS	0.69	0.55		VI		0.75	0.61
		HS	0.785	0.645			HS	0.920	0.780
	VIII	0.92	0.78			VIII	HS	1.06	0.92
	VI	LS	0.69	0.55	Fe^{3+}	IV	HS	0.63	0.49
Cr^{5+}	IV		0.485	0.345		V		0.72	0.58
	VI		0.63	0.49		VI	LS	0.69	0.55
	VIII		0.71	0.57			HS	0.785	0.645
Cr^{6+}	IV		0.40	0.26		VIII	HS	0.92	0.78
	VI		0.58	0.44	Fe^{4+}	VI		0.725	0.585
Cs^{1+}	VI		1.81	1.67	Fe^{6+}	IV		0.39	0.25
	VIII		1.88	1.74	Fr^{1+}	VI		1.94	1.80
	IX		1.92	1.78	Ga^{3+}	IV		0.61	0.47
	X		1.95	1.81		V		0.69	0.55
	XI		1.99	1.85		VI		0.760	0.620
	XII		2.02	1.88		VII		1.14	1.00
Cu^{1+}	II		0.60	0.46	Gd^{3+}	VI		1.078	0.938
	IV		0.74	0.60		VIII		1.193	1.053
	VI		0.91	0.77		IX		1.247	1.107
Cu^{2+}	IVsp		0.71	0.57	Ge^{2+}	VI		0.87	0.73
	V		0.79	0.65	Ge^{4+}	IV		0.530	0.390
	VI		0.87	0.73		VI		0.670	0.530
Cu^{3+}	VI	LS	0.68	0.54	H^{1+}	I		−0.24	−0.38
D^{1+}	II		0.04	0.10		II		−0.04	−0.08
Dy^{2+}	VI		1.21	1.07	Hf^{4+}	IV		0.72	0.58
	VII		1.27	1.13		VI		0.85	0.71
	VIII		1.33	1.19		VII		0.90	0.76
Dy^{3+}	VI		1.052	0.912		VIII		0.97	0.83
	VII		1.11	0.97	Hg^{1+}	III		1.11	0.97
	VIII		1.167	1.027		VI		1.33	1.19
	IX		1.223	1.083	Hg^{2+}	II		0.83	0.69
Er^{3+}	VI		1.030	0.890		IV		1.10	0.96
	VII		1.085	0.945		VI		1.16	1.02
	VIII		1.144	1.004		VIII		1.28	1.14
Eu^{2+}	VI		1.31	1.17	Ho^{3+}	VI		1.041	0.901
	VII		1.34	1.20		VIII		1.155	1.015
	VIII		1.39	1.25		IX		1.212	1.072
	IX		1.44	1.30		X		1.26	1.12
	X		1.49	1.35	I^{1-}	VI		2.06	2.20

Table 1.11 (continued)

Ion	CN	SP	CR	IR	Ion	CN	SP	CR	IR
I⁵⁺	IIIpy		0.58	0.44	Mo⁵⁺	IV		0.60	0.46
	VI		1.09	0.95		VI		0.75	0.61
I⁷⁺	IV		0.56	0.42	Mo⁶⁺	IV		0.55	0.41
	VI		0.67	0.53		V		0.64	0.50
In³⁺	IV		0.76	0.62		VI		0.73	0.59
	VI		0.940	0.800		VII		0.87	0.73
	VIII		1.06	0.92	N³⁻	IV		1.32	1.46
Ir³⁺	VI		0.82	0.68	N³⁺	VI		0.30	0.16
Ir⁴⁺	VI		0.765	0.625	N⁵⁺	III		0.044	0.104
Ir⁵⁺	VI		0.71	0.57		IV		0.27	0.13
K¹⁺	IV		1.51	1.37	Na¹⁺	IV		1.13	0.99
	VII		1.60	1.46		V		1.14	1.00
	VIII		1.65	1.51		VI	1.16	1.02	
	IX		1.69	1.55		VII	1.26	1.12	
	X		1.73	1.59		VIII	1.32	1.18	
	XII		1.78	1.64		IX	1.38	1.24	
La³⁺	VI		1.172	1.032		XII	1.53	1.39	
	VII		1.24	1.10	Nb²⁺	VI		0.86	0.72
	VIII		1.300	1.160	Nb⁴⁺	VI		0.82	0.68
	IX		1.356	1.216		VIII		0.93	0.79
	X		1.41	1.27	Nb⁵⁺	IV		0.62	0.48
	XII		1.50	1.36		VI		0.78	0.64
Li¹⁺	IV		0.730	0.590		VII		0.83	0.69
	VI		0.90	0.76		VIII		0.88	0.74
	VIII		1.06	0.91	Nd²⁺	VIII		1.43	1.29
Lu³⁺	VI		1.001	0.861		IX		1.49	1.35
	VIII		1.117	0.977	Nd³⁺	VI		1.123	0.983
	IX		1.172	1.032		VIII		1.249	1.109
Mg²⁺	IV		0.71	0.57		IX		1.303	1.163
	V		0.80	0.66		XII		1.41	1.27
	VI		0.860	0.720	Ni²⁺	IV		0.69	0.55
	VIII		1.03	0.89		IVsp		0.63	0.49
Mn²⁺	IV	HS	0.80	0.66		V		0.77	0.63
	V	HS	0.89	0.75		VI		0.830	0.690
	VI	LS	0.81	0.67	Ni³⁺	VI	LS	0.70	0.56
		HS	0.970	0.830			HS	0.74	0.60
	VII	HS	1.04	0.90	Np²⁺	VI		1.24	1.10
	VIII		1.10	0.96	Np³⁺	VI		1.15	1.01
Mn³⁺	V		0.72	0.58	Np⁴⁺	VI		1.01	0.87
	VI	LS	0.72	0.58		VIII		1.12	0.98
		HS	0.785	0.645	Np⁵⁺	VI		0.89	0.75
Mn⁴⁺	IV		0.53	0.39	Np⁶⁺	VI		0.86	0.72
	VI		0.670	0.530	Np⁷⁺	VI		0.85	0.71
Mn⁵⁺	IV		0.47	0.33	O²⁻	II		1.21	1.35
Mn⁶⁺	IV		0.395	0.255		III		1.22	1.36
Mn⁷⁺	IV		0.39	0.25		IV		1.24	1.38
	VI		0.60	0.46		VI		1.26	1.40
Mo³⁺	VI		0.83	0.69		VIII		1.28	1.42
Mo⁴⁺	VI		0.790	0.650	OH¹⁻	II		1.18	1.32

continued

Table 1.11 (continued)

Ion	CN	SP	CR	IR	Ion	CN	SP	CR	IR
	III		1.20	1.34	Pu^{3+}	VI		1.14	1.00
	IV		1.21	1.35	Pu^{4+}	VI		0.94	0.80
	VI		1.23	1.37		VIII		1.10	0.96
Os^{4+}	VI		0.770	0.630	Ra^{2+}	VIII		1.62	1.48
Os^{5+}	VI		0.715	0.575		XII		1.84	1.70
Os^{6+}	V		0.63	0.49	Rb^{1+}	VI		1.66	1.52
	VI		0.685	0.545		VII		1.70	1.56
Os^{7+}	VI		0.665	0.525		VIII		1.75	1.61
Os^{8+}	IV		0.53	0.39		IX		1.77	1.63
P^{3+}	VI		0.58	0.44		X		1.80	1.66
P^{5+}	IV		0.31	0.17		XI		1.83	1.69
	V		0.43	0.29		XII		1.86	1.72
	VI		0.52	0.38		XIV		1.97	1.93
Pa^{3+}	VI		1.18	1.04	Re^{4+}	VI		0.77	0.63
Pa^{4+}	VI		1.04	0.90	Re^{5+}	VI		0.72	0.58
	VIII		1.15	1.01	Re^{7+}	IV		0.52	0.38
Pa^{5+}	VI		0.92	0.78		VI		0.67	0.53
	IX		1.09	0.95	Rh^{3+}	VI		0.805	0.665
Pb^{2+}	IVpy		1.12	0.98	Rh^{4+}	VI		0.74	0.60
	VI		1.33	1.19	Rh^{5+}	VI		0.69	0.55
	VII		1.37	1.23	Ru^{3+}	VI		0.82	0.68
	VIII		1.43	1.29	Ru^{4+}	VI		0.760	0.620
	IX		1.49	1.35	Ru^{5+}	VI		0.705	0.565
	X		1.54	1.40	Ru^{7+}	IV		0.52	0.38
	XI		1.59	1.45	Ru^{8+}	IV		0.50	0.36
	XII		1.63	1.49	S^{2-}	VI		1.70	1.84
Pb^{4+}	IV		0.79	0.65	S^{4+}	VI		0.51	0.37
	V		0.87	0.73	S^{6+}	IV		0.26	0.12
	VI		0.915	0.775		VI		0.43	0.29
	VIII		1.08	0.94	Sb^{3+}	IVpy		0.90	0.76
Pd^{1+}	II		0.73	0.59		V		0.94	0.80
Pd^{2+}	IVsp		0.78	0.64		VI		0.90	0.76
	VI		1.00	0.86	Sb^{5+}	VI		0.74	0.60
Pd^{3+}	VI		0.90	0.76	Sc^{3+}	VI		0.885	0.745
Pd^{4+}	VI		0.755	0.615		VIII		1.010	0.870
Pm^{3+}	VI		1.11	0.97	Se^{2-}	VI		1.84	1.98
	VIII		1.233	1.093	Se^{4+}	VI		0.64	0.50
	IX		1.284	1.144	Se^{6+}	IV		0.42	0.28
Po^{4+}	VI		1.08	0.94		VI		0.56	0.42
	VIII		1.22	1.08	Si^{4+}	IV		0.40	0.26
Po^{6+}	VI		0.81	0.67		VI		0.540	0.400
Pr^{3+}	VI		1.13	0.99	Sm^{2+}	VII		1.36	1.22
	VIII		1.266	1.126		VIII		1.41	1.27
	IX		1.319	1.179		IX		1.46	1.32
Pr^{4+}	VI		0.99	0.85	Sm^{3+}	VI		1.098	0.958
	VIII		1.10	0.96		VII		1.16	1.02
Pt^{2+}	IVsp		0.74	0.60		VIII		1.219	1.079
	VI		0.94	0.80		IX		1.272	1.132
Pt^{4+}	VI		0.765	0.625		XII		1.38	1.24
Pt^{5+}	VI		0.71	0.57	Sn^{4+}	IV		0.69	0.55

Table 1.11 (continued)

Ion	CN	SP	CR	IR	Ion	CN	SP	CR	IR
	V		0.76	0.62	Tm³⁺	VI		1.020	0.880
	VI		0.830	0.690		VIII		1.134	0.994
	VII		0.89	0.75		IX		1.192	1.052
	VIII		0.95	0.81	U³⁺	VI		1.165	1.025
Sr²⁺	VI		1.32	1.18	U⁴⁺	VI		1.03	0.89
	VII		1.35	1.21		VII		1.09	0.95
	VIII		1.40	1.26		IX		1.19	1.05
	IX		1.45	1.31		XII		1.31	1.17
	X		1.50	1.36	U⁵⁺	VI		0.90	0.76
	XII		1.58	1.44		VII		0.98	0.84
Ta³⁺	VI		0.86	0.72	U⁶⁺	II		0.59	0.45
Ta⁴⁺	VI		0.82	0.68		IV		0.66	0.52
Ta⁵⁺	VI		0.78	0.64		VI		0.87	0.73
	VIII		0.88	0.74		VII		0.95	0.81
Tb³⁺	VI		1.063	0.923		VIII		1.00	0.86
	VII		1.12	0.98	V²⁺	VI		0.93	0.79
	VIII		1.180	1.040	V³⁺	VI		0.780	0.640
Tb⁴⁺	VI		0.90	0.76	V⁴⁺	V		0.67	0.53
	VIII		1.02	0.88		VI		0.72	0.58
Tc⁴⁺	VI		0.785	0.645		VIII		0.86	0.72
Tc⁵⁺	VI		0.74	0.60	V⁵⁺	IV		0.495	0.355
Tc⁷⁺	IV		0.51	0.37		V		0.60	0.46
	VI		0.70	0.56		VI		0.68	0.54
Te²⁻	VI		2.07	2.21	W⁴⁺	VI		0.80	0.66
Te⁴⁺	III		0.66	0.52	W⁵⁺	VI		0.76	0.62
	IV		0.80	0.66	W⁶⁺	IV		0.56	0.42
	VI		1.11	0.97		V		0.65	0.51
Te⁶⁺	IV		0.57	0.43		VI		0.74	0.60
	VI		0.70	0.56	Xe⁸⁺	IV		0.54	0.40
Th⁴⁺	VI		1.08	0.94		VI		0.62	0.48
	VIII		1.19	1.05	Y³⁺	VI		1.040	0.900
	IX		1.23	1.09		VII		1.10	0.96
	X		1.27	1.13		VIII		1.159	1.019
	XI		1.32	1.18		IX		1.215	1.075
	XII		1.35	1.21	Yb³⁺	VI		1.008	0.868
Ti²⁺	VI		1.00	0.86		VII		1.065	0.925
Ti³⁺	VI		0.810	0.670		VIII		1.125	0.985
Ti⁴⁺	IV		0.56	0.42		IX		1.182	1.042
	V		0.65	0.51	Zn²⁺	IV		0.74	0.60
	VI		0.745	0.605		V		0.82	0.68
	VIII		0.88	0.74		VI		0.880	0.740
Tl¹⁺	VI		1.64	1.50		VIII		1.04	0.90
	VIII		1.73	1.59	Zr⁴⁺	IV		0.73	0.59
	XII		1.84	1.70		V		0.80	0.66
Tl³⁺	IV		0.89	0.75		VI		0.86	0.72
	VI		1.025	0.885		VII		0.92	0.78
	VIII		1.12	0.98		VIII		0.98	0.84
Tm²⁺	VI		1.17	1.03		IX		1.03	0.89
	VII		1.23	1.09					

than 1000 crystal structures, Shannon and Prewitt (1969) made two distinct assumptions:

1. The oxygen radius in VI-fold coordination is effectively 1.40 Å, and is thus identical to the value proposed by Pauling and Ahrens.
2. The oxygen radius in VI-fold coordination is 1.26 Å. This value is consistent with the theoretical estimates of Fumi and Tosi (1964) for F^-, given the assigned difference between the ionic radii of F^- and O^{2-}.

Shannon and Prewitt (1969) named the ionic radii derived from the first assumption "effective ionic radii" (IRs) and those derived from the second assumption "crystal radii," thus stating that they are more appropriate to real structures.

Shannon's (1976) tabulation, reported in the preceding pages, is a simple updating of the previous tabulation of Shannon and Prewitt (1969), and obeys the following assumptions:

1. The additivity of cationic and anionic radii in reproducing the interionic distance (eq. 1.24) is valid, provided that coordination number, electron spin, degree of covalence, repulsive forces, and polyhedral distortion (eq. 1.49) are all taken into account.
2. According to the above limitations, effective ionic radii and crystal radii are independent of the type of structure.
3. Both cationic and anionic radii vary with coordination number.
4. With the same anion, unit-cell volumes in isostructural series are proportional (but not necessarily in a linear fashion) to the cationic volume.
5. Covalence effects on the shortening of metal-oxygen and metal-fluorine distances are not comparable.
6. The mean interatomic distance for a coordination polyhedron within a given structure varies in a reproducible fashion with the degree of polyhedral distortion and with coordination number.

We focus attention on the fact that the crystal radii (CRs) for the various cations listed in table 1.11 are simply equivalent to the effective ionic radii (IRs) augmented by 0.14 Å. Wittaker and Muntus (1970) observed that the CR radii of Shannon and Prewitt (1969) conform better than IR radii to the "radius ratio principle"* and proposed a tabulation with intermediate values, consistent with the above principle (defined by the authors as "ionic radii for geochemistry"), as particularly useful for silicates. It was not considered necessary to reproduce the

---

* The so-called "radius ratio principle" establishes that, for a cation/anion radius ratio lower than 0.414, the coordination of the complex is 4. The coordination numbers rise to 6 for ratios between 0.414 and 0.732 and to 8 for ratios higher than 0.732. Actually the various compounds conform to this principle only qualitatively. Tossell (1980) has shown that, if Ahrens's ionic radii are adopted, only 60% of compounds conform to the "radius ratio principle."

tabulation here, because the actual incertitude of the interpolated values is approximately equal to the CR-IR gap.

1.11 Forces Interacting Between Two Atoms at Defined Distances

1.11.1 Coulombic Attraction

The attractive force exerted by a cation of pointlike charge Z_+ on an anion of pointlike charge Z_- at distance r is given by

$$F_C = \frac{Z_+ Z_- e^2}{r^2}. \qquad (1.51)$$

where e is the electronic charge. The coulombic potential at distance r is obtained by integration of equation 1.51 in dr from infinite distance to r:

$$\Phi_C = \int -F_C \, dr = \frac{Z_+ Z_- e^2}{r}. \qquad (1.52)$$

The potential energy progressively decreases with decreasing distance between the two ions. Thus, theoretically, the two ions could collapse if opposite forces did not oppose their excessive proximity.

1.11.2 Short-Range Repulsion

First of all, let us note that short-range coulombic repulsion exists, due to the homologous nature of the electronic charges of the two atoms at short distances from each other. However, the main repulsive force is quantum-mechanistic and is a manifestation of Heisenberg's indetermination principle. The more electrons are confined in a limited region of space, the more their angular momentum and hence their kinetic energy increases (of section 1.18).

The increase of energy with decreasing distance is exponential. Born first suggested a potential in the form

$$\Phi_R = \frac{b}{r^n}, \qquad (1.53)$$

where b is a constant and n (Born exponent) is an integer (between 5 and 12) that depends on the nature of interacting ions. Equation 1.53 is then replaced by

$$\Phi_R = b \exp\left(\frac{-r}{\rho}\right), \tag{1.54}$$

which expresses the exponential dependency of the radial functions better. In equation 1.54, b and ρ are two constants that usually are defined, respectively, as "repulsion factor" and "hardness factor."

1.11.3 Dispersive Forces

If electron clouds of cations and anions are deformable in an electric field and thus have polarizabilities of α_+ and α_-, the value of the dipole moment induced in the cations by the presence of anions (μ_{-+}) and induced in the anions by the presence of cations (μ_{+-}) are, respectively, proportional to α_+/r^3 and α_-/r^3. The electric field acting on the cations is proportional to $-\alpha_+\overline{\mu}^2_{-+}/r^6$, where $\overline{\mu}^2_{-+}$ is the mean square dipole moment of the cations.

Polarization has been formally represented as the movement of a charge Z_e through a displacement distance d (cf. Section 1.8). The displacement is due to field **F**, and the drop in potential energy caused by movement in the field is Fd. If the force acting on charge Z_e by virtue of the parent nucleus is

$$F = -K_F d \tag{1.55}$$

(see eq. 1.14), the resultant fall in potential energy will be

$$-K_F d^2 + \frac{1}{2} K_F d^2 = -\frac{1}{2} K_F d^2 - \frac{1}{2} \alpha \mathbf{F}^2. \tag{1.56}$$

The interaction of the two dipoles can thus be associated with a potential energy proportional to the square of the electric field **F**. Moreover, **F** itself is proportional to r^{-3}, so that the functional dependency of the dispersive energy over the distance will be proportional to r^{-6}.

Accounting for the instantaneous higher moments of the charge distributions of the atoms leads to an inverse eighth-power functional form (dipole-quadrupole interactions). The bulk dispersive potential is represented as shown by Mayer (1933):

$$\Phi_D = \frac{dd_{+-}}{r^6} + \frac{dq_{+-}}{r^8}. \tag{1.57}$$

The dipole-dipole (dd_{+-}) and dipole-quadrupole (dq_{+-}) coefficients can be derived from free-ion polarizability α and the *mean excitation* $\overline{E}$ by applying

$$dd_{+-} = \frac{3}{2}\alpha_+\alpha_-\left(\frac{\overline{E}_+\overline{E}_-}{\overline{E}_+ + \overline{E}_-}\right) \tag{1.58}$$

and

$$dq_{+-} = \frac{27}{8}\left(\frac{\alpha_+\alpha_-}{e^2}\right)\left(\frac{\overline{E}_+\overline{E}_-}{\overline{E}_+ + \overline{E}_-}\right)\left(\frac{\alpha_+\overline{E}_+}{n_+} + \frac{\alpha_-\overline{E}_-}{n_-}\right) \tag{1.59}$$

(Mayer, 1933; Boswara and Franklyn, 1968), where n_+ and n_- are *effective electrons* (see Tosi, 1964, for an extended discussion of the significance of the various terms).

1.12   Lattice Energy of an Ionic Crystal

The lattice energy of an ionic crystal is the amount of energy required at absolute zero temperature to convert one mole of crystalline component into constituent ions in a gaseous state at infinite distance. It is composed of the various forms of energies, as shown above. The calculation is in fact somewhat more complex because of the presence of various ions of alternating charges in a regular tridimensional network.

1.12.1   Madelung Constant

To approach the complexity of a real tridimensional structure, let us first consider the case of a monodimensional array of alternating positive and negative charges, each at distance r from its first neighbor. We will assume for the sake of simplicity that the dispersive potential is negligible. The total potential is therefore

$$\Phi_{\text{total}} = \Phi_C + \Phi_R. \tag{1.60}$$

The potential at the (positive) central charge in figure 1.11 is that generated by the presence of two opposite charges at distance r, by two homologous charges at distance $2r$, by two opposite charges at distance $3r$, and so on. The coulombic potential is then represented by a serial expansion of the type

$$\Phi_C = -\frac{2e^2}{r} + \frac{2e^2}{2r} - \frac{2e^2}{3r} + \frac{2e^2}{4r} - \cdots = -\frac{2e^2}{r}\left(1 - \frac{1}{2} + \frac{1}{3} - \frac{1}{4} + \cdots\right). \tag{1.61}$$

The term in parentheses in equation 1.61, which in this particular case corresponds to the natural logarithm of 2 ($\ln 2 = 0.69$), is known as the *Madelung constant* ($\mathcal{M}$) and has a characteristic value for each structure.

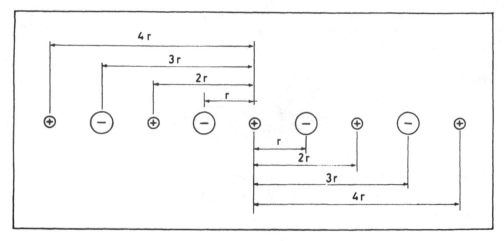

Figure 1.11 Monodimensional array of alternating charges.

Let us now consider the coulombic potential at a positive central charge in a regular tridimensional network of alternating charges. The cubic compound NaCl is exemplified in this respect. Taking the central charge on a Na^+ ion, we find six Cl^- ions as first neighbors at distance r_0, corresponding to $\frac{1}{2}$ the cell edge a_0 (cf. figure 1.2). The second neighbors are 12 Na^+ ions at distance $r_0\sqrt{2}$, the third neighbors are eight Cl^- ions at distance $r_0\sqrt{3}$, the fourth neighbors are six Na^+ ions at distance $2r_0$, the fifth neighbors are 24 Cl^- ions at distance $r_0\sqrt{5}$, and so on. The coulombic potential acting on the cation is

$$\Phi_{C,Na+} = -\frac{Z_{+-}^2 e^2}{r_0}\left(6 - \frac{12}{\sqrt{2}} + \frac{8}{\sqrt{3}} - \frac{6}{\sqrt{4}} + \frac{24}{\sqrt{5}} - \cdots\right). \tag{1.62}$$

where Z_{+-} is the largest common factor of the pointlike charges. If we now repeat the calculation adopting Cl^- as the central charge, we obtain a coulombic potential that (for this particular structure) is identical to that acting at the cationic site. The bulk coulombic potential is given by the summation of the cationic and anionic parts divided by 2:

$$\Phi_C = \frac{\Phi_{C,Na+} + \Phi_{C,Cl-}}{2} = -\frac{Z_{+-}^2 e^2 \mathcal{M}}{r_0}. \tag{1.63}$$

Let us now go back to the monodimensional array of figure 1.11. The repulsive potential at the central charge is composed of

$$\Phi_R = 2b\left[\exp\left(\frac{-r}{\rho}\right) + \exp\left(\frac{-2r}{\rho}\right) + \exp\left(\frac{-3r}{\rho}\right) + \cdots\right] \approx 2b\exp\left(\frac{-r}{\rho}\right) \tag{1.64}$$

(the effect of repulsive forces vanishes rapidly with increasing distance and, for practical purposes, extended effects beyond the first neighbors at first sight can be neglected). We then have

$$\Phi_{\text{total}} = \frac{-1.38 Z_{+-}^2 e^2}{r} + 2b \exp\left(\frac{-r}{\rho}\right). \tag{1.65}$$

The ions in the array arrange themselves in order to minimize the bulk energy of the substance. The equilibrium condition is analytically expressed by equating to zero the first derivative of the potential energy over distance:

$$\left(\frac{\partial \Phi_{\text{total}}}{\partial r}\right)_{r=r_0} = 0. \tag{1.66}$$

Applying the above condition and solving over b, we obtain

$$\Phi_{\text{total}(r=r_0)} = \frac{-1.38 Z_{+-}^2 e^2}{r_0}\left(1 - \frac{\rho}{r_0}\right). \tag{1.67}$$

Equation 1.67 is known as *Born-Landé equation*.

If we now consider the effect of the crystal lattice, by introducing Madelung constant $\mathcal{M}$ and multiplying by Avogadro's number N_0 (i.e., the number of atoms per weight formula unit), we get

$$\Phi_{\text{total}(r=r_0)} = \frac{-\mathcal{M} N_0 Z_{+-}^2 e^2}{r_0}\left(1 - \frac{\rho}{r_0}\right). \tag{1.68}$$

Equation 1.68 is the *Born-Mayer equation*.

1.12.2 Zero-Point Energy

The lattice energy of a crystalline substance U with purely ionic bonds and negligible polarization effects is given by equation 1.68, changed in sign and with the subtraction of an energetic term known as "zero-point energy":

$$U = -\Phi_{\text{total}} - E_0. \tag{1.69}$$

To understand the significance of this last term, we must go back to the monodimensional harmonic oscillator and quantum mechanics.

Atoms in a crystal are in fact in vibrational motion even at zero temperature.

Under these thermal conditions, vibrational quantum number n is at the lowest possible value. When we substitute the value obtained by the vibrational frequency for force constant K_F of the harmonic oscillator (eq. 1.14):

$$K_F = 4\pi^2 m v_0^2,\qquad(1.70)$$

equation 1.17 becomes:

$$E_{(n)} = \left(n + \frac{1}{2}\right) h v_0,\qquad(1.71)$$

which, with $n = 0$, gives the zero-point energy—i.e., the vibrational energy present even at the zero point:

$$E_0 = \frac{1}{2} h v_0.\qquad(1.72)$$

However, for a real crystal at zero temperature it is impossible to group all vibrational motions on the lowest single vibrational mode and, if the crystal behaves as a Debye solid (see later), zero-point energy is expressed as:

$$E_0 = \frac{9}{4} N_0 h v_{max},\qquad(1.73)$$

where v_{max} is the highest vibrational mode occupied in the crystal.

1.12.3 Ladd and Lee's Equation

Accounting for all the various forms of energy outlined above (including dispersive forces) and solving for b at equilibrium conditions we obtain the equation of Ladd and Lee (1958):

$$U = \frac{\mathcal{M} N_0 Z_{+-}^2 e^2}{r_0}\left(1 - \frac{\rho}{r_0}\right) + \frac{N_0 dd_{+-}}{r_0^6}\left(1 - \frac{6\rho}{r_0}\right) + \frac{N_0 dq_{+-}}{r_0^8}\left(1 - \frac{8\rho}{r_0}\right)$$
$$- \frac{9}{4} N_0 h v_{max}.\qquad(1.74)$$

1.12.4 Quantitative Calculations: an Example

Adopting the equation of Catti (1981), the lattice energy of an ionic crystal may be expressed by means of the double summations

Table 1.12 Lattice energies of olivine end-members; values in kJ/mole (Ottonello, 1987). E_{BHF} = energy of the thermochemical cycle; E_C = coulombic energy; E_R = repulsive energy; E_{DD} and E_{DQ} = dispersive energies.

Compound	E_{BHF}	$\overline{H}_{f,T_r,P_r}$	$-U$	E_C	E_R	E_{DD}	E_{DQ}
Mg_2SiO_4	18456.4	−2175.7	−20632.1	−24131.3	3835.3	−259.4	−76.0
Fe_2SiO_4	19268.4	−1481.6	−20750.3	−23934.6	3779.3	−442.1	−152.9
Ca_2SiO_4	17608.5	−2316.2	−19925.1	−23105.4	3546.9	−282.4	−84.3
Mn_2SiO_4	18825.4	−1728.1	−20553.5	−23734.8	3641.3	−348.5	−111.6
Co_2SiO_4	19443.9	−1406.7	−20850.6	−24040.8	3855.4	−490.5	−174.6
Ni_2SiO_4	19609.9	−1405.2	−21015.1	−24219.8	4027.4	−596.4	−226.2
Zn_2SiO_4	19326.0	−1390.4	−20716.4	−24068.0	3875.2	−393.9	−129.6

$$U = \sum_{i=1}^{n} \sum_{j=i}^{n} Z_i Z_j C_{ij}^{EL} + \sum_{i=1}^{n} \sum_{j=i}^{n} b_{ij} C_{ij}^{R} + \sum_{i=1}^{n} \sum_{j=i}^{n} dd_{ij} C_{ij}^{DD} + \sum_{i=1}^{n} \sum_{j=i}^{n} dq_{ij} C_{ij}^{DQ} \quad (1.75)$$

extended to all ions in the lattice.

Coefficients b_{ij}, dd_{ij}, and dq_{ij} in equation 1.75 have the same significance as coefficient b in equation 1.54 and dd_{+-} and dq_{+-} in equation 1.57. Coefficients C_{ij}^{EL}, C_{ij}^{R}, C_{ij}^{DD}, and C_{ij}^{DQ} depend on crystal symmetry and on interionic distances and are obtained by means of standard mathematical routines that cannot be detailed here (see appendix A in Tosi, 1964 for exhaustive treatment). Table 1.12 lists as examples the lattice energy values obtained through application of equation 1.75 to orthorhombic silicates (olivine, spatial group *Pbnm*). Values are expressed in kilojoules per mole. We note that lattice energy is mainly (about 90%) composed of the attractive coulombic term. Repulsive energy is about 15% of bulk energy and is of opposite sign. Dispersive energy is greatly subordinate (1–3%).

Figure 1.12 shows the potential well for Ni_2SiO_4 obtained by expanding and/or compressing isotropically the bond distances among all the ions in the structure, with respect to the equilibrium position. The stable state condition of the solid compound is at the minimum of the potential well (Φ_0 at r_0; see eq. 1.66). We note that the energy modification is asymmetric with respect to the minimum of the potential well (dashed line in figure 1.12), due to the different functional dependencies of r on the various forms of energy. As we will see later, this asymmetry (which is responsible for the thermal expansion of the substance) may be translated into an *anharmonicity* of vibrational motions of the ions in the structure.

1.12.5 Semiempirical Calculations: the Kapustinsky Method

As already noted, the estimate of the lattice energy of a crystalline compound requires detailed knowledge of its structure. When the structure is not known

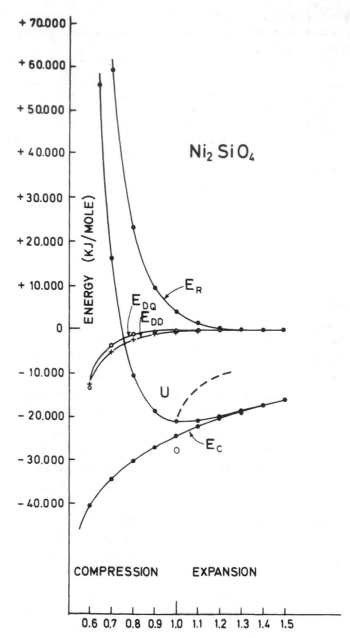

Figure 1.12 Potential well for Ni_2SiO_4 olivine. Abscissa values are fractional values of isotropic compression and/or expansion.

with sufficient precision, or when calculation of the Madelung constant is particularly time-consuming, alternative semiempirical calculations can be carried out. The *Kapustinsky method* is perhaps the most popular among the semiquantitative approaches. The difference between the Madelung constant of a given crystal and the Madelung constant of the same compound with the structure of sodium

Table 1.13 Thermochemical radii for polyanionic complexes (Greenwood, 1970).

Complex	r'	Complex	r'	Complex	r'	Complex	r'
BF_4	2.28	BeF_4	2.45	ClO_4	2.36	IO_4	2.49
ClO_4	2.36	SO_4^{2-}	2.30	SeO_4^{2-}	2.43	TeO_4^{2-}	2.54
CrO_4^2	2.40	MoO_4^{2-}	2.54	PO_4^{3-}	2.38	AsO_4^{3-}	2.48
SbO_4^3	2.60	BiO_4^{3-}	2.68	SiO_4^{4-}	2.40	OH	1.40
SH	1.95	SeH	2.15	CN	1.82	CNO	1.59
CNS	1.95	NH^{2-}	1.30	NO_2	1.55	HCO_3	1.58
$CH_3CO_2^-$	1.58	HCO_3	1.63	CO_3^{2-}	1.85	NO_3	1.89
ClO_3	2.00	BrO_3	1.91	IO_3	1.89	O_3^{2-}	1.80

chlorine is compensated by the difference between the actual interionic distances in the crystal in question and those obtained from the ionic radii in VI-fold coordination with oxygen. Thus derived is a set of "thermochemical radii" that represent the energy counterpart of a given ion (or polyanion) appropriate to any structure. The Kapustinsky equation reads as follows:

$$U = 287.2 \frac{v_F Z^2_{+-}}{r'_+ + r'_-}\left(1 - \frac{0.345}{r'_+ + r'_-}\right), \qquad \left(\frac{\text{Kcal}}{\text{mole}}\right) \qquad (1.76)$$

where v_F is the number of ions per formula unit and r'_+ and r'_- are the thermochemical radii. Table 1.13 lists typical thermochemical radii for the most important polyanionic groups. Coupling the thermochemical radius of the polyanionic group SiO_4^{4-} with the ionic radius of Mg^{2+} or Fe^{2+} in VI-fold coordination with oxygen, the lattice energy of Mg_2SiO_4 or Fe_2SiO_4 is readily derived by application of equation 1.76.

An extended form of the Kapustinsky equation has been proposed by Saxena (1977). Its formulation is a compromise between equation 1.76 and the Ladd and Lee equation (eq. 1.74):

$$U = a' + \frac{b'}{r'_+ + r'_-}\left(1 - \frac{0.345}{r'_+ + r'_-}\right) + \frac{c'}{\left(r'_+ + r'_-\right)^6}\left(1 - \frac{6 \cdot 0.345}{r'_+ + r'_-}\right)$$
$$+ \frac{d'}{\left(r'_+ + r'_-\right)^8}\left(1 - \frac{8 \cdot 0.345}{r'_+ + r'_-}\right) \qquad (1.77)$$

The third and fourth terms on the right side of equation 1.77 take into account the effect of dispersive potential. Constants a', b', c', and d' in equation 1.77 have no direct physical meaning and are derived by interpolation procedures carried out on the main families of crystalline compounds.

1.13 Born-Haber-Fayans Thermochemical Cycle

In calculating the lattice energy of a crystal, we adopt an arbitrary reference condition (two isolated ions in the gaseous state and at infinite distance) to which we assign a zero potential. It is worth stressing that this condition is not equivalent to the standard state commonly adopted in thermochemical calculations, which is normally that of *element at stable state at reference P,T.*

The relationships between the two different states and between the "enthalpy of formation from the elements at standard state" (H_f^0) and the "lattice energy" (U) are easily understood by referring to the Born-Haber-Fayans thermochemical cycle. In this cycle, the formation of a crystalline compound from isolated atoms in the gaseous state is visualized as a stepwise process connecting the various transformations. Let us follow the condensation process of a crystal MX formed from a metal M and a gaseous molecule X_2:

1. First we transform metal M into a gaseous atom $M_{(gas)}$ and the gaseous molecule $\frac{1}{2}X_2$ into a gaseous atom $X_{(gas)}$:

$$M_{(metal)} \xrightarrow{\;E_S\;} M_{(gas)} \tag{1.78}$$

$$\frac{1}{2}X_{2(gas)} \xrightarrow{\;\frac{1}{2}E_D\;} X_{(gas)}. \tag{1.79}$$

The energies required by these processes are composed of the sublimation energy of the metal (E_S) and the dissociation energy of the gaseous molecule ($\frac{1}{2}E_D$).

2. We now transform the gaseous atoms into gaseous ions:

$$M_{(gas)} \xrightarrow{\;I_M\;} M_{(gas)}^+. \tag{1.80}$$

$$X_{(gas)} \xrightarrow{\;-E_X\;} X_{(gas)}^-. \tag{1.81}$$

The required energies are, respectively, those of first ionization (I_M) and electron affinity ($-E_X$).

3. To transform the two gaseous ions into a crystalline molecule, we must expend energy corresponding to the *lattice energy* of the crystalline substance (i.e., the lattice energy with its sign changed):

$$M_{(gas)}^+ + X_{(gas)}^- \xrightarrow{\;-U\;} MX_{(crystal)}. \tag{1.82}$$

4. We then close the cycle by transforming crystal MX into its component elements in the standard state. To do this, we must furnish energy corresponding to the enthalpy of formation from the elements:

$$MX_{(crystal)} \xrightarrow{-H^0_f} M_{(metal)} + \frac{1}{2}X_{2(gas)}. \qquad (1.83)$$

The complete cycle is visualized in the following scheme:

$$
\begin{array}{ccc}
M_{(metal)} + \dfrac{1}{2}X_{2(gas)} & \xrightarrow{E_S + \frac{1}{2}E_D} & M_{(gas)} + X_{(gas)} \\[2mm]
\uparrow {-H^0_f} & & \downarrow I_M - E_X \\[2mm]
MX_{(crystal)} & \xleftarrow{-U} & M^+_{(gas)} + X^-_{(gas)}
\end{array}
\qquad (1.84)
$$

By applying the Hess additivity rule to the involved energy terms, we have:

$$H^0_f = E_S + \frac{1}{2}E_D + I_M - E_X - U. \qquad (1.85)$$

Equation 1.85 can be generalized for polyatomic molecules as follows:

$$H^0_{f,MX_n} = E_S + \frac{n}{2}E_D + \sum_1^n I_M - nE_X - U + (1+n)RT, \qquad (1.86)$$

where $\sum_1^n I_M$ is the summation of the first n ionization potentials of M.

Lastly, note that equation 1.86 includes the expansion work $[(1+n)RT]$ on the gaseous ions from absolute zero to standard state temperature (this energy term should be included in equations 1.80 and 1.81, but the energy variation associated with this term is negligible, being well within the experimental incertitude involved in magnitudes I_M, E_S, E_D, E_X, and U).

The tabulated values of E_{BHF} for olivine end-members (table 1.12) correspond to the summation of the first four terms on the right side of equation 1.86. Table 1.14 lists the various energy terms of the Born-Haber-Fayans thermochemical

Table 1.14 Born-Haber-Fayans energy terms for alkali halides. Values in k/mole (adapted from Grenwood, 1970).

Compound	$-H^0_f$	E_s	$\frac{1}{2}E_D$	$\sum_1^n I_M$	E_X	U
LiCl	408.8	138.9	121.3	520.1	351.5	837.6
NaCl	410.9	100.4	121.3	495.8	360.7	767.8
KCl	436.0	81.2	121.3	418.8	354.8	702.4
RbCl	430.5	77.8	121.3	402.9	354.8	677.8
CsCl	433.0	70.3	121.3	375.7	359.8	640.6

cycle for alkali halides. We emphasize that the electron affinity of chlorine should be constant in all compounds and that observed differences must be imputed to experimental approximation.

1.14   Internal Energy of an Ionic Crystal and Temperature

The internal energy of a crystal at zero temperature is equivalent to its lattice energy. When heat is applied to an ionic crystal, its internal energy is increased by vibrational effects. If ionic bonds are considered in light of the harmonic approximation, this increase in energy can be viewed as a statistical increase in the amplitude of vibrational motion (v_{max} in eq. 1.73). As already noted (see also section 3.1), the ionic bond does not conform perfectly to the harmonic approximation, because anharmonic vibrational terms are present (i.e., the crystal has a definite thermal expansion). Although the treatment of anharmonicity in vibrational studies results in a complex approach (see Kittel, 1971 for a detailed account of this problem), generally speaking the variation of potential energy of a pair of ions displaced from their equilibrium position by thermal agitation can be represented by a polynomial expansion in x of the following type (Deganello, 1978):

$$U_{(x)} = cx^2 - gx^3 - fx^4,\tag{1.87}$$

where c, g, and f are positive. The term $-gx^3$ reflects the asymmetricity of repulsive effects, and $-fx^4$ accounts for the attenuation of vibrational motions at high amplitude. With increasing temperature, gx^3 and fx^4 become more and more significant. Thus, generally, thermal expansion is not a linear function in T.

1.14.1   Internal Energy and Thermal Expansion

The relationship between the increase in internal energy, due to the first term on the right in equation 1.87, and thermal expansion is easily understood if we consider the form of the potential well for a crystal (cf. figure 1.12). If we heat the crystal and inhibit any thermal expansion, we observe a sudden increase in energy, caused mainly by repulsive effects between neighboring ions: the increase in energy is identical to that observed on the compressional side of the potential well in figure 1.12 (harmonic condition). Indeed, Debye formulated a principle stating that linear thermal expansion is simply related to heat capacity at constant volume (C_V). By means of the Born repulsive model (eq. 1.53), it can be shown that, for a cubic crystal with a single vibrational frequency, the Debye principle can be translated as follows:

$$\alpha_l = \frac{C_V}{2U}\left(\frac{n+2}{n}\right),\tag{1.88}$$

where α_l is the linear thermal expansion coefficient and n is the Born exponent (see Das et al., 1963 for an explanatory application of eq. 1.88).

1.14.2 Linear and Volumetric Thermal Expansion of Crystalline Solids

The thermal expansion of a crystal with increasing temperature is generally described by a second-order polynomial in T of type

$$V_T = V_0 + aT + bT^2,\tag{1.89}$$

where V_0 is the volume at reference condition and a and b are experimentally determined factors, varying from substance to substance. Equation 1.89 is purely functional and cannot be used in thermochemical calculations that require a generalized equation. This is provided by linear isobaric thermal expansion coefficient α_l and volumetric isobaric thermal expansion coefficient α_V, defined as follows:

$$\alpha_l = \frac{1}{d}\left(\frac{\partial d}{\partial T}\right)_{P,X}\tag{1.90}$$

$$\alpha_V = \frac{1}{V}\left(\frac{\partial V}{\partial T}\right)_{P,X}.\tag{1.91}$$

We have already seen that α_l can be related to U by means of the Debye principle. The usefulness of α_V will be evident later, when we treat the thermodynamic properties of crystalline substances at high T.

Another coefficient generally used in thermal expansion calculations is the *mean polyhedral linear expansion coefficient* between temperatures T_1 and T_2 $(\overline{\alpha}_{T_1-T_2})$:

$$\overline{\alpha}_{T_1-T_2} = \frac{2}{d_1+d_2}\left(\frac{d_2-d_1}{T_2-T_1}\right).\tag{1.92}$$

Examining the mean polyhedral linear thermal expansion coefficients of a large number of crystalline compounds, Hazen and Prewitt (1977) formulated two important generalizations:

Table 1.15 Relationship among mean polyhedral linear thermal expansion ($\bar{\alpha}_{20-1000}$), Pauling's bond strength, and ionicity factor (adapted from Hazen and Finger, 1982).

Cation	Anion	ν_c	$\dfrac{S_i^2 Z_+ Z_-}{\nu_c}$	$\bar{\alpha}_{20-1000}$	Cation	Anion	ν_c	$\dfrac{S_i^2 Z_+ Z_-}{\nu_c}$	$\bar{\alpha}_{20-1000}$
Re^{6+}	O^{2-}	6	1	1	Li^+	O^{2-}	4	1/4	22
Bi^{5+}	O^{2-}	6	5/6	0	Li^+	O^{2-}	6	1/6	20
Si^{4+}	O^{2-}	4	1	0	K^+	O^{2-}	6	1/6	21
Ti^{4+}	O^{2-}	6	2/3	8	Na^+	O^{2-}	6	1/6	17
Zr^{4+}	O^{2-}	7	4/7	8	Na^+	O^{2-}	7	1/7	35
Hf^{4+}	O^{2-}	7	4/7	8	Na^+	O^{2-}	8	1/8	22
U^{4+}	O^{2-}	8	1/2	10	Na^+	O^{2-}	9	1/9	18
Th^{4+}	O^{2-}	8	1/2	9	Li^+	F^-	6	0.125	46
Ce^{4+}	O^{2-}	8	1/2	13	Na^+	Cl^-	6	0.125	51
Al^{3+}	O^{2-}	4	3/4	1	K^+	Cl^-	6	0.125	46
Al^{3+}	O^{2-}	5	3/5	6	K^+	Br^-	6	0.125	49
Al^{3+}	O^{2-}	6	1/2	9	K^+	I^-	6	0.125	47
V^{3+}	O^{2-}	6	1/2	13	Na^+	F^-	6	0.125	45
Fe^{3+}	O^{2-}	6	1/2	9	Rb^+	Br^-	6	0.125	44
Cr^{3+}	O^{2-}	6	1/2	8	Ca^{2+}	F^-	8	0.19	21
Ti^{3+}	O^{2-}	6	1/2	8	Cs^+	Br^-	8	0.094	68
Bi^{3+}	O^{2-}	6	1/2	9	Mn^{2+}	S^{2-}	6	0.27	18
Be^{2+}	O^{2-}	4	1/2	9	Pb^{2+}	S^{2-}	6	0.27	22
Zn^{2+}	O^{2-}	4	1/2	7	Pb^{2+}	Te^{2-}	6	0.27	20
Ni^{2+}	O^{2-}	6	1/3	14	Pb^{2+}	Se^{2-}	6	0.27	21
Co^{2+}	O^{2-}	6	1/3	14	Zn^{2+}	S^{2-}	4	0.40	9
Fe^{2+}	O^{2-}	6	1/3	13	Zn^{2+}	Te^{2-}	4	0.40	10
Mn^{2+}	O^{2-}	6	1/3	15	Zn^{2+}	Se^{2-}	4	0.40	9
Cd^{2+}	O^{2-}	6	1/3	13	Al^{3+}	As^{3-}	4	0.56	5
Mg^{2+}	O^{2-}	6	1/3	14	Ga^{3+}	As^{3-}	4	0.56	7
Mg^{2+}	O^{2-}	8	1/4	13	B^{3+}	N^{4-}	4	0.45	13
Ca^{2+}	O^{2-}	6	1/3	15	Ti^{4+}	N^{4-}	6	0.53	9
Ca^{2+}	O^{2-}	7	2/7	16	U^{4+}	N^{4-}	6	0.53	9
Ca^{2+}	O^{2-}	8	1/4	14	Nb^{4+}	C^{4-}	6	0.53	7
Sr^{2+}	O^{2-}	6	1/3	14	Ta^{4+}	C^{4-}	6	0.53	7
Ba^{2+}	O^{2-}	6	1/3	15	Ti^{4+}	C^{4-}	6	0.53	8
Ba^{2+}	O^{2-}	9	2/9	15	Zr^{4+}	C^{4-}	6	0.53	7
Pb^{2+}	O^{2-}	12	1/6	23	C^{4+}	C^{4-}	4	0.80	3.5

1. For each polyhedron, the mean polyhedral linear thermal expansion is independent of the type of bonding of the polyhedron with the other structural units of the crystal.
2. All oxygen-based polyhedra with the same Pauling bond strength (cationic charge/coordination number; see section 1.10.3) have the same mean polyhedral linear expansion.

These generalizations led the authors to propose the following empirical relation, valid for oxygen-based polyhedra:

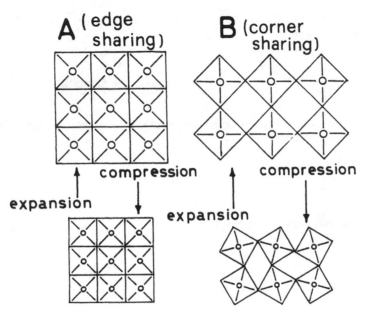

Figure 1.13 Expansion and/or compression effects on crystal structures (planar section). A = shared edge; B = shared corner. From R. M. Hazen and L. W. Finger, *Comparative Crystal Chemistry,* copyright © 1982 by John Wiley and Sons. Reprinted by permission of John Wiley & Sons.

$$\bar{\alpha}_{20-1000} = 32.9 \left(0.75 - \frac{Z_+}{v_c} \right) \times 10^{-6} \, °\text{C}^{-1} . \tag{1.93}$$

Equation 1.93 was later superseded by a more general equation proposed by Hazen and Finger (1979):

$$\bar{\alpha}_{20-1000} = \left(\frac{4v_C}{S_i^2 Z_+ Z_-} \right) \times 10^{-6} \, °\text{C}^{-1} . \tag{1.94}$$

The term S_i^2 is defined by the authors as an "ionicity factor" and assumes a value of 0.5 for oxides and silicates, 0.75 for halides, 0.40 for calcogenides, 0.25 for phosphides and arsenides, and 0.2 for nitrides and carbides (Z_- is the anion charge). Equation 1.94 is based on the thermal expansion data listed in table 1.15.

It must be emphasized that the bulk thermal expansion of a crystal is not simply related to the mean expansion of the various polyhedra within the structure, because progressive readjustments of polyhedral orientations are often dictated by geometric constraints. As an example, figure 1.13 shows the different behaviors of crystals with polyhedra sharing an edge (case A) or a corner (case B). In case B, the thermal expansion is accompanied by a substantial modification of the angular value of the shared corner. The same type of reasoning is valid for compressional effects.

1.15 Internal Energy and Pressure

The equation of state of a cubic crystal under hydrostatic pressure is given by

$$P = -\left(\frac{\partial A}{\partial V}\right)_{T,X}, \tag{1.95}$$

where A is the Helmholtz free energy, composed of internal energy U plus a vibrational term (A_{vib}):

$$A_{(V,T,X)} = U_{(V,X)} + A_{\text{vib}(V,T,X)}. \tag{1.96}$$

Adopting the harmonic approximation, the vibrational term is not explicitly dependent on volume, and equation 1.95 may be reexpressed as follows:

$$P = -\left(\frac{\partial U}{\partial P}\right)_{T,X} \tag{1.97}$$

We now introduce two new parameters that describe the changes in interionic distances and volume with pressure: *isothermal linear compression coefficient* β_l, and *isothermal volumetric compression coefficient* β_V:

$$\beta_l = -\frac{1}{d}\left(\frac{\partial d}{\partial P}\right)_{T,X} \tag{1.98}$$

$$\beta_V = -\frac{1}{V}\left(\frac{\partial V}{\partial P}\right)_{T,X} \tag{1.99}$$

The analogy with thermal expansion coefficients is evident when one compares equations 1.98 and 1.99 with equations 1.90 and 1.91, respectively. For compressibility, as for thermal expansion, a mean coefficient that defines the volumetric variation ($V_2 - V_1$) for a finite pressure range ($T_2 - T_1$) may be introduced. This coefficient, the "mean volumetric isothermal compressibility," is given by

$$\bar{\beta}_V = -\frac{2}{V_1 + V_2}\left(\frac{V_2 - V_1}{P_2 - P_1}\right). \tag{1.100}$$

The reciprocal of β_V, expressed in units of pressure, is called "bulk modulus" (K):

$$K = \frac{1}{\beta_V}.$$ (1.101)

Combining equations 1.97 and 1.99, we obtain

$$\beta_V = -\frac{1}{V}\left(\frac{\partial^2 U}{\partial V^2}\right)_{T,X}^{-1}.$$ (1.102)

As already noted, equation 1.102 is valid when vibrational energy does not depend explicitly on volume. In a more refined treatment of the problem, the vibrational energy of the crystal can be treated as purely dependent on T. This assumption brings us to the *Hildebrand equation of state* and to its derivative on V:

$$\frac{\partial U}{\partial V} = -P + T\frac{\alpha_V}{\beta_V}$$ (1.103)

$$V\left(\frac{\partial^2 U}{\partial V^2}\right) = \frac{1}{\beta_V} + \frac{T}{\beta_V^2}\left[\left(\frac{\partial \beta_V}{\partial T^2}\right)_{P,X} + \frac{\alpha_V}{\beta_V}\left(\frac{\partial \beta_V}{\partial P}\right)_{T,X}\right].$$ (1.104)

Equation 1.104 is obtained by application of thermodynamic relations (see chapter 2):

$$\left(\frac{\partial S}{\partial V}\right)_T = \frac{\alpha_V}{\beta_V}$$ (1.105)

$$V\left(\frac{\partial \alpha_V}{\partial V}\right)_T = \frac{1}{\beta_V}\left(\frac{\partial \beta_V}{\partial P}\right)_T.$$ (1.106)

An even more precise treatment, based on the assumption that the vibrational Helmholtz free energy of the crystal, divided by temperature, is a simple function of the ratio between T and a characteristic temperature dependent on the volume of the crystal, leads to the *Mie-Gruneisen equation of state* (see Tosi, 1964 for exhaustive treatment):

$$\frac{\partial U}{\partial V} = -P + \frac{A_{\text{vib}}\alpha_V}{C_V\beta_V}.$$ (1.107)

The ratio $\alpha_V/C_V\beta_V$ multiplied by the reciprocal of the density of the substance ($1/\rho$) is called the *Gruneisen thermodynamic parameter* (γ_G):

$$\gamma_G = \frac{\alpha_V}{C_V \beta_V} \frac{1}{\rho}.$$ (1.108)

In a more or less complex fashion, equations 1.97, 1.103, and 1.107 describe the response of the crystal, in terms of energy, to modification of the intensive variable P. We note that, in all cases, the differential forms combine internal energy and volume and are thus analytical representations of the potential diagram seen above for Ni_2SiO_4 in figure 1.12.

For simple compounds such as NaCl, where a single interionic distance (r) is sufficient to describe the structure, if we adopt the simplified form of equation 1.97 and combine the partial derivatives of U and V on r, we obtain

$$\beta_V = \frac{\left(\dfrac{\partial V}{\partial r}\right)^3}{V\left(\dfrac{\partial V}{\partial r}\dfrac{\partial^2 U}{\partial r^2} - \dfrac{\partial U}{\partial r}\dfrac{\partial^2 V}{\partial r^2}\right)}.$$ (1.109)

The error introduced by neglecting the term $T\,\alpha_V/\beta_V$ (cf. eq. 1.97 and 1.103) is, in the case of alkali halides, about 0.5 to 1.5% of the value assigned to U (Ladd, 1979).

For cubic structures with more than one interionic distance—as, for instance, spinels (multiple oxides of type AB_2O_4 with A and B cations in tetrahedral and octahedral coordination with oxygen, respectively)—it is still possible to use equation 1.109, but the partial derivatives must be operated on the cell edge, which is, in turn, a function of the various interionic distances (Ottonello, 1986).

1.15.1 Pressure-Volume Empirical Relationships for Crystalline Solids

As early as 1923, Bridgman had observed that the compressibility of metals is proportional to their molar volume raised to $\frac{4}{3}$. Anderson and coworkers (cf. Anderson, 1972, and references therein) extended Bridgman's empirical relationship to various isostructural classes of compounds, noting that, within a single isostructural class, the following relation is valid:

$$\overline{\beta}_V V = \text{constant}.$$ (1.110)

Hazen and Finger (1979) extended equation 1.110 to mean polyhedral compressibility $\overline{\beta}_{V_P}$ (mean compressibility of a given coordination polyhedron within a crystal structure), suggesting that it is related to the charge of ions in the polyhedron through an ionicity factor, analogous to what we have already seen for thermal expansion—i.e.,

Table 1.16 Polyhedral compressibility moduli in selected natural and synthetic compounds (adapted from Hazen and Finger, 1982). Data are expressed in megabars (1Mbar = 10^6 bar). Str = type of structure. $K = 1/\overline{\beta}_{V_p}$.

Polyhedron	Str	K	Polyhedron	Str	K	Polyhedron	Str	K
Ni-O	1	1.96	Ca-O	10	0.85	Mn-F	5	0.9
Co-O	1	1.85	Na-O	12	0.32	Ti-C	1	1.9
Fe-O	1	1.53	Zr-O	13	2.8	U-C	1	1.6
Mn-O	1	1.43	Si-O	13	>2.5	Zr-C	1	1.9
Ca-O	1	1.10	Li-F	1	0.66	C-C	16	5.8
Sr-O	1	0.91	Na-F	1	0.45	C-C	17	5.9
Ba-O	1	0.69	K-F	1	0.293	Ca-S	1	0.43
Be-O	2	2.5	Rb-F	1	0.273	Sr-S	1	0.40
Zn-O	2	1.4	Li-Cl	1	0.315	Ba-S	1	0.35
U-O	3	2.3	Na-Cl	1	0.240	Pb-S	1	0.48
Th-O	3	1.93	K-Cl	1	0.180	Cd-S	2	0.61
Al-O	4	2.4	Rb-Cl	1	0.160	Zn-S	2	0.77
Fe-O	4	2.3	Li-Br	1	0.257	Zn-S	15	0.76
Cr-O	4	2.3	Na-Br	1	0.200	Sn-S	18	1.2
V-O	4	1.8	K-Br	1	0.152	Zn-Se	15	0.60
Si-O	5	3.2	Rb-Br	1	0.138	Ca-Se	1	0.49
Ge-O	5	2.7	Li-I	1	0.188	Sr-Se	1	0.43
Ti-O	5	2.2	Na-I	1	0.161	Ba-Se	1	0.36
Ru-O	5	2.7	K-I	1	0.124	Pb-Se	1	0.34
Mn-O	5	2.8	Rb-I	1	0.111	Cd-Se	2	0.54
Sn-O	5	2.3	Cs-Cl	14	0.182	Ca-Te	1	0.42
Si-O	6	>5	Cs-Br	14	0.155	Sr-Te	1	0.334
Ge-O	6	>4	Cs-I	14	0.129	Ba-Te	1	0.305
Mg-O	7	1.3	Th-Cl	14	0.236	Pb-Te	1	0.41
Al-O	7	2.2	Th-Br	14	0.225	Sn-Te	1	0.42
Si-O	7	3	Cu-Cl	15	0.40	Cd-Te	15	0.42
Ca-O	7	1.15	Ag-I	15	0.243	Hg-Te	15	0.44
Al-O	7	2.2	Ca-F	3	0.86	Zn-Te	15	0.51
Si-O	7	3	Ba-F	3	0.57	Ga-Sb	1	0.56
Mg-O	8	1.5	Pb-F	3	0.61	In-Sb	1	0.47
Ni-O	9	1.5	Sr-F	3	0.70	Ga-As	1	0.75
Si-O	9	>2.5	Mg-F	5	1.0	In-As	1	0.58

1 = NaCl   2 = Zincite   3 = Fluorite   4 = Corundum   5 = Rutile   6 = Quartz   7 = Garnet
8 = Olivine   9 = Spinel   10 = Clinopyroxene   11 = Mica   12 = Feldspar   13 = Zircon   14 = CsCl
15 = Cubic ZnS   16 = Diamond   17 = Graphite   18 = CdI

$$\overline{\beta}_{V_P} = \frac{0.044 d^3}{S_i^2 Z_+ Z_-} \; (\text{Mbar}^{-1}).$$

(1.111)

S_i^2 in equation 1.111 has the same significance and values already seen in equation 1.94, and d is the bond distance.

Experimental values of mean polyhedral compressibility modulus in various compounds are listed in table 1.16.

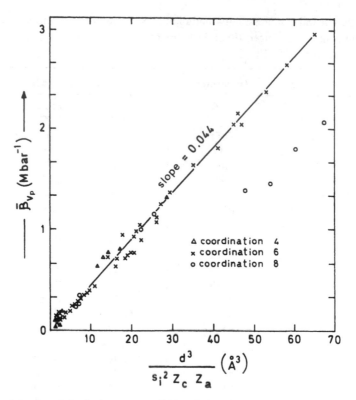

Figure 1.14 Mean polyhedral compressibility for different coordination states. Interpolant is equation 1.111. From R. M. Hazen and L. W. Finger, *Comparative Crystal Chemistry*, copyright © 1982 by John Wiley and Sons. Reprinted by permission of John Wiley & Sons.

The reproducibility of equation 1.111 may be deduced from figure 1.14, based on the tabulated values. Consistent deviations are noted only for some VIII-fold coordinated complexes.

1.15.2 Effect of Pressure on Compressibility

The compressibility of crystals varies with pressure. The simplest functional form that can be adopted to describe the P-V relations for a solid is a polynomial expansion in P, as we have already seen for temperature (cf. eq. 1.89):

$$V_P = V^0 + aP + bP^2. \tag{1.112}$$

An expansion of the series to the third term is not sufficiently accurate to reproduce the actual behavior of solids and it is preferable to use more physically sound functional forms, such as the first-order equation of state of Murnaghan:

$$P = \frac{K_0}{K'}\left[\left(\frac{V_0}{V}\right)^{K'} - 1\right] = P_0 \tag{1.113}$$

or the first-order Birch form:

$$P = \frac{3}{2} K_0 \left[\left(\frac{V_0}{V}\right)^{7/3} - \left(\frac{V_0}{V}\right)^{5/3}\right]\left\{1 - \frac{3}{4}(4 - K')\left[\left(\frac{V_0}{V}\right)^{2/3} - 1\right]\right\} + P_0, \tag{1.114}$$

where K_0 is the bulk modulus at zero pressure, V_0 is the molar volume at zero pressure, and K' is the first derivative of the bulk modulus in dP.

Equations 1.113 and 1.114 are the most frequently used in geochemical studies. However, the most physically sound equation of state for solids is perhaps the Born-Mayer form:

$$P = 3K_0\left\{\left(\frac{V_0}{V}\right)^{2/3}\exp\left\{\left(-\frac{r_0}{\rho}\right)\left[\left(\frac{V_0}{V}\right)^{-1/3} - 1\right]\right\} - \left(\frac{V_0}{V}\right)^{4/3}\right\}$$

$$\times\left[\left(\frac{r_0}{\rho}\right) - 2\right]^{-2} + P_0. \tag{1.115}$$

Although not normally adopted in geochemistry, this form best takes into account the functional form of short-range repulsive forces (cf. eq. 1.54).

Second-order equations of state (i.e., equations involving the second derivative of the bulk modulus in P) have also been proposed (see for this purpose the synoptic table in Kim et al., 1976). However these high-order equations are of limited application in geochemistry because of the still large incertitude involved in high-P compressional studies.

1.16 Crystal Field Theory

For better comprehension of the crystal field theory, we must reconsider to some extent the wavelike behavior of electrons.

1.16.1 Recalls on Atomic Orbitals

The wave function of the hydrogen atom in the ground state ($n = 1$) is given by

$$\psi_1 = \left(\frac{1}{\pi a_0^3} \right)^{\frac{1}{2}} \exp\left(-\frac{r}{a_0} \right), \tag{1.116}$$

where

$$a_0 = \frac{h^2}{4\pi^2 me^2}. \tag{1.117}$$

In equation 1.116, a_0 is Bohr's radius for the hydrogen ion in the ground state ($a_0 = 0.529$ Å). The probability distribution relative to equation 1.116 is

$$\rho_1 = \psi_1^2 = \frac{1}{\pi a_0^3} \exp\left(\frac{-2r}{a_0} \right) \tag{1.118}$$

and is spherically symmetrical with respect to the center of coordinates, located in the nucleus. The probability functions for the excited states of the hydrogen atom have a more complex spatial dependency.

We have already seen (section 1.1.2) that the wave equation (eq. 1.2, 1.4) is function of both time and spatial coordinates. However, the wave function may be rewritten in the form

$$\psi = \psi_0 \exp(2\pi i \, vt), \tag{1.119}$$

where i is the imaginary number ($i = \sqrt{-1}$). In equation 1.119, ψ_0 is a function only of spatial coordinates, whereas the exponential part represents the periodic variation with time.

The so-called "conjugate of ψ," written $\bar{\psi}$, is

$$\bar{\psi} = \psi_0 \exp(-2\pi i \, vt). \tag{1.120}$$

Then

$$\bar{\psi}\psi = \psi_0^2. \tag{1.121}$$

The product in equation 1.121, which is independent of time, represents the density of distribution—i.e., "the probability of finding a given particle in the region specified by the spatial coordinates of ψ." The application of wave equation 1.10 to hypothetical atoms consisting of a nucleus of charge Z and a single electron progressively occupying the various orbitals (*hydrogenoid functions*) requires transformation to polar coordinates (system sketched in figure 1.15). The corresponding transformation of equation 1.10 is

$$\frac{1}{r^2}\frac{\partial}{\partial r}\left(r^2\frac{\partial\psi}{\partial r}\right)+\frac{1}{r^2\sin\theta}\frac{\partial}{\partial\theta}\left(\sin\theta\frac{\partial\psi}{\partial\theta}\right)$$
$$+\frac{1}{r^2\sin^2\theta}\frac{\partial^2\psi}{\partial\phi^2}+\frac{8\pi^2m(E-\Phi)}{h^2}\psi=0.\tag{1.122}$$

As already shown (Section 1.2.1), the energy eigenvalues that satisfy wave equation 1.10 are given by equation 1.22:

$$E_{(n)}=-\frac{2\pi^2me^4Z^2}{n^2h^2}.\tag{1.123}$$

In equation 1.123, n must be a positive integer.
    Posing

$$n=n'+l+1\tag{1.124}$$

(Hinshelwood, 1951), it is evident that l cannot be greater than $n-1$. It is also clear that $E_{(n)}$ does not depend on the total value of n but on the separate values of n' and l. The correspondence between n and the principal quantum number is thus obvious: the energy levels follow the prediction of Bohr's model, assuming distinct values according to quantum numbers n, l, and m_l, which lead to configurations s, p, and d. Equation 1.122 is then satisfied by a product of separate functions, one dependent on l and a function of angular coordinates θ and ϕ, and the other dependent on n, l, and a function of radial distance r:

$$\psi=\psi_{l(\theta,\phi)}\,\psi_{n,l_{(r)}}.\tag{1.125}$$

In quantum states with $n>1$, quantum number l assumes different values; for instance, for $n=3$, $l=0,1,2$. When l is equal to or greater than 1, several independent wave functions exist $(2l+1)$. An electron of the second level, sublevel p, can occupy three $2p$ orbitals of the same energy, described by three distinct wave functions. These orbitals, which, in the absence of external perturbation, rigorously have the same energy, are called "atomic degenerate orbitals" (ADs).
    The analytical expressions of the various wave functions, for the ground state $(n=1)$ and for orbitals $2p$ and $3d$, are listed in figure 1.15. The same figure also shows the "form" (actually, the contour surface) attained by the five ADs of sublevel $3d$ and, for comparative purposes, the three ADs of sublevel $2p$ and the AD $1s$ for the ground state $(n=1)$. The orientation of the atomic orbitals depends on angular factor ψ_l and not on principal quantum number n. Orbitals of the same l thus have the same orientation, regardless of the value of n.

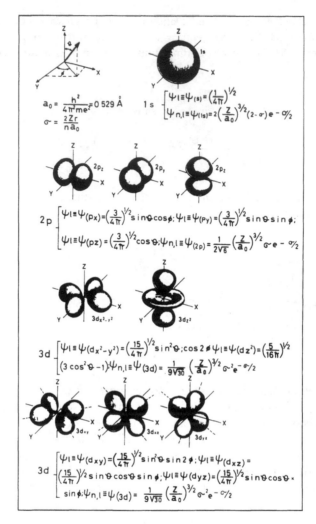

Figure 1.15 $1s$, $2p$, and $3d$ degenerate atomic orbitals (ADs) and corresponding wave functions (cf. eq. 1.125). Upper left: spherical coordinate system. Adapted from Harvey and Porter (1976).

1.16.2 Crystal Field Effect

Let us now consider a coordination polyhedron, in which the cation (with its nucleus at the center of the spherical polar coordinates) is surrounded by six anionic ligands, located two-by-two on the polar axes ("octahedral field," cf. figure 1.16). If we move the six ligands from infinite distance toward the nucleus of the central cation, we generate interaction forces of the type described in section 1.11. In particular, short-range repulsion between the electron clouds of the anionic ligands and the electrons of the d orbitals of the central atom gives rise to a generalized increase in orbital energies. However, the repulsion is more marked for orbitals oriented along the Cartesian axes (d_{z^2} and $d_{x^2-y^2}$; cf. figures 1.15 and 1.16) and less marked for orbitals transversally oriented with respect to the

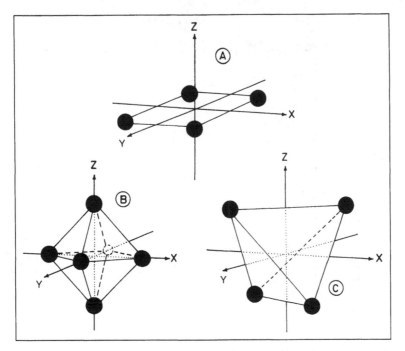

Figure 1.16 Positions of anionic ligands for various configurational states: A = square-planar coordination; B = octahedral coordination; C = tetrahedral coordination. Cation nuclei are located in center of Cartesian coordinates.

Cartesian axes (d_{xy}, d_{xz}, and d_{yz}). The *degeneracy* (equal energy) of the five d orbitals is thus partly eliminated. The energy gap between the first and second groups of orbitals (defined by convention d_{γ}, d_{ε} or e_g, t_{2g}, respectively) is known as "crystal field splitting" and is usually denoted by the symbol Δ (or 10Dq).

According to electrostatic theory, the magnitude of crystal field splitting is given by

$$\Delta = \frac{5\,e\,\mu a^4}{d^6},$$ (1.126)

where e is the electronic charge, μ is the dipole moment of the ligand (cf. section 1.8), a is the mean radial distance of d orbitals, and d is the cation-to-anion distance. With respect to a spherical field, where the degeneracy of d orbitals is maintained, the increase in energy of orbitals e_g is $\frac{3}{5}\Delta$, and the decrease in energy of orbitals t_{2g} is $\frac{3}{5}\Delta$. Orbitals e_g and t_{2g} are therefore called, respectively, "destabilized" and "stabilized."*

* The crystal field effect is due primarily to repulsive effects between electron clouds. As we have already seen, the repulsive energy is of opposite sign with respect to coulombic attraction and the dispersive forces that maintain crystal cohesion. An increase in repulsive energy may thus be interpreted as actual destabilization of the compound.

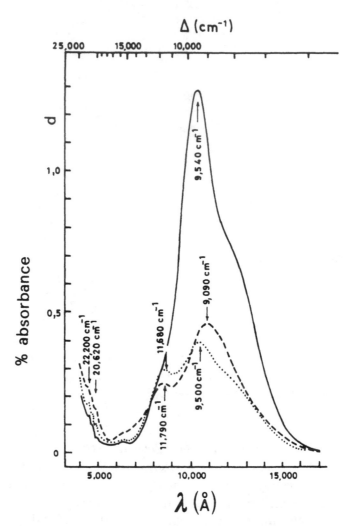

Figure 1.17 Absorption spectrum of a forsteritic olivine under polarized light. Ordinate axis represents optical density (relative absorption intensity, I/I_0). From R. G. Burns (1970), *Mineralogical Applications of Crystal Field Theory*. Reprinted with the permission of Cambridge University Press.

With the cation and the coordination number held constant, the magnitude of Δ depends on the type of ligand and increases according to the spectrochemical series

$$Br^- < Cl^- < F^- < OH^- < O^{2-} < H_2O < EDTA^{4-}$$

$$< NH^{2-} < NO^- < CN^- .$$

(1.127)

A "weak field" is usually defined as one generated by ligands at the beginning of the spectrochemical series; a "strong field" is caused by ligands at the end of the series.

The value of Δ may be deduced by absorption of light measurements on the various complexes. The frequency of the maximum of the first absorption band corresponds to the transference of an electron from a t_{2g} orbital to an e_g orbital and directly gives the value of Δ expressed in cm^{-1} (see eq. 1.128, below). As an example, figure 1.17 shows the absorption spectrum, under polarized light, of an olivine of composition $(Mg_{0.88}Fe_{0.12})_2SiO_4$ (88% forsterite, 12% fayalite). The position of the absorption maximum varies slightly as a function of the orientation of the polarized light with respect to optical axes, as an effect of crystal symmetry. The frequency of the absorption maximum is related to the Δ parameter according to

$$v_a = \frac{\Delta}{h} \tag{1.128}$$

where h is Planck's constant.

1.16.3 Octahedral Site Preference Energy

Figure 1.18 shows energy levels for d orbitals in crystal fields of differing symmetry. The splitting operated by the octahedral field is much higher than that of the tetrahedral field ($\Delta_t = \frac{4}{9}\Delta_O$) and lower than the effect imposed by the square planar field. If we define as "crystal field stabilization energy" the decrease in

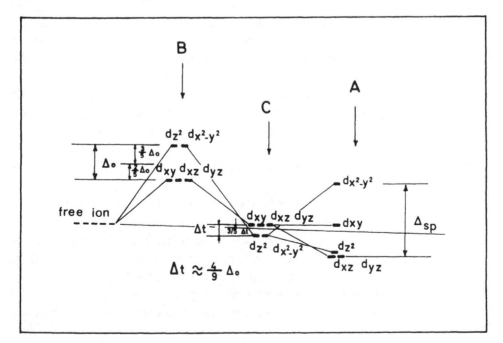

Figure 1.18 Crystal field splitting for d orbitals: A = square-planar field; B = octahedral field; C = tetrahedral field.

Table 1.17 Electronic configuration and crystal field stabilization energies for $3d$ electrons in transition elements. (Ar) = argon core: $1s^2 2s^2 2p^6 3s^2 3p^6$.

Ion	Electronic Configuration	Spin	Octahedral Field	Tetrahedral Field	OSPE
Sc^{3+}; Ti^{4+}	(Ar)$3d^0$		0	0	0
Ti^{3+}	(Ar)$3d^1$	↑	$\frac{2}{5}$	$\frac{12}{45}$	$\frac{6}{45}$
V^{3+}	(Ar)$3d^2$	↑ ↑	$\frac{4}{5}$	$\frac{24}{45}$	$\frac{12}{45}$
Cr^{3+}	(Ar)$3d^3$	↑ ↑ ↑	$\frac{6}{5}$	$\frac{16}{45}$	$\frac{38}{45}$
Mn^{2+}; Fe^{3+}	(Ar)$3d^5$	↑ ↑ ↑ ↑ ↑	0	0	0
Fe^{2+}	(Ar)$3d^6$	↑↓ ↑ ↑ ↑ ↑	$\frac{2}{5}$	$\frac{12}{45}$	$\frac{6}{45}$
Co^{2+}	(Ar)$3d^7$	↑↓ ↑↓ ↑ ↑ ↑	$\frac{4}{5}$	$\frac{24}{45}$	$\frac{12}{45}$
Ni^{2+}	(Ar)$3d^8$	↑↓ ↑↓ ↑↓ ↑ ↑	$\frac{6}{5}$	$\frac{16}{45}$	$\frac{38}{45}$
Cu^{2+}	(Ar)$3d^9$	↑↓ ↑↓ ↑↓ ↑↓ ↑	$\frac{3}{5}$	$\frac{8}{45}$	$\frac{19}{45}$
Zn^{2+}	(Ar)$3d^{10}$	↑↓ ↑↓ ↑↓ ↑↓ ↑↓	0	0	0

energy with respect to a spherical field and compare the values for the tetrahedral and octahedral fields, we see that, for all configurations (besides $3d^0$, $3d^5$, and $3d^{10}$, for which crystal field stabilization is obviously zero), the octahedral field has a higher stabilization effect, called "Octahedral Site Preference Energy" (OSPE) (cf. table 1.17).

In 1957, almost contemporaneously, McClure (1957) and Dunitz and Orgel (1957) proposed precise calculations of crystal field stabilization energies in tetrahedral and octahedral fields, and the corresponding OSPE values. Their data are summarized in table 1.18. The Δ_o values reported by McClure were actually derived for hydrated ions (although the OH^- groups are near O^{2-} in spectrochemical series 1.127), whereas those listed by Dunitz and Orgel were obtained for crystalline oxides and glasses from spectroscopic data on absorption of light. As noted by Dunitz and Orgel, the Δ_o values for O^{2-} and OH^- ligands are identical for bivalent ions but very different for trivalent ions (8 to 12% lower for O^{2-} with respect to OH^- ligands). In both works, the Δ_t value was not experimentally obtained but was assumed to correspond to $\frac{4}{9}\Delta_0$, in agreement with theory. The estimated OSPE values thus have a high level of incertitude. In some instances, for trivalent ions, Dunitz and Orgel imposed the "strong field" condition (cf. section 1.16.2).

Both McClure (1957) and Dunitz and Orgel (1957) derived their OSPE values with the aim of explaining the intracrystalline distribution of transition elements observed in multiple oxides of cubic structure (spinels). According to these authors, a more or less high OSPE value was crucial in intracrystalline partitioning. Their hypothesis was later taken up again by several authors. In particular, Burns (1970) furnished several examples of natural compounds, calculating the crystal field stabilization energy terms from adsorption spectra under polarized light (cf. figure 1.17).

Nevertheless, the application of crystal field theory to natural compounds

Table 1.18 Crystal field stabilization energies for $3d$ transition elements according to McClure (1957) (1) and Dunitz and Orgel (1957) (2).

Ion	Δ_o (cm^{-1})	Δ_t (cm^{-1})	Octahedral Stabilization Energy (kJ/mole)	Tetrahedral Stabilization Energy (kJ/mole)	OSPE (kJ/mole)
Ti^{3+} (1)	20,300	9000	96.7	64.4	32.3
Ti^{3+} (2)	18,300	8130	87.4	58.6	28.8
V^{3+} (1)	18,000	8400	128.4	120.0	8.4
V^{3+} (2)	16,700	7400	160.2*	106.7	53.5*
V^{2+} (1)	11,800	5200	168.2	36.4	131.8
Cr^{3+} (1)	17,600	7800	251.0	55.6	195.4
Cr^{3+} (2)	15,700	6980	224.7	66.9*	157.8*
Cr^{2+} (1)	14,000	6200	100.4	29.3	71.1
Mn^{3+} (1)	21,000	9300	150.2	44.4	105.8
Mn^{3+} (2)	18,900	8400	135.6	40.2	95.4
Mn^{2+} (1)	7,500	3300	0	0	0
Fe^{3+} (1)	14,000	6200	0	0	0
Fe^{2+} (1)	10,000	4400	47.7	31.4	16.3
Fe^{2+} (2)	10,400	4620	49.8	33.1	16.7
Co^{3+} (1)	–	7800	188.3	108.8	79.5
Co^{2+} (1)	10,000	4400	71.5	62.7	8.8
Co^{2+} (2)	9,700	4310	92.9*	61.9	31.0*
Ni^{2+} (1)	8,600	3800	122.6	27.2	95.4
Ni^{2+} (2)	8,500	3780	122.2	36.0*	86.2*
Cu^{2+} (1)	13,000	5800	92.9	27.6	65.3
Cu^{2+} (2)	12,600	5600	90.4	26.8	63.6

*Valid for strong field conditions.

must be carried out with extreme caution. OSPE is only a tiny fraction of the cohesive energy of the crystal. For spinels in particular, it can be shown that the transition from normal structure (type $[A]_T[BB]_OO_4$, where $[A]_T$ is a bivalent ion in a tetrahedral site and $[B]_O$ is a trivalent ion in an octahedral site) to inverse structure ($[B]_T[AB]_OO_4$) corresponds in terms of lattice energy to less than 200 kJ/mole and also depends on other energy effects. As an example, table 1.19 compares lattice energy terms ($-U = E_C + E_R$), the modulus of the bulk energy gap between normal and inverse conditions ($|\Delta U|$), and the modulus of the crystal field stabilization energy gap ($|\Delta CFSE|$) for some spinel compounds. In the light of the contents of table 1.19 and of the above discussion, it is not surprising that CFSE predictions (without detailed consideration of all forms of static energies) are denied in most cases by experimental observations.

The CFSE contribution to lattice energy is almost insignificant for meta- and orthosilicates in which normal and distorted octahedral coordinations are present ($M1$ and $M2$ sites, respectively). As shown in table 1.20, the CFSE gap between normal and distorted octahedral fields is in fact only a few kJ/mole.

Table 1.19 Lattice energy and CFSE for selected spinel compounds. Values of U, E_c, and E_r from Ottonello (1986). CFSE values calculated with values proposed by McClure (1957). Data in kJ/mole; x = degree of inversion.

Compound	x	U	E_C	E_R	$|\Delta U|$	$|\Delta CFSE|$
$CoFe_2O_4$	1	18,712	−21,730	3018	270	8.8
$CoCr_2O_4$	0	18,918	−22,090	3172	300	186.6
$CuFe_2O_4$	1	18,805	−21,697	2892	150	65.3
$FeAl_2O_4$	0	19,029	−22,844	3815	400	16.3
$FeFe_2O_4$	1	18,655	−21,664	3009	220	16.3
$FeCr_2O_4$	0	18,810	−22,029	3219	260	179.1
$NiCr_2O_4$	0	18,948	−22,116	3168	200	100.0
$NiFe_2O_4$	0	18,793	−21,829	3036	220	95.4

Table 1.20 CFSE for $M1$ and $M2$ sites in some silicates. Data from Burns (1970) and Wood (1974), expressed in kJ/mole.

Compound	Ion	Site $M1$	Site $M2$	$M1 - M2$
Fayalite	Fe^{2+}	57.7	53.6	4.1
Forsterite	Fe^{2+}	61.7	54.8	6.9
Bronzite	Fe^{2+}	48.1	49.0	−0.9
Ferrosilite	Fe^{2+}	46.0	47.3	−1.3
Ni-Mg olivine	Ni^{2+}	114.2	107.5	6.7

1.16.4 Field Strength and Spin Condition

Hund's rule establishes that all orbitals of a given sublevel must have been occupied by single electrons with parallel spins before two electrons of opposite spins can occupy a single sublevel. Let us consider, for instance, a metallic ion with d^5 configuration. Based on Hund's rule, the first three electrons occupy the t_{2g} degenerate orbitals with parallel spins. The remaining two electrons then have two possibilities:

1. to occupy the two e_g degenerate orbitals with parallel spins or
2. to fill two of the three t_{2g} degenerate orbitals with opposite spins.

These two configurations are energetically determined by the influence of the crystal field and are denoted, respectively, as conditions of "high" or "low" spin. In the presence of a weak octahedral field (generated by ligands at the beginning of the spectrochemical series), the value of Δ_o is low and the resulting electron configuration is that sketched in figure 1.19A. In the presence of a strong field (ligands at the end of the spectrochemical series), crystal field splitting is high and electrons tend to occupy the t_{2g} stabilized orbitals, opposing the spin and

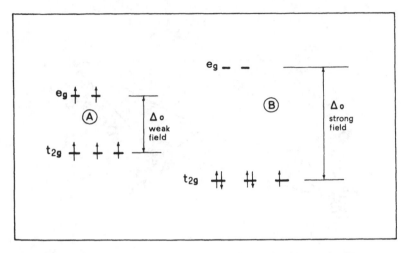

Figure 1.19 Electron configurations under high spin (A) and low spin (B), corresponding, respectively, to weak and strong field conditions.

leaving the destabilized e_g orbitals free (figure 1.19B). The magnetic properties of the compound (which depend on the electron spin condition) are thus influenced by the crystal field. It should be recalled here that crystal field theory was originally developed mainly to account for the magnetic properties of crystalline compounds (Bethe, 1929).

1.17 Ligand Field Theory

1.17.1 Concepts of Molecular Orbitals

Like atomic orbitals (AOs), molecular orbitals (MOs) are conveniently described by quantum mechanics theory. Nevertheless, the approach is more complex, because the interaction involves not simply one proton and one electron, as in the case of AOs, but several protons and electrons. For instance, in the simple case of two hydrogen atoms combined in a diatomic molecule, the bulk coulombic energy generated by the various interactions is given by four attractive effects (proton-electron) and two repulsive effects (proton-proton and electron-electron; cf. figure 1.20):

$$\Phi_C = -e^2 \left(\frac{1}{d_{P_A e_A}} + \frac{1}{d_{P_B e_B}} + \frac{1}{d_{P_A e_B}} + \frac{1}{d_{P_B e_A}} \right) + e^2 \left(\frac{1}{d_{P_A P_B}} + \frac{1}{d_{e_A e_B}} \right). \quad (1.129)$$

To obtain the wave function appropriate to the case examined in figure 1.20, we must introduce Φ_C in the general equation (eq. 1.10). However, the complexity of the problem does not allow an analytical solution, and the wave function is thus derived by application

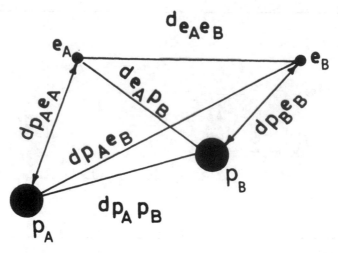

Figure 1.20 Coulombic interactions in a diatomic molecule composed of hydrogen ions; p_A and e_A are the proton and electron of atom A, and p_B and e_B are those of atom B.

of the "variational principle" (Hirschfelder et al., 1954). The general equation is rewritten in the form

$$\left(\Phi - \frac{h^2}{8\pi^2 m} \nabla^2 \right) \psi = E\psi. \tag{1.130}$$

The term in parentheses in equation 1.130 is defined as the "quantum-mechanic hamiltonian" ($\mathcal{H}$; cf. appendix 2):

$$\mathcal{H}\psi = E\psi. \tag{1.131}$$

Multiplying both terms in equation 1.131 by ψ, extracting E, and integrating over the Cartesian coordinates, we obtain the energy of molecular orbitals (MOs):

$$E = \frac{\int \psi \mathcal{H} \psi \, d\tau}{\int \psi^2 \, d\tau}, \tag{1.132}$$

where

$$d\tau = dx \cdot dy \cdot dz. \tag{1.133}$$

Equation 1.132 has the property of never giving an energy that is lower than the true energy (*resonance principle;* cf. Pauling, 1960). This property allows us to assign to the MO wave functions that are obtained by linear combination of the AO functions of the separate atoms (Linear Combination of Atomic Orbitals, or LCAO, method), by progressive adjustment of the combinatory parameters, up to achievement of the lowest energy.

The LCAO approximation may be expressed as:

$$\Psi_{MO} = \psi_A + \lambda\psi_B, \tag{1.134}$$

where ψ_A and ψ_B are wave functions of the AOs of atoms A and B, respectively. The λ constant denotes the bond prevalence and is an expression of the polarity of MO. For instance, with $\lambda = 0$, the MO of the AB molecule coincides with the AO of atom A; the electron of B is completely displaced on A and the bond is purely ionic. With $\lambda = 1$, the MO is a weighted mean of the AOs of A and B, the electrons rotate along a bicentric orbit around the two nuclei, and the bond is perfectly covalent.

If we apply the variational procedure to hydrogen atoms at ground state, we find that the lowest orbitals in order of increasing energy are

$$\Psi_g = \psi_{A(1s)} + \psi_{B(1s)} \qquad (\lambda = 1) \tag{1.135}$$

and

$$\Psi_u = \psi_{A(1s)} - \psi_{B(1s)} \qquad (\lambda = -1) \tag{1.136}$$

(g = gerade = even parity; u = ungerade = odd parity).

Orbital Ψ_g has lower energy with respect to the two constituent AOs and thus forms a stable molecule (cf. section 1.3). This orbital is thus called a "bonding orbital" (or "attractive state"). Orbital Ψ_u has higher energy with respect to the two AOs and is called an "antibonding orbital" (or "repulsive state") (cf. figure 1.21). The above discussion is valid for any kind of molecule—i.e., "the combination of any couple of AOs leads always to the formation of one bonding and one antibonding MO".

The two MOs of a diatomic molecule are classified as σ, π, δ, etc., according to whether the projection of angular momentum along the molecular axis is 0, ±1, ±2, etc. (in units of $h/2\pi$). Moreover, the antibonding condition is indicated by an asterisk; for

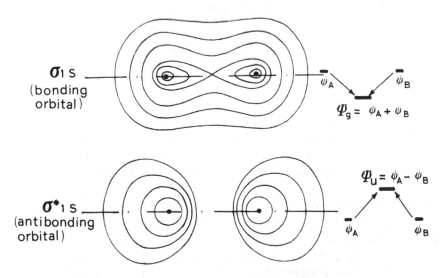

Figure 1.21 Bonding and antibonding conditions for a hydrogen molecule. Adapted from Harvey and Porter (1976).

instance, σ_{1s}^* is the antibonding MO with rotational symmetry of the hydrogen molecule (cf. figure 1.21).

The choice of the two AOs that can efficiently combine to form the MO may be based on three principles:

1. The energies of the two AOs ψ_A and ψ_B must be of similar magnitude.
2. The two AOs must overlap as much as possible.
3. The two AOs must have the right symmetry with respect to the internuclear axis.

Based on the above principles, only some combinations among the AO of the cation and those of the ligands are possible. Plausible combinations are selected as follows:

1. Once the symmetry of the complex is known, the AO of the central cation that best conforms to the studied case is selected—i.e., for an octahedral complex, we select the AO directed along the Cartesian axes $[ns, np_x, np_y, np_z, (n-1)\,d_{z^2}, (n-1)d_{x^2-y^2}]$.
2. For each ligand, the six AOs appropriate to the symmetry of the central cation (denoted by symbols $\phi_1, \phi_2, \phi_3, \phi_4, \phi_5, \phi_6$) are selected.
3. By applying the LCAO method and the principle of lowest energy, the selected AOs are combined to form the Ψ function of the corresponding MO. Figure 1.22A shows an example of combination of AO d_{xy} with $\pi-$bonding orbitals, and figure 1.22B shows a probable combination between $d_{x^2-y^2}$ orbitals of the central cation with $\sigma-$bonding orbitals. The function corresponding to the latter case is

$$\Psi_{x^2-y^2} = \alpha d_{x^2-y^2} + \left(\frac{1-\alpha^2}{4}\right)^{1/2} (\phi_1 + \phi_2 - \phi_3 - \phi_4) \qquad (1.137)$$

and the antibonding condition is expressed by

$$\Psi_{x^2-y^2}^* = (1-\alpha^2)^{1/2} d_{x^2-y^2} - \frac{\alpha}{2}(\phi_1 + \phi_2 - \phi_3 - \phi_4) \qquad (1.138)$$

(a complete tabulation of LCAO-MO wave functions can be found in classical physical chemistry textbooks). In equations 1.137 and 1.138, the α term denotes the degree of overlapping of the AOs of the anionic ligands with those of the central metal cation. This term is analogous to the λ factor in equation 1.134. With $\alpha^2 = 0$, the bond is purely ionic; with $\alpha^2 = \frac{1}{2}$, it is purely covalent.

The energy levels for the σ orbitals of the octahedral complex are sketched in figure 1.23A. For each orbital, both bonding (Ψ) and antibonding (Ψ^*) conditions are represented. Also, the energies of the $\Psi_{z^2}^*$ and $\Psi_{x^2-y^2}^*$ MOs are identical so that, in analogy with AO nomenclature, they are defined as "double degenerate" (the Ψ_{pz}, Ψ_{py}, and Ψ_{pz} MOs and their corresponding antibonding conditions are thus "triple degenerate"). Orbitals d_{xy}, d_{xz}, and d_{yz} do not form any σ bonds and thus maintain their original energies.

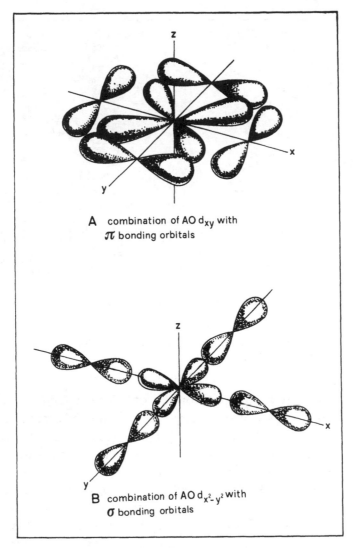

A combination of AO d_{xy} with
π bonding orbitals

B combination of AO $d_{x^2-y^2}$ with
σ bonding orbitals

Figure 1.22 Examples of π-bonding orbitals (A) and σ-bonding orbitals (B) for d electrons.

1.17.2 Ligand Field Effect

Let us now imagine locating the electrons on the various sublevels according to the Aufbau principle, applying Hund's rule to degenerate orbitals. As an octahedral complex has 12 electrons pertaining to the ligands, they occupy two-by-two the Ψ_{z^2}, $\Psi_{x^2-y^2}$, Ψ_s, Ψ_{px}, Ψ_{py}, and Ψ_{pz} bonding MOs; consequently six d electrons (d_{xy}, d_{xz}, and d_{yz}) remain on their original AOs but four are displaced on antibonding degenerate orbitals $\Psi^*_{z^2}$ and $\Psi^*_{x^2-y^2}$. We therefore reach the same condition described by crystal field theory: three (triply) degenerate t_{2g} orbitals (AOs d_{xy}, d_{xz}, and d_{yz}) and two (doubly) degenerate e_g orbitals (antibonding MOs $\Psi^*_{z^2}$ and $\Psi^*_{x^2-y^2}$).

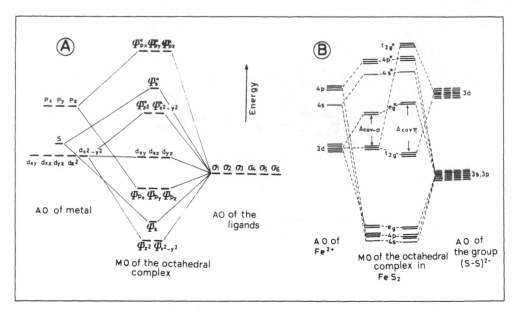

Figure 1.23 General energy level diagram for an octahedral complex (A) and energy levels of pyrite (B). Part (B) from Burns and Vaughan (1970). Reprinted with permission of The Mineralogical Society of America.

The analogy between the two theories is only formal. Crystal field theory is a purely electrostatic approach that does not take into consideration the formation of MOs and the nature of the bond. According to crystal field theory, optical and magnetic properties are ascribed to crystal field splitting between two AOs, whereas in ligand field theory energy splitting occurs between AOs (d_{xy}, d_{xz}, and d_{yz}) and antibonding MOs $\Psi^*_{z^2}$ and $\Psi^*_{x^2-y^2}$. Lastly, note that the σ condition implies combinations between AOs that are oriented in the same direction, and crystal field theory assigns the highest energy to such configurations.

The fact that the ligand field theory can account for the fractional character of the bond through the α parameter (cf. eq. 1.137 and 1.138) leads to important applications for covalent compounds, for which electrostatic theory is not adequate.

The most classical example of application is that of sulfur compounds. Figure 1.23B shows the energy level scheme developed by Burns and Vaughan (1970) for pyrite (FeS_2).

The structure of FeS_2 is similar to that of NaCl, with Fe^{2+} occupying the position of Na^+ and that of Cl^- occupied by the mean point of a (S-S) group. Each iron atom is surrounded by six sulfur atoms, pertaining to six distinct (S-S) groups.

The energy level diagram of figure 1.23B is similar to that of figure 1.23A (although the formal representation of degeneracy is different), but two distinct conditions of ligand field separation, corresponding to σ and π symmetry, are considered in this case. The σ orbitals of the ligands are composed of the six hybrid orbitals $3s$, $3p$ of sulfur (one for each ligand at the corners of the coordination polyhedron). The $3d$ orbitals of Fe^{2+} do not form MOs and remain unchanged on the AO of the metal cation. Burns and Vaughan (1970) define $\Delta_{cov-\sigma}$ as the crystal field separation between the $3d$ AO of Fe^{2+} and the t_{2g} antibonding MOs, implying that the bond is purely covalent (we recall for this purpose that the condition of perfect covalence is reached when $\alpha^2 = \frac{1}{2}$; see, for instance, eq. 1.138). These authors also suggest that, in FeS_2 compounds, bonding and antibonding MOs with π symmetry may be formed between the $3d$ orbitals of the $(S-S)^{2-}$ group and those of Fe^{2+}. In this case, energy splitting should be higher than for the σ symmetry and therefore this should be the type of bonding present in pyrite. When comparing figures 1.23A and B, it must be noted that although the two representations of degenerate orbitals are different (on the same line in part A and piled in part B), they have rigorously the same significance. Besides the quoted work of Burns and Vaughan (1970) on transition element disulfides (CuS_2, NiS_2, CoS_2, FeS_2), ligand field theory has been successfully applied to thiospinels (AB_2S_4 compounds with cubic structures) with a full account of several of their physical properties (hardness, reflectivity, mutual solubility, density; Vaughan et al., 1971).

1.17.3 Jahn-Teller Effect

"All nonlinear nuclear configurations are unstable for an orbitally degenerate electronic state." This principle, formulated on theoretical grounds by Jahn and Teller (1937) by application of group theory to the electronic states of polyatomic molecules, has profound consequences for the significance of local symmetry in crystal lattices. This principle establishes that the perfect symmetric condition of a regular coordination polyhedron is always unstable with respect to a distorted state, whenever the central cation (or the cation-ligand MO) has orbital degeneracy. Jahn and Teller (1937) explicitly stated that the effect of instability applies to degenerate electrons that contribute appreciably to molecular binding. Inner shells are thus excluded from the effect of nuclear displacement.

Too often the Jahn-Teller principle is presented in a reversed fashion—i.e., "the distortion of a coordination polyhedron eliminates orbital degeneracy." This formulation renders the principle ineffective, because the distortion is at first sight attributed to external forces, whereas in fact it is the result of the intrinsic instability of the symmetric condition.

The Jahn-Teller principle finds applications both in the framework of crystal field theory and in the evaluation of energy levels through the LCAO-MO approach. In both cases, practical applications are restricted to $3d$ transition elements.

Octahedrally coordinated transition ions with odd numbers of $3d$ electrons in the destabilized e_g orbitals (i.e., d^4, d^9, and low-spin d^7 configurations) are subject

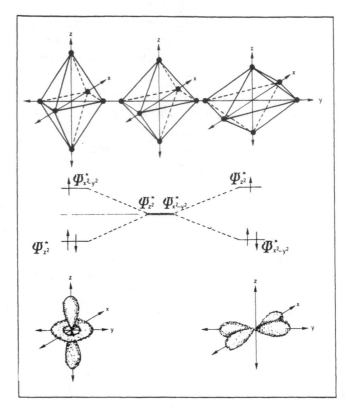

Figure 1.24 Energy splitting of antibonding MOs in tetragonally distorted octahedral sites: (left) elongation along axis z; (right) compression along axis z.

to marked Jahn-Teller distortion. The same is true for d^3, d^4, d^8, and d^9 configurations in tetrahedral coordination. Let us consider, for instance, Cu^{2+} in octahedral coordination with oxygen ligands. Cu^{2+} has a $(t_{2g})^6(e_g)^3$ configuration. If the octahedron is regular, the e_g orbitals are degenerate, but if it is elongated in direction z, the short-range repulsion between the d_{z^2} AO of the central cation and the AO of oxygen is lower in direction z and higher on the x-y plane.

Orbitals $d_{x^2-y^2}$ are thus destabilized with respect to d_{z^2}, and two electrons occupy d_{z^2} in a low-spin configuration. The opposite is observed in the case of compressional distortion along axis z. However, this "crystal field" interpretation is more appropriately translated in terms of ligand field LCAO-MO theory by stating that ligand field splitting occurs between antibonding orbitals Ψ^*. For a d^9 complex with three electrons in the antibonding MOs, the alternative configurations $(\Psi_{z^2}^*)^2 (\Psi_{x^2-y^2}^*)^1$ and $(\Psi_{z^2}^*)^1 (\Psi_{x^2-y^2}^*)^2$ may be attained. The first configuration leads to a higher antibonding character (higher energy) in direction z (left side of figure 1.24); the second leads to higher energy along x and y (right side of figure 1.24). Higher energy is equivalent to a weaker bond and hence to a greater bond distance: the Jahn-Teller effect thus results in distortion of the coordination polyhedron.

In geochemistry, the Jahn-Teller effect is relevant for metals Fe^{2+} and Cu^{2+} in octahedral complexes and for Cr^{3+} and Ni^{2+} in tetrahedral complexes. Other transition ions (e.g., Cr^{2+} and Co^{2+}) require unusual oxidation or low-spin conditions that can be reached only under extreme pressure.

Distortion of coordination polyhedra is commonly observed in silicates. Because distortion of the coordination field implies elimination of degeneracy, it is important to evaluate to what extent the nondegenerate condition contributes to the stability of the phase. Systematic MO calculations in this direction have been carried out by Wood and Strens (1972) on the basis of the AOM (Angular Overlap Model) integrals of Ballhausen (1954).

1.18 The Gordon-Kim Modified Electron Gas (MEG) Model

The ionic models discussed in section 1.12 involve some sort of empiricism in the evaluation of repulsive and dispersive potentials. They thus need accurate parameterization based on experimental values. They are useful in predicting interaction energies *within a family of isostructural compounds,* but cannot safely be adopted for predictive purposes outside the parameterized chemical system or in cases involving structural changes (i.e., phase transition studies).

However, the energies of crystalline substances can be evaluated from first principles with reasonable precision. This goal, which is now in sight as a result of impressive improvements in automated computing capabilities, will lead to significant advances in theoretical computational geochemistry, as excellently outlined by Tossell and Vaughan (1992).

We will limit ourselves here to introducing the simplest of the quantum mechanics procedures: the Modified Electron Gas (MEG) treatment of Gordon and Kim (1971). This procedure is on the borderline between the classical atomistic approach and quantum mechanics *ab initio* calculations that determine energy by applying the variational principle. A short introduction to MEG treatment should thus be of help in filling the conceptual gap between the two theories.

MEG treatment is based on the following main assumptions:

1. Each atom is represented by its electronic charge density. No rearrangement or distortion of separate atomic densities takes place when atoms are brought together.
2. In the overlap region between two atoms (A and B), densities are additive and the total electron density is

$$\rho = \rho_A + \rho_B. \tag{1.139}$$

3. Total *interaction energy* is represented by a coulombic term (V_C) calculated between all charges (both electrons and nuclei), plus an electron gas term (V_{eg}) related to local density in the overlap region:

$$V_{total} = V_C + V_{eg}. \tag{1.140}$$

4. Only the outer regions of the atoms in which atomic densities overlap contribute significantly to the interaction.
5. The electron densities of the separate atoms are described by Hartree-Fock wave functions approximated by analytic extended (Slater-type) basis sets.

The coulombic energy between interacting atoms A and B at distance R is the summation of four terms:

$$\Phi_C = \frac{Z_A Z_B}{R} + \int \int \frac{[\rho_A(r_1) + \rho_B(r_1)][\rho_A(r_2) + \rho_B(r_2)]}{r_{12}} dr_1\, dr_2$$
$$-Z_A \int \frac{[\rho_A(r_1) + \rho_B(r_1)]}{r_{1A}} dr_1 - Z_B \int \frac{\rho_A(r_1) + \rho_B(r_1)}{r_{1B}} dr_1, \tag{1.141}$$

where Z_A and Z_B are nuclear charges of atoms A and B, r_{12} is the distance between two electrons, and r_{1A} and r_{1B} are the distances between electrons and nuclei.

The first term on the right in equation 1.141 is repulsion between the two nuclei, the second term represents electron-electron repulsion, and the third and fourth terms are attraction effects between electrons and nuclei. If we subtract the coulombic energies of separate atoms from equation 1.141:

$$\Phi_{C,A} = \frac{1}{2} \int \left[\frac{\rho_A(r_1)\rho_A(r_2)}{r_{12}} \right] dr_1\, dr_2 - Z_A \int \left[\frac{\rho_A(r_1)}{r_{1A}} \right] dr_1 \tag{1.142}$$

$$\Phi_{C,B} = \frac{1}{2} \int \left[\frac{\rho_B(r_1)\rho_B(r_2)}{r_{12}} \right] dr_1\, dr_2 - Z_B \int \left[\frac{\rho_B(r_1)}{r_{1B}} \right] dr_1, \tag{1.143}$$

we obtain the coulombic interaction V_C:

$$V_C = \frac{Z_A Z_B}{R} + \int \int \frac{\rho_A(r_1)\rho_B(r_2)}{r_{12}} dr_1\, dr_2 - Z_B \int \frac{\rho_A(r_1)}{r_{1B}} dr_1$$
$$-Z_A \int \frac{\rho_B(r_1)}{r_{2A}} dr_2, \tag{1.144}$$

All integrals in equation 1.144 may be evaluated analytically but, in the case of multielectron atoms, the calculations are quite complex and involve many-term summations. Gordon and Kim (1971) thus suggest a solution by two-dimensional (Gauss-Laguerre or Gauss-Legendre) numerical quadrature after reduction of equation 1.144 into a single integrand that, in the case of two interacting univalent ions, takes the form

$$V_C = -R^{-1} + \int \int \rho_A(r_1)\rho_B(r_2)$$

$$\left[\frac{Z_A Z_B + 1}{(Z_A - 1)(Z_B + 1)} R^{-1} + r_{12}^{-1} - \frac{Z_B}{Z_B + 1} r_{1B}^{-1} - \frac{Z_A}{Z_A - 1} r_{2A}^{-1} \right] dr_1 dr_2. \quad (1.145)$$

The form of equation 1.145 is obtained by introducing

$$\int \rho_A(r)\, dr = Z_A - 1 \quad (1.146)$$

and

$$\int \rho_B(r)\, dr = Z_B - 1. \quad (1.147)$$

Similar forms can be obtained for coulombic interaction between neutral atoms or between atoms and ions by appropriate consideration of the integrated electron densities of the two species (in the case of interaction between two neutral atoms, the right sides of equations 1.146 and 1.147 are, respectively, equivalent to the nuclear charges of the two species: Z_A and Z_B).

The energy of the electron gas is composed of two terms, one Hartree-Fock term (Φ_{HF}) and one correlation term (Φ_{corr}). The Hartree-Fock term comprises the zero-point kinetic energy density and the exchange contribution (first and second terms on the right in equation 1.148, respectively):

$$\Phi_{HF}(\rho) = \frac{3}{10}(3\pi^2)^{2/3}\rho^{2/3} - \frac{4}{3}\left(\frac{3}{\pi}\right)^{1/3}\rho^{1/3}. \quad (1.148)$$

The correlation term (expressed in Hartree) is accurately reproduced at high electron density ($r_s \geq 10$) by

$$\Phi_{corr} = 0.0311 \ln(r_s) - 0.048 + 0.009 r_s \ln(r_s) - 0.018 r_s \quad (1.149)$$

and at low electron density ($r_s \leq 0.7$) by

$$\Phi_{corr} = -0.438 r_s^{-1} + 1.325 r_s^{-3/2} - 1.47 r_s^{-2} - 0.4 r_s^{-5/2}, \quad (1.150)$$

where r_s is related to electron density through

$$r_s = \sqrt[3]{\frac{3}{4\pi a_0^3 \rho}},$$

(1.151)

in which a_0 is the Bohr radius.

In the intermediate electron density range, Gordon and Kim (1971) used the interpolant expression:

$$\Phi_{corr} = -0.06156 + 0.01898 \ln(r_s)$$

(1.152)

(alternative equations have been proposed, following the Gordon-Kim introductory paper; see Clugston, 1978 for a discussion).

The electron gas functional is given by the summation of the two terms:

$$\Phi_{eg} = \Phi_{HF} + \Phi_{corr}$$

(1.153)

and the electron gas contribution to the interatomic interaction is

$$V_{eg} = \int dr \left\{ \left[\rho_A(r_A) + \rho_B(r_B) \right] \Phi_{eg}(\rho_A + \rho_B) - \rho_A(r_A) \Phi_{eg}(\rho_A) - \rho_B(r_B) \Phi_{eg}(\rho_B) \right\}$$

(1.154)

Because electron densities are positive-definite, the contribution of the electron gas to interaction energy takes the same sign as the electron gas functional (see Clugston, 1978). Hence, the kinetic energy term is repulsive (cf. eq. 1.148 and 1.154), exchange energy is attractive (cf. eq. 1.148 and 1.154), and the correlation term is also attractive (cf. eq. 1.149, 1.150, and 1.152).

Since its appearance in 1971, the MEG model has been improved to various extents. The most effective modification was perhaps that suggested by Rae (1973), which introduced a correction factor to lower the exchange energy and added dispersive series (these modifications were partly embodied in the MEG calculation routine by Green and Gordon, 1974; see Clugston, 1978 for a careful treatment). The main advantage of the Gordon-Kim MEG treatment with respect to more sophisticated *ab initio* or density functional theories is undoubtedly its computational economy.

The MEG model has been extensively used to determine lattice energies and interionic equilibrium distances in ionic solids (oxides, hydroxides, and fluorides: Mackrodt and Stewart, 1979; Tossell, 1981) and defect formation energies (Mackrodt and Stewart, 1979). Table 1.21 compares the lattice energies and cell edges of various oxides obtained by MEG treatment with experimental values.

Table 1.21 MEG values of lattice energy and lattice parameter for various oxides, compared with experimental values. Source of data: Mackrod and Stewart (1979). U_L is expressed in kJ/mole; a_0 (in Å) corresponds to the cell edge for cubic substances, whereas it is the lattice parameter in the a plane for Al_2O_3, Fe_2O_3, and Ga_2O_3 and it is the lattice parameter parallel to the sixfold axis of the hexagonal unit cell in rutiles $CaTiO_3$ and $BaTiO_3$.

Oxide	U_L			a_0	
	MEG	Experimental		MEG	Experimental
Li_2O	3,049	2,904		2.30	2.31
BeO	4,767	4,526	4,603	2.72	2.70
MgO	3,932	3,937	3,898	2.18	2.11
α-Al_2O_3	15,622	15,477		13.2	13.0
CaO	3,474	3,570	3,483	2.46	2.41
TiO_2	11,811	12,158		4.52	4.59
MnO	3,618	3,815		2.36	2.22
α-Fe_2O_3	14,349	15,082		14.3	13.7
Ga_2O_3	14,445	15,188	15,613	14.3	13.4
SrO	3,271	3,310	3,261	2.62	2.58
CdO	3,348	3,879		2.58	2.35
SnO_2	10,962	11,869	11,367	4.87	4.74
BaO	3,088	3,126	3,078	2.78	2.76
CeO_2	10,392	10,344		2.72	2.71
PbO_2	10,585	11,753	11,570	5.05	4.95
ThO_2	9,794	9,939	10,103	2.87	2.80
UO_2	9,881	10,296		2.84	2.73
$CaTiO_3$	15,352	15,622		3.90	3.84
$BaTiO_3$	14,908	15,265		4.04	4.01
$MgAl_2O_4$	19,646	19,366		4.12	4.04

Although the MEG model is essentially ionic in nature, it may also be used to evaluate interactions in partially covalent compounds by appropriate choices of the wave functions representing interacting species. This has been exemplified by Tossell (1985) in a comparative study in which MEG treatment was coupled with an *ab initio* Self Consistent Field-Molecular Orbital procedure. In this way, Tossell (1985) evaluated the interaction of CO_3^{2-} with Mg^{2+} in magnesite ($MgCO_3$). Representing the CO_3^{2-} ion by a 4–31G wave function, holding its geometry constant with a carbon-to-oxygen distance of 1.27 Å, and calculating its interaction energy at various Mg-C distances, Tossell was able to reproduce the experimental structure with reasonable precision. Calculations also showed that the covalently bonded CO_3^{2-} anion has a smaller repulsive interaction with Mg^{2+} than that of the overlapping free ions $C^{4+}(O^{2-})_3$. This results from the fact that the electron densities along the C-O internuclear axis are higher near the midregion of the C-O bond than the sum of overlapping spherical free-ion den-

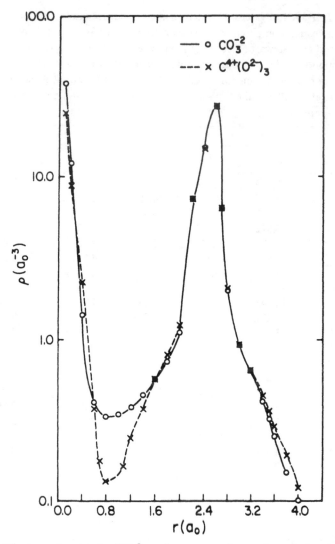

Figure 1.25 Electron density in CO_3^{2-} minus sum of superimposed spherical free-ion densities along C-O internuclear axis. Reprinted from Tossell (1985), with permission of Elsevier Science B.V.

sities, whereas they are lower outside the oxygen nucleus (figure 1.25). Because electron density falls in the peripheral zone of the CO_3^{2-} group, short-range repulsion falls with approaching cations.*

* Incidentally it must be noted that the interaction energy of Mg^{2+} with CO_3^{2-} does not correspond to the lattice energy of the crystalline substance, but is about $^1/_{10}$ of its value. Most of the cohesive energy is due to the covalent bonds within the CO_3^{2-} group.

1.19 Polarization Energy

We have already introduced the concept of ionic polarizability (section 1.8) and discussed to some extent the nature of dispersive potential as a function of the individual ionic polarizability of interacting ions (section 1.11.3). We will now treat another type of polarization effect that is important in evaluation of defect energies (chapter 4).

Let us imagine subtracting a $-e$ charge at a given point in the crystal lattice: at that point, effective charge $+e$ exerts polarization energy of the type

$$E_P = -\frac{1}{2} e\Phi_e, \tag{1.155}$$

where Φ_e is the electric potential induced by the polarized medium at charge $+e$ taken at the origin. If the medium is composed of i ions, each at distance r_i and with polarizability α_i, equation 1.155 will be expressed as

$$E_P = -\frac{1}{2} e \sum_i \frac{\mu_i}{r_i^2} = -\frac{1}{2} e \sum_i \alpha_i \frac{F_i}{r^2}, \tag{1.156}$$

where F_i is the component directed along r_i (toward effective charge $+e$) of total electric field Φ_e on the ith ion and μ_i is the corresponding dipole moment. Applying differential operators (cf. appendix 2), we can write

$$F_i = \left(\frac{e}{r_i^2} + F_j \frac{\gamma_i}{r_i} \right), \tag{1.157}$$

where γ_i is the modulus of dipole ij at distance r_i from charge $+e$ and F_j is the summation of the gradients of the electric potential from the various j ions toward i—i.e.,

$$F_j = \sum_j \nabla_i \Phi_{e,j} \tag{1.158}$$

and

$$\nabla_i \Phi_{e,j} = \left(\frac{\partial \Phi_{e,j}}{\partial x}, \frac{\partial \Phi_{e,j}}{\partial y}, \frac{\partial \Phi_{e,j}}{\partial z} \right). \tag{1.159}$$

The solution of the system shown in equations 1.156 to 1.159 leads to a large number of equations (cf. Rittner et al., 1949; Hutner et al., 1949), and simplifying

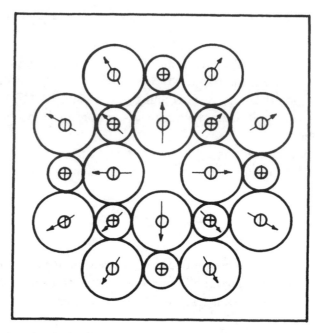

Figure 1.26 Induced polarization effects arising from a cationic vacancy in a crystal (negative net charge). Arrows mark induced dipoles at various lattice positions. From Lasaga (1981c). Reprinted with permission of The Mineralogical Society of America.

assumptions must be introduced to solve the calculations properly. The generally adopted method is that of Mott and Littleton (1938), which solves the polarization energy between two ions (i and j) through the system

$$E_P = -\frac{1}{2} e\Phi_e \tag{1.160}$$

$$\Phi_e = -ea^3 \left(M_i \sum_i \frac{1}{r^4} + M_j \sum_j \frac{1}{r^4} \right) \tag{1.161}$$

$$M_i = \frac{2\alpha_i}{\alpha_i + \alpha_j} \frac{1}{4\pi} \left(1 - \frac{1}{\varepsilon_o} \right) \tag{1.162}$$

$$M_j = \frac{2\alpha_j}{\alpha_i + \alpha_j} \frac{1}{4\pi} \left(1 - \frac{1}{\varepsilon_o} \right), \tag{1.163}$$

where a^3 is the volume occupied by the pair of ions and ε_0 is the dielectric constant of the medium.

Table 1.22 Optical (α^e) and static (α^T) polarizabilities for silicates. Calculated values obtained by additivity principle starting from polarizability of oxide constituents (adapted from Lasaga and Cygan, 1982).

Mineral	Formula	α^e Calculated	α^e Experimental	α^T Calculated	α^T Experimental
Forsterite	Mg_2SiO_4	6.44	6.35	11.48	11.55
Fayalite	Fe_2SiO_4	8.80	8.33	12.62	12.12
Orthoclase	$KAlSi_3O_8$	13.31	13.11	23.83	25.98
Enstatite	$MgSiO_3$	4.65	4.56	8.16	7.96
Diopside	$CaMgSi_2O_6$	10.48	9.85	18.20	18.78
Tephroite	Mn_2SiO_4	8.54	8.26	13.76	14.45
Phenacite	Be_2SiO_4	5.48	5.46	9.14	8.51
Willemite	Zn_2SiO_4	8.62	8.08	13.78	14.35
Monticellite	$CaMgSiO_4$	7.62	7.52	13.36	14.56
Rhodonite	$MnSiO_3$	5.70	5.62	9.30	10.45
Zircon	$ZrSiO_4$	7.50	7.55	12.46	11.67
Edembergite	$CaFeSi_2O_6$	11.66	10.95	18.77	22.89
Giadeite	$NaAlSi_2O_6$	9.61	8.84	17.35	17.96
Leucite	$KAlSi_2O_6$	10.45	10.46	18.99	23.09
Nefeline	$NaAlSiO_4$	6.75	6.71	12.51	13.46
Microcline	$KAlSi_3O_8$	13.31	13.23	23.83	26.61
Albite	$NaAlSi_3O_8$	12.47	12.25	22.19	26.37
Anorthite	$CaAl_2Si_2O_8$	12.88	13.35	22.58	26.49
Almandine	$Fe_3Al_2Si_3O_{12}$	21.68	20.07	33.89	23.93
Spessartine	$Mn_3Al_2Si_3O_{12}$	21.29	20.02	35.60	32.20
Sillimanite	Al_2SiO_5	7.05	7.36	12.54	15.22
Beryl	$Be_3Al_2Si_6O_{18}$	25.28	26.32	43.19	53.02
Sphene	$CaTiSiO_5$	10.96	10.42	17.22	19.18
Wollastonite	$CaSiO_3$	5.83	5.62	10.04	11.35

The Mott-Littleton approach has various degrees of approximation (up to 4). In the zero-order approximation, polarization is regarded as the result of a series of induced dipoles at the various lattice sites (cf. figure 1.26). In I-, II-, III-, and IV-order approximations, Mott and Littleton performed precise calculations, respectively, for the I, II, III, and IV series of cells surrounding the origin, treating the remaining part of the crystal in the same manner as the zero-order approximation (see Rittner et al., 1949a,b for a detailed account of the method). Adopting the zero-order Mott-Littleton approximation, Lasaga (1980) proposed polarization calculations for silicates. For the Mg_2SiO_4 component, the system shown in equations 1.160 to 1.163 is rewritten as follows:

$$E_P = -\frac{1}{2} e\Phi_e \qquad (1.164)$$

$$\Phi_e = -e\frac{V_{cell}}{4}\left(M_{Mg}\sum_{Mg}\frac{1}{r^4} + M_{Si}\sum_{Si}\frac{1}{r^4} + M_O\sum_O\frac{1}{r^4}\right) \qquad (1.165)$$

$$M_i = \frac{\alpha_i^T}{2\alpha_{Mg}^T + \alpha_{Si}^T + 4\alpha_O^T} \frac{1}{4\pi} \left(1 - \frac{1}{\varepsilon_0^T}\right). \tag{1.166}$$

The α^T terms in equation 1.166 represent total ionic polarizability, composed of electronic polarizability α^e plus an additional factor α^d, defined as a "displacement term," due to the fact that the charges are not influenced by an oscillating electric field (as in the case of experimental optical measurements) but are in a "static field" (Lasaga, 1980):

$$\alpha^T = \alpha^e + \alpha^d. \tag{1.167}$$

Based on this concept, the dielectric constant is modified by application of the Clausius-Mosotti relation:

$$\frac{\varepsilon_0^T - 1}{\varepsilon_0^T + 2} = \frac{\pi}{3V_{cell}} (2\alpha_{Mg}^T + \alpha_{Si}^T + 4\alpha_O^T). \tag{1.168}$$

Table 1.22 lists the values of optical polarizability α^e and static polarizability α^T, calculated by Lasaga and Cygan (1982) for various silicates on the basis of the additivity principle

$$\alpha_{Mg_2SiO_4}^T = 2\alpha_{MgO}^T + \alpha_{SiO_2}^T. \tag{1.169}$$

Stemming from a comparison between calculated and experimental methods, the proposed calculations appear quite precise and thus open up new perspectives in polarization calculations for natural crystalline phases. As we will see in chapter 4, polarization energy is of fundamental importance in the evaluation of defect equilibria and consequent properties.

CHAPTER TWO

Concepts of Chemical Thermodynamics

2.1 General Concepts and Definitions

In geology it is customary to consider systems in which the intensive variables pressure (P) and temperature (T) are characteristic of the ambient and, therefore, are prefixed and constant. In these conditions, the Gibbs free energy of the system (G) is at minimum at equilibrium. The treatments presented in this chapter are based on this fundamental principle. Let us first introduce in an elementary fashion some fundamental definitions.

SYSTEM – "Whatever part of the real world which is the subject of a thermodynamic discussion is generally defined *system*" (Lewis and Randall, 1970). As far as we are concerned, a *system* may be a group of atoms, or minerals, or rocks. The limit of the system can be defined in any way one wishes (for instance, an outcrop, a hand specimen, or a thin section). We generally define a system in such a way that all phases within it are in thermodynamic equilibrium. The modifications that take place in a system may or may not imply interactions with matter external to the system itself, so we consider a system that does not exchange matter or energy with the exterior as "isolated," a system that is able to exchange energy but not matter with the exterior as "closed," and a system that exchanges both matter and energy as "open." A system is also called "heterogeneous" if it is composed of several *phases.*

PHASE – A phase is defined as a fraction (region) of a system that has peculiar and distinguishable chemical and physical properties. For instance, a "magmatic" system may be composed of mineral solid phases, fluids (H_2O, CO_2, etc.), and melts.

COMPONENT – Each phase in a system is in turn composed of chemical *components*. The choice of the various components is, under certain provisos, arbitrary. For instance, the olivine solid mixture $(Mg, Fe)_2SiO_4$ may be considered to be composed of fayalite (Fe_2SiO_4) and forsterite (Mg_2SiO_4) end-members, of oxides

(MgO; FeO; SiO_2), or of elements (Mg; Fe; O_2). As we will see in section 2.6, the choice of the various components is no longer arbitrary whenever the "phase rule" is applied.

EQUILIBRIUM – This term is used to indicate a condition in which the various components are subjected to dynamic and reversible exchanges among various phases within the heterogeneous system. *Stable equilibrium,* dictated by the principle of minimization of Gibbs free energy, must be distinguished from *metastable equilibrium* or *apparent equilibrium* conditions, which are often encountered in geological systems. The classic example of metastable equilibrium is the secretion of carbonate shells by marine organisms, composed of the denser high-*P* stable aragonite phase of the $CaCO_3$ component (see table 2.1 and figure 2.6). In the *P-T* conditions of synthesis, the calcite polymorph, which has lower Gibbs free energy, should be formed instead of aragonite. Another example of metastable equilibrium is silica solubility in water: at room *P-T* conditions the stable phase is quartz; however, because this phase does not nucleate easily, metastable equilibrium between amorphous silica and a saturated solution is observed. There is *apparent equilibrium* in a system whenever two or more of its phases do not apparently interact to any measurable extent and it is impossible to determine whether stable equilibrium exists. The *local equilibrium* in a system represents a zone of complete reversibility of exchanges (or "zero affinity"; see section 2.12) restricted to a region sufficiently large for compositional fluctuations to be negligible.

SOLUTION and MIXTURE – There is some confusion between these two terms in geological literature. According to the I.U.P.A.C. (International Union for Pure and Applied Chemistry), the term *mixture* must be adopted whenever "all components are treated in the same manner", whereas *solution* is reserved for cases in which it is necessary to distinguish a "solute" from a "solvent." This distinction in terminology will be more evident after the introduction of the concept of "standard state." It is nevertheless already evident that we cannot treat an aqueous solution of NaCl as a mixture, because the solute (NaCl) in its stable (crystalline) state has a completely different aggregation state from that of the solvent (H_2O) and, because NaCl is a strong electrolyte (see section 8.2), we cannot even imagine pure aqueous NaCl.

2.2 Gibbs Free Energy and Chemical Potential

To understand fully the chemical reaction processes that take place in rock assemblages, it is necessary to introduce the concept of *chemical potential.* Much the same as in a gravitational potential, in which an object tends to fall from a high to a low altitude, in a chemical potential field the reaction or "flow direction of components" always tends to proceed from a high to a low chemical potential region.

The Gibbs free energy of a system (G_{total}) is composed of all the Gibbs free energies of the various phases in the system, defined by the chemical potentials of their components. Algebraically, this is expressed as

$$G_{total} = \sum_i \mu_i n_i ,$$

(2.1)

where n_i is the number of moles of the ith component, μ_i is its chemical potential, and summation is extended to all components.

For a system at constant T and P,

$$\left(\frac{\partial G_{total}}{\partial n_i} \right)_{P,T} = \mu_i .$$

(2.2)

As already stated, chemical reactions proceed spontaneously in the direction of lowering of chemical potentials. Based on equation 2.1, minimization of chemical potentials corresponds to minimization of the Gibbs free energy of the system.

2.3 Gibbs Free Energy of a Phase as a Function of the Chemical Potentials of Its Components

The Gibbs free energy of a phase is a measurable property that depends on pressure, temperature, structure, and composition. Because the chemical potential is the partial derivative of the Gibbs free energy with respect to composition, it graphically represents the tangent to the Gibbs free energy curve at any given position of the compositional field. For instance, figure 2.1 shows the chemical

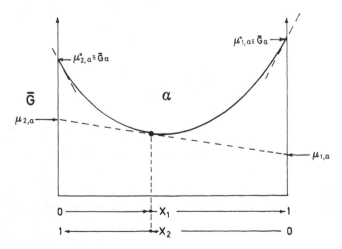

Figure 2.1 Relationships between chemical potentials of two components and Gibbs free energy of a binary phase.

potential of component 1 in a solid mixture α composed of two components (1 and 2) and analytically defined by

$$\mu_{1,\alpha} = \frac{\partial G_\alpha}{\partial n_1}. \tag{2.3}$$

Analogously, for component 2,

$$\mu_{2,\alpha} = \frac{\partial G_\alpha}{\partial n_2}. \tag{2.4}$$

Reasoning in terms of *molar* Gibbs free energy ($\overline{G}$, the energy of one mole of the given phase) and considering a homogeneous system composed of one phase with two components, because

$$X_1 + X_2 = 1, \tag{2.5}$$

where X_1 and X_2 are the mole fractions of the components in the system, we have

$$\frac{d\overline{G}_\alpha}{dX_1} = \frac{\mu_{1,\alpha} - \overline{G}_\alpha}{1 - X_1} \quad \text{and} \quad \frac{d\overline{G}_\alpha}{dX_2} = \frac{\mu_{2,\alpha} - \overline{G}_\alpha}{1 - X_2}, \tag{2.6}$$

from which we obtain

$$\mu_{1,\alpha} = \overline{G}_\alpha + \left(1 - X_1\right)\frac{d\overline{G}_\alpha}{dX_1}. \tag{2.7}$$

and

$$\mu_{2,\alpha} = \overline{G}_\alpha + X_1\frac{d\overline{G}_\alpha}{d\left(1 - X_1\right)}. \tag{2.8}$$

It is also evident from figure 2.1 that the molar Gibbs free energy of a pure phase composed of a single component is equivalent to the chemical potential of the component itself.

Let us now consider two coexisting phases α and β, each having components 1 and 2 and the Gibbs free energy relationships outlined in figure 2.2. The following four identities apply:

$$\mu_{1,\alpha} = \overline{G}_\alpha + \left(1 - X_{1,\alpha}\right)\frac{d\overline{G}_\alpha}{dX_{1,\alpha}} \tag{2.9}$$

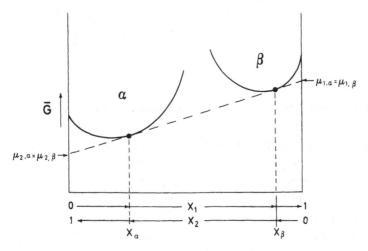

Figure 2.2 Relationships between chemical potential and Gibbs free energy in a two-phase system with partial miscibility.

$$\mu_{2,\alpha} = \overline{G}_\alpha + X_{1,\alpha} \frac{d\overline{G}_\alpha}{d(1 - X_{1,\alpha})} \tag{2.10}$$

$$\mu_{1,\beta} = \overline{G}_\beta + (1 - X_{1,\beta}) \frac{d\overline{G}_\beta}{dX_{1,\beta}} \tag{2.11}$$

$$\mu_{2,\beta} = \overline{G}_\beta + X_{1,\beta} \frac{d\overline{G}_\beta}{d(1 - X_{1,\beta})}. \tag{2.12}$$

It is clear from figure 2.2 that the chemical potentials of the two phases at equilibrium are equal:

$$\mu_{1,\alpha} = \mu_{1,\beta} \tag{2.13}$$

$$\mu_{2,\alpha} = \mu_{2,\beta}. \tag{2.14}$$

We thus write

$$\overline{G}_\alpha + (1 - X_{1,\alpha}) \frac{d\overline{G}_\alpha}{dX_{1,\alpha}} = \overline{G}_\beta + (1 - X_{1,\beta}) \frac{d\overline{G}_\beta}{dX_{1,\beta}} \tag{2.15}$$

and

$$\overline{G}_\alpha + X_{1,\alpha} \frac{d\overline{G}_\alpha}{d(1 - X_{1,\alpha})} = \overline{G}_\beta + X_{1,\beta} \frac{d\overline{G}_\beta}{d(1 - X_{1,\beta})}. \tag{2.16}$$

Equations 2.15 and 2.16 are the general equilibrium relations for two coexisting phases.

We can now examine phase stability in the compositional field: for amounts of component 2 in α higher than X_α the tangent to the Gibbs free energy curve of phase α never intercepts the loop of phase β (in other words, the potentials of the two components in the two phases are never identical); only phase α is stable in such conditions. Analogously for $X_1 > X_\beta$, only phase β is stable, whereas stable coexistence of the two phases is the case for $X_\beta < X_1 < X_\alpha$ and $X_\alpha < X_2 < X_\beta$. Although equations 2.15 and 2.16 imply that only two phases may form in the binary system, this constraint is generally not valid, and the conformation of the Gibbs free energy minimum curve in the G-X space becomes more complex as the number of phases that may nucleate in the system increases. Although there is no analytical solution to this problem, the graphical outline of the equilibrium is quite simple. Let us consider, for instance, the isobaric-isothermal G-X plot depicted in figure 2.3: four phases with partial miscibility of components may form in the system (i.e., α, β, γ, and δ), but the principle of minimization of the Gibbs free energy and the equality condition of potentials must be obeyed over the entire compositional range.

The conformation of the $G = f(X)$ curve that satisfies these conditions can be determined by draping a rope under the $\overline{G}$-X loops of the single phases and pulling it up on the ends, as shown in figure 2.3A. The resulting minimum Gibbs

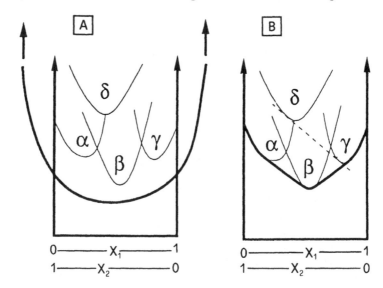

Figure 2.3 Conformation of the minimum Gibbs free energy curve in the binary compositional field (modified from Connolly, 1992).

free energy curve may be conceived as composed of two kinds of regions (figure 2.3B): *linear regions,* in which the curve spans the $\overline{G}$-X loops of the single phases, and *nonlinear regions,* in which the curve overlaps the $\overline{G}$-X loop of a single phase. Figure 2.3 highlights three important facts (Connolly, 1992):

1. Linear regions (constant slope, hence constant potentials) define the composition interval over which a two-phase assemblage is stable.
2. Because the minimum Gibbs free energy curve of the system is never convex, the chemical potential of any component will always increase with the increase of its molar proportion in the system.
3. Because P and T are arbitrarily fixed and the conformations of the $\overline{G}$-X loops of the various phases are mutually independent, no more than two phases may be stable in any part of the compositional space.

These arguments may be extended to an n-compositional space: the straight paths will be replaced by planes in a three-component system, by hyperplanes in a four-component system, and so on. Moreover, in an n-component system at fixed P and T, no more than n phases may be stable at the same P-T-X condition, and their compositions are uniquely determined at any combination of P, T, and X.

2.4 Relationships Between Gibbs Free Energy and Other Thermodynamic Magnitudes

The internal energy of a system is a "state function"—i.e., the variation of the internal energy of the system when passing from state A to state B does not depend on the reaction path but only on initial and final state conditions (cf. appendix 2).

For a crystalline phase, we have already seen that internal energy corresponds to lattice energy at the zero point. At higher T, the internal energy of the phase increases as a result of the increase in vibrational motion of all the atoms in the lattice.

We have also seen (in chapter 1) that enthalpy and lattice energy are related through the Born-Haber-Fayans thermochemical cycle, on the basis of the energy additivity principle of Hess. The enthalpy or "heat content" of a phase (H) is composed of the internal energy U at the T of interest and the PV product:

$$H = U + PV. \tag{2.17}$$

Equation 2.17 is general and can be extended to a heterogeneous system.

The Gibbs free energy of a phase (or a system) is composed of its enthalpy minus the TS product:

$$G = H - TS. \tag{2.18}$$

The concept of the entropy (S) of a system is too often defined cryptically, for the following reasons:

1. The definition is an operative concept.
2. The operative concept is essentially based on the relationship between *work* and *energy*, and accounts for the reversibility of processes in thermodynamic cycles (in relation to the development of thermal engines in the 19th century, or "steam age": cf. Kittel, 1989).

The first satisfactory definition of entropy, which is quite recent, is that of Kittel (1989): "entropy is the natural logarithm of the quantum states accessible to a system." As we will see , this definition is easily understood in light of Boltzmann's relation between configurational entropy and permutability. The definition is clearly "nonoperative" (because the number of quantum states accessible to a system cannot be calculated). Nevertheless, the entropy of a phase may be experimentally measured with good precision (with a calorimeter, for instance), and we do not need any "operative" definition. Kittel's definition has the merit to having put an end to all sorts of nebulous definitions that confused causes with effects. The fundamental P-V-T relation between state functions in a closed system is represented by the exact differential (cf. appendix 2)

$$dU = T \, dS - P \, dV. \tag{2.19}$$

Differentiating equation 2.17, we obtain

$$dH = dU + P \, dV + V \, dP, \tag{2.20}$$

which, combined with equation 2.19, gives

$$dH = T \, dS + V \, dP. \tag{2.21}$$

Differentiating equation 2.18 and combining it with equation 2.21, we also obtain

$$dG = -S \, dT + V \, dP. \tag{2.22}$$

The differential equations above are valid for reversible processes taking place in closed systems in which there is no flow of matter among the various phases.

The bulk differential of the Gibbs free energy for a compositionally variable phase is

$$dG = \left(\frac{\partial G}{\partial T}\right)_{P,n_j} dT + \left(\frac{\partial G}{\partial P}\right)_{T,n_j} dP + \left(\frac{\partial G}{\partial n_i}\right)_{T,P,n_j} dn_i, \qquad (2.23)$$

which gives

$$dG = -S\,dT + V\,dP + \sum_i \mu_i\,dn_i. \qquad (2.24)$$

In equation 2.24, contrary to equation 2.22, the Gibbs free energy of the phase is a function not only of the intensive variables T and P, but also of composition. Equation 2.24 is thus of more general validity and can also be used in open systems or whenever there is flow of components among the various phases in the system. Like the exact differential dG, we can reexpress exact differentials dH and dU as

$$dH = T\,dS + V\,dP + \left(\frac{\partial H}{\partial n_i}\right)_{T,P,n_j} dn_i \qquad (2.25)$$

and

$$dU = T\,dS - P\,dV + \left(\frac{\partial U}{\partial n_i}\right)_{T,P,n_j} dn_i. \qquad (2.26)$$

The "Gibbs equations" 2.24, 2.25, and 2.26 (Gibbs, 1906) are exact differentials, composed of the summations of the partial derivatives over all the variables of which derived magnitude is a function (cf. appendix 2).

The partial derivatives not explicit in equations 2.25 and 2.26 are

$$\left(\frac{\partial G}{\partial T}\right)_{P,n_j} = -S \qquad (2.27)$$

and

$$\left(\frac{\partial G}{\partial P}\right)_{T,n_j} = V. \qquad (2.28)$$

The need to make explicit the compositional derivations in equations 2.25 and 2.26 is dictated by the fact that, in the Gibbs (1906) sense, all partial derivatives over composition are called "chemical potentials" and are symbolized by μ, regardless of the kind of magnitude that undergoes partial derivation (i.e., not only the partial derivative of Gibbs free energy G, as in equation 2.23, for instance, but also $\partial H/\partial n_i$ and $\partial U/\partial n_i$). This fact gives rise to some confusion in several textbooks.

Deriving the G/T ratio in dT, we obtain

$$\left[\frac{\partial(G/T)}{\partial T}\right]_{P,n_j} = -\frac{H}{T^2}. \tag{2.29}$$

Moreover, if we derive molar enthalpy $\overline{H}$ (the enthalpy of one mole of the substance) with respect to T, we have

$$\left(\frac{\partial \overline{H}}{\partial T}\right)_{P,n_j} = C_P, \tag{2.30}$$

where C_P is the specific heat of the substance at constant P—i.e., "the amount of heat that is needed to increase by 1 kelvin one mole of the substance at the P and T of interest."

2.5 Partial Molar Properties

We have defined the "chemical potential of a component" as the partial derivative of the Gibbs free energy of the system (or, for a homogeneous system, of the phase) with respect to the number of moles of the component at constant P and T—i.e.,

$$\mu_i = \left(\frac{\partial G}{\partial n_i}\right)_{P,T}. \tag{2.31}$$

As we have already seen, chemical potential varies with phase composition. If we derive this chemical potential with respect to T and P at constant composition, we obtain

$$\left(\frac{\partial \mu_i}{\partial T}\right)_{P,n_i} = -\left(\frac{\partial S}{\partial n_i}\right)_{P,T} = -s_i \tag{2.32}$$

$$\left(\frac{\partial \mu_i}{\partial P}\right)_{T,n_i} = \left(\frac{\partial V}{\partial n_i}\right)_{P,T} = v_i \tag{2.33}$$

$$\left[\frac{\partial\left(\mu_i/T\right)}{\partial T}\right]_{P,n_i} = -\frac{h_i}{T^2}, \tag{2.34}$$

where magnitudes s_i, v_i, and h_i are defined as the "*partial molar properties*" of the ith component in the phase. These magnitudes should not be confused with molar properties $\overline{S}$, $\overline{V}$, and $\overline{H}$, with which they nevertheless coincide in the condition of a pure component in a pure phase (we have also seen that, in the same condition $\mu_i \equiv \overline{G}$).

"Partial molar properties," like molar properties, are intensive magnitudes (i.e., they do not depend on the extent of the system), and the same relationships that are valid for molar properties hold for them as well. For instance,

$$\mu_i = h_i - Ts_i. \tag{2.35}$$

As we will see , partial molar properties are of general application in the thermodynamics of mixtures and solutions.

2.6 Gibbs Phase Rule

The *phase rule* formulated by Gibbs establishes the degree of freedom (*variance*) as a function of the number of phases and of the numbers of components and of intensive (P, T) and condition variables (gravitational field, magnetic field, electric field, etc.). This rule is intuitively assimilated to the degree of determination of linear systems in elementary algebra: the variance of a system of variables correlated by linear equations is equivalent to the number of independent variables less the number of correlation equations. If we consider an algebraic system with two variables (x and y) that are not mutually dependent (i.e., there are no correlation equations linking the two variables), the variance of the system is 2; if $y = f(x)$, the variance is 1 because a fixed value of y corresponds to a given value of x. If there are two correlation equations between x and y, the variance is 0 (the system is completely determined).

Let us now consider a heterogeneous thermodynamic system at equilibrium. If there are Φ phases in the system, it can easily be seen that $\Phi - 1$ equations of type 2.15 and 2.16 apply for each component in the system. Hence, if there are n components, the number of equations will be $n(\Phi - 1)$. Moreover, the following mass-balance equation holds for each phase:

$$\sum_i X_{i,\alpha} = 1, \tag{2.36}$$

where $X_{i,\alpha}$ is the molar fraction of the ith component in phase α.

The number of *condition equations* is therefore

$$n(\Phi - 1) + \Phi. \tag{2.37}$$

The system variables are composed of $n\Phi$ compositional terms plus ambient variables that are usually two in number: temperature and pressure (hydrostatic and/or lithostatic-isotropic pressure). The *variance* (V) of the system is readily obtained by subtracting the number of condition equations from the total number of variables ($n\Phi + 2$):

$$V = (n\Phi + 2) - \left[n(\Phi - 1) + \Phi\right] = n - \Phi + 2. \tag{2.38}$$

When applying the Gibbs phase rule, it must be remembered that the choice of components is not arbitrary: *the number of components is the minimum number compatible with the compositional limits of the system.*
Consider, for instance, the polymorphic reaction

$$\underset{\text{calcite}}{CaCO_3} \Leftrightarrow \underset{\text{aragonite}}{CaCO_3}. \tag{2.39}$$

There are two phases in reaction and, apparently, three components, corresponding to atoms Ca, C, and O. However, the compositional limits of the system are such that

$$Ca = C \tag{2.40}$$

and

$$O = 3C. \tag{2.41}$$

Because conditions 2.40 and 2.41 hold for all phases in the system, the minimum number of components necessary to describe phase chemistry is 1.
Consider again the exchange reaction

$$\underset{\text{pyroxene}}{2\,MgSiO_3} + \underset{\text{olivine}}{Fe_2SiO_4} \Leftrightarrow \underset{\text{pyroxene}}{2\,FeSiO_3} + \underset{\text{olivine}}{Mg_2SiO_4}. \tag{2.42}$$

Because in *all* phases

$$O = Mg + Fe + 2 \tag{2.43}$$

and

$$Si = 1, \qquad (2.44)$$

there are only two independently variable components (Fe and Mg).

2.7 Stability of Polymorphs

Polymorphism occurs whenever a given component exists under different aggregation states as a function of P and T. The stable state requires that the chemical potential of the component (hence, the Gibbs free energy of the phase for a phase composed of a single component) be at minimum at equilibrium. Figure 2.4 shows examples of G-T plots for Al_2SiO_5 in various P conditions.

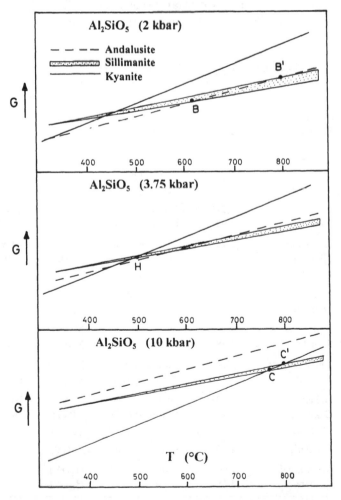

Figure 2.4 *G-T* plots for Al_2SiO_5 in various P conditions (schematic).

Al_2SiO_5 occurs in nature in the solid state in three different aggregation forms: *kyanite,* which is triclinic, dense, and stable at high *P*, and *sillimanite* and *andalusite,* which are orthorhombic, less dense, and stable at lower *P* (see table 2.1). If we examine the *G-T* diagram for *P* = 10 kbar in figure 2.4, we note that the kyanite polymorph has the lowest Gibbs free energy over a wide *T* range (*T* < 780 °C). At *T* > 780 °C, sillimanite has the lowest energy. At *P* = 2 kbar, the *T* range in which kyanite has the lowest energy is restricted to *T* < 320 °C, andalusite has the lowest energy for 320 °C < T < 600 °C, and sillimanite has the lowest energy at *T* > 600 °C. At *P* = 3.75 kbar, we see that the Gibbs free energy is equal for all three polymorphs at *T* ≅ 500 °C (504 ± 20 °C according to Holdaway and Mukhopadhyay, 1993).

By combining the various observations obtained from the *G-T* diagrams in different *P* conditions, we can build up a *P-T* diagram plotting the stability fields of the various polymorphs, as shown in figure 2.5. The solid dots in figures 2.4 and 2.5 mark the phase transition limits and the triple point, and conform to the experimental results of Richardson et al. (1969) (A, R, B′, C′) and Holdaway (1971) (A, H, B, C). The dashed zone defines the uncertainty field in the

Table 2.1 Examples of polymorphism in geology.

Compound	Phase	Density, g/cm^3	Symmetry	Elemental Coordination		
SiO_2	Trydimite	2.27	Hexagonal	Si(IV)		
	Cristobalite	2.35	Cubic	Si(IV)		
	Quartz	2.654	Trigonal	Si(IV)		
	Cohesite	3.01	Monoclinic	Si(IV)		
	Stishovite	4.28	Tetragonal	Si(VI)		
$CaCO_3$	Calcite	2.715	Trigonal	C(III)	Ca(VI)	
	Aragonite	2.94	Orthorhombic	C(III)	Ca(IV)	
Al_2SiO_5	Andalusite	3.15	Orthorhombic	Al(IV)	Al(V)	Si(IV)
	Sillimanite	3.25	Orthorhombic	Al(IV)	Al(IV)	Si(IV)
	Kyanite	3.6	Triclinic	Al(IV)	Al(VI)	Si(IV)
C	Graphite	2.25	Hexagonal	C(III)		
	Diamond	3.51	Cubic	C(IV)		
Mg_2SiO_4	Forsterite	3.23	Orthorhombic	Mg(VI)	Mg(VI)	Si(IV)
	Spinel	3.58	Cubic	Mg(IV)	Mg(VI)	Si(IV)
$MgSiO_3$	Clinoenstatite	3.19	Monoclinic	Mg(VI)	Si(IV)	
	Enstatite	3.20	Orthorhombic	Mg(VI)	Si(IV)	
$Al_2O_3H_2O$	Bohemite	3.02	Orthorhombic	Al(VI)	H(II)	
	Diaspore	3.4	Orthorhombic	Al(VI)	H(II)	
ZnS	Wurtzite	4.0	Hexagonal	Zn(IV)		
	Sphalerite	4.09	Cubic	Zn(IV)		
HgS	Metacinnabar	7.60	Trigonal	Hg(IV)		
	Cinnabar	8.18	Cubic	Hg(VI)		

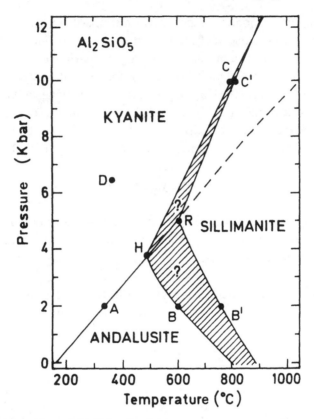

Figure 2.5 Polymorphs of Al₂SiO₅. Dashed area covers uncertainty range among the various experiments. Marked dots B, B', C, C', and H are those plotted in figure 2.4.

andalusite-kyanite and kyanite-sillimanite transitions, arising from the comparison of the two sets of data. In figure 2.4 (dashed area), the uncertainty is regarded as an error progression involving the value of the Gibbs free energy of sillimanite with increasing T. In fact, as discussed by Newton (1987), the bias must be imputed to the formation of metastable fibrolite in the andalusite-sillimanite transition in the experimental runs of Richardson et al. (1969), and to difficulty in reaching equilibrium for the kyanite-sillimanite transition.*

If we now recall the phase rule, it is evident that, at the P-T conditions represented by point D in figure 2.5, slight variations in the P or T values will not induce any change in the structural state of the phase (there are one phase and one component: the variance is 2). At point A in the same figure, any change in one of the two intensive variables will induce a phase transition. To maintain the coexistence of kyanite and andalusite, a dP increment consistent with the slope of the univariant curve (there are two phases and one component: the variance

* A recent study by Holdaway and Mukhopadhyay (1993) essentially confirms the stability diagram of Holdaway (1971). However, it is of interest to show how even slight errors in the assigned Gibbs free energy of a phase drastically affect the stability fields of polymorphs.

is then 1) would correspond to each dT variation. At the triple point R (or H), all three polymorphs coexist stably, but any minimal change in the P-T conditions results in the disappearance of one or two phases (variance is 0).

Figure 2.6 shows P-T stability diagrams for several components exhibiting polymorphism in geology (the Co_2SiO_4 orthosilicate, which is not a major constituent of rock-forming minerals, is nevertheless emblematic of phase transitions observed in the earth's mantle; cf. section 5.2.3).

Stability fields conform qualitatively to the following general rules:

1. With increasing T, low-density structural forms are tendentially more stable.
2. Increasing P increases the stability of high-density polymorphs.
3. High-T phase transitions frequently proceed in the direction of increasing crystalline symmetry.

The significance of the first two rules will be more evident in sections 2.10 and 3.7, when we discuss calculation of the Gibbs free energy of a phase at various P-T conditions.

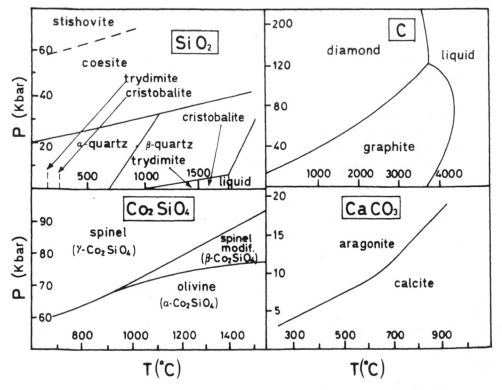

Figure 2.6 Selected cases of polymorphism in geology (simplified).

2.8 Solid State Phase Transitions

We saw in the preceding section that, at solid state in nature, a given component may assume various structural states as a function of P and T variables. In given P-T conditions, there is a given stable form. If one of the intensive variables is changed, at a certain moment the univariant equilibrium of phase transition is reached, and we then observe a solid state transformation leading to the new aggregation state that is stable in the new P-T conditions. The Gibbs free energy modification connected with the solid state phase transition is zero at equilibrium, but the partial derivatives of the Gibbs free energy with respect to the P and T variables are not zero.

We now distinguish solid state transformations as "first-order transitions" or "lambda transitions." The latter class groups all high-order solid state transformations (second-, third-, and fourth-order transformations; see Denbigh, 1971 for exhaustive treatment). We define *first-order transitions* as all solid state transformations that involve discontinuities in enthalpy, entropy, volume, heat capacity, compressibility, and thermal expansion at the transition point. These transitions require substantial modifications in atomic bonding. An example of first-order transition is the solid state transformation (see also figure 2.6)

$$\underset{\text{graphite}}{C} \quad \Leftrightarrow \quad \underset{\text{diamond}}{C.} \tag{2.45}$$

This reaction requires breaking of at least one bond and modification of the coordination state of carbon (cf. table 2.1). During phase transformation, an intermediate configuration with high static potential is reached. In order to reach this condition it is necessary to furnish the substance with a great amount of energy (mainly in thermal form). First-order transitions, as a result of this energy requirement, are difficult to complete, and metastable persistence of the high-energy polymorph may often be observed. Diamond, for instance, at room P-T conditions is a metastable form of carbon; aragonite, as already stated, at room P-T conditions is a metastable form of $CaCO_3$, and so on.

In *lambda transitions,* no discontinuity in enthalpy or entropy as a function of T and/or P at the transition zone is observed. However, heat capacity, thermal expansion, and compressibility show typical perturbations in the lambda zone, and T (or P) dependencies before and after transition are very different.

Alpha-beta (α-β) transitions of the condensed forms of SiO_2 quartz, trydimite, and cristobalite may all be regarded as lambda transformations. Their kinetics are higher than those of quartz-trydimite, quartz-cristobalite, and quartz-coesite, which are first-order transformations. Figure 2.7 plots in detail the evolution of enthalpy, entropy, heat capacity, and volume at the transition zone

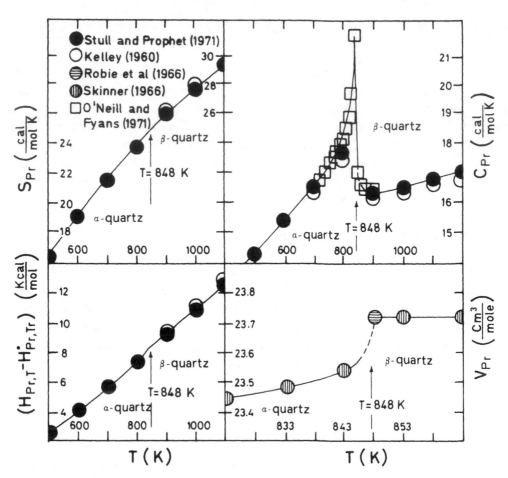

Figure 2.7 Enthalpy, entropy, heat capacity, and volume modification in α-quartz/ β-quartz transition region. Reprinted from H. C. Helgeson, J. Delany, and D. K. Bird, *American Journal of Science,* 278A, 1–229, with permission.

between α-quartz and β-quartz. The lambda point at 1 bar for this transition occurs at $T = 848$ K. (The α-quartz/β-quartz transition in fact implies minimal variations in enthalpy and entropy at the transition point and is therefore interpreted by many authors as an overlapping first-order-plus-lambda transition. We will discuss this transition more extensively in section 5.8.)

As we have already stated, the Gibbs free energy modification connected with a polymorphic transition is zero. The univariant equilibrium along the transition curve is described by the Clapeyron equation:

$$\frac{dP}{dT} = \frac{\Delta S^0_{\text{transition}}}{\Delta V^0_{\text{transition}}} = K_{\text{transition}}, \qquad (2.46)$$

where $\Delta S^0_{\text{transition}}$ and $\Delta V^0_{\text{transition}}$ are, respectively, the standard molar entropy and standard molar volume of transition, and $K_{\text{transition}}$ is the Clapeyron slope of the solid-solid transition. For a first-order transition at $P < 10$ kbar, $K_{transition}$ is almost constant at all temperatures. With several substitutions from equation 2.46, we derive

$$\Delta Cp^0_{transition} = 2K_{transition}T\left(\frac{\partial \Delta V^0_{transition}}{\partial T}\right)_P + K^2_{transition}T\left(\frac{\partial \Delta V^0_{transition}}{\partial P}\right)_T. \quad (2.47)$$

Equation 2.47 describes the interdependence of thermal expansion, compressibility, and heat capacity of a first-order transition and furnishes a precise tool for the evaluation of the internal consistency of experimental data in solid state transition studies (see Helgeson et al., 1978 for a careful application of eq. 2.47).

2.8.1 Phase Transitions and Landau Theory

The application of Landau theory to rock-forming minerals has been promoted by Ekhard Salje and his coworkers in an attempt to achieve better quantification of complex transition phenomena (mainly in feldspars, but also in pyroxenes and spinels; Salje et al., 1985; Salje, 1985, 1988; Carpenter and Salje, 1994a,b; Carpenter, 1988).

If "excess specific heat," defined as "that part of the specific heat that is due to the structural phase transition" (cf. Salje, 1988), is denoted by $\Delta C_{P,excess,trans}$, the excess entropy associated with transition within the upper and lower limits of C_P perturbation is given by

$$\Delta S_{excess,trans} = \int_{T_l}^{T_u} \frac{\Delta C_{P,excess,trans}}{T} dT, \quad (2.48)$$

and the excess enthalpy is given by

$$\Delta H_{excess,trans} = \int_{T_l}^{T_u} \Delta C_{P,excess,trans} dT. \quad (2.49)$$

The excess thermodynamic properties correlated with phase transitions are conveniently described in terms of a macroscopic order parameter Q. Formal relations between Q and the excess thermodynamic properties associated with a transition are conveniently derived by expanding the Gibbs free energy of transition in terms of a Landau potential:

$$\Delta G_{excess,trans,T,P,\mu,Q} = G(\nabla Q)^2 - \mathcal{H}Q + AQ^2 + BQ^4 + CQ^6, \quad (2.50)$$

(Landau and Lifshitz, 1980), where $\mathcal{H}$ is the conjugate field of the order parameter, coefficient A is a function of temperature:

$$A = a(T - T_c),\tag{2.51}$$

and the remaining coefficients are not explicitly dependent on T. T_c in equation 2.51 is the critical temperature (e.g., the Curie temperature in the case of magnetic transitions).

In the case where transitions involve low-symmetry phases, general equation 2.50 may be recast into the unidimensional form

$$\Delta G_{\text{excess,trans}} = \frac{1}{2}a(T - T_c)Q^2 + \frac{1}{4}BQ^4 + \frac{1}{6}CQ^6\tag{2.52}$$

(Carpenter, 1988). Because the excess Gibbs free energy of transition must always be at a minimum with respect to the macroscopic order parameter Q, i.e.:

$$\left(\frac{\partial \Delta G_{\text{excess,trans}}}{\partial Q}\right)_{T,P} = 0,\tag{2.53}$$

the various types of transitions are evaluated on the basis of condition 2.53 by setting $Q = 0$ at 0 K. The type of transition depends essentially on the signs of B and C.

For a second-order transition, $B > 0$ and $C = 0$, and condition 2.53 leads to

$$\Delta G_{\text{excess,trans}} = \frac{1}{2}a(T - T_c)Q^2 + \frac{1}{4}BQ^4\tag{2.54}$$

$$\Delta S_{\text{excess,trans}} = -\frac{\partial \Delta G_{\text{excess,trans}}}{\partial T} = -\frac{1}{2}aQ^2\tag{2.55}$$

$$\Delta H_{\text{excess,trans}} = \Delta G_{\text{excess,trans}} - T\,\Delta S_{\text{excess,trans}} = -\frac{1}{2}aT_cQ^2 + \frac{1}{4}BQ^4\tag{2.56}$$

$$\Delta C_{P,\text{excess,trans}} = \frac{a}{2B}T, \quad T \le T_c\tag{2.57}$$

$$T_c = \frac{B}{a}.$$ (2.58)

The equilibrium temperature of transition is coincident with the critical temperature:

$$T_{\text{trans}} = T_c.$$ (2.59)

For a tricritical transition (i.e., a high-order transition crossing a first-order phase boundary), $B = 0$ and $C > 0$, and condition 2.53 leads to

$$\Delta G_{\text{excess,trans}} = \frac{1}{2} a (T - T_c) Q^2 + \frac{1}{6} C Q^6$$ (2.60)

$$\Delta S_{\text{excess,trans}} = -\frac{1}{2} a Q^2$$ (2.61)

$$\Delta H_{\text{excess,trans}} = -\frac{1}{2} a T_c Q^2 + \frac{1}{6} C Q^6$$ (2.62)

$$\Delta C_{P,\text{excess,trans}} = \frac{aT}{4\sqrt{T_c}} (T_c - T)^{-1/2}, \quad T < T_c$$ (2.63)

$$T_c = \frac{C}{a}$$ (2.64)

$$T_{\text{trans}} = T_c.$$ (2.65)

For a first-order transition, $B < 0$ and $C > 0$, and condition 2.53 leads to

$$\Delta G_{\text{excess,trans}} = \frac{1}{2} a (T - T_c) Q^2 + \frac{1}{4} B Q^4 + \frac{1}{6} C Q^6$$ (2.66)

$$\Delta S_{\text{excess,trans}} = -\frac{1}{3} a Q^2 \left\{ 1 + \left[1 - \frac{3}{4} \left(\frac{T - T_c}{T_{\text{trans}} - T_c} \right) \right] \right\}$$ (2.67)

$$\Delta H_{\text{excess,trans}} = -\frac{1}{2} a T_c Q^2 + \frac{1}{4} B Q^4 + \frac{1}{6} C Q^6$$ (2.68)

$$T_{trans} = T_c + \frac{3 B^2}{16aC} . \tag{2.69}$$

The latent heat of the transformation L is

$$L = \frac{1}{2} a Q_{trans}^2 T_{trans}, \tag{2.70}$$

where Q_{trans} is the order parameter at T_{trans}.

Figure 2.8 shows the behavior of the macroscopic order parameter Q as a function of T/T_c (T/T_{trans} for first-order transitions).

For a second-order transition,

$$Q = \left(1 - \frac{T}{T_c} \right)^{1/2}, \quad T < T_c . \tag{2.71}$$

For a tricritical transition,

$$Q = \left(1 - \frac{T}{T_c} \right)^{1/4}, \quad T < T_c . \tag{2.72}$$

For a first-order transition, the macroscopic order parameter shows a jump at T_{trans} from $Q = 0$ to $Q = Q_{trans}$—i.e., below T_{trans},

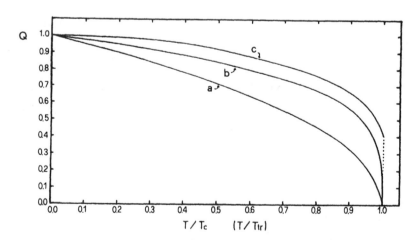

Figure 2.8 The order parameter Q as a function of T/T_c for second-order (a) and tricritical (b) transitions, and as a function of T/T_{trans} for a first-order transition (c) with $T_c = 0.99 T_{trans}$ and $Q_{trans} = 0.4$. From Carpenter (1988). Reprinted by permission of Kluwer Academic Publishers.

$$Q^2 = \frac{2}{3} Q_{\text{trans}}^2 \left\{ 1 + \left[1 - \frac{3}{4} \left(\frac{T - T_c}{T_{\text{trans}} - T_c} \right) \right]^{1/2} \right\},$$ (2.73)

and at T_{trans},

$$Q_{\text{trans}} = \pm \left[\frac{4a(T_{\text{trans}} - T_c)}{b} \right]^{1/2}.$$ (2.74)

Curve c in figure 2.8, which is valid for a first-order transition, has been calculated by Carpenter (1988) for a relatively small difference between T_{trans} and T_c ($T_c = 0.99 T_{\text{trans}}$) and a relatively high Q_{trans} ($Q_{\text{trans}} = 0.4$), resulting in a small jump at T_{trans}.

Obviously, if we know experimentally the behavior of the macroscopic ordering parameter with T, we may determine the corresponding coefficients of the Landau expansion (eq. 2.52). However, things are not so easy when different transitions are superimposed (such as, for instance, the displacive and order-disorder transitions in feldspars). In these cases the Landau potential is a summation of terms corresponding to the different reactions plus a coupling factor associated with the common elastic strain.

We will see detailed application of Landau theory to complex superimposed transition phenomena when we treat the energetics of feldspars in chapter 5.

2.9 Standard State and Thermodynamic Activity

From what we have seen up to this point, we can perform "relative" evaluations of the modification of Gibbs free energy with composition (section 2.2) and with state conditions of the system (section 2.4). However, in order to calculate thermodynamic equilibrium, we need a rigorous energy scale—one that refers to a firmly established "starting level" of the chemical potential. This level, or *standard state,* which is kept constant as calculations develop, allows us to evaluate quantitatively the change in chemical potential that takes place, for instance, in a solid state transition such as those discussed in the preceding section. For the same purpose, we introduce the concept of *thermodynamic activity,* which reflects the difference in chemical potential for a given component in a given phase at fixed P-T conditions with respect to chemical potential at standard state.

The algebraic formulation of the relationship among chemical potential μ_i, standard state chemical potential μ_i^0, and thermodynamic activity a_i is

$$\mu_{i,\alpha} = \mu_{i,\alpha}^0 + RT \ln a_{i,\alpha},$$ (2.75)

where $\mu_{i,\alpha}$ is the chemical potential of the ith component in phase α and R is the gas constant. The reversed expression for thermodynamic activity is therefore

$$a_{i,\alpha} = \exp\left(\frac{\mu_{i,\alpha} - \mu_{i,\alpha}^0}{RT}\right). \qquad (2.76)$$

It must be emphasized that the choice of a particular standard state of reference has no influence on the result of equilibrium calculations and is only a matter of convenience. It is generally convenient to adopt as standard state for solid components the condition of "pure component in the pure phase at the P and T of interest" or, alternatively, the condition of "pure component in the pure phase at $P = 1$ bar and $T = 298.15$ K."

2.9.1 Standard State of Pure Component in Pure Phase at P and T of Interest

We can identify this reference condition by means of the notation

$$\mu_{i,\alpha}^0 = \mu_{i,\alpha(X_i=1)}. \qquad (2.77)$$

The relationship between the thermodynamic activity and molar fraction of a given component i in mixture α is described, over the whole compositional range, by

$$a_{i,\alpha} = X_{i,\alpha}\gamma_{i,\alpha}, \qquad (2.78)$$

where $\gamma_{i,\alpha}$ is the *rational activity coefficient,* which is a function of P, T, and composition and embodies interaction effects among the component of interest and all the other components in the mixture. It is obvious from equation 2.75 that, at standard state, the activity of the component is equal to 1. If the adopted standard state is that of "pure component," we also have

$$X_{i,\alpha} = 1 \Rightarrow \gamma_{i,\alpha} = 1. \qquad (2.79)$$

Combining equation 2.78 with equation 2.75, we derive the general activity-concentration relation that is valid over the entire compositional range:

$$\mu_{i,\alpha} = \mu_{i,\alpha}^0 + RT \ln X_{i,\alpha} + RT \ln \gamma_{i,\alpha}. \qquad (2.80)$$

Based on equation 2.80, we can distinguish three different compositional fields:

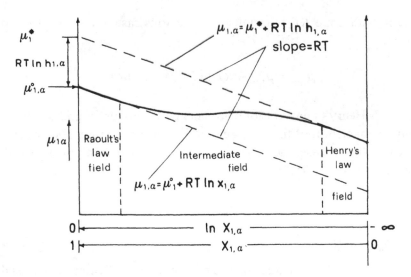

Figure 2.9 Activity–molar fraction relationship for component 1 in mixture α.

1. Compositional range of validity of Raoult's law
2. Compositional range of validity of Henry's law
3. Intermediate field

These fields are shown in a semilogarithmic plot in figure 2.9, relative to component 1 in mixture α.

Compositional Range of Validity of Raoult's Law

This field is extended from the condition of "pure component" ($X_{1,\alpha} = 1$) to a given concentration value, and is characterized by a straight-line correlation with slope RT on the semilogarithmic plot:

$$\mu_{1,\alpha} = \mu^0_{1,\alpha} + RT \ln X_{1,\alpha}. \tag{2.81}$$

Chemical interactions among component 1 and all other components in the mixture are virtually absent, because the mixture is composed essentially of component 1 itself ($\gamma_{1,\alpha} = 1$). If we compare equations 2.81 and 2.75, we see that, in this compositional field, we have

$$\alpha_{1,\alpha} = X_{1,\alpha}. \tag{2.82}$$

Compositional Range of Validity of Henry's Law

This field ranges from a given (generally low) value of the molar concentration of component 1 down to infinite dilution ($X_{1,\alpha} = 0$). In this case, too, the relationship between chemical potential and the natural logarithm of the molar fraction is straight, with slope RT, but, unlike the previous case, it also involves an intercept term:

$$\mu_{1,\alpha} = \mu_{1,\alpha}^0 + RT \ln X_{1,\alpha} + RT \ln h_{1,\alpha}, \qquad (2.83)$$

where $h_{1,\alpha}$ is Henry's constant for component 1 in mixture α and depends on the P, T, and composition of the mixture, but not on the amount of component 1 in the mixture. In this case, the thermodynamic activity of component 1 is defined, in terms of equation 2.78, by

$$\gamma_{1,\alpha} = h_{1,\alpha}. \qquad (2.84)$$

Intermediate Field

Equation 2.78 is still valid, and the chemical potential of component 1 is defined by equation 2.80, with γ varying with P, T, and composition (see figure 2.9).

2.9.2 Standard State of Infinite Dilution of Component

This standard state is adopted for *solutions* in which, as already stated, the condition of *solute* is distinguished from that of *solvent*. Moreover, it is (implicitly) adopted in trace element geochemistry (see chapter 10).

If we embody solute-solvent interactions (observed at low solute concentrations) in the standard state condition, i.e.:

$$\mu_{1,\alpha}^* = \mu_{1,\alpha}^0 + RT \ln h_{1,\alpha}, \qquad (2.85)$$

in the field of Henry's law we have

$$\mu_{1,\alpha} = \mu_{1,\alpha}^* + RT \ln X_{1,\alpha}, \qquad (2.86)$$

where $\mu_{1,\alpha}^*$ is the chemical potential for Henry's standard state. In this case, too, as shown in figure 2.9, we observe in the field of Henry's law a straight-line relation with slope RT between chemical potential and the natural logarithm of the molar concentration, but the intercept is now $\mu_{1,\alpha}^*$. Again, we introduce an activity coefficient varying with $X_{1,\alpha}$ for the intermediate concentration field, and we have

$$\mu_{1,\alpha} = \mu_{1,\alpha}^* + RT \ln a_{1,\alpha} \qquad (2.87)$$

$$a_{1,\alpha} = X_{1,\alpha} \gamma_{1,\alpha}^*, \qquad (2.88)$$

$$\gamma_{1,\alpha}^* = f(X_{1,\alpha}). \tag{2.89}$$

We will see in chapter 8 that, in the case of aqueous solutions, it is convenient to adopt the condition of "hypothetical 1-molal solution at $P = 1$ bar and $T = 298.15$ K" as the standard state. Most experimental data on aqueous solutions conform to this reference condition. In this case, the resulting activity coefficient is defined as the "practical activity coefficient" and must not be confused with the rational activity coefficient of general relation 2.80.

2.10 Calculation of Gibbs Free Energy of a Pure Phase in Various *P-T* Conditions

From the differential equations (expressed in terms of *molar* properties)

$$d\overline{H} = \left(\frac{\partial \overline{H}}{\partial T} \right)_P dT + \left(\frac{\partial \overline{H}}{\partial P} \right)_T dP \tag{2.90}$$

and

$$d\overline{S} = \left(\frac{\partial \overline{S}}{\partial T} \right)_P dT + \left(\frac{\partial \overline{S}}{\partial P} \right)_T dP, \tag{2.91}$$

where

$$\left(\frac{\partial \overline{H}}{\partial P} \right)_T = \overline{V} - T \left(\frac{\partial \overline{V}}{\partial T} \right)_P \tag{2.92}$$

and

$$\left(\frac{\partial \overline{S}}{\partial P} \right)_T = - \left(\frac{\partial \overline{V}}{\partial T} \right)_P, \tag{2.93}$$

recalling equation 2.30 and integrating, we obtain

$$\overline{H}_{T,P} = \overline{H}_{T_r,P_r}^0 + \int_{T_r}^{T} C_P \, dT + \int_{P_r}^{P} \left[\overline{V} - T \left(\frac{\partial \overline{V}}{\partial T} \right)_P \right] dP \tag{2.94}$$

and

$$\overline{S}_{T,P} = \overline{S}^0_{T_r,P_r} + \int_{T_r}^{T} C_P \frac{dT}{T} - \int_{P_r}^{P} \left(\frac{\partial \overline{V}}{\partial T}\right)_P dP. \tag{2.95}$$

Because equation 2.18 holds at all P-T conditions, we have

$$\overline{G}_{T,P} = \overline{H}_{T,P} - T\overline{S}_{T,P}. \tag{2.96}$$

The molar Gibbs free energy of the phase at P and T of interest is obtained by combining equations 2.94, 2.95, and 2.96 for known values of molar enthalpy and molar entropy at T_r and P_r reference conditions ($\overline{H}^0_{T_r,P_r}$ and $\overline{S}^0_{T_r,P_r}$ respectively).

2.11 Gibbs-Duhem Equation

The general form of the Gibbs-Duhem equation for an n-component system can be expressed as

$$X_1 \, d \ln a_1 + X_2 d \ln a_2 + \cdots + X_n \, d \ln a_n = 0. \tag{2.97}$$

For a binary system, we have

$$X_1 \, d \ln a_1 + X_2 \, d \ln a_2 = 0, \tag{2.98}$$

which, by application of the properties of the exact differentials (see appendix 2), may be expressed as

$$X_1 \frac{\partial \ln a_1}{\partial X_2} + X_2 \frac{\partial \ln a_2}{\partial X_2} = 0. \tag{2.99}$$

Based on equation 2.99, the corresponding activity trend of component 2 may be deduced, whenever the activity trend of component 1 in the mixture with the increase in the molar fraction of component 2 is known.

An analogous expression holds for the activity coefficients:

$$X_1 \frac{\partial \ln \gamma_1}{\partial X_2} + X_2 \frac{\partial \ln \gamma_2}{\partial X_2} = 0. \tag{2.100}$$

In integrated form, equation 2.100 becomes

$$\ln \gamma_1 = -\int_0^{X_2} \frac{X_2}{1 - X_2} \frac{\partial \ln \gamma_2}{\partial X_2} \, dX_2, \tag{2.101}$$

which, with integration by parts, may be transformed as shown by Wagner (1952):

$$\ln \gamma_1 = -\frac{X_2}{1 - X_2} \ln \gamma_2 + \int_0^{X_2} \frac{\ln \gamma_2}{(1 - X_2)^2} \, dX_2. \tag{2.102}$$

Evaluation of the integral in equation 2.102 may be performed graphically or by polynomial interpolation.

In the case of ternary or higher-order mixtures, solution of the Gibbs-Duhem equation is again based on application of the properties of the exact differentials (Lewis and Randall, 1970):

$$\mu_1 \, dn_1 + \mu_2 \, dn_2 + \cdots + \mu_n \, dn_n = dG. \tag{2.103}$$

The solution of the problem requires integration procedures along pseudo-binary lines that result in the combination of integral forms of the type in equation 2.101. In this context it is unnecessary to proceed further with detailed treatment, for which reference may be made to Lewis and Randall (1970).

2.12 Local Equilibrium, Affinity to Equilibrium, and Detailed Balancing

The "principle of local equilibrium in disequilibrium processes" (Prigogine, 1955) establishes that every natural system contains a macroscopic region combining a number of molecules sufficiently large for microscopic fluctuations to be negligible, within which chemical exchanges are reversible and equilibrium is complete. At local equilibrium, there is no flux of components among coexisting phases; moreover, the spatial extent of the local equilibrium zone is time-dependent (this fact renders the principle of local equilibrium a formidable tool in the investigation of geological systems). The degree of attainment of local equilibrium may be evaluated by introducing the concept of *thermodynamic affinity:* because a reaction proceeds in the sense of minimization of the Gibbs free energy of the system, for a generic reaction j taking place at constant temperature and pressure, thermodynamic affinity (A_j) is defined as

$$A_j = -\left(\frac{\partial G}{\partial \xi_j} \right)_{P,T} = -RT \ln \frac{Q_j}{K_j}, \tag{2.104}$$

where ξ_j is the progress variable for the jth reaction, K_j is the equilibrium constant, and Q_j is the observed activity product of the (macroscopic) reaction:

$$Q_j = \prod_i a_{i,j}^{v_{i,j}}, \qquad (2.105)$$

where $a_{i,j}$ is the activity of the ith term in the jth reaction, and $v_{i,j}$ is the stoichiometric coefficient of the ith term in the jth reaction (positive for products and negative for reactants).

If the equilibrium condition of a system is perturbed by modifications in intensive variables P and T, or by exchanges of energy or matter with the exterior, the system will tendentially achieve a new Gibbs free energy minimum through modification of the aggregation states in its various regions (*phases*) or through modification of phase compositions by flow of components. In a heterogeneous system, several reactions will generally concur to restore equilibrium. The overall chemical affinity to equilibrium of a heterogeneous system (A_{system}) is related to the individual affinities of the various reactions through

$$A_{\text{system}} = \sum_j \sigma_j A_j, \qquad (2.106)$$

where σ_j is the relative rate of the jth reaction with respect to the overall reaction progress variable (ξ) (Temkin, 1963):

$$\sigma_j = \frac{d\xi_j}{d\xi} \qquad (2.107)$$

Because, in the zone of local equilibrium,

$$A_{\text{system}} = \sum_j A_j = 0, \qquad (2.108)$$

and because activities can always be translated into compositional variables (molalities in fluids, or molar concentrations in condensed phases), we may represent local equilibria as discrete points in a compositionally continuous trend of locally reversible but overall irreversible exchanges (figure 2.10); i.e., for a given component i, in terms of total number of moles n_i,

$$n_{i,\text{end}} - n_{i,\text{start}} = \int_{\xi=0}^{\xi=1} \left(\sum_j \frac{\partial n_{i,j}}{\partial \xi_j} \right) d\xi. \qquad (2.109)$$

As exemplified in figure 2.10, in the attainment of new equilibrium a single reaction may proceed in opposite directions (i.e., forward and backward). In hydrolytic equilibria, the dissolution process is conventionally defined as moving in

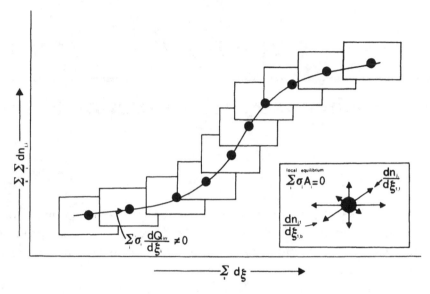

Figure 2.10 Schematic representation of concept of local equilibrium in disequilibrium processes. Reprinted from P. F. Sciuto and G. Ottonello, *Geochimica et Cosmochimica Acta,* 59, 2207–2214, copyright © 1995, with kind permission from Elsevier Science Ltd., The Boulevard, Langford Lane, Kidlington OX5 1 GB, UK.

a forward direction and the precipitation process as moving in a backward or "*reverse*" direction—i.e., in the case of silica,

$$\text{forward}(k_+)\rightarrow$$

$$SiO_{2(solid)} + 2\,H_2O \quad \rightarrow \quad H_4SiO^0_{4(aqueous)} \cdot \tag{2.110}$$

$$\leftarrow \text{backward}(k_-)$$

Because at equilibrium there is no flow of components, the rate constants of forward reaction (k_+) and reverse reaction (k_-) must counterbalance. Moreover, because affinity to equilibrium is zero, activity product Q must be equivalent to equilibrium constant K. For a generic reaction j at equilibrium, we have

$$\frac{k_+}{k_-} = K_j^n \tag{2.111}$$

where n is a positive coefficient related to the orders of forward and backward reactions (see section 8.21.1) and to stoichiometry (see Dembigh, 1971, for an appropriate treatment of the actual significance of n). Equation 2.111, which is the fundamental link between thermodynamics and kinetics, establishes that local equilibrium is not associated with cessation of processes but rather with a dynamic balance between flows moving in opposite directions.

CHAPTER THREE

Thermochemistry of Crystalline Solids

Systematic analysis of the mixing properties of crystalline compounds, coupled with impressive experimental efforts in the last three decades, has produced a qualitative leap forward in our comprehension of the chemical-physical processes operating in the earth's lithosphere. Excellent works of systematic revision and theoretical framing (to which we refer for detailed treatment) have also appeared (Thompson, 1969; Saxena, 1973; Helgeson et al., 1978; Ganguly and Saxena, 1987; Navrotsky, 1994; Anderson, 1995). In this chapter, only those concepts that are operatively most significant in crystal energetics will be outlined, after an elementary digression on the quantum-mechanic interpretation of entropy and heat capacity. The quantum-mechanic interpretation of the vibrational motion of atoms clarifies the real significance of the macroscopic thermodynamic properties of crystalline substances and allows us to rationalize the plethora of spectroscopic data for thermochemical interpretation. Interested readers will find adequate treatments in specialized volumes such as that edited by Kieffer and Navrotsky (1985).

3.1 Entropy, Heat Capacity, and Vibrational Motion of Atoms in Crystals

Let us consider a diatomic molecule and assume that it behaves as a harmonic oscillator with two masses, m_1 and m_2, connected by an "ideal" (constant-force) spring. At equilibrium, the two masses are at a distance X_0; by extending or compressing the distance by an amount X, a force F will be generated between the two masses, described by Hooke's law (cf. equation 1.14):

$$F = -K_F X. \tag{3.1}$$

The indefinite integration in dX of equation 3.1 gives the potential energy of the harmonic oscillator (cf. equation 1.16):

$$\Phi_{(X)} = \frac{K_F X^2}{2}.$$ (3.2)

Force constant K_F thus represents the curvature of the potential with respect to distance:

$$K_F = \frac{\partial^2 \Phi_{(X)}}{\partial X^2}.$$ (3.3)

Most diatomic molecules have a force constant in the range 10^2 to 10^3 N · m^{-1}. A common tool for the calculation of K_F in diatomic molecules (often extended to couples of atoms in polyatomic molecules) is "Badger's rule":

$$K_F = \frac{186}{\left(X_0 - d_{ij}\right)^3} \left(N \cdot m^{-1}\right),$$ (3.4)

where d_{ij} is a constant relative to i and j atoms. The equation of motion for the harmonic oscillator is, in the case of diatomic molecules,

$$m^* \frac{d^2 X}{dt^2} = -K_F X,$$ (3.5)

where m^* is the reduced mass (typically 10^{-26} to 10^{-27} kg)

$$m^* = \frac{m_1 m_2}{m_1 + m_2}.$$ (3.6)

The frequency of the harmonic oscillator (cycles per unit of time) is given by

$$v_0 = \frac{1}{2\pi} \sqrt{\frac{K_F}{m^*}}$$ (3.7)

and is typically about 10^{-12} to 10^{-14} s^{-1} (McMillan, 1985).

The Schrödinger equation, as we have already seen (section 1.17.1), can be written in the general form

$$\mathcal{H}\psi = E\psi,$$ (3.8)

where $\mathcal{H}$ is the quantum-mechanic hamiltonian

$$\mathcal{H} = \Phi - \frac{h^2}{8\pi^2 m} \nabla^2. \tag{3.9}$$

Introducing the potential of the harmonic oscillator (eq. 3.2) in the monodimensional equivalent of equation 3.9 (i.e., the Schrödinger equation for one-dimensional stationary states; see eq. 1.9), we obtain

$$\frac{d^2 \psi_{(X)}}{dX^2} + \frac{8\pi^2 m^*}{h^2} \left(E - \frac{K_F X^2}{2} \right) \psi_{(X)} = 0. \tag{3.10}$$

The admitted energy levels correspond to the eigenvalues of equation 3.10 for the various quantum numbers n (see section 1.1.2):

$$E_{(n)} = \left(\frac{K_F h^2}{4\pi^2 m^*} \right)^{1/2} \left(n + \frac{1}{2} \right) \tag{3.11}$$

and may be expressed as functions of the characteristic oscillation frequencies of the harmonic oscillator (v_0) by substituting equation 3.7 in equation 3.11:

$$E_{(n)} = h v_0 \left(n + \frac{1}{2} \right). \tag{3.12}$$

We thus obtain a set of energy levels separated by an energy gap $\Delta E = h v_0$ and bounded by the potential function 3.2 (figure 3.1).

Based on equation 3.12, the lowest vibrational mode ($n = 0$) also has a certain amount of energy, known as *zero-point energy:*

$$E_0 = \frac{1}{2} h v_0. \tag{3.13}$$

This energy contributes to the lattice energy of the substance at $T = 0$, in the terms outlined in section 1.12.2.

The vibrational functions, corresponding to the energy eigenvalues of equation 3.11 or "vibrational eigenfunctions," are given by

$$\psi_{(n)} = C_{(n)} H_{(n,y)} \exp\left(-\frac{y^2}{X} \right), \tag{3.14}$$

where

$$y = \left(\frac{4\pi^2 m^* K_F}{h^2}\right) X.$$ (3.15)

$C_{(n)}$ is the "normalization factor":

$$C_{(n)} = \left(2^n n! \sqrt{\pi}\right)^{-1/2}$$ (3.16)

and $H_{(n,y)}$ is the Hermite polynomial:

$$
\begin{aligned}
H_{(0,y)} &= 1 \\
H_{(1,y)} &= 2y \\
H_{(2,y)} &= 4y^2 - 2 \\
H_{(3,y)} &= 8y^3 - 12y.
\end{aligned}
$$ (3.17)

The vibrational motion of atoms in diatomic molecules and, by extension, in crystals cannot be fully assimilated to harmonic oscillators, because the potential well is asymmetric with respect to X_0. This asymmetry is due to the fact that the short-range repulsive potential increases exponentially with the decrease of interionic distances, while coulombic terms vary with $1/X$ (see, for instance, figures 1.13 and 3.2). To simulate adequately the asymmetry of the potential well, empirical *asymmetry terms* such as the *Morse potential* are introduced:

$$\Phi_{(X)} = D_e \left[1 - \exp(-aX)\right]^2$$ (3.18)

where D_e is the depth of the potential well and

$$a = \pi v_0 \sqrt{\frac{2m^*}{D_e}}.$$ (3.19)

The anharmonicity of atomic vibration in crystals may also be approximated by adding higher-order terms to the potential energy function of equation 3.2:

$$\Phi_{(X)} = \frac{K_F}{2} X^2 + \frac{K_F}{3} X^3 + \cdots.$$ (3.20)

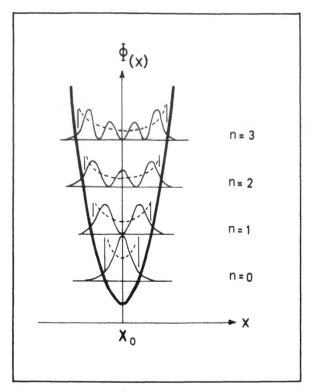

Figure 3.1 Energy levels and wave functions of harmonic oscillator. Heavy line: bounding potential (3.2). Light solid lines: quantum-mechanic probability density distributions for various quantum vibrational numbers ($\psi^2_{(n)}$, see section 1.16.1). Dashed lines: classical probability distribution; maximum classical probability is observed in the zone of inversion of motion where velocity is zero. From McMillan (1985). Reprinted with permission of The Mineralogical Society of America.

The energy eigenvalues for equation 3.20 (McMillan, 1985) are

$$E_{(n)} = h v_e \left[\left(n + \frac{1}{2} \right) - X_e \left(n + \frac{1}{2} \right)^2 + Y_e \left(n + \frac{1}{2} \right)^3 + \cdots \right], \qquad (3.21)$$

where v_e replaces v_0 in the presence of anharmonicity and X_e and Y_e are *anharmonicity constants*.

Zero-point energy, in the presence of anharmonicity (eq. 3.20), is

$$E_0 = h v_e \left(\frac{1}{2} - \frac{X_e}{4} + \frac{Y_e}{8} \right) \qquad (3.22)$$

and is lower with respect to the zero-point energy of the harmonic oscillator (X_e and Y_e are generally positive with $1 >> X_e >> Y_e$; cf. McMillan, 1985).

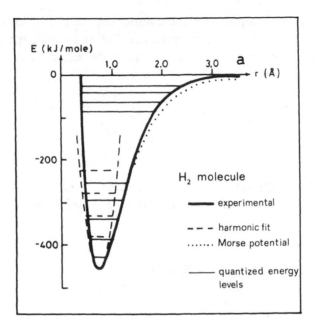

Figure 3.2 The experimental potential of H_2 molecule (heavy line) compared with harmonic fit (dashed line) and Morse potential (dotted line). From McMillan (1985). Reprinted with permission of The Mineralogical Society of America.

Figure 3.2 shows the experimental potential well for the H_2 molecule, compared with the harmonic fit and the Morse potential. Horizontal lines represent quantized energy levels. Note that, as vibrational quantum number n increases, the energy gap between neighboring levels diminishes and the equilibrium distance increases, due to the anharmonicity of the potential well. The latter fact is responsible for the thermal expansion of the substance.

At high T, when the spacing of vibrational energy levels is low with respect to thermal energy, crystalline solids begin to show the classical behavior predicted by kinetic theory, and the heat capacity of the substance at constant volume (C_V) approaches the theoretical limit imposed by free motion of all atoms along three directions, in a compound with n moles of atoms per formula unit (*limit of Dulong and Petit*):

$$C_V = 3nR. \tag{3.23}$$

At low T, thermal energy is not enough to ensure statistical occupancy of all energy levels accessible to each atom, and the heat capacity (either C_V or C_P) approaches zero as absolute zero is approached.

The relationship among heat capacity, entropy, and temperature in crystalline solids may be understood on the basis of two fundamental concepts: the *Boltzmann factor* and the *partition function* (or "summation over the states," from the German term *Zustandsumme*). Consider a system in which energy levels E_0, E_1,

$E_2, E_3, \ldots, E_n$ are occupied and the components in the various excited states may be recognized. The ratio of the atoms of a given component in excited state E_i, (N_i), with respect to all the other atoms of the same component at ground state E_0, (N_{v_0}), is given by the Boltzmann factor:

$$\frac{N_i}{N_{v_0}} = \exp\left(-\frac{\Delta E_i}{kT}\right), \tag{3.24}$$

where ΔE_i is the energy gap between excited state E_i and E_0. The partition function is the summation of the relative statistical population over all accessible quantum states:

$$Q = \sum_{i=0}^{\infty} \exp\left(-\frac{\Delta E_i}{kT}\right) \tag{3.25}$$

and total energy is given by

$$U = \sum_{i=0}^{\infty} N_i E_i. \tag{3.26}$$

The constant-volume heat capacity of the substance (C_V) represents the variation of total energy U with T in the harmonic approximation (X is constant over T), and the integration of C_V over T gives the (harmonic) entropy of the substance:

$$C_V = \left(\frac{\partial U}{\partial T}\right); \qquad S_{(T)} = \int_0^T C_V \frac{dT}{T}. \tag{3.27}$$

If a single energy level exists in the ground state, then clearly the partition function tends toward 1 as T decreases toward zero (*nondegenerate ground state*). However, because several (n) configurations may occur at ground state (*degenerate ground state*), the system has a nonzero entropy even at the zero point:

$$S_0 = R \ln(n). \tag{3.28}$$

The extension of the quantum-mechanic interpretation of the vibrational motion of atoms to a crystal lattice is obtained by extrapolating the properties of the diatomic molecule. In this case there are $3\mathcal{N}$ independent harmonic oscillators ($\mathcal{N}$ is here the number of atoms in the primitive unit cell—e.g., fayalite has four formula units Fe_2SiO_4 in the unit cell, which gives $3\mathcal{N} = 3 \times 4 \times 7 = 84$ vibrational normal modes*), each with a characteristic vibrational frequency:

$$v_t = \lambda_t^{1/2}. \tag{3.29}$$

The energy of the system is then established, in terms of characteristic frequencies and of a set of vibrational quantum numbers for the various levels:

$$E = \left(n_1 + \frac{1}{2}\right)hv_1 + \left(n_2 + \frac{1}{2}\right)hv_2 + \left(n_3 + \frac{1}{2}\right)hv_3 + \cdots. \tag{3.30}$$

The energy of the nondegenerate ground state is given by the summation of the various oscillators over the characteristic frequency:

$$E_0 = \frac{1}{2}h\sum_{t=1}^{3\mathcal{N}} v_t. \tag{3.31}$$

Planck showed that the mean energy of a great number of oscillators, each with a characteristic angular frequency $\omega_t = 2\pi v_t$, is given by

$$\langle\langle E \rangle\rangle = \frac{\hbar\omega_t}{\left(e^{\hbar\omega_t/kT} - 1\right)}, \tag{3.32}$$

where

$$\hbar = \frac{h}{2\pi}. \tag{3.33}$$

Einstein (1907) applied equation 3.32 (originally conceived by Planck for the quantization of electromagnetic energy) to the quantization of particle energy, describing the internal energy of a solid composed of N_0 atoms as

$$U = \frac{3N_0}{\left(e^{\hbar\omega_t/kT} - 1\right)}. \tag{3.34}$$

C_V is then obtained from equation 3.34 by partial derivation in dT:

* The three translational and three rotational degrees of freedom become *normal modes* when the "unit cell molecule" is embedded in the crystal.

$$C_V = \left(\frac{\partial U}{\partial T}\right) = \frac{3N_0 k X^2 e^X}{\left(e^X - 1\right)^2} = 3RE_{(X)} \tag{3.35}$$

where

$$X = \frac{h\nu_t}{kT} = \frac{\hbar\omega_t}{kT} \tag{3.36}$$

and

$$E_{(X)} = \frac{X^2 e^X}{\left(e^X - 1\right)^2} = \frac{X^2}{4\sin^2\left(\dfrac{X}{2}\right)} . \tag{3.37}$$

Equation 3.37, known as the Einstein function, is tabulated for various X-values (see, for instance, Kieffer, 1985). In the Einstein function, the characteristic frequency ω_t (and the corresponding characteristic temperature θ_E; see, for instance, eq. 3.40) has an arbitrary value that optimizes equation 3.35 on the basis of high-T experimental data. Extrapolation of equation 3.35 at low temperature results in notable discrepancies from experimental values. These discrepancies found a reasonable explanation after the studies of Debye (1912) and Born and Von Karman (1913).

In the Debye model, the Brillouin zone (see section 3.3) is replaced by a sphere of the same volume in the reciprocal space (cf. eq. 3.57):

$$\frac{4}{3}\pi K_{\text{max}}^3 = \frac{\left(2\pi\right)^3}{V_L} . \tag{3.38}$$

Maximum wave vector K_{max} has maximum angular frequency ω_D:

$$\omega_D = \mathcal{V}_M K_{\text{max}}, \tag{3.39}$$

where $\mathcal{V}_M$ is the mean velocity of acoustic waves in the crystal.

The characteristic frequency ω_D leads to the definition of a characteristic temperature for the solid, known as the *Debye temperature* (whose nature is analogous to that of θ_E):

$$\theta_D = \frac{\hbar\omega_D}{k} . \tag{3.40}$$

The Debye temperature of the solid defines the form of the vibrational spectrum in the acoustic zone (low frequency) and is related to the molar volume of the solid V and to the mean velocity of acoustic waves $\mathcal{V}_M$ through

$$\theta_D = \frac{\hbar}{k}\left(\frac{6\pi^2 N_0}{ZV}\right)^{1/3} \mathcal{V}_M, \tag{3.41}$$

where Z is the number of formula units per unit cell.

The heat capacity at constant volume in the Debye model is given by

$$C_v = 9nR\left(\frac{T}{\theta_D}\right)^3 \int_0^{\theta_D/T} \frac{e^X}{\left(e^X - 1\right)^3} X^4\, dX = 3nRD\left(\frac{\theta_D}{T}\right), \tag{3.42}$$

where $D(\theta_D/T)$ is a tabulated function known as *Debye heat capacity* (see table 3 in Kieffer, 1985).

3.2 Heat Capacity at Constant P and Maier-Kelley Functions

As we have already seen, the heat capacity is the amount of heat that must be furnished to raise the temperature of a given substance by 1 K at the T of interest. If constant pressure is maintained during heat transfer, heat capacity is defined as "heat capacity at constant P" (C_P). As already seen in chapter 2, C_P is the partial derivative of the enthalpy of the substance at constant P and composition—i.e.,

$$C_P = \left(\frac{\partial H}{\partial T}\right)_{P,n}. \tag{3.43}$$

If heat transfer takes place at constant volume, the magnitude is defined as "heat capacity at constant volume" (C_V) and is equivalent, as we have seen, to the partial derivative of the internal energy of the substance at constant volume and composition:

$$C_V = \left(\frac{\partial U}{\partial T}\right)_{V,n}. \tag{3.44}$$

C_P and C_V are related through

$$C_P = C_V + \frac{\alpha_V^2}{\beta_V} TV, \tag{3.45}$$

where α_V and β_V are, respectively, the *isobaric thermal expansion coefficient* and the *isothermal compressibility coefficient* of the substance (see also section 1.15).

We have seen in chapter 2 that the heat capacity at constant P is of fundamental importance in the calculation of the Gibbs free energy, performed by starting from the standard state enthalpy and entropy values

$$\overline{H}_{T,P_r} = \overline{H}^0_{T_r,P_r} + \int_{T_r}^{T} C_P \, dT, \tag{3.46}$$

$$\overline{S}_{T,P_r} = \overline{S}^0_{T_r,P_r} + \int_{T_r}^{T} C_P \frac{dT}{T}, \tag{3.47}$$

and

$$\overline{G}_{T,P_r} = \overline{H}_{T,P_r} - T\overline{S}_{T,P_r}. \tag{3.48}$$

With increasing T, C_P, like C_V, approaches a limit imposed by the Dulong and Petit rule:

$$C_V = 3nR \tag{3.49}$$

and

$$C_P = 3nR + \frac{\alpha_V^2}{\beta_V} TV, \tag{3.50}$$

where n is the number of atoms in the compound formula.

The C_P value of crystalline substances is obtained directly from calorimetric measures operated at various T. C_P dependency on T is commonly represented by polynomial expansions of the type:

$$C_P = K_1 + K_2 T^A + K_3 T^B + \cdots + K_n T^N, \tag{3.51}$$

where $K_1, K_2, K_3, \ldots, K_n$ are interpolation constants and $A, B, \ldots, N$ are fixed exponents. Already in 1932, Maier and Kelley had proposed an interpolation of the type

$$C_P = K_1 + K_2T + K_3T^{-2}. \tag{3.52}$$

Since then, several other polynomial expansions have been proposed. Haas and Fisher (1976), for instance, proposed an interpolation with five terms in the form

$$C_P = K_1 + K_2T + K_3T^{-2} + K_4T^2 + K_5T^{-1/2}. \tag{3.53.1}$$

The thermodynamic tables of Robie et al. (1978) use the Haas-Fisher polynomial (eq. 3.53.1). Helgeson et al. (1978) use the Maier-Kelley expansion, changing sign at the third term:

$$C_P = K_1 + K_2T - K_3T^{-2}. \tag{3.53.2}$$

Barin and Knacke (1973) and Barin et al. (1977) adopt a four-term polynomial:

$$C_P = K_1 + K_2T + K_3T^{-2} + K_4T^2. \tag{3.53.3}$$

More recently, Berman and Brown (1985) proposed an interpolation of the type

$$C_P = K_1 + K_2T^{-1/2} + K_3T^{-2} + K_4T^{-3}, \tag{3.54}$$

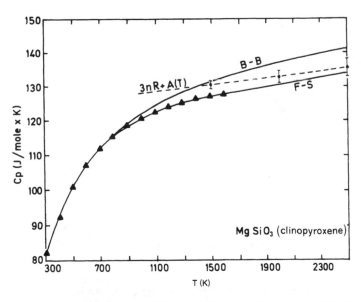

Figure 3.3 Interpolation properties of Maier-Kelley functions, exemplified by compound clinoenstatite. Experimental data from Robie et al. (1978). B-B = Berman-Brown polynomial; F-S = Fei-Saxena polynomial. Dashed line: Dulong and Petit limit. Reprinted from Y. Fei and S. K. Saxena, *Geochimica et Cosmochimica Acta,* 51, 251–254, copyright © 1987, with kind permission from Elsevier Science Ltd., The Boulevard, Langford Lane, Kidlington OX5 1GB, UK.

where $K_2 \leq 0$ and $K_3 \leq 0$. Polynomial 3.54 seems to obey the limits imposed by vibrational theory better than the preceding forms (cf. eq. 3.50).

Even more recently, Fei and Saxena (1987) proposed the more complex functional form

$$C_P = 3nR\left(1 + K_1 T^{-1} + K_2 T^{-2} + K_3 T^{-3}\right) + \left(A + BT\right) + C'_P, \qquad (3.55)$$

where A and B are coefficients related, respectively, to thermal expansion and compressibility, and C'_P is the deviation from the theoretical Dulong and Petit limit due to anharmonicity, internal disorder, and electronic contributions. Figure 3.3 shows how the Berman-Brown and Saxena-Fei equations conform with the experimental data of Robie et al. (1978) for clinoenstatite ($MgSiO_3$).

As has been repeatedly observed by various authors (see, for instance, Robie et al., 1978; Haas et al., 1981; Robinson and Haas, 1983), Maier-Kelley and Haas-

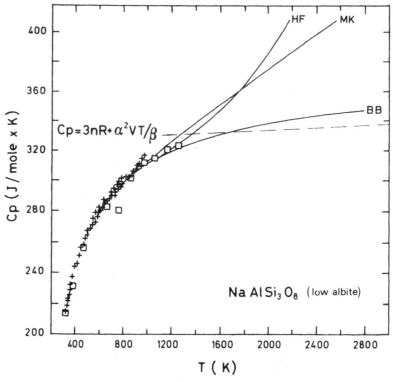

Figure 3.4 Comparison of interpolation properties of Maier-Kelley (MK), Haas-Fisher (HF), and Berman-Brown (BB) functions for low albite. Dashed line: Dulong and Petit limit (eq. 3.50). From R. G. Berman and T. H. Brown, Heat capacity of minerals in the system $Na_2O-K_2O-CaO-MgO-FeO-Fe_2O_3-Al_2O_3-SiO_2-TiO_2-H_2O-CO_2$: representation, estimation, and high temperature extrapolation, *Contributions to Mineralogy and Petrology*, 89, 168–183, figure 3, copyright © 1985 by Springer Verlag. Reprinted with the permission of Springer-Verlag GmbH & Co. KG.

Fisher polynomials can be used only within the T range for which they were created. Extrapolation beyond the T limits of validity normally implies substantial error progression in high-T entropy and enthalpy calculations. For instance, figure 3.4 compares Maier-Kelley, Haas-Fisher, and Berman-Brown polynomials for low albite. As can be seen, the first two interpolants, if extended to high T, definitely exceed the Dulong and Petit limit. The Berman-Brown interpolant also passes this limit, but the bias is less dramatic.

3.3 Entropy and Heat Capacity from Vibrational Spectra: Kieffer's Model

It has already been stated that, theoretically, N atoms in a crystal have $3N$ possible vibrational modes. Obviously, if we knew the energy associated with each vibrational mode at all T and could sum the energy terms in the manner discussed in section 3.1, we could define the internal energy of the crystal as a function of T, C_V could then be obtained by application of equation 3.27, and (harmonic) entropy could also be derived by integration of C_V in dT/T.

Although the number of atoms in a crystal is extremely high, we can imagine the crystal as generated by a spatial reproduction of the asymmetric unit by means of symmetry operations. The calculation can thus be restricted to a particular portion of space, defined as the *Brillouin zone* (Brillouin, 1953).

If we consider a primitive Bravais lattice with cell edges defined by vectors $\mathbf{a}_1$, $\mathbf{a}_2$, and $\mathbf{a}_3$, the corresponding reciprocal lattice is defined by reciprocal vectors $\mathbf{b}_1$, $\mathbf{b}_2$, and $\mathbf{b}_3$ so that

$$\mathbf{b}_1 = \frac{(2\pi)\mathbf{a}_2 \times \mathbf{a}_3}{|\mathbf{a}_1 \cdot \mathbf{a}_2 \times \mathbf{a}_3|}; \qquad \mathbf{b}_2 = \frac{(2\pi)\mathbf{a}_3 \times \mathbf{a}_1}{|\mathbf{a}_2 \cdot \mathbf{a}_3 \times \mathbf{a}_1|}; \qquad \mathbf{b}_3 = \frac{(2\pi)\mathbf{a}_1 \times \mathbf{a}_2}{|\mathbf{a}_3 \cdot \mathbf{a}_1 \times \mathbf{a}_2|}. \tag{3.56}$$

The volume of Brillouin zone V_B corresponds to $(2\pi)^3$ times the reciprocal of cell volume V_L commonly adopted in crystallography, and the origin of the coordinates is at the center of the cell (and not at one of the corners):

$$V_B = \frac{(2\pi)^3}{V_L}. \tag{3.57}$$

Wave vectors $\mathbf{K}_{(\eta)}$ are defined in the reciprocal space according to

$$\mathbf{K}_{(\eta)} = \eta_1 \mathbf{b}_1 + \eta_2 \mathbf{b}_2 + \eta_3 \mathbf{b}_3, \tag{3.58}$$

where η_1, η_2, and η_3 are integers (Kieffer, 1979a).

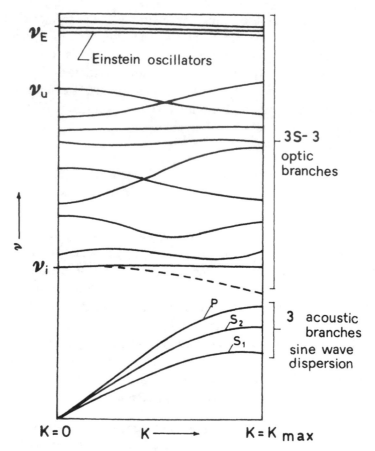

Figure 3.5 Phonon dispersion diagram for a complete unit cell with $3\mathcal{N}$ degrees of freedom. From Kieffer (1985). Reprinted with permission of The Mineralogical Society of America.

N is thus replaced by the number of atoms in the corresponding primitive unit cell ($\mathcal{N}$; see section 3.1). We now (theoretically) have two possibilities:

1. We can compute from first principles all possible vibrational modes for $3\mathcal{N}$ oscillators in the cell unit, solving the Schrödinger equation with appropriate atomic (and/or molecular) wave functions.
2. We can deduce the vibrational modes of atoms from experimental data on the interaction of radiation of various frequencies (photons and phonons) with the crystalline matter.

The first approach is still computationally prohibitive because of the chemical-structural complexity of the crystalline materials of our interest. The second type of approach is also rather complex, however, and necessitates several operative assumptions.

Figure 3.5 shows the path of the vibrational frequencies of the crystal along a given direction in the reciprocal lattice. For each value of **K** there are $3\mathcal{N}$ degrees

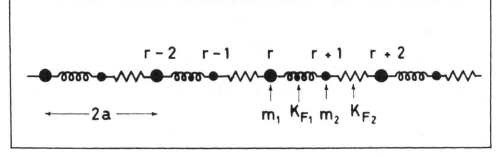

Figure 3.6 Diatomic chain with two force constants K_{F_1} and K_{F_2} and different atomic masses ($m_1 > m_2$). Reprinted with permission from Kieffer (1979a), *Review of Geophysics and Space Physics,* 17, 1–19, copyright © 1979 by the American Geophysical Union.

of freedom, and the vectors are defined in the Brillouin zone from $\mathbf{K} = 0$ to $\mathbf{K} = \mathbf{K}_{max}$. In this *phonon dispersion diagram,* each vibrational mode, for a given value of wave vector $\mathbf{K}$, pertains to a branch represented by a curve in the figure. Three *acoustic branches* (which approximate zero frequency at $\mathbf{K} = 0$) and $3\mathcal{N} - 3$ *optical branches* (the three at the highest frequency being *Einstein oscillators*) can be recognized.*

The nature of the wave vectors sketched in figure 3.5 is not dissimilar from what can be deduced for a harmonic oscillator with two force constants K_{F_1} and K_{F_2}.

Consider a diatomic chain in which the atoms, of distinct masses m_1 and m_2, are positioned at distance a (figure 3.6). The repetition distance of the chain is $2a$, and the Brillouin zone falls between $-\pi/2a$ and $\pi/2a$. If only the interactions between first neighbors are significant, the equation of motion for atom r at position μ is given by

$$F_r = K_{F_1}\left(\mu_{r-1} - \mu_r\right) + K_{F_2}\left(\mu_{r-1} - \mu_r\right) = m_1 \frac{d^2\mu_r}{dt^2} \tag{3.59}$$

(Kieffer, 1985) and, for the atom at position $r + 1$,

$$F_{r+1} = K_{F_1}\left(\mu_r - \mu_{r+1}\right) + K_{F_2}\left(\mu_{r+2} - \mu_{r+1}\right) = m_2 \frac{d^2\mu_{r+1}}{dt^2}. \tag{3.60}$$

Because we are in a single dimension, we can adopt a scalar notation (i.e., K instead of $\mathbf{K}$).

---

* The motion of atoms in the lattice can be depicted as a wave propagation (phonon). By "dispersion" we mean the variation in the wave frequency as reciprocal space is traversed. The propagation of sound waves is similar to the translation of all atoms of the unit cell in the same direction; hence the set of translational modes is commonly defined as an "acoustic branch." The remaining vibrational modes are defined as "optical branches," because they are capable of interaction with light (see McMillan, 1985, and Tossell and Vaughan, 1992, for more exhaustive explanations).

The solution of equations 3.59 and 3.60 is a wave equation of amplitudes ξ_1 and ξ_2:

$$\mu_r = \xi_2 \exp(irKa - i\omega t) \tag{3.61}$$

$$\mu_{r+1} = \xi_1 \exp\left[i(r+1)Ka - i\omega t\right], \tag{3.62}$$

where ω is the angular frequency

$$\omega = 2\pi v \tag{3.63}$$

The system comprised of equations 3.62 and 3.63 has solutions only when the determinant of coefficients ξ_1 and ξ_2 (*secular determinant*) is zero. The coefficient matrix, or *dynamic matrix*, is

$$\begin{vmatrix} m_1\omega^2 - K_{F_1} - K_{F_2} & K_{F_1}\exp(iKa) + K_{F_2}\exp(-iKa) \\ K_{F_1}\exp(-iKa) + K_{F_2}\exp(iKa) & m_2\omega^2 - K_{F_1} - K_{F_2} \end{vmatrix}. \tag{3.64}$$

The general solution of matrix 3.64 is

$$\omega^2 = \frac{K_{F_1} + K_{F_2}}{2m^*}\left\{1 \pm \left[1 - \frac{16K_{F_1}K_{F_2}m^{*2}}{\left(K_{F_1} + K_{F_2}\right)^2 m_1 m_2}\sin^2\left(\frac{Ka}{2}\right)\right]^{1/2}\right\}, \tag{3.65}$$

where m^* is the reduced mass (see section 3.1 and eq. 3.6).

For low values of K, $\sin^2 Ka \cong K^2a^2$, and equation 3.65 has roots

$$\omega_0 \approx \left(K_{F_1} + K_{F_2}\right)\left(m_1 + m_2\right)^{-1}K^2a^2 \tag{3.66}$$

and

$$\omega_1 \approx \left(K_{F_1} + K_{F_2}\right)\left(\frac{1}{m_1} + \frac{1}{m_2}\right). \tag{3.67}$$

The first solution represents acoustic vibrational modes, and the second represents optical vibrational modes.

A typical vibrational spectrum of a crystalline phase appears as a section of the dispersion diagram along the ordinate axis. Figure 3.7 shows a generalized

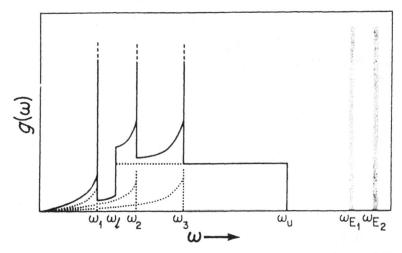

Figure 3.7 Schematic representation of a vibrational spectrum of a crystalline phase. Dotted curves: acoustic branches and optical continuum. Solid line: total spectrum. ω_{E1} and ω_{E2}: Einstein oscillators. Reprinted with permission from Kieffer (1979c), *Review of Geophysics and Space Physics*, 17, 35–39, copyright © 1979 by the American Geophysical Union.

diagram of the density of states: in this case the ordinate axis represents the number of frequencies between ω and $\omega + d\omega$. Acoustic modes generally prevail at low frequencies, the optical continuum occurs at intermediate frequencies, and Einstein's oscillators are found at high frequencies. According to Kieffer (1985), the density of states for acoustic branches may be represented by three sinusoidal functions:

$$g_{(\omega)} = \sum_{i=1}^{3} 3N_0 \left(\frac{2}{\pi}\right)^3 \frac{1}{Z} \frac{\left[\arcsin\left(\omega/\omega_i\right)\right]^2}{\left(\omega_i^2 - \omega\right)^{1/2}} + g_{0(\omega_i)}, \qquad (3.68)$$

where ω_i is the maximum frequency of each acoustic branch (see Kieffer, 1979c). Posing $X = \hbar\omega/kT$ (cf. eq. 3.36), equation 3.68 (*dispersed acoustic function;* cf. Kieffer, 1985) can be rewritten as

$$S_{(X_i)} = \left(\frac{2}{\pi}\right)^3 \int_0^{X_i} \frac{\left[\arcsin\left(X/X_i\right)\right]^2 X^2 e^2 \, dX}{\left(X_i^2 - X^2\right)^{1/2} \left(e^X - 1\right)^2}. \qquad (3.69)$$

In the optical continuum, frequency distribution is approximately constant (see figure 3.7):

$$g_{(\omega)} = g_0 \quad for \quad \omega > \omega_3. \tag{3.70}$$

The corresponding vibrational function in the optical continuum frequency field has the form

$$\mathcal{K}_{(X_l, X_u)} = \int_{X_l}^{X_u} \frac{X^2 e^X \, dX}{\left(X_u - X_l\right)^X \left(e^X - 1\right)^2} \tag{3.71}$$

(Kieffer, 1979c, 1985), where X_l and X_u correspond, respectively, to the lower and upper bounds.

For the Einstein oscillators,

$$g_{(\omega)} = g_0 + g_E \quad for \quad \omega = \omega_E, \tag{3.72}$$

and the corresponding vibrational function is

$$\mathcal{E}_{(X_E)} = \frac{X_E^2 e^{X_E}}{\left(e^{X_E} - 1\right)^2} \tag{3.73}$$

(cf. eq. 3.37). The molar heat capacity at constant V, normalized to a monatomic equivalent ($C_V{}^*$), based on vibrational functions 3.69, 3.71, and 3.73, is

$$C_V^* = \frac{3N_0 k}{\mathcal{N}} \sum_{i=1}^{3} S_{(X_i)} + 3N_0 k\left(1 - \frac{1}{\mathcal{N}} - q\right)\mathcal{K}_{(X_l, X_u)} + 3N_0 kq\mathcal{E}_{(X_E)}, \tag{3.74}$$

where $\mathcal{N} = nZ$ is the number of atoms in the primitive cell and q is the proportion of vibrational modes due to Einstein oscillators.

Because $N_0 k = R$, the unnormalized heat capacity (C_V) is

$$C_V = \frac{3R}{Z} \sum_{i=1}^{3} S_{(X_i)} + 3nR\left(1 - \frac{1}{\mathcal{N}} - q\right)\mathcal{K}_{(X_l, X_u)} + 3nRq E_{(X_E)}. \tag{3.75}$$

Because the dispersed acoustic function 3.69, the optic continuum function 3.71, and the Einstein function 3.73 may be tabulated for the limiting values of undimensionalized frequencies (see tables 1, 2, 3 in Kieffer, 1979c), the evaluation of C_V reduces to the appropriate choice of lower and upper cutoff frequencies for the optic continuum (i.e., X_l and X_u limits of integration in eq. 3.71), of the three zone edges for the dispersed acoustic function (i.e., three values of the X_i upper limit of integration in eq. 3.69), and of the frequency bands representing Einstein oscillators (i.e., X_E in eq. 3.73).

The nondimensionalized frequencies are related to linear and angular frequencies by equation 3.36. The conversion factor from linear frequencies in cm^{-1} to undimensionalized frequencies is $ch/k = 1.4387864$ cm (where c is the speed of light in vacuum). Acoustic branches for the various phases of interest may be derived from acoustic velocities through the guidelines outlined by Kieffer (1980). Vibrational modes at higher frequency may be derived by infrared (IR) and Raman spectra. Note incidentally that the tabulated values of the dispersed sine function in Kieffer (1979c) are 3 times the real ones (i.e., the listed values must be divided by 3 to obtain the appropriate $S_{(X_l)}$ value for each acoustic branch; see also Kieffer, 1985).

To appreciate the predictive properties of Kieffer's model, it is sufficient to compare calculated and experimental entropy values for several phases of geochemical interest in table 3.1, which also lists entropy values obtained through application of Debye's and Einstein's models. One advantage of Kieffer's model with respect to the two preceding formulations is its wider T range of applicability (Debye's model is appropriate to low frequencies and hence to low T, whereas Einstein's model is appropriate to high frequencies and hence to high T).

The bias observed between experimental measurements and Kieffer's model predictions is due to the relative paucity of experimental data concerning cutoff frequencies of acoustic branches, and also to the assumption that the frequencies of the lower optical branches are constant with **K** and equivalent to those detected by Raman and IR spectra (corresponding only to vibrational modes at **K** = 0). Indeed, several of these vibrational modes, and often the most important ones, are inactive under Raman and IR radiation (Gramaccioli, personal communication). The limits of the Kieffer model and other hybrid models with respect to nonempirical computational procedures based on the equation of motion of the Born–Von Karman approach have been discussed by Ghose et al. (1992).

3.4 Empirical Estimates of Heat Capacity at Constant Pressure

For many crystalline compounds, C_P values are experimentally known and tabulated at various T conditions. In several cases, the data are also presented in the interpolated form of a heat capacity function through Maier-Kelley-type polynomials.

Whenever the heat capacity C_P or heat capacity function $C_P = f(T)$ for a given substance is not known, reasonable estimates can be afforded through additivity algorithms. The simplest procedure is decomposition of the chemical for-

Table 3.1 Entropy values obtained by application of Einstein (E), Debye (D), and Kieffer (K) models, compared with experimental data at three different temperatures. Data are expressed in J/(mole × K). Values in parentheses are Debye temperatures (θ_D) and Einstein temperatures (θ_E) adopted in the respective models (from Kieffer, 1985).

Mineral/T(K)	Experimental Value	E	D	K
Halite		(210)	(306)	
298.15	34.8	34.2	33.3	35.1
700	55.0	55.1	54.0	55.9
1000	63.1	63.9	62.8	64.7
Periclase		(567)	(942)	
298.15	13.2	12.4	9.9	13.2
700	31.2	30.9	26.9	31.2
1000	39.2	39.4	35.3	39.7
Brucite		(1184)	(697)	
298.15	12.6	2.4	15.2	14.4
400	17.5	5.4	21.2	19.4
Corundum		(705)	(1026)	(model 1)
298.15	10.0	8.6	8.6	10.3
700	26.8	25.8	25.0	27.3
1000	34.7	34.2	33.3	35.6
Spinel		(680)	(879)	(model 2)
298.15	11.3	9.2	11.0	11.3
700	28.5	26.6	28.5	28.5
1000	36.9	35.0	36.9	36.9
Quartz		(815)	(567)	
298.15	13.7	6.4	19.3	13.9
700	29.1	22.5	38.9	29.7
α-Cristobalite		(819)	(519)	
298.15	14.3	6.4	21.2	13.8
400	19.1	11.0	27.8	18.5
Cohesite		(779)	(676)	(model 1)
298.15	13.5	7.0	15.8	14.0
700	29.5	23.5	34.7	30.2
1000	37.4	31.8	43.3	38.2
Stishovite		(757)	(921)	(model 1)
298.15	9.0	7.5	10.2	10.6
700	24.5	24.2	27.5	26.9
1000	32.0	32.5	35.2	35.1
Rutile		(580)	(781)	(model 1)
298.15	16.6	11.9	13.1	16.6
700	34.0	30.3	31.3	34.5
1000	42.1	38.8	39.8	42.9

Table 3.1 *(continued)*

Mineral/T(K)	Experimental Value	E	D	K
Albite		(753)	(472)	
298.15	15.9	7.6	23.3	15.9
700	32.3	24.3	43.4	32.2
1000	40.2	32.6	52.1	40.2
Microcline		(746)	(460)	
298.15	16.5	7.7	23.9	16.2
700	32.7	24.5	44.0	32.4
1000	40.8	32.8	52.7	40.5
Anorthite		(757)	(518)	
298.15	15.3	7.5	21.3	14.8
700	32.3	24.2	41.1	31.0
1000	40.7	32.5	49.8	39.2
Clinoenstatite		(691)	(719)	
298.15	13.4	8.9	14.6	13.6
700	29.9	26.2	33.2	30.6
1000	37.9	34.6	41.8	38.9
Orthoenstatite		(705)	(719)	
298.15	13.1	8.6	14.6	13.1
700	–	25.8	33.2	30.1
1000	–	34.2	41.8	38.3
Diopside		(680)	(654)	
298.15	14.2	9.2	16.4	14.5
700	31.2	26.6	35.5	31.5
1000	39.5	35.0	44.1	39.8
Jadeite		(711)	(724)	
298.15	13.2	8.4	14.5	13.7
700	29.7	25.6	33.1	30.5
1000	37.5	34.0	41.6	38.7
Tremolite		(763)	(547)	
298.15	13.3	7.4	20.1	14.1
700	30.1	24.0	39.8	30.8
1000	38.3	32.3	48.5	38.9
Muscovite		(860)	(520)	
298.15	14.6	5.7	21.2	13.8
700	31.0	21.3	41.0	30.0
1000	39.2	29.4	49.7	38.0
Talc		(882)	(525)	
298.15	12.4	5.4	21.0	12.5
700	28.6	20.8	40.8	28.5
Calcite		(766)	(468)	
298.15	18.3	7.3	23.5	19.2
700	35.3	23.9	43.6	35.3
1000	43.3	32.2	52.3	43.3

continued

Table 3.1 *(continued)*

Mineral/T(K)	Experimental Value	E	D	K
Zircon		(694)	(601)	
298.15	14.0	8.82	18.1	14.4
700	31.0	26.2	37.5	31.2
1000	39.3	34.5	46.2	39.5
Forsterite		(670)	(747)	
298.15	13.4	9.4	13.9	14.0
700	30.2	27.0	32.3	30.2
1000	38.1	35.4	40.9	39.5
Pyrope		(697)	(794)	
298.15	13.3	8.8	12.8	13.6
700	28.2	26.1	30.9	30.5
1000	36.5	34.4	39.4	38.7
Grossular		(704)	(821)	
298.15	12.7	8.6	12.2	12.6
700	29.7	25.8	30.1	29.5
1000	37.9	34.2	38.6	37.7
Kyanite		(757)	(916)	
298.15	10.4	7.5	10.3	10.8
700	26.6	24.2	27.6	27.2
1000	34.2	32.5	36.0	35.4
Andalusite		(754)	(761)	
298.15	11.5	7.5	13.5	12.0
700	27.5	24.3	32.0	28.4
1000	35.1	32.6	40.4	36.5
Sillimanite		(789)	(782)	
298.15	12.0	6.9	13.0	12.3
700	28.3	23.2	31.3	28.5
1000	36.1	31.5	39.8	36.6

mula into the oxide constituents, followed by stoichiometric summation of the C_P values of such constituents. The procedure is generalized by the expression

$$C^0_{P_r,T_r,i} = \sum_j v_{j,i} C^0_{P_r,T_r,j},$$ (3.76)

where $C^0_{P_r,T_r,i}$ is the standard state heat capacity for mineral i, $C^0_{P_r,T_r,j}$ is the standard state heat capacity for the jth constituent oxide, and $v_{j,i}$ is the stoichiometric number of moles of the jth oxide in mineral i.

For instance, general equation 3.76 applied to forsterite gives

$$C^0_{P_r,T_r,\text{Mg}_2\text{SiO}_4} = 2C^0_{P_r,T_r,\text{MgO}} + C^0_{P_r,T_r,\text{SiO}_2}.$$ (3.77)

Table 3.2 Coefficients of Berman-Brown polynomials for the commonest oxide constituents of rock-forming minerals. The resulting C_P is in J/(mole $\times$ K) (from Berman and Brown, 1985).

Oxide	K_1	$K_2 \times 10^{-2}$	$K_3 \times 10^{-5}$	$K_4 \times 10^{-7}$
Na_2O	95.148	0.0	−51.0405	83.3648
K_2O	105.140	−5.7735	0.0	0.0
CaO	60.395	−2.3629	0.0	−9.3493
MgO	58.196	−1.6114	−14.0458	11.2673
FeO	77.036	−5.8471	0.0	0.5558
Fe_2O_3	168.211	−9.7572	0.0	−17.3034
TiO_2	85.059	−2.2072	−22.5138	22.4979
SiO_2	87.781	−5.0259	−25.2856	36.3707
Al_2O_3	155.390	−8.5229	−46.9130	64.0084
H_2O^a	106.330	−12.4322	0.0	9.0628
H_2O^b	87.617	−7.5814	0.0	0.5291
CO_2	119.626	−15.0627	0.0	17.3869

[a]"Structural" water. [b]"Zeolitic" water.

Because equation 3.76 is valid at the various T conditions, by application of the additivity of polynomials, it follows that the coefficients of Maier-Kelley-type functions for mineral i can also be derived from the corresponding coefficients of the constituent oxides—i.e.,

$$K_{1,i} = \sum_j v_{j,i} K_{1,j} \tag{3.78.1}$$

$$K_{2,i} = \sum_j v_{j,i} K_{2,j} \tag{3.78.2}$$

$$K_{n,i} = \sum_j v_{j,i} K_{n,j}. \tag{3.78.3}$$

Table 3.2 lists the optimal values of the interpolation coefficients estimated by Berman and Brown (1987) for the most common oxide constituents of rock-forming minerals. These coefficients, through equations 3.78.1, 3.78.2, and 3.78.3, allow the formulation of polynomials of the same type as equation 3.54, whose precision is within 2% of experimental C_P values in the T range of applicability. However, the tabulated coefficients cannot be applied to phases with lambda transitions (see section 2.8).

Because the heat capacities of crystalline solids at various T are related to the vibrational modes of the constituent atoms (cf. section 3.1), they may be expected to show a functional relationship with the coordination states of the various atoms in the crystal lattice. It was this kind of reasoning that led Robinson and

Table 3.3 Coefficients of "structural components" valid for polynomial 3.79. Coefficient K_6 appears only at integration for calculation of entropy. The resulting C_P is in J/(mole $\times$ K) (from Robinson and Haas, 1983).

Structural component	K_1	K_2	K_3	K_6	K_4	K_5
Al$_2$O$_3$-IV	156.985	6.4774E-3	0.00	−992.000	0.0000	−1372.21
Al$_2$O$_3$-V	205.756	−7.82311E-3	0.00	−1349.63	0.0000	−2084.06
Al$_2$O$_3$-VI	222.740	−8.20451E-3	0.00	−1507.24	0.0000	−2464.56
CaO-VI	78.8255	−1.91875E-3	0.00	−480.538	0.0000	−622.865
CaO-VII	78.8255	−1.91875E-3	0.00	−471.709	0.0000	−622.865
CaO-VIII	83.6079	−2.97891E-3	19,661.5	−515.167	0.0000	−716.401
Fe$_2$O$_3$-IV	318.412	−4.89380E-3	417,088	0.00000	2.6E-5	−3307.95
Fe$_2$O$_3$-VI	318.412	−4.89380E-3	417,088	0.00000	2.6E-5	−3307.95
FeO-VI	81.1612	0.00000000	0.00	−485.209	0.0000	651.941
F	13.9627	1.28265E-2	0.00	−52.2387	0.0000	0.00000
H$_2$O	56.9125	0.00000000	0.00	−300.702	0.0000	−263.847
OH	129.124	−6.01221E-3	632,070	−886.693	0.0000	−1645.32
K$_2$O-VIII	7.71711	5.27163E-2	0.00	108.368	0.0000	656.875
K$_2$O-VI	42.4609	1.70942E-2	0.00	−125.937	0.0000	171.435
MgO-IV	43.0846	7.44796E-4	0.00	−206.902	0.0000	0.00000
MgO-VI	89.9331	−3.19321E-3	0.00	−588.796	0.0000	−872.529
MgO-VIII	47.8300	0.00000000	−810,599	−245.321	0.0000	0.00000
Na$_2$O-VI	58.0738	1.24598E-2	0.00	−226.355	0.0000	−45.8234
Na$_2$O-VII	58.0738	1.24598E-2	0.00	−251.172	0.0000	−45.8234
Na$_2$O-VIII	58.0738	1.24598E-2	0.00	−259.204	0.0000	−45.8234
SiO$_2$-IV	109.383	−2.77591E-3	0.00	−704.147	0.0000	−1083.05

Haas (1983) to formulate a set of coefficients valid for "structural components" that take into account the structural states of the various elements in the phase of interest (table 3.3). The coefficients are then applied to a modified form of the Haas-Fisher polynomial—i.e.,

$$C_P = K_1 + 2K_2 T + K_3 T^{-2} + K_4 T^2 + K_5 T^{-1/2}. \qquad (3.79)$$

Using the structural coefficients in table 3.3 implies knowledge of the conditions of internal disorder of the various cations in the structure. In most cases (especially for pure components), this is not a problem. For instance, table 3.4 shows the decomposition into "structural components" of the *acmite* molecule, in which all Na$^+$ is in VIII-fold coordination with oxygen, Fe^{3+} is in VI-fold coordination, and Si^{4+} is in tetrahedral coordination. In some circumstances, however, the coordination states of the various elements are not known with sufficient precision, or, even worse, vary with T. A typical example is spinel (MgAl$_2$O$_4$), which, at $T < 600$ °C, has inversion $X = 0.07$ (i.e., 7% of the tetrahedrally coordinated sites are occupied by Al^{3+}) and, at $T = 1300$ °C, has inversion $X = 0.21$ (21% of tetrahedral sites occupied by Al^{3+}). Clearly the calculation of

Table 3.4 Coefficients of modified Haas-Fisher polynomial 3.79 for *acmite* component of pyroxene ($NaFeSi_2O_6$), after application of structural coefficients in table 3.3. The resulting C_P is in J/(mole × K) (adapted from Robinson and Haas, 1983).

Component	Moles	Partial Contribution to the Coefficients				
		K_1	K_2	K_3	K_4	K_5
Na_2O-VIII	0.5	29.0370	6.22990E-3	0.0	0.0	−0.02291
Fe_2O_3-VI	0.5	159.206	2.44690E-3	208,544	1.2856E-5	−1653.98
SiO_2-IV	2.0	218.766	−5.55182E-3	0.0	0.0	−2166.10
$NaFeSi_2O_6$	1.0	407.009	3.12498E-3	208,544	1.2856E-5	−3842.99

the Haas-Fisher coefficients of the C_P function through the structural components is different for the two different occupancy states. The resulting Haas-Fisher polynomials thus have only local validity and cannot be extended to the whole T range of stability of the phase (note, however, that, according to Berman and Brown, 1985, the structural effect on C_P at high T is quite limited, affecting less than 2% of the bulk amount).

A method alternative to the simple additive procedures described above is outlined by Helgeson et al. (1978). An exchange reaction is written, involving the compound of interest and an isostructural compound for which the C_P function is known with sufficient precision. It is then assumed that the ΔC_P of the reaction is negligible. Consider for instance the exchange reaction between the wollastonite and clinoenstatite components of pyroxene:

$$CaSiO_3 + MgO \Leftrightarrow MgSiO_3 + CaO . \qquad (3.80)$$

$$\text{wollastonite} \qquad\qquad \text{clinoenstatite}$$

The heat capacities of $MgSiO_3$, CaO, and MgO being known, we derive the heat capacity of $CaSiO_3$ by applying

$$C_{P\ CaSiO_3} = C_{P\ MgSiO_3} + C_{P\ CaO} - C_{P\ MgO} . \qquad (3.81)$$

According to Helgeson et al. (1978), the method is quite precise and leads to estimates better than those provided by simple additivity procedures.

Table 3.5 compares estimates made by the different methods: the procedure of Helgeson et al. (1978) effectively leads to the best estimates. According to Robinson and Haas (1983), however, the result of this procedure is path-dependent, because it is conditioned by the selected exchange components: different exchange reactions (with different reference components) can lead to appreciably different results for the same component.

Table 3.5 Comparison of simple additivity of oxide constituents (column II) and exchange method of Helgeson et al. (1978) (column I), as methods of estimating heat capacity for crystalline components. Experimental values are shown for comparison in column III. Lower part of table: adopted exchange reactions (for which it is assumed that $\Delta C_{P \text{ reaction}} = 0$). Data in J/(mole × K) (adapted from Helgeson et al., 1978).

Mineral	Formula	I	II	III
Wollastonite	$CaSiO_3$	84.1	87.4	85.4
Clinoenstatite	$MgSiO_3$	80.3	82.4	79.1
Forsterite	Mg_2SiO_4	116.7	120.9	118.0
Fayalite	Fe_2SiO_4	141.8	143.9	133.1
Diopside	$CaMgSi_2O_6$	164.4	169.5	156.9
Akermanite	$Ca_2MgSi_2O_7$	208.8	212.5	212.1
Merwinite	$Ca_3MgSi_2O_8$	253.6	255.2	252.3
Low albite	$NaAlSi_3O_8$	208.8	207.5	205.0
Microcline	$KalSi_3O_8$	212.5	215.1	202.9

$$CaSiO_3 + MgO \Leftrightarrow MgSiO_3 + CaO$$
$$MgSiO_3 + CaO \Leftrightarrow CaSiO_3 + MgO$$
$$Mg_2SiO_4 + 2CaO \Leftrightarrow \alpha\text{-}Ca_2SiO_4 + 2MgO$$
$$Fe_2SiO_4 + 2MgO \Leftrightarrow Mg_2SiO_4 + 2FeO$$
$$CaMgSi_2O_6 \Leftrightarrow MgSiO_3 + CaSiO_3$$
$$Ca_2MgSi_2O_7 + \alpha\text{-}Al_2O_3 \Leftrightarrow Ca_2Al_2SiO_7 + MgO + SiO_2$$
$$Ca_3MgSi_2O_8 \Leftrightarrow Ca_2MgSi_2O_7 + CaO$$
$$NaAlSi_3O_8 + CaO + 0.5\alpha\text{-}Al_2O_3 \Leftrightarrow CaAl_2Si_2O_8 + SiO_2 + 0.5Na_2O$$
$$KalSi_3O_8 + 0.5Na_2O \Leftrightarrow NaAlSi_3O_8 + 0.5K_2O$$

3.5 Empirical Estimates of Standard State Entropy

As shown by Helgeson et al. (1978), satisfactory estimates of standard state molar entropy for crystalline solids can be obtained through reversible exchange reactions involving the compound of interest and an isostructural solid (as for heat capacity, but with a volume correction). Consider the generalized exchange reaction

$$M_v X + v\, M^i O \Leftrightarrow M_v^i X + v\, MO, \tag{3.82}$$

where $M_v X$ and $M_v^i X$ are isostructural solids, and the molar entropies of $M_v X$, $M^i O$, and MO are known; X is the common anionic radical, and v is the number of moles of oxide components in the isostructural solids.

Posing

$$S_S^0 = \Delta S_{\text{reaction}} + \overline{S}_{M_v X}^0 = \overline{S}_{M_v^i X}^0 + v\overline{S}_{MO}^0 - v\overline{S}_{M^i O}^0 \tag{3.83}$$

and

$$V_S^0 = \Delta V_{\text{reaction}} + \overline{V}_{M_v X}^0 = \overline{V}_{M_v^i X}^0 + v\overline{V}_{MO}^0 - v\overline{V}_{M^iO}^0, \tag{3.84}$$

where $\overline{S}^0$ and $\overline{V}^0$ are, respectively, standard molar entropies and standard molar volumes, the standard molar entropy of the compound of interest is given by

$$\overline{S}_{M_v^i X}^0 = \frac{S_S^0 \left(V_S^0 + \overline{V}_{M_v^i X}^0 \right)}{2V_S^0}. \tag{3.85}$$

For nonisostructural solids, satisfactory estimates can be obtained with the method proposed by Fyfe et al. (1958):

$$\overline{S}_{M_v^i X}^0 = \sum_j v_{j,i}\overline{S}_j^0 + K\left(\overline{V}_{M_v^i X}^0 - \sum_j v_{j,i}\overline{V}_j^0 \right), \tag{3.86}$$

where $v_{j,i}$ is the number of moles of the jth oxide per formula unit of compound i, $\overline{V}_j^0$ is the standard molar volume of the jth oxide, and $\overline{S}_j^0$ is the standard molar entropy of the jth oxide. Constant K is related to the partial derivative of entropy with respect to volume, corresponding to the ratio between isobaric thermal expansion α and isothermal compressibility β:

$$\left(\frac{\partial S}{\partial V} \right)_T = \left(\frac{\partial P}{\partial T} \right)_V = \frac{\alpha}{\beta}. \tag{3.87}$$

Holland (1989) reconsidered the significance of constant K in light of Einstein's model for the heat capacity of solids (see eq. 3.35 and 3.45):

$$\left(\frac{\partial S}{\partial V} \right)_{298} = \frac{nRX^2}{\overline{V}_{M_v^i X}^0 \left(e^X - 1 \right)\left(1 - e^{-X} \right)}, \tag{3.88}$$

where n is the number of atoms in the formula unit and X is the nondimensionalized frequency (see eq. 3.36). As shown by Holland (1989), the mean value of $(\partial S/\partial V)_{298}$, obtained by application of equation 3.88 to a large number of compounds, is 1.07 ± 0.11 J $\cdot$ K^{-1} $\cdot$ cm^{-3}. An analogous development using Debye's model (more appropriate to low-T conditions; see section 3.1) leads to a mean partial derivative $(\partial S/\partial V)_{298}$ of 0.93 ± 0.10 J $\cdot$ K^{-1} $\cdot$ cm^{-3}. Based on the above evidence, Holland (1989) proposed a set of $\overline{S}_j^0 - \overline{V}_j^0$ finite differences for

Table 3.6 Comparison of predictive capacities of various equations in estimating standard molar entropy ($T = 298.15$ K, $P = 1$ bar). Column I = simple summation of standard molar entropies of constituent oxides. Column II = equation 3.86. Column III = equation 3.86 with procedure of Holland (1989). Column IV = equation 3.85. Values are in J/(mole × K). Lower part of table: exchange reactions adopted with equation 3.85 (from Helgeson et al., 1978) and $\bar{S}_j^!$ finite differences for structural oxides (Holland, 1989).

Mineral	Formula Unit	Experimental	I	II	III	IV
Clinoenstatite	$MgSiO_3$	67.8	70.3	61.5	67.2	67.8
Wollastonite	$CaSiO_3$	82.0	81.2	82.4	79.3	83.7
Diopside	$CaMgSi_2O_6$	143.1	149.4	131.0	144.2	143.9
Giadeite	$NaAlSi_2O_6$	133.5	145.6	120.1	134.8	136.0
Tephroite	Mn_2SiO_4	163.2	160.7	159.4	162.1*	162.3
Ca-olivine	Ca_2SiO_4	120.5	120.9	128.0	120.5	120.9
Phenakite	Be_2SiO_4	64.4	69.5	64.4	–	69.0
Willemite	Zn_2SiO_4	131.4	128.4	131.4	–	131.8
Akermanite	$Ca_2MgSi_2O_7$	209.2	189.1	195.8	208.0	209.2
Merwinite	$Ca_3MgSi_2O_8$	253.1	228.9	227.2	252.2	250.6
Low albite	$NaAlSi_3O_8$	207.1	187.0	203.8	206.8	203.3
Nefeline	$NaAlSiO_4$	124.3	104.2	120.1	125.8	125.9
Tremolite	$Ca_2Mg_5Si_8O_{22}(OH)_2$	548.9	584.9	554.8	553.2	551.5
Caolinite	$Al_2Si_2O_5(OH)_4$	202.9	213.8	216.7	188.4	201.7

$MgSiO_3 + MnO \Leftrightarrow MnSiO_3 + MgO$
$CaSiO_3 + MnO \Leftrightarrow MnSiO_3 + CaO$
$CaMgSi_2O_6 \Leftrightarrow MgSiO_3 + CaO + SiO_2$
$NaAlSi_2O_6 + CaO + MgO \Leftrightarrow CaMgSi_2O_6 + 0.5Na_2O + 0.5Al_2O_3$
$Mn_2SiO_4 + 2MgO \Leftrightarrow Mg_2SiO_4 + 2MnO$
$Ca_2SiO_4 + 2BeO \Leftrightarrow Be_2SiO_4 + 2CaO$
$Be_2SiO_4 + 2MgO \Leftrightarrow Mg_2SiO_4 + 2BeO$
$Zn_2SiO_4 + 2MgO \Leftrightarrow Mg_2SiO_4 + 2ZnO$
$Ca_2MgSi_2O_7 + Al_2O_3 \Leftrightarrow Ca_2Al_2SiO_7 + MgO + SiO_2$
$Ca_3MgSi_2O_8 + CaO \Leftrightarrow 2Ca_2SiO_4 + MgO$
$NaAlSi_3O_8 + 0.5K_2O \Leftrightarrow KAlSi_3O_8 + 0.5Na_2O$
$NaAlSiO_4 + 0.5K_2O \Leftrightarrow KAlSiO_4 + 0.5Na_2O$
$Ca_2Mg_5Si_8O_{22}(OH)_2 \Leftrightarrow Mg_3Si_4O_{10}(OH)_2 + 2CaMgSi_2O_6$
$Al_2Si_2O_5(OH)_4 + 2SiO_2 \Leftrightarrow Al_2Si_4O_{10}(OH)_2 + H_2O$

Component	$\bar{S}_j^!$	Component	$\bar{S}_j^!$	Component	$\bar{S}_j^!$
[4]SiO_2	17.45	[8]CaO	27.37	[8]Na_2O	56.32
[6]Al_2O_3	22.60	[8-10]CaO	34.37	[9-12]Na_2O	79.49
[4]Al_2O_3	28.89	[gt]CaO	17.86	[a]K_2O	79.55
[6]MgO	15.75	[4-8]FeO	30.78	[b]K_2O	87.96
[4]MgO	18.77	[gt]FeO	36.50	$H_2O(a)$	15.71
[7]MgO	21.06	[6]MnO	33.41	$H_2O(b)$	7.44
[gt]MgO	26.06	[6]TiO_2	32.63		
[6]CaO	21.94	[6]Fe_2O_3	50.24		

[6] = VI-fold coordination

[a] = framework positions

[b] = interlayer positions and cavities

[gt] = valid for garnets

$H_2O(a)$ = diaspore, gibbsite, serpentine, dioctahedral micas

$H_2O(b)$ = chlorites, talc, trioctahedral micas, amphiboles

*29.8 J· K^{-1} · $mole^{-1}$ added to account for electronic contribution to configurational disorder.

structural oxides, to be used in conjunction with equation 3.86 at $K = 1$. These terms, in units of $J \cdot K^{-1} \cdot mole^{-1}$, are reported in the lower part of table 3.6.

Standard molar entropy estimates obtained by means of equation 3.85 are generally within 1% deviation of the calorimetric entropy for the majority of silicates, provided that the application does not involve ferrous compounds. Estimates obtained by applying equation 3.86 are generally less precise. When applied to ferrous compounds, either equation (3.85 or 3.86) leads to values that are generally higher than calorimetric ones. According to Helgeson et al. (1978), the bias is around 10 J per mole of FeO in the compound. According to Burns (1970) and Burns and Fyfe (1967), the lower entropy of ferrous compounds is due to the nonspherical symmetry of Fe^{2+} ions in oxides and silicates. More precisely, the entropy differences are due to the different electronic contributions of the d orbitals in the various coordination states (see equations 4 to 7 in Wood, 1980). To account for such contributions, appropriate correction factors for nonspherical ions can be introduced in equation 3.85. With the method of Holland (1989), equation 3.86 reduces to $\overline{S}^0_{M_i^\nu X} = \Sigma_j \, v_{j,i} \overline{S}^1_j + \overline{V}^0_{M_i^\nu X}$ where the $\overline{S}^1_j$ terms are finite differences $\overline{S}^0_j - \overline{V}^0_j$ for structural oxides. The precision of the method is comparable to that of equation 3.85 (average absolute deviation = $1.41 \, J \cdot K^{-1} \cdot mole^{-1}$ over 60 values).

Comparative evaluation of the predictive properties of equations 3.85 and 3.86 is given in table 3.6. Column I lists values obtained by simple summation of the entropies of the constituent oxides. This method (sometimes observed in the literature) should be avoided, because it is a source of significant errors. The lower part of table 3.6 lists the adopted exchange reactions and the $\overline{S}^1_j$ terms of Holland's model.

3.6 Estimates of Enthalpy and Gibbs Free Energy of Formation from Elements by Additivity: the ΔO^{2-} Method

Both the enthalpy of formation from the elements and the Gibbs free energy of formation from the elements of various crystalline compounds at the standard state of 298.15 K, 1 bar, can be predicted with satisfactory approximation through the linear additivity procedures developed by Yves Tardy and colleagues (Tardy and Garrels, 1976, 1977; Tardy and Gartner, 1977; Tardy and Viellard, 1977; Tardy, 1979; Gartner, 1979; Viellard, 1982). Tardy's method is based on the definition of the ΔO^{2-} parameter, corresponding to the enthalpy ($\Delta_H O^{2-}$) or the Gibbs free energy ($\Delta_G O^{2-}$) of formation of a generic oxide $M_{2/z}O_{(crystal)}$ from its aqueous ion:

$$\Delta_H O^{2-} \, M^{z+} = \Delta H^0_f \, M_{2/z}O_{\text{(crystal)}} - \frac{2}{z} \Delta H^0_f \, M^{z+}_{\text{(aqueous)}} \qquad (3.89)$$

$$\Delta_G O^{2-} \, M^{z+} = \Delta G^0_f \, M_{2/z}O_{\text{(crystal)}} - \frac{2}{z} \Delta G^0_f \, M^{z+}_{\text{(aqueous)}} . \qquad (3.90)$$

Table 3.7 Values of $\Delta_G O^{2-}$ for various cations. Data in kJ/mole (from Tardy and Garrels, 1976).

Ion	$\Delta_G O^{2-}$	Ion	$\Delta_G O^{2-}$	Ion	$\Delta_G O^{2-}$
H^+	-223.63	Fe^{2+}	-154.22	Y^{3+}	-143.09
Li^+	$+27.36$	Co^{2+}	-151.46	La^{3+}	-108.66
Na^+	$+147.19$	Ni^{2+}	-166.48	Ce^{3+}	-118.87
K^+	$+245.68$	Cu^{2+}	-192.17	Eu^{3+}	-142.38
Rb^+	$+273.63$	Zn^{2+}	-174.47	Ti^{3+}	-244.22
Cs^+	$+289.62$	Cd^{2+}	-150.50	V^{3+}	-210.33
Cu^+	-246.77	Hg^{2+}	-222.84	Cr^{3+}	-205.31
Ag^+	-165.06	Pd^{2+}	-250.62	Mn^{3+}	-241.42
Tl^+	-82.51	Ag^{2+}	-257.32	Fe^{3+}	-236.02
Be^{2+}	-225.10	Sn^{2+}	-231.04	Au^{3+}	-234.60
Mg^{2+}	-114.64	Pb^{2+}	-165.02	Ge^{4+}	-246.98
Ca^{2+}	-50.50	Al^{3+}	-203.89	Sn^{4+}	-263.26
Sr^{2+}	-2.51	Ga^{3+}	-228.74	Pb^{4+}	-260.75
Ba^{2+}	$+32.22$	In^{3+}	-207.78	Zr^{4+}	-259.24
Ra^{2+}	$+71.13$	Tl^{3+}	-246.98	Hf^{4+}	-259.41
V^{2+}	-168.62	Bi^{3+}	-225.81	U^{4+}	-226.35
Mn^{2+}	-135.56	Sc^{3+}	-205.64	Th^{4+}	-222.59

The ΔO^{2-} parameter is thus in linear relation with the solution energy of a given crystalline oxide $M_{2/z}O_{(crystal)}$ in water, through

$$2\,H^+ + M_{2/z}O_{(crystal)} \Leftrightarrow \frac{2}{z}M^{z+}_{(aqueous)} + H_2O_{(liquid)} \tag{3.91}$$

$$\Delta G^0_{reaction} = -\Delta_G O^{2-} + \Delta G^0_f\,H_2O_{(liquid)} \tag{3.92}$$

$$\Delta H^0_{reaction} = -\Delta_H O^{2-} + \Delta H^0_f\,H_2O_{(liquid)}. \tag{3.93}$$

Table 3.7 lists the $\Delta_G O^{2-}$ parameters derived by Tardy and Garrels (1976) on the basis of the Gibbs free energy of formation of oxides, hydroxides, and aqueous ions. The ΔO^{2-} parameters allow us to establish linear proportionality between the enthalpy (or Gibbs free energy) of formation of the compound from the constituent oxides within a given class of solids. If we define $\Delta H_{compound}$ and $\Delta G_{compound}$ as, respectively, the enthalpy of formation and the Gibbs free energy of formation of a given double oxide, starting from the oxide components $M_{2/z_1+}O$ and $N_{2/z_2}O$, the general proportionality relations between these two magnitudes and the ΔO^{2-} parameters of the corresponding cations have the form

$$\Delta H_{compound} = -\alpha_H\,\frac{n_1 n_2}{n_1 + n_2}\left(\Delta_H O^{2-}\,M^{z_1^+} + \Delta_H O^{2-}\,N^{z_2^+}\right) \tag{3.94}$$

and

$$\Delta G_{compound} = -\alpha_G \frac{n_1 n_2}{n_1 + n_2} \left(\Delta_G O^{2-} \; M^{z_i^+} + \Delta_G O^{2-} \; N^{z_i^+} \right) \qquad (3.95)$$

where n_1 and n_2 are the numbers of oxygen ions linked, respectively, to the M^{z_1+} and N^{z_2+} cations, and α_H and α_G are the correlation parameters for a given class of compounds (double oxides). Adding to the $\Delta H_{compound}$ (or $\Delta G_{compound}$) the enthalpy (or Gibbs free energy) of formation from the elements of the oxides in question, the enthalpy (or Gibbs free energy) of formation from the elements of the compound of interest is readily derived.

For instance, let us consider the compound clinoenstatite. Its enthalpy of formation from the constituent oxides is

$$\Delta H_{MgSiO_3} = H^0_{f,MgSiO_3} - H^0_{f,MgO} - H^0_{f,SiO_2}. \qquad (3.96.1)$$

$$\text{With} \; \frac{n_1 n_2}{n_1 + n_2} = \frac{2}{3}; \qquad \alpha_H = -1.3095 \qquad (3.96.2)$$

$$\text{and} \; \Delta_H O^{2-} \; Mg^{2+} = -139.95; \qquad \Delta_H O^{2-} \; Si^{4+} = -193.33, \qquad (3.96.3)$$

we obtain

$$\Delta H_{MgSiO_3} = -46.60; \qquad H^0_{f,MgSiO_3} = -46.60 - 601.49 - 910.70$$
$$(3.97)$$
$$= -1558.79.$$

The experimental value is -1547.45 (kJ/mole). The estimate thus has an approximation of about 10 kJ/mole.

As figure 3.8 shows, equations 3.94 and 3.95 appear as straight lines of slope α_H (or α_G in the case of Gibbs free energy) in a binary plot $\Delta_{compound} = f(\Delta O^{2-})$. The intercept value on the abscissa, corresponding to zero on the ordinate, represents the $\Delta_H O^{2-}$ of the "reference cation" based on equation 3.94. Thus, in the case of metasilicates (figure 3.8B), we have $\Delta_H O^{2-} Si^{4+} = -193.93$ kJ/mole (see also equations 3.96.3 and 3.97), and, for orthosilicates (figure 3.8A), $\Delta_H O^{2-} Si^{4+} = -204.53$ kJ/mole.

Table 3.8 lists values of α_H and $\Delta_H O^{2-}_{reference \; cation}$ for silicates and aluminates. The tabulated parameters, as simple interpolation factors, have good correlation coefficients, confirming the quality of the regression.

Table 3.9 lists general regression parameters between $\Delta_G O^{2-}$ and the Gibbs free energy of formation from constituent oxides, for various classes of substances. The values conform to the general equation

$$\Delta G_{compound} = a \times \Delta_G O^{2-} \; M^{z+} + b. \qquad (3.98)$$

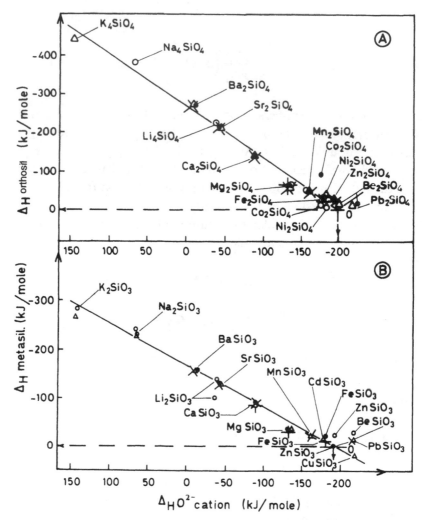

Figure 3.8 Relationship between $\Delta H_{compound}$ and $\Delta_H O^{2-}$ in orthosilicates (A) and metasilicates (B). The $\Delta_H O^{2-}$ value of the reference cation is found at a value of zero on the ordinate. Reprinted from Viellard (1982), Sciences Geologiques, Memoir n°69, Université Louis Pasteur, with kind permission of the Director of Publication.

Table 3.8 Regression parameters $\Delta_H O^{2-}$ for silicates and aluminates. Values in kJ/mole (from Tardy and Garrels, 1976, 1977; Tardy and Viellard, 1977; Tardy and Gartner, 1977).

Compound	α_H	$\Delta_H O^{2-}_{\text{ref. cation}}$	Points	R^2	Precision
$MSiO_3$	−1.3095	−193.93	13	0.9987	±9.84
M_2SiO_4	−1.305	−204.53	14	0.9961	25.3
$M_3Si_2O_7$	−1.213	−201.92	3	0.9999	8.4
MSi_2O_5	−1.258	−181.26	4	0.9993	6.2
M_3SiO_5	−1.297	−193.91	3	0.9703	73.5
MAl_2O_4	−0.797	−224.52	12	0.9374	22.3

Table 3.9 General regression parameters between $\Delta_H O^{2-}$ and $\Delta G_{compound}$. Values in kJ/mole (from Tardy and Garrels, 1976, 1977; Tardy and Viellard, 1977; Tardy and Gartner, 1977).

Compound	a	b	Points	R^2	Precision
$M(OH)_x$	−0.206	−46.02	104	0.975	10.46
$M_x SiO_3$	−0.673	−126.90		0.998	11.72
$M_x SiO_4$	−1.01	−189.12		0.985	16.74
$M_x PO_4$	−1.285	−418.48	18	0.972	49.37
$M_x SO_4$	−0.541	−159.79	37	0.982	57.74
$M_x NO_3$	−0.541	−159.79	20	0.984	31.80
$M_x CO_3$	−0.771	−199.41	32	0.987	35.15

3.7 Gibbs Free Energy of a Phase at High P and T, Based on the Functional Forms of Heat Capacity, Thermal Expansion, and Compressibility

We may first assume that isothermal compressibility β_V and isobaric thermal expansion coefficient α_V are independent, respectively, of T and P. Equations 1.91 and 1.99, integrated on T and P, respectively, give

$$\overline{V}_{T_r,P} = \overline{V}^0_{T_r,P_r} \, \exp\left[-\beta_V \left(P - P_r\right)\right] \tag{3.99}$$

and

$$\overline{V}_{T,P_r} = \overline{V}^0_{T_r,P_r} \, \exp\left[\alpha_V \left(T - T_r\right)\right] . \tag{3.100}$$

The final state of volume does not depend on the integration path (see appendix 2), and thus

$$\overline{V}_{T,P} = \overline{V}^0_{T_r,P_r} \, \exp\left[\alpha_V \left(T - T_r\right)\right] \exp\left[-\beta_V \left(P - P_r\right)\right] . \tag{3.101}$$

The ΔP values for which the change of volume is significant are quite high, and reference pressure P_r (normally $P_r = 1$) can be ignored in the development of calculations. The integrals on pressure in equations 2.94 and 2.95 take the forms

$$\int_{P_r}^{P} \left[\overline{V} - T\left(\frac{\partial \overline{V}}{\partial T}\right)_P\right] dP = P\overline{V}^0_{T_r,P_r} \exp\left[\alpha_V \left(T - T_r\right)\right]\left(1 - \frac{1}{2}\beta_V P\right)\left(1 - \alpha_V T\right)$$

$$\tag{3.102}$$

and

$$\int_{P_r}^{P} \left(\frac{\partial V}{\partial T}\right)_P dP = \alpha_V P \overline{V}^0_{T_r,P_r} \exp\left[\alpha_V\left(T - T_r\right)\right]\left(1 - \frac{1}{2}\beta_V P\right). \qquad (3.103)$$

We have seen (section 3.2) that heat capacity at constant P can be expressed in a functional form representing its T-dependency. Adopting, for instance, the Haas-Fisher polynomial:

$$C_P = K_1 + K_2 T + K_3 T^{-2} + K_4 T^2 + K_5 T^{-1/2}, \qquad (3.104)$$

the integrals on T in equations 2.94 and 2.95 take the forms

$$\int_{T_r}^{T} C_P \, dT = K_1\left(T - T_r\right) + \frac{K_2}{2}\left(T^2 - T_r^2\right) - K_3\left(T^{-1} - T_r^{-1}\right)$$
$$+ \frac{K_4}{3}\left(T^3 - T_r^3\right) + 2K_5\left(T^{1/2} - T_r^{1/2}\right) \qquad (3.105)$$

and

$$\int_{T_r}^{T} C_P \frac{dT}{T} = K_1 \ln\left(\frac{T}{T_r}\right) + K_2\left(T - T_r\right) - \frac{K_3}{2}\left(T^{-2} - T_r^{-2}\right)$$
$$+ \frac{K_4}{2}\left(T^2 - T_r^2\right) - 2K_5\left(T^{-1/2} - T_r^{-1/2}\right). \qquad (3.106)$$

The enthalpy of the phase at the P and T of interest is derived from the standard state values of enthalpy and volume ($\overline{H}^0_{T_r,P_r}$ and $\overline{V}^0_{T_r,P_r}$) through

$$\overline{H}_{T,P} = \overline{H}^0_{T_r,P_r} + \int_{T_r}^{T} C_P \, dT$$
$$+ P\overline{V}^0_{T_r,P_r} \exp\left[\alpha_V\left(T - T_r\right)\right]\left(1 - \frac{1}{2}\beta_V P\right)\left(1 - \alpha_V T\right). \qquad (3.107)$$

Analogously, for entropy we have

$$\overline{S}_{T,P} = \overline{S}^0_{T_r,P_r} + \int_{T_r}^{T} C_P \frac{dT}{T} + \alpha_V P \overline{V}^0_{T_r,P_r} \exp\left[\alpha_V\left(T - T_r\right)\right]\left(1 - \frac{1}{2}\beta_V P\right) \qquad (3.108)$$

and for Gibbs free energy we have

$$\overline{G}_{T,P} = \overline{G}^0_{T_r,P_r} - \overline{S}^0_{T_r,P_r}(T - T_r) + \int_{T_r}^{T} C_P \, dT - T \int_{T_r}^{T} C_P \frac{dT}{T}$$

$$+ P\overline{V}^0_{T_r,P_r} \exp\left[1 - \alpha_V(T - T_r) - \frac{1}{2}\beta_V P\right].$$

(3.109)

3.8 Solid Mixture Models

We will now consider in some detail the mixture models that can be applied to rock-forming minerals. These models are of fundamental importance for the calculation of equilibria in heterogeneous systems. In the examples given here we will refer generally to binary mixtures forming isomorphous compounds over the entire compositional field. Each of these mixtures also has a well-defined crystal structure, with two or more energetically distinguishable structural sites. Extension of the calculations to ternary mixtures will also be given. The discussion will be developed in terms of *molar properties* (i.e., referring to one mole of substance). The term "molar" will be omitted throughout the section, for the sake of simplicity.

3.8.1 Ideal Mixture

An "ideal" binary crystalline mixture obeys the equations

$$H_{\text{mixing}} = 0 \tag{3.110}$$

and

$$S_{\text{mixing}} = -R\left(X_A \ln X_A + X_B \ln X_B\right), \tag{3.111}$$

where X_A and X_B are the molar fractions of components A and B in a binary mixture. The entropy of mixing described by equation 3.111 derives from the *Boltzmann equation:*

$$S_{\text{configuration}} = k \ln Q, \tag{3.112}$$

where k is the *Boltzmann constant:*

$$k = \frac{R}{N_0} \tag{3.113}$$

and Q is "permutability"—i.e., the total number of possible configurational arrangements in the mixture. Q can be calculated through the combinatorial formula

$$Q = \frac{N_0!}{(N_0 A)!(N_0 B)!}. \tag{3.114}$$

Equations 3.112 and 3.114 give

$$S_{\text{configuration}} = k \ln\left[\frac{N_0!}{(N_0 A)!(N_0 B)!}\right] = k\left[\ln N_0! - \ln(N_0 A)! - \ln(N_0 B)!\right]. \tag{3.115}$$

By application of *Stirling's approximation* for large numbers (note that $N_0 = 6.022 \times 10^{23}$):

$$\ln N_0! = N_0 \ln N_0 - N_0, \tag{3.116}$$

as

$$N_0 = N_0 A + N_0 B, \tag{3.117}$$

we have

$$S_{\text{mixing}} = k\left[N_0 \ln N_0 - N_0 A \ln(N_0 A) - N_0 B \ln(N_0 B)\right]$$
$$= -kN_0\left[\frac{N_0 A}{N_0 A + N_0 B} \ln\left(\frac{N_0 A}{N_0 A + N_0 B}\right) + \frac{N_0 B}{N_0 A + N_0 B} \ln\left(\frac{N_0 B}{N_0 A + N_0 B}\right)\right]. \tag{3.118}$$

Because

$$\frac{N_0 A}{N_0 A + N_0 B} = X_A \quad \text{and} \quad \frac{N_0 B}{N_0 A + N_0 B} = X_B, \tag{3.119}$$

equation 3.111 is readily derived. Moreover, because

$$G_{\text{mixing}} = H_{\text{mixing}} - TS_{\text{mixing}}, \tag{3.120}$$

then

$$G_{\text{mixture}} = X_A \mu_A^0 + X_B \mu_B^0 + G_{\text{mixing}} \tag{3.121}$$

and, for an ideal mixture,

$$G_{\text{mixture}} = X_A \mu_A^0 + X_B \mu_B^0 + RT\left(X_A \ln X_A + X_B \ln X_B\right). \tag{3.122}$$

Equations 3.121 and 3.122 distinguish the bulk Gibbs free energy of the mixture (G_{mixture}) from the Gibbs free energy term involved in the mixing procedure (G_{mixing}).

Let us now consider in detail the mixing process of two generic components AN and BN, where A and B are cations and N represents common anionic radicals (for instance, the anionic group SiO_4^{4-}):

$$AN + BN \rightarrow (A, B)\, N \tag{3.123}$$

$$G_{\text{mixing}} = \Delta G_{123}. \tag{3.124}$$

If mixture (A,B)N is ideal, mixing will take place without any heat loss or heat production. Moreover, the two cations will be fully interchangeable: in other words, if they occur in the same amounts in the mixture, we will have an equal opportunity of finding A or B over the same structural position. The Gibbs free energy term involved in the mixing process is

$$
\begin{aligned}
G_{\text{mixing}} = \Delta G_{123} &= RT\left(X_A \ln X_A + X_B \ln X_B\right) \\
&= RT\left[X_A \ln X_A + \left(1 - X_A\right)\ln\left(1 - X_A\right)\right].
\end{aligned}
\tag{3.125}
$$

Adopting as standard state the condition of "pure component at T and P of interest," for whatever composition in the binary field, we have

$$\mu_A - \mu_A^0 = \left(\frac{\partial \Delta G_{123}}{\partial X_A}\right)_{P,T} = RT \ln X_A \tag{3.126}$$

and

$$\mu_B - \mu_B^0 = \left(\frac{\partial \Delta G_{123}}{\partial X_B}\right)_{P,T} = RT \ln X_B = RT \ln\left(1 - X_A\right). \tag{3.127}$$

Recalling now the relationship linking chemical potential to thermodynamic activity (eq. 2.75 in section 2.9), for all compositions in the binary field we will obtain

$$a_A = X_A \quad and \quad a_B = X_B = \left(1 - X_A\right). \tag{3.128}$$

The concepts above can be extended to multicomponent mixtures ($i > 2$), representing the Gibbs free energy of the mixture as

$$G_{\text{mixture}} = \sum_i \mu_i^0 X_i + G_{\text{mixing}}. \tag{3.129}$$

The Gibbs free energy of mixing can also be expressed as

$$G_{\text{mixing}} = \sum_i \left(\mu_i - \mu_i^0\right) X_i \tag{3.130}$$

and for an "ideal" multicomponent mixture we can write

$$G_{\text{mixing}} = G_{\text{ideal mixing}} = RT \sum_i X_i \ln X_i. \tag{3.131}$$

Figure 3.9A shows that the Gibbs free energy of ideal mixing is symmetric-concave in shape, with a maximum depth directly dependent on T. Note that the ideal Gibbs free energy of mixing term is always present *even in nonideal mixtures* (see figure 3.9B, C, and D, and section 3.11).

3.8.2 Generalities on Nonideal Mixtures

Maintaining the same standard state previously adopted for the ideal case, for nonideal mixtures we have

$$\mu_i - \mu_i^0 = \left(\frac{\partial \Delta G_{123}}{\partial X_i}\right)_{P,T} = \left(\frac{\partial G_{\text{mixing}}}{\partial X_i}\right)_{P,T} = RT \ln a_i, \tag{3.132}$$

where $a_i = X_i \gamma_i$ and γ_i is the activity coefficient.

The Gibbs free energy of mixing is given by

$$G_{\text{mixing}} = RT \sum_i X_i \ln a_i = RT \sum_i X_i \ln X_i + RT \sum_i X_i \ln \gamma_i$$

$$= G_{\text{ideal mixing}} + G_{\text{excess mixing}}, \tag{3.133}$$

where $G_{\text{excess mixing}}$ is the *excess Gibbs free energy of mixing* term.

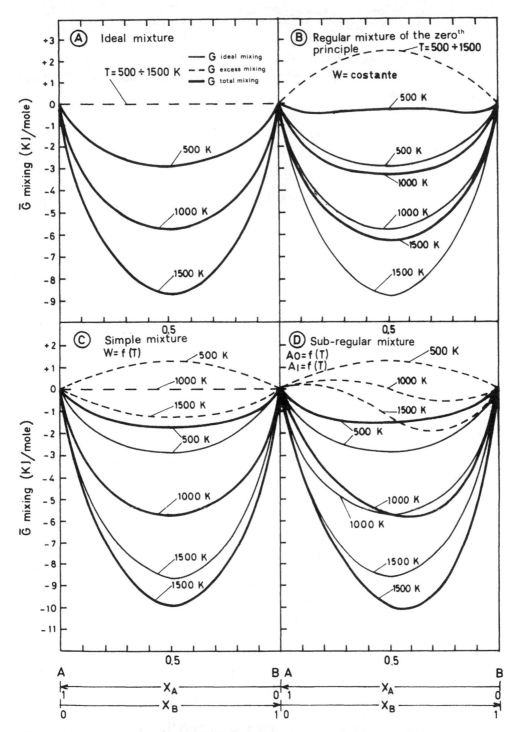

Figure 3.9 Conformation of Gibbs free energy curve in various types of binary mixtures. (A) Ideal mixture of components A and B. Standard state adopted is that of pure component at T and P of interest. (B) Regular mixture with complete configurational disorder: $W = 10$ kJ/mole for $500 < T(K) < 1500$. (C) Simple mixture: $W = 10 - 0.01 \times T(K)$ (kJ/mole). (D) Subregular mixture: $A_0 = 10 - 0.01 \times T$ (kJ/mole); $A_1 = 5 - 0.01 \times T$ (kJ/mole). Adopting corresponding Margules notation, an equivalent interaction is obtained with $W_{BA} = 15 - 0.02 \times T$ (kJ/mole); $W_{AB} = 5$ (kJ/mole).

It is clear from equation 3.133 that the thermodynamic properties of any *nonideal* mixture are *always* related to the corresponding properties of an ideal mixture through *excess terms:*

$$S_{\text{excess mixing}} = -\left(\frac{\partial G_{\text{excess mixing}}}{\partial T}\right) =$$

$$= -RT\left(X_A \frac{\partial \ln \gamma_A}{\partial T} + X_B \frac{\partial \ln \gamma_B}{\partial T}\right) - R\left(X_A \ln \gamma_A + X_B \ln \gamma_B\right) \qquad (3.134)$$

$$H_{\text{excess mixing}} = -T^2\left(\frac{\partial G_{\text{excess mixing}}/T}{\partial T}\right)$$

$$= -RT^2\left(X_A \frac{\partial \ln \gamma_A}{\partial T} + X_B \frac{\partial \ln \gamma_B}{\partial T}\right) \qquad (3.135)$$

$$V_{\text{excess mixing}} = \left(\frac{\partial G_{\text{excess mixing}}}{\partial P}\right) = RT\left(X_A \frac{\partial \ln \gamma_A}{\partial P} + X_B \frac{\partial \ln \gamma_B}{\partial P}\right), \quad (3.136)$$

where γ_A and γ_B are the activity coefficients of components AN and BN, respectively, in the (A,B)N mixture.

Let us now review the main types of nonideal mixtures.

3.8.3 Athermal Mixtures

In this type of nonideal mixture, deviations from ideality are solely due to entropic terms:

$$H_{\text{excess mixing}} = 0 \qquad (3.137)$$

$$G_{\text{excess mixing}} = H_{\text{excess mixing}}. \qquad (3.138)$$

3.8.4 Regular Mixtures with Complete Configurational Disorder (Zeroth Approximation)

Deviation from ideality is entirely due to enthalpic terms. The entropy of mixing is that of an ideal mixture:

$$S_{\text{excess mixing}} = 0 \qquad (3.139)$$

$$G_{\text{excess mixing}} = -TS_{\text{excess mixing}} \cdot \quad (3.140)$$

Excess enthalpy is parameterized through *interaction parameter W*:

$$G_{\text{excess mixing}} = X_A X_B W \quad (3.141)$$

$$W = N_0 w \quad (3.142)$$

$$w = \frac{2w_{AB} - w_{AA} - w_{BB}}{2} \cdot \quad (3.143)$$

The terms w_{AB}, w_{AA}, and w_{BB} are, respectively, the ionic interactions between A-B, A-A, and B-B atoms in the (A,B)N mixture. Note that, in this model, *W is not* dependent on *T* and *P* (see also figure 3.9B). The condition of complete disorder is often defined as "approximation of the Zeroth principle."

3.8.5 Simple Mixtures

Interaction parameter W^I *does* depend on *T* and *P*, and the excess Gibbs free energy of mixing is described as in the preceding model:

$$G_{\text{excess mixing}} = X_A X_B W^I. \quad (3.144)$$

T-dependency implies an entropic contribution to the nonideality of the mixing process:

$$-S_{\text{excess mixing}} = X_A X_B \frac{\partial W^I}{\partial T} \quad (3.145)$$

$$H_{\text{excess mixing}} = X_A X_B \left(W^I - T \frac{\partial W^I}{\partial T} \right). \quad (3.146)$$

The activity coefficients may be related to the molar concentration through the expressions

$$\ln \gamma_A = \frac{W^I}{RT} X_B^2 \quad (3.147)$$

and

$$\ln \gamma_B = \frac{W^I}{RT} X_A^2. \quad (3.148)$$

It can easily be deduced from equations 3.147 and 3.148 that the activity-composition relationships for the two components are symmetric in the compositional field (see also figure 3.9C).

3.8.6 Quasi-Chemical Model

The interaction between components in a mixture is regarded as the result of short-range electrostatic and repulsive interactions among the various ions. Ion A tends to be coordinated by A ions, if this condition minimizes the electrostatic-plus-repulsive potential of the structure, or, alternatively, by B ions, if this is the more favorable configuration in terms of energy.

The molar fractions of components in mixture $X_A + X_B$ are converted into *associative fractions* ϕ_A and ϕ_B through *contact factors* q_A and q_B:

$$\phi_A = 1 - \phi_B = \frac{X_A q_A}{X_A q_A + X_B q_B} \tag{3.149}$$

(see the concept of "associate" in section 4.8).

The tendency toward configurational disorder is expressed by parameter ω_q:

$$\omega_q = \left[1 + 4\phi_A \phi_B (\lambda - 1)\right]^{1/2}, \tag{3.150}$$

where

$$\lambda = \exp\left(\frac{2W^{II}}{ZRT}\right). \tag{3.151}$$

(Green, 1970). W^{II} in equation 3.151 is the interaction parameter of the quasi-chemical model and Z is the coordination number.

The activity coefficients of components in mixtures are expressed by

$$\gamma_A = \left[1 + \frac{\phi_B (\omega_q - 1)}{\phi_A (\omega_q + 1)}\right]^{Z q_A / 2} \tag{3.152}$$

and

$$\gamma_B = \left[1 + \frac{\phi_A (\omega_q - 1)}{\phi_B (\omega_q - 1)}\right]^{Z q_B / 2} \tag{3.153}$$

The excess Gibbs free energy of mixing is given by

$$G_{\text{excess mixing}} = \frac{Z}{2} RT \left\{ X_A q_A \ln \left[1 + \frac{\phi_B (\omega_q - 1)}{\phi_A (\omega_q + 1)} \right] + X_B q_B \ln \left[1 + \frac{\phi_A (\omega_q - 1)}{\phi_B (\omega_q + 1)} \right] \right\}.$$

(3.154)

Guggenheim (1952) named W^{II} "cooperative energy", intuitively associating this term with an exchange reaction of the type

$$AA + BB \Leftrightarrow AB + AB.$$

(3.155)

It can easily be seen that, when $W^{II} = 0$ in equation 3.151, then $\lambda = 1$ and also $\omega_q = 1$ (cf. eq. 3.150); it follows that activity coefficients γ_A and γ_B are 1 at all concentrations (cf. eq. 3.152 and 3.153). Nevertheless, if $2W^{II}/RT > 0$, then ω_q is greater than 1; in this condition configuration AA-BB is more stable than AB-AB. There is then a tendency toward *clustering* of ions A-A and B-B, which is phenomenologically preliminary to *unmixing* (the activity coefficients are correspondingly greater than 1). If $2W^{II}/RT < 0$, then ω_q is less than 1, and there is a tendency toward mixing.

The quasi-chemical model was derived by Guggenheim for application to organic fluid mixtures. Applying it to crystalline solids is not immediate, because it necessitates conceptual modifications of operative parameters, such as the above-mentioned "contact factor." Empirical methods of derivation of the above parameters, based on structural data, are available in the literature (Green, 1970; Saxena, 1972). We will not treat this model, because it is of scanty application in geochemistry. More exhaustive treatment can be found in Guggenheim (1952) and Ganguly and Saxena (1987).

3.8.7 Multisite and Reciprocal Mixtures

Let us consider a multisite mixture of type $(A, B, C, \ldots)_{v_1} (M, N, O, \ldots)_{v_2} Z$, where cations of types A, B, C and M, N, O, respectively, occupy energetically distinct sites present in one mole of substance in the stoichiometric amounts v_1 and v_2, and Z is the common anionic group. Applying the permutability concept to each distinct site and assuming random mixing and the absence of interactions on sites, the activity of component $A_{v_1} M_{v_2} Z$ in the mixture may be expressed as

$$a_{A_{v_1} M_{v_2} Z} = X_A^{v_1} X_M^{v_2}.$$

(3.156)

Equation 3.156 follows the general treatment proposed by Temkin (1945) for fused salts.

If site-interaction energy terms are nonnegligible, equation 3.156 must be modified to account for *site-activity coefficients,* as follows:

$$a_{A_{v_1}M_{v_2}Z} = X_A^{v_1}\gamma_A^{v_1} X_M^{v_2}\gamma_M^{v_2}.$$

(3.157)

If, however, site interactions of atoms on energetically equivalent sites are equal and the standard molar volumes of mixing components are not dissimilar (i.e., within 5 to 10% difference), equation 3.157 may be simplified and the activity of any component i in the mixture (a_i) may be expressed in a generalized fashion as shown by Helgeson et al. (1978):

$$a_i = k_i \prod_s \prod_j X_{j,s}^{v_{s,j,i}},$$

(3.158)

where $v_{s,j,i}$ is the number of sth energetically equivalent sites occupied by the jth atom in one mole of the ith component, $X_{j,s}$ is the molar fraction of atom j on site s, and k_i is a proportionality constant imposed by the limit

$$\lim_{X_i \to 1} a_i = 1.$$

(3.159)

The constant k_i may be also expressed in a generalized fashion, as shown by Helgeson et al. (1978):

$$k_i = \prod_s \prod_j \left(\frac{v_{s,j,i}}{v_{s,i}}\right)^{-v_{s,j,i}},$$

(3.160)

where $v_{s,i}$ is the number of sites of type s in the ith component:

$$v_{s,i} = \sum_j v_{s,j,i}.$$

(3.161)

Equation 3.158 is consistent with

$$G_{\text{mixing}} = RT \sum_i X_i \left(\ln k_i + \sum_s \sum_j v_{s,j,i} \ln X_{j,s} \right)$$

(3.162)

(Kerrick and Darken, 1975; Helgeson et al., 1978) and is strictly valid only for isovolumetric mixtures.

It must be noted that, when only one kind of atom occupies the s sites in the end-member formula, as in our compound $A_{v_1}M_{v_2}Z$, for instance, then $v_{s,j,i} \equiv v_{s,i}$ and k_i reduces to 1. In silicate end-members of geochemical interest, we often have two types of atoms occupying the identical sublattice. For example, the end-member *clintonite* of trioctahedral brittle micas has formula $Ca(Mg_2Al)(SiAl_3)O_{10}(OH)_2$, with Al and Mg occupying three energetically equivalent octahedral sites and Si and Al occupying four energetically equivalent tetrahedral sites (cf. section 5.6). We thus have, for octahedral sites, $v_{s,i} = 3$, $v_{s,j,i} = 2$ for Mg, and $v_{s,j,i} = 1$ for Al. For tetrahedral sites, $v_{s,i} = 4$, $v_{s,j,i} = 1$ for Si, and $v_{s,j,i} = 3$ for Al. The resulting proportionality constant is

$$k_i = \left(\frac{2}{3}\right)^{-2}\left(\frac{1}{3}\right)^{-1}\left(\frac{1}{4}\right)^{-1}\left(\frac{3}{4}\right)^{-3} = 64. \tag{3.163}$$

In some cases, attribution of coefficients $v_{s,i}$ and $v_{s,j,i}$ is not so straightforward as it may appear at first glance. This is the case, for instance, for K-feldspar ($KAlSi_3O_8$), with Al and Si occupying $T1$ and $T2$ tetrahedral sites. Assuming, for the sake of simplicity, that $T1$ and $T2$ sites are energetically equivalent (which is not the case; see section 5.7.2), we have $v_{s,i} = 4$, $v_{s,j,i} = 1$ for Al, and $v_{s,j,i} = 3$ for Si, so that

$$k_i = \left(\frac{1}{4}\right)^{-1}\left(\frac{3}{4}\right)^{-3} = 9.481. \tag{3.164}$$

Adopting topologic symmetry, with two $T1$ sites energetically different from the two $T2$ sites, and assuming Al and Si to be randomly distributed in $T1$ and $T2$, we obtain

$$k_i = \left(\frac{0.5}{2}\right)^{-0.5}\left(\frac{1.5}{2}\right)^{-1.5}\left(\frac{0.5}{2}\right)^{-0.5}\left(\frac{1.5}{2}\right)^{-1.5} = 6.158. \tag{3.165}$$

Assuming all T sites to be energetically different (which happens at low temperatures, when $T1_o$, $T1_m$, $T2_o$, and $T2_m$ nonenergetically equivalent sites exist and Al is stabilized in site $T1_o$; see section 5.7.2), we have

$$k_i = \left(\frac{1}{1}\right)^{-1}\left(\frac{1}{1}\right)^{-1}\left(\frac{1}{1}\right)^{-1}\left(\frac{1}{1}\right)^{-1} = 1. \tag{3.166}$$

Because activity is related to molar concentration through the rational activity coefficient, from equation 3.157 we may derive

$$\gamma_{A_{v_1}M_{v_2}Z} = \gamma_A^{v_1}\gamma_M^{v_2} \tag{3.167.1}$$

$$X_{A_{v_1}M_{v_2}Z} = X_A^{v_1}X_M^{v_2} \tag{3.167.2}$$

$$a_{A_{\nu_1} M_{\nu_2} Z} = a_A^{\nu_1} a_M^{\nu_2}. \tag{3.167.3}$$

Equations 3.167.1, 3.167.2, and 3.167.3 find their analogs in the theory of electrolyte solutions, as we will see in detail in section 8.4.

Let us now consider again the mixture $(A, B, C, \ldots)_{\nu_1}(M, N, O, \ldots)_{\nu_2}Z$. We can identify four components, $A_{\nu_1} M_{\nu_2} Z$, $A_{\nu_1} N_{\nu_2} Z$, $B_{\nu_1} M_{\nu_2} Z$, and $B_{\nu_1} N_{\nu_2} Z$, energetically related by the *reciprocal* reaction

$$A_{\nu_1} M_{\nu_2} Z + B_{\nu_1} N_{\nu_2} Z \Leftrightarrow B_{\nu_1} M_{\nu_2} Z + A_{\nu_1} N_{\nu_2} Z. \tag{3.168}$$

According to equation 3.168, the energy properties of the four components are not mutually independent but, once the Gibbs free energy is fixed for three of them, the energy of the fourth is constrained by the reciprocal energy term $\Delta G_{168} \equiv \Delta G_{reciprocal}$. Accepting this way of reasoning, the relationship between rational activity coefficient and site-activity coefficients is no longer given by equation 3.167.1 but, according to Flood et al. (1954), involves a reciprocal term ($\gamma_{reciprocal}$) whose nature is analogous to the *cooperative energy* term expressed by Guggenheim (1952) in the quasi-chemical model (see preceding section):

$$a_{A_{\nu_1} M_{\nu_2} Z} = \left(X_A^{\nu_1} \gamma_A^{\nu_1} X_M^{\nu_2} \gamma_M^{\nu_2} \right) \gamma_{reciprocal} \tag{3.169}$$

$$\gamma_{reciprocal} = \exp\left[\left(1 - X_A \right)\left(1 - X_M \right) \frac{\Delta \overline{G}_{reciprocal}^0}{RT} \right], \tag{3.170}$$

where $\Delta \overline{G}_{reciprocal}^0$ is the Gibbs free energy change of reciprocal reaction 3.168.

3.9 General Equations of Excess Functions for Nonideal Binary Mixtures

So far, we have seen that deviation from ideal behavior may affect one or more thermodynamic magnitudes (e.g., enthalpy, entropy, volume). In some cases, we are able to associate macroscopic interactions with real (microscopic) interactions of the various ions in the mixture (for instance, coulombic and repulsive interactions in the quasi-chemical approximation). In practice, it may happen that none of the models discussed above is able to explain, with reasonable approximation, the macroscopic behavior of mixtures, as experimentally observed. In such cases (or whenever the numeric value of the energy term for a given substance is more important than actual comprehension of the mixing process), we adopt general (and more flexible) equations for the excess functions.

Let us consider a generic mixture of components 1 and 2 in the binary compositional field $X_1 = 0 \rightarrow 1$, $X_2 = 1 \rightarrow 0$. The excess Gibbs free energy of mixing is expressed as a polynomial function of the composition:

$$G_{\text{excess mixing}} = X_1 X_2 \left[A_0 + A_1 (X_1 - X_2) + A_2 (X_1 - X_2)^2 + \cdots \right], \quad (3.171)$$

and the activity coefficients are defined by

$$RT \ln \gamma_1 = G_{\text{excess mixing}} + X_2 \left(\frac{\partial G_{\text{excess mixing}}}{\partial X_1} \right) = \quad (3.172)$$

$$= X_2^2 \left[A_0 + A_1 (3X_1 - X_2) + A_2 (X_1 - X_2)(5X_1 - X_2) + \cdots \right]$$

$$RT \ln \gamma_2 = X_1^2 \left[A_0 - A_1 (3X_2 - X_1) + A_2 (X_2 - X_1)(5X_2 - X_1) + \cdots \right]. \quad (3.173)$$

Guggenheim's polynomial expansion (equation 3.171; Guggenheim, 1937) and the two Redlich-Kister equations (3.172 and 3.173; Redlich and Kister, 1948) are of general applicability for any type of mixture:

1. If $A_0 = A_1 = A_2 = 0$, the excess Gibbs free energy of mixing is zero throughout the compositional field and the mixture is *ideal.*
2. If $A_0 \neq 0$ and $A_1 = A_2 = 0$, the mixture is *regular:*

$$A_0 \equiv W \Rightarrow G_{\text{excess mixing}} = X_1 X_2 W. \quad (3.174)$$

3. If all the odd coefficients in the polynomial expansions (i.e., A_1, A_3, . . .) are zero, the mixture is *symmetric* (i.e., γ_1 and γ_2 assume identical values for symmetric compositions in the binary field; see, for instance, figure 3.9C).
4. If $A_0 \neq 0$ and $A_1 \neq 0$, the excess functions do not exhibit symmetrical properties over the compositional field. In this case, the mixture is defined as *subregular* (see, for instance, figure 3.9D).

The subregular model approximates several silicate mixtures with sufficient precision, as we will see in chapter 5. For a subregular mixture, we have

$$G_{\text{excess mixing}} = X_1 X_2 \left[A_0 + A_1 (X_1 - X_2) \right] \quad (3.175)$$

$$RT \ln \gamma_1 = X_2^2 \left[A_0 + A_1 (3X_1 - X_2) \right] \quad (3.176)$$

$$RT \ln \gamma_2 = X_1^2 \left[A_0 - A_1 (3X_2 - X_1) \right]. \quad (3.177)$$

Van Laar parameters A_0 and A_1 can be translated into the corresponding Margules subregular parameters W_{12} and W_{21}, posing

$$W_{12} = A_0 - A_1 ; \qquad W_{21} = A_0 + A_1 \qquad (3.178)$$

and

$$A_0 = \frac{W_{21} + W_{12}}{2} ; \qquad A_1 = \frac{W_{21} - W_{12}}{2} . \qquad (3.179)$$

The Margules subregular model then becomes

$$G_{\text{excess mixing}} = X_1 X_2 \left(W_{12} X_2 + W_{21} X_1 \right) \qquad (3.180)$$

$$RT \ln \gamma_1 = X_2^2 \left[W_{12} + 2X_1 \left(W_{21} - W_{12} \right) \right] \qquad (3.181)$$

$$RT \ln \gamma_2 = X_1^2 \left[W_{21} + 2X_2 \left(W_{12} - W_{21} \right) \right] . \qquad (3.182)$$

Figure 3.9D shows the form of the curve of the excess Gibbs free energy of mixing obtained with Van Laar parameters variable with T: the mixture is subregular—i.e., asymmetric over the binary compositional field.

3.10 Generalizations for Ternary Mixtures

So far, we have seen several ways of calculating the Gibbs free energy of a two-component mixture. To extend calculations to ternary and higher-order mixtures, we use empirical combinatory extensions of the binary properties. We summarize here only some of the most popular approaches. An extended comparative appraisal of the properties of ternary and higher-order mixtures can be found in Barron (1976), Grover (1977), Hillert (1980), Bertrand et al. (1983), Acree (1984), and Fei et al. (1986).

Because the ideal Gibbs free energy of mixing contribution is readily generalized to n-component systems (cf. eq. 3.131), the discussion involves only excess terms.

3.10.1 Wohl Model

According to Wohl (1953), the excess Gibbs free energy of mixing for a three-component mixture is represented by

$$G_{\text{excess mixing}} = X_1 X_2 \left(W_{12} X_2 + W_{21} X_1 \right)$$

$$+ X_2 X_3 \left(W_{23} X_3 + W_{32} X_2 \right) + X_1 X_3 \left(W_{13} X_3 + W_{31} X_1 \right)$$

$$+ X_1 X_2 X_3 \left[\frac{1}{2} \left(W_{12} + W_{21} + W_{13} + W_{31} + W_{23} + W_{32} \right) - C \right],$$

(3.183)

where X_1, X_2, and X_3 are the molar fractions of components 1, 2, and 3 in the mixture, W_{12}, W_{21}, W_{13}, W_{31}, W_{23}, and W_{32} are binary interaction parameters between the components (see eq. 3.180), and C is a ternary interaction constant whose value must be determined experimentally.

The activity coefficient of component 1 in a ternary mixture is defined by

$$RT \ln \gamma_1 = X_2^2 \left[W_{12} + 2X_1 \left(W_{21} - W_{12} \right) \right] + X_3^2 \left[W_{13} + 2X_1 \left(W_{31} - W_{13} \right) \right] +$$

$$X_2 X_3 \left[\frac{1}{2} \left(W_{12} + W_{21} + W_{13} + W_{31} - W_{23} - W_{32} \right) + X_1 \left(W_{21} - W_{12} + W_{31} - W_{13} \right) + \right.$$

$$\left. \left(X_2 - X_3 \right) \left(W_{23} - W_{32} \right) - \left(1 - 2X_1 \right) C \right].$$

(3.184)

(To obtain the corresponding function for component 2, substitute 2 for 1, 3 for 2, and 1 for 3, and for component 3, substitute 3 for 1, 1 for 2, and 2 for 3.)

3.10.2 Hillert Model

According to the *Hillert model* (Hillert, 1980), the excess Gibbs free energy of mixing of a ternary mixture is given by

$$G_{\text{excess mixing}} = \left(\frac{X_2}{1 - X_1} \right) \left\{ X_1^* X_2^* \left[A_{0_{12}} + A_{1_{12}} \left(X_1^* - X_2^* \right) \right] \right\} + \left(\frac{X_3}{1 - X_1} \right) \times$$

$$\left\{ X_1^* X_3^* \left[A_{0_{13}} + A_{1_{13}} \left(X_1^* - X_3^* \right) \right] \right\}$$

(3.185)

$$+ X_2 X_3 \left[A_{0_{23}} + A_{1_{23}} \left(V_{23} - V_{32} \right) \right],$$

where

$$V_{23} = \frac{1}{2} \left(1 + X_2 - X_3 \right)$$

(3.186)

and

$$V_{32} = \frac{1}{2}\left(1 + X_3 - X_2\right).$$

(3.187)

The A_{012}, A_{013}, A_{023}, A_{112}, A_{113}, and A_{123} parameters in equation 3.185 are analogous to the Van Laar parameters in eq. 3.175 and represent the binary interactions between components 1 and 2, 1 and 3, and 2 and 3, respectively. X_1, X_2, and X_3 are the molar fractions of components 1, 2, and 3 in the mixture, and X_1^*, X_2^*, and X_3^* are the corresponding scaled fractions in the binary field.

3.10.3 Kohler Model

The *Kohler model* is a general model based on linear combination of the binary interactions among the components in a mixture, calculated as if they were present in binary combination (relative proportions) and then normalized to the actual molar concentrations in the multicomponent system. The generalized expression for the excess Gibbs free energy is

$$G_{\text{excess mixing}} = \sum_{i \neq j} \left(X_i + X_j\right)^2 \cdot G_{ij,\text{binary}},$$

(3.188)

where $G_{ij,\text{binary}}$ is the excess Gibbs free energy calculated for i and j components as if they were in binary combination—i.e., for a two-component subregular Margules notation:

$$G_{ij,\text{binary}} = X_i^* X_j^* \left(W_{ij} X_j^* + W_{ji} X_i^*\right)$$

(3.189)

$$X_i^* = \frac{X_i}{X_i + X_j}$$

(3.190)

$$X_j^* = \frac{X_j}{X_i + X_j}.$$

(3.191)

Consider, for instance, a three-component system (1, 2, 3). We have

$$G_{\text{excess mixing}} = \left(X_1^* + X_2^*\right)^2 G_{12,\text{binary}} + \left(X_1^* + X_3^*\right)^2 G_{13,\text{binary}}$$

$$+ \left(X_2^* + X_3^*\right)^2 G_{23,\text{binary}}$$

(3.192)

$$G_{12,\text{binary}} = X_1^* X_2^* \left(W_{12} X_2^* + W_{21} X_1^*\right)$$

(3.193)

$$X_1^* = \frac{X_1}{X_1 + X_2}; \qquad X_2^* = \frac{X_2}{X_1 + X_2}, \qquad (3.194,195)$$

and so on.

The application of multicomponent mixing models to the crystalline state is a dangerous exercise. For instance, Fei et al. (1986) have shown that, adopting the same binary interaction parameters between components in a mixture, extensions to the ternary field through the Wohl and Hillert models do not lead to the same results. Moreover, from a microscopic point of view, we cannot expect ternary interactions to be linear combinations of binary terms simply because the interionic potential field in a crystal is not a linear expression of the single properties of the interacting ions. We have seen (in chapter 1) how the lattice energy of a crystalline substance is a complex function of individual ionic properties (effective charge, repulsive radius, hardness factor, polarizability, etc.) and of structure (coordination, site dimension, etc.)—even in a simple pair-potential static approach (however, see also relationships between electron density and coulombic, kinetic, and correlation energy terms in section 1.18). Undoubtedly, multicomponent mixing models cannot be expected to have heuristic properties when applied to the crystalline state, but must be simply adopted as parametric models to describe mixing behavior (experimentally observed or calculated from first principles) in multicomponent space in an easy-to-handle fashion (see, as an example, Ottonello, 1992).

3.11   Solvus and Spinodal Decomposition in Binary Mixtures

In chapter 2 we defined equilibrium conditions for coexisting phases:

1. Equality of chemical potentials of the various components in all coexisting phases
2. Minimization of the Gibbs free energy of the system

Let us now imagine that we are dealing with a regular mixture (A,B)N with an interaction parameter $W = +20$ kJ/mole. The Gibbs free energy of mixing at various temperatures will be

$$
\begin{aligned}
G_{\text{mixing}} &= G_{\text{ideal mixing}} + G_{\text{excess mixing}} \\
&= RT\big(X_A \ln X_A + X_B \ln X_B\big) + X_A X_B W.
\end{aligned}
\qquad (3.196)
$$

The Gibbs free energy of mixing curves will have the form shown in figure 3.10A. By application of the above principles valid at equilibrium conditions, we deduce that the minimum Gibbs free energy of the system, at low T, will be

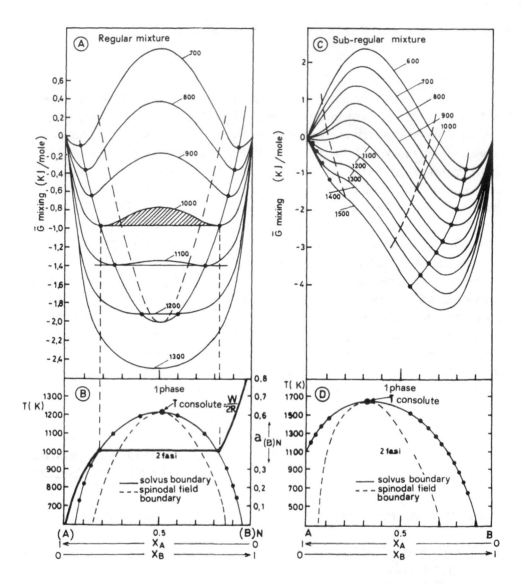

Figure 3.10 Solvus and spinodal decomposition fields in regular (B) and subregular (D) mixtures. Gibbs free energy of mixing curves are plotted at various T conditions in upper part of figure (A and C, respectively). The critical temperature of unmixing (or "consolute temperature") is the highest T at which unmixing takes place and, in a regular mixture (B), is reached at the point of symmetry.

reached by the coexistence of two phases: one rich in component $(A)N$, and the other rich in $(B)N$. The loci of tangency points (or "binodes") at the various values of T will define a compositional field, in a T-X space, within which the two phases will coexist stably at equilibrium—the "solvus field of the solid mixture" (figure 3.10B). In this case, the solvus is symmetrical with respect to composition.

Instead, if we have a solid mixture whose excess Gibbs free energy of mixing is approximated by a subregular Margules model with $W_{AB} = 5$ kJ/mole and $W_{BA} = 35$ kJ/mole, the Gibbs free energy of mixing at the various values of T is

$$
\begin{aligned}
G_{\text{mixing}} &= G_{\text{ideal mixing}} + G_{\text{excess mixing}} \\
&= RT\left(X_A \ln X_A + X_B \ln X_B\right) + X_A X_B \left(W_{AB} X_B + W_{BA} X_A\right)
\end{aligned}
\tag{3.197}
$$

(figure 3.10C). The resulting solvus field is asymmetric (figure 3.10D).

Figure 3.10 also shows the fields of "spinodal decomposition"* defined by the loci of the points of inflection in the Gibbs free energy of mixing curves. These points obey the following general conditions:

$$
\left(\frac{\partial^2 G_{\text{mixing}}}{\partial X^2}\right) = 0
\tag{3.198}
$$

and

$$
\left(\frac{\partial^3 G_{\text{mixing}}}{\partial X^3}\right) = 0.
\tag{3.199}
$$

Recalling that

$$
G_{\text{mixing}} = G_{\text{ideal mixing}} + G_{\text{excess mixing}},
\tag{3.200}
$$

we can write

$$
\frac{\partial^2 G_{\text{excess mixing}}}{\partial X^2} = -\frac{\partial^2 G_{\text{ideal mixing}}}{\partial X^2} = -\frac{RT}{X(1 - X)}
\tag{3.201}
$$

and

$$
\frac{\partial^3 G_{\text{excess mixing}}}{\partial X^3} = -\frac{\partial^3 G_{\text{ideal mixing}}}{\partial X^3} = -\frac{RT(2X - 1)}{X^2(1 - X)^2}.
\tag{3.202}
$$

* The term "spinode" was proposed by van der Waals, who formulated it in analogy with the shape of a thorn ("spine") given by the intersection of the tangent plane to the $\mu = f(P)$ function for gaseous phases at the critical point (see figure 1 in van der Waals, 1890; see also Cahn, 1968, for an extended discussion of the etymology of the term).

The *critical unmixing* condition is obtained from equations 3.201 and 3.202 by substituting for the left-side terms the algebraic form of the excess function appropriate to the case under consideration. For a *simple mixture,* we have, for instance (cf. eq. 3.144),

$$\frac{\partial^2 G_{\text{excess mixing}}}{\partial X^2} = -2W^I \tag{3.203}$$

and

$$\frac{\partial^3 G_{\text{excess mixing}}}{\partial X^3} = 0, \tag{3.204}$$

which, combined with equations 3.201 and 3.202, give

$$2W^I = \frac{RT}{X(1-X)} \tag{3.205}$$

and

$$\frac{RT(2X-1)}{X^2(1-X)^2} = 0. \tag{3.206}$$

Because the simple mixture has symmetric properties, the system defined by equations 3.205 and 3.206 can be easily solved. Setting $X = 0.5$, we have

$$T = T_{\text{consolute}} = \frac{W^I}{2R}. \tag{3.207}$$

If the mixture is subregular, definition of the limits of spinodal decomposition is more complex. For a subregular Margules model (figure 3.10C and D), we have

$$\frac{\partial^2 G_{\text{excess mixing}}}{\partial X^2} = -4\left(W_{21} - \frac{1}{2}W_{12}\right) + 6X(W_{21} - W_{12}) \tag{3.208}$$

and

$$\frac{\partial^3 G_{\text{excess mixing}}}{\partial X^3} = 6(W_{21} - W_{12}) \tag{3.209}$$

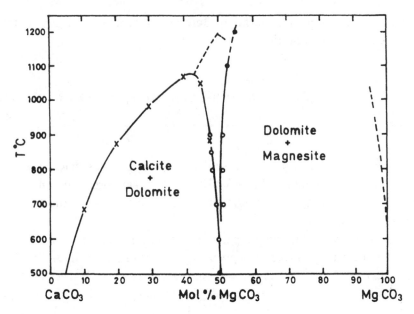

Figure 3.11 Miscibility gaps in compositional field $CaCO_3$–$MgCO_3$. Reprinted from J. R. Goldsmith and H. C. Heard (1961), *Journal of Geology*, 69, 45–74, copyright © 1961 by The University of Chicago, with permission of The University of Chicago Press.

and, for a two-coefficient Van Laar model,

$$\frac{\partial^2 G_{\text{excess mixing}}}{\partial X^2} = -2A_0 + 6A_1\left(2X - 1\right) \tag{3.210}$$

and

$$\frac{\partial^3 G_{\text{excess mixing}}}{\partial X^3} = 12A_1. \tag{3.211}$$

The combination of equations 3.208, 3.209 or 3.210, and 3.211 with general equations 3.201 and 3.202 gives rise to transcendental equations that must be solved using iterative procedures. In the treated cases, we have seen a single solvus field that occupies a limited portion of the compositional field (figure 3.10B), and an asymmetric solvus field that extends to the condition of pure component (figure 3.10D). In nature, there are also cases in which two or more solvi may be observed over the same binary range. One example is the $CaCO_3$–$MgCO_3$ system (figure 3.11). Actually, within the compositional field, we observe second-order phase transitions (see section 2.8): the spatial group *dolomite* ($Ca_{0.5}Mg_{0.5}CO_3$; group $R\bar{3}$) is a subgroup of *calcite* ($CaCO_3$; group $R\bar{3}c$), and the definition "miscibility gap" or even "pseudo-solvus" instead of "solvus field" would be preferable (Navrotsky and Loucks, 1977). We note that the consolute temperatures for the two miscibility gaps are markedly different; moreover, the gap at low Mg content is strongly asymmetric.

3.12 Spinodal Decomposition and Critical Nucleation Radius

Let us again consider a solid mixture (A,B)N with a solvus field similar to the one outlined in the T-X plot in figure 3.10B, and let us analyze in detail the "form" of the Gibbs free energy of mixing curve in the zone between the two binodes (shaded area in figure 3.10A).

We can identify the binodal tangency point that marks the equality of potentials for the coexisting phases at a given composition (X_b in figure 3.12A). We also note that the Gibbs free energy of mixing curve, after the binode, increases its slope up to an inflection point called the *spinode* (X_s). Beyond the spinode, the slope of the curve decreases progressively. If we now analyze the geometrical relationships between composition and Gibbs free energy of mixing of two generic points intermediate between binode and spinode ($X^I_{(1)}$ and $X^{II}_{(1)}$ in figure 3.12A), we see that *the algebraic summation of the Gibbs free energies at the two compositional points is always higher than the Gibbs free energy of the mean composition $X_{(1)m}$*. Based on the Gibbs free energy minimization principle, a generic mixture of composition $X_{(1)m}$ apparently will be unable to unmix in two phases, because any compositional fluctuation in the neighborhood of $X_{(1)m}$ would result in an increase in the Gibbs free energy of the system. This apparently contrasts with the statements of the preceding paragraphs—i.e., within the solvus field, the minimum Gibbs free energy condition is reached with splitting in two phases. Indeed, if we analyze the bulk Gibbs free energy of the system when unmixing is completed, we see that it is always lower than the corresponding energy of the homogeneous phase. The increase in energy related to compositional fluctuations in the zone between binode X_b and spinode X_s must be conceived of as an "energy threshold of activation of the unmixing process" that must be overcome if the process is to terminate. The spinodal field defined by the inflection points in the Gibbs free energy of mixing curve thus represents a "kinetic limit" and not a phase boundary in the proper sense of the term.

The Gibbs free energy variation connected with compositional fluctuations in the vicinity of $X_{(1)m}$ (figure 3.12B) peaks near the so-called "critical nucleation radius" (r_c) and then decreases progressively with the increase of (compositional) distance. Let us analyze the significance of r_c from a microscopic point of view. A generic mixture (A,B)N in which AN is present at (macroscopic) concentration $X_{(1)m}$ and BN is at a concentration of $(1 - X_{(1)m})$ may exhibit compositional fluctuations around $X_{(1)m}$, on the microscopic scale. If these fluctuations are such that ΔX_{AN} at $X_{(1)m}$ is higher than ΔX_{r_c}, the energy threshold of the process will

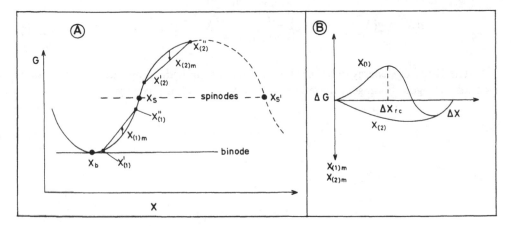

Figure 3.12 Energy relationships between solvus and spinodal decompositions. (A) Portion of Gibbs free energy of mixing curve in zone between binodal (X_b) and spinodal (X_s) points. (B) Gibbs free energy variation as a consequence of compositional fluctuations around intermediate points $X_{(1)m}$ and $X_{(2)m}$.

be overcome; a nucleus of composition $(A_{X_{(1)m}} + \Delta X)N$ with $\Delta X > \Delta X_{r_c}$ will be formed and "homogeneous nucleation" will take place. The process will be one of "heterogeneous nucleation" if the compositional step necessary to overcome the critical nucleation is favored by the presence of extended defects, such as limits of grains and/or dislocations. If we consider the equation relating composition and Gibbs free energy in the compositional field between the two spinodes, any compositional fluctuation in the vicinity of $X_{(2)m}$ will result in a decrease in the Gibbs free energy of the system (figure 3.12A and B): the mixture is therefore intrinsically unstable in this compositional range, and will tend to decompose spontaneously. The compositional region between the solvus and the spinodal fields is thus one of *metastability,* whereas the region within the spinodes is one of *intrinsic instability.*

For analytical comprehension of the kinetics of spinodal decomposition processes, we must be able to evaluate the Gibbs free energy of a binary mixture of nonuniform composition. According to Cahn and Hilliard (1958), this energy can be expressed by the linear approximation

$$G = \int_V \left[g(X) + K(\nabla X)^2 \right] dV, \tag{3.212}$$

where the first term, $g(X)$, represents the Gibbs free energy of an infinitesimal volume of mixture with uniform composition, and the second term, $K(\nabla X)^2$, defines a local energy gradient that is always positive because K is inherently positive. The expansion of $g(X)$ in the vicinity of a mean composition X_0 is given by the following expansion truncated at the third term:

$$g(X) = g(X_0) + (X - X_0)\left(\frac{dg}{dX}\right)_{X_0} + \frac{1}{2}(X - X_0)^2\left(\frac{d^2g}{dX^2}\right)_{X_0} + \cdots. \qquad (3.213)$$

The Gibbs free energy difference between the system with compositional fluctuations and an analogous homogeneous system of composition X_0 is given by

$$\Delta G = \int_V \left[\Delta g(X) + K(\nabla X)^2\right] dV, \qquad (3.214)$$

where

$$\Delta g(X) = g(X) - g(X_0). \qquad (3.215)$$

Compositional fluctuations may be represented through a Fourier expansion, and are sufficiently approximated (for our interests) by the first sinusoidal term

$$X - X_0 = A \sin \beta_z, \qquad (3.216)$$

where β_z is the scalar component of the wave vector β along the decomposition direction z, and A is a constant term. The Gibbs free energy variation due to periodic fluctuation is

$$\Delta G = \frac{A^2}{4}\left[\left(\frac{d^2g}{dX^2}\right) + 2K\beta^2\right]. \qquad (3.217)$$

Because K is inherently positive, the stability conditions of the mixture with compositional fluctuations can be deduced by the value of the differential term d^2g/dX^2. The spinodal locus is defined by $d^2g/dX^2 = 0$; if $d^2g/dX^2 > 0$, then $\Delta G > 0$, and the mixture is stable with respect to infinitesimal fluctuations (i.e., region between binodes and spinodes, point $X_{(1)}$ in figure 3.12A). However, if $d^2g/dX^2 < 0$, then $\Delta G < 0$ for compositional fluctuations of wavelength overcoming the critical value

$$\lambda_{\text{critical}} = \frac{2\pi}{\beta_{\text{critical}}} = \left[-\frac{8\pi^2 K}{\left(d^2g/dX^2\right)_{X_0}}\right]^{1/2}. \qquad (3.218)$$

Because there is no energy barrier opposed to compositional fluctuations for compositions within the spinodal field, the decomposition mechanism *is entirely a diffusional process.* Unmixing effectively begins with a minimal compositional fluctuation that extends over the entire volume of the system.

The flow of matter as a function of the local chemical potential along a given direction is described by the Cahn (1968) equation

$$J_A - J_B = M\left[\frac{d(\mu_A - \mu_B)}{dX}\right], \qquad (3.219)$$

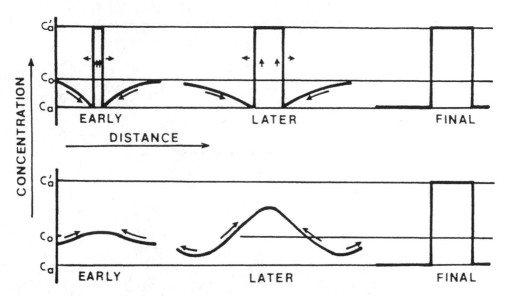

Figure 3.13 Schematic evolution of concentration profiles in decomposition processes. Upper part: nucleation and growth. Lower part: spinodal decomposition. Reprinted from Cahn (1968), with kind permission of ASM International, Materials Park, Ohio.

where μ_A and μ_B are the chemical potentials of components A and B, J_A and J_B are their fluxes, and M is the atomic mobility per unit volume. The diffusion equation along z is

$$\frac{dX}{dt} = -\frac{d\bar{J}}{dz} = M\left(\frac{d^2g}{dX^2}\right)\left(\frac{d^2X}{dz^2}\right) - 2MK\left(\frac{d^4X}{dz^4}\right), \tag{3.220}$$

where $\bar{J}$ is the interdiffusional flux ($\bar{J} = J_A = -J_B$) (note that eq. 3.220 can be generalized to all directions by substituting the ∇^2X operator for the second derivative in dz; see eq. 8 in Cahn, 1968).

Because mobility term M is always positive, the diffusion coefficient $M(d^2g/dX^2)$ depends on the sign of the second derivative of the Gibbs free energy. Within the spinodal field, (d^2g/dX^2) is negative and the diffusion coefficient is thus also negative. This means that there is "uphill" diffusion: the atoms of a given species migrate from low to high concentration zones. The final result is a *clustering* of atoms of the same species. Nevertheless, if we are in the compositional zone between spinode and binode, (d^2g/dX^2) is positive, the diffusion coefficient is then also positive, and diffusion takes place from high to low concentration zones (*nucleation and growth*). The two different accretion processes are shown in figure 3.13 as a function of time (and distance).

Let us now consider the process from a thermal point of view (figure 3.14) and imagine freezing the system from point p to point p' (corresponding to temperature T'): we begin to see unmixing phenomena in p'. Actually it may happen that the mixture remains metastably homogeneous (i.e., the critical nucleation radius is not overcome) down to a point p'' (temperature T'') at which we effectively observe decomposition processes taking place (at p'' the system enters the

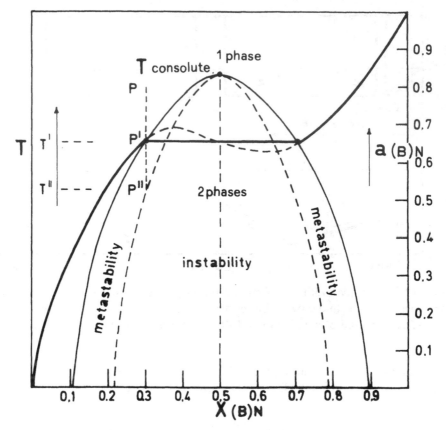

Figure 3.14 Stability relations in a binary mixture (A,B)N as a function of temperature. Heavy, solid line: activity trend for component (B)N in the case of binodal decomposition. Dashed line: activity trend in the case of spinodal decomposition.

spinodal field). In terms of thermodynamic activity of components, we observe a progressive increase in the activity of component (B)N in mixture, with an increase in its molar concentration, in the metastability region between the solvus and spinodal decomposition fields, and a progressive decrease (due to decomposition) within the spinodal field. There is therefore a maximum and a minimum in the transition zone between *metastability* and *instability*. Nevertheless, if decomposition already takes place in p' (at the limits of the solvus field), the activity of component (B)N remains constant throughout the compositional range lying—at that temperature—within the solvus field (figure 3.14).

3.13   Chemical Solvus and Coherent Solvus

The exsolution process in either binodal or spinodal decomposition may lead to coherent or incoherent interfaces between the unmixed phases (figure 3.15). The

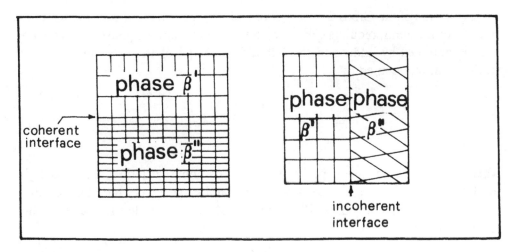

Figure 3.15 Types of interfaces between exsolving phases.

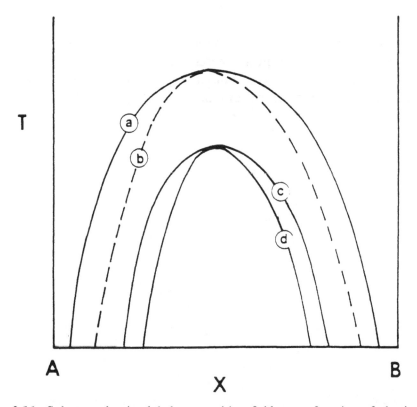

Figure 3.16 Solvus and spinodal decomposition fields as a function of elastic strain. (a) "Strain-free" or "chemical" solvus; (b) strain-free spinodal; (c) "coherent" solvus; (d) coherent spinodal. From Ganguly and Saxena (1992). Reprinted with permission of Springer-Verlag, New York.

representation of the Gibbs free energy of a binary mixture, as a continuous function of composition, requires coherence between the unmixed phases. This coherence is obtained by adding to the Gibbs free energy of the mixture an elastic strain term, represented by

$$\Delta g_{elastic} = \frac{\eta^2 E_y}{(1-v)} (X - X_0)^2,$$ (3.221)

where η is linear expansion per unit of compositional variation, E_y is Young's modulus for X_0, and v is Poisson's ratio. The Gibbs free energy modification due to periodic fluctuation (eq. 3.217) is thus modified to include this elastic strain term:

$$\Delta G = \frac{A^2}{4} \left[\left(\frac{d^2 g}{dX^2} \right) + 2K\beta^2 + \frac{\eta^2 E_y}{(1-v)} \right].$$ (3.222)

The energy of elastic strain modifies the Gibbs free energy curve of the mixture, and the general result is that, in the presence of elastic strain, both solvus and spinodal decomposition fields are translated, pressure and composition being equal, to a lower temperature, as shown in figure 3.16.

CHAPTER FOUR

Some Concepts of Defect Chemistry

When investigating the thermochemical properties of crystalline substances, we, more or less consciously, adopt an idealization of the phase of interest: we imagine a crystal as being composed of a regular network of atoms, with interionic distances dictated by the symmetry operations of the spatial group. In this idealized picture, a crystal is an unperturbed tridimensional replication of the asymmetric unit. Actually, this hypothetical homogeneity-structural continuity is, in real phases, interrupted by local irregularities called *lattice defects*. Lattice defects are furthermore subdivided into two main groups: *extended defects* and *point defects*.

4.1 Extended Defects

Extended defects are primarily composed of linear dislocations, shear planes, and intergrowth phenomena. Figure 4.1A and B, for example, show two types of linear dislocation: an *edge dislocation* and a *screw dislocation*.

Extended defects interrupt the continuity of the crystal, generating crystal subgrains whose dimensions depend, in a complex fashion, on the density of extended defects per unit area. Table 4.1 gives examples of reported dislocation densities and subgrain dimensions in olivine crystals from the San Carlos peridotite nodules (Australia). Assuming a mean dislocation density within 1.2×10^5 and 6×10^5 cm^{-2}, Kirby and Wegner (1978) deduced that a directional strain pressure of 35 to 75 bar acted on the crystals prior to their transport to the surface by the enclosing lavas.

The presence of extended defects does not significantly affect the energetic properties of crystalline compounds, because the concentration of extended defects necessary to create significant energy modifications is too high. For instance, Wriedt and Darken (1965) observed deviations in the solubility of nitrogen in steels induced by an extremely high dislocation density: 10^{11} cm^{-2} was the limit at which the solubility modification could be measured. Such densities are not normally observed in natural phases, whose dislocation densities normally range between 10^4 and 10^9. Kohlstedt and Vander Sande (1973) observed dislocation

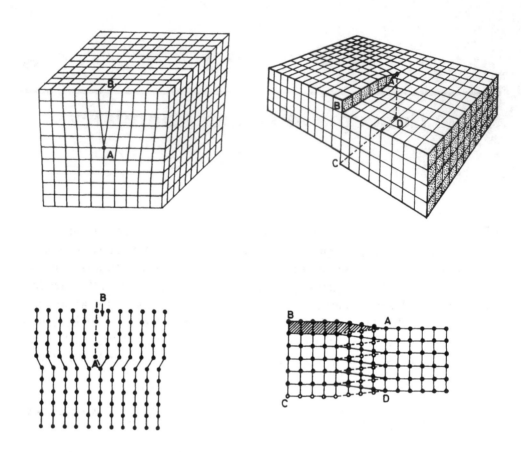

A edge dislocation   B screw dislocation

Figure 4.1 Examples of extended defects in crystals: edge dislocation (A) and screw dislocation (B).

Table 4.1 Dislocation densities (cm^{-2}) and mean subgrain dimensions (cm) in San Carlos olivines (from Kirby and Wegner, 1978). n.d. = not determined.

Sample	Mean Subgrain Dimension			Dislocation Density	
	[100]	[010]	[001]	Surficial	Internal
(1)	0.07 ± 0.04	–	0.22 ± 0.08	5.85 × 10^6	5 ± 4 × 10^5
(2)	0.03 ± 0.02	–	0.05 ± 0.02	2.74 × 10^6	1.2 ± 0.4 × 10^5
(3)	0.26 ± 0.02	–	0.04 ± 0.03	n.d.	n.d.
(4)	0.02 ± 0.01	–	0.02 ± 0.01	n.d.	3 ± 2 × 10^5
(5)	0.02 ± 0.01	0.02 ± 0.01	–	n.d.	6 ± 3 × 10^5
(6)	–	0.02 ± 0.01	0.02 ± 0.01	n.d.	1.3 ± 0.3 × 10^5

densities of 10^6 to 10^7 cm^{-2} in natural orthopyroxenes, and dislocation densities of 10^5 to 10^9 cm^{-2} have been observed in olivine (Kohlstedt and Goetze, 1974; Buiskool Toxopeus and Boland, 1976; Kirby and Wegner, 1978).

Anomalously high extended defect concentrations may be achieved in crystals by submitting them to directional stress. For example, Willaime and Gaudais (1977) induced dislocation densities of 10^9 cm^{-2} in sanidine crystals (KAlSi$_3$O$_8$; triclinic), and Ardell et al. (1973) reached a dislocation density of 10^{12} cm^{-2} in quartz with the same method.

Extended defects such as dislocations *cannot exist in the crystal at thermodynamic equilibrium* (Friedel, 1964), and their energy contribution must be evaluated through the elastic theory. Calculations in this sense (Holder and Granato, 1969) lead to energies between 1 and 22 eV per atomic unit length. At a defect energy of about 20 eV and a dislocation density of 10^9 cm^{-2}, the resulting energy is about 50 J/mole. This is lower than experimental error in most experimental measurements of crystal energetics, and may virtually be neglected in calculations of both mixing properties and intercrystalline equilibria.

4.2 Point Defects

Point defects exist in crystals *at equilibrium conditions.* They primarily affect some chemical-physical features of crystalline substances, such as electronic and ionic conductivity, diffusivity, light absorption and light emission, etc., and, in subordinate fashion, their thermodynamic properties. As already noted, our idealized picture of a crystal has little to do with reality. As extended defects interrupt the symmetry of crystals, point defects modify their chemistry significantly, inducing defect equilibria with coexisting phases and modifying stoichiometry. The disorder induced by the presence of point defects is grouped into two main categories: *intrinsic disorder* and *extrinsic disorder.*

4.3 Intrinsic Disorder

Intrinsic disorder is observed in conditions of perfect stoichiometry of the crystal. It is related to two main defect equilibria: *Schottky defects* and *Frenkel defects.*

Let us consider a generic oxide MO; in conditions of perfect stoichiometry, one mole of MO is composed of one mole of M and half a mole of O$_2$. If a cation is subtracted from a cationic position, creating a doubly ionized *cationic vacancy* (V_M''), then, in order to maintain crystal electroneutrality, one oxygen ion must be subtracted from a lattice site pertaining to the anion, thus creating an *anionic vacancy* ($V_O^{\cdot\cdot}$). Using the Kröger-Vink notation, generally adopted to describe point defect equilibria, the process, defined as "Schottky defectuality," is represented as follows:

$$MO \rightarrow V_M^{\parallel} + V_O^{\cdot\cdot} \qquad (4.1)$$

or

$$M_M^{x} + O_O^{x} \rightarrow V_M^{\parallel} + V_O^{\cdot\cdot} \qquad (4.2)$$

or

$$0 \rightarrow V_M^{\parallel} + V_O^{\cdot\cdot}, \qquad (4.3)$$

where the superscript ^{x} indicates neutrality relative to the ideal crystal, $^{\parallel}$ stands for negative excess charge taken by the cation vacancy, $^{\cdot}$ stands for a positive excess charge taken by the oxygen vacancy, and the subscripts indicate the lattice position of the defect.

Let us now imagine moving a generic cation M from its normal position M (M_M^{x}) to an interstitial position i, leaving the previously occupied cationic site empty:

$$M_M^{x} \rightarrow M_i^{\cdot\cdot} + V_M^{\parallel}. \qquad (4.4)$$

The equilibrium thus established is a "Frenkel defect." In both the Schottky and Frenkel equilibria, the stoichiometry of the crystal is unaltered (figure 4.2). Assuming that the thermodynamic activity of the various species obeys Raoult's law, thus corresponding to their molar concentrations (denoted hereafter by square brackets), the constant of the Schottky process is reduced to

$$K_S = \left[V_M^{\parallel} \right]\left[V_O^{\cdot\cdot} \right], \qquad (4.5)$$

and that of the Frenkel defect is

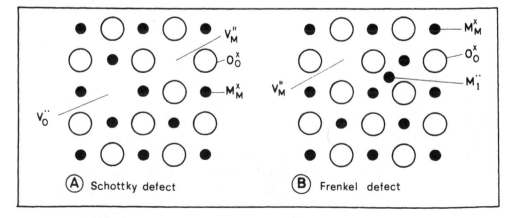

Figure 4.2   Intrinsic Schottky (A) and Frenkel (B) defects in a MO crystal.

$$K_F = \left[M_i^{\cdot\cdot} \right]\left[V_M^{\parallel} \right] \tag{4.6}$$

(because the defect concentrations are extremely low—i.e., generally lower than 10^{-4} in terms of molar fractions—the activity of ionic species at their normal positions is assumed to be 1).

4.4 Extrinsic Disorder

Let us imagine immersing the crystal MO in an atmosphere in which the partial pressure of oxygen (P_{O_2}) is lower than the partial pressure of intrinsic stability (i.e., the partial pressure of oxygen at which crystal MO is perfectly stoichiometric: $P_{O_2}^*$). We observe two distinct defect processes:

$$O_O^x \rightarrow \frac{1}{2} O_{2(gas)} + V_O^{\cdot\cdot} + 2e^{|} \tag{4.7}$$

and

$$M_M^x + O_O^x \rightarrow \frac{1}{2} O_{2(gas)} + M_i^{\cdot\cdot} + 2e^{|}. \tag{4.8}$$

The first process produces doubly ionized positive oxygen vacancies and electrons ($e^{|}$) and the second produces doubly ionized positive cation interstitials and electrons. The equilibrium constants of the two processes are given by

$$K_7 = \left[V_O^{\cdot\cdot} \right]\left[e^{|} \right]^2 P_{O_2}^{1/2} \tag{4.9}$$

and

$$K_8 = \left[M_i^{\cdot\cdot} \right]\left[e^{|} \right]^2 P_{O_2}^{1/2}. \tag{4.10}$$

Because the bulk crystal is electroneutral, we can apply the "electroneutrality condition." For process 4.7 we write

$$2\left[V_O^{\cdot\cdot} \right] = \left[e^{|} \right] \tag{4.11}$$

(i.e., the molar concentration of electrons must be twice the molar concentration of doubly ionized positive vacancies). From equations 4.9 and 4.11 we obtain

$$\left[e^{|} \right] = 2^{1/3} K_7^{1/3} P_{O_2}^{-1/6} \tag{4.12}$$

and

$$\left[V_O^{..}\right] = \frac{K_7^{1/3}}{4^{1/3}} P_{O_2}^{-1/6}.$$

(4.13)

Applying the electroneutrality condition to the equilibrium 4.8, we have

$$2\left[M_i^{..}\right] = \left[e^|\right],$$

(4.14)

which, combined with equation 4.10, gives

$$\left[e^|\right] = 2^{1/3} K_8^{1/3} P_{O_2}^{-1/6}$$

(4.15)

and

$$\left[M_i^{..}\right] = \frac{K_8^{1/3}}{4^{1/3}} P_{O_2}^{-1/6}.$$

(4.16)

At P_{O_2} higher than $P_{O_2}^*$, we observe the formation of doubly ionized cationic vacancies and positive electronic vacancies (usually defined as *electron holes* by symbol $h^.$):

$$\frac{1}{2} O_{2_{(gas)}} \rightarrow O_O^x + V_M^{||} + 2h^..$$

(4.17)

The constant of equilibrium 4.17 is

$$K_{17} = \left[V_M^{||}\right]\left[h^.\right]^2 P_{O_2}^{-1/2}$$

(4.18)

and defect concentrations are given by

$$\left[h^.\right] = 2^{1/3} K_{17}^{1/3} P_{O_2}^{1/6}$$

(4.19)

and

$$\left[V_M^{||}\right] = \frac{K_{17}^{1/3}}{4^{1/3}} P_{O_2}^{1/6}.$$

(4.20)

Figure 4.3 plots defect concentrations as a function of the partial pressure of oxygen, ranging from low P_{O_2} to intrinsic pressure $P_{O_2}^*$ to high P_{O_2}.

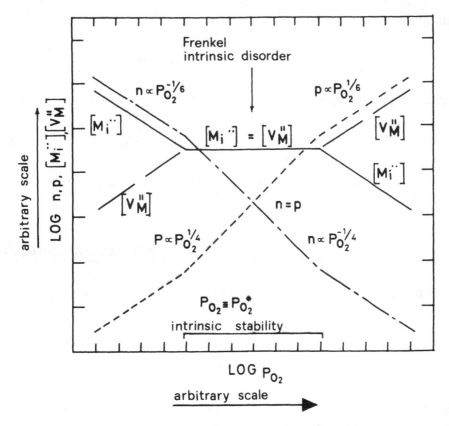

Figure 4.3 Defect concentration as a function of partial pressure of oxygen. Magnesium Frenkel defects are assumed to predominate at stoichiometric composition. At low P_{O_2}, electronic conductivity is of type n (negative charge transport) and proportional to $P_{O_2}^{-1/6}$. At high P_{O_2}, electronic conductivity is of type p (positive charge transport) and proportional to $P_{O_2}^{1/6}$. From R. Morlotti and G. Ottonello, Solution of rare earth elements in silicate solid phases: Henry's law revisited in light of defect chemistry, *Physics and Chemistry of Minerals*, 8, 87–97, figure 1, copyright © 1982 by Springer Verlag. Reprinted with the permission of Springer-Verlag GmbH & Co. KG.

Diagrams such as that presented in figure 4.3 are commonly used in the interpretation of the defects in crystalline substances. Based on the equations above, the defect concentration depends on P_{O_2} in an exponential fashion (proportional to $P_{O_2}^{-1/6}$ and $P_{O_2}^{1/6}$ in low and high P_{O_2} ranges, respectively; because the plot is logarithmic, the exponent becomes the slope of the function). Outside the intrinsic stability range, the stoichiometry of the crystal, as we have seen, is altered by exchanges with the surrounding atmosphere. However, the above defect scheme is an oversimplification of the complex equilibria actually taking place in heterogeneous systems. Indeed, defect processes are not simply ruled by the partial pressure of oxygen, but by the chemistry of all phases present in the system. For example, in the presence of a silica-rich phase, forsterite (Mg_2SiO_4) can incorporate SiO_2 in excess with respect to its normal stoichiometry through the defect equilibrium

$$2\,SiO_2 \Leftrightarrow 2\,V_{Mg}^{\parallel} + Si_i^{\cdots} + Si_{Si}^{x} + 4\,O_O^{x} \tag{4.21}$$

or through the equilibrium

$$SiO_2 \Leftrightarrow 2\,V_{Mg}^{\parallel} + Si_{Si}^{x} + 2\,V_O^{\cdot\cdot} + 2\,O_O^{x} \tag{4.22}$$

(cf. Smith and Stocker, 1975). Based on the experimental data of Pluschkell and Engell (1968), the silica excess in this way reaches 2.4% at 1400 °C.

4.5   Point Impurities

Let us imagine equilibrating a fayalite crystal (Fe_2SiO_4) in an atmosphere sufficiently oxidizing to allow a defect equilibrium of the following type to proceed toward the right:

$$2\,Fe_{Fe}^{x} \rightarrow V_{Fe}^{\parallel} + 2\,Fe_{Fe}^{\cdot} \tag{4.23}$$

or consider that a limited amount of component $Sm_4(SiO_4)_3$ may dissolve into a forsteritic olivine according to the process

$$Sm_4(SiO_4)_3 + 4\,Mg_{Mg}^{x} + 2\,Si_{Si}^{x} + 8\,O_O^{x} \Leftrightarrow 4\,Sm_{Mg}^{\cdot} + 2\,V_M^{\parallel} + SiO_2 \tag{4.24}$$

(Morlotti and Ottonello, 1984). In both cases, the crystal maintains its electroneutrality, but locally we observe the presence of cations whose charge (3+ in this case) is not that of the normal constituent (2+) and which, in the case of equilibrium 4.24, do not correspond to a major component. These "point impurities," normally present in the crystal at equilibrium, significantly affect the chemical-physical properties of the phase, acting, for instance, as activators in light emission processes or as centers of heterogeneous nucleation (see section 3.12). Processes such as 4.24 also regulate the solubility of trace components in rock-forming minerals; the implications of these processes in trace element geochemistry are discussed in chapter 10.

4.6   Energy of Formation of Point Defects

We have seen that defect concentrations in crystals may be described with the aid of electroneutral equilibria involving species of differing charges. We have also seen that defect concentrations are related to the partial pressure of oxygen in the surrounding atmosphere and to the chemistry of coexisting phases in a heterogeneous system. Combining equations of the same types as equations 4.12, 4.13, 4.15, 4.16, 4.19 and 4.20 with experimental data on the actual defect population at a given temperature, we may derive the values of the constants of the various processes and, through them, evaluate the Gibbs free energy involved in each process.

Let us imagine, for instance, that we "measured" the defect population of a generic oxide MO composed essentially of Schottky defects (such a measure could be obtained, for instance, by thermogravimetry at various T and P_{O_2} conditions or by electronic and ionic conductivity measurements at various values of T and P_{O_2}). We can then write

$$0 \rightarrow V_M^{\parallel} + V_O^{\cdot\cdot} \tag{4.25}$$

and

$$K_S = \left[V_M^{\parallel}\right]\left[V_O^{\cdot\cdot}\right], \tag{4.26}$$

where $[V_M^{\parallel}] = [V_O^{\cdot\cdot}]$ is the concentration of vacant sites at temperature T. The Gibbs free energy variation connected with the defect process is*

$$\Delta G_S = -kT \ln K_S, \tag{4.27}$$

where k is the Boltzmann constant ($k = R/N_0$).

An alternative (and probably more precise) method for evaluating defect energies is based on the calculation of lattice energy potentials.

Consider a Schottky equilibrium taking place in a forsterite crystal and involving Mg^{2+} cations and O^{2-} anions, displacing them from their lattice positions and transporting them to the surface of the crystal:

$$Mg_{Mg}^{x} + O_O^{x} \rightarrow V_M^{\parallel} + V_O^{\cdot\cdot} + MgO_{(surface)} \ . \tag{4.28}$$

To subtract the cation Mg^{2+} from its lattice position in the crystal and to bring it to the surface, we must work against the static potentials (coulombic plus repulsive plus dispersive) at the Mg site. In terms of energy, this work corresponds to half the lattice contribution of Mg^{2+} (in the Mg site of interest—i.e., $M1$ or $M2$; see section 5.2) to the bulk static energy of the phase (see also section 1.12):

$$\Delta E_{U,i} = -\frac{1}{2}(E_{C,i} + E_{R,i} + E_{D,i}). \tag{4.29}$$

---

* Because the energy is defined at the atom scale, the standard state is implicit in the definition and the zero superscript is hereafter omitted.

Subtracting a charge in a given lattice position also contributes to the defect energy with an induced polarization term whose significance was described in section 1.19:

$$\Delta E_P = -\frac{1}{2} e \Phi_e. \tag{4.30}$$

Table 4.2 lists defect energies calculated with the method described above in fayalite and forsterite crystals. Note that the energies obtained are on the magnitude of some eV—i.e., substantially lower than the energies connected with extended defects. Note also that, ionic species being equal, the defect energies depend on

Table 4.2 Defect energies in forsterite and fayalite based on lattice energy calculations. $I =$ ionization potential; $E =$ electron affinity; $E_d =$ dissociation energy for O_2; $\Delta H =$ enthalpy of defect process (adapted from Ottonello et al., 1990).

Process	ΔE_C	ΔE_R	ΔE_D	$\Delta CFSE$	ΔE_P	I	E	E_d	ΔH
				FAYALITE					
(1)	22.50	−4.89	0.89	0.31	−15.31	-	-	-	3.50
(2)	24.60	−4.81	0.87	0.28	−14.84	-	-	-	6.10
(3)	−27.34	6.14	−1.07	-	12.44	-	7.46	2.79	0.42
(4)	−27.32	7.11	−1.15	-	12.65	-	7.46	2.79	1.54
(5)	−52.20	12.8	−2.06	-	24.68	-	14.92	5.58	3.72
(6)	11.60	5.59	−0.87	0.28	−42.94	30.64	-	-	4.30
(7)	15.80	5.75	−0.72	0.22	−42.00	30.64	-	-	9.77
(8)	−18.67	−6.70	1.01	0.62	−7.66	30.64	-	-	−0.76
(9)	−20.17	−6.59	0.99	0.56	−7.42	30.64	-	-	−1.99
(10)	−1.07	-	-	-	-	-	-	-	−1.07
(11)	−1.03	-	-	-	-	-	-	-	−1.03
				FORSTERITE					
(12)	23.44	−4.59	0.11	-	−17.14	-	-	-	1.82
(13)	23.44	−6.29	2.08	-	−17.14	-	-	-	2.09
(14)	25.16	−4.46	0.13	-	−16.46	-	-	-	4.37
(15)	55.39	−12.71	1.48	-	−26.22	-	−7.46	−2.79	7.69
(16)	54.73	−14.51	1.54	-	−26.22	-	−7.46	−2.79	4.89
(17)	52.73	−13.10	1.50	-	−26.00	-	−7.46	−2.79	4.89

(1) $Fe_{M1}^{\times} \rightarrow V_{M1}^{||} + Fe_{(surface)}^{2+}$

(2) $Fe_{M2}^{\times} \rightarrow V_{M2}^{||} + Fe_{(surface)}^{2+}$

(3) $\frac{1}{2} O_{2(gas)} \rightarrow O_{(O1.surface)}^{2-}$

(4) $\frac{1}{2} O_{2(gas)} \rightarrow O_{(O2.surface)}^{2-}$

(5) $\frac{1}{2} O_{2(gas)} \rightarrow O_{(O3.surface)}^{2-}$

(6) $Fe_{M1}^{\times} + h^{\cdot} \rightarrow V_{M1}^{||} + Fe_{Si}^{|} + Si_{(surface)}^{4+}$

(7) $Fe_{M2}^{\times} + h^{\cdot} \rightarrow V_{M2}^{||} + Fe_{Si}^{|} + Si_{(surface)}^{4+}$

(8) $Fe_{M1}^{\times} + h^{\cdot} \rightarrow Fe_{M1}^{\cdot}$

(9) $Fe_{M2}^{\times} + h^{\cdot} \rightarrow Fe_{M2}^{\cdot}$

(10) $Fe_{M1}^{\cdot} + Fe_{Si}^{|} \rightarrow \{Fe_{M1}Fe_{Si}^{|}\}^{\times}$

(11) $Fe_{M2}^{\cdot} + Fe_{Si}^{|} \rightarrow \{Fe_{M2}Fe_{Si}^{|}\}^{\times}$

(12) $Mg_{M1}^{\times} \rightarrow V_{M1}^{||} + Mg_{(surface)}^{2+}$

(13) $Fe_{M1}^{\times} \rightarrow V_{M1}^{||} + Fe_{(surface)}^{2+}$

(14) $Mg_{M2}^{\times} \rightarrow V_{M2}^{||} + Mg_{(surface)}^{2+}$

(15) $O_{O1}^{\times} + 2e^{|} \rightarrow V_{O1}^{||} + \frac{1}{2}O_{2(gas)}$

(16) $O_{O2}^{\times} + 2e^{|} \rightarrow V_{O2}^{||} + \frac{1}{2}O_{2(gas)}$

(17) $O_{O3}^{\times} + 2e^{|} \rightarrow V_{O3}^{||} + \frac{1}{2}O_{2(gas)}$

Table 4.3 Energy of Schottky and Frenkel equilibria in halides, oxides, and sulfides. (1) Kröger (1964); (2) Barr and Liliard (1971); (3) Greenwood (1970).

Compound	Process	ΔH_D, Calculated (eV)	ΔH_D, Observed (eV)	K_0, Observed
AgBr (1)	$0 \to V'_{Ag} + V^{\cdot}_{Br}$	-	0.9-2.2	8000-20,000
AgBr (1)	$Ag^{x}_{Ag} \to V'_{Ag} + Ag^{\cdot}_i$	-	0.87-1.27	157-1500
AgCl (1)	$Ag^{x}_{Ag} \to V'_{Ag} + Ag^{\cdot}_i$	-	1.08-1.69	36-2300
AgI (1)	$Ag^{x}_{Ag} \to V'_{Ag} + Ag^{\cdot}_i$	0.6-0.8	0.69	-
CsCl (2)	$0 \to V'_{Cs} + V^{\cdot}_{Cl}$	-	1.87	-
CsBr (2)	$0 \to V'_{Cs} + V^{\cdot}_{Br}$	-	2.00	-
CsI (2)	$0 \to V'_{Cs} + V^{\cdot}_{I}$	-	1.91	-
CdS (1)	$0 \to V'_{Cd} + V^{\cdot}_{S}$	-	2.7	-
CdTe (1)	$Cd^{x}_{Cd} \to V'_{Cd} + Cd^{\cdot}_i$	-	1.4	0.00017
KBr (1)	$0 \to V'_{K} + V^{\cdot}_{Br}$	1.92	1.96-2.0	-
KCl (1)	$0 \to V'_{K} + V^{\cdot}_{Cl}$	2.18-2.21	2.1-2.4	45
KI (2)	$0 \to V'_{K} + V^{\cdot}_{I}$	-	1.61	-
LiBr (1)	$0 \to V'_{Li} + V^{\cdot}_{Br}$	-	1.80	815
LiCl (1)	$0 \to V'_{Li} + V^{\cdot}_{Cl}$	-	2.12	1700
LiF (1)	$0 \to V'_{Li} + V^{\cdot}_{F}$	-	2.68	500
LiI (1)	$0 \to V'_{Li} + V^{\cdot}_{I}$	-	1.34	500
NaBr (1)	$0 \to V'_{Na} + V^{\cdot}_{Br}$	-	1.66-1.68	-
NaCl (1)	$0 \to V'_{Na} + V^{\cdot}_{Cl}$	1.92-2.12	2.02-2.12	5.4
PbS (1)	$0 \to V'_{Pb} + V^{\cdot}_{S}$	-	1.75	0.27
ZnS (1)	$0 \to V''_{Zn} + V^{\cdot\cdot}_{S}$	4-6	-	-
MgO (3)	$0 \to V''_{Mg} + V^{\cdot\cdot}_{O}$	-	4.77	-
SrO (1)	$0 \to V''_{Sr} + V^{\cdot\cdot}_{O}$	3-5	-	-
BaO (1)	$0 \to V''_{Ba} + V^{\cdot\cdot}_{O}$	4.2	3.7	-
CaO (3)	$0 \to V''_{Ca} + V^{\cdot\cdot}_{O}$	3-5	4.34	-
UO$_2$ (1)	$O^{x}_{O} \to V^{\cdot\cdot}_{O} + O''_{i}$	-	3.44	-

the lattice positions of the defects themselves. For instance, creation of a cationic vacancy on site M (V''_{M1}) requires less energy than that necessary to create V''_{M2}. Also, the amount of energy required to create an oxygen vacancy on site $O3$ ($V^{\cdot\cdot}_{O3}$) is lower than that necessary to create $V^{\cdot\cdot}_{O1}$ or $V^{\cdot\cdot}_{O2}$.

Table 4.2 lists energy values assigned to point impurities such as Fe_{M1} and Fe_{M2}. In the calculation of their energies, an ionization term necessary to subtract one electron from Fe^{2+} has also been added (ionization potential). With defect notation, this process can be expressed as

$$Fe^{x}_{M1} + h^{\cdot} \to Fe^{\cdot}_{M1}. \qquad (4.31)$$

An alternative way of calculating defect energies on the basis of static potentials is that outlined by Fumi and Tosi (1957) for alkali halides, in which the energy of the Schottky process is seen as an algebraic summation of three terms:

$$Na^{x}_{Na} \to V'_{Na} + Na^{+}_{(gas)} \qquad (+4.80\,\text{eV}) \qquad (4.32)$$

$$Cl_{Cl}^{x} \rightarrow V_{Cl}^{\cdot} + Cl_{(gas)}^{-} \qquad (+5.14\,eV) \qquad (4.33)$$

$$Na_{(gas)}^{+} + Cl_{(gas)}^{-} \rightarrow Na_{Na}^{x} + Cl_{Cl}^{x} \qquad (-7.82\,eV) \qquad (4.34)$$

$$0 \rightarrow V_{Na}^{|} + V_{Cl}^{\cdot} \qquad (+2.12\,eV) \qquad (4.35)$$

Note that in this case the energies of processes 4.32 and 4.33 are doubled with respect to the energy necessary to bring the ion to the surface of the crystal; moreover, the energy of process 4.34 is equal to the lattice energy of the crystal.

For comparative purposes, table 4.3 lists defect energies (enthalpies) of Schottky and Frenkel processes in halides, oxides, and sulfides. The constant K_0 appearing in the table is the "preexponential factor" (see section 4.7) raised to a power of 1/2.

4.7 Defect Concentration as a Function of Temperature and Pressure

As we have seen, a given enthalpy term can be associated with each defect process (tables 4.2 and 4.3). The defect process generally also involves an entropic term, which may be quantified through

$$\exp\left(\frac{\Delta S_D}{k}\right) = g_1 g_2 f, \qquad (4.36)$$

where g_1 and g_2 are the electronic degeneracies of the atoms participating in the defect process and f is a vibrational term whose nature will be discussed later.

Because a defect entropy (ΔS_D), a defect enthalpy (ΔH_D), and a defect volume (ΔV_D) exist, the Gibbs free energy of the defect process is also defined:

$$\Delta G_D = \Delta H_D - T\Delta S_D + P\Delta V_D. \qquad (4.37)$$

The enthalpic term ΔH_D is the dominant one in equation 4.37 and determines the variation of the defectual concentration with T. For instance, for a Schottky defect we can write

$$K_S = K_0 \exp\left(-\frac{\Delta H_S}{kT}\right), \qquad (4.38)$$

in which the so-called "preexponential factor" K_0 is given by

$$K_0 = \exp\left(\frac{\Delta S_S}{k}\right), \qquad (4.39)$$

where ΔS_S is the amount of entropy associated with the Schottky process.

If we assume, at first approximation, that ΔH_S does not change with T, it is obvious that the increase in defect concentration with T is simply exponential. Table 4.4, for instance, lists Schottky defect concentrations calculated in this way at various T by Lasaga (1981c) for NaCl and MgO, assuming defect enthalpies of 2.20 and 4.34 eV, respectively.

The approximation discussed above is in fact rather rough, and there are actually two general cases.

1. A linear variation of the Gibbs free energy of the defect process with T, of type

$$\Delta G_D = \Delta G_D^0 + \beta' T, \qquad (4.40)$$

in which β' is the slope of the function. In this case, an Arrhenius plot of the constant of the defect process results in a straight line:

$$\ln K_D \propto \left(\frac{1}{T}\right). \qquad (4.41)$$

The preexponential factor also depends on T according to

$$K_0 = g_1 g_2 f \exp\left(\frac{-\beta'}{k}\right). \qquad (4.42)$$

Table 4.4 Schottky defect concentrations in NaCl and MgO at various T (from Lasaga, 1981c).

T (K)	T (°C)	NaCl		MgO	
		$[V_{Na}^{\mid}]$	$[V_{Na}^{\mid}]/cm^3$	$[V_{Mg}^{\mid\mid}]$	$[V_{Mg}^{\mid\mid}]/cm^3$
0	−273	0	0	0	0
25	298	2.5×10^{-19}	5.6×10^3	2.1×10^{-37}	1.1×10^{-14}
200	473	1.9×10^{-12}	4.2×10^{10}	7.9×10^{-24}	4.2×10^{-1}
400	673	5.9×10^{-9}	1.3×10^{14}	5.8×10^{-17}	3.1×10^6
600	873	4.5×10^{-7}	1.0×10^{16}	3.0×10^{-13}	1.6×10^{10}
800	1073	6.8×10^{-6}	1.5×10^{17}	6.5×10^{-11}	3.5×10^{12}
1000	1273	Melt		2.6×10^{-9}	1.4×10^{14}
1500	1773	Melt		6.9×10^{-7}	3.7×10^{16}
2000	2273	Melt		1.6×10^{-5}	8.6×10^{17}

2. A linear variation with T of the enthalpy of the process:

$$\Delta H_D = \Delta H_D^0 + \beta T. \tag{4.43}$$

In this case, the constant of the process does not lead to a straight function in an Arrhenius plot and the preexponential factor varies with T according to

$$K_0 = g_1 g_2 f T^{\beta/k}. \tag{4.44}$$

Generally, nevertheless, the defect concentration tendentially increases with T.

Reexpressing the constant of the Schottky process as

$$\ln K_S = -\frac{\Delta G_S}{kT} \tag{4.45}$$

and deriving in P, we obtain

$$\left(\frac{\partial \ln K_S}{\partial P}\right)_{T,X} = -\frac{1}{kT}\left(\frac{\partial \Delta G_S}{\partial P}\right)_{T,X} = -\frac{\Delta V_S}{kT}, \tag{4.46}$$

where ΔV_S is the volume variation associated with the formation of the defect. The value of ΔV_S is generally positive for Schottky defects in ionic crystals, because the lattice around the vacant sites undergoes relaxation due to static effects. The corresponding ΔV_F is generally lower for Frenkel defects. It follows that the P effect on the defect population is more marked for Schottky disorder than for Frenkel disorder. Because ΔV_S is inherently positive, the defect population generally decreases with increasing P.

4.8 Associative Processes

Point defects of opposite charges tend to be reciprocally attracted, forming the so-called *associate defects*.

Let us consider the general case of two point defects A and B forming an associate defect (AB) according to the process

$$A + B \rightarrow (AB). \tag{4.47}$$

If we imagine that both defects A and B and associate defect (AB) are distributed in conditions of complete configurational disorder, and if we substitute thermodynamic activity for the molar concentration, we have

$$K_{A,B} = \frac{[(AB)]}{[A][B]}. \tag{4.48}$$

The Gibbs free energy modification due to associative processes (disregarding volumetric effects) is

$$\Delta G_{AB} = -kT \ln K_{AB} = \Delta H_{AB} - T\Delta S_{AB}. \tag{4.49}$$

The enthalpy change related to associative process (ΔH_{AB}) is due essentially to coulombic interactions and, subordinately, to polarization, repulsion, covalent bonding, elastic interactions, and vibrational effects. The latter two causes are generally negligible and may have some effects only at low T.

The coulombic energy of associative processes is conveniently described by

$$\Delta H_{AB,\text{coulombic}} = -\frac{q^2}{\varepsilon_s r}, \tag{4.50}$$

where q is the charge of interacting defects, r is the distance of shortest approach, and ε_s is the static dielectric constant of the crystal.* Evaluation of the coulombic interaction through equation 4.50 usually leads to overestimation of the energy of the associative process (see table 4.5).

The entropy variation connected with associative processes (ΔS_{AB}) may be ascribed to two different contributions: configurational and vibrational. The configurational term is given by

$$\Delta S_{AB,\text{conf}} = k \ln\left(\frac{Z}{\sigma}\right), \tag{4.51}$$

where Z is the number of ways in which the associate may be formed and σ is a symmetry factor. For an associate defect (AB), Z is the number of equivalent sites that can be occupied by B at the shortest distance from A (or by A at the shortest distance from B). For the same associate defect (AB), $\sigma = 1$, but becomes 2 for an associate dimer (AA), 3 for a trimer (AAA), and so on.

The vibrational entropy ($\Delta S_{AB,\text{vibr}}$) contributing to the bulk entropy of defect process arises from the variation of the vibrational spectrum of the crystal in the neighborhood of the associate:

---

* Actually, as pointed out by Kröger (1964), at small distances the use of the static dielectric constant based on a continuum approach to interaction energy is not justified, because one would rather adopt the permittivity of a vacuum. However, using the static dielectric constant leads to underestimation of the binding energy by 10 to 15%, counterbalancing the fact that short-range repulsive forces are neglected.

$$\Delta S_{AB,\text{vibr}} = k \ln f \tag{4.52}$$

$$f = \prod_j \frac{v_{0,j}}{v'_j}, \tag{4.53}$$

where $v_{0,j}$ and v'_j are vibrational frequencies related, respectively, to free and associated defects (see Kröger, 1964, for a more detailed treatment).

For a pair of Schottky defects with, respectively, x and y first neighbors and v' and v'' vibrational frequencies, vibrational term f has the form

$$f = \left(\frac{v_0}{v'}\right)^x \left(\frac{v_0}{v''}\right)^y. \tag{4.54}$$

For instance, for a Schottky defect pair in NaCl, we have

$$\frac{v_0}{v'} = \frac{v_0}{v''} = 2 \tag{4.55}$$

$$x = y = 6, \tag{4.56}$$

from which we obtain $f = 4096$. If the vibrational frequency of the defects increases as a result of association, then $f > 1$, and the vibrational entropy also increases. The opposite happens when $f < 1$.

In the case of a Frenkel defect pair, if v_i is the frequency of interstitial atom i and v'_i is the vibrational frequency of the Z first neighbors surrounding it, we have

$$f = \left(\frac{v_0}{v_i}\right)^z \left(\frac{v_0}{v'_i}\right)^z \left(\frac{v_0}{v'}\right)^x. \tag{4.57}$$

Combining equations 4.49, 4.51, and 4.52, we obtain

$$K_{AB} = \frac{Zf}{\sigma} \exp\left(-\frac{\Delta H_{AB}}{kT}\right). \tag{4.58}$$

For the simple associate defect (AB), $\sigma = 1$ and the relative concentrations of associate defect (AB) and free defects A and B are ruled by

$$\left[(AB)\right] + \left[A\right] = \left[A\right]_{\text{total}} \tag{4.59}$$

and

$$\left[(AB)\right] + \left[B\right] = \left[B\right]_{total}. \tag{4.60}$$

Introducing fractions

$$\beta_{AB} = \frac{\left[(AB)\right]}{\left[A\right]_{total}}; \qquad \beta_A = \frac{\left[A\right]}{\left[A\right]_{total}}; \qquad \beta_B = \frac{\left[B\right]}{\left[B\right]_{total}}, \tag{4.61}$$

we have

$$\frac{\beta_{AB}}{\beta_A \beta_B} = \left[A\right]_{total} Zf \exp\left(-\frac{\Delta H_{AB}}{kT}\right). \tag{4.62}$$

For a generic defect concentration c, such as

$$c = \left[A\right]_{total} = \left[B\right]_{total}, \tag{4.63}$$

at the point where half the defects are associated, we have

$$\frac{\beta_{AB}}{\beta_A \beta_B} = 1 \tag{4.64}$$

and

$$\left(\frac{\Delta H_{AB}}{kT}\right)_{0.5} = \ln c + \ln Zf. \tag{4.65}$$

Figure 4.4 plots relative concentrations of associate defect (AB) and free defects A and B calculated as functions of $\Delta H_{AB}/kT$ for various values of c, with $Z = 4$ and $f = 1$. Note that when the enthalpy change connected with the association process is low (and/or T is high) all defects are practically free, whereas for high (negative) values of $\Delta H_{AB}/kT$ all defects are associated. The dashed line in figure 4.4 represents the equality condition outlined by equation 4.65.

Table 4.5 lists the energies of associative processes in halides and simple sulfides. As already mentioned, the enthalpy modification resulting from coulombic interactions is generally higher with respect to the values resulting from a detailed calculation. Table 4.5 also lists association energies involving point impurities. We will see in chapter 10, when treating the stabilization of trace elements in crystals, that association energy plays an important role in trace element distribution processes.

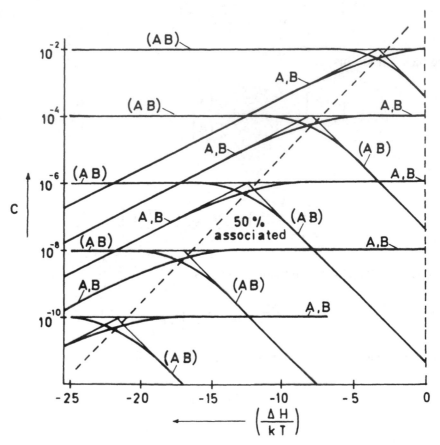

Figure 4.4 Relative concentrations of free and associate defects for various values of total defect concentration c, with $Z = 4$ and $f = 1$ (from Kröger, 1964; redrawn). Dashed line: equality conditions in populations of free and associate defects.

4.9 Stability of a Crystalline Compound in the Presence of Defect Equilibria: Fayalite as an Example

We have already noted that the concept of "stoichiometric crystal" is an extreme idealization of an effectively more complex reality. In the presence of extrinsic disorder, stoichiometry varies as a function of the chemistry of the coexisting phases and of T and P. To clarify this concept better, the procedure developed by Nakamura and Schmalzried (1983) to describe fayalite may be briefly recalled.

We define $\bar{\xi}$ and $\bar{\eta}$ as nonstoichiometry parameters describing the chemistry of fayalite according to

$$\bar{\xi} = \frac{n_{Si}}{n_{Si} + n_{Fe}} - \frac{1}{3} \qquad (4.66)$$

Table 4.5 Energies of associative processes in halides and simple sulfides. Values in eV (from Kröger, 1964; modified).

Compound	Associative Process	$-\Delta H_{AB}$, Coulombic Calculated	$-\Delta H_{AB}$, Total Calculated	$-\Delta H_{AB}$, Experimental	K_0
AgBr	$Cd^{\cdot}_{Ag} + V'_{Ag} \rightarrow (Cd^{\cdot}_{Ag}V'_{Ag})^{\times}$	0.22	-	0.16	-
AgCl	$Cd^{\cdot}_{Ag} + V'_{Ag} \rightarrow (Cd^{\cdot}_{Ag}V'_{Ag})^{\times}$	0.12	-	0.18	-
AgCl	$V'_{Ag} + V^{\cdot}_{Cl} \rightarrow (V'_{Ag}V^{\cdot}_{Cl})^{\times}$	-	-	0.43	-
CaF$_2$	$Y^{\cdot}_{Ca} + F'_{i} \rightarrow (Y^{\cdot}_{Ca}F'_{i})^{\times}$	-	-	1.43	-
CaF$_2$	$Na'_{Ca} + V^{\cdot}_{F} \rightarrow (Na'_{Ca}V^{\cdot}_{F})^{\times}$	-	-	0.07	-
KBr	$Ca^{\cdot}_{K} + V'_{K} \rightarrow (Ca^{\cdot}_{K}V'_{K})^{\times}$	-	-	0.56	-
KCl	$Ca^{\cdot}_{K} + V'_{K} \rightarrow (Ca^{\cdot}_{K}V'_{K})^{\times}$	0.69	0.32	0.52	0.6
KCl	$Cd^{\cdot}_{K} + V'_{K} \rightarrow (Cd^{\cdot}_{K}V'_{K})^{\times}$	0.69	0.32	-	-
KCl	$Sr^{\cdot}_{K} + V'_{K} \rightarrow (Sr^{\cdot}_{K}V'_{K})^{\times}$	0.69	0.39	0.21-0.42	-
KCl	$V'_{K} + V^{\cdot}_{Cl} \rightarrow (V'_{K}V^{\cdot}_{Cl})^{\times}$	0.98	0.58–0.72	-	-
LiF	$Mg^{\cdot}_{Li} + V'_{Li} \rightarrow (Mg^{\cdot}_{Li}V'_{Li})^{\times}$	-	-	0.7	-
NaCl	$Ca^{\cdot}_{Na} + V'_{Na} \rightarrow (Ca^{\cdot}_{Na}V'_{Na})^{\times}$	0.6	0.38	0.67	0.04
NaCl	$Cd^{\cdot}_{Na} + V'_{Na} \rightarrow (Cd^{\cdot}_{Na}V'_{Na})^{\times}$	0.6	0.38–0.44	0.3–0.34	-
NaCl	$Sr^{\cdot}_{Na} + V'_{Na} \rightarrow (Sr^{\cdot}_{Na}V'_{Na})^{\times}$	0.6	0.45	-	-
NaCl	$Mn^{\cdot}_{Na} + V'_{Na} \rightarrow (Mn^{\cdot}_{Na}V'_{Na})^{\times}$	-	-	0.39–0.7	1.3
NaCl	$V'_{Na} + V^{\cdot}_{Cl} \rightarrow (V'_{Na}V^{\cdot}_{Cl})^{\times}$	0.91	0.44–0.60	-	-
ZnS	$V''_{Zn} + V^{\cdot\cdot}_{S} \rightarrow (V''_{Zn}V^{\cdot\cdot}_{S})^{\times}$	-	0.3–1.0	-	-

and

$$\eta = \frac{n_O}{n_{Si} + n_{Fe}} - \frac{4}{3}, \tag{4.67}$$

where the n-terms are the moles of the various elements. Obviously, in the absence of extrinsic disorder, parameters $\bar{\xi}$ and $\bar{\eta}$ take on a value of zero. Figure 4.5A shows the Gibbs phase triangle for the Fe-Si-O system. The (stoichiometric) compound Fe_2SiO_4, coexisting with SiO_2 and Fe, with Fe and FeO, or with FeO and Fe_3O_4, is represented by a dot. Actually this scheme is not of general validity, and, in the presence of extrinsic disorder, the *compositional field of intrinsic stability* is expanded, as shown in figure 4.5B.

Based on thermogravimetric experiments on the compound Fe_2SiO_4 at various T and P_{O_2} conditions, Nakamura and Schmalzried (1983) established that the extrinsic disorder of fayalite is conveniently represented by the equilibrium

$$10\,Fe^{\times}_{Fe} + Si^{\times}_{Si} + 2\,O_{2(gas)} \Leftrightarrow 3V''_{Fe} + 6Fe^{\cdot}_{Fe} + \left(Fe'_{Si}Fe'_{Fe}\right)^{\times} + FeSiO_4 . \tag{4.68}$$

The defect scheme shown in equation 4.68 was later confirmed by electromotive force measurements with galvanic cells (Simons, 1986) and by diffusivity mea-

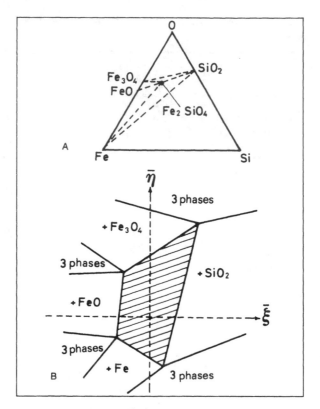

Figure 4.5 Nonstoichiometry of compound Fe_2SiO_4 (fayalite) in the Fe-Si-O system. (A) Usual representation of Fe_2SiO_4 as a discrete point in Gibbs space. (B) Defect equilibria expand intrinsic stability field of phase through variation in nonstoichiometry parameters. From A. Nakamura and H. Schmalzried, On the nonstoichiometry and point defects in olivine, *Physics and Chemistry of Minerals,* 10, 27–37, figure 1, copyright © 1983 by Springer Verlag. Reprinted with the permission of Springer-Verlag GmbH & Co. KG.

surements (Brinkmann and Laqua, 1985) and is in reasonable agreement with the defect energies reported in table 4.2.

As shown in figure 4.5B, extrinsic disorder (nonstoichiometry) may vary considerably, as a function of solid paragenesis, between the low P_{O_2} stability limit defined by equilibrium

$$2\,Fe + SiO_2 + O_{2(gas)} \Leftrightarrow Fe_2SiO_4 \tag{4.69}$$

and the high P_{O_2} stability limit defined by

$$3\,Fe_2SiO_4 + O_{2(gas)} \Leftrightarrow 2\,Fe_3O_4 + 3\,SiO_2 . \tag{4.70}$$

Table 4.6 shows the energy of extrinsic disorder calculated for the solid mixture $(Fe, Mg)_2SiO_4$ at $T = 1200$ °C, based on the defect scheme of equation 4.68 and on the defect energies of table 4.2.

We note that the defect energy contribution associated with extrinsic disorder varies considerably as a function of the partial pressure of oxygen of the system. These energy amounts may significantly affect the intracrystalline disorder, with marked consequences on thermobarometric estimates based on intracrystalline distribution. As we will see in detail in chapter 10, most of the apparent complexities affecting trace element distribution may also be solved by accurate evaluation of the defect state of the phases.

4.10 Diffusion in Crystals: Atomistic Approach

Diffusion processes in crystals are markedly conditioned by point defects. The presence of vacancies (or interstitials) allows ions to jump from one site to a vacant neighboring position, leaving behind a vacancy: in this way an *elemental diffusion process* is created (together with complementary *vacancy migration*) as a summation of single displacement steps. During its movement from the occupied to the vacant site, the ion performs a saddle-shaped trajectory, of the type exemplified in figure 4.6, to overcome the repulsive barriers of neighboring ions. Work against the repulsive potential requires a certain amount of energy (mainly accessible in thermal form). This energy, which is the only energy required during displacement, is defined as "migration energy" (actually, not only the repulsive potential but also all the static energy forms present in the crystal field play a role in determining the amount of migration energy, although the repulsive term is the dominant one).

Table 4.7 lists the *migration enthalpies* of some halides and simple oxides

Table 4.6 Energy contributions deriving from extrinsic disorder of mixture $(Fe,Mg)_2SiO_4$ at $T = 1200$ °C. Values expressed in J/mole; P_{O_2} is in bar (from Ottonello et al., 1990).

$X_{Fe_2SiO_4}$	(1)		(2)	
	P_{O_2}	ΔG_D	P_{O_2}	ΔG_D
0.2	1.96E-14	−6.21	2.65E-4	−401.80
0.3	4.41E-14	−10.35	2.32E-5	−377.72
0.4	7.85E-14	−16.66	4.14E-6	−397.71
0.5	1.23E-13	−26.37	1.08E-6	−445.34
0.6	1.77E-13	−41.26	3.63E-7	−517.86
0.7	2.40E-13	−64.06	1.44E-7	−617.18

(1) = maximum reduction limit (equilibrium 4.69)

(2) = maximum oxidation limit (equilibrium 4.70)

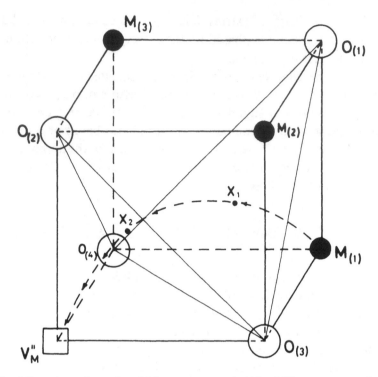

Figure 4.6 Migration of a cation M from an occupied site $M_M^\times$ to a vacant site $V_M^{\parallel}$ in an ionic crystal MO. Trajectory passes through centers of equilateral triangles O_1-O_4-O_3 (X_1) and O_2-O_4-O_3 (X_2).

for *vacancy migration* and *interstitial migration*. Note that the required energies are subordinate with respect to the energies of defect formation (cf. tables 4.3 and 4.7).

Let us now consider the crystal MO. If the diffusion takes place by migration of cationic vacancies, the number of atoms that undergo the process depends on the vacancy concentration $[V_M^{\parallel}]$ and the thermal state of single atoms M (the jump takes place only whenever atom M in the neighborhood of the vacancy has sufficient energy to perform it). The diffusion coefficient associated with the vacancy migration process is given by

$$D = \frac{d^2 v_0}{3} \left[V_M^{\parallel} \right] \exp\left(-\frac{H_{V_M^{\parallel}}}{kT} \right) \qquad (4.71)$$

(Jost, 1960; Lasaga, 1981c), where d is the interionic distance between vacant and occupied sites, and v_0 is a specific vibrational frequency of the crystal (Einstein frequency, or a fraction of the Debye frequency; see Lasaga, 1981c).

A similar relation holds for anionic vacancies and for diffusion through interstitial migration (replace $[V_M^{\parallel}]$ and $H_{V_M^{\parallel}}$ by $[V_O^{\cdot\cdot}]$ and $H_{V_O^{\cdot\cdot}}$, and so on).

Because at high T the vacancy concentration depends exponentially on T, i.e.

Table 4.7 Migration enthalpy in alkali halides and simple oxides. Values in eV. Data from Barr and Liliard (1971) (1) and Greenwood (1970) (2).

Compound	H_{V_M}	H_{V_X}	H_{M_i}	H_{X_i}
LiF	0.70(1)	0.65(1)	-	-
CaF$_2$	-	-	-	1.65(1)
SrF$_2$	-	-	-	1.00(1)
BaF$_2$	-	-	-	0.78(1)
LiCl	0.39(1)	-	-	-
NaCl	0.71(1)	1.00(1)	-	-
AgBr	-	-	0.17(1)	-
KCl	0.71(1)	1.00(1)	-	-
KBr	0.65(1)	0.91(1)	-	-
NaBr	0.78(1)	1.17(1)	-	-
KI	1.21(1)	-	-	-
AgCl	-	-	0.13(1)	-
PbCl$_2$	-	0.35(1)	1.56(2)	-
PbBr$_2$	-	0.30(1)	-	-
MgO	3.38(1)	2.69(2)	-	-
CaO	3.51(1)	-	-	-
PbO	2.91(1)	-	-	-
BeO	2.43(1)	-	-	-
UO$_2$	1.21(2)	-	-	-
ZrO$_2$	1.08(2)	-	-	-

H_{V_M} = migration enthalpy of cationic vacancy
H_{V_X} = migration enthalpy of anionic vacancy
H_{M_i} = migration enthalpy of cationic interstitial
H_{X_i} = migration enthalpy of anionic interstitial

$$\left[V_M^{\parallel} \right] = K_S^{1/2} = \exp\left(-\frac{\Delta H_S}{2kT} \right) \exp\left(\frac{\Delta S_S}{2k} \right), \tag{4.72}$$

equation 4.71 can be reexpressed as

$$D = D_0 \exp\left(-\frac{Q_D}{kT} \right), \tag{4.73}$$

where Q_D represents the "bulk activation energy of the diffusion process," composed of the enthalpy of vacancy formation plus the enthalpy of vacancy migration:

$$Q_D = \frac{\Delta H_S}{2} + H_{V_M^{\parallel}}, \tag{4.74}$$

and D_0 groups the preexponential terms:

$$D_0 = \frac{d^2 v_0}{3} \exp\left(\frac{\Delta S_S}{2k}\right). \tag{4.75}$$

Equation 4.75 finds its application in the region of intrinsic disorder (a similar equation can be developed for Frenkel defects), where Schottky and Frenkel defects are dominant with respect to point impurities and nonstoichiometry.

With decreasing temperature, as we have seen, the intrinsic defect population decreases exponentially and, at low T, extrinsic disorder becomes dominant. Moreover, extrinsic disorder for oxygen-based minerals (such as silicates and oxides) is significantly affected by the partial pressure of oxygen in the system (see section 4.4) and, in the region of intrinsic pressure, by the concentration of point impurities. In this new region, term Q_D does not embody the enthalpy of defect formation, but simply the enthalpy of migration of the defect—i.e.,

$$Q_D = H_{V_M^{\parallel}}. \tag{4.76}$$

In an Arrhenius plot of diffusivity, the transition from condition 4.74 to condition 4.76 corresponds to a change of slope, which in turn corresponds to transition from intrinsically dominated defects (high T) to extrinsically dominated defects (low T). Comparison of the slopes of the function in the two regions also allows evaluation of the respective values of ΔH_S and $H_{V_M^{\parallel}}$ (or ΔH_F and $H_{M_i^{\cdot\cdot}}$). For instance, figure 4.7 shows diffusivity measurements in $(Mg, Fe)_2SiO_4$ olivine mixtures, after Buening and Buseck (1973). We see a change in the slope at about 1125 °C, ascribable to the above transition. This transition is also clearly detectable in Co-Mg interdiffusion experiments (Morioka, 1980; $T = 1300$ °C) and in Fe-Mg interdiffusion experiments (Misener, 1974; $T = 1100$ °C). T values being equal, Buening and Buseck (1973) also observed a marked anisotropy of diffusion as a function of crystallographic direction: diffusion was higher along axis c, indicating preferential migration by vacancy diffusion on $M1$ sites, in agreement with the energies of table 4.2 (the same anisotropy was observed in Ni diffusion by Clark and Long, 1971 and Morioka, 1980 for Co-Mg interdiffusion, and by Misener, 1974 for Fe-Mg interdiffusion).

Lastly, diffusivity appears to increase progressively with the amount of Fe in the system, in agreement with the defect scheme outlined in section 4.9 (see equation 4.68).

Based on the data of Buening and Buseck (1973), the energy of diffusion activation along axis c is 2.735 eV (corresponding to 264 kJ/mole) in the intrinsic high-T region, and 1.378 eV (corresponding to 133 kJ/mole) in the extrinsic low-T region.

Applying equation 4.75, we obtain an enthalpy of formation of Schottky defects of 2.64 eV, which is somewhat higher than the value derived by Ottonello et al. (1990) with static potential calculations (see table 4.2):

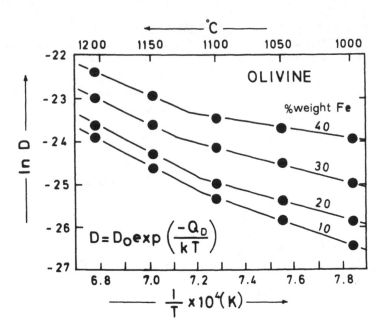

Figure 4.7 Arrhenius plot of cationic diffusivity in olivines. Reprinted with permission from Buening and Buseck (1973), *Journal of Geophysical Research,* 78, 6852–6862, copyright © 1973 by the American Geophysical Union.

$$\mathrm{Fe}^{\times}_{M1} \rightarrow \mathrm{V}^{\parallel}_{M1} + \mathrm{Fe}^{2+}_{(surface)} \qquad \left(+2.09\,\mathrm{eV}\right). \qquad (4.77)$$

(To evaluate the energy of the Schottky process, based on the energies of table 4.2, the concomitant process of anion vacancy formation and the recombination energies of cation and anion on the surface of the crystal must be added to process 4.77.)

Not all crystalline compounds exhibit slope modifications similar to those shown for olivine in figure 4.7. For instance, figure 4.8 shows diffusivity plots for feldspars according to the revision of Yund (1983). The temperature ranges investigated by various authors are sufficiently wide, but in any case we do not observe any slope modifications imputable to modified defect regimes.

However, it must be emphasized that interpretation of elemental diffusion in feldspars is complicated by the structural state of the polymorphs, which vary in a complex fashion with temperature, chemistry and re-equilibration kinetics. These complexities also account for the controversies existing in the literature regarding diffusion energy in these phases (see also, incidentally, figure 4.8). Elemental diffusivity data for rock-forming silicates are listed in table 4.8.

Whenever diffusivity rates are not experimentally known and cannot be estimated by static potential calculations, approximate values can be obtained by empirical methods. The most popular of these methods establishes a linear relationship between the enthalpy of the Schottky process (and enthalpy of migration) and the melting temperature of the substance (expressed in K):

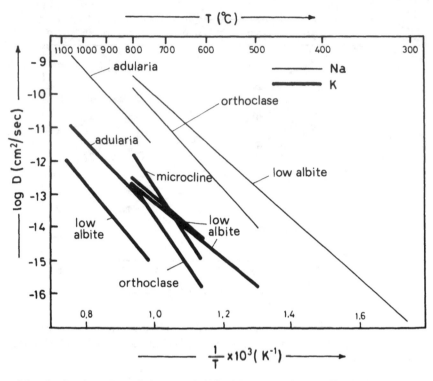

Figure 4.8 Arrhenius plot of elemental diffusivity in feldspars. From Yund (1983). Reprinted with permission of The Mineralogical Society of America.

$$\Delta H_S \approx 2.14 \times 10^{-3}\, T_f \qquad \left(\text{eV}\right) \tag{4.78}$$

and

$$\Delta H_{V_M} \approx 0.014 T_f \qquad \left(\text{eV}\right). \tag{4.79}$$

Equations 4.78 and 4.79 are valid for ionic solids. The analogous relations for metals are

$$\Delta H_S \approx 7.5 \times 10^{-4} T_f \qquad \left(\text{eV}\right) \tag{4.80}$$

and

$$\Delta H_{V_M} \approx 6 \times 10^{-4} T_f \qquad \left(\text{eV}\right). \tag{4.81}$$

The significance of empirical equations 4.78, 4.79, 4.80, and 4.81 has not yet been clarified in a satisfactory fashion. The enthalpy of Frenkel processes in metals can be related to the Debye temperature of the solid through

Table 4.8 Elemental diffusivity in silicates (from Lasaga, 1981c; Morioka and Nagasawa, 1991).

Compound	Element	T-Range, (°C)	D_0, (cm²s⁻¹)	Q_D, (eV)	Reference
Mg_2SiO_4	Fe_{M1}	1125–1200	5.6×10^{-2}	2.73	Buening and Buseck (1973)
Mg_2SiO_4	Fe_{M1}	1100–1125	5.9×10^{-7}	1.38	Buening and Buseck (1973)
Mg_2SiO_4	Fe_{M2}	900–1000	1.1×10^{-2}	2.61	Misener (1974)
Mg_2SiO_4	Ni	1250–1450	2.2×10^2	4.29	Morioka (1981)
Mg_2SiO_4	Ni	1149–1234	1.1×10^{-5}	2.00	Clark and Long (1971)
Mg_2SiO_4	Co	1150–1400	2.8×10^{-1}	3.24	Morioka (1980, 1981)
Mg_2SiO_4	Ca	1200–1400	7.3×10^1	4.31	Morioka (1981)
Mg_2SiO_4	Mg_{M1}	1300–1400	1.5×10^3	4.60	Morioka (1981)
Mg_2SiO_4	Mg_{M2}	1000–1150	1.8×10^{-8}	1.48	Sockel and Hallwig (1977)
Mg_2SiO_4	O_{O1}	1150–1600	1×10^{-4}	3.32	Jaoul et al. (1980)
Mg_2SiO_4	O_{O2}	1275–1628	3.5×10^{-3}	3.86	Reddy et al. (1980)
Mg_2SiO_4	O_{O3}	1472–1734	2.9×10^{-2}	4.31	Ando et al. (1981)
Mg_2SiO_4	Si	1300–1700	1.5×10^{-6}	3.94	Jaoul et al. (1981)
Fe_2SiO_4	Mg_{M1}	900–1100	1.5×10^{-2}	2.16	Misener (1974)
Fe_2SiO_4	Mg_{M1}	1125–1200	2.9×10^{-2}	2.55	Buening and Buseck (1973)
Fe_2SiO_4	Mg_{M1}	1000–1125	7.6×10^{-7}	1.28	Buening and Buseck (1973)
Co_2SiO_4	Si	1200–1300	8.9×10^{-3}	3.44	Schmaizried (1978)
$KAlSi_3O_8$	O	350–700	4.5×10^{-8}	1.11	Giletti et al. (1978)
Adularia	O	400–700	5.3×10^{-7}	1.28	Yund and Anderson (1974)
Orthoclase	Na	500–800	8.9	2.29	Foland (1974)
Orthoclase	K	500–800	16.1	2.96	Foland (1974)
Orthoclase	Rb	500–800	38	3.17	Foland (1974)
Orthoclase	Sr	800–870	6×10^{-4}	1.79	Misra and Venkatasubramanian (1977)
Microcline	O	400–700	2.8×10^{-6}	1.28	Yund and Anderson (1974)
Microcline	K	600–800	133.8	3.04	Lin and Yund (1972)
Microcline	Sr	800–870	5×10^{-4}	1.67	Misra and Venkatasubramanian (1977)
$NaAlSi_3O_8$	O	350–800	2.3×10^{-9}	0.92	Giletti at al. (1978)
$NaAlSi_3O_8$	O	600–800	2.5×10^{-5}	1.61	Anderson and Kasper (1975)
$NaAlSi_3O_8$	Na	600–940	5×10^{-4}	1.52	Bailey (1971)
$NaAlSi_3O_8$	Na	200–600	2.3×10^{-6}	0.82	Lin and Yund (1972)
Biotite	K	550–700	2.7×10^{-10}	0.91	Hofman and Giletti (1970)
Anorthite	O	350–800	1.4×10^{-7}	1.14	Giletti et al. (1978)

$$\theta_D = 34.3\left(\frac{\Delta H_F}{MV^{2/3}}\right)^{1/2}, \qquad (4.82)$$

where ΔH_F is in cal/mole. M is the molar weight and V is the molar volume of the crystal (Mukherjee, 1965). Equation 4.82 is related to the preceding empirical formulations through the *Lindemann equation*:

$$T_f \propto \theta_D. \qquad (4.83)$$

Detailed experimental studies on rock-forming minerals (see Chakraborty and Ganguly, 1991, and references therein) have shown that the pressure effect on elemental diffusivity cannot be neglected in petrologic studies, in which highly variable pressure regimes are commonly involved.

This effect can be analytically expressed by the partial derivative

$$\frac{\partial \ln D}{\partial P} = -\frac{\Delta V^+}{kT},$$

(4.84)

where ΔV^+ is the *activation volume of diffusion*—i.e., "the volume change associated with raising the system from a potential energy minimum to the activated state" (Ganguly and Saxena, 1987). Equation 4.84 implies that the activation energy of the diffusion process is a function of pressure:

$$Q_{D(P)} = Q_{D(P_r)} + \Delta V^+ (P - P_r),$$

(4.85)

where P_r is the P condition of reference.

In light of equation 4.85, and adopting $P_r = 1$, equation 4.73 is more conveniently rewritten in the form

$$D_{(P)} = D_0 \exp\left[\frac{-Q_D + (P - 1)\Delta V^+}{kT}\right].$$

(4.86)

Table 4.9 gives a summary of diffusion data for divalent cations in aluminous garnets, with the relative activation volume. Although estimation of activation volume is still largely uncertain, its evaluation is essential when dealing with the wide baric regimes encountered in petrologic studies. As shown in the third column of table 4.9, the presence of ΔV^+ implies substantial modifications of Q_D on the kbar scale of pressure.

4.11   Diffusion and Interdiffusion

The diffusion coefficient for a given ion in a crystal is determined, as we have seen, by the atomistic properties of the ion in the structural sites where the vacancy (or interstitial) participating in the migration process is created (see eq. 4.71). The units of diffusion (and/or "self-diffusion") are usually cm^2sec^{-1}. *Fick's first law* relates the diffusion of a given ion A (J_A) to the concentration gradient along a given direction X:

$$J_A = -D_A \left(\frac{\partial C_A}{\partial X}\right)_{T,P},$$

(4.87)

Table 4.9 Summary of self-diffusion data for divalent cations in aluminosilicate garnets (after Chakraborty and Ganguly, 1991; energy data converted to eV scale, P in kbar). See Chakraborty and Ganguly, 1991, for references.

Cation	D_0 (cm²s⁻¹)	ΔV^+ (cm³mole⁻¹)	$Q_{(P)}$(eV)
Mn^{2+}	5.15×10^{-4}	6.04 ± 2.93	$2.626 + 0.0063P$
Fe^{2+}	6.36×10^{-4}	5.63 ± 2.87	$2.854 + 0.0059P$
Mg^{2+a}	1.11×10^{-3}	5.27 ± 2.96	$2.948 + 0.0055P$
Mg^{2+b}	2.79×10^{-4}	3.22 ± 2.49	$2.838 + 0.0034P$

ᵃT range, 1100–1475 °C.
ᵇT range, 750–1475 °C.

where $(\partial C_A/\partial X)_{T,P}$ is the concentration gradient along X at temperature T and pressure P, and J_A is the flux of A in the same direction. Equation 4.87 can be applied only at *stationary state*—i.e., whenever the concentration gradient does not change with time. If it does change with time, its modification is described by *Fick's second law:*

$$\left(\frac{\partial C_A}{\partial t}\right)_{T,P} = \frac{\partial}{\partial X}\left[D_A\left(\frac{\partial C_A}{\partial X}\right)\right]. \tag{4.88}$$

If there is simultaneous diffusion of more than one component in the crystal, the flux of A in direction X depends on the individual diffusivities of all diffusing components (Darken, 1948), and the individual diffusivity coefficient in equations 4.87 and 4.88 is replaced by interdiffusion coefficient $\tilde{D}$; i.e., for the simultaneous diffusion of two ions A and B,

$$J_A = -\tilde{D}\left(\frac{\partial C_A}{\partial X}\right)_{T,P}; \qquad J_B = -\tilde{D}\left(\frac{\partial C_B}{\partial X}\right)_{T,P} \tag{4.89}$$

and

$$\left(\frac{\partial C_A}{\partial t}\right)_{T,P} = \frac{\partial}{\partial X}\left[\tilde{D}\left(\frac{\partial C_A}{\partial X}\right)\right]; \qquad \left(\frac{\partial C_B}{\partial t}\right)_{T,P} = \frac{\partial}{\partial X}\left[\tilde{D}\left(\frac{\partial C_B}{\partial X}\right)\right]. \tag{4.90}$$

Let us now consider the diffusion of two neutral species A and B in two portions of metal slag welded at one interface (figure 4.9A). If the diffusivity of A is higher than that of B, the portion on the right in figure 4.9A will tendentially grow, whereas the portion on the left will become smaller (figure 4.9B). However, if we keep the metal slag in a fixed position, the interface will be displaced to the left (*Kirkendall effect,* figure 4.9C). The interdiffusion coefficient is related to individual diffusivities according to

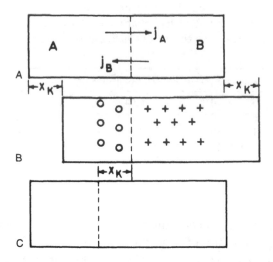

Figure 4.9 Kirkendall effect in a slag for metals A and B, with $D_A > D_B$. (A) Initial configuration of diffusion couple. Position of interface fixed (dashed line) during diffusion (accretion of right part and shrinking of left part of slag block). (C) Position of slag block fixed during diffusion (interface moves leftward). From P. Haasen, *Physical Metallurgy,* copyright © 1978 by Cambridge University Press. Reproduced with modifications by permission of Cambridge University Press.

$$\tilde{D} = X_A D_B + X_B D_A. \tag{4.91}$$

If, however, the diffusing species are electrically charged, the net flux at the interface is effectively zero, to maintain the charge neutrality. Particularly, if A and B are of the same charge, the flux of A will equal that of B:

$$\sum_i J_i = 0. \tag{4.92}$$

Imposing condition 4.92 on Fick's first law results in

$$\tilde{D} = \frac{D_A D_B}{X_A D_A + X_B D_B} \tag{4.93}$$

(see Ganguly and Saxena, 1987 for a more detailed derivation of equations 4.91 and 4.93).

Equation 4.90 is of particular interest when treating diffusion in nonmetallic solids, where diffusion is primarily limited to ionic species. Note that, based on equations 4.91 and 4.93, $\tilde{D} \rightarrow D_B$ when $X_B \rightarrow 0$. Thus, in a diluted binary mixture, the interdiffusion coefficient assumes the same value as the diffusion coefficient for the minor component.

In interdiffusion studies involving ions of the same charge, the dependence of the single-ion diffusivity on concentration is obtained by means of annealing experiments involving single oriented crystals of different compositions kept in contact along a smooth surface. The interface coincides with the *Boltzmann-Matano plane*—i.e., "a plane such that the gain of a diffusing species on one side equals its loss on the other side" (Crank, 1975):

$$\int_0^{C_0} X \, dC = 0, \tag{4.94}$$

where C_0 is the initial concentration in the diffusing couple and X is the distance from the Boltzmann-Matano plane.

Figure 4.10 shows profiles of Mn^{2+} and Mg^{2+} distribution in the

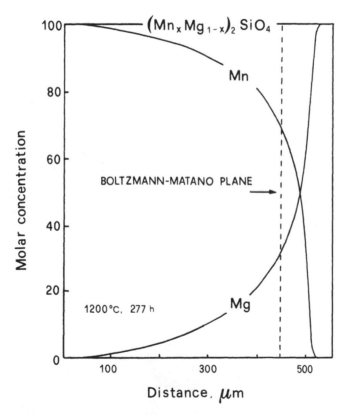

Figure 4.10 Distribution profiles of Mn^{2+} and Mg^{2+} along axis c in the olivine mixture $(Mn_xMg_{1-x})_2SiO_4$ after couple annealing at 1200 °C for 277 hours. From Morioka and Nagasawa (1991). Reprinted with permission of Springer-Verlag, New York.

$(Mn_xMg_{1-x})_2SiO_4$ olivine mixture obtained by Morioka (1981). The concentration profiles are symmetric, indicating that the diffusivities are independent of the concentration of the diffusing ion. On the contrary, possible asymmetry in the diffusion profiles indicate that concentration depends significantly on the diffusing cations. In this case, the interdiffusion coefficient can be obtained by the Boltzmann-Matano equation:

$$\tilde{D}_{C=C_1} = -\frac{1}{2t}\frac{\partial X}{\partial C}\int_0^{C_1} X\, dC,$$
(4.95)

where $\tilde{D}_{C=C_1}$ is the interdiffusion coefficient at concentration C_1 intermediate between infinite dilution and initial concentration C_0 ($0 \leq C_1 \leq C_0$), and t is the duration of the diffusion run.

CHAPTER FIVE

Silicates

Qualitatively and quantitatively, compounds of silicon and oxygen are the class of substances of greatest importance in the earth's crust and mantle, in regard to both mass and variety of structural forms.

In silicon and oxygen compounds, Si^{4+} has coordination IV or VI. The SiO_4^{4-} tetrahedron is energetically quite stable, with mixed ionic-covalent bonds.

The tetrahedral coordination of the $[SiO_4]^{4-}$ group could be regarded as a direct consequence of the capability of the silicon atom to form four hybrid orbitals sp^3, each directed toward one of the apexes of a regular tetrahedron, as shown in figure 5.1. The energy gap between $3s$ and $3p$ orbitals at $Z = 14$ is limited, and thus the combination is at first sight energetically plausible. Detailed quantum mechanics calculations confirm that the tetrahedral symmetry corresponds to the minimum energy for Si-O bonds.

"Hybrid orbitals" may be considered as "perfectioned AOs," adopted in the calculation of localized MOs in polyatomic molecules, with the LCAO method (cf. section 1.17.1). In the case of hybrid orbitals sp^3, the four linear combinations of s and p orbitals (Te_1, Te_2, Te_3, Te_4) that lead to tetrahedral symmetry are

$$Te_1 = \frac{1}{2}\left(s + p_x + p_y + p_z\right) \tag{5.1}$$

$$Te_2 = \frac{1}{2}\left(s + p_x - p_y - p_z\right) \tag{5.2}$$

$$Te_3 = \frac{1}{2}\left(s - p_x + p_y - p_z\right) \tag{5.3}$$

$$Te_4 = \frac{1}{2}\left(s - p_x - p_y + p_z\right). \tag{5.4}$$

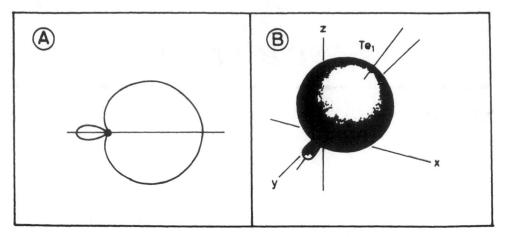

Figure 5.1 Tetrahedral hybrid orbitals sp^3. (A) Section along development axis. (B) Spatial form and orientation of hybrid orbital Te_1 (from Harvey and Porter, 1976; redrawn and reproduced with permission).

However, things are more complex when we look at the MO energy diagram (figure 5.2) developed by Tossell et al. (1973) for the $[SiO_4]^{4-}$ cluster on the basis of a density functional molecular method (MS-SCF $X\alpha$: Multiple Scattering Self Consistent Field). The orbital with the lowest binding energy is $1t_1$, which is essentially the nonbonding AO $2p$ of oxygen, followed in order of increasing energy by the two $1e$ and $5t_2$ MOs with π symmetry (≈ 5 eV) formed by combination of the AOs $3s$ and $3p$ of silicon with the AO $2p$ of oxygen, and the $4t_2$ MO (≈ 7 eV) also formed by combination of the AO $3p$ of silicon with the AO $2p$ of oxygen but with σ symmetry (high-energy orbitals in the inner valence region are predominantly AO $2s$ in character).

Determination of electron density maps for the α-quartz polymorph establishes that the charge transfer between silicon and oxygen is not complete and that a residual charge of $+1.0$ (± 0.1) electron units (e.u.) remains localized on silicon, whereas a charge of -0.5 (± 0.1) e.u. is localized on each oxygen atom. The interpretation of this fact in terms of the bond ionicity is not as univocal as it may appear at first glance.

According to Pauling (1980), electron density maps confirm the fractional ionic character ($f_i = 51\%$) deduced from electronegativity (see eq. 1.32 in section 1.7). Pauling's reasoning is as follows: the length of the Si-O bond in all compounds in which the $[SiO_4]^{4-}$ group shares the oxygen atoms with other tetrahedrally coordinated ions is 1.610 Å (± 0.003; cf. Baur, 1978). Because the observed Si-O distance is about 0.20 Å lower than the sum of the single-bond radii and there are four Si-O bonds, the bond number 1.55 is obtained and covalence is 6.20 e.u. The charge transfer from oxygen to silicon is thus 2.2 valence electrons (i.e., 6.2 to 4). Based on the net charge of $+1.0$ e.u. on silicon, the fractional ionic character is therefore

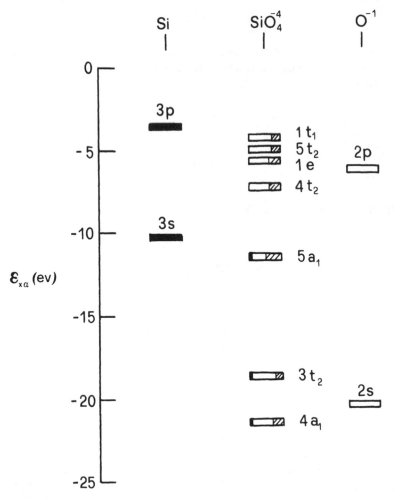

Figure 5.2 MO energy levels for the $[SiO_4]^{4-}$ cluster based on MS-SCF $X\alpha$ calculations. Reprinted from Tossell et al. (1973), with permission of Elsevier Science B.V.

$$f_i = \frac{1 + 2.2}{6.2} = 0.5161 \approx 51.6\%, \tag{5.5}$$

which is in good agreement with the value of 51% deduced from electronegativity.

However, starting from the same experimental evidence, Stewart et al. (1980) adopt the following definition of fractional ionic character:

$$f_i = \frac{100 \times \text{residual charge}}{\text{formal charge}} = \frac{100 \times 1}{4} = 25. \tag{5.6}$$

According to Stewart et al. (1980), the fractional ionic character thus defined is in agreement with the value deduced from electronegativity theory for molecular orbitals. The energy of a generic atom A is defined as

$$W_A = a_A + b_A n_A + c_A n_A^2, \tag{5.7}$$

where a_A, b_A, and c_A are constants valid for atom A, and n_A is the number of electrons in the orbital: 0, 1, or 2. Constants b_A and c_A are related to ionization potential I and to the electron affinity E:

$$b_A = I_A \tag{5.8}$$

$$c_A = \frac{1}{4}\left(E_A - I_A\right). \tag{5.9}$$

The definition of Iczkowski and Margrave (1961) for electronegativity (see section 1.7) gives

$$\chi_A = \frac{\partial W_A}{\partial n} = b_A + 2c_A n_A. \tag{5.10}$$

According to the treatment of Hinze et al. (1963), the charge transfer obeys the equality principle outlined in equation 1.31 (section 1.7). The transfer of fractional charges n_A and n_B for two atoms A and B results in a stable bond with a fractional ionic character:

$$f_i = \frac{1}{2}\left(\frac{\chi_A - \chi_B}{c_A - c_B}\right), \tag{5.11}$$

where χ_A and χ_B are the initial electronegativities of the AOs that form the bond. In the case of α-quartz, applying the electronegativity values of orbitals χ_A and χ_B and coefficients c_A and c_B, both known for states sp and sp^2 and evaluable for hybrid orbitals sp^3 on the basis of bond angles (for the Si-O-Si bond between neighboring tetrahedral groups, the bond angle is 143.68°; cf. Le Page and Donnay, 1976; Hinze and Jaffe, 1962), Stewart et al. (1980) obtain

$$f_i = \frac{1}{2}\left(\frac{2.25 - 5.80}{-2.26 - 4.74}\right) = 0.2536, \tag{5.12}$$

which is in good agreement with the definition of electronegativity expressed by equation 5.6.

Based on the controversies outlined above, discussion on the degree of ionicity in natural silicates appears highly conjectural and superfluous, because there is no agreement even for building group $[SiO_4]^{4-}$. Therefore, no mention will be made hereafter of the degree of ionicity (or covalence) in these compounds.

Disregarding the type of bond, the tetrahedral group $[SiO_4]^{4-}$ is usually represented in the conventional ways outlined in figure 5.3, with the oxygen atoms at the corners and silicon at the center of the tetrahedron, and the structure of silicates is normally described as a function of the relative arrangements of the

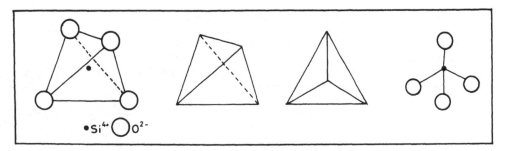

Figure 5.3 Conventional ways of representing group [SiO$_4$]$^{4-}$.

Table 5.1 Structural arrangements of groups [SiO$_4$]$^{4-}$ and [SiO$_6$]$^{8-}$; x = known, 0 = not known (from Liebau, 1982).

Arrangement of Groups	[SiO$_4$]$^{4-}$ (Tetrahedral)	[SiO$_6$]$^{8-}$ (Octahedral)
Isolated	x	x
Corner sharing	x	x
Edge sharing	x	x
Face sharing	0	0

[SiO$_4$]$^{4-}$ groups. Besides coordination IV, silicon also assumes coordination VI with oxygen. Although the number of phases containing [SiO$_6$]$^{8-}$ groups is limited (about 30), the importance of octahedral coordination becomes more and more evident with the advance of studies on high-P compounds. Both groups, [SiO$_4$]$^{4-}$ and [SiO$_6$]$^{8-}$, exhibit various structural arrangements (see table 5.1).

For [SiO$_4$]$^{4-}$ groups, corner sharing and presence in isolated groups are the most common structural arrangements (only fibrous SiO$_2$ seems to exhibit an edge-sharing configuration, although this arrangement is rather suspect; S. Merlino, personal communication). For [SiO$_6$]$^{8-}$ groups, the edge-sharing configuration (as in stishovite, for instance) is as common as the corner-sharing arrangement.

Several elements (generally those with high field strength) exhibit isomorphous substitution of silicon in the tetrahedron (Be^{2+}, Al^{3+}, Fe^{3+}, Ga^{3+}, B^{3+}, Ge^{4+}, Ti^{4+}, P^{5+}). The most important diadochy is that of aluminum, which forms pentavalent tetrahedral groups [AlO$_4$]$^{5-}$. Tetrahedral groups with isomorphous substitution at the center of the tetrahedron are usually designated [TO$_4$].

The chemical classification of silicates is based on their "multiplicity" and type of structural arrangement. It partly differs from the classification commonly adopted in mineralogy (table 5.2).

In mineralogy, monosilicates are called "nesosilicates" and all other oligosilicates are called "sorosilicates"; moreover, the term "polysilicates" is replaced by "inosilicates."

The *multiplicity* of the groups establishes the number m of [TO$_4$] groups that

Table 5.2 Chemical classification of silicates.

Oligosilicates: monosilicates, disilicates, trisilicates, tetrasilicates
Cyclosilicates: monocyclosilicates, dicyclosilicates
Polysilicates: monopolysilicates, dipolysilicates, tripolysilicates, tetrapolysilicates, pentapolysilicates
Phyllosilicates: monophyllosilicates, diphyllosilicates
Tectosilicates

Table 5.3 Kostov's classification of silicates. A = axial; P = planar; I = isometric (from Liebau, 1982).

| | Classes | | |
Cations	$X_{Si} > 4$	$3 > X_{Si} > 1$	$X_{Si} < 1$
Be, Al, Mg, (Fe)	A-P-I	A-P-I	A-P-I
Zr, Ti, (Sn), Nb	A-P-I	A-P-I	A-P-I
Ca, (REE), Mn, Ba	A-P-I	A-P-I	A-P-I
Zn, Cu, Pb, U	A-P-I	A-P-I	A-P-I

link to form multiple tetrahedra. Multiplicities $m = 2, 3, 4, 8, 9$, and 10 have been observed. When several $[TO_4]$ groups are linked to form rings, the number of mutually linked rings is defined as multiplicity: the maximum multiplicity observed in this case is 2.

If $[TO_4]$ groups are linked to form chains, multiplicity is the number of linked chains; the observed multiplicities in this case are $m = 2, 3, 4$, and 5. Analogously, multiplicity is defined as the number of mutually linked sheets. If several $[TO_4]$ groups are linked in a chain, *periodicity* (p) is the number of groups that defines the structural motive (i.e., after p groups, the chain is obtained by simple translation). *Dimensionality* of $[TO_4]$ groups is the degree of condensation of a crystal structure. The condensation of an infinite number of tetrahedra may lead to infinite unidimensional chains ($d = 1$), bidimensional sheets ($d = 2$), and tridimensional lattices ($d = 3$). The dimensionality of an isolated $[TO_4]$ tetrahedron is zero ($d = 0$).

The alternative classification of Kostov (1975) is based on the "degree of silicification" (X_{Si}), expressed by the ratio

$$X_{Si} = \frac{Si + Al}{M'}, \qquad (5.13)$$

where $M' = M_2^+; M^{2+}; M_{0.666}^{3+}; M_{0.5}^{4+}$.

The so-called degree of silicification is inversely proportional to the Si:O atomic ratio and represents the degree of condensation of the $[TO_4]$ groups. Three *classes* with distinct silicification degrees are thus distinguished, and each *class* is further subdivided into *groups*, as a function of the geochemical affinity of the elements, and into *subclasses*, as a function of the morphology of the substance (table 5.3). We refer readers to Liebau (1982) for a more exhaustive discussion on classification schemes for silicates.

5.2 Olivines

Olivines are *monosilicates* ("nesosilicates" in the mineralogical classification) and crystallize in the orthorhombic system (spatial group *Pbnm*). The structural arrangement is shown in figure 5.4: isolated $[SiO_4]^{4-}$ groups (*nesos* means "island") are mutually linked by divalent cations. This arrangement gives rise to two nonequivalent octahedral sites, usually designated $M1$ and $M2$, in which the cations are located, respectively, between the bases of tetrahedra ($M1$ sites) and between a basis and an edge of two neighboring tetrahedra ($M2$ sites). As figure 5.4A shows, tetrahedral apexes are alternate in opposite directions; the dashed line in the figure contours the elementary cell. The formula unit usually adopted for olivine is M_2SiO_4, with M as generic divalent cation. The elementary cell is composed of four formula units. Figure 5.4B details the conformation of the two octahedral sites. The $M1$ octahedron shares six of its 12 edges with the neighboring polyhedra (four with octahedra and two with tetrahedra), whereas the $M2$ octahedron shares only three edges (two with octahedra and one with a tetrahedron). Because shared edges are usually shorter than nonshared edges, to minimize repulsion between neighboring cations (Pauling's third rule), the octahedral symmetry *in both sites* is generally distorted (although the degree of distortion is more evident for the $M2$ site).

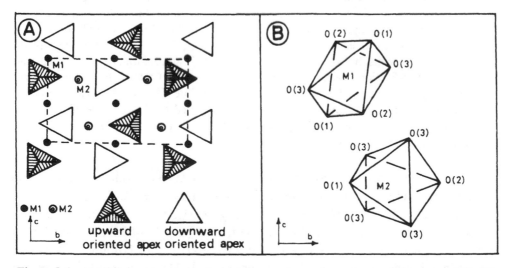

Figure 5.4 (A) Olivine structure: dashed line contours elementary cell (axes c and b). (B) Details of coordination state and distortion of $M1$-$M2$ sites; O(1), O(2), O(3) oxygens occupy nonequivalent positions; distortion from perfect octahedral symmetry is more marked for $M2$ site.

5.2.1 Chemistry and Occurrence

The chemistry of natural olivines is conveniently described by the four main components Mg_2SiO_4 (forsterite), Fe_2SiO_4 (fayalite), Mn_2SiO_4 (tephroite), and Ca_2SiO_4 (Ca-olivine), whose molar fractions generally sum up to more than 99% of the total composition. To these major components must be added the minor terms Ni_2SiO_4, Co_2SiO_4, Zn_2SiO_4, and Cr_2SiO_4.

Table 5.4 lists analyses of some natural specimens. In some instances, there is a silicon deficiency with respect to the stoichiometric formula and diadochy of Al^{3+}, Ti^{4+}, and Fe^{3+} on the tetrahedral sites (cases 1, 2, 3, and 4). In other cases, the amount of silicon is near the theoretical value and only divalent cations are virtually present in the $M1$ and $M2$ positions (cases 5 and 6).

Olivine is the main mineral, in terms of mass and volume, of the earth's upper mantle (peridotites), where it is present as a mixture of the main components forsterite and fayalite, in molar proportions of 85 to 95% and 15 to 5%, respectively. The intrinsic stability limit of the mixture is assumed to coincide with the seismic discontinuity observed at a depth of about 400 km, where olivine undergoes transition to the β- and γ-spinel polymorphs (Ringwood, 1969). Olivine is also a primary phase of mafic rocks (basalts, gabbros, dolerites), where it occurs with amounts of forsterite ranging from Fo_{80} to Fo_{50}. Its importance (in terms of relative abundance) decreases considerably in acidic rocks (granites, syenites, rhyolites, trachytes), where fayalite prevails. Monticellite ($CaMgSi_2O_6$) is normally found as a constituent in Si-Mg marbles, where it is a marker of the sanidine facies (and probably also of the granulitic facies) of thermal regional metamorphism (Brown, 1970). Terms of the tephroite-fayalite mixture are commonly found in Fe-Mn ore deposits associated with skarns and in metamorphosed manganiferous sediments.

The Ni_2SiO_4 content in natural olivines rarely exceeds 1% in mole (weight % of NiO < 0.5), and is usually directly correlated to the forsterite content (see, for instance, Davis and Smith, 1993). The inverse correlation is observed for Mn_2SiO_4 (weight % of MnO < 4). Figure 5.5 shows the relative distributions of Mn_2SiO_4, Ni_2SiO_4, and Ca_2SiO_4 in effusive and plutonic rocks, based on the synthesis of Simkin and Smith (1970): marked control of the mineralization environment can be seen for Ca_2SiO_4. This effect is less evident for the manganous term and is virtually absent for Ni. The relationship between amount of Ca_2SiO_4 and crystallization conditions may be imputed to the pressure effect on the solvus field existing for the join Mg_2SiO_4-$CaMgSiO_4$ (cf. Finnerty, 1977; Finnerty and Boyd, 1978; see section 5.2.5).

Table 5.5 lists silicate components that are known to crystallize with the olivine structure, according to the compilation of Ganguly (1977). The pure components Zn_2SiO_4 and Cr_2SiO_4 have never been synthesized in the laboratory, but are known to form the mixtures $(Mg,Zn)_2SiO_4$ and $(Mg,Cr)_2SiO_4$, in which they reach maximum molar concentrations of 24% (Sarver and Hummel, 1962) and 5% (Matsui and Syono, 1968; Ghose and Wan, 1974), respectively. It must be noted that the presence of chromium as a minor component of terrestrial olivines is restricted to kimberlitic rocks, with a maximum content of 0.15 weight % Cr_2O_3, although greater abundances (>0.4% Cr_2O_3) have been observed in lunar

Table 5.4 Olivine major element compositions (in weight %). Samples occur in different types of rocks: (1) = forsterite from a metamorphosed limestone: (2) = hortonolite from an olivine gabbro: (3) = fayalite from a pantelleritic obsidian; (4) = fayalite from an Fe-gabbro; (5) = forsterite from a cumulitic peridotite; (6) = forsterite from a tectonitic peridotite. Samples (1) to (4) from Deer et al. (1983); sample (5) from Ottonello et al. (1979); sample (6) from Piccardo and Ottonello (1978).

Oxide	Samples					
	(1)	(2)	(3)	(4)	(5)	(6)
SiO_2	41.07	34.94	30.56	30.15	39.52	40.00
TiO_2	0.05	0.43	0.72	0.20	-	-
Al_2O_3	0.56	0.91	0.09	0.07	-	-
Fe_2O_3	0.65	1.46	0.10	0.43	-	-
FeO	3.78	40.37	60.81	65.02	13.27	11.11
MnO	0.23	0.68	3.43	1.01	-	-
MgO	54.06	20.32	3.47	1.05	46.36	47.55
CaO	-	0.81	1.13	2.18	-	-
NiO	-	-	-	-	0.07	0.46
$H_2O^{(+)}$	0.05	0.09	-	-	-	-
Total	100.45	99.11	100.31	100.11	99.22	99.12

Cations	Ionic Fractions on a Four-Oxygens Basis					
Si	0.979	0.990	0.996	1.002	0.992	0.997
Al	0.016	0.032	0.004	0.003	-	-
Ti	0.001	0.009	0.018	0.005	-	-
Fe^{3+}	0.012	0.032	0.004	0.011	-	-
Total IV	1.008	1.063	1.022	1.021	0.992	0.997
Mg	1.920	0.881	0.169	0.052	1.735	1.766
Fe^{2+}	0.075	0.983	1.659	1.808	0.279	0.232
Ni	-	-	-	-	0.001	0.009
Mn	0.005	0.017	0.094	0.028	-	-
Ca	-	0.025	0.039	0.078	-	-
Total VI	2.000	1.906	1.961	1.966	2.015	2.017

specimens. Because chromium is present in nature in the valence states Cr^{2+} and Cr^{3+}, but occurs in olivine only (or predominantly) in the Cr^{2+} form (Burns, 1975; Sutton et al. 1993), the paucity of Cr_2SiO_4 in terrestrial olivines must be ascribed to the oxidation state of the system, which is generally too high to allow the formation of an olivine compound with stoichiometry Cr_2SiO_4. Lastly, it must be noted that the synthesis of compounds of type $LiLnSiO_4$ or $NaLnSiO_4$ (in which Ln is a generic rare earth with atomic number not lower than that of Ho) apparently contrasts with natural evidence. Although microprobe analyses report

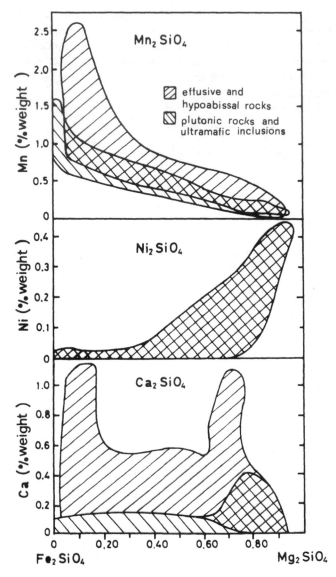

Figure 5.5 Relative distributions of minor components Mn₂SiO₄, Ni₂SiO₄, and Ca₂SiO₄ in fayalite-forsterite mixtures (weight %).

sufficient amounts of Na₂O in natural olivines, the rare earth element (REE) abundances are generally at ultratrace levels (i.e., REE < 1 ppm; see, for instance, Ottonello, 1980; Frey, 1982) and the relative distribution of the various terms of the series does not reveal any crystal chemical control.

5.2.2 Structural Properties

Table 5.6 lists the structural properties of the various olivine components according to the compilation of Smyth and Bish (1988). Note the conspicuous vari-

Table 5.5 Olivine compounds, known and synthesized.

Component	Name	Component
Mg_2SiO_4	Forsterite	$CaCoSiO_4$
Fe_2SiO_4	Fayalite	$LiScSiO_4$
Ca_2SiO_4	Larnite	$LiYSiO_4$
Mn_2SiO_4	Tephroite	$LiLnSiO_4$ (Ln = Ho-Lu)
Co_2SiO_4	Co-olivine	$LiInSiO_4$
Ni_2SiO_4	Liebembergite	$NaYSiO_4$
$CaMgSiO_4$	Monticellite	$NaLnSiO_4$ (Ln = Ho-Lu)
$CaFeSiO_4$	Kirschsteinite	
$CaMnSiO_4$	Glaucocroite	

Table 5.6 Structure of olivine compounds (from Smyth and Bish, 1988). Molar volume is expressed in cm^3/mole; molar weight in g/mole; density in g/cm^3; cell edges and cell volume are in Å and $Å^3$, respectively.

Formula units per unit cell: 4
System: Orthorhombic
Class: *mmm*
Spatial group: *Pbnm*

Phase	Forsterite	Fayalite	Monticellite	Kirschsteinite
Component	Mg_2SiO_4	Fe_2SiO_4	$MgCaSiO_4$	$FeCaSiO_4$
Molar weight	40.708	203.778	156.476	188.011
Density	3.227	4.402	3.040	3.965
Molar volume	43.603	46.290	51.472	47.415
Cell parameters:				
a	4.7534	4.8195	4.822	4.844
b	10.1902	10.4788	11.108	10.577
c	5.9783	6.0873	6.382	6.146
Volume	289.58	307.42	41.84	314.89
Reference	Fujino et al. (1981)	Fujino et al. (1981)	Onken (1965)	Brown (1970)

Phase	Larnite	Tephroite	Co-olivine	Liebembergite
Component	Ca_2SiO_4	Mn_2SiO_4	Co_2SiO_4	Ni_2SiO_4
Molar weight	172.244	201.960	209.950	209.503
Density	2.969	4.127	4.719	4.921
Molar volume	58.020	48.939	44.493	42.574
Cell parameters:				
a	5.078	4.9023	4.7811	4.726
b	11.225	10.5964	10.2998	10.118
c	6.760	6.2567	6.0004	5.913
Volume	385.32	325.02	295.49	282.75
Reference	Czaya (1971)	Fujino et al. (1981)	Brown (1970)	Lager and Meagher (1978)

ation of density between larnite (2.969 g/cm^3) and liebembergite (4.921 g/cm^3), which results from the fact that a low molecular weight is associated to the highest molar volume (larnite) and vice versa (liebembergite).

Figure 5.6 plots the molar volumes of various olivine mixtures against the cubes of the ionic radii of the major cations in VI-fold coordination with oxygen, according to the compilation of Brown (1982). The figure has been partially modified to correct the volume of Ca_2SiO_4 on the basis of the data of Czaya (1971). Based on figure 5.6, olivine compounds seem to obey the simple proportionality established by Vegard's rule (with the exception of the pure component larnite):

$$V_{cell} \propto r^3. \tag{5.14}$$

Brown (1970) calculated the following equation, which is valid for olivine compounds (ionic radii of Shannon and Prewitt, 1969):

$$V_{cell} = 188.32r^3 + 220.17 \qquad (\text{Å}^3). \tag{5.15}$$

Equation 5.15 is based on 13 values, has a correlation coefficient R2 = 0.99, and does not include the cell volume of the pure compound Ca_2SiO_4 (see figure 5.6). It can be opportunely modified to give molar volumes according to

$$\overline{V}^0 = \left(188.32r^3 + 220.17\right)\frac{N_0 \times 10^{-24}}{v} \qquad (\text{cm}^3/\text{mole}), \tag{5.16}$$

where N_0 is Avogadro's number and v is the number of formula units per unit cell.

Equations 5.15 and 5.16 give only approximate values for the volumes of olivine compounds. More accurate study of binary mixtures outlines important deviations from ideal behavior, which result in slight curves in cell parameter vs. composition plots, as shown in figure 5.7. The greatest deviations regard cell edges and are particularly evident for the $(Mg,Ni)_2SiO_4$ mixture.

The volume properties of crystalline mixtures must be related to the crystal chemical properties of the various cations that occupy the nonequivalent lattice sites in variable proportions. This is particularly true for olivines, in which the relatively rigid $[SiO_4]^{4-}$ groups are isolated by $M1$ and $M2$ sites with distorted octahedral symmetry. To link the various interionic distances to the properties of cations, the concept of *ionic radius* is insufficient; it is preferable to adopt the concept of *crystal radius* (Tosi, 1964; see section 1.9). This concept, as we have already noted, is associated with the radial extension of the ion in conjunction with its neighboring atoms. Experimental electron density maps for olivines (Fujino et al., 1981) delineate well-defined minima (cf. figure 1.7) marking the maximum radial extension (r_{max}) of the neighboring ions:

$$r_{max} = \min 4\pi r^2 \rho(r), \tag{5.17}$$

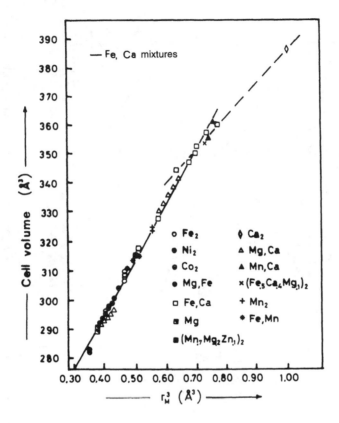

Figure 5.6 Relationship between cell volume and cubed ionic radius of main cation in VI-fold coordination with oxygen. From Brown (1982). Reprinted with permission of The Mineralogical Society of America.

where $\rho(r)$ is electron density at distance r. Equation 5.17 may be operatively translated to the form

$$r_{max} = \frac{C_1}{Z_{eff}} = \frac{C_1}{Z_n - S_e},$$

(5.18)

where C_1 is a constant determined by the electronic configuration of the inner shells, and *effective nuclear charge* Z_{eff} is the actual nuclear charge minus screening constant S_e (cf. section 1.5).

The energy of the atomic orbitals in the hydrogenoid approximation is proportional to the square of the effective charge, and is (numerically) equivalent to the ionization potential

$$E_{(n)} = -\frac{R_y h c z^2}{n^2} = \frac{Z_{eff}^2 C_2}{n^2},$$

(5.19)

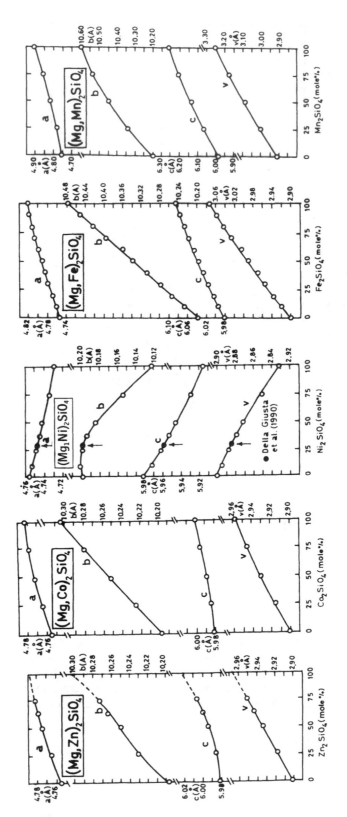

Figure 5.7 Molar volumes and cell edges in binary olivine mixtures (from Akimoto et al., 1976; modified). Note conspicuous deviations from linearity in cell edges of $(Mg,Ni)_2SiO_4$ mixture, satisfactorily reproduced by site-occupancy calculations based on parameters in table 5.7.

where R_y is Rydberg's constant, h is Planck's constant, c is the speed of light, and z is the ionic charge.

In light of equations 5.18 and 5.19, a relationship can be derived between ionization potential (I) and crystal radius (Della Giusta et al., 1990):

$$r_{max} = \left[\frac{C_1}{\left(C_2^{-1/2} n \right)} \right] I^{-1/2}.$$

(5.20)

Equation 5.20 is general and differs from the proportionality of Ahrens, which is valid for ions of homologous charge:

$$r \propto I^{-2}.$$

(5.21)

Because the effective charge may also be related to free ion polarizability α_i^f:

$$\alpha_i^f = 4n^4 \, a_0^3 \, Z_{eff}^{-4},$$

(5.22)

where a_0 is Bohr's radius (0.529 Å), the crystal radius may be related to polarizability through

$$r_{max} = \left[\frac{C_1}{\left(1.4142 n \, a_0^{3/4} \right)} \right] \alpha_i^{f^{1/4}}.$$

(5.23)

(Della Giusta et al., 1990). Experimental observations show that equations 5.20 and 5.23 both hold.

In olivine, 24 independent interatomic distances must be evaluated in order to define the structure. For all 24 distances (D_{ij}) plus cell edges and volume, regressions of the following type may be proposed:

$$D_{ij} = \sum_s \sum_i \omega_{ij,s} X_{i,s} + \xi_j.$$

(5.24)

In equation 5.24, $\omega_{ij,s}$ and ξ_j are adjustable coefficients valid for a given distance j in a given site s, and $X_{i,s}$ are the atomic fractions of i ions in site s.

Equation 5.24 is quite precise. The maximum bias between observed and predicted distances for a population of 55 samples is in fact 0.005 Å, practically coinciding with the experimental error. The coefficients of equation 5.24 may be directly derived from the ionization potential and ion polarizability, based on equations 5.20 and 5.23:

$$\omega_{ij,s} = K_{1j,s} \left(\sum_{m=1}^z I_m \right)_i^{-1/2} + K_{2j,s} \, \alpha_i^{f^{1/4}} + K_{3j,s}.$$

(5.25)

In equation 5.25, $[\sum_{m=1}^z I_m]_i$ is the summation of the ionization potentials of ion i up to valence z, and $K_{1j,s}$, $K_{2j,s}$ and $K_{3j,s}$ are numerical constants valid for site s. The values of

Table 5.7 Constants of equation relating coefficients $\omega_{ij,s}$ to ionization potential and polarizability of cations in *M1* and *M2* sites in olivine. R is the correlation coefficient (from Della Giusta et al., 1990).

Distance	K_1	K_2	K_3	Site of Reference	R
O2A-O3A	1.4060	0.0720	1.8584	M1	0.936
M1-O1B	4.9132	0.1563	−0.7328	M1	0.986
M1-O2A	6.1811	0.2281	−0.1680	M1	0.991
M1-O3A	7.6123	0.2705	−0.3521	M1	0.986
Mean M1-O	6.2303	0.2204	−0.8553	M1	0.991
O1B-O2A	8.8115	0.3572	0.3179	M1	0.988
O1B-O3A	6.1686	0.1827	0.9582	M1	0.966
O1B-O3C	11.552	0.4195	0.752	M1	0.993
O2A-O3C	16.151	0.5740	−0.6596	M1	0.905
O1B-O2B	6.9951	0.1934	1.0069	M1	0.991
M2-O1A	8.4392	0.3092	−1.4253	M2	0.982
M2-O3C	7.3525	0.2576	−0.7356	M2	0.968
M2-O3D	5.6366	0.1738	−0.7300	M2	0.967
M2-O3E	6.4770	0.1854	−1.0222	M2	0.987
Mean M2-O	6.6693	0.2149	−1.0299	M2	0.994
O1A-O3E	16.395	0.5200	−3.6118	M2	0.998
O3D-O3F	11.072	0.3221	−1.7030	M2	0.999
O3E-O3F	10.732	0.3251	−2.3447	M2	0.990
O2C-O3D	11.591	0.4004	−2.4462	M2	0.981
O2C-O3E	6.3085	0.2119	−1.4248	M2	0.964
a_0	7.3633	0.2328	2.1574	M1	0.986
a_0	1.7883	0.0634	0.1920	M2	0.957
b_0	25.462	0.7845	−0.7220	M2	0.994
c_0	11.430	0.2874	1.3087	M1	0.991
c_0	11.322	0.3398	−0.8387	M2	0.997
V	1255.84	38.3462	−3.7834	M1	0.992
V	1466.70	44.3668	−499.44	M2	0.995

these constants for the principal interatomic distances (excluding the $[SiO_4]^{4-}$ tetrahedron) are listed in table 5.7.

Table 5.8 lists cell edges and fractional atomic coordinates obtained for an olivine mixture of composition $Mg_{1.4}Ni_{0.6}SiO_4$ with 53.2% of the Ni distributed in *M1* and 46.8% in *M2*. The values obtained by calculation almost coincide with natural observation (see also figure 5.7).

5.2.3 Phase Stability as a Function of P and T

With increasing T and constant P, olivines expand their octahedral sites *M1* and *M2*, whereas the tetrahedral group $[SiO_4]^{4-}$ remains virtually unaffected. Table 5.9 lists linear thermal expansion coefficients for *M1-O* and *M2-O* polyhedra in some olivine compounds, according to Lager and Meagher (1978). These values are generally higher than those obtained by the method of Hazen and Prewitt (1977) (see eq. 1.93 and section 1.14.2). In particular, thermal expansion is anom-

Table 5.8 Measured (*m*) structural parameters for mixture $Mg_{1.4}Ni_{0.6}SiO_4$ compared with results of calculations (*c*) based on equation 5.24 and observed site occupancies (Ottonello et al., 1989)

	a_0		b_0		c_0		V	
	m	*c*	*m*	*c*	*m*	*c*	*m*	*c*
	4.7458	4.7473	10.1986	10.1968	5.9563	5.9559	288.35	288.33

	x		y		z	
	m	*c*	*m*	*c*	*m*	*c*
*M*1	0.0	0.0	0.0	0.0	0.0	0.0
*M*2	0.9900	0.9902	0.2761	0.2760	0.2500	0.2500
Si	0.4257	0.4258	0.0936	0.0936	0.2500	0.2500
O(1)	0.7660	0.7665	0.0924	0.0926	0.2500	0.2500
O(2)	0.2202	0.2202	0.4461	0.4460	0.2500	0.2500
O(3)	0.2763	0.2764	0.1626	0.1627	0.0326	0.0322

Table 5.9 Linear thermal expansion coefficients in olivines ($°C^{-1}$).

Compound	$\alpha_{l_{M1-O}}$ ($\times 10^5$)	$\alpha_{l_{M2-O}}$ ($\times 10^5$)
Ni_2SiO_4	1.54	1.31
Mg_2SiO_4	1.93	1.72
$Mg_{0.69}Fe_{0.31}SiO_4$	1.17	1.22
Fe_2SiO_4	1.26	1.50
$CaMgSiO_4$	1.74	1.33
$CaMnSiO_4$	1.54	1.52
Mean value	1.53	1.43
Equation 1.93	1.38	1.38

alously high for Mg^{2+}-bearing polyhedra. With increasing *T*, the rigidity of $[SiO_4]^{4-}$ groups and the marked expandibility of octahedral sites result in increased distortion of *M*1-O and *M*2-O polyhedra. However, at whatever *T*, the *M*1 site remains less distorted than the *M*2 site.

Octahedral distortion prevents calculation of the volumetric thermal expansion coefficient of the phase stemming from the linear polyhedral thermal expansion values. However, volumetric thermal expansions may be obtained by least-squares regression of the observed *PVT* relations. Table 5.10 lists, for instance, the coefficients of a thermal expansion relation of the type

$$\alpha_V = \alpha_0 + \alpha_1 T + \alpha_2 T^{-2}. \tag{5.26}$$

These coefficients are valid for a pressure of 1 bar, within the stability field of the phase.

Table 5.10 Isobaric thermal expansion of olivine compounds. Regressions for Mg_2SiO_4 and Fe_2SiO_4 from Fei and Saxena (1986).

Compound	$\alpha_0 \times 10^4$	$\alpha_1 \times 10^8$	α_2	Reference
Mg_2SiO_4	0.2660	0.8736	-0.2487	Suzuki et al. (1981)
Fe_2SiO_4	0.2711	0.6885	-0.5767	Suzuki et al. (1981)
$CaMgSiO_4$	0.3160	0.7008	-0.1469	Skinner (1966)
$CaFeSiO_4$	0.2260	1.4456	-0.1652	Skinner (1966)

Incidentally, it must be noted that, although most olivine compounds are stable (at $P = 1$ bar) up to the melting temperature, in some compounds polymorphic transitions are observed at temperatures below the melting point (for instance, Ca_2SiO_4 has a transition to a monoclinic polymorph at 1120 K). Forsterite at 1 bar melts at $T_f = 2163$ K, whereas fayalite melts at a much lower temperature ($T_f = 1490$ K). The significance of this difference, as shown by Hazen (1977), must be ascribed to different polyhedral thermal expansions for polyhedra containing Mg^{2+} and Fe^{2+} cations. Hazen (1977) pointed out that, if the curve of volumes of the members of the $(Mg,Fe)_2SiO_4$ mixture is extrapolated at various T conditions (V_T) up to the melting point of each compound, the same limiting volume is reached ($V_{T_f} = 319$ Å^3; cf. figure 5.8). According to Hazen this volume represents a sort of critical condition, beyond which the thermal expansion of M1-O and M2-O polyhedra is not compatible (notwithstanding distortion) with the rigidity of the groups: the bonds break and the compound melts. This interpretation is extremely convincing and its rationalization of the melting behavior of olivines should be tentatively extended to other silicate phases. We will examine the thermodynamic significance of solid-liquid equilibria in the melting process of the mixture $(Mg,Fe)_2SiO_4$ (section 6.5), observing that the phase relations give rise to Roozeboom diagrams of type I (section 7.1.3). We will also see that the observed phase boundaries may be ascribed to essentially ideal behavior, in both the solid and liquid aggregation states of the mixture.

Concerning the effect of pressure on the structure of olivine, experimental studies show that compression affects only M1-O and M2-O polyhedra (the $[SiO_4]^{4-}$ group is virtually incompressible). Pressure thus has the same (although inverse) effect as temperature. It must be noted, however, that, unlike temperature, the increase of pressure do not seem to cause further distortion of octahedral sites (Brown, 1982). This is quite reasonable: octahedral distortion is largely a result of the fact that one (in the M2 site) and two (in the M1 site) of the octahedral edges are shared with tetrahedra, and are thus shorter. Because the $[SiO_4]^{4-}$ tetrahedron is substantially rigid, increased T causes increased dissimilarity of the polyhedral edges, whereas increased P renders the octahedral edges more similar in length (S. Merlino, personal communication).

The polyhedral compressibility of the various olivine compounds conforms satisfactorily to the generalizations of Anderson (1972) and Hazen and Finger

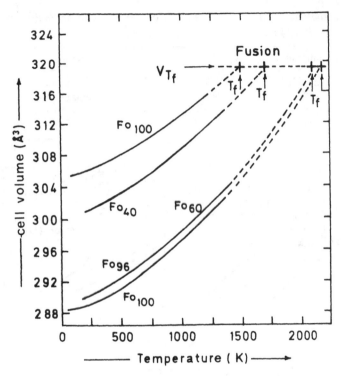

Figure 5.8 Relationship between cell volume and temperature for compounds of mixture $(Mg,Fe)_2SiO_4$. Extrapolation to melting point (T_f) identifies the same limiting volume $(V_{T_f} = 319 \text{ Å}^3)$ for the various terms of the mixture. From Hazen (1977). Reprinted with permission of The Mineralogical Society of America.

(1979) (cf. section 1.15.1 and eq. 1.110 and 1.111): there is an inverse linear relation between polyhedral compressibility and polyhedral volume.

Table 5.11 lists experimental bulk moduli for some olivine end-members.

Combining thermal expansion and compressibility data for compounds in the $(Mg,Fe)_2SiO_4$ mixture, Hazen (1977) proposed the following equation of state for olivines of the series Mg_2SiO_4-Fe_2SiO_4:

$$V_{cell} + \left(290 + 0.17X_{Fa} - 0.006T + 0.000006T^2\right)\left(1 - \frac{P}{1350 - 0.16T}\right), \quad (5.27)$$

where T is in °C, P is in bar, and V is in Å^3.

We have already noted that the $(Mg,Fe)_2SiO_4$ mixture at high P undergoes polymorphic transitions to the forms β (modified spinel) and γ (spinel) by progressive thickening, which leads first to the $[SiO_4]^{4-}$ tetrahedra sharing a corner, thus forming sorosilicate groups $[Si_2O_7]^{6-}$ (β form; Moore and Smith, 1970), and then to the maximum close packing (eutaxis) of oxygen ions (cubic form γ).

Figure 5.9A summarizes the phase relations observed by Akimoto et al.

Table 5.11 Bulk modulus K_0 (Mbar) and its first derivative on P (K') for some olivine end-members.

Compound	K_0 (Mbar)	K'	Reference
Mg_2SiO_4	1.379	5.0	Sumino et al. (1977)
Fe_2SiO_4	1.600	5.0	Sumino (1979)
Mn_2SiO_4	1.310	5.0	Sumino (1979)
Co_2SiO_4	1.502	5.0	Sumino (1979)

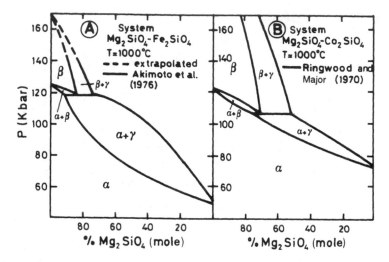

Figure 5.9 Phase stability relations in binary systems Mg_2SiO_4-Fe_2SiO_4 (A) and Mg_2SiO_4-Co_2SiO_4 (B) at $T = 1000$ °C.

(1976) at $T = 1000$°C in the binary system Mg_2SiO_4-Fe_2SiO_4. The phase diagram is essentially identical to that previously proposed by Ringwood and Major (1970). Three two-phase fields are observed: $\alpha + \gamma$, $\alpha + \beta$, and $\beta + \gamma$. Phase β is stable only for molar Fe_2SiO_4 amounts under 30%. The peculiar "rabbit ears" conformation of the binary loops is also observed in the Mg_2SiO_4-Co_2SiO_4 system (figure 5.9B; Ringwood and Major, 1970), but in this case the "ears" are more developed and phase stability is extended to 50% in moles of the component Co_2SiO_4. The transition from β (orthorhombic) to γ (cubic) is first-order. Assuming that upper-mantle olivine has a composition $Fo_{89}Fa_{11}$, Ringwood and Major (1970) deduced that the phase transition to forms β and γ involves a portion of mantle about 27 km thick. With a geothermal gradient of 30 bar/°C for the β–γ transition, the discontinuity in the velocity of seismic waves, observed at about 400 km depth, may be ascribed to this structural rearrangement. The pressure in the transition zone is 132 Kbar and the temperature 1600°C (the discontinuity is observed at 370–420 km depth and p-waves increase their velocity by 0.6–1 km/s).

5.2.4 Thermodynamic Properties of End-Members

Table 5.12 reports a compilation of thermochemical data for the various olivine components (compound Zn_2SiO_4 is fictitious, because it is never observed in nature in the condition of pure component in the olivine form). Besides standard state enthalpy of formation from the elements ($T_r = 298.15$ K; $P_r = 1$ bar; pure component), the table also lists the values of bulk lattice energy and its constituents (coulombic, repulsive, dispersive). Note that enthalpy of formation from elements at standard state may be derived directly from bulk lattice energy, through the Born-Haber-Fayans thermochemical cycle (see section 1.13).

Standard state entropy values and Maier-Kelley coefficients of heat capacity at constant P, with respective T limits of validity, are listed for the same components in table 5.13. The adopted polynomial expansion is the Haas-Fisher form:

Table 5.12 Thermochemical data for various olivine end-members ($T_r = 298.15$ K; $P_r = 1$ bar). Listed values in kJ/mole (from Ottonello, 1987). H^0_{f,T_r,P_r} = standard state enthalpy of formation from the elements; U_L = lattice energy; E_{BHF} = energy of Born-Haber-Fayans cycle; E_C = coulombic energy; E_{DD} = dipole-dipole interactions; E_{DQ} = dipole-quadrupole interactions; E_R = repulsive energy.

Compound	E_{BHF}	H^0_{f,T_r,P_r}	U_L	E_C	E_{DD}	E_{DQ}	E_R
Mg_2SiO_4	18,456.4	−2175.7[1]	20,632.1	−24,131.3	−259.4	−76.8	3835.4
		−2170.4[2]	20,626.8				3840.6
		−2176.9[3]	20,633.3				3834.0
		−2163.6[4]	20,620.0				3847.3
		−2171.9[8]	20,628.3				3839.2
Fe_2SiO_4	19,268.7	−1481.6[1]	20,750.3	−23,934.6	−442.1	−152.9	3779.3
		−1479.4[2]	20,748.0				3781.6
		−1479.9[3]	20,748.5				3781.1
		−1478.8[8]	20,747.5				3782.1
Ca_2SiO_4	17,608.5	−2316.2[2]	19,925.1	−23,105.4	−282.4	−84.3	3546.9
Mn_2SiO_4	18,825.4	−1728.1[2]	20,553.5	−23,734.8	−348.5	−111.6	3641.3
		−1727.2[3]	20,552.6				3642.2
		−1732.7[8]	20,558.1				3636.7
Co_2SiO_4	19,443.9	−1406.7[3]	20,850.6	−24,040.8	−480.5	−174.7	3855.4
Ni_2SiO_4	19,609.9	−1405.2[3]	21,015.1	−24,219.8	−596.5	−226.2	4027.4
		−1393.1[5]	21,004.0				4038.5
		−1405.8[6]	21,015.7				4026.7
Zn_2SiO_4	19,326.0	−1390.4[7]	20,716.4	−24,068.0	−393.9	−129.6	3875.2

(1) Helgeson et al. (1978)

(2) Robie et al. (1978)

(3) Barin and Knacke (1973); Barin et al. (1977)

(4) JANAF (1974–1975)

(5) Levitskii et al. (1975)

(6) Mah and Paneratz (1976)

(7) Ottonello (1987) extrapolated value

(8) Holland and Powell (1990)

Table 5.13 Selected standard state entropy ($S^0_{T_r,P_r}$; $T_r = 298.15\ K$, $P_r = 1$ bar) and Maier-Kelley coefficients of heat capacity function. References as in table 5.12. Data in J/(mole $\times$ K).

Compound	$S^0_{T_r,P_r}$	K_1	$K_2 \times 10^3$	$K_3 \times 10^{-5}$	$K_4 \times 10^5$	K_5	T Limit	Reference
Mg_2SiO_4	95.19	149.83	27.36	−35.648	0	0	1800	(1)
Fe_2SiO_4	148.32	152.76	39.162	−28.033	0	0	1800	(1)
Ca_2SiO_4	120.50	132.57	52.51	−19.049	0	0	1200	(2)
Mn_2SiO_4	163.20	512.52	182.73	46.026	5.2058	−6640.4	1800	(2)
Co_2SiO_4	158.57	157.40	22.05	−26.694	0	0	1690	(3)
Ni_2SiO_4	110.04	163.18	19.748	−24.309	0.88282	0	1818	(3)
Zn_2SiO_4	122.04	−7.921	111.04	−70.715	−1.7405	3108.3	1800	(7)

$$C_P = K_1 + K_2 T + K_3 T^{-2} + K_4 T^2 + K_5 T^{-1/2}. \tag{5.28}$$

The polynomial expansion in equation 5.28 allows the calculation of the high-T enthalpy and entropy of the compound of interest, from standard state values, through the equations

$$H^0_{T_r,P_r} \equiv H^0_{f,T_r,P_r} \tag{5.29}$$

$$H_{T,P_r} = H^0_{T_r,P_r} + \int_{T_r}^{T} C_P\, dT \tag{5.30}$$

$$S_{T,P_r} = S^0_{T_r,P_r} + \int_{T_r}^{T} C_P\, \frac{dT}{T} \tag{5.31}$$

(see section 3.7 and eq. 3.105 and 3.106 for the form of integrals 5.30 and 5.31).
    We recall also that the Gibbs free energy of the compound at the T and P of interest may be obtained by applying

$$G_{T,P} = H_{T,P_r} - TS_{T,P_r} + \int_{P_r}^{P} V_{T,P}\, dP. \tag{5.32}$$

To solve equation 5.32, we must first calculate the volume of the compound at the T of interest and $P = P_r$. Adopting polynomial expansions of the type shown in equation 5.26, we have

$$V_{T,P_r} = V^0_{T_r,P_r}\left(1 + \int_{T_r}^{T} \alpha_{(T)}\, dT\right)$$

$$= V^0_{T_r,P_r}\left[1 + \alpha_0(T - T_r) + \frac{\alpha_1}{2}(T^2 - T_r^2) - \alpha_2(T^{-1} - T_r^{-1})\right]. \tag{5.33}$$

Integration on P depends on the form of the equation of state. For instance, adopting the Murnaghan form:

$$V_{T,P} = V_{T,P_r}\left(1 + \frac{K'}{K_0}P\right)^{-1/K'},$$ (5.34)

we have

$$\int_{P_r}^{P} V_{T,P}\, dP = \int_{1}^{P} V_{T,P}\, dP$$

$$= V_{T,P_r}\left\{\left(\frac{K_0}{K'-1}\right)\left[\left(1 + \frac{K'}{K_0}P\right)^{(K'-1)/K_0} - 1\right]\right\}.$$ (5.35)

We have already seen that the binary mixture $(Mg,Fe)_2SiO_4$ undergoes polymorphic transition with increasing pressure. The univariant curves in figure 5.9 define the loci of equality of potential for the components in the two phases at equilibrium. The Gibbs free energy of polymorph $\alpha-\beta$ (or polymorphs $\alpha-\gamma$ and $\beta-\gamma$) is thus identical at each point on the univariant curve. On the basis of this principle, Fei and Saxena (1986) proposed a list of thermochemical values that (within the limits of experimental approximation) reproduce P/X stability fields ($T = 1000$ °C) for the system Mg_2SiO_4-Fe_2SiO_4. These values are reported in table 5.14.

When adopting the values in table 5.14, note that they refer to half a mole of

Table 5.14 Compilation of thermodynamic data for compounds β-Mg_2SiO_4, β-Fe_2SiO_4, γ-Mg_2SiO_4 and γ-Fe_2SiO_4, after Fei and Saxena (1986). $T_r = 298.15$; $P_r = 1$ bar. Note that each listed value refers to one-half mole of compound (see first column).

Compound	H^0_{f,T_r,P_r} kJ/mole	$S^0_{T_r,P_r}$ J/(mole × K)	$V^0_{T_r,P_r}$ cm³/mole	C_P J/(mole × K)			
				K_1	$K_3 \times 10^{-7}$	$K_5 \times 10^{-3}$	$K_6 \times 10^{-8}$
β-$MgSi_{0.5}O_2$	-1072.450	44.53	20.270	108.218	0	-0.7362	-2.7895
β-$FeSi_{0.5}O_2$	-736.473	70.93	21.575	136.332	0	-1.3198	2.2412
γ-$MgSi_{0.5}O_2$	-1069.300	42.75	19.675	100.096	-0.066	-0.5184	-2.4582
γ-$FeSi_{0.5}O_2$	-738.334	68.44	21.010	130.986	0	-1.2041	1.3419

Compound	Thermal Expansion (K^{-1})			Compressibility (Mbar)	
	$\alpha_0 \times 10^4$	$\alpha_1 \times 10^8$	α_2	K_0	K'
β-$MgSi_{0.5}O_2$	0.2711	0.6885	-0.5767	1.600	4.00
β-$FeSi_{0.5}O_2$	0.2319	0.7117	-0.2430	1.660	4.00
γ-$MgSi_{0.5}O_2$	0.2367	0.5298	-0.5702	2.130	4.00
γ-$FeSi_{0.5}O_2$	0.2455	0.3591	-0.3703	1.970	4.00

orthosilicate component (i.e., $MgSi_{0.5}O_2$, $FeSi_{0.5}O_2$) and must thus be doubled to be consistent with the values in tables 5.12 and 5.13. Note also that Fei and Saxena 1986) use the Berman-Brown polynomial for C_P, in the form

$$C_P = K_1 + K_3 T^{-2} + K_5 T^{-1/2} + K_6 T^{-3} \qquad (5.36)$$

(see section 3.2). It is also important to note that β-Fe_2SiO_4 is fictitious because it never occurs in nature as a pure compound (cf. figure 5.9A).

To reproduce the binary loops of the system Mg_2SiO_4-Fe_2SiO_4 accurately, Fei and Saxena (1986) use a (symmetric) regular mixture model for α-β-γ phases with the following interaction parameters (cf. section 3.8.4): $W_\alpha = -8.314$ kJ/mole; $W_\beta = -9.977$ kJ/mole; $W_\gamma = -11.210$ kJ/mole. These parameters also refer to one-half mole of substance and must thus be multiplied by 2 to be consistent with the usual stoichiometry.

Alternative thermochemical tabulations to reproduce the P/X stability fields of the system Mg_2SiO_4-Fe_2SiO_4 may be found in Bina and Wood (1987) and Navrotsky and Akaogi (1984).

5.2.5 Mixing Properties and Intracrystalline Disorder

If we consider the seven components in tables 5.12 and 5.13 as representative of the chemistry of natural olivines, it is clear that 21 regular binary interaction parameters (disregarding ternary and higher-order terms) are necessary to describe their mixing properties, through a combinatory approach of the Wohl or Kohler type (cf. section 3.10). In reality, the binary joins for which interactions have been sufficiently well characterized are much fewer. They are briefly described below.

(Mg,Fe)$_2$SiO$_4$ Mixture

Mixing properties of the binary join Mg_2SiO_4-Fe_2SiO_4 have been extensively investigated by various authors. There is general consensus on the fact that the mixture is almost ideal and that the excess Gibbs free energy of mixing is near zero. Electrochemical measurements by Nafziger and Muan (1967) and Kitayama and Katsura (1968) at $T = 1200$ °C and $T = 1204$ °C, respectively ($P = 1$ bar), suggest slight deviation from ideality, which can be modeled with an interaction parameter of about 1.5 to 2 kJ/mole (see also Williams, 1971; Driessens, 1968; Sack, 1980). Calorimetric measurements by Sahama and Torgeson (1949) indicate zero heat of mixing at 25 °C. However, more recent measurements by Wood and Kleppa (1981) show the existence of positive enthalpy of mixing at $T = 970$ K and $P = 1$ bar, reproducible through a subregular mixing model of the form

$$H_{mixing} = 2\left(W_{H,FeMg} X_{Fe} X_{Mg} + W_{H,MgFe} X_{Mg} X_{Fe}\right), \qquad (5.37)$$

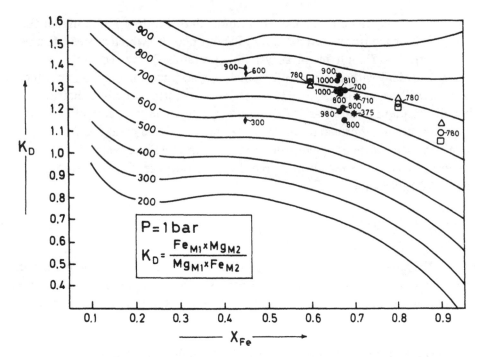

Figure 5.10 Experimentally observed intracrystalline disorder in $(Mg,Fe)_2SiO_4$ mixture, compared with theoretical distribution curves generated by interionic potential calculations. ● = Aikawa et al. (1985); ◆ = Smyth and Hazen (1973); ✳ = Brown and Prewitt (1973); □,○,△ = Ottonello et al. (1990). From G. Ottonello, F. Princivalle, and A. Della Giusta, Temperature, composition and f_{O_2} effects on intersite distribution of Mg and Fe^{2+} in olivines, *Physics and Chemistry of Minerals*, 17, 301–12, copyright © 1990 by Springer Verlag. Reprinted with the permission of Springer-Verlag GmbH & Co. KG.

where $W_{H,FeMg} = 1$ kcal/mole; $W_{H,MgFe} = 2$ kcal/mole, and X_{Mg} and X_{Fe} are mole fractions of magnesian and ferrous components in the mixture.

As regards intracrystalline disorder on $M1$ and $M2$ sites, there is agreement on the fact that Fe^{2+} and Mg^{2+} are almost randomly distributed between the two sites. In figure 5.10, the intracrystalline distribution coefficient

$$K_D = \frac{Fe_{M1} Mg_{M2}}{Fe_{M2} Mg_{M1}},$$
(5.38)

experimentally observed for various terms of the mixture at various T conditions and $P = 1$ bar, is plotted against the molar fraction of the ferrous end-member. The same figure also shows distribution curves calculated on the basis of an interionic potential model. These curves define the conditions of internal disorder in which the Gibbs free energy of the mixture is minimized and thus represent stable equilibrium. There is satisfactory agreement between high-T experimental results and theoretical calculations. It is interesting to note that the interionic potential calculations predict K_D values lower than 1 at low T and higher than 1 at high T. Starting from a temperature of about 600 °C (which represents a probable kinetic low-T limit in intracrystalline exchanges), we see progressive "increase of order" with increasing T, which is also apparent in the experimental study of Aikawa et al. (1985). However, because intracrystalline distribution is virtually random at all P and T conditions of interest, the entropy of mixing, within reasonable approximation, may approach the ideal case (i.e., $S_{excess\,mixing}$ = 0). We may therefore adopt for the mixture of interest a regular (or subregular) mixture model of the zeroth principle. For this purpose, Chatterjee (1987) suggests a Margules model:

$$G_{excess\,mixing} = X_{Fe}X_{Mg}\left(W_{FeMg}X_{Mg} + W_{MgFe}X_{Fe}\right), \tag{5.39}$$

where $W_{FeMg} = W_{MgFe} = +9$ kJ/mole. This model agrees with the calorimetric data of Wood and Kleppa (1981) but contrasts with the interaction parameter adopted by Fei and Saxena (1986) on the basis of the polymorphic transitions observed in the same system (cf. section 5.2.4).

Recent experiments by Akamatsu et al. (1993) show that pressure has a significant effect on Fe-Mg intracrystalline distribution ($\partial K_D/\partial P \cong 2$ Mbar^{-1}). At the upper pressure limit of stability of olivine ($P \approx 0.12$ Mbar, corresponding to a depth of about 400 km; p-wave seismic discontinuity, cf. section 5.2.3), the increase in K_D with respect to the $P =$ 1 bar condition, T and composition being equal, is about 0.24. For a Fo$_{80}$ olivine, this K_D variation is similar to that induced by a T increase of about 250 °C (cf. figure 5.11).

(Mg,Ca)$_2$SiO$_4$ Mixture

The formation of the intermediate compound CaMgSiO$_4$ (monticellite), with limited solubility toward the end-member components Mg$_2$SiO$_4$ and Ca$_2$SiO$_4$, is well known in this system. Warner and Luth (1973) experimentally investigated the miscibility between CaMgSiO$_4$ and Mg$_2$SiO$_4$ in the T range 800 to 1300 °C at $P = 2.5$ and 10 kbar. The miscibility gap was parameterized by the authors with a simple mixture model in which the interaction parameter (kJ/mole) varies with T (K) and P (kbar) according to

$$W_{T,P} = 85.094 - 56.735\left(\frac{T}{1000}\right) + 15.175\left(\frac{T}{1000}\right)^2 - 1.487P + 1.292\left(\frac{PT}{1000}\right). \tag{5.40}$$

The miscibility gap extends over practically the whole compositional field from CaMgSiO$_4$ to Mg$_2$SiO$_4$, and the maximum amount of monticellite miscible in Mg$_2$SiO$_4$ is lower than 5% in molar proportions. Based on equation 5.40, the miscibility gap should expand with increasing pressure. Figure 5.11 shows the Gibbs free energy of mixing curves calculated at $T = 600$ °C and $P = 1$ bar, with a static interionic potential approach. Based on the principles of minimization of Gibbs free energy at equilibrium and of equality of chemical potentials at equilib-

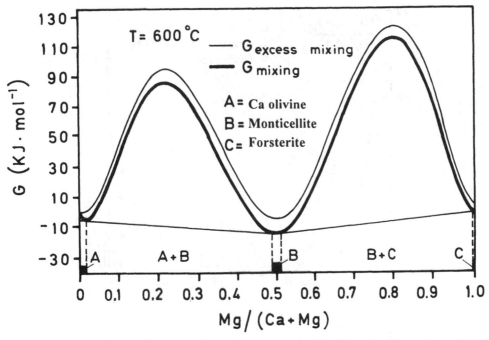

Figure 5.11 Gibbs free energy of mixing in binary join Mg_2SiO_4-Ca_2SiO_4 at $T = 600\ °C$ and $P = 1$ bar, calculated with a static interionic potential approach. Reprinted from G. Ottonello, *Geochimica et Cosmochimica Acta*, 3119–3135, copyright © 1987, with kind permission from Elsevier Science Ltd., The Boulevard, Langford Lane, Kidlington OX5 1GB, UK.

rium, the plotted curves define two solvi (cf. sections 2.3 and 3.11): one extending from $(Mg_{0.02}Ca_{0.98})_2SiO_4$ to $(Mg_{0.49}Ca_{0.51})_2SiO_4$ and the other extending from $(Mg_{0.51}Ca_{0.49})_2SiO_4$ to pure forsterite. According to the model, monticellite is thus virtually insoluble in forsterite.

$(Mg,Ni)_2SiO_4$ Mixture

According to the experiments of Campbell and Roeder (1968), at $T = 1400\ °C$ the $(Mg,Ni)_2SiO_4$ mixture is virtually ideal. More recent measurements by Seifert and O'Neill (1987) at $T = 1000\ °C$ seem to confirm this hypothesis. Because the intracrystalline distribution of Mg^{2+} and Ni^{2+} on $M1$ and $M2$ sites is definitely nonrandom (Ni^{2+} is preferentially stabilized in $M1$), the presumed ideality implies that negative enthalpic terms of mixing counterbalance the positive excess entropy arising from the configurational expression

$$S_{\text{configuration}} = -R\left(\sum_i X_{i,M1} \ln X_{i,M1} + \sum_i X_{i,M2} \ln X_{i,M2} \right) \quad (5.41)$$

—i.e., $S_{\text{configuration}}$ does not reach the maximum value of equipartition (11.526 J/(mole × K)) and consequently an excess term in S appears.

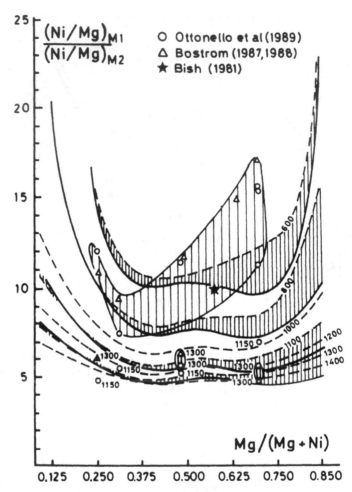

Figure 5.12 Comparison between calculated and experimentally observed intracrystalline disorder in $(Mg,Ni)_2SiO_4$ mixture at various T, and $P = 1$ bar. Theoretical distribution trends calculated with two values of hardness factor ($\rho = 0.20$ and 0.21). From Ottonello et al. (1989). Reprinted with permission of The Mineralogical Society of America.

Figure 5.12 shows how Ni^{2+} and Mg^{2+} ions are distributed between $M1$ and $M2$ sites in the $(Mg, Ni)_2SiO_4$ mixture at various T ($P = 1$ bar). Experimentally observed distributions are compared with the results of static interionic potential calculations carried out at two different values of the hardness factor ρ (cf. section 1.11.2).

Lastly, as far as volume is concerned, the mixture $(Mg,Ni)_2SiO_4$ exhibits conspicuous deviations from ideality (see section 5.2.2): at high pressure, these deviations presumably induce unmixing phenomena, not observed at $P = 1$ bar.

For this purpose, it is of interest to recall that electrochemical galvanic cell measurements indicate additional complexities of the system: Ottonello and Morlotti (1987) ob-

Table 5.15 Activity of component Ni_2SiO_4 in $(Mg, Ni)_2SiO_4$ mixture: (1) results of interionic potential model (Ottonello et al., 1989); (2) Campbell and Roeder, (1968).

	$T = 1400\,°C, P = 1$ bar		$T = 1000\,°C,$ $P = 1$ bar	
$X_{Ni_2SiO_4}$	(1)	(2)	$X_{Ni_2SiO_4}$	(1)
0.18	0.19	0.18	0.18	0.18
0.22	0.25	0.26	0.21	0.23
0.35	0.37	0.39	0.33	0.36
0.40	0.41	0.42	0.42	0.41
0.52	0.53	0.53	0.51	0.51
0.72	0.70	0.75	0.63	0.62
0.89	0.94	0.93	0.65	0.63
			0.71	0.66
			0.79	0.76
			0.81	0.80
			0.90	0.98

served two isoactivity regions in the binary join, one at approximately $X_{Ni} = 0.2$ and the other between 0.4 and 0.6. Ottonello and Morlotti attributed this complex thermodynamic feature to the formation of the intermediate compound $(Ni_{0.25}Mg_{0.75})_2SiO_4$, with limited solubility toward the end-member components Mg_2SiO_4 and Ni_2SiO_4. Actually, as shown by Campbell and Roeder (1968), in the presence of excess silica, an olivine mixture of this limiting composition reacts with SiO_2 to form pyroxene. Because the galvanic cell compartment contains $(Mg,Ni)_2SiO_4 + SiO_2 + Ni$, the two isoactivity regions indicate the coexistence of a (reacted) pyroxene-type $(Mg,Ni)SiO_3$ phase with a $(Mg,Ni)_2SiO_4$ mixture. The activity data derived from electromotive force (EMF) measurements cannot thus be considered representative of the $(Mg,Ni)_2SiO_4$ mixture, and the most reliable activity data are those of Campbell and Roeder (1968), which can be reproduced by appropriate parameterization of the repulsive potentials in the mixture, as shown in table 5.15.

(Mg,Mn)$_2$SiO$_4$ Mixture

According to Glasser (1960), the $(Mg,Mn)_2SiO_4$ mixture is almost ideal in the temperature range $T = 1300$ to $1600\,°C$ ($P = 1$ bar). The experiments of Maresch et al. (1978) at $T = 700$ to $1100\,°C$ ($P = 2$ kbar) indicate the presence of unmixed phases in the composition range Fo_{40} to Fo_{75}. According to lattice energy calculations, there are two solvi at $P = 1$ bar, extending from intermediate composition toward the two end-members (figure 6 in Ottonello, 1987). As regards intracrystalline disorder, Mn^{2+} appears preferentially stabilized in $M2$ at all investigated T conditions (Francis and Ribbe, 1980; Urusov et al., 1984).

(Fe,Ca)$_2$SiO$_4$ Mixture

This mixture shows several analogies with the Mg_2SiO_4-Ca_2SiO_4 system. Also in this case, an intermediate compound (kirschsteinite, $FeCaSiO_4$) forms, with limited solubility toward the two end-members. Mukhopadhyay and Lindsley (1983) show that, at $T = 800$ °C, the solvus field in the Ca-rich side of the join extends from $(Fe_{0.08}Ca_{0.92})_2SiO_4$ to $(Fe_{0.42}Ca_{0.58})_2SiO_4$. Bowen et al. (1933) also observed an extended miscibility gap between $FeCaSiO_4$ and Fe_2SiO_4. These mixing properties are confirmed by lattice energy calculations (Ottonello, 1987). In $(Fe,Ca)_2SiO_4$, as in the $(Mg,Ca)_2SiO_4$ mixture, Ca^{2+} is stabilized at all T conditions in the $M2$ site with more distorted octahedral symmetry.

(Fe,Mn)₂SiO₄ Mixture

According to Schwerdtfeger and Muan (1966), this mixture is essentially ideal at $T = 1150$ °C, $P = 1$ bar, although the intracrystalline distribution measurements of Annersten et al. (1984) and Brown (1970) show that Fe^{2+} is preferentially distributed on the $M1$ site.

(Co,Fe)₂SiO₄ Mixture

According to Masse et al. (1966), this mixture shows slight positive deviations from ideality of activity values at $T = 1180$ °C, $P = 1$ bar. There are no experimental data on intracrystalline distribution.

(Mg,Fe,Ca)₂SiO₄ Mixture

Only the quadrilateral portion of the Mg_2SiO_4-Ca_2SiO_4-Fe_2SiO_4 system with $X_{Ca_2SiO_4} \leq 0.5$ is of relevant interest in geochemistry. Mixing behavior in the olivine quadrilateral was investigated by Davidson and Mukhopadhyay (1984). As figure 5.13 shows, the solvus between Ca-rich and Ca-poor compositional terms extends within the quadrilateral, progressively shrinking toward the Fe-Ca join.

Multicomponent mixing behavior has been parameterized by Davidson and Mukhopadhyay (1984) with a *nonconvergent site-disorder* model. The interacting components were first expanded in a set of six site-ordered end-members: $Fe_{M1}Mg_{M2}SiO_4$, $Mg_{M1}Fe_{M2}SiO_4$, $Fe_{M1}Fe_{M2}SiO_4$, $Mg_{M1}Mg_{M2}SiO_4$, $Fe_{M1}Ca_{M2}SiO_4$, and $Mg_{M1}Ca_{M2}SiO_4$. It was then assumed for simplicity that mixing of Fe^{2+}-Mg^{2+} is ideal in both $M1$ and $M2$ sites and that mixing of Fe^{2+}-Ca^{2+} and Mg^{2+}-Ca^{2+} is nonideal in the $M2$ site. The authors then followed the guidelines of Thompson (1969, 1970) and expanded the Gibbs free energy of mixing of the mixture into two distinct contributions arising from configurational entropy and enthalpic interaction terms:

$$G_{mixture} = G_{end-members} + G_{mixing} \tag{5.42}$$

$$G_{mixing} = H_{mixing} - TS_{mixing} \tag{5.43}$$

$$G_{end-members} = \mu^0_{MgCa} X_{Mg,M1} X_{Ca,M2} + \mu^0_{FeCa} X_{Fe,M1} X_{Ca,M2}$$
$$+ \mu^0_{MgMg} X_{Mg,M1} X_{Mg,M2} + \mu^0_{FeFe} X_{Fe,M1} X_{Fe,M2} \tag{5.44}$$
$$+ \mu^0_{MgFe} X_{Mg,M1} X_{Fe,M2} + \mu^0_{FeMg} X_{Fe,M1} X_{Mg,M2}$$

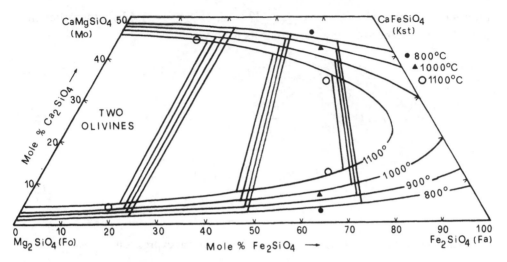

Figure 5.13 Mixing properties of $(Mg,Fe,Ca)_2SiO_4$ olivines at $P = 1$ kbar and various temperatures, after Davidson and Mukhopadhyay (1984). Tie lines connect coexisting unmixed phases according to the proposed mixture model. Experimentally determined compositions are also shown. From P. M. Davidson and D. K. Mukhopadhyay, Ca-Fe-Mg olivines: phase relations and a solution model, *Contributions to Mineralogy and Petrology*, 86, 256–263, figure 3, copyright © 1984 by Springer Verlag. Reprinted with the permission of Springer-Verlag GmbH & Co. KG.

$$H_{mixing} = W_{MgCa}X_{Mg,M2}X_{Ca,M2} + W_{FeCa}X_{Fe,M2}X_{Ca,M2} \qquad (5.45)$$

$$\begin{aligned} S_{mixing} = -R(&X_{Mg,M1}\ln X_{Mg,M1} + X_{Fe,M1}\ln X_{Fe,M1} \\ &+ X_{Mg,M2}\ln X_{Mg,M2} + X_{Fe,M2}\ln X_{Fe,M2} + X_{Ca,M2}\ln X_{Ca,M2}) \end{aligned} \qquad (5.46)$$

The mixing properties were then solved by minimizing the Gibbs free energy of the mixture with respect to ordering parameter t:

$$t = X_{Fe,M2} - X_{Fe,M1} \qquad (5.47)$$

$$\begin{aligned} \frac{\partial G_{mixture}}{\partial t} = 0 = &-RT\ln\left(\frac{X_{Fe,M1}X_{Mg,M2}}{X_{Mg,M1}X_{Fe,M2}}\right) - X_{Ca,M2}\left(W_{MgCa} - W_{FeCa}\right) \\ &+ \Delta G_*^0\left(0.5X_{Ca,M2} + t\right) + F^0 - \Delta G_E^0\left(1 - 0.5X_{Ca,M2}\right) \end{aligned} \qquad (5.48)$$

and by applying the concept of equality of chemical potential for components in coexisting (unmixed) phases at equilibrium (tie lines in figure 5.13; cf. also sections 2.3 and 3.11).

Five excess terms appear in equation 5.48: two of them are the usual regular interaction parameters (i.e., $W_{\text{MgCa}} = 32.9$ kJ, $W_{\text{FeCa}} = 21.4$ kJ); the other three are related to standard-state potentials as follows:

$$\Delta G_E^0 = \mu^0_{\text{Fe}_{M1}\text{Mg}_{M2}} - \mu^0_{\text{Mg}_{M1}\text{Fe}_{M2}} = -0.84 \,\text{kJ} \tag{5.49}$$

$$\Delta G_*^0 = \mu^0_{\text{Mg}_{M1}\text{Fe}_{M2}} - \mu^0_{\text{Fe}_{M1}\text{Mg}_{M2}} = 7.0 \pm 3.9 \,\text{kJ} \tag{5.50}$$

$$F^0 = 2\left(\mu^0_{\text{Mg}_{M1}\text{Ca}_{M2}} - \mu^0_{\text{Fe}_{M1}\text{Ca}_{M2}} \right) + \mu^0_{\text{Fe}_{M1}\text{Fe}_{M2}}$$
$$- \mu^0_{\text{Mg}_{M1}\text{Mg}_{M2}} = 12.7 \pm 1.6 \,\text{kJ}. \tag{5.51}$$

Note that equation 5.42 is analogous to the canonical expression

$$G_{\text{mixture}} = \sum_i \mu_i^0 X_i + G_{\text{ideal mixing}} + G_{\text{excess mixing}}. \tag{5.52}$$

However, Davidson and Mukhopadhyay (1984) did not use the macroscopic fractions X_i but thermodynamic fractions arising from pair probabilities and consistent with the six–end-member identification (see section 3.8.7). Moreover, nonrandom configurational contributions were accounted for in the entropy term (which is not ideal, because Ca is absent in the $M1$ site). In Thompson's (1969, 1970) notation, entropic effects are treated in the same manner, and the standard state Gibbs free energies of the pure end-members and all nonconfigurational effects are grouped in a *vibrational* Gibbs energy term (G^*):

$$G = -TS^{\text{IC}} + G^*. \tag{5.53}$$

The word "vibrational," is in this case somewhat misleading, because enthalpic interactions are essentially static in nature and are *purely* static whenever the zeroth approximation is appropriate (see Della Giusta and Ottonello, 1993). Lastly, it must be noted that the excess term F^0 represents a reciprocal energy arising from the noncoplanarity of the four quadrilateral end-members (cf. section 3.8.7). The model of Davidson and Mukhopadhyay (1984) was reconsidered by Hirschmann (1991), who proposed an analogous treatment for $(\text{Ni},\text{Mg},\text{Fe})_2\text{SiO}_4$ mixtures.

5.3 Garnets

The structure of garnet was determined for the first time by Menzer (1926). The general formula unit for garnet may be expressed as $X_3 Y_2 Z_3 O_{12}$. Garnet has a body-centered-cubic cell (spatial group $Ia3d$) and 8 formula units per unit cell. Cations X, Y, and Z occupy nonequivalent sites with, respectively, dodecahedral, octahedral, and tetrahedral symmetry. The main IV-fold coordinated cation is Si^{4+}. Figure 5.14 shows that in garnet, too, the $[\text{TO}_4]$ groups are isolated from

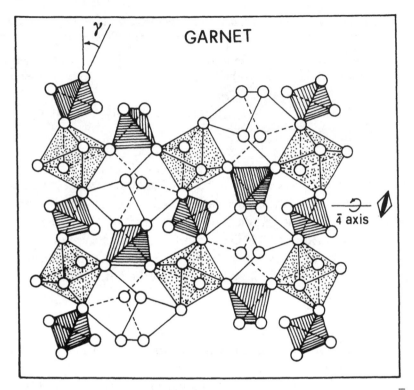

Figure 5.14 Structure of garnet. Positional angle defines degree of rotation on $\bar{4}$ axis.

one other (nesosilicates). The octahedra are also isolated, and are linked by single apexes to tetrahedra. Dodecahedra share edges either with tetrahedra and octahedra or with other dodecahedra. With respect to other silicates, the structure of garnet is very dense as regards oxygen packing. Assuming a spherical oxygen ion with an ionic radius of 1.38 Å, in noncalcic garnets oxygen occupies 70% of the volume of the unit cell—a value near the theoretical value of maximum packing (*eutaxis;* 74%; cf. Meagher, 1982).

5.3.1 Chemistry and Occurrence

The variety of symmetries in the garnet structure (coordinations 4, 6, and 8) allows considerable compositional range. Table 5.16 lists the elements commonly present in positions X, Y, and Z. The diadochy of Al, Ti^{4+}, and Fe^{3+} in the tetrahedral site has been confirmed by Mössbauer spectroscopy on natural Fe-Ti-bearing garnets (Schwartz and Burns, 1978), and the presence of phosphorus in these sites, observed in upper mantle garnet, is attributable, according to Bishop et al. (1976), to coupled substitutions of the type

$$Ca_X + Si_Z \Leftrightarrow Na_X + P_Z. \tag{5.54}$$

Table 5.16   Elements present at major, minor, and trace levels in the garnet phase.

Abundance	X-site (Dodecahedral Symmetry)	Y-site (Octahedral Symmetry)	Z-site (Tetrahedral Symmetry)
Major	Ca, Mn, Fe^{2+}, Mg	Al, Cr^{3+}, Fe^{3+}	Si
Minor	Zn, Y, Na	Ti^{3+}, Ti^{4+}, V^{3+}, Fe^{2+}, Zr, Sn,	Al, Ti^{4+}, Fe^{3+}, P
Trace	Li, Be, B, F, Sc, Cu, Ga,Ge, Sr, Nb, Ag, Cd, In, REE		

Table 5.17 lists analyses of several types of natural garnet, after Deer et al. (1983). The calculation of ionic fractions on the basis of 24 oxygens shows effective diadochy of Al on Si in tetrahedral site Z, whereas Ti^{4+} and Fe^{3+} appear to be stabilized in octahedral site Y. Deviation from ideal closure in sites Y and X (6 and 4, respectively) may be attributed in some instances to erroneous estimation of ferric iron (samples 4, 5, and 6), but in others it is difficult to interpret (samples 1, 2, and 3).

As suggested by Winchell (1933), the main garnet components may be subdivided into two groups with complete miscibility of their members: the calcic group of *ugrandites,* comprising uvarovite ($Ca_3Cr_2Si_3O_{12}$), grossular ($Ca_3Al_2Si_3O_{12}$), and andradite ($Ca_3Fe_2Si_3O_{12}$); and the group of *pyralspites,* including pyrope ($Mg_3Al_2Si_3O_{12}$), almandine ($Fe_3Al_2Si_3O_{12}$), and spessartine ($Mn_3Al_2Si_3O_{12}$). Six other less common compositional terms (which, however, become more and more important in phase stability studies at high P) may be associated with these six end-members. The names and formula units of these six terms are listed in table 5.18 (note that all the natural garnets listed in table 5.17 have compositions that may be reexpressed as molar proportions of the various compositional terms in table 5.18).

Table 5.19 summarizes the main occurrences of the various compositional terms. Because garnets are solid mixtures, the listed components merely indicate the main compositional terms of the phase.

5.3.2   Structural Properties

Using multiple regression techniques, Novak and Gibbs (1971) obtained linear equations that give cell edge and oxygen positional parameters as functions of the radii of cations occupying sites X, Y, and Z. The equation of Novak and Gibbs for cell edge is

$$a = 9.04 + 1.61r_X + 1.89r_Y \qquad (\text{Å}). \tag{5.55}$$

Equation 5.55 was later modified by Novak and Colville (1975) to include the dimension of site Z:

Table 5.17 Compositions (in weight %) of some natural garnets (from Deer et al., 1983): (1) almandine from quartz-biotite gneiss; (2) andradite from metamorphosed andesite; (3) grossular from garnet-bearing gneiss; (4) pyrope from eclogite; (5) spessartine from a cornubianitic rock; (6) uvarovite from a garnet-tremolite-pyrrothite vein.

Constituent Oxide	Samples					
	(1)	(2)	(3)	(4)	(5)	(6)
SiO_2	38.03	37.03	38.69	41.52	35.84	35.88
TiO_2	-	0.04	0.55	trace	0.03	-
Al_2O_3	22.05	8.92	18.17	23.01	20.83	1.13
Cr_2O_3	-	-	-	0.22	-	27.04
Fe_2O_3	0.88	18.34	5.70	1.22	0.65	2.46
FeO	29.17	2.25	3.78	12.86	1.78	-
MnO	1.57	1.09	0.64	0.33	33.37	0.03
MgO	6.49	0.83	0.76	16.64	2.48	0.04
CaO	1.80	30.26	31.76	4.71	5.00	33.31
$H_2O^{(+)}$	-	0.48	0.13	} 0.16	-	} 0.18
$H_2O^{(-)}$	-	0.16	0.06		-	
Total	99.99	99.40	100.24	100.67	99.98	100.07

Ionic fractions on 24-oxygens basis

	(1)	(2)	(3)	(4)	(5)	(6)
Si	5.951	6.043	5.966	5.999	5.808	5.964
Al	0.049	-	0.034	0.001	0.192	0.036
Total site Z	6.000	6.043	6.000	6.000	6.000	6.000
Al	4.019	1.716	3.268	3.912	3.786	0.186
Cr	0.102	-	-	0.026	-	3.554
Fe^{3+}	-	2.253	0.662	0.132	0.080	0.308
Ti	-	0.005	0.064	-	0.004	-
Total site Y	4.121	3.974	3.994	4.070	3.870	4.048
Mg	1.513	0.201	0.175	3.580	0.598	0.010
Fe^{2+}	3.818	0.307	0.487	1.552	0.241	-
Mn	0.203	0.151	0.083	0.040	4.581	0.004
Ca	0.302	5.292	5.248	0.730	0.868	5.934
Total site X	5.836	5.951	9.993	5.902	6.288	5.948

Molar proportions of main components (%)

	(1)	(2)	(3)	(4)	(5)	(6)
Almandine	65.4	5.2	8.1	26.1	4.0	-
Andradite	2.7	56.9	18.2	3.3	2.2	7.7
Grossular	2.5	32.0	69.4	8.9	11.2	2.4
Pyrope	25.8	3.4	2.9	60.3	7.9	0.2
Spessartine	3.6	2.5	1.4	0.7	74.7	0.1
Uvarovite	-	-	-	0.7	-	89.6

Table 5.18 Chemistry and terminology of main components of the garnet phase.

Name	Formula Unit	References on Structure
Pyrope	$Mg_3Al_2Si_3O_{12}$	Novak and Gibbs (1971)
Almandine	$Fe_3Al_2Si_3O_{12}$	Skinner (1966), Geiger et al. (1987)
Spessartine	$Mn_3Al_2Si_3O_{12}$	Skinner (1966)
Grossular	$Ca_3Al_2Si_3O_{12}$	Novak and Gibbs (1971)
Uvarovite	$Ca_3Cr_2Si_3O_{12}$	Huckenholtz and Knittel (1975)
Andradite	$Ca_3Fe_2Si_3O_{12}$	Novak and Gibbs (1971)
Knorringite	$Mg_3Cr_2Si_3O_{12}$	Irefune et al. (1982)
Calderite	$Mn_3Fe_2Si_3O_{12}$	Lattard and Schreyer (1983)
Skiagite	$Fe_3Fe_2Si_3O_{12}$	Karpinskaya et al. (1983), Woodland and O'Neill (1993)
Khoharite	$Mg_3Fe_2Si_3O_{12}$	McConnell (1966)
Mn-Cr garnet	$Mn_3Cr_2Si_3O_{12}$	Fursenko (1981)
Fe-Cr garnet	$Fe_3Cr_2Si_3O_{12}$	Fursenko (1981)

Table 5.19 Occurrence of garnets in various types of rocks.

Mineral	Occurrence
Almandine $Fe_3Al_2Si_3O_{12}$	Eclogites (alm-pir), gneisses, metamorphic schists, granites, pegmatites, granulites (mineral index)
Grossular $Ca_3Al_2Si_3O_{12}$	Typical of contact-metamorphosed limestones (skarns, marbles, cornubianites)
Andradite $Ca_3Fe_2Si_3O_{12}$	In contact-metamorphic zones, in association with magnetite and Fe-rich silicates (feldspathoid-bearing igneous rocks)
Uvarovite $Ca_3Cr_2Si_3O_{12}$	Mainly in chromite deposits
Spessartine $Mn_3Al_2Si_3O_{12}$	Pegmatites and other igneous acidic rocks, metamorphic rocks and Mn-rich clay deposits
Pyrope $Mg_3Al_2Si_3O_{12}$	Basic and ultrabasic rocks (peridotites, serpentinites), kimberlitic xenoliths, rodingites, eclogites

$$a = 8.44 + 1.71r_X + 1.78r_Y + 2.17r_Z \qquad (\text{Å}). \qquad (5.56)$$

Both equations are based on the ionic radii of Shannon and Prewitt (1969). The method of Novak and Gibbs (1971) was then reexamined by Hawthorne (1981b), and finally extended by Basso (1985) to hydrogarnets. Basso's (1985) equations are shown in table 5.20. They determine cell edge, fractional atomic coordinates of oxygen, and mean metal-to-oxygen bond distances with satisfactory precision. For even better precision, it is good practice to submit the obtained parameters to a further optimization procedure, based on crystallographic constraints (i.e., *Distance-Least-Squares,* or DLS, treatment).

Table 5.21 shows structural data on various garnet end-members, compared

Table 5.20 Equations for structural simulation of garnet (Basso, 1985).

Fractional coordinates of oxygen

$X_O = 0.0258r_X + 0.0093r_Y - 0.0462r_Z + 0.0171$

$Y_O = -0.0261r_X + 0.0261r_Y + 0.0310r_Z + 0.0514$

$Z_O = -0.0085r_X + 0.0323r_Y - 0.0237r_Z + 0.6501$

Metal-to-oxygen distances

$X1 - O = 0.558r_X + 0.298r_Y + 0.244r_Z + 1.447$

$X2 - O = 0.702r_X + 0.083r_Y + 1.673$

$Y - O = 0.154r_X + 0.754r_Y + 0.132r_Z + 1.316$

$Z - O = 1.026r_X + 1.366$

Cell edge

$$
a = \frac{8}{5}\left\{ \frac{3(X1 - O)^2}{2} + 2(X2 - O)^2 + \frac{7(Y - O)^2}{5} + \frac{3(Z - O)^2}{2} + \right.
$$
$$
+ \left\{ \left[\frac{3(X1 - O)^2}{2} + 2(X2 - O)^2 + \frac{7(Y - O)^2}{5} + \frac{3(Z - O)^2}{2} \right]^2 \right.
$$
$$
-20\left[\left(X1 - O \right)^2 - \left(Z - O \right)^2 \right]^2
$$
$$
-\frac{5}{4}\left[\left(X1 - O \right)^2 - 4\left(X2 - O \right)^2 + 2\left(Y - O \right)^2 + \left(Z - O \right)^2 \right]^2
$$
$$
\left. \left. -5\left[\left(X1 - O \right)^2 - 2\left(Y - O \right)^2 - \left(Z - O \right)^2 \right]^2 \right]^{1/2} \right\}^{1/2} \right\}
$$

with the values obtained by the method of Basso (1985), followed by the DLS procedure of Baerlocher et al. (1977). The optimization procedure determinates cell parameters for the six components not pertaining to the ugrandite and pyralspite series.

On the basis of equations of type 5.55 and of analogous expressions for positional parameters, Novak and Gibbs (1971) proposed several guidelines to establish whether a "garnet" compound may form and remain stable, according to the dimensions of cations occupying sites X, Y, and Z. These guidelines consider the following three factors.

1. The maximum distance compatible with Si-O bonds
2. The O-O distance of the nonshared octahedral edge
3. The mean dimensions of sites X and Y compared with the simple summations of $r_X + r_{O^{2-}}$ and $r_Y + r_{O^{2-}}$, respectively.

On the basis of these principles, Novak and Gibbs (1971) proposed a map of compatible r_X and r_Y dimensions for the formation of a garnet phase. This map (figure 5.15) shows that the commonly observed components plot in the low-volume zone of the compatibility area. Theoretically, compounds with higher volumes may form and remain stable at appropriate P and T conditions.

Table 5.21 Structural data on garnet components

Component	Formula Unit	Cell Edge (Å) Observed	Cell Edge (Å) Calculated	Calculated Positional Parameter of Oxygen X_O	Y_O	Z_O
Pyrope	$Mg_3Al_2Si_3O_{12}$	11.459	11.4591	0.03291	0.05033	0.65364
Almandine	$Fe_3Al_2Si_3O_{12}$	11.526	11.5263	0.03960	0.04925	0.65330
Spessartine	$Mn_3Al_2Si_3O_{12}$	11.621	11.6210	0.03545	0.04770	0.65281
Grossular	$Ca_3Al_2Si_3O_{12}$	11.845	11.8453	0.03902	0.04399	0.65163
Uvarovite	$Ca_3Cr_2Si_3O_{12}$	11.996	11.9962	0.03982	0.04664	0.65463
Andradite	$Ca_3Fe_2Si_3O_{12}$	12.058	12.0583	0.04016	0.04772	0.65588
Knorringite	$Mg_3Cr_2Si_3O_{12}$	-	11.6040	0.03387	0.05290	0.65670
Calderite	$Mn_3Fe_2Si_3O_{12}$	-	11.8288	0.03672	0.05137	0.65715
Skiagite	$Fe_3Fe_2Si_3O_{12}$	11.728	11.7320	0.03529	0.05288	0.65769
Khoharite	$Mg_3Fe_2Si_3O_{12}$	-	11.6637	0.03427	0.05395	0.65806
Mn-Cr garnet	$Mn_3Cr_2Si_3O_{12}$	-	11.7683	0.03635	0.05030	0.65588
Fe-Cr garnet	$Fe_3Cr_2Si_3O_{12}$	-	11.6720	0.03489	0.05183	0.65640

Component	Cell Volume (Å^3)	Molar Volume (cm^3/mole)	Molar Weight (g/mole)	Density (g/cm^3)
Pyrope	1504.67	113.287	403.122	3.559
Almandine	1531.21	115.284	496.668	4.308
Spessartine	1569.39	118.159	495.021	4.189
Grossular	1661.90	125.124	450.450	3.600
Uvarovite	1726.27	129.971	500.480	3.851
Andradite	1753.18	131.997	508.180	3.850
Knorringite	1562.51	117.641	453.152	3.852
Calderite	1655.09	124.612	552.751	4.436
Skiagite	1614.79	121.578	554.398	4.560
Khoharite	1586.75	119.466	460.852	3.858
Mn-Cr garnet	1629.83	122.710	545.051	4.442
Fe-Cr garnet	1590.14	119.722	546.698	4.566

5.3.3 Compressibility and Thermal Expansion

The experimental data of Meagher (1975) on pyrope and grossular, and of Rakai (1975) on spessartine and andradite, show that Si-O distances do not vary appreciably with increasing temperature and that thermal expansion affects both octahedral and dodecahedral sites in an almost linear fashion. The different thermal behaviors of the various symmetries lead to rotation of the $[SiO_4]$ groups on the $\overline{4}$ axis (*positional angle;* cf. figure 5.14). With increasing T we also observe progressive increases in the shared O-O octahedral distances, whereas nonshared O-O edges remain virtually unchanged.

Because of the importance of garnet in upper-mantle parageneses, the compressibility of this phase has been the subject of many experimental studies. Hazen and Finger (1978), studying the compressibility of pyrope, observed that the

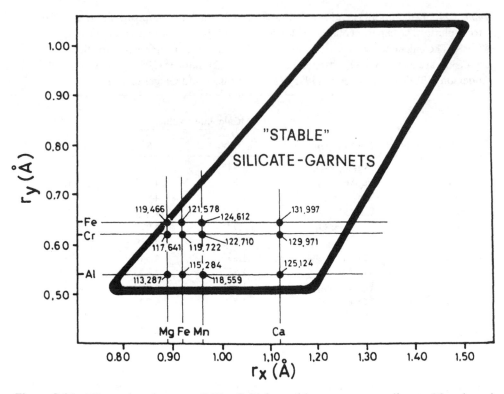

Figure 5.15 Dimensional compatibility field for stable garnets according to Novak and Gibbs (1971). Points correspond to 12 main components listed in table 5.21. Molar volumes are also indicated.

[MgO$_8$] dodecahedron is more compressible than the [AlO$_6$] octahedron, and that the [SiO$_4$] group, unlike that of olivine, is also compressible. Thermal expansion and compressibility are thus not perfectly antipathetic in this case. Notwithstanding experimental studies on the compressibilities of the various compositional terms, complete agreement among authors does not exist (Isaak and Graham, 1976; Sato et al., 1978; Leitner et al., 1980; Babuska et al., 1978; Hazen and Finger, 1978; Bass, 1986). Calculation of garnet compressibility and thermal expansion based on structure simulations at various pressures and temperatures gives bulk moduli that are within the range of experimental values and may be extended to not previously investigated compounds. Values (listed in table 5.22) are relative to the Birch-Murnaghan equation of state. The same table also compares calculated thermal expansion values with the experimental values of Skinner (1956), Suzuki and Anderson (1983), Isaak et al. (1992), and Armbruster and Geiger (1993) reconducted to the usual polynomial expansion in T:

$$\alpha_{V,T} = \frac{1}{V}\left(\frac{\partial V}{\partial T}\right)_P = \alpha_0 + \alpha_1 T + \alpha_2 T^2. \qquad (5.57)$$

Table 5.22 Bulk modulus and thermal expansion for various compositional terms of the garnet phase. The experimental range for bulk moduli is relative to a common value of $K' = 4$. The calculated values are from Ottonello et al. (1996). References for thermal expansion are as follows: (1) Ottonello et al. (1996); (2) Skinner (1956); (3) Suzuki and Anderson (1983); (4) Isaak et al. (1992); (5) Armbruster and Geiger (1993).

Compound	Formula Unit	Calculated Values		Experimental Range	
		K_0 (Mbar)	K'	K_0	($K' = 4$)
Pyrope	$Mg_3Al_2Si_3O_{12}$	1.776	4.783	1.66–1.79	
Almandine	$Fe_3Al_2Si_3O_{12}$	1.837	4.795	1.75–1.80	
Spessartine	$Mn_3Al_2Si_3O_{12}$	1.797	4.833	1.72–1.75	
Grossular	$Ca_3Al_2Si_3O_{12}$	1.673	4.887	1.39–1.69	
Uvarovite	$Ca_3Cr_2Si_3O_{12}$	1.743	4.977	1.43–1.65	
Andradite	$Ca_3Fe_2Si_3O_{12}$	1.726	4.919	1.38–1.59	
Knorringite	$Mg_3Cr_2Si_3O_{12}$	1.876	4.842	-	
Calderite	$Mn_3Fe_2Si_3O_{12}$	1.818	4.823	-	
Skiagite	$Fe_3Fe_2Si_3O_{12}$	1.867	4.762	-	
Khoharite	$Mg_3Fe_2Si_3O_{12}$	1.857	4.799	-	
Mn-Cr garnet	$Mn_3Cr_2Si_3O_{12}$	1.852	4.894	-	
Fe-Cr garnet	$Fe_3Cr_2Si_3O_{12}$	1.902	4.831	-	

Compound	Thermal Expansion			
	$\alpha_0 \times 10^5 \ K^{-1}$	$\alpha_1 \times 10^9 \ K^{-2}$	$\alpha_2 \ K$	Reference
Pyrope	2.431	5.649	−0.2205	(1)
	2.311	5.956	−0.4538	(2)
	2.244	9.264	−0.1761	(3)
Almandine	1.781	11.558	−0.2000	(1)
	1.776	12.140	−0.5071	(2)
Spessartine	2.962	2.398	−0.8122	(1)
	2.927	2.726	−1.1560	(2)
Grossular	2.224	5.272	−0.4119	(1)
	1.951	8.089	−0.4972	(2)
	2.610	3.390	−0.7185	(4)
Uvarovite	2.232	5.761	−0.2329	(1)
Andradite	2.208	6.019	−0.2005	(1)
	2.103	6.839	−0.2245	(2)
	1.869	9.909	−0.1486	(5)
Knorringite	2.894	2.9642	−0.7748	(1)
Calderite	2.898	3.7316	−0.5455	(1)
Skiagite	2.336	7.0286	−0.2950	(1)
Khoharite	2.626	5.3687	−0.4521	(1)
Mn-Cr garnet	2.862	4.0015	−0.5160	(1)
Fe-Cr garnet	2.584	3.7865	−0.4282	(1)

Leitner et al. (1980) showed that the bulk moduli of terms of pyrope-almandine-grossular and andradite-grossular mixtures vary with chemical composition in a perfectly linear fashion (within the range of experimental incertitude). This behavior is also implicit in the generalization of Anderson (1972).

5.3.4 Thermodynamic Properties of End-Members

Table 5.23 lists thermochemical data for various garnet endmembers. The proposed enthalpy values for garnet reported in the literature are affected by large discrepancies that reach 60 kJ/mole in the case of almandine and 45 kJ/mole for andradite (see table 7 in Anovitz et al., 1993; and table 4 in Zhang and Saxena, 1991). The calculated lattice energy values shown in table 5.23 are consistent with the selected values of standard state enthalpy of formation from the elements, through the Born-Haber-Fayans thermochemical cycle (cf. section 1.13). The calculated repulsive energies are also consistent with the Huggins-Mayer formulation (see Tosi, 1964) for a common value of the hardness parameter ($\rho = 0.48$). Table 5.23 also lists enthalpy of formation values for unknown compounds calculated by Ottonello et al. (1996) with an interionic potential model. These values are affected by the same range of incertitude observed for ugranditic and pyralspitic terms.

Table 5.24 lists selected data concerning entropy and isobaric heat capacity, covering andradite, grossular, pyrope, and almandine terms, compared with results of calculations based on the Kieffer model (Ottonello et al., 1996).

Table 5.23 Thermochemical data for various endmembers of the garnet series. E_{CFS} = crystal field stabilization energy. The other column heads are defined as in table 5.12. Data in kJ/mole.

Compound	E_{BHF}	H^0_{f,T_r,P_r}	U_L	E_C	E_R	E_{DD}	E_{DQ}	E_{CFS}
Pyrope	59,269.2	−6290.8[1]	65,560.0	−77,060.6	12,287.8	−599.8	−187.4	0.0
Almandine	60,487.6	−5261.3[2]	65,611.0	−76,916.7	12,328.7	−769.6	−253.4	−138.0
Spessartine	59,822.8	−5686.9[3]	65,509.8	−76,704.9	12,097.9	−684.2	−218.6	0.0
Grossular	57,997.4	−6632.8[4]	64,630.2	−76,148.6	12,360.9	−645.3	−197.2	0.0
Uvarovite	58,356.4	−6034.6[5]	63,919.0	−75,446.1	12,396.7	−660.6	−209.1	−472.0
Andradite	58,489.9	−5758.9[6]	64,248.8	−75,160.5	11,779.0	−660.5	−206.8	0.0
Knorringite	59,628.2	−5576.1[7]	64,732.4	−76,294.3	12,390.4	−623.9	−204.5	−472.0
Calderite	60,315.3	−4719.8[7]	65,035.1	−75,661.1	11,556.8	−701.7	−229.1	0.0
Skiagite	60,980.1	−4276.2[7]	65,118.4	−75,856.7	11,783.8	−783.5	−262.0	−138.0
Khoharite	59,761.7	−5299.5[7]	65,061.2	−75,981.6	11,750.0	−626.5	−203.1	0.0
Mn-Cr garnet	60,181.8	−5007.1[7]	64,716.9	−75,965.8	12,182.0	−701.5	−231.7	−472.0
Fe-Cr garnet	60,846.7	−4555.3[7]	64,792.0	−76,162.5	12,420.4	−784.6	−265.3	−610.0

References: [1]Charlu et al. (1975); [2]Anovitz et al. (1993); [3]Boeglin (1981); [4]Charlu et al. (1978); [5]Kiseleva et al. (1977); [6]Helgeson et al. (1978); [7]Ottonello et al. (1996)

Table 5.24 Entropy and heat capacity function values for the various garnet end-members compared with calorimetric values.

| Compound | $S^0_{T_r, P_r}$ | $C_P = K_1 + K_2T + K_3T^{-2} + K_5T^{-1/2} + K_4T^2$ ||||| T Range | References |
		$K_1 \times 10^{-2}$	$K_2 \times 10$	$K_3 \times 10^{-5}$	$K_5 \times 10^{-3}$	$K_4 \times 10^5$		
Pyrope	266.283	8.7293	−1.4750	−4.7243	−8.6643	4.1084	298.15-m.p.	(1)
	266.27	8.025	−0.958	−23.52	−7.2877	0	350–1000	(2,3)
		8.72988	−1.3744	0.045	−8.7943	3.3408	820–1300	(4)
Almandine	342.562	8.7094	−1.6067	−1.6953	−8.5176	5.2364	298.15-m.p.	(1)
	342.6	2.9274	3.4652	−72.099	0.7163	15.317	400–1000	(5)
Spessartine	331.499	8.7486	−1.4680	−4.7045	−8.6060	4.0201	298.15-m.p.	(1)
Grossular	260.125	8.7296	−1.4846	−4.6095	−8.6330	3.9643	298.15-m.p.	(1)
	260.12							(2)
Uvarovite	286.81	8.7266	−1.4730	−3.1499	−8.5940	4.0649	298.15-m.p.	(1)
Andradite	316.425	8.7479	−1.4858	3.4215	−8.6066	4.1040	298.15-m.p.	(1)
	316.4	8.0924	−0.7025	−6.789	−7.403	0	300–1000	(6)
Knorringite	305.17	8.7557	−1.4656	−4.5706	−8.6212	4.0758	298.15-m.p.	(1)
Calderite	414.588	8.7683	−1.4606	6.9140	−8.5977	4.1656	298.15-m.p.	(1)
Skiagite	369.871	8.7074	−1.4476	−4.0045	−8.6526	4.3207	298.15-m.p.	(1)
Khoharite	344.417	8.7616	−1.4715	5.2712	−8.6516	4.2446	298.15-m.p.	(1)
Mn-Cr garnet	367.791	8.7736	−1.4637	5.2303	−8.6415	4.2123	298.15-m.p.	(1)
Fe-Cr garnet	357.955	8.7447	−1.4813	−1.5757	−8.5953	4.0796	298.15-m.p.	(1)

(1) Ottonello et al. (1996); (2) Haselton and Westrum (1980); (3) Newton et al. (1977); (4) Tequi et al. (1991); (5) Anovitz et al. (1993); (6) Robie et al. (1987)

In the Kieffer model (Kieffer 1979a, b, and c; cf. section 3.3), the isochoric heat capacity is given by

$$C_V = \frac{3R}{Z} \sum_{i=1}^{3} S_{(X_i)} + 3nR \left(1 - \frac{1}{N} - q \right) \mathcal{K}_{(X_l, X_u)} + 3nRq\mathcal{E}_{(X_E)}, \tag{5.58}$$

where $S_{(X_i)}$ is the dispersed acoustic function, $\mathcal{K}_{(X_l, X_u)}$ represents the optic continuum, and $\mathcal{E}_{(X_E)}$ is the Einstein function (see eq. 3.69, 3.71, and 3.73 in section 3.3). The expression for third-law harmonic entropy is:

$$
\begin{aligned}
S = &\frac{3R}{Z} \left(\frac{2}{\pi} \right)^3 \sum_{i=1}^{3} \int_0^{X_i} \frac{\left[\arcsin(X/X_i) \right]^2 X \, dX}{\left(X_i^2 - X^2 \right)^{1/2} (e^X - 1)} \\
&- \frac{3R}{Z} \left(\frac{2}{\pi} \right)^3 \sum_{i=1}^{3} \int_0^{X_i} \frac{\left[\arcsin(X/X_i) \right]^2}{\left(X_i^2 - X^2 \right)^{1/2}} \ln(1 - e^{-X}) \, dx \\
&+ \frac{3R \left(1 - \dfrac{3}{3s} - q \right)}{X_u - X_l} \int_{X_l}^{X_u} \frac{X \, dX}{(e^X - 1)} - \frac{3R \left(1 - \dfrac{3}{3s} - q \right)}{X_u - X_l} \int_{X_l}^{X_u} \ln(1 - e^{-X_E}) \, dX \\
&+ \frac{3RqX_E}{(e^{X_E} - 1)} - 3Rq \ln(1 - e^{-X_E}),
\end{aligned}
\tag{5.59}
$$

Table 5.25 Parameters of the Kieffer model for calculation of C_V and harmonic entropy for the 12 garnet end-members in the system $(Mg, Fe, Ca, Mn)_3(Al, Fe, Cr)_2Si_3O_{12}$. $\omega^*_{l,K_{max}}$ is the lower cutoff frequency of the optic continuum optimized on the basis of the experimental value of entropy at $T = 298.15$ °C. θ_D is the Debye temperature of the substance. S_m and S_{an} are, respectively, the magnetic spin and anharmonicity contributions to third-law entropy expressed in J/(mole × K). See section 3.3 for the significance of the various frequencies (cm^{-1}) (from Ottonello et al., 1996).

Compound	θ_D	ω_1	ω_2	ω_3	$\omega_{l,K=0}$	$\omega_{l,K_{max}}$	$\omega^*_{l,K_{max}}$	ω_u	m_1/m_2	$E1$	$E2$	S_m	S_{an}
Pyrope	793	73	73	130	149	115.7	124.0	638	1.519	975	875	0	3.316
Almandine	731	68	68	122	117	103.2	78.4	635	3.491	975	875	40.146	2.280
Spessartine	743	69	69	121	113	99.6	-	631	3.434	965	865	44.694	2.845
Grossular	821	74	74	126	180	152.2	141.9	619	2.505	950	850	0	2.295
Uvarovite	733	71	71	124	170	143.7	-	609	2.505	930	830	23.053	3.098
Andradite	723	70	70	123	132	111.6	121.2	589	2.505	925	825	29.796	3.174
Knorringite	701	64	64	118	139	107.9	-	628	1.519	955	855	23.053	2.931
Calderite	642	59	59	108	65	57.2	-	600	3.434	940	840	74.490	3.875
Skiagite	631	58	58	108	69	60.8	-	695	3.491	950	850	69.942	3.316
Koharite	690	64	64	116	101	78.4	-	608	1.519	950	850	29.796	3.406
Mn-Cr garnet	652	60	60	109	103	90.6	-	621	3.434	945	845	67.747	3.904
Fe-Cr garnet	640	59	59	110	107	94.3	-	625	3.491	955	855	63.199	3.331

where

$$X = \frac{\hbar\omega}{kT}, \qquad (5.60)$$

ω is angular frequency, $\hbar$ is Planck's constant ($h/2\pi$), R is the Gas constant, $N = nZ$ is the number of atoms in the primitive cell, and q is the total proportion of vibrational modes ascribed to Einstein's oscillators.

The relevant parameters of the model for the various end-members adopted by Otto-nello et al. (1996) are listed in table 5.25. The ω_l limit is the most critical parameter for the estimation of macroscopic thermodynamic functions. Values of ω_l at $K = 0$ (center of the Brillouin zone) based on infrared spectra may be reduced to $\omega_{l,K_{max}}$ at $K = K_{max}$ by applying

$$\omega_{l,K_{max}} = \omega_{l,K=0}\left(1 + \frac{m_2}{m_1}\right)^{-1/2}, \qquad (5.61)$$

where m_1 and m_2 are reduced masses of intervening ions (oxygen and "heavy" cation, respectively, in our case; see Kieffer, 1980). $\omega_{l,K_{max}}$ values for pyrope, almandine, spessartine, grossular, and andradite are those of Kieffer (1980). For uvarovite, a value of 170 cm^{-1} for $\omega_{l,K=0}$ was estimated from the spectrum of Moore et al. (1971), and the corresponding $\omega_{l,K_{max}}$ was obtained by application of equation 5.61.

The model parameters relative to the uncommon garnet end-members were obtained by Bokreta (1992) through linear regression involving the atomistic properties of the intervening ions. Equation 5.59 and the various vibrational function terms in eq. 5.58 were calculated by gaussian integration. Values on nondimensionalized frequencies $\omega_{l,K_{max}}$ were adjusted, stemming from the initial guess values with a trial-and-error procedure, matching the calculated S with experimental third-law entropy values at $T = 298.15$ K and $P = 1$ bar, after correction for a magnetic spin entropy contribution of the form

$$S_m = R\ln(2q+1), \tag{5.62}$$

where q is $\frac{3}{2}$ for Cr^{3+}, $\frac{5}{2}$ for Mn^{3+} and Fe^{2+}, 2 for Fe^{3+}, and correction for anharmonicity, according to

$$S_{an} = \int_0^T V\alpha^2 K_0\, dT. \tag{5.63}$$

The procedure of Ottonello et al. (1996a) is operationally different from that indicated by Kieffer (1980), which suggested that the calculated C_V be fitted to the calorimetric C_P at 100 K after subtraction of the anharmonicity contribution

$$C_P = C_V + TV\alpha^2 K_0. \tag{5.64}$$

Figure 5.16A shows, for example, how calculated C_P values for pyrope compare with low-T calorimetric measurements of Haselton and Westrum (1980). Adopting the initial guess value of $\omega_{l,K_{max}}$ (115.7 cm^{-1}; see table 5.25) leads to an entropy estimate at $T_r = 298.15$ of 272.15 J/(mol × K) (268.834 harmonic plus 3.316 anharmonic). The calorimetric value being 266.27, the initial guess value of $\omega_{l,K_{max}}$ is too low and must be raised to 124.0 cm^{-1} to match the calorimetric evidence. As we see in the upper part of figure 5.16 also the resulting C_P in the low-T region compares better with calorimetric values after adjustment of the nondimensionalized frequency at the limit of the Brillouin zone. In the lower part of figure 5.16B, model predictions for pyrope are compared with high-T calorimetric data of Newton et al. (1977) and Tequi et al. (1991), indicating a reasonable agreement with the latter.

5.3.5 Mixing Properties

The existence of unmixing processes in the garnet phase was long masked by its isotropic nature, which inhibits mineralogical investigation under polarized light, and was first observed by Cressey (1978) by transmission electron microscopy on a garnet of composition $(Mg_{0.21}Fe_{0.49}Ca_{0.29}Mn_{0.01})_3Al_2Si_3O_{12}$. Indeed, the existence of positive interactions among garnet components in binary mixtures implies the opening of solvus fields at given values of the consolute temperature (T_c; see section 3.11). For instance, if we adopt an interaction parameter of 3.0 kcal/mole for the $(Mg,Mn)_3Al_2Si_3O_{12}$ mixture, as proposed by Chakraborty and Ganguly (1992), the consolute temperature is

$$T_c = \frac{W}{2R} = \frac{3000}{2 \times 1.98726} = 756(K) = 483(°C). \tag{5.65}$$

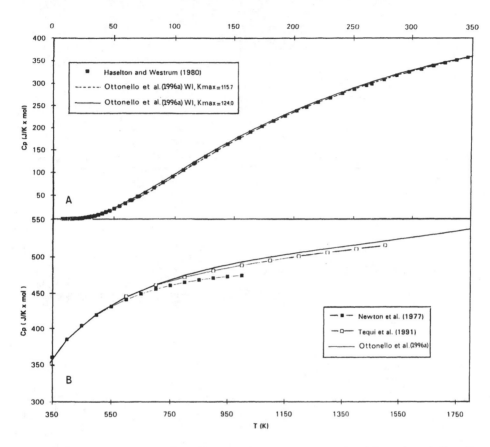

Figure 5.16 Heat capacity of pyrope at constant *P*, as determined from the Kieffer vibrational model, compared with low-*T* (upper part of figure) and high-*T* (lower part of figure) experimental evidences. From Ottonello et al. (1996). Reprinted with permission of The Mineralogical Society of America.

This temperature may be somewhat enhanced at high-pressure regimes in light of the observed positive deviations from Vegard's law (Ganguly et al., 1993).

In spite of a conspicuous effort spent in deciphering the interaction properties of binary garnet mixtures, agreement in the literature is still scanty. Figure 5.17 shows, for example, enthalpic interactions in binary aluminiferous garnet mixtures, according to various authors, represented as subregular Margules models and compared with calorimetric values and results of interionic potential model calculations. The corresponding subregular Margules interaction parameters are listed in table 5.26.

Agreement is also poor concerning entropy and volume excess terms. Because divalent cations (Mg, Ca, Fe, Mn) occupy only dodecahedral sites whereas octahedral sites are reserved for trivalent cations (Cr, Fe^{3+}, Al), each cation has only one site at its disposal and permutability is fixed by stoichiometry (cf. section 3.8.1). As regards the occupancy on tetrahedral positions, we have already seen that analyses of natural specimens show silicon deficiencies, compensated by Al^{3+}

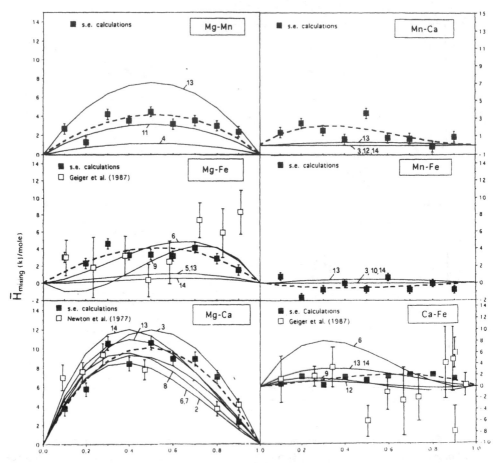

Figure 5.17 Enthalpic interactions in the various binary joins of aluminiferous garnets. □: calorimetric data; ■: results of interionic potential calculations. The corresponding subregular Margules interaction parameters are listed in table 5.26 (from Ottonello et al., in prep.).

in tetrahedral coordination (cf. table 5.17). This fact should be carefully evaluated in studies of the mixing properties of aluminiferous garnets. Unfortunately, agreement among the various experimental sources is so poor that it prevents appreciation of the effect of Al-Si substitutions in natural mixtures (cf. tables 5.26 and 5.27). Hence, the existence of excess entropy terms in mixtures must be provisionally ascribed either to substantial short-range ordering on sites or to vibrational effects.

Particularly, for the join $(Ca,Fe)_3Al_2Si_3O_{12}$, the recent data of Koziol (1990) confirm that excess entropy of mixing is virtually absent (cf. table 5.27). Existing experimental data on the $Mg_3Al_2Si_3O_{12}$-$Ca_3Al_2Si_3O_{12}$ system show that the excess entropy of mixing terms are probably symmetric and coupled to a slightly positive excess volume. Newton and Wood (1980) suggested that the molar volume of the

Table 5.26 Subregular Margules parameters for enthalpic interactions in aluminiferous garnets (from Ottonello et al., in prep.). Values in kJ/mole on a three-cation basis. The subregular Margules model is $H_{mixing,\,ij} = X_i X_j (W_{ij} X_i X_j + W_{ji} X_j X_i)$ with (1) $Mg_3Al_2Si_3O_{12}$, (2) $Mn_3Al_2Si_3O_{12}$, (3) $Ca_3Al_2Si_3O_{12}$, and (4) $Fe_3Al_2Si_3O_{12}$.

W_{H12}	W_{H21}	W_{H13}	W_{H31}	W_{H14}	W_{H41}	W_{H23}	W_{H32}	W_{H24}	W_{H42}	W_{H34}	W_{H43}	Reference
-	-	93.6	93.6	-	-	-	-	-	-	-	-	(1)
-	-	25.1	47.9	-	-	-	-	-	-	-	-	(2)
-	-	48.1	48.1	-	-	0	0	0	0	-	-	(3)
4.5	4.5	-	-	-	-	-	-	-	-	-	-	(4)
-	-	-	-	6.4	2.1	-	-	-	-	-	-	(5)
-	-	12.6	50.8	31.4	2.5	-	-	-	-	−7.9	58.0	(6)
-	-	12.6	50.8	-	-	-	-	-	-	-	-	(7)
-	-	15.0	56.5	-	-	-	-	-	-	-	-	(8)
-	-	-	-	36.2	−15.8	-	-	-	-	−9.1	13.7	(9)
-	-	-	-	-	-	-	-	0	0	-	-	(10)
12.6	12.6	-	-	-	-	-	-	-	-	-	-	(11)
-	-	-	-	-	-	0	0	-	-	−3.3	7.8	(12)
30.3	30.3	25.9	59.3	6.4	2.1	1.4	1.4	1.9	1.9	2.6	20.3	(13)
0	0	21.6	69.2	3.7	0.2	0	0	0	0	2.6	20.3	(14)
46.3	46.3	25.9	25.9	6.4	2.1	0	0	2.3	2.3	2.6	20.3	(15)
17.7	15.1	38.6	42.4	18.3	14.1	−0.4	14.6	−1.2	−3.2	12.6	2.0	(16)

(1) Hensen et al., 1975; phase equilibration; (2) Newton et al., 1977; calorimetric; (3) Ganguly and Kennedy, 1974; phase equilibration; (4) Wood et al., 1994; phase equilibration; (5) Hackler and Wood, 1989; phase equilibration; (6) Ganguly and Saxena, 1984; multivariate analysis; (7) Haselton and Newton, 1980; calorimetric; (8) Wood, 1988; phase equilibration; (9) Geiger et al., 1987; calorimetric; (10) Powenceby et al., 1987; phase equilibration; (11) Chakraborty and Ganguly, 1992; diffusion experiments; (12) Koziol, 1990; phase equilibration; (13) Ganguly and Cheng, 1994; multivariate analysis; (14) Berman, 1990; multivariate analysis; (15) Ganguly, 1995; personal communication; (16) Ottonello et al. (in prep.); structure-energy calculations.

$(Mg,Ca)_3Al_2Si_3O_{12}$ and $(Fe,Ca)_3Al_2Si_3O_{12}$ mixtures has a sigmoidal path (positive excess volume over most of the binary join, coupled with slight negative deviations observed in the vicinity of the Mg or Fe end-member). However according to Dempsey (1980), this could be indicative of a displacive transition to space group $I2_13$. Indeed, if we compare the experimental results of Newton and Wood (1980) and Ganguly et al. (1993) for the same binary join, the assignment of negative excess volume becomes highly conjectural. The experiments of Koziol (1990) on the $(Ca,Mn)_3Al_2Si_3O_{12}$ join indicate that the binary mixture is virtually ideal. According to Woodland and O'Neill (1993), molar volumes of mixing are also ideal for the $Fe_3(Al,Fe)_2Si_3O_{12}$ join. Proposed subregular entropic and volumetric Margules parameters for aluminiferous garnets are presented in table 5.27.

As regards the calcic garnets, there is general consensus that the $Ca_3(Al,Fe^{3+})_2Si_3O_{12}$ mixture is essentially ideal (Holdaway, 1972; Ganguly, 1976; Perchuk and Aranovich, 1979; Bird and Helgeson, 1980). For the $Ca_3(Cr,Al)_2Si_3O_{12}$ mixture, the calorimetric measurements of Wood and Kleppa (1984) indicate negative enthalpy of mixing terms, which can be modeled with an enthalpic interaction parameter W_H falling between 0 and −3 kcal/mole. According to

Table 5.27 Entropic (W_S, cal/mole × K) and volumetric (W_V; cm³/mole) terms of binary subregular interaction parameters for garnets in the system (Fe, Mg, Ca, Mn)$_3$Al$_2$Si$_3$O$_{12}$.

	$W_{S.12}$	$W_{S.21}$	$W_{V.12}$	$W_{V.21}$	Reference
Mg(1)-Ca(2)	4.5	4.5	-	-	(1)
	11.19	0	3.0	0	(2)
	2.51	2.51	0.58	0.12	(4)
	-	-	1.73	0.36	(5)
Fe(1)-Ca(2)	4.5	4.5	-	-	(1)
	0	0	-	-	(1)
	0	0	3.0	0	(2)
	0	0	0	0	(3)
	1.21	1.21	0.3	0.3	(4)
Fe(1)-Mg(2)	0	0	-	-	(1)
	0	0	−0.09	0.9	(2)
	0	0	0.03	0.2	(4)
Ca(1)-Mn(2)	0	0	-	-	(4)
Mg(1)-Mn(2)	3.73	3.73	0	0	(4)
Fe(1)-Mn(2)	0	0	0	0	(4)

(1) Ganguly and Saxena (1984); (2) Newton et al. (1986); (3) Koziol (1990); (4) Ganguly (personal communication); (5) Ganguly et al. (1993).

Table 5.28 Macroscopic interaction parameters for garnets in the three four-component systems (Mg, Mn, Ca, Fe)$_3$Al$_2$Si$_3$O$_{12}$, (Mg, Mn, Ca, Fe)$_3$Cr$_2$Si$_3$O$_{12}$, and (Mg, Mn, Ca,Fe)$_3$Fe$_2$Si$_3$O$_{12}$ (Wohl's 1946 formulation). Terms in parentheses are Berman's (1990) thernary interaction parameters (from Ottonello et al., in prep.). X cation: (1) = Mg; (2) = Mn; (3) = Ca; (4) = Fe^{2+}.

	Y Cation			
	Al^{3+}		Cr^{3+}	Fe^{3+}
W_{12}	17.673		14.758	−0.773
W_{21}	15.112		11.480	5.236
W_{13}	38.624		32.928	38.111
W_{31}	42.418		61.085	46.801
W_{14}	18.316		11.741	27.985
W_{41}	14.142		18.767	6.726
W_{23}	−0.392		7.904	19.090
W_{32}	14.549		21.798	−1.067
W_{24}	−1.156		−12.002	12.765
W_{42}	−3.193		6.984	−14.429
W_{34}	12.605		11.107	30.821
W_{43}	1.963		−3.189	−4.458
W_{123}	−23.254	(58.825)	21.389	−13.313
W_{124}	11.904	(45.424)	24.347	−65.766
W_{134}	19.892	(11.470)	−12.967	−27.540
W_{234}	2.654	(1.975)	−4.676	104.290
W_{1234}	0	(0)	0	0

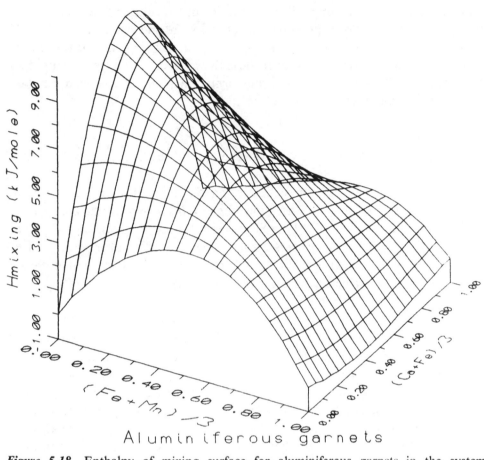

Figure 5.18 Enthalpy of mixing surface for aluminiferous garnets in the system $(Mg,Mn,Ca,Fe)_3Al_2Si_3O_{12}$ based on the Wohl's model parameters in table 5.28 (from Otto-nello et al., in prep.).

Mattioli and Bishop (1984), the mixture is of regular type but has a positive bulk interaction parameter W between 0.6 and 3 kcal/mole. Extension to multicomponent mixtures was attempted by Berman (1990), Ganguly (personal communication), and Ottonello et al. (in prep.). Berman (1990) applied the model of Berman and Brown (1984), originally conceived for silicate melts, to aluminiferous garnets and deduced the magnitude of ternary interaction parameters by applying

$$W_{i,j,k} = \frac{\left(W_{i,i,j} + W_{i,j,j} + W_{i,i,k} + W_{i,k,k} + W_{j,j,k} + W_{j,k,k}\right)}{2} - C_{i,j,k} \quad (5.66)$$

and neglecting ternary term $C_{i,j,k}$ (note that eq. 5.66 corresponds to the part on the right in square brackets in Wohl's equation 3.183). Ganguly (personal communication) applied the formulation of Cheng and Ganguly (1994) and found the ternary interaction terms to be insignificant. Wohl's model parameters for enthalpic interactions in aluminiferous, ferriferous, and chromiferous garnets obtained by static interionic potential calculations are shown in table 5.28. Although the three systems investigated are strictly quaternary, there is no need for a quaternary interaction term because the enthalpy of mixing surface (figure 5.18) does not exhibit an upward-convex conformation (contrary to what has been observed for clinopyroxenes; see section 5.4.5).

5.4 Pyroxenes

Pyroxenes are *polysilicates* ("inosilicates" in the mineralogical classification), crystallizing, respectively, in the monoclinic (*clinopyroxenes;* spatial groups *C2/c, P2$_1$/c, P2/n*) and orthorhombic systems (*orthopyroxenes;* spatial groups *Pbca, Pbcn*). Chain periodicity p is 2 and defines the length of the c edge of the unit cell (about 5.3 Å). Multiplicity m is 1 (cf. section 5.1). The peculiar characteristic of pyroxene structures is the bonding of a tetrahedral group [SiO$_4$] with two neighboring groups, through bridging oxygens, to form infinite silicate chains of composition [SiO$_3$]$_n$. Figure 5.19A shows the geometry of these chains, projected along the various axes and in perspective. Figure 5.19B shows more generally the relative positions of the chains in the structure: each chain has an apical and a basal front. Crystallographic sites located between two neighboring basal fronts are larger (coordination VI or VIII with oxygen) and are generally defined as *M*2 positions. Sites between two neighboring apical fronts are smaller (VI-fold coordination) and are usually defined as *M*1 positions. Neighboring chains may be displaced relative to one another by tilting of the tetrahedra on the bridging oxygens, thus allowing incorporation of different kinds of metal cations in *M*2 sites, as a function of the degree of distortion. This is the origin of the polymorphic transitions encountered in nature under various *P, T,* and *X* conditions.

The most representative spatial group of clinopyroxenes is *C2/c,* this structure having been resolved as long ago as 1928 by Warren and Bragg. It is characterized by sheets of single tetrahedral chains (parallel to axis c), alternating with octahedral sheets with VI-fold or VIII-fold coordinated cations. The *M*1 octahedron is quite regular, whereas *M*2 has distorted symmetry coordinated by eight oxygens. The spatial group *P2$_1$/c* (pigeonite, clinoenstatite, clinoferrosilite) differs from *C2/c* in having two types of tetrahedral chains distinct in symmetry (A and B). In the polymorphic *P2$_1$/c-C2/c* transition (e.g., clinoferrosilite-ferrosilite) one observes doubling of cell edge a and orthogonalization of angle β between edges a and b. The *Pbca* structure was resolved by Warren and Modell (1930), who noted its close correspondence with the *C2/c* structure (from which it differs essentially by doubling of edge a of the unit cell and by the presence of two symmetrically distinct tetrahedral chains). This structure contains octahedral sheets with alternating orientations ($\pm$), composed of *M*1 and *M*2 sites with, respectively, regular and distorted symmetry.

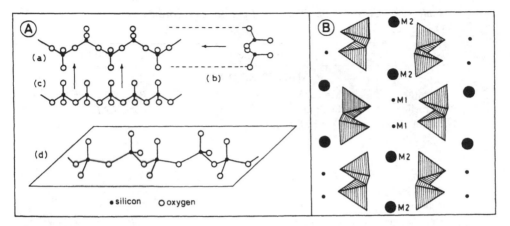

Figure 5.19 (A) Schematic representation of $[SiO_3]_n$ chains of pyroxenes: (a) projection on plane (100): (b) projection along axis Z; (c) projection along axis Y; (d) perspective representation. (B) Relative arrangement of tetrahedral chains in pyroxene structure (seen in their terminal parts with intercalation of $M1$ and $M2$ positions. From Putnis and McConnell (1980). Reproduced with modifications by permission of Blackwell Scientific Publications, Oxford, Great Britain.

5.4.1 Chemistry and Occurrence

The general formula for pyroxenes is XYZ_2O_6 where X is a generic cation in site $M2$ (VIII-fold coordination in clinopyroxene and VI-fold in orthopyroxene), Y is a generic cation in $M1$, and Z is a tetrahedrally coordinated cation. The most important diadochy in tetrahedral positions is that of Al^{3+}, replacing Si^{4+}, although diadochy of Fe^{3+} has been observed under particular T and P conditions (Huckenholtz et al., 1969). Al^{3+} and Fe^{3+} also occur in site $M1$, together with Cr^{3+} and Ti^{4+}. Mg^{2+}, Fe^{2+}, and Mn^{2+} occupy both $M1$ and $M2$ sites, whereas Li^+, Na^+, and Ca^{2+} occupy $M2$ positions exclusively.

According to Cameron and Papike (1982), pyroxenes contain Cr^{2+} and Ti^{3+} in rocks equilibrated at low f_{O_2} (lunar specimens, meteorites). However, spectroscopic evidence is ambiguous and insufficient for a safe attribution (Rossman, 1982). Some authors (Bocchio et al., 1979; Ghose et al., 1986; Griffin and Mottana, 1982) report the presence of Mn^{3+} in $M1$ sites in clinopyroxene. Davoli (1987) reexamined this hypothesis, proposing precise structural criteria to detect the presence of Mn^{3+} in the monoclinic phase (the ratio Mn^{3+}/Mn^{2+} may be a potential f_{O_2} barometer).

The chemical classification of pyroxene components disregards to some extent the represented structural groups, being based essentially on the chemistry of major cations. As an example, table 5.29 reports the classification of Deer et al. (1978), which subdivides pyroxene components into five groups (Mg-Fe, Ca, Ca-Na, Na, and Li pyroxenes). In this classification, the term "enstatite" identifies a component crystallizing in different structural classes (*Pbca*, *P2₁/c*, *Pbcn*).

Table 5.29 Chemical classification of pyroxenes according to Deer et al. (1978).

Name	Stoichiometry	Structural Class
	Mg-Fe PYROXENES	
Enstatite	$Mg_2Si_2O_6$	*Pbca, P2_1/c, Pbcn*
Ferrosilite	$Fe_2Si_2O_6$	*Pbca, P2_1/c*
Orthopyroxene	$(Mg,Fe,Ca)_2Si_2O_6$	*Pbca*
Pigeonite	$(Mg,Fe,Ca)_2Si_2O_6$	*P2_1/c, C2/c*
	Ca PYROXENES	
Augite	$(Ca,R^{2+})(R^{2+},R^{3+},Ti^{4+})(Si,Al)_2O_6$	*C2/c*
Diopside	$CaMgSi_2O_6$	*C2/c*
Hedembergite	$CaFeSi_2O_6$	*C2/c*
Johannsenite	$CaMnSi_2O_6$	*C2/c*
	Ca-Na PYROXENES	
Omphacite	$(Ca,Na)(R^{2+},Al)Si_2O_6$	*C2/c, P2/n, P2*
Augitic aegirine	$(Ca,Na)(R^{2+},Fe^{3+})Si_2O_6$	*C2/c*
	Na PYROXENES	
Jadeite	$NaAlSi_2O_6$	*C2/c*
Acmite	$NaFeSi_2O_6$	*C2/c*
Ureyite	$NaCrSi_2O_6$	*C2/c*
	Li PYROXENES	
Spodumene	$LiAlSi_2O_6$	*C2, C2/c*

R^{2+} in table 5.29 represents divalent cations (Mn^{2+}, Fe^{2+}, Mg^{2+}), and R^{3+} represents trivalent cations (Fe^{3+}, Cr^{3+}, Al^{3+}). Thus, from the crystal chemical point of view, the terms "augite," "omphacite," and "augitic aegirine" do not identify precise stoichiometry. Analogously, the general term "orthopyroxene," according to Deer et al. (1978), represents a mixture of magnesian and ferroan components, and the term "pigeonite" identifies an Mg-Fe-Ca mixture.

In geochemistry it is necessary to describe the composition of pyroxene by end-members that are compositionally simple and stoichiometrically well defined. It is also opportune to distinguish the various structural classes, because *P-T* stability and reactivity vary greatly with type of polymorph. This inevitably requires the formulation of "fictitious" components (i.e., components that have never been synthesized as pure phases in the structural form of interest, but that are present as members of pyroxene mixtures). For example, table 5.30 gives the list of pyroxene geochemical components proposed by Ganguly and Saxena (1987) (partly modified here).

In compiling table 5.30, the "orthopyrope" component ($Mg_{1.5}Al_{1.0}Si_{1.5}O_6$) of Ganguly and Saxena was discarded, because it may be obtained as a mixture of Mg-Tschermak and enstatite. Moreover, Cr- and Li-bearing terms (ureyite and spodumene) are added to the *C2/c* class and a Co-bearing term to the *Pbca* class. The manganous term ($Mn_2Si_2O_6$; Tokonami et al., 1979) has been added to the *P2_1/c* class. It must be noted that natural pigeonites are mixtures of compositional

Table 5.30 Pyroxene components that may be adopted in geochemistry. x = actually existing; o = fictive.

Component	Stoichiometry	System	Structural Class	Occurrence
Clinoenstatite	$Mg_2Si_2O_6$	Monoclinic	$P2_1/c$	x
Clinoferrosilite	$Fe_2Si_2O_6$	Monoclinic	$P2_1/c$	x
Mn clinopyroxene	$Mn_2Si_2O_6$	Monoclinic	$P2_1/c$	x
Diopside	$CaMgSi_2O_6$	Monoclinic	$C2/c$	x
Ca-Tschermak	$CaAl_2SiO_6$	Monoclinic	$C2/c$	x
Jadeite	$NaAlSi_2O_6$	Monoclinic	$C2/c$	x
Hedembergite	$CaFeSi_2O_6$	Monoclinic	$C2/c$	x
Al clinopyroxene	$Al_2Al_2O_6$	Monoclinic	$C2/c$	o
Wollastonite	$Ca_2Si_2O_6$	Monoclinic	$C2/c$	x
Acmite	$NaFeSi_2O_6$	Monoclinic	$C2/c$	x
Ureyite	$NaCrSi_2O_6$	Monoclinic	$C2/c$	x
Spodumene	$LiAlSi_2O_6$	Monoclinic	$C2/c$	x
Enstatite	$Mg_2Si_2O_6$	Orthorhombic	$Pbca$	x
Ferrosilite	$Fe_2Si_2O_6$	Orthorhombic	$Pbca$	x
Orthodiopside	$CaMgSi_2O_6$	Orthorhombic	$Pbca$	o
Mg-Tschermak	$MgAl_2Si_2O_6$	Orthorhombic	$Pbca$	o
Orthohedembergite	$CaFeSi_2O_6$	Orthorhombic	$Pbca$	o
Al orthopyroxene	$Al_2Al_2O_6$	Orthorhombic	$Pbca$	o
Co orthopyroxene	$Co_2Si_2O_6$	Orthorhombic	$Pbca$	x

terms $Mg_2Si_2O_6$, $Fe_2Si_2O_6$, $CaMg_2Si_2O_6$, and $CaFe_2Si_2O_6$, crystallizing in the spatial group $P2_1/c$ of the monoclinic system. Two additional terms with structure $P2_1/c$ ($CaMg_2Si_2O_6$ and $CaFe_2Si_2O_6$) must then be added for complete representation of pigeonites. The same is true for components $Mg_2Si_2O_6$ and $Fe_2Si_2O_6$ of the $C2/c$ class. In geochemistry, nevertheless, it is usual to assume that the energetic properties of the two monoclinic classes differ solely by a coherence term. The energy of coherence (which may be assimilated to a *strain energy,* cf. section 3.13) is embodied in the Gibbs free energy of mixing terms, thus avoiding the formulation of a redundant number of fictious components.

Pyroxenes (as feldspars) occupy a position that is "chemically central in the realm of rock compositions" (Robinson, 1982). They are therefore found ubiquitously and virtually in any kind of paragenesis developing in the P-T-X conditions typical of earth's mantle and crust. Table 5.31 lists the main occurrences.

Table 5.32 lists chemical compositions of natural pyroxenes. Pyroxene chemistry is quite complex and analyses must follow precise stoichiometric rules based on site occupancies. Cameron and Papike (1982) follow four rules in the selection of "acceptable" analyses among the plethora of values reported in the literature:

1. The sum of Si and tetrahedrally coordinated Al must not be higher than 2.02 or lower than 1.98.
2. The sum of $M1$ cations (Al, Fe^{3+}, Cr^{3+}, Ti^{4+}, Mg, Fe^{2+}, Mn) must be higher than 0.98.

Table 5.31 Occurrence of main compositional terms of pyroxenes in various types of rocks.

Term	Stoichiometry	Occurrence
ORTHORHOMBIC PYROXENES		
Enstatite-bronzite	$(Mg,Fe)_2Si_2O_6$	Gabbros; Mg-rich ultramafic rocks; serpentinites; meteorites
Hyperstene	$(Mg,Fe)_2Si_2O_6$ $0.3 < Fe_2Si_2O_6 < 0.5$	Charnockites; norites; Fe-rich ultramafic rocks; andesites; rocks with high-P metamorphism
MONOCLINIC PYROXENES		
Diopside-hedembergite	$(Ca,Mg,Fe)_2Si_2O_6$	Cr-rich ultramafic rocks; contact metamorphism (cipollines, cornubianites)
Pigeonite	$(Mg,Fe,Ca)_2Si_2O_6$ $Ca_2Si_2O_6 < 0.12$	Basalts and dolerites; kimberlites; meteorites
Augite	$(Ca,Mg,Fe,Al)_2(Si,Al)_2O_6$	Basalts, andesites, tephrites, limburgites gabbros, dolerites, ultramafic rocks. Sometimes in granites
Aegirine	$NaFeSi_2O_6$	Nepheline syenites; hypersodic granites
Omphacite	$(Ca,Mg,Na,Al)_2(Al,Si)_2O_6$	Eclogites
Spodumene	$LiAlSi_2O_6$	Pegmatites

3. The sum of $M2$ cations must range between 0.98 and 1.02.
4. The residual of charges in the equilibrium $X + Y = Z$ with $X = Na_{M2}$, $Y = Al_{M1} + Cr_{M1} + Fe^{3+}_{M1} + 2Ti^{4+}_{M1}$, and $Z = Al_T$, must be lower than 0.030.

Based on these four rules, Cameron and Papike (1982) selected 175 analyses out of 405 reported by Deer et al. (1978) and discarded 230 compositions. This selection is extremely rigorous and does not take into account either the possible stabilization of Fe^{3+} in tetrahedral sites or the existence of extrinsic disorder (cf. chapter 4). Robinson (1982) showed that, by accepting a limited amount of cationic vacancies in $M2$ sites and assuming possible stabilization of Fe^{3+} in tetrahedral positions, 117 additional analyses out of the 230 discarded by Cameron and Papike (1982) may be selected.

Samples 6 and 7 in table 5.32 are from the Zabargad peridotite (Red Sea) and are representative of the chemistry of upper mantle pyroxenes (Bonatti et al., 1986). The absence of Fe_2O_3 in these samples is due to the fact that microprobe analyses do not discriminate the oxidation state of iron, which is thus always expressed as FeO. It must be noted here that the observed stoichiometry (based on four oxygen ions) is quite consistent with the theoretical formula and that no Fe^{3+} is required to balance the negative charges of oxygen.

Table 5.32 Compositions (in weight %) of natural pyroxenes (samples 1–5 from Deer et al., 1983; samples 6 and 7 from Bonatti et al., 1986): (1) enstatite from a pyroxenite; (2) ferrosilite from a thermometamorphic iron band; (3) hedembergite; (4) chromian augite from a gabbroic rock of the Bushveld complex; (5) aegirine from a riebeckite-albite granitoid; (6) diopside from a mantle peridotite; (7) enstatite from a mantle peridotite.

Oxide	Samples						
	(1)	(2)	(3)	(4)	(5)	(6)	(7)
SiO_2	57.73	45.95	48.34	52.92	51.92	50.87	54.03
TiO_2	0.04	0.10	0.08	0.50	0.77	0.61	0.15
Al_2O_3	0.95	0.90	0.30	2.80	1.85	6.47	3.97
Fe_2O_3	0.42	0.31	1.50	0.85	31.44	-	-
Cr_2O_3	0.46	-	-	0.88	-	0.68	0.26
FeO	3.57	41.65	22.94	5.57	0.75	2.62	7.11
MnO	0.08	5.02	3.70	0.15	-	0.08	0.18
NiO	0.35	-	-	0.10	-	0.01	0.02
MgO	36.13	3.49	1.06	16.40	-	15.13	33.13
CaO	0.23	1.43	21.30	19.97	-	21.79	0.67
Na_2O	-	-	0.14	0.35	12.86	1.23	0.01
K_2O	-	-	0.03	0.01	0.19	0.07	0.05
$H_2O^{(+)}$	0.52	0.65	0.46	0.10	0.17	-	-
$H_2O^{(-)}$	0.04	0.09	-	0.07	-	-	-
Total	100.52	99.59	99.85	100.67	99.95	99.56	99.58
Si	1.972	1.972	1.988	1.929	1.986	1.858	1.886
Al	0.028	0.028	0.012	0.071	0.014	0.142	0.114
Total T	2.00	2.00	2.00	2.00	2.00	2.00	2.00
Al	0.010	0.018	0.006	0.049	0.069	0.136	0.049
Ti	0.001	0.003	0.002	0.014	0.022	0.017	0.004
Fe^{3+}	0.010	0.010	0.046	0.024	0.905	-	-
Cr	0.012	-	-	0.026	-	0.020	0.007
Mg	1.839	0.223	0.065	0.891	-	0.824	1.724
Ni	0.010	-	-	0.003	-	0.000	0.001
Fe^{2+}	0.102	1.495	0.789	0.170	0.024	0.080	0.208
Mn	0.002	0.183	0.129	0.005	-	0.002	0.005
Ca	0.008	0.066	0.939	0.780	-	0.853	0.025
Na	-	-	0.011	0.024	0.953	0.087	0.001
K	-	-	0.001	0.000	0.008	0.003	0.002
Total $M1 + M2$	1.99	2.00	1.99	1.99	1.98	2.02	2.02
Mg/(Mg + Fe + Ca)	93.9	12.4	3.3	47.6	-	46.9	88.1
Fe/(Mg + Fe + Ca)	5.7	83.9	49.0	10.7	-	4.6	10.6
Ca/(Mg + Fe + Ca)	0.4	3.7	47.7	41.7	-	48.5	1.3

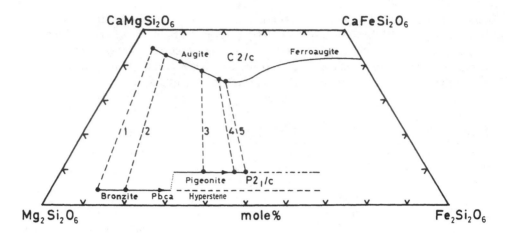

Figure 5.20 Pyroxene quadrilateral. Solid lines: compositionally continuous series; dashed tie lines connect compositional terms coexisting on opposite sides of miscibility gap. From G. M. Brown (1967). Mineralogy of basaltic rocks, in *Basalts*, H. H. Hess and A. Poldervaart, eds., copyright © 1967 by John Wiley and Sons. Reproduced with modifications by permission of John Wiley & Sons.

Most natural pyroxenes fall in the compositional field $Mg_2Si_2O_6$-$CaMg_2Si_2O_6$-$CaFe_2Si_2O_6$-$Fe_2Si_2O_6$ which is usually called the "pyroxene quadrilateral" (fig. 5.20). This system contains one compositionally continuous series of monoclinic *C2/c* pyroxenes (augites, Fe-augites) and another compositionally continuous series of orthorhombic *Pbca* pyroxenes (bronzite-hyperstene). The two series are separated by a miscibility gap whose nature will be discussed later. An intermediate series with structure *P2₁/c* (pigeonites) is also stable in the system.

5.4.2 Structural Properties

Table 5.33 lists the structural characters of various pure components of pyroxene crystallizing in the stable state, according to the synthesis of Smyth and Bish (1988). Note the doubling of cell edge *a* in the transition from monoclinic to orthorhombic structure, composition being equal (i.e., clinoenstatite-enstatite; clinoferrosilite-ferrosilite). This effect may be represented by the equation

$$a_{opx} = 2 \times a_{cpx} \sin \beta, \tag{5.67}$$

where β is the angle between edges *a* and *c*.

It should be stressed that doubling of cell edge *a* in the transition from $P2_1/c$ (or *C2/c*) to *Pbca* structure does not imply doubling of the molar volume, because the number of unit formulas per unit cell (*Z*) is also doubled (see table 5.33).

Table 5.33 Structure of pure pyroxene components in stable state. Molar volume in cm³/mole; molar weight in g/mole; cell edges in Å; cell volume in Å³. Adapted from Smyth and Bish (1988).

Phase	Diopside	Hedembergite	Jadeite	Acmite
Formula unit	$CaMgSi_2O_6$	$CaFeSi_2O_6$	$NaAlSi_2O_6$	$NaCrSi_2O_6$
Molar weight	216.560	248.095	202.140	231.005
Density	3.279	3.656	3.341	3.576
Molar volume	66.039	67.867	60.508	64.606
Z	4	4	4	4
System	Monoclinic	Monoclinic	Monoclinic	Monoclinic
Class	*2/m*	*2/m*	*2/m*	*2/m*
Space group	*C2/c*	*C2/c*	*C2/c*	*C2/c*
Cell parameters:				
a	9.746	9.845	9.423	9.658
b	8.899	9.024	8.564	8.795
c	5.251	5.245	5.223	5.294
β	105.63	104.70	107.56	107.42
Cell volume	438.58	450.72	401.85	429.06
Reference	Cameron et al. (1973 b)	Cameron et al. (1973 b)	Cameron et al. (1973 b)	Clark et al. (1969)

Phase	Ureyite	Spodumene	Ca-Tschermak
Formula unit	$NaCrSi_2O_6$	$LiAlSi_2O_6$	$CaAlAlSiO_6$
Molar weight	227.154	186.089	218.125
Density	3.592	3.176	3.438
Molar volume	63.239	58.596	63.445
Z	4	4	4
System	Monoclinic	Monoclinic	Monoclinic
Class	*2/m*	*2/m*	*2/m*
Space group	*C2/c*	*C2/c*	*C2/c*
Cell parameters:			
a	9.579	9.461	9.609
b	8.722	8.395	8.652
c	5.267	5.218	5.274
β	107.37	110.09	106.06
Cell volume	419.98	389.15	421.35
Reference	Cameron et al. (1973 b)	Sasaki et al. (1980)	Okamura et al. (1974)

Phase	Enstatite	Ferrosilite	Co-orthopyroxene
Formula unit	$Mg_2Si_2O_6$	$Fe_2Si_2O_6$	$Co_2Si_2O_6$
Molar weight	200.792	263.862	270.035
Density	3.204	4.002	4.222
Molar volume	62.676	65.941	63.963
Z	8	8	8
System	Orthorhombic	Orthorhombic	Orthorhombic

continued

Table 5.33 *(continued)*

Phase	Enstatite	Ferrosilite	Co-orthopyroxene
Class	*mmm*	*mmm*	*mmm*
Space group	*Pbca*	*Pbca*	*Pbca*
Cell parameters:			
a	8.2271	8.4271	8.296
b	8.819	9.076	8.923
c	5.179	5.237	5.204
β	90	90	90
Cell volume	832.49	875.85	849.58
Reference	Sasaki et al. (1982)	Sasaki et al. (1982)	Sasaki et al. (1982)

Phase	Clinoenstatite	Clinoferrosilite	Mn-clinopyroxene
Formula unit	$Mg_2Si_2O_6$	$Fe_2Si_2O_6$	$Mn_2Si_2O_6$
Molar weight	200.792	263.862	262.044
Density	3.188	4.005	3.819
Molar volume	62.994	65.892	68.608
Z	4	4	4
System	Monoclinic	Monoclinic	Monoclinic
Class	*2/m*	*2/m*	*2/m*
Space group	$P2_1/c$	$P2_1/c$	$P2_1/c$
Cell parameters:			
a	9.626	9.7085	9.864
b	8.825	9.0872	9.179
c	5.188	5.2284	5.298
β	108.33	108.43	108.22
Cell volume	418.36	437.60	455.64
Reference	Morimoto et al. (1960)	Burnham (1967)	Tokonami et al. (1979)

Concerning the volume properties of the various polymorphs with respect to their chemical compositions, existing information on *Pbca* and $P2_1/c$ structures indicates linear dependence of mean $M1$ and $M2$ site dimensions on the ionic radii of the occupying cations (algebraic mean). As figure 5.21 shows, this linearity is more marked for $M1$ than for $M2$, which has distorted symmetry.

Figure 5.22 shows cell edges and volumes of synthetic pyroxenes in the $Mg_2Si_2O_6$-$CaMg_2Si_2O_6$-$CaFe_2Si_2O_6$-$Fe_2Si_2O_6$ quadrilateral, according to Turnock et al. (1973). The various parameters do not appear to evolve in linear fashion with the chemistry of the system, and the relationship is particularly complex for edge c.

The various lattice site dimensions in $C2/c$ clinopyroxenes depend in a complex fashion on the bulk chemistry of the phase. Nevertheless, the various interionic distances (and cell edges and volumes) may be expressed as functions of cationic occupancies of lattice

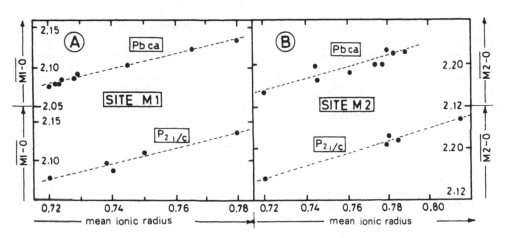

Figure 5.21 Mean cation-to-oxygen distances in sites $M1$ (A) and $M2$ (B) plotted against ionic radii of occupying cations (algebraic mean). Adopted radii are those of Shannon (1976). Data from Cameron and Papike (1981).

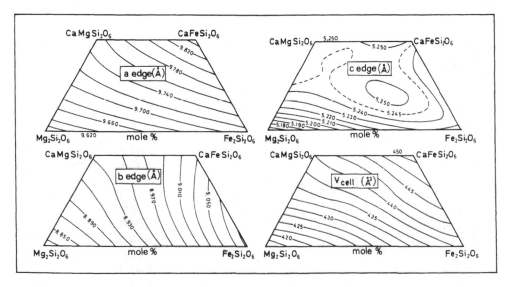

Figure 5.22 Cell parameters of synthetic pyroxenes in the $Mg_2Si_2O_6$-$CaMg_2Si_2O_6$-$CaFe_2Si_2O_6$-$Fe_2Si_2O_6$ quadrilateral. From Turnock et al. (1973). Reprinted with permission of The Mineralogical Society of America.

sites, through multiple regression parameters analogous to those proposed by Lumpkin and Ribbe (1983) and Ottonello (1987) for olivines. These parameters (table 5.34) allow estimations of a precision near that of experimental error. The mean discrepancy between estimated and experimental values is 0.002 Å for T-O, $M1$-O, and $M2$-O distances and 0.004 Å for O-O distances. To obtain, for instance, distance T-O1 we apply

$$D_{T-O1} = 0.4894r_Z - 0.1205r_X - 0.0090r_Y + 1.4479, \qquad (5.68)$$

Table 5.34 Multiple regression parameters relating site population to interionic distances in *C2/c* pyroxenes.

Distance	$Z(T)$	$X(M2)$	$Y(M1)$	Constant
T-O1	0.4894	−0.1205	−0.0090	1.4479
T-O2	0.6058	0.0042	0.0009	1.2680
T-O3*A*1	0.4130	0.1645	0.0368	1.2879
T-O3*A*2	0.5045	0.2151	0.0736	1.1807
Mean *T*-O	0.5033	0.0659	0.0256	1.2959
O1*A*1-O2*A*2	0.9590	−0.1717	−0.0538	2.4170
O1*A*1-O3*A*1	0.9994	0.2413	0.1001	1.8700
O2*A*1-O3*A*1	0.8560	0.0922	0.0656	2.0755
O1*A*1-O3*A*2	1.0174	0.2637	0.0604	1.8977
O2*A*1-O3*A*2	0.5151	0.0241	0.0581	2.2260
O3*A*1-O3*A*2	0.4842	0.1592	−0.0053	2.2888
*M*1-O1*A*2	0.1981	0.8682	0.2433	1.0684
*M*1-O1*A*1	0.2117	0.6872	−0.1261	1.6636
*M*1-O2	−0.1021	0.8627	−0.0973	1.5851
Mean *M*1-O	0.3488	0.8976	0.0719	1.1705
O1*A*1-O1*B*1	−0.1164	0.2724	−0.1609	2.8378
O1*B*1-O1*A*2	0.2976	2.0568	0.0461	1.1266
O1*A*1-O1*A*2	0.6910	0.6797	−0.0893	2.3024
O1*A*1-O2*C*1	0.0447	1.5730	−0.1913	2.0718
O1*B*2-O2*C*1	−0.3369	1.0620	0.7987	1.4958
O1*A*2-O2*C*1	−0.0050	0.9138	−0.5247	2.8250
O2*C*1-O2*D*1	0.2763	0.7692	0.0378	2.2430
*M*2-O1	0.5574	0.1390	0.8411	1.0302
*M*2-O2	0.6639	0.0160	1.3030	0.5274
*M*2-O3*C*1	0.2968	0.9035	−0.7956	2.6450
*M*2-O3*C*2	−1.4111	−0.0657	−0.3851	3.9561
Mean *M*2-O	0.0251	0.2471	0.2406	2.0416
O1*A*1-O2*C*2	0.0588	−0.4499	0.7313	2.6130
O1*A*1-O3*C*2	−0.9130	−0.0194	0.5252	3.5457
O2*C*2-O3*C*2	−0.5931	−0.6688	0.9047	3.9398
O2*C*2-O3*D*2	0.3201	0.4011	0.9837	1.8535
O3*C*1-O3*D*2	−0.2204	0.8072	0.1395	2.2979
O3*C*1-O3*D*1	2.1568	1.5370	0.8506	0.1710
a	1.4187	1.8303	0.3395	7.3108
b	0.3150	1.8880	0.0721	7.3158
c	0.7793	0.1775	0.1483	4.5631
β	−13.2743	−12.5773	−8.3044	131.2997
Volume	171.9523	217.2755	51.4963	136.2097

where

$$r_Z = r_{Al^{IV}} X_{Al,T} + r_{Si} X_{Si,T} \qquad (5.69)$$

$$r_X = r_{Mg} X_{Mg,M2} + r_{Ca} X_{Ca,M2} \qquad (5.70)$$

$$r_Y = r_{Al^{VI}} X_{Al,M1} + r_{Fe^{3+}} X_{Fe^{3+},M1} + r_{Cr^{3+}} X_{Cr^{3+},M1} + \cdots. \qquad (5.71)$$

The ionic radii in equations 5.69 to 5.71 ($r_{Al^{IV}}$, $r_{Al^{VI}}$, r_{Si}, . . .) are those of Shannon (1976) in the appropriate coordination state with oxygen and $X_{Al,T}$, $X_{Si,T}$, . . . are the atomic occupancies in the various sites.

5.4.3 Effect of Intensive Variables on Structure and Phase Stability

The structural behavior of pyroxenes with the increase of T (pressure and chemical composition being equal) is to some extent similar to what we saw previously for olivine and garnet. The tetrahedral sites are relatively insensitive to temperature, whereas $M1$ and $M2$ sites expand appreciably with T. Site expansion is a function of the occupying cations, and site $M2$ generally exhibits more marked thermal expansion with respect to $M1$, as a result of the lower mean charge of its site population and the higher coordination number (VI-VIII as opposed to VI). Because the "inert" tetrahedra are linked to form chains (unlike monosilicates) and share corners with two sites with different polyhedral thermal expansions, with increasing T there is progressive structural decoupling, partially adsorbed by rotation of the tetrahedra at the point of junction (on axis c; cf. figure 5.23) or by minimal rotation out of the chain plane. At high T, the chains are stretched to a greater length, as shown in figure 5.23, as a consequence of the increased angle θ.

Experimental studies on $C2/c$ pyroxenes (Cameron et al., 1973b; Finger and Ohashi, 1976) at temperatures up to $T = 1000\ °C$ show that the structure remains stable without transition phenomena, and that the increase of the angle θ is 2 to 3°. Analogous studies on $Pbca$ pyroxenes (hyperstene, ferrosilite; Smyth, 1973) show phase stability in the T range 25 to 1000 °C but a more conspicuous increase in the angle θ (i.e., 10° for chain A and 15° for chain B in ferrosilite; 5° for chain A and 10° for chain B in hyperstene). However, phase transitions have been observed in $P2_1/c$ pyroxenes. Brown et al. (1972) observed a transition to the form

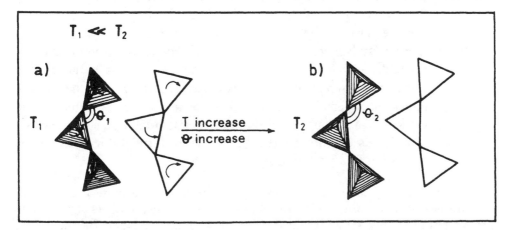

Figure 5.23 Extension of tetrahedral chains of pyroxenes as a function of temperature.

Table 5.35 Molar volume (cm³/mole), isobaric thermal expansion (K⁻¹), and isothermal bulk modulus (Mbar) of pyroxene end-members according to Saxena (1989)

Phase	Component	V_{T_r,P_r}	$\alpha_0 \times 10^4$	$\alpha_1 \times 10^8$	α_2	K_0	K'
			Thermal Expansion			Compressibility	
		C2/c + P2₁/c CLINOPYROXENES					
Clinoenstatile	$Mg_2Si_2O_6$	62.91	0.20303	1.28049	0.0010	1.10	4.20
Clinoferrosilite	$Fe_2Si_2O_6$	66.28	0.23326	−0.10413	−0.0059	1.00	4.20
Diopside	$CaMgSi_2O_6$	66.10	0.18314	1.41273	0.0011	1.00	4.20
Ca-Tschermak	$CaAl_2SiO_6$	63.56	0.2470	0.00000	0.0000	1.20	4.20
Jadeite	$NaAlSi_2O_6$	60.40	0.13096	2.23308	0.17442	1.33	4.20
Hedembergate	$CaFeSi_2O_6$	68.26	0.18314	1.41273	0.0011	1.00	4.20
Al-elinopyroxene	$Al_2Si_2O_6$	52.4	0.20303	1.28050	0.0010	1.10	4.20
Wollastonite	$Ca_2Si_2O_6$	79.86	0.1000(?)	-	-	1.00	4.00
		Pbca ORTHOPYROXENES					
Enstatite	$Mg_2Si_2O_6$	62.66	0.20303	1.28049	0.0010	1.07	4.20
Ferrosilite	$Fe_2Si_2O_6$	65.90	0.23326	−0.10413	−0.0059	1.01	4.20
Orthodiopside	$CaMgSi_2O_6$	69.09	0.18314	1.41273	0.0011	1.07	4.20
Orthohedembergite	$CaFeSi_2O_6$	69.22	0.18314	1.41273	0.0011	1.075	4.20
Al-orthopyroxene	$Al_2Si_2O_6$	53.56	0.20303	1.28050	0.0010	1.00	4.20
		Pbcn ORTHOPYROXENES					
Protoenstatite	$Mg_2Si_2O_6$	64.88	(?)	-	-	1.10	4.00

C2/c at about 960 °C in pigeonite, and Smyth (1974) observed a first-order transition in the low-*T* clinopyroxene at $T = 725$ °C.

Pressure has an antithetical effect on the interionic distances of pyroxenes with respect to temperature, and the observed compressibilities of *M2*, *M1*, and *T* sites obey the generalizations of Anderson (1972) and Hazen and Prewitt (1977).

The properties above indicate that the responses of the various pyroxene polymorphs to modifications of *P* and *T* intensive variables are different. It is essentially this fact that determines the opening of miscibility gaps during the cooling stage from "magmatic" *P-T* regimes. To rationalize this phenomenon in thermodynamic terms, we must distinguish the structural properties of the various compositional terms crystallizing in the various spatial groups with appropriate volume factors. Table 5.35 lists the values of molar volume, isobaric thermal expansion, and isothermal bulk modulus for pyroxene polymorphs in the thermodynamic database of Saxena (1989). Note that, for several components, Saxena (1989) adopts identical thermal expansion parameters for the structural classes *C2/c* and *Pbca*. This assumption renders the calculations straightforward in phase transition studies but is not compulsory.

Let us examine in detail the phase stability relationships of the four end-members of the pyroxene quadrilateral. Figure 5.24A shows the *P* and *T* stability ranges of the various polymorphs of the $Mg_2Si_2O_6$ compound, according to Lindsley (1982). The stable form

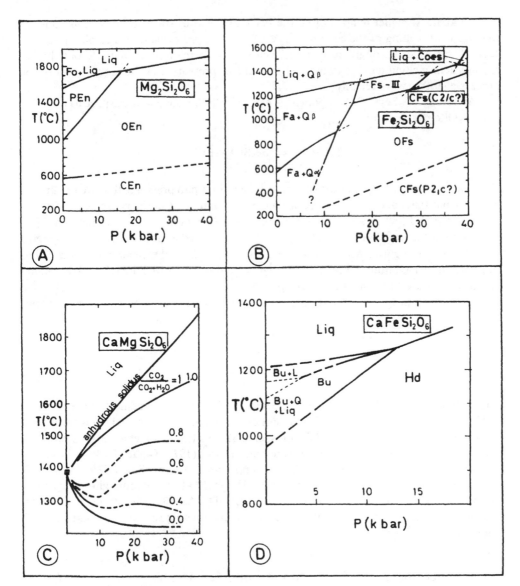

Figure 5.24 Phase stability relations for end-members of pyroxene quadrilateral. Melting curves refer to anhydrous conditions. Solidus curves for CaMgSi$_2$O$_6$ in saturated vapor phase conditions are also shown for various CO$_2$/H$_2$O ratios in the vapor phase. Dashed lines are extrapolated. From Lindsley (1982). Reprinted with permission of The Mineralogical Society of America.

at low T is $C2/c$ (clinoenstatite; CEn in figure 5.24A). At a given T between 566 °C (1 bar) and about 720 °C (40 kbar), there is a first-order transition to the orthorhombic form $Pbca$ (orthoenstatite; OEn in figure 5.24A). At $P = 1$ bar and $T = 985$ °C, there is transition from orthoenstatite ($Pbca$) to protoenstatite ($Pbcn$; PEn in figure 5.24A). The thermobaric gradient for this polymorphic transition is quite high and, with increasing P, the T stability range of the $Pbcn$ polymorph decreases progressively and vanishes at about

$P = 15$ kbar. At higher temperatures, the *Pbca* form is thus stable at all T conditions, up to congruent melting of the compound. At $P = 1$ bar, protoenstatite is stable up to $T = 1557$ °C, at which incongruent melting occurs, with formation of forsterite plus liquid. Incongruent melting has been observed only at low pressure, but the high-P limit of the process is not known precisely.

At low P, the $Fe_2Si_2O_6$ component is extrinsically unstable with respect to α-quartz plus fayalite paragenesis:

$$Fe_2Si_2O_6 \Leftrightarrow Fe_2SiO_4 + SiO_2 . \tag{5.72}$$

This reaction proceeds spontaneously toward the right, up to pressures that are not known precisely but that are estimated to be around 5 to 8 kbar for T in the range 200 to 400 °C. At higher P and low T, a monoclinic phase is stable (clinoferrosilite; probably in $P2_1/c$ form; CFs in figure 5.24B), and at high P and T the orthorhombic form *Pbca* is stable (OFs in the same figure). According to Lindsley (1982), below the melting point the orthorhombic phase has another transition to a pyroxenoid phase (ferrosilite III; Fs-III in figure 5.24B) or to the monoclinic $C2/c$ polymorph.

At high P, $CaMgSi_2O_6$ undergoes congruent melting (figure 5.24C). The melting temperature at $P = 1$ bar is about 1390 °C and (not shown in figure) the melting process at low P is incongruent with the formation of forsterite:

$$CaMgSi_2O_6 \Leftrightarrow Mg_2SiO_4 + melt . \tag{5.73}$$

The T stability of $CaMgSi_2O_6$ is greatly affected by fluid partial pressure. In the presence of H_2O-rich fluids, the thermobaric gradient of the melting reaction (which is positive in anhydrous conditions, as observed for most crystalline silicates; cf. chapter 6) is reversed, according to the experiments of Rosenhauer and Eggler (1975) (figure 5.24C).

The stable form of the $CaFeSi_2O_6$ compound is monoclinic $C2/c$ in a wide P-T range (hedembergite; Hd in figure 5.24D). At $P < 13$ kbar, hedembergite has a transition to the triclinic form bustamite (Bu in figure 5.24D). At low P, bustamite melts incongruently to produce silica plus melt. Nevertheless, the melting reaction is congruent at P higher than 2 to 3 kbar.

5.4.4 Thermodynamic Properties of End-Members

We have already noted the necessity in geochemistry for complete definition of the thermodynamic properties of all compositional terms required to describe the chemistry of natural pyroxenes in their various structural forms ($C2/c$; $P2_1/c$; *Pbca*; *Pbcn*). Because the pyroxene quadrilateral is usually described as a quaternary system (it is in fact a portion of the ternary system $Mg_2Si_2O_6$-$Ca_2Si_2O_6$-$Fe_2Si_2O_6$) and because four polymorphs form in the system, we should theoretically attribute 16 thermodynamic parameters for each magnitude to describe quantitatively phase stability relations at the compositional limits (plus mixing properties inside the quadrilateral; see section 5.4.5). Actually, the *Pbcn* structure is important only for the magnesian end-member protoenstatite. Moreover, $C2/c$ and $P2_1/c$ are often treated as a single polymorph, under the assumption that no apparent discontinuities are observable in the Gibbs free energy curve of the crystalline mixtures as a function of their chemical composition (see, for instance, Buseck et al., 1982). Table 5.36 lists the thermodynamic data of end-members at standard state of pure component according to Saxena (1989). These data are internally consistent and, combined with the appropriate mixing properties (see section 5.4.5), quantitatively describe the phase relations observed within the quadrilateral at the various P and T conditions of interest (the database of Saxena, 1989 is given preference here with respect to the more recent revision of Saxena et al., 1993). For comparative purposes, the same table also lists entropy, enthalpy of formation from the elements, and the heat capacity function adopted by Berman (1988) and by Holland and Powell (1990).

Table 5.37 lists the various terms of the lattice energy of some clinopyroxene components. Data refer to the $C2/c$ spatial group. Components $Mg_2Si_2O_6$ and $Fe_2Si_2O_6$, which crystallize in $P2_1/c$, must thus be considered "fictitious."

5.4.5 Mixing Properties and Subsolidus Phase Relations in the Pyroxene Quadrilateral $Mg_2Si_2O_6$-$CaMgSi_2O_6$-$CaFeSi_2O_6$-$Fe_2Si_2O_6$

During crystallization of basaltic magmas, orthopyroxene ($Pbca$), pigeonite ($P2_1/c$), and/or augite ($C2/c$) form. Figure 5.25A reports the solidus relations drawn by Huebner (1982) on the basis of existing petrologic data. As outlined by that author, figure 5.25A cannot actually be considered as a phase diagram, because the P and T values are not specified and the natural systems in which pyroxene phases form by crystallization are chemically heterogeneous. With these provisions, figure 5.25A furnishes a realistic picture of the solidus stability relations of the various pyroxene polymorphs. There is a miscibility gap between orthopyroxene and augite, extending from the iron-free join to an Fe/(Fe + Mg) molar ratio of 0.45. For more ferroan compositions, augite coexists stably with pigeonite and the miscibility gap shrinks progressively with increasing iron content in the system. Three pyroxene phases coexist stably in the transition zone between the two miscibility gaps (i.e., 0.40 < Fe/(Fe + Mg) < 0.60). This three-phase field is larger for pyroxenes of stratified complexes, as shown in figure 5.25B [0.25 < Fe/(Fe + Mg) < 0.70]. The coexistence of three pyroxenes in magmatic systems at solidus conditions is imputed to the heterogeneous reaction (Ross and Huebner, 1975):

$$\begin{array}{cccc} \text{orthopyroxene} + \text{liquid} & \Leftrightarrow & \text{pigeonite} + \text{augite}. \\ Pbca & & P2_1/c & C2/c \end{array} \quad (5.74)$$

In this interpretation, the persistence of the orthorhombic polymorph must be imputed to metastability associated to low reaction kinetics. At subsolidus temperatures, characteristic of most metamorphic regimes (including the granu-

Table 5.36 Thermodynamic properties of pure pyroxene components in their various structural forms according to Saxena (1989) (1), Berman (1988) (2), and Holland and Powell (1990) (3) database. $S^0_{T_r,P_r}$ = standard state entropy of pure component at T_r = 298.15 K and P_r = 1 bar (J/mole); H^0_{f,T_r,P_r} = enthalpy of formation from elements at same standard state conditions. Isobaric heat capacity function C_P is $C_P = K_1 + K_2T + K_3T^{-2} + K_4T^2 + K_5T^{-3} + K_6T^{-1/2} + K_7T^{-1}$.

Phase	Component	Reference	H^0_{f,T_r,P_r}	$S^0_{T_r,P_r}$	K_1	$K_2 \times 10^4$	$\dfrac{K_3}{10^6}$	K_4	$\dfrac{K_5}{10^8}$	$\dfrac{K_6}{10^2}$	$\dfrac{K_7}{10^4}$
					$C2/c + P2_1/c$ CLINOPYROXENES							
Clinoenstatite	$Mg_2Si_2O_6$	(1)	−3088.0	134.5	288.90	37.64	−2.700	0	9.224	0	−3.876
		(2)	−3091.9	132.7	279.92	0	8.8004	0	10.7142	−9.94	0
		(3)	−3083.2	135.0	356.2	−29.9	−0.5969	0	0	−31.8503	0
Diopside	$CaMgSi_2O_6$	(1)	−3201.8	143.1	221.21	328	−6.586	0	0	0	0
		(2)	−3200.6	142.5	305.41	0	−7.166	0	9.2184	−16.049	0
		(3)	−3200.2	142.7	314.5	0.41	−2.7459	0	0	−20.201	0
Ca-Tschermak	$CaAl_2SiO_6$	(1)	−3306.0	139.3	280.08	26.85	−5.344	0	9.611	0	−2.711
		(2)	−3298.8	140.8	310.7	0	−0.4827	0	1.666	−16.577	0
		(3)	−3305.6	138.0	347.6	−69.74	−1.7816	0	0	−27.575	0
Jadeite	$NaAlSi_2O_6$	(1)	−3029.9	134.7	265.66	138.6	3.294	0	−4.93	0	−3.826
		(2)	−3025.1	108.6	311.29	0	−5.3503	0	6.6257	−20.051	0
		(3)	−3029.9	133.5	301.1	101.43	−2.2393	0	0	−20.551	0
Hedembergite	$CaFeSi_2O_6$	(1)	−2838.8	170.3	229.33	341.84	−6.28	0	0	0	0
		(3)	−2843.4	175.0	310.4	125.70	−1.8460	0	0	−20.400	0

Phase	Component	Reference	H^0_{f,T_r,P_r}	$S^0_{T_r,P_r}$	K_1	$K_2 \times 10^4$	$\dfrac{K_3}{10^6}$	K_4	$\dfrac{K_5}{10^8}$	$\dfrac{K_6}{10^2}$	$\dfrac{K_7}{10^4}$
Al-clinopyroxene	$Al_2Al_2O_6$	(1)	−3289.8	116.0	160.72	740	0	0	0	0	0
Wollastonite	$Ca_2Si_2O_6$	(1)	−3269.5	162.1	280.42	36.	262.81	0	−2.708	0	−3.816
		(2)	−3263.0	163.6	298.14	0	−7.3186	0	9.687	−13.806	0
		(3)	−3266.3	163.4	165.1	−18.41	−0.7933	0	0	−11.998	0
		Pbca ORTHOPYROXENES									
Enstatite	$Mg_2Si_2O_6$	(1)	−3092.6	132.5	288.9	37.64	−2.70	0	9.224	0	−3.876
		(2)	−3091.1	132.3	333.16	0	−4.5412	0	5.583	−2.4012	0
		(3)	−3089.4	132.5	356.2	−29.90	−0.5969	0	0	−31.853	0
Ferrosilite	$Fe_2Si_2O_6$	(1)	−2390.4	189.1	263.78	1.2428	−9.894	0	0.285	0	0.04638
		(2)	−2388.8	191.8	338.12	0	−4.1942	0	5.850	−23.86	0
		(3)	−2388.2	192.0	357.4	−27.56	−0.7111	0	0	−29.926	0
Orthodiopside	$CaMgSi_2O_6$	(1)	−3182.1	143.1	221.21	328	−6.586	0	0	0	0
		(3)	−3192.8	148.5	314.5	0.41	−2.7459	0	0	−20.201	0
Orthohedembergite	$CaFeSi_2O_6$	(1)	−2833.6	169.8	229.33	341.84	−6.28	0	0	0	0
Al-orthopyroxene	$Al_2Al_2O_6$	(1)	−3282.0	124.8	160.72	740	0	0	0	0	0
		Pbcn ORTHOPYROXENES									
Protoenstatite	$Mg_2Si_2O_6$	(1)	−3086.2	137.6	246.436	132.44	−24.06	0	39.18	0	1.1197
		(2)	−3087.9	134.9	333.16	0	−4.5412	0	5.583	−24.012	0

Table 5.37 Lattice energy terms for $C2/c$ pyroxenes. Values in kJ/mole. E_{BHF} = energy of Born-Haber-Fayans thermochemical cycle; U_L = lattice energy; E_C = coulombic energy; E_R = repulsive energy; E_{DD} = dipole-dipole interactions; E_{DQ} = dipole quadrupole interactions; H_{f,T_r,P_r}^0 = enthalpy of formation from the elements at 298.15 K, 1bar reference conditions

Phase	Component	E_{BHF}	H_{f,T_r,P_r}^0	U_L	E_C	E_{DD}	E_{DQ}	E_R
Clinoenstatite	$Mg_2Si_2O_6$	30,543.6	−3087.1[a]	33,639.1	−39,215.7	−398.0	−122.5	6097.1
			−3094.2[i]	33,637.8				6098.4
			−3088.0[b]	33,631.6				6104.6
Clinoferrosilite	$Fe_2Si_2O_6$	31,355.9	−2387.2[c]	33,743.1	−39,254.6	−590.0	−203.8	6305.4
			−2361.9[d]	33,717.8				6330.7
Diopside	$CaMgSi_2O_6$	30,119.6	−3210.8[a]	33,330.4	−38,878.6	−437.7	−132.6	6118.5
			−3205.5[ef]	33,325.1				6123.8
			−3203.2[g]	33,322.8				6126.1
			−3201.8[hi]	33,321.4				6127.5
Hedembergite	$CaFeSi_2O_6$	30,525.8	−2849.2[j]	33,375.0	38,766.0	−533.9	−174.5	6099.3
			−2838.2[gh]	33,364.6				6109.7
Ca-Tschermak	$CaAl_2SiO_6$	28,301.7	−3306.0[n]	31,604.4	−37,152.9	−464.4	−141.9	6157.8
			−3280.2[ng]	31,578.6				6183.6
Jadeite	$NaAlSi_2O_6$	31,928.5	−3030.7[c]	35,280.3	−41,115.0	−459.0	−145.5	6439.2
			−3029.9[k]	35,279.5				6440.0
			−3029.4[a]	35,279.0				6440.5
			−3023.8[i]	35,272.9				6446.1
			−3011.5[g]	35,261.1				6458.4
			−3010.8[f]	35,260.4				6459.0
Acmite	$NaFeSi_2O_6$	32,174.8	−2593.7[l]	34,768.5	−40,588.7	−468.3	−150.2	6438.6
	$CaMnSi_2O_6$	30,304.2	−2947.2[m]	33,251.4	−38,472.7	−466.5	−144.8	5832.7
Ureyite	$NaCrSi_2O_6$	32,108.1	−2900.7[d]	35,008.7	−40,728.4	−470.4	−150.2	6340.2
Ca-Ti-Tschermak	$CaTiAl_2O_6$	27,244.1	−3553.1[d]	30,797.2	−36,066.6	−490.1	−153.7	5913.1

[a]Robie et al. (1978)
[b]Saxena and Chatterjee (1986)
[c]Saxena et al. (1986)
[d]Ottonello et al. (1992)
[e]Wagman et al. (1981)

[f]Naumov et al. (1971)
[g]Helgeson et al. (1978)
[h]Saxena (1989)
[i]Berman (1988)
[j]Stull and Prophet (1971)

[k]Hemingway et al. (1981)
[l]Calculated from the enthalpy of formation from the oxides of Viellard (1982)
[m]Newton and McCready (1948)
[n]Chatterjee (1989)

lite facies), pigeonite disappears and a unique miscibility gap extends from the orthorhombic *Pbca* polymorph to the $C2/c$ monoclinic form (figure 5.25C). The gap covers a wide compositional range, from the iron-free join up to $Fe/(Fe + Mg) \cong 0.80$. For higher relative amounts of Fe, pyroxene is extrinsically unstable with respect to the olivine plus liquid paragenesis (Huebner, 1982).

In order to assess appropriately the above phase stability relations, we must rely on experimental evidence concerning the limiting binary join and, particularly, attention must focus on the $Mg_2Si_2O_6$-$CaMgSi_2O_6$ and $Fe_2Si_2O_6$-$CaFeSi_2O_6$ joins. Figure 5.26A shows phase relations in the system $Mg_2Si_2O_6$-$CaMgSi_2O_6$ at $P = 15$ kbar, based on the experiments of Lindsley and Dixon (1976) (LD in figure 5.26A) and Schweitzer (1977) (S). A wide miscibility gap is evident between

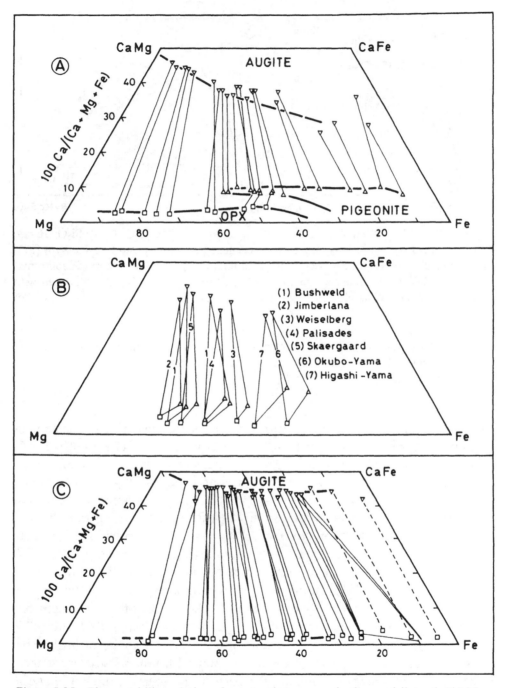

Figure 5.25 Phase stability relations for natural pyroxenes in the quadrilateral. (A) Magmatic pyroxenes (not reequilibrated at low *T*). (B) Three-phase region for pyroxenes of stratified complexes and magmatic series. (C) Pyroxenes reequilibrated in subsolidus conditions. □ = *Pbca* orthopyroxene; ∇ = augite (*C2/c*); Δ = pigeonite (*P2₁/c*); ◊ = olivine. From Huebner (1982). Reprinted with permission of The Mineralogical Society of America.

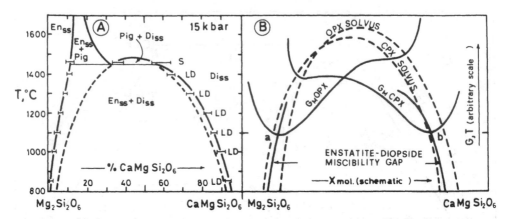

Figure 5.26 (A) Phase stability relations along binary join $Mg_2Si_2O_6$-$CaMgSi_2O_6$ at $P =$ 15 kbar (anhydrous conditions) based on experiments of Lindsley and Dixon (1976) (LD) and Schweitzer (1977) (S). En = enstatite solid mixture (*Pbca*); Di = diopside solid mixture (*C2/c*); Pig = pigeonite solid mixture (*P2₁/c*). From Lindsley (1982). Reprinted with permission of The Mineralogical Society of America. (B) Schematic representation of Gibbs free energy properties responsible for opening of a miscibility gap between orthorhombic (opx) and monoclinic (cpx) forms (Saxena, 1973). Coherence is assumed between *P2₁/c* and *C2/c* polymorphs, which are thus represented by a single Gibbs free energy curve. Note distinction between "solvus field" (intrinsic instability) and "miscibility gap" (extrinsic instability).

the orthorhombic (opx) and monoclinic (cpx) polymorphs. This gap is complicated by the appearance at $T = 1450$ °C of pigeonite. At T higher than 1450 °C, two distinct miscibility gaps occur (*Pbca-P2₁/c* and *Pbca-C2/c*). With increasing P, the gap widens and the stability field of pigeonite progressively disappears ($P =$ 25 kbar). As shown in part B of figure 5.26, the opx-cpx gap may be conceived as the result of the combined intrinsic instabilities of the two polymorphs: i.e., the "miscibility gap" between cpx and opx covers the two "solvi" valid, respectively, for the opx and cpx phases. A third phase (pigeonite) appears at the T-X zone of intersection of the two solvus limbs on the Ca-poor side of the binary join. Lindsley et al. (1981), on the basis of all existing experimental data on the $Mg_2Si_2O_6$-$CaMgSi_2O_6$ join, proposed a model valid for the pressure range 1 bar $< P < 40$ kbar. This model does not distinguish spatial groups $P2₁/c$ and $C2/c$ and treats the equilibria as stability relations between a generic orthorhombic phase and a generic monoclinic phase. In this model, the Gibbs free energy curves for the monoclinic mixture thus imply the existence of a certain amount of *elastic strain energy* (to maintain coherence) the nature of which was described in section 3.13.

To solve the energy model, Lindsley et al. (1981) based their calculations on the principle of equality of chemical potentials of components in mixture for phases at equilibrium (cf. section 2.3):

$$\mu_{Mg_2Si_2O_6,cpx} = \mu_{Mg_2Si_2O_6,opx} \tag{5.75}$$

$$\mu_{CaMgSi_2O_6,cpx} = \mu_{CaMgSi_2O_6,opx} \tag{5.76}$$

applied at every point in the P-T-X space of interest. Application of the usual relations between chemical potential and thermodynamic activity to identities 5.75 and 5.76 gives

$$\mu_{Mg_2Si_2O_6,cpx} = \mu^0_{Mg_2Si_2O_6,cpx} + RT \ln a_{Mg_2Si_2O_6,cpx} \tag{5.77}$$

$$\mu_{Mg_2Si_2O_6,opx} = \mu^0_{Mg_2Si_2O_6,opx} + RT \ln a_{Mg_2Si_2O_6,opx} \tag{5.78}$$

$$\mu_{CaMgSi_2O_6,cpx} = \mu^0_{CaMgSi_2O_6,cpx} + RT \ln a_{CaMgSi_2O_6,cpx} \tag{5.79}$$

$$\mu_{CaMgSi_2O_6,opx} = \mu^0_{CaMgSi_2O_6,opx} + RT \ln a_{CaMgSi_2O_6,opx} \tag{5.80}$$

$$\Delta\mu^0_{Mg_2Si_2O_6} = RT \ln \frac{a_{Mg_2Si_2O_6,opx}}{a_{Mg_2Si_2O_6,cpx}} \tag{5.81}$$

$$\Delta\mu^0_{CaMgSi_2O_6} = RT \ln \frac{a_{CaMgSi_2O_6,opx}}{a_{CaMgSi_2O_6,cpx}} . \tag{5.82}$$

Adopting for both components the standard state of pure component at the T_r and P_r of reference, as for a pure component:

$$\mu^0 \equiv \overline{G}_{T_r,P_r} , \tag{5.83}$$

Table 5.38 Model of Lindsley et al. (1981) for binary system $Mg_2Si_2O_6$-$CaMgSi_2O_6$.

Reaction (*a*) $\quad Mg_2Si_2O_6$ (opx) $\Leftrightarrow Mg_2Si_2O_6$ (cpx)	
Reaction (*b*) $\quad CaMgSi_2O_6$ (opx) $\Leftrightarrow CaMgSi_2O_6$ (cpx)	
$\Delta H_{a,T_r,P_r} = 3.561$ kJ/mole	Mixing model:
$\Delta S_{a,T_r,P_r} = 1.91$ J/(mole $\times$ K) $\qquad$ opx:	$\begin{cases} W = 25 \text{ kJ/mole} \\ \text{(regular mixture)} \end{cases}$
$\Delta V_{a,T_r,P_r} = 0.0355$ J/(bar $\times$ mole)	
$\Delta H_{b,T_r,P_r} = -21.178$ kJ/mole $\qquad$ cpx:	$\begin{cases} W_{H,12} = 25.484 \text{ kJ/mole} \\ W_{H,21} = 31.216 \text{ kJ/mole} \\ W_{V,21} = 0.0812 \text{ J/bar} \times \text{mole} \\ W_{V,12} = -0.006 \text{ J/bar} \times \text{mole} \\ W_{12} = W_{H,12} + PW_{V,12} \end{cases}$
$\Delta S_{b,T_r,P_r} = -8.16$ J/(mole $\times$ K)	
$\Delta V_{b,T_r,P_r} = -0.0908$ J/(bar $\times$ mole)	with $1 = Mg_2Si_2O_6$, $2 = CaMgSi_2O_6$

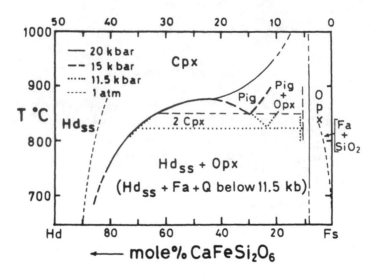

Figure 5.27 Phase stability relations along binary join CaFeSi$_2$O$_6$-Fe$_2$Si$_2$O$_6$ under various P conditions. From Lindsley (1982). Reprinted with permission of The Mineralogical Society of America.

Table 5.39 Model of Lindsley (1981) for binary system Fe$_2$Si$_2$O$_6$-CaFeSi$_2$O$_6$.

Reaction (a) Fe$_2$Si$_2$O$_6$ (opx) $\Leftrightarrow$ Fe$_2$Si$_2$O$_6$ (cpx)	
Reaction (b) CaFeSi$_2$O$_6$ (opx) $\Leftrightarrow$ CaFeSi$_2$O$_6$ (cpx)	

$\Delta H_{a,T_r,P_r} = 1.843$ kJ/mole   Mixing model:

$\Delta S_{a,T_r,P_r} = 1.68$ J/mole × K   opx: $\begin{cases} W = 15 \text{ kJ/mole} \\ \text{(regular mixture)} \end{cases}$

$\Delta V_{a,T_r,P_r} = 0.0369$ J/(bar × mole)

$\Delta H_{b,T_r,P_r} = -5.261$ kJ/mole   cpx: $\begin{cases} W_{H,12} = 16.941 \text{ kJ/mole} \\ W_{H,21} = 20.697 \text{ kJ/mole} \\ W_{V,12} = 0.0059 \text{ J/bar × mole} \\ W_{V,21} = -0.00235 \text{ J/bar × mole} \end{cases}$

$\Delta S_{b,T_r,P_r} = 0.53$ J/(mole × K) $W_{12} = W_{H,12} + PW_{V,12}$

$\Delta V_{b,T_r,P_r} = -0.0951$ J/(bar × mole)   with 1 = Fe$_2$Si$_2$O$_6$, 2 = CaFeSi$_2$O$_6$

we also have

$$\Delta\mu^0 \equiv \Delta\overline{G}_{T_r,P_r}. \tag{5.84}$$

The $\Delta\overline{G}_{T_r,P_r}$ of reaction between pure components may be split into enthalpic, entropic, and volumetric parts:

$$\Delta\overline{G}_{T_r,P_r} = \Delta\overline{H}_{T_r,P_r} - T_r\Delta\overline{S}_{T_r,P_r} + P_r\Delta\overline{V}_{T_r,P_r} \tag{5.85}$$

The parameters of the thermodynamic model of Lindsley et al. (1981) are reported in table 5.38. The mixing properties described are those of a *regular mixture* for orthopyroxene and of a *subregular mixture* (asymmetric with Mar-

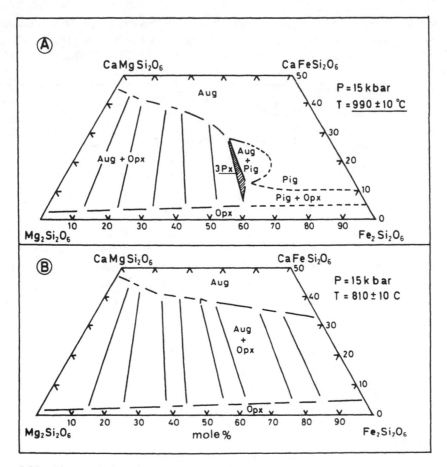

Figure 5.28 Phase relations in pyroxene quadrilateral for $P = 15$ kbar at two subsolidus temperatures. From Lindsley (1982). Reprinted with permission of The Mineralogical Society of America.

gules notation) for clinopyroxene. Moreover, for the monoclinic phase, the interaction is split into enthalpic and volumetric terms, with no excess entropic term of mixing. (In the interpretation of Lindsley et al. (1981), because Ca never occurs in $M1$ sites, the configurational entropy is that of ideal mixing of Ca and Mg in $M2$ sites, with no excess terms.)

Figure 5.27 shows phase stability limits in the (pseudo)binary system $CaFeSi_2O_6$-$Fe_2Si_2O_6$ in various P conditions, as reported by Lindsley (1982). At high P (i.e., above 11–12 kbar), the stability relations are similar to those previously seen for the $CaMgSi_2O_6$-$Mg_2Si_2O_6$ join: there is a miscibility gap between monoclinic (cpx) and orthorhombic (opx) phases. Moreover, within a limited P range, pigeonite forms. In the iron-rich side of the join, the stability diagram shows additional complexities resulting from the *extrinsic* instability of ferrosilite with respect to the fayalite plus liquid paragenesis (see also figure 5.24D).

For the binary join $CaFeSi_2O_6$-$Fe_2Si_2O_6$, Lindsley (1981) proposed a thermo-

Table 5.40 Binary interaction parameters for pyroxenes. Parameters refer to an ionic mixing model in which n is the number of sites over which permutability is calculated—i.e., $G_{mixing} = nRT (X_1 \ln X_1 + X_2 \ln X_2)$. Data in J/mole ($H$), J/(mole $\times$ K) (S), and J/(bar $\times$ mole) (V), respectively.

| Mixture | | Margules Paramters | | | | |
		H	S	V	n	Reference
CLINOPYROXENES ($C2/c$; $P2_1/c$)						
$Mg_2Si_2O_6$-$Fe_2Si_2O_6$	$W_{12} =$	−960.5	0.041	0.0557	2	(1)
	$W_{21} =$	1190.1	−0.081	−0.0180		
	$W_{12} =$	−1558.6	24.389	0.0282	1	(1)
	$W_{21} =$	2518.6	24.651	0.0226		
$Mg_2Si_2O_6$-$CaMgSi_2O_6$	$W_{12} =$	25,484	0.0	0.0812	1	(2)
	$W_{21} =$	31,216	0.0	−0.0061		
$Mg_2Si_2O_6$-$CaFeSi_2O_6$	$W_{12} =$	52,971	15.987	0.0416	2	(1)
	$W_{21} =$	29,085	3.538	0.0294		
	$W_{12} =$	93,300	45.0	0.0	1	(3)
	$W_{21} =$	20,000	−28.0	0.0		
$Fe_2Si_2O_6$-$CaMgSi_2O_6$	$W_{12} =$	46,604	14.996	0.019	2	(1)
	$W_{21} =$	22,590	3.940	0.072		
	$W_{12} =$	15,000	0.0	0.0	1	(3)
	$W_{21} =$	24,000	0.0	0.0		
$Fe_2Si_2O_6$-$CaFeSi_2O_6$	$W_{12} =$	16,941	0.0	0.0059	1	(4)
	$W_{21} =$	20,697	0.0	−0.023		
$CaMgSi_2O_6$-$CaFeSi_2O_6$	$W_{12} =$	13,984	−0.001	0.0068	1	(1)
	$W_{21} =$	17,958	0.004	−0.0084		
	$W_{12} =$	12,000	0.0	0.0	1	(3)
	$W_{21} =$	12,000	0.0	0.0		
$Mg_2Si_2O_6$-$NaAlSi_2O_6$	$W_{12} =$	30,000	0.0	0.0	1	(5)
	$W_{21} =$	30,000	0.0	0.0		
$CaFeSi_2O_6$-$NaAlSi_2O_6$	$W_{12} =$	25,200	17.05	0.0	1	(5)
	$W_{21} =$	0.0	1.85	0.0	1	(5)
$CaMgSi_2O_6$-$CaAl_2SiO_6$	$W_{12} =$	30,918	18.78	0.146	1	(6)
	$W_{21} =$	35,060	−44.72	0.469		

| Mixture | | Redlich-Kister Parameters | | | | |
		H	S	V	n	Reference
CLINOPYROXENES ($C2/c$; $P2_1/c$)						
$CaMgSi_2O_6$-$NaAlSi_2O_6$	$A_0 =$	12,600	9.45	0.0	1	(7)
	$A_1 =$	12,600	7.60	0.0		
	$A_2 =$	−21,400	−16.2	0.0		
	$A_0 =$	19,800	−4.4	0.0	1	(8)
	$A_1 =$	9600	9.1	0.0		
	$A_2 =$	−8200	−16.4	0.0		

Table 5.40 *(continued)*

Mixture		Redlich-Kister Parameters				
		H	S	V	n	Reference
NaAlSi$_2$O$_6$-CaAl$_2$SiO$_6$	$A_0 =$	7000	13.0	0.0	1	(8)
	$A_1 =$	9300	0.0	0.0		
	$A_2 =$	−1100	0.0	0.0		

Ideal Mixtures

Mixture	Reference
NaAlSi$_2$O$_6$-NaFeSi$_2$O$_6$	(9)
CaMgSi$_2$O$_6$-NaFeSi$_2$O$_6$	(10)
Mg$_2$Si$_2$O$_6$-Fe$_2$Si$_2$O$_6$	(3)

Mixture		Margules Parameters				
		H	S	V	n	Reference
ORTHOPYROXENES (*Pbca*)						
Mg$_2$Si$_2$O$_6$-Fe$_2$Si$_2$O$_6$	$W_{12} =$	13,100	15.0	0.0	1	(1) $T < 873$ K
	$W_{21} =$	3370	5.0	0.0		
Mg$_2$Si$_2$O$_6$-CaMgSi$_2$O$_6$	$W_{12} =$	25,000	0.0	0.0	1	(2)
	$W_{21} =$	25,000	0.0	0.0		
Mg$_2$Si$_2$O$_6$-CaFeSi$_2$O$_6$	$W_{12} =$	60,000	0.0	0.0	1	(3)
	$W_{21} =$	30,000	0.0	0.0		
Fe$_2$Si$_2$O$_6$-CaMgSi$_2$O$_6$	$W_{12} =$	16,000	0.0	0.0	1	(3)
	$W_{21} =$	20,000	0.0	0.0		
Fe$_2$Si$_2$O$_6$-CaFeSi$_2$O$_6$	$W_{12} =$	30,000	0.0	0.0	1	(4)
	$W_{21} =$	33,000	0.0	0.0		
CaMgSi$_2$O$_6$-CaFe$_2$Si$_2$O$_6$	$W_{12} =$	5000	0.0	0.0	1	(3)
	$W_{21} =$	7000	0.0	0.0		
Ideal Mixtures						
Mg$_2$Si$_2$O$_6$-Fe$_2$Si$_2$O$_6$		$T > 873$ K				(3)

References:

(1) Ottonello et al. (1992); based on interionic potential calculations

(2) Lindsley et al. (1981); experimental

(3) Saxena et al. (1986); based on phase equilibria

(4) Lindsley et al. (1981); experimental

(5) Chatterjee (1989); based on phase equilibria

(6) Gasparik and Lindsley (1980); estimate on the basis of various experiments

(7) Gasparik (1985); estimate on the basis of various experiments

(8) Cohen (1986); estimate on the basis of various experiments

(9) Ganguly and Saxena (1987); estimate on the basis of various experiments

(10) Ganguly (1973); based on crystal-chemical arguments

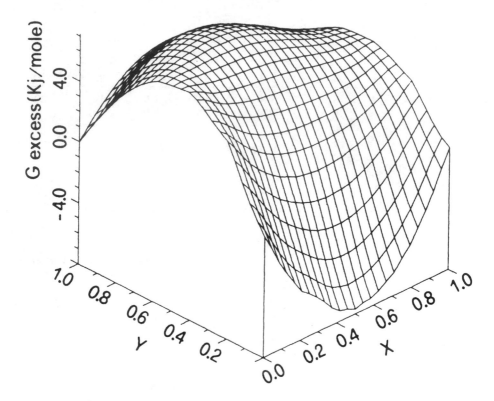

Figure 5.29 Excess Gibbs free energy surface at $T = 1000$ °C and $P = 1$ bar obtained with Wohl model parameters of table 5.41 and consistent with results of static interionic potential calculations in $C2/c$ pyroxene quadrilateral. From G. Ottonello, *Contributions to Mineralogy and Petrology,* Interactions and mixing properties in the (C2/c) clinopyroxene quadrilateral, 111, 53–60, figure 6, copyright © 1992 by Springer Verlag. Reprinted with permission of Springer-Verlag GmbH & Co. KG.

dynamic model similar to that of Lindsley et al. (1981) for the $CaMgSi_2O_6$-$Mg_2Si_2O_6$ join. The model parameters are listed in table 5.39.

Figure 5.28 shows stability relations in the pyroxene quadrilateral, based on experimental evidence, at $P = 15$ kbar. The miscibility gap contours depend markedly on the mixing properties observed along the binary joins $CaMgSi_2O_6$-$Mg_2Si_2O_6$ and $CaFeSi_2O_6$-$Fe_2Si_2O_6$ (cf. figures 5.26 and 5.27). Figure 5.28 confirms petrological observations on natural pyroxenes (figure 5.25)—i.e., at high T the miscibility gap between monoclinic ($C2/c$) and orthorhombic ($Pbca$) phases is small, with the formation, at intermediate compositions, of pigeonite ($P2_1/c$), which in turn causes the formation of two separate miscibility gaps ($C2/c$-$P2_1/c$ and $P2_1/c$-$Pbca$); at low T, pigeonite disappears and the single miscibility gap $Pbca$-$C2/c$ expands throughout the compositional field.

Table 5.40 lists some of the binary interaction parameters that may be used to describe the mixing properties of monoclinic and orthorhombic pyroxenes. Readers are referred to sections 3.8 and 3.9 for the meanings of these parameters. Some of the interaction parameters in table 5.40 are deduced from experimental observations; part are assumed or calculated by interionic potential models (Ottonello et al., 1992) and still require confirmation.

We have seen from the comparison of figures 5.25 and 5.28 that the properties of the $MgSi_2O_6$-$CaMgSi_2O_6$-$CaFeSi_2O_6$-$Fe_2Si_2O_6$ quadrilateral in synthetic systems can be extended to natural pyroxenes as well. It may thus be deduced that the effect on the phase energetics of minor components in natural mixtures, such as $NaAlSi_2O_6$, $Al_2Al_2O_6$, $CaAl_2SiO_6$, $NaFeSi_2O_6$, $NaCrSi_2O_6$, and $LiAlSi_2O_6$, are limited. However, some of these components may predominate in peculiar parageneses (see table 5.31). In this case, their interaction properties must be carefully evaluated in order to define the energy of the mixture. A more extensive discussion on the binary interaction properties of components outside the pyroxene quadrilateral may be found in Ganguly and Saxena (1987).

Concerning multicomponent interactions in the pyroxene quadrilateral, figure 5.29 shows the excess Gibbs free energy surface generated at $T = 1000$ °C and $P = 1$ bar by a Wohl model simulation of calculated interaction energies in $C2/c$ pyroxenes, on the basis of interionic static potentials (Ottonello, 1992). Excess energy terms are slightly asymmetric in the compositional field and reach a maximum at Fe/(Fe + Mg) $\approx$ 0.4 and Ca/(Fe + Mg) $\approx$ 0.6.

The adopted Wohl model formulation is

$$
\begin{aligned}
G_{\text{excess mixing}} = &\; X_1 X_2 \left(W_{12} X_2 + W_{21} X_1 \right) + X_1 X_3 \left(W_{13} X_3 + W_{31} X_1 \right) \\
&+ X_1 X_4 \left(W_{14} X_4 + W_{41} X_1 \right) + X_2 X_4 \left(W_{24} X_4 + W_{42} X_2 \right) \\
&+ X_3 X_4 \left(W_{34} X_4 + W_{43} X_3 \right) \\
&+ X_1 X_2 X_3 \left[\frac{1}{2} \left(W_{12} + W_{21} + W_{13} + W_{31} + W_{23} + W_{32} \right) - W_{123} \right] \\
&+ X_1 X_2 X_4 \left[\frac{1}{2} \left(W_{12} + W_{21} + W_{14} + W_{41} + W_{24} + W_{42} \right) - W_{124} \right] \\
&+ X_1 X_3 X_4 \left[\frac{1}{2} \left(W_{13} + W_{31} + W_{14} + W_{41} + W_{34} + W_{43} \right) - W_{134} \right] \\
&+ X_2 X_3 X_4 \left[\frac{1}{2} \left(W_{23} + W_{32} + W_{24} + W_{42} + W_{34} + W_{43} \right) - W_{234} \right] \\
&+ X_1 X_2 X_3 X_4 \left[\frac{1}{2} \left(W_{12} + W_{21} + W_{13} + W_{31} + W_{14} + W_{41} + W_{23} \right. \right. \\
&\qquad\qquad\quad \left. \left. + W_{32} + W_{24} + W_{42} + W_{34} + W_{43} \right) - W_{1234} \right].
\end{aligned}
\tag{5.86}
$$

The excess terms refer to a strict macroscopic formulation in which the ideal Gibbs free energy of mixing term is given by

$$
G_{\text{ideal mixing}} = -T S_{\text{ideal mixing}} = RT \sum_i X_i \ln(X_i),
\tag{5.87}
$$

Table 5.41 Macroscopic interaction parameters for $C2/c$ quadrilateral pyroxenes. Resulting excess Gibbs free energies are in J/mole (from Ottonello, 1992).

$W_{12} = -1558.6 - 24.398T + 0.0282P$
$W_{21} = 2518.64 - 24.651T + 0.0226P$
$W_{13} = 23,663.1 - 0.016T + 0.0208P$
$W_{31} = 32,467.3 + 0.007T + 0.0066P$
$W_{14} = 50,774.2 - 38.907T + 0.0771P$
$W_{41} = 27,841.2 - 24.609T + 0.0510P$
$W_{23} = 49,128.0 - 41.017T - 0.0040P$
$W_{32} = 17,326.7 - 23.620T + 0.1375P$
$W_{34} = 13,984.2 + 0.001T + 0.0068P$
$W_{43} = 17,958.0 - 0.004T - 0.0084P$
$W_{24} = 16,109.3 - 0.033T + 0.0453P$
$W_{42} = 20,099.8 + 0.034T + 0.0530P$
$W_{123} = -810.1 - 26.123T - 0.0226P$
$W_{124} = -87,866.4 - 20.612T + 0.0188P$
$W_{134} = -39,017.2 - 34.475T - 0.0935P$
$W_{234} = 55,293.4 - 37.629T - 0.1014P$
$W_{1234} = 74,713.2 - 316.028T + 0.5061P$

(1) = $Mg_2Si_2O_6$; (2) = $Fe_2Si_2O_6$; (3) = $CaMgSi_2O_6$; (4) = $CaFeSi_2O_6$; T is in K, P in bar

where X_i represents molar fractions of the components in the mixture. Using a strictly macroscopic formulation leads to negative entropic interactions along binary joins in which intracrystalline disorder occurs in two separate sites (i.e., $M1$ and $M2$: cf. table 5.41). A nonnegligible quaternary interaction term results (W_{1234} in eq. 5.86), because the system is treated as quaternary whereas it is in fact ternary. The complete set of interaction parameters of the model is listed in table 5.41.

5.4.6 Exsolutive Processes

Owing to interaction energies among the various compositional terms in mixture (table 5.40), the principle of minimimization of the Gibbs free energy of the system at equilibrium does not allow the phase to remain homogeneous with decreasing temperature. Extension and completion of exsolutive processes depend on the chemistry of the (initially) homogeneous phase and on the kinetics of the cooling event. The information gained through the study of exsolutive processes is thus essentially kinetic.

In pyroxenes, exsolutive processes proceed either by nucleation and growth or by spinodal decomposition (see sections 3.11, 3.12, and 3.13). Figure 5.30B shows the spinodal field calculated by Saxena (1983) for $Ca_{0.5}Mg_{0.5}SiO_3$ (diopside) and $MgSiO_3$ (clinoenstatite) in a binary mixture, by application of the subregular Margules model of Lindsley et al. (1981):

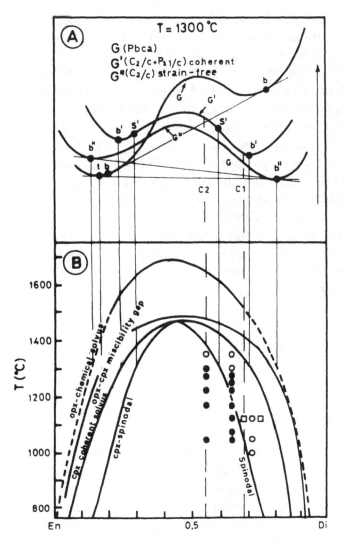

Figure 5.30 (A) Simplified Gibbs free energy curves for various polymorphs along enstatite-diopside join at $T = 1300$ °C. (B) Resulting solvus, spinodal field, and miscibility gap compared with experimental data of McCallister and Yund (1977) on pyroxene un-mixing kinetics (part B from Ganguly and Saxena (1992). Reprinted with permission of Springer-Verlag, New York).

$$W_{12} = 25,484 - 0.0812P \qquad \text{(J/mole)} \qquad (5.88)$$

$$W_{21} = 31,216 - 0.061P \qquad \text{(J/mole)} \qquad (5.89)$$

where P is expressed in bar.

The spinodal field, calculated at $P = 1$ bar on the basis of equations 3.201 and 3.202 (section 3.11) conforms quite well to the experimental observations of

McCallister and Yund (1977), obtained by heating clinopyroxenes to various T and observing the effects of thermal annealing by transmission electron microscopy (TEM). Filled symbols in figure 5.30B indicate unmixing evidences after heating for 19 to 186 hours. Open circles indicate homogeneous composition (no evidence of unmixing after prolonged heating), and open squares indicate unmixing after prolonged thermal treatment (i.e., more than 186 hours). Spinodal decomposition takes place spontaneously (19 to 186 hours, as a function of T) for initial compositions inside the calculated spinodal field, whereas metastable persistence of the homogeneous phase is observed outside the field.

To understand fully the nature of the unmixing process, we must consider the form of the Gibbs free energy curves for $T = 1300$ °C for the various polymorphs. Figure 5.30A shows the Gibbs free energy curves for the orthorhombic phase (*Pbca*; G). For the monoclinic polymorph, two separate curves are drawn: one for a monoclinic phase grouping the two classes *C2/c* and *P2$_1$/c* (G' through a *coherence energy* term (see section 3.13), and one in the absence of coherence energy (*C2/c*; G'', *strain-free*).

Let us now consider a composition C1 intermediate between binodal point b' and spinodal point s' on the G' curve in figure 5.30A. Unmixing requires marked compositional fluctuation, to overcome the critical nucleation radius of the phase (cf. section 3.12). The result is a *coherent solvus unmixing* (or, in the absence of a coherence energy term, a *chemical solvus unmixing*). If the compositional fluctuation is sufficiently large, two phases (*Pbca* and *C2/c*) form: pigeonite plus diopside in the first case and orthopyroxene plus diopside in the second. It must be noted that the *intrinsic instability* field is located by the tangent to the minima of the Gibbs free energy curve of the <u>single</u> phase *C2/c* (G'') and that the minimum Gibbs free energy is indeed achieved here by coexistence (i.e., tangency of the minima) of the two phases *Pbca* (G) and *C2/c* (G'', i.e., *extrinsic instability*), all other conditions being metastable with respect to this arrangement.

Marked compositional fluctuation requires high ionic diffusivity within the phase, and we know that elemental diffusivities decrease exponentially with decreasing T. The probability of attainment (and completion) of the process is thus greater at high T than at low T. The process is also enhanced by the presence of nucleation centers (*heterogeneous nucleation*) composed of extended defects (dislocations, grain boundaries, etc.). If the cooling rate of the system is too high with respect to ionic diffusivity, the process does not take place and the phase remains *metastably homogeneous* (as experimentally observed for composition C1; cf. figure 5.30B).

Let us now consider an initial composition C2. We are within the two spinodal points on the Gibbs free energy coherent curve G': decomposition takes place spontaneously throughout the entire phase, with modulated compositional fluctuations whose amplitudes increase as the process advances. We always obtain pigeonite plus diopside, although theoretically the orthopyroxene plus diopside paragenesis, also in this case, corresponds to the minimum energy of the system.

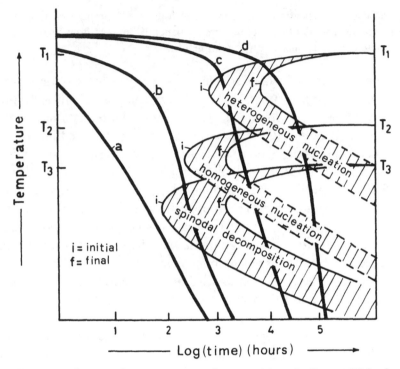

Figure 5.31 *TTT* diagram for a pyroxene of composition similar to C2 in figure 5.30. Curves *a*, *b*, *c*, and *d*: cooling rates of a lava flow at increasing depths from surface (*a*) to 1 m inside the body (*d*), based on calculations of Provost and Bottinga (1974). From Buseck et al. (1982). Reprinted with permission of The Mineralogical Society of America.

Exsolution kinetics are appropriately described with the aid of *TTT* (Time-Temperature-Transformation) diagrams, in which the cooling rates of the system cross the kinetic boundaries of the beginning and end of the process. For example, figure 5.31 is a *TTT* diagram for a pyroxene with a composition like that of C2 in figure 5.30 (Buseck et al., 1982). Curves *a*, *b*, *c*, and *d* are cooling rates calculated by Provost and Bottinga (1974) for a basaltic lava: curve *a* is the cooling rate at the surface of the lava flow, and curves *b*, *c*, and *d* are those at increasing depths beneath the surface. These curves encounter the kinetic curves of the beginning and end of heterogeneous nucleation (cpx-opx miscibility gap or chemical solvus in figure 5.30B), homogeneous nucleation (coherent solvus), and spinodal decomposition (spinodal field) at various values of time and temperature. Note that the three processes take place under different temperature conditions, times being equal.

Let us examine the significance of figure 5.31 in some detail: cooling rate *a* does not cross any transformation curve; a pyroxene of composition similar to C2 at the surface of the lava flow thus remains *metastably homogeneous* throughout the entire cooling process. Cooling rate *b* encounters the starting curve for spinodal decomposition but not the curve of the end of the process; the cooling rate of the system is still too high to allow completion of the exsolutive process. Cooling rate *c* encounters the starting curve for heterogeneous nucleation but not its end curve; the process starts but does not reach completion. At lower temperatures, cooling rate curve *c* crosses the *TT* field of homogeneous nucleation;

a second process of nucleation and growth develops until completion. Cooling rate d, which corresponds to a depth of about 1 meter beneath the lava flow surface, brings the process of heterogeneous nucleation to completion.

TEM investigation of pyroxenes identifies the type of exsolutive process as a function of phase microstructure. Spinodal decomposition initially produces characteristic "tweed microstructures" [compositional modulations oriented in a crossed fashion along the (100) and (001) planes] or modulated microstructures with compositional bands of sinusoidal form whose wavelength depends on the degree of advancement of the process [plane (001)]. The process then evolves with the formation of tiny intercalated lamellae ($\approx$ 1000 Å), internally coherent but of contrasting composition (i.e., at opposite sides of the spinodal field). Heterogeneous nucleation often develops at grain boundary limits and is characterized by incoherent interfaces between phases of different structures. Recognition of homogeneous nucleation is not so immediate.

Independent of the type of exsolutive process, pyroxene exolutions usually develop on the (001) and (100) planes, probably because coherence energy is minimal along the axis perpendicular to these planes (development of compositional fluctuations is maximum along the direction that minimizes the value of G; cf. eq. 3.212; see also Morimoto and Tokonami, 1969, for an example of calculation).

For a more detailed discussion on exsolutive processes in pyroxenes, there is an excellent synthesis in Buseck et al. (1982).

5.5 Amphiboles

Amphiboles are *dipolysilicates* (multiplicity = 2; periodicity = 3; dimensionality = 1) and have several structural analogies with pyroxenes, from which they differ primarily by virtue of a doubling of tetrahedral chains $[T_4O_{11}]_n$. Like those of pyroxenes, their cationic structural positions are characterized by the relative arrangement of the apical and basal parts of tetrahedral groups $[SiO_4]$ (compare figures 5.19 and 5.32): sites occurring between two basal parts of neighboring tetrahedral chains are called $M4$ (like $M2$ sites in pyroxenes), and sites between apical parts of neighboring chains (more restricted) are called $M1$, $M2$, and $M3$ (like $M1$ sites in pyroxenes). Sites $M1$, $M2$, and $M3$ are octahedrally coordinated, whereas sites $M4$ may be either VI-fold or VIII-fold coordinated. The double-chain arrangement leads to the formation of another nonequivalent position A, located between two rings formed by the basal parts of opposing neighboring tetrahedral chains (figure 5.32). OH^- groups or halogens (F^-, Cl^-) are also located at the centers of hexagonal rings at apical levels of tetrahedral chains.

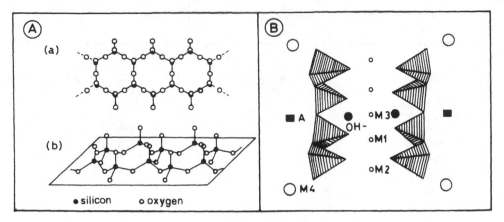

Figure 5.32 Structural features of amphiboles. (A) Double chain $[T_4O_{11}]_n$ seen along axis *c* (a) and in perspective (b). (B) Double chain seen from terminal part and various cationic positions (compare with figure 5.19, for analogies with the pyroxene structure).

5.5.1 Chemistry and Occurrence

The general formula for amphiboles may be expressed as

$$W_{0-1}X_2Y_5Z_8\,O_{22}\left(OH, F, Cl\right)_2, \tag{5.90}$$

where *W* is in site *A*, generally occupied by alkalis (Na, K) or by cationic vacancies $V_A^!$; *X* is in site *M4*, generally occupied by univalent and divalent cations; *Y* groups the *M1*, *M2*, and *M3* positions; and *Z* groups two nonequivalent tetrahedral positions (*T1* and *T2*). Table 5.42 lists the cationic occupancies of major elements on the various nonequivalent sites of the structure.

Concerning minor and trace elements, in joesmithite Be^{2+} is found in site *T1*. Moreover, Ge-amphiboles have been synthesized in which Ge^{4+} completely replaces Si^{4+} in sites *T1* and *T2* (Grebenshchikov et al., 1974). Transition elements Co, Ni, and Cr^{3+} appear to occupy sites *M1*, *M2*, *M3*, and *M4*, whereas Zn is stabilized essentially in sites *M1*, *M2*, and *M3* (Klein and Ito, 1968).

The chemical classification of amphiboles groups the various terms according to their chemical compositions independent of the represented structural classes.

Table 5.43 furnishes a simplified classification, with the various members composed of mixtures of Mg, Fe^{2+} and Al, Fe^{3+} terms: for instance, barroisite is a generic monoclinic (*C2/m*) mixture $CaNa(Mg, Fe)_3(Al, Fe^{3+})_2Si_7AlO_{22}(OH)_2$ composed of the end-members aluminous barroisite $[CaNaMg_3Al_2Si_7Al\,O_{22}(OH)_2]$, ferrous-aluminous barroisite $[CaNaFe_3Al_2Si_7AlO_{22}(OH)_2]$, ferric barroisite $[CaNaMg_3Fe_2^{3+}Si_7AlO_{22}(OH)_2]$ and ferrous-ferric barroisite $[CaNaFe_3Fe_2^{3+}Si_7AlO_{22}(OH)_2]$.

The structure of amphibole admits a large number of isomorphous substitutions, and the group shows a marked heterogeneity in natural specimens. Table

Table 5.42 Cationic occupancies in amphiboles. (?) = dubious assignment

Cation	W	X	Y			Z	
	A	M4	M1	M2	M3	T1	T2
Si						×	×
Al			×	×	×	×	×
Fe^{3+}			×	×	×	×(?)	×(?)
Ti			×	×	×	×(?)	×(?)
Fe^{2+}		×	×	×	×		
Mn		×	×	×	×		
Mg		×	×	×	×		
Ca		×					
Li	×	×	×	×	×		
Na	×	×					
K	×						

Table 5.43 Amphibole terminology and chemical classification. After Hawthorne (1981a), with modifications. Note that X and Y here do not correspond to identification 5.90.

Name	Stoichiometry	Structural Class
	Fe-Mg-Mn AMPHIBOLES	
Anthophyllite	$Na_X(Mg,Mn,Fe)_{7-Y}Al_Y(Al_{X+Y},Si_{8-X-Y})O_{22}(OH,F,Cl)_2$ X + Y < 1.00	Pnma
Gedrite	$Na_X(Mg,Mn,Fe)_{7-Y}Al_Y(Al_{X+Y},Si_{8-X-Y})O_{22}(OH,F,Cl)_2$ X + Y ≥ 1.00	Pnma
Holmquistite	$Li_2(Mg,Mn,Fe)_3(Fe^{3+},Al)_2Si_8O_{22}(OH,F,Cl)_2$	Pnma; C2/m
Cummingtonite	$(Mg,Mn,Fe)_7Si_8O_{22}(OH,F,Cl)_2$	$P2_1/m$; C2/m
	CALCIC AMPHIBOLES	
Actinolite	$Ca_2(Mg,Fe)_5Si_8O_{22}(OH)_2$	C2/m
Edenite	$NaCa_2(Mg,Fe)_5Si_7AlO_{22}(OH)_2$	C2/m
Pargasite	$NaCa_2(Mg,Fe)_4AlSi_6Al_2O_{22}(OH)_2$	C2/m
Hastingsite	$NaCa_2(Mg,Fe)_4Fe^{3+}Si_6Al_2O_{22}(OH)_2$	C2/m
Tschermakite	$Ca_2(Mg,Fe)_3(Al,Fe^{3+})_2Si_6Al_2O_{22}(OH)_2$	C2/m
Aluminous hornblende	$Ca_2(Mg,Fe)_4AlSi_7AlO_{22}(OH)_2$	C2/m
Kaersutite	$NaCa_2(Mg,Fe)_4TiSi_6Al_2(O,OH)_{24}$	C2/m
	SODIC-CALCIC AMPHIBOLES	
Richterite	$NaCaNa(Mg,Fe)_5Si_8O_{22}(OH)_2$	C2/m
Winchite	$CaNa(Mg,Fe)_4(Al,Fe^{3+})Si_8O_{22}(OH)_2$	C2/m
Barroisite	$CaNa(Mg,Fe)_3(Al,Fe^{3+})_2Si_7AlO_{22}(OH)_2$	C2/m
Katophorite	$NaCaNa(Mg,Fe)_4(Al,Fe^{3+})Si_7AlO_{22}(OH)_2$	C2/m
Taramite	$NaCaNa(Mg,Fe)_3(Al,Fe^{3+})_2Si_6Al_2O_{22}(OH)_2$	C2/m
	ALKALINE AMPHIBOLES	
Glaucophane	$Na_2(Mg,Fe)_3Al_2Si_8O_{22}(OH)_2$	C2/m
Riebeckite	$Na_2(Mg,Fe)_3Fe_2^{3+}Si_8O_{22}(OH)_2$	C2/m
Eckermannite	$NaNa_2(Mg,Fe)_4AlSi_8O_{22}(OH)_2$	C2/m
Arfvedsonite	$NaNa_2(Mg,Fe)_4Fe^{3+}Si_8O_{22}(OH)_2$	C2/m
Kozulite	$NaNa_2Mn_4(Al,Fe^{3+})Si_8O_{22}(OH)_2$	C2/m

Table 5.44 Compositions (in weight %) of some natural amphiboles (from Deer et al., 1983): (1) = anthophyllite from a serpentinite; (2) gedrite from a gedrite-kyanite-garnet paragenesis; (3) cummingtonite from a oligoclase-biotite schist; (4) common hornblende from a tonalite; (5) pargasite from a metamorphic limestone; (6) basaltic hornblende from a latite; (7) glaucophane from a glaucophane schist.

Oxide	Samples						
	(1)	(2)	(3)	(4)	(5)	(6)	(7)
SiO_2	58.48	44.89	51.53	44.99	48.10	45.17	57.73
TiO_2	0.03	0.67	0.31	1.46	0.10	2.11	-
Al_2O_3	0.57	17.91	5.02	11.21	11.05	7.68	12.04
Fe_2O_3	0.58	0.67	0.82	3.33	0.67	14.30	1.16
FeO	7.85	13.31	16.91	13.17	1.65	2.81	5.41
MnO	0.27	0.37	0.22	0.31	-	0.41	-
MgO	29.25	18.09	20.84	10.41	20.60	13.44	13.02
CaO	0.14	0.40	1.34	12.11	12.50	11.18	1.04
Na_2O	0.08	1.45	0.65	0.97	2.54	1.35	6.98
K_2O	0.02	0.05	0.00	0.76	1.24	1.09	0.68
$H_2O^{(+)}$	2.60	2.02	2.15	1.48	0.71	0.19	2.27
$H_2O^{(-)}$	0.20	0.00	0.64	0.04	0.11	0.06	-
F^-	-	-	-	-	1.90	0.35	-
	100.20	99.87	100.43	100.41	101.17	100.14	100.33
$O^{2-} + F^- + Cl^-$	-	-	-	-	0.80	0.14	-
Total	100.20	99.87	100.43	100.41	100.37	100.00	100.33
Ionic fractions on the basis of 24 ($O^= + OH^- + F^-$)							
Si	7.885	6.325	7.364	6.669	6.760	6.728	7.789
Al	0.090	1.675	0.636	1.331	1.240	1.272	0.211
Total T	7.98	8.00	8.00	8.00	8.00	8.00	8.00
Al	-	1.301	0.209	0.629	0.592	0.076	-
Ti	0.003	0.071	0.033	0.163	0.011	0.236	0.118
Fe^{3+}	0.058	0.070	0.087	0.370	0.071	1.602	2.618
Mg	5.876	3.799	4.438	2.300	4.315	2.984	0.611
Fe^{2+}	0.885	1.569	2.022	1.633	0.194	0.350	-
Mn	0.031	0.044	0.027	0.039	-	0.051	-
Na	0.020	0.396	0.180	0.278	0.693	0.390	1.826
Ca	0.020	0.060	0.205	1.923	1.882	1.784	0.150
K	0.004	0.008	-	0.144	0.221	0.208	0.117
Total $X + Y + W$	6.90	7.32	7.20	7.48	7.98	7.68	7.14
OH^-	2.338	1.898	2.049	1.462	0.667	0.190	2.043
F^-	-	-	-	-	0.845	0.165	-

5.44 lists chemical compositions of natural amphiboles (Deer et al., 1983), together with the corresponding structural formulae on a 24-anion basis (O^{2-}, OH^-, F^-, Cl^-). Because of the complex chemistry of natural amphiboles, calculation of the structural formula from a complete set of chemical analyses should begin with an evaluation of the relative amounts of H_2O, F, and Cl. If the analysis

Table 5.45 Occurrence of amphiboles in rocks. Note that X and Y here do not correspond to site identification 5.90.

ORTHORHOMBIC AMPHIBOLES

Name: *anthophyllite*
Formula: $Na_X(Mg,Mn,Fe)_{7-Y}Al_Y(Al_{X+Y},Si_{8-X-Y})(OH,F,Cl)_2$ $X + Y < 1.00$
Occurrence: crystalline schists, peridotites, serpentinites

Name: *gedrite*
Formula: $Na_X(Mg,Mn,Fe)_{7-Y}Al_Y(Al_{X+Y},Si_{8-X-Y})(OH,F,Cl)_2$ $X + Y \geq 1.00$
Occurrence: amphibolites, eclogites, cordierite-bearing and garnet-bearing gneisses

MONOCLINIC AMPHIBOLES

Name: *cummingtonite*
Formula: $(Mg,Mn,Fe)_7Si_8O_{22}(OH,F,Cl)_2$
Occurrences: cornubianites, low-metamorphic schists, micaschists, (iron ores)

Name: *actinote* (tremolite, actinolite)
Formula: $Ca_2(Mg,Fe)_5Si_8O_{22}(OH)_2$
Occurrence: crystalline schists, calcic cornubianites, cipollines

Name: *common hornblende*
Formula: $Na_X Ca_2(Mg,Fe)_{5-X}Al_Y(Al_{X+Y},Si_{8-X-Y})O_{22}(OH,F,Cl)_2$
Occurrence: calc-alkaline granites, sienites, diorites and corresponding effusive terms, lamprophires, amphibolites, amphibole-bearing hornfels.

Name: *pargasite*
Formula: $NaCa_2(Mg,Fe)_4AlSi_6Al_2O_{22}(OH,F,Cl)_2$
Occurrence: paragneisses, methamorphosed limestones, amphibole peridotites

Name: *basaltic hornblende*
Formula: $Na_X Ca_2(Mg,Fe)_{5-X}(Al,Fe^{3+})_Y(Al_{X+Y},Si_{8-X-Y})O_{22}(OH,F,Cl)_2$
Occurrence: basic volcanites, gabbros, and ultramafic rocks

Name: *glaucophane*
Formula: $Na_2(Mg,Fe)_3Al_2Si_8O_{22}(OH,F,Cl)_2$
Occurrence: glaucophanites, eclogites, prasinites

Name: *arfvedsonite*
Formula: $NaNa_2(Mg,Fe)_4Fe^{3+}Si_8O_{22}(OH,F,Cl)_2$
Occurrence: hypersodic granites, nepheline sienites, metamorphic rocks

---

is very accurate, one can begin directly with a structural calculation based on 24 anions (O^{2-}, OH^-, F^-, Cl^-). However, if there are any doubts about the internal precision of the data set, it is preferable to assume that the bulk charges of OH^-, F^-, and Cl^- amount to 2 and that the cationic charges are balanced by these two negative charges plus 22 oxygens (O^{2-}).

According to Robinson et al. (1982), the correct procedure for assigning site occupancies is as follows.

Sites T (T1 + T2)

Add Si and Al, followed by Fe^{3+} and Ti^{4+} up to 8. If Si is higher than 8, there is an analytical problem. If Si + Al + Fe^{3+} + Ti^{4+} does not reach 8, there is again an analytical problem.

Sites Y (M1 + M2 + M3)

Add the remaining parts of Al, Ti, and Fe^{3+} with Cr^{3+}, Mg, Fe^{2+}, Zn, Mn, Ca, Li, and Na up to 5.0. If Ca and Na are necessary to reach 5.0, this may indicate an analytical problem, although the presence of Ca in limited amounts in Y sites is reasonable in some sodic amphiboles (Hawthorne, 1976).

Site X (M4)

Add the remaining parts of Mg, Fe^{2+}, Zn, Mn, Ca, Li, and Na in this precise order up to 2.0. If the sum is substantially lower than 2.0, there is an analytical problem or, alternatively, a high number of cationic vacancies on the site.

Site W

Add the remaining part of Ca with Li, Na, and K up to 1.0. If some Ca still remains after this operation, the analysis is suspect. There are also problems if the sum exceeds 1.0. Crystal-chemical data indicate that Na + K occupancies generally vary between 0 and 1, and it is precisely this variability that sometimes renders structural reconstruction problematic.

Table 5.45 lists the main occurrences of the various compositional terms. The name of the mineral reflects the dominant component in the mixture.

5.5.2 Structural Properties

The structural classes occurring in amphiboles are those of the monoclinic ($C2/m$, $P2_1/m$, $P2a$) and orthorhombic ($Pnma$, $Pnmn$) systems. The most common space groups are $C2/c$ and $P2_1/m$ (monoclinic) and $Pnma$ (orthorhombic; cf. table 5.43). Table 5.46 lists structural data for some compositional terms belonging to two of the spatial groups of interest ($C2/m$ and $Pnma$). As in pyroxenes, note doubling of cell edge a and orthogonalization of cell angle β in the transition from monoclinic to orthorhombic symmetry. Note also that densities are on average lower than those of corresponding pyroxene compounds (cf. tables 5.33 and 5.46).

Evaluation of the effects of chemistry on volume properties of amphiboles is complicated by the high number of sites in which isomorphous substitutions take place. Limiting ourselves to the main structural classes $C2/m$, $Pnma$, and $P2_1/m$, we present the following evaluations.

C2/m Amphiboles

There are three sites with pseudo-octahedral coordination ($M1$, $M2$, and $M3$) whose mean dimensions exhibit straight correlation with the ionic radii of cations occupying these lattice sites. The relationship is complicated by the fact that, for sites $M1$ and $M3$, two of the coordinating anions may be OH^-, F^-, or Cl^- instead of O^{2-}. Figure 5.33A shows the relationship between mean $\langle M\text{-}O \rangle$ distance and

Table 5.46 Structural data on amphibole end-members (from Smyth and Bish, 1988). Molar volume in cm^3/mole; molar weight in g/mole; cell edges in Å; cell volume in Å^3.

Compound		Cummingtonite	Pargasite	Tremolite
Formula	W	-	$Na_{0.63}K_{0.30}$	$Na_{0.38}K_{0.12}$
	X	$Fe_{1.7}Mg_{0.3}$	Ca_2	$Ca_{1.8}Mg_{0.2}$
	Y	$Fe_{0.6}Mg_{4.4}$	$Fe_{1.10}Mg_{3.25}Al_{0.55}$	Mg_5
	Z	Si_8	$Al_{1.86}Si_{6.14}$	$Al_{0.23}Si_{7.8}$
Molar weight		853.72	870.30	824.57
Density		3.142	3.165	3.010
Molar volume		271.68	274.94	273.93
Z		2	2	2
System		Monoclinic	Monoclinic	Monoclinic
Class		*2/m*	*2/m*	*2/m*
Group		*C2/m*	*C2/m*	*C2/m*
Cell parameters:				
a		9.51	9.910	9.863
b		18.19	18.022	18.048
c		5.33	5.312	5.285
β		101.92	105.78	104.79
Volume		902.14	912.96	909.60
Reference		Fisher (1966)	Robinson et al. (1973)	Hawthorne and Grundy (1976)

Compound		Glaucophane	Anthophyllite	Gedrite
Formula	W	$Na_{0.04}$	-	$Na_{0.05}Ca_{0.03}$
	X	$Na_{1.80}Ca_{0.20}$	$\left. \begin{matrix} X \\ Y \end{matrix} \right\} Fe_{1.47}Mg_{5.53}$	$\left. \begin{matrix} X \\ Y \end{matrix} \right\} Fe_{1.1}Mg_{4.5}Al_{1.2}$
	Y	$Fe_1Mg_{2.38}Al_{1.58}Ti_{0.06}$		
	Z	$Al_{0.08}Si_{7.92}$	Si_8	$Al_{1.8}Si_{6.3}$
Molar weight		822.03	827.23	827.42
Density		3.135	3.111	3.184
Molar volume		262.25	265.88	259.84
Z		2	4	4
System		Monoclinic	Orthorhombic	Orthorhombic
Class		*2/m*	*mmm*	*mmm*
Group		*C2/m*	*Pmna*	*Pmna*
Cell parameters:				
a		9.541	18.560	18.531
b		17.740	18.013	17.741
c		5.295	5.2818	5.249
β		103.667	90.0	90.0
Volume		870.83	1765.82	1725.65
Reference		Papike and Clark (1968)	Finger (1970)	Papike and Ross (1970)

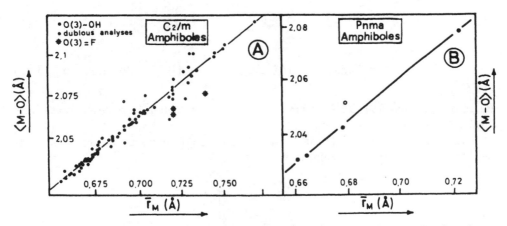

Figure 5.33 Correlations between algebraic mean of ionic sizes of intervening ions in M positions $(\bar{r}_M)$ and mean dimension of M sites ($M1$, $M2$, $M3$) in $C2/m$ (A) and $Pnma$ (B) amphiboles. From Hawthorn (1981a). Reprinted with permission of The Mineralogical Society of America.

mean ionic radius of intervening cations. This may be described by the two equations

$$\langle M - O \rangle = 1.527 + 0.764 \bar{r}_M \qquad (\text{Å}) \qquad (5.91)$$

and

$$\langle M - O \rangle = 1.017 + 0.817 \bar{r}_M + 0.354 r_{O3} \qquad (\text{Å}), \qquad (5.92)$$

where $\bar{r}_M$ is the mean radius (algebraic mean) of cations occupying sites $M1$, $M2$, and $M3$, and equation 5.92 explicitly takes into account the dimensions of coordinating O3 anions (r_{O3}).

Concerning tetrahedral sites, several authors agree that a straight correlation exists between the amount of Al in the tetrahedral site and the mean site dimension. From the equations proposed by Robinson et al. (1973) we may derive

$$\langle T1 - O \rangle = 0.0013 + 1.6184 X_{Al,T1} \qquad (\text{Å}) \qquad (5.93)$$

$$\langle T2 - O \rangle = 0.0008 + 1.6292 X_{Al,T2} \qquad (\text{Å}), \qquad (5.94)$$

where $X_{Al,T1}$ is the occupancy of Al^{3+} cations in tetrahedral site $T1$. Equations 5.93 and 5.94 have a correlation coefficient R2 = 0.98. Similar correlations may be found in Ungaretti (1980). Concerning site $M4$, Ungaretti et al. (1981) showed that $M4$-site occupancy markedly affects cell edges with linear proportionalities of the type

$$a/c = 1.8653 - 0.0258 X_{\text{Na},M4} \qquad \left(R2 = 0.994\right) \qquad (5.95)$$

$$b + c = 23.3338 - 0.1647 X_{\text{Na},M4} \qquad \left(R2 = 0.992\right). \qquad (5.96)$$

Based on equations 5.91 to 5.96, we may expect $C2/m$ amphibole mixtures to have nearly ideal behavior in regard to volume properties.

Any deviations from ideal behavior may be attributable to a high site distortion, represented in linear terms by parameter Δ:

$$\Delta = \sum_{i=1}^{n} \left[\frac{\left(\dfrac{l_i - l_m}{l_m}\right)^2}{n} \right] \times 10^4, \qquad (5.97)$$

where l_i is the individual length of a single bond in the polyhedron, l_m is the mean length of polyhedral bonds, and n is the number of bonds in the polyhedron. Site distortion may also be described in angular terms by parameter σ^2:

$$\sigma^2 = \sum_{i=1}^{n} \frac{\left(\Theta_i - \Theta_m\right)^2}{n-1}, \qquad (5.98)$$

where Θ_i is the individual bond angle and Θ_m is the *ideal* bond angle.

As table 5.47 shows, the observed distortions of the various sites are in some cases quite high and there is also considerable variability in the degrees of polyhedral distortion of the various compositional terms.

Hawthorne (1976) showed that the distortion parameter of site $M2$ is a function of the formal charge on site $M4$. Because site $M2$ shows the most severe distortion, the simple relationships connecting site $M4$ occupancy and cell edges may be complicated to some extent by the distortion induced on site $M2$.

Detailed investigations of the volumes of binary $C2/m$ mixtures with Fe-Mg, Na-K, and Fe^{3+}-Al coupled substitutions invariably show substantial ideality of mixing. The systems studied comprise:

$NaCa_2Mg_4AlSi_6Al_2O_{22}(OH)_2$-$NaCa_2Fe_4AlSi_6Al_2O_{22}(OH)_2$
(pargasite–ferro-pargasite; Charles, 1980)

$Na_2Mg_3Al_2Si_8O_{22}(OH)_2$-$Na_2Fe_3Fe_2Si_8O_{22}(OH)_2$
(glaucophane-riebeckite; Borg, 1967)

$Na_2CaMg_5Si_8O_{22}(OH,F)_2$-$KNaCaMg_5Si_8O_{22}(OH,F)_2$
(sodium–magnesio-richterite–potassium-magnesio-richterite; Huebner and Papike, 1970)

$Na_2CaMg_5Si_8O_{22}(OH,F)_2$-$Na_2CaFe_5Si_8O_{22}(OH,F)_2$
(magnesio-richterite–ferro-richterite; Charles, 1974)

$Na_2Mg_3Fe_2Si_8O_{22}(OH)_2$-$Na_2Fe_3Al_2Si_8O_{22}(OH)_2$
(magnesio-riebeckite–ferro-glaucophane; Borg, 1967)

For the first four systems, the experimental data cover the entire compositional field, whereas for magnesio-riebeckite–ferro-glaucophane, the data are limited to the field $X_{Na_2Mg_3Fe_2Si_8O_{22}(OH)_2} = 1.0 \rightarrow 0.5$. Within the studied compositional range and within experimental approximation, the volume of the mixture is ideal.

Pnma Amphiboles

In orthorhombic amphiboles, as in monoclinic $C2/m$ amphiboles, site dimensions are linearly dependent on the ionic radii of occupying cations (figure 5.33B). However, no empirical equations exist to derive M-O mean dimensions from cationic occupancy, because of the paucity of experimental information. The same is true for sites $M4$, $T1$, $T2$, and A.

Concerning cell volume, the experimental data of Popp et al. (1976) on the system $Mg_7Si_8O_{22}(OH)_2$-$Fe_7Si_8O_{22}(OH)_2$ indicate substantial ideality of mixing.

$P2_1/m$ Amphiboles

Volume properties in the system $Mg_7Si_8O_{22}(OH)_2$-$Fe_7Si_8O_{22}(OH)_2$ (cummingtonite-grunerite) show well-defined linearity in the compositional range $X_{Fe_7Si_8O_{22}(OH)_2} = 0.60 \rightarrow 1.00$ (Klein, 1964; Viswanathan and Ghose, 1965); Vegard's law probably holds over the entire compositional field.

Investigations of T effects on amphibole volume are complicated by dehydroxylation. Experimental data show that loss of OH^- groups is favored by the presence of Fe^{2+}, based on equilibrium:

$$Fe^{2+} + OH^- \Leftrightarrow Fe^{3+} + O^{2-} + \frac{1}{2}H_2 \qquad (5.99)$$

(as already suggested by Barnes, 1930).

Instead, in magnesian end-members, the configuration (MgMgMg)OH remains stable up to high temperatures (Ernst and Wai, 1970). Detailed studies on tremolite and other chemically similar amphiboles (Cameron et al., 1973a, b) show maximum expansion along cell edge a and minimum expansion on cell edge c (in this respect, amphiboles differ from pyroxenes, which show maximum expansion along edge b; cf. Sueno et al., 1973).

High-T structural refinements indicate that mean polyhedral expansions

Table 5.47 Polyhedral distortion parameters for *C2/m* amphiboles
(from Hawthorne, 1981a).

Phase	T1		T2		M1	
	Δ	σ^2	Δ	σ^2	Δ	σ^2
Grunerite	0.29	0.8	0.78	15.8	2.25	36.0
Glaucophane	0.05	0.6	1.88	17.4	0.21	69.5
Tremolite	0.54	5.0	4.14	19.9	0.15	35.6
Fluor-richterite	2.06	14.3	5.60	22.4	0.62	43.1
Fluor-tremolite	0.38	6.2	3.45	19.3	0.01	46.9
Potassic pargasite	0.59	6.6	1.42	19.1	1.25	50.5
Ferro-tschermakite	0.63	4.8	0.61	17.0	4.00	57.3
Potassic ossi-kaersutite	0.29	5.8	1.16	18.9	14.35	47.9
Potassic arfvedsonite	1.05	11.4	5.54	21.4	0.51	44.6
Potassic ferri-taramite	1.25	3.6	1.53	13.8	2.45	45.8
Fluor-riebeckite	0.02	4.4	2.02	12.7	0.29	47.9
Ferro-glaucophane	0.06	0.7	2.04	14.8	0.39	70.5

Phase	M2		M3	
	Δ	σ^2	Δ	σ^2
Grunerite	2.79	43.2	0.11	60.9
Glaucophane	15.77	35.0	0.34	85.0
Tremolite	5.52	22.9	0.09	43.6
Fluor-richterite	12.54	33.1	0.97	43.3
Fluor-tremolite	5.76	21.9	1.03	47.9
Potassic pargasite	5.98	24.3	0.01	75.7
Ferro-tschermakite	5.00	19.6	0.33	106.0
Potassic ossi-kaersutite	7.06	31.2	0.05	56.8
Potassic arfvedsonite	17.27	44.5	0.60	67.6
Potassic ferri-taramite	8.53	28.7	0.08	82.3
Fluor-riebeckite	15.48	39.1	0.26	68.7
Ferro-glaucophane	14.36	31.9	1.14	97.4

differ markedly according to coordination number (figure 5.34). In this phase, too, Si tetrahedra are almost inert to *T* modifications, whereas octahedral sites show maximum thermal expansion. Concerning site *M4*, the structural response to *T* modifications is linear for VIII-fold coordination but not for VI-fold coordination. The contrasting behavior of octahedral and tetrahedral sites results in decoupling between tetrahedral and octahedral chain moduli, partially compensated by rotation of tetrahedral groups (this type of distortion directly affects the dimensions of sites *A*; see figure 5.35).

Heating experiments on $P2_1/m$ amphiboles (cummingtonite) show transition to the *C2/m* polymorph at low *T* (50 to 100 °C; Prewitt et al., 1970; Sueno et al., 1972) in analogy to the behavior of pigeonitic pyroxenes ($P2_1/c$), which show transition to the form *C2/c* (cf. section 5.4.3). Phase stability limits are markedly conditioned by the chemistry of the system and, particularly, by the partial pres-

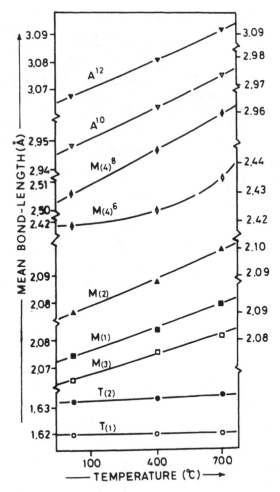

Figure 5.34 Effect of temperature on mean dimensions of various coordination polyhedra in tremolite (*C2/m*). From Hawthorn (1981a). Reprinted with permission of The Mineralogical Society of America.

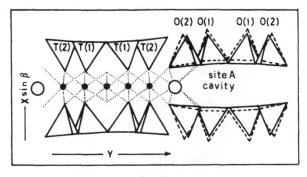

Figure 5.35 Structure of tremolite projected on (001) plane. Dashed lines on $[T_4O_{11}]_n$ tetrahedral chains: displacement (highly exaggerated for illustrative purposes) due to thermal decoupling. From Sueno et al. (1973). Reprinted with permission of The Mineralogical Society of America.

Table 5.48 Thermal expansion and compressibility of some amphibole end-members according to Saxena et al. (1993) (1) and Holland and Powell (1990) (2).

Phase	Stoichiometry	$\alpha_0 \times 10^5$	$\alpha_1 \times 10^9$	$\alpha_2 \times 10^3$	α_3	Reference
Anthophyllite	$Mg_7Si_8O_{22}(OH)_2$	6.44229	9.4909	4.715	−4.97167	(1)
		2.7882	0	0	0	(2)
Fe-anthophyllite	$Fe_7Si_8O_{22}(OH)_2$	2.8345	0	0	0	(2)
Cummingtonite	$Mg_7Si_8O_{22}(OH)_2$	2.7956	0	0	0	(2)
Grunerite	$Fe_7Si_8O_{22}(OH)_2$	10.67	24.8	−30.67	1.6558	(1)
		2.8417	0	0	0	(2)
Tremolite	$Ca_2Mg_5Si_8O_{22}(OH)_2$	7.46263	5.2307	−13.168	−0.56802	(1)
Fe-tremolite	$Ca_2Fe_5Si_8O_{22}(OH)_2$	3.1117	0	0	0	(2)
Glaucophane	$Na_2Mg_3Al_2Si_8O_{22}(OH)_2$	4.416	9.727	−15.43	2.226	(1)
		2.6871	0	0	0	(2)
Fe-glaucophane	$Na_2Fe_3Al_2Si_8O_{22}(OH)_2$	3.0462	0	0	0	(2)
Edenite	$NaCa_2Mg_5AlSi_7O_{22}(OH)_2$	3.1008	0	0	0	(2)
Hornblende	$Ca_2Mg_4Al_2Si_7O_{22}(OH)_2$	3.0752	0	0	0	(2)
Fe-hornblende	$Ca_2Fe_4Al_2Si_7O_{22}(OH)_2$	3.1183	0	0	0	(2)
Magnesio-riebeckite	$Na_2Mg_3Al_2Si_8O_{22}(OH)_2$	2.6907	0	0	0	(2)
Pargasite	$NaCa_2Mg_4Al_3Si_6O_{22}(OH)_2$	3.0871	0	0	0	(2)

Phase	Stoichiometry	$\beta_0 \times 10^6$	$\beta_1 \times 10^{10}$	$\beta_2 \times 10^{14}$	$\beta_3 \times 10^{17}$	Reference
Anthophyllite	$Mg_7Si_8O_{22}(OH)_2$	1.86582	4.3803	2.3045	3.27547	(1)
		1.2434	0	0	0	(2)
Fe-anthophyllite	$Fe_7Si_8O_{22}(OH)_2$	1.1481	0	0	0	(2)
Cummingtonite	$Mg_7Si_8O_{22}(OH)_2$	1.2467	0	0	0	(2)
Grunerite	$Fe_7Si_8O_{22}(OH)_2$	1.447	5.042	−25.05	12.35	(1)
		1.1511	0	0	0	(2)
Tremolite	$Ca_2Mg_5Si_8O_{22}(OH)_2$	1.8257	0.48749	23.491	−2.8108	(1)
Fe-tremolite	$Ca_2Fe_5si_8O_{22}(OH)_2$	1.3083	0	0	0	(2)
Glaucophane	$Na_2Mg_3Al_2Si_8O_{22}(OH)_2$	1.4563	1.1979	8.47	0.42052	(1)
		1.1132	0	0	0	(2)
Fe-glaucophane	$Na_2Fe_3Al_2Si_8O_{22}(OH)_2$	1.1658	0	0	0	(2)
Edenite	$NaCa_2Mg_5AlSi_7O_{22}(OH)_2$	1.1074	0	0	0	(2)
Hornblende	$Ca_2Mg_4Al_2Si_7O_{22}(OH)_2$	1.2968	0	0	0	(2)
Fe-hornblende	$Ca_2Fe_4Al_2Si_7O_{22}(OH)_2$	1.2903	0	0	0	(2)
Magnesio-riebeckite	$Na_2Mg_3Al_2Si_8O_{22}(OH)_2$	1.1058	0	0	0	(2)
Pargasite	$NaCa_2Mg_4Al_3Si_6O_{22}(OH)_2$	1.1025	0	0	0	(2)

sure of fluids (f_{H_2O}) and by redox state (f_{O_2}). An appropriate discussion of such limits is beyond the scope of this text, and readers are referred to the work of Gilbert et al. (1982).

Table 5.48 summarizes thermal expansion and compressibility data for amphibole end-members according to the databases of Holland and Powell (1990) and Saxena et al. (1993). Isobaric thermal expansion (α, K^{-1}) and isothermal compressibility (β, bar^{-1}) may be retrieved from the listed coefficients by applying the polynomial expansions

$$\alpha_{(T)} = \alpha_0 + \alpha_1 \times T + \alpha_2 \times T^{-1} + \alpha_3 \times T^{-2} \qquad (5.100)$$

and

$$\beta_{(T)} = \beta_0 + \beta_1 \times T + \beta_2 \times T^2 + \beta_3 \times T^3. \qquad (5.101)$$

The P-derivative of the bulk modulus (K'; cf. section 1.15.2) is 4.0 in all cases.

5.5.3 Thermodynamic Properties of End-Members

Thermodynamic data on amphibole end-members are mostly derived from phase equilibria, because of the difficulty of synthesizing pure components for direct calorimetric investigation. Table 5.49 lists existing values for some pure components. Most data come from the internally consistent set of Holland and Powell (1990).

5.5.4 Mixing Properties and Intracrystalline Disorder

The chemistry of many natural amphiboles is conveniently described by a compositional quadrilateral like that seen for pyroxenes. The quadrilateral components are anthophyllite [$Mg_7Si_8O_{22}(OH)_2$], grunerite [$Fe_7Si_8O_{22}(OH)_2$], tremolite [$Ca_2Mg_5Si_8O_{22}(OH)_2$], and ferro-tremolite [$Ca_2Fe_5Si_8O_{22}(OH)_2$] (figure 5.36).

There is complete miscibility among the calcic terms with $C2/m$ structures. These terms correspond to $C2/c$ calcic pyroxenes (augites). Orthoamphiboles with $Pnma$ structures (anthophyllite series) are equivalent to $Pbca$ orthopyroxenes, although with a much more restricted Fe^{2+}/Mg compositional range. Cummingtonites, with $C2/m$ and $P2_1/m$ structures, are analogous to pigeonites with $C2/c$ or $P2_1/c$ structures. As for pyroxenes, amphiboles have an extended miscibility gap between calcic monoclinic terms and Ca-poor orthorhombic terms. Moreover, the three-pyroxene paragenesis (augite-orthopyroxene-pigeonite; cf. figures 5.25B and 5.28A) is similar to the three-amphibole paragenesis (hornblende-anthophyllite-cummingtonite) observed in metamorphic terrains. Notwithstanding the above analogies, the chemistry of amphiboles is much more complex than that of their pyroxene counterparts, and the study of mixing behavior is complicated by the possibility of isomorphous substitutions over a large number of nonequivalent sites (A, $M4$, $M1$, $M2$, $M3$, $T1$, $T2$).

Besides the amphibole quadrilateral discussed above, a potentially useful interpretative analysis is based on the $M4$-site population projected on the compositional triangle Na – Ca – (Fe^{2+}+Mg+Mn). A plot of this type groups metamorphic amphiboles into three main assemblages (mafic, calcic, and sodic), as shown in figure 5.37. The same figure also shows the miscibility gaps responsible for this compositional splitting, together with the tie lines connecting coexisting amphiboles at the borders of the miscibility gaps.

Table 5.49 Thermodynamic data of amphibole end-members, $T_r = 298.15$ K; $P_r = 1$ bar; H^0_{f,T_r,P_r} in kJ/mole; $S^0_{T_r,P_r}$ in J/(mole × K); $V^0_{T_r,P_r}$ in cm³/mole. (1) Helgeson et al. (1978); (2) Berman (1988); (3) Saxena et al. (1993); (4) Holland and Powell (1990). Heat capacity equation: $C_P = K_1 + K_2 T + K_3 T^{-2} + K_4 T^2 + K_5 T^{-3} + K_6 T^{-1/2} + K_7 T^{-1}$.

Phase	Stoichiometry	Structure	H^0_{f,T_r,P_r}	$S^0_{T_r,P_r}$	$V^0_{T_r,P_r}$	Reference
Anthophyllite	$Mg_7Si_8O_{22}(OH)_2$	Pnma	−12,071.0	535.19	265.44	(3)
			−12,069	535.20	265.60	(2)
			−12,064.14	537.0	265.40	(4)
Fe-anthophyllite	$Fe_7Si_8O_{22}(OH)_2$	Pnma	−9625.86	729.0	278.70	(4)
Cummingtonite	$Mg_7Si_8O_{22}(OH)_2$	$P2_1/m$	−12,075.73	542.0	264.70	(4)
Grunerite	$Fe_7Si_8O_{22}(OH)_2$	$P2_1/m$	−9631.5	714.6	278.73	(3)
				683.2	256.67	(1)
			−9614.14	734.0	278.0	(4)
Tremolite	$Ca_2Mg_5Si_8O_{22}(OH)_2$	C2/m	−12,305.6	551.2	272.68	(2)
			−12,302.50	550.0	272.70	(4)
Fluor-tremolite	$Ca_2Mg_5Si_8O_{22}(F)_2$	C2/m		541.4	270.45	(1)
Fe-tremolite	$Ca_2Fe_5Si_8O_{22}(OH)_2$	C2/m	−10,848.9	684.1	282.80	(1)
			−10,527.10	705.0	282.80	(4)
Edenite	$NaCa_2Mg_5AlSi_7O_{22}(OH)_2$	C2/m	−12,580.53	599.0	270.90	(4)
Hornblende	$Ca_2Mg_4Al_2Si_7O_{22}(OH)_2$	C2/m	−12,420.29	551.0	269.90	(4)
Fe-hornblende	$Ca_2Fe_4Al_2Si_7O_{22}(OH)_2$	C2/m	−10,999.97	690.0	279.00	(4)
Mg-riebeckite	$Na_2Mg_3Al_2Si_8O_{22}(OH)_2$	C2/m	−11,087.79	602.0	271.30	(4)
Pargasite	$NaCa_2Mg_4Al_3Si_6O_{22}(OH)_2$	C2/m	−12,719.83	591.0	272.10	(4)
Fe-pargasite	$NaCa_2Fe_4AlAl_2Si_6O_{22}(OH)_2$	C2/m		776.1	279.89	(1)
Glaucophane	$Na_2Mg_3Al_2Si_8O_{22}(OH)_2$	C2/m		543.9	269.70	(1)
			−11,963.86	535.0	260.50	(4)
Fe-glaucophane	$Na_2Fe_3Al_2Si_8O_{22}(OH)_2$	C2/m	−10,901.10	624.0	265.90	(4)
Richterite	$Na_2CaMg_5Si_8O_{22}(OH)_2$	C2/m		575.7	272.80	(1)

Phase	K_1	$K_2 \times 10^3$	$K_3 \times 10^{-5}$	K_4	$K_5 \times 10^{-7}$	$K_6 \times 10^{-2}$	$K_7 \times 10^{-3}$	Reference
Anthophyllite	1128.7	38.86	50.43	0	−76.42	0	−151	(3)
	1219.31	0	−347.661	0	440.09	−57.665	0	(2)
	1277.3	25.825	−97.046	0	0	−90.747	0	(4)
Fe-anthophyllite	1383.1	30.669	−42.247	0	0	−112.576	0	(4)
Cummingtonite	1277.3	25.825	−97.046	0	0	−90.747	0	(4)
Grunerite	763.704	356.55	−178.27	0	28.902	0	5.5255	(3)
	831.9	254.93	−165.48	0	0	0	0	(1)
	1383.1	30.669	−42.247	0	0	−112.576	0	(4)
Tremolite	1229.36	0	−320.899	0	420.881	−64.019	0	(2)
	1214.4	26.528	−123.62	0	0	−73.885	0	(4)
Fluor-tremolite	768.8	231.17	−182.97	0	0	0	0	(1)
Fe-tremolite	828.1	246.65	−172.26	0	0	0	0	(1)
	1290.0	29.991	−84.475	0	0	−89.47	0	(4)
Edenite	1264.9	24.090	−125.60	0	0	−77.040	0	(4)
Hornblende	1229.6	25.438	−121.635	0	0	−77.503	0	(4)
Fe-hornblende	1290.0	28.209	−90.319	0	0	−89.971	0	(4)
Mg-riebeckite	1701.5	−115.65	70.216	0	0	−185.336	0	(4)
Pargasite	1280.2	22.997	−123.595	0	0	−80.658	0	(4)
Fe-pargasite	893.6	179.87	−197.86	0	0	0	0	(1)
Glaucophane	796.0	249.49	−209.24	0	0	0	0	(1)
	1717.5	−121.07	70.75	0	0	−192.72	0	(4)
Fe-glaucophane	1762.9	−118.992	94.237	0	0	−202.071	0	(4)
Richterite	815.0	255.64	−193.09	0	0	0	0	(1)

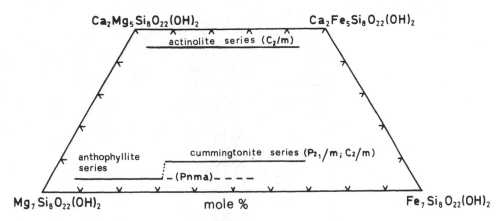

Figure 5.36 Amphibole quadrilateral.

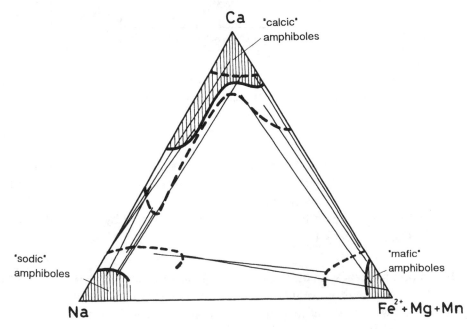

Figure 5.37 Locations of metamorphic amphiboles in compositional triangle Na – Ca – (Fe^{2+}+Mg+Mn) (atom proportions). Miscibility gaps are simplified. Solid lines are microprobe analyses; dashed lines are wet chemical analyses. From Robinson et al. (1982). Reprinted with permission of The Mineralogical Society of America.

According to Ghose (1982), there are three general conditions (rules) leading to immiscibility between two amphibole components:

1. Site $M4$ is occupied in one of the two components by a cation with a large ionic radius (e.g., Ca^{2+}) and by cations of much shorter radii (e.g., Mg^{2+} and Fe^{2+}) in the other component (hornblende-cummingtonite or actinolite-cummingtonite).

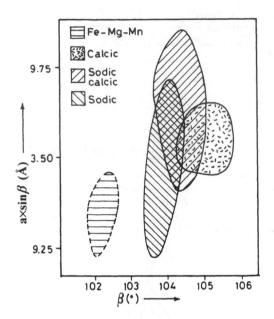

Figure 5.38 Structural parameters of amphiboles belonging to various chemical groups. From Hawthorn (1981a). Reprinted with permission of The Mineralogical Society of America.

2. Site A is partially or fully occupied in one component and completely empty in the other (hornblende-actinolite, hornblende-cummingtonite, gedrite-anthophyllite, richterite-riebeckite).
3. Site $M4$ is occupied by Ca^{2+} in one component and by Na^+ in the other (actinolite-riebeckite).

Based on these rules, mixing properties appear to be strongly conditioned by the charges and dimensions of intervening cations in positions $M4$ and A. Indeed, figure 5.38 shows that structures and dimensions of monoclinic amphiboles are substantially affected by occupancies on site $M4$, which appear to condition not only cell edge a but also angle β: the four fields outlined in figure 5.38 are emblematic confirmations of general rules 1 and 3 proposed by Ghose (1982).

According to Robinson et al. (1982), a convenient way of rationalizing mixing properties of amphiboles in the system SiO_2-Al_2O_3-Fe_2O_3-TiO_2-MgO-FeO-CaO-Na_2O-H_2O is to consider tremolite [$\square Ca_2Mg_5Si_8O_{22}(OH)_2$] (where $\square$ symbolizes a cationic vacancy on site A) as a base component. Tremolite is then transformed into the other components of interest by a set of independent exchange reactions, as summarized in table 5.50. The resulting components are 8 (including tremolite). In this compositional field there are 28 binary joins, and calculation of mixing properties in the multicomponent system is thus prohibitive. Even limiting

Table 5.50 *C2/m* amphibole end-member formulation of Robinson et al. (1982).

Base Component (Tremolite)	Exchange	Resulting Component
□ $Ca_2Mg_5Si_8O_{22}(OH)_2$	$Fe \Leftrightarrow Mg$	□ $Ca_2Fe_5Si_8O_{22}(OH)_2$ Ferro-actinolite
□ $Ca_2Mg_5Si_8O_{22}(OH)_2$	□ $Si \Leftrightarrow Na^AAl^{IV}$	□ $NaCa_2Mg_5AlSi_7O_{22}(OH)_2$ Edenite
□ $Ca_2Mg_5Si_8O_{22}(OH)_2$	$MgSi \Leftrightarrow Al^{VI}Al^{IV}$	□ $Ca_2Mg_3Al_2Si_6O_{22}(OH)_2$ Tschermak
□ $Ca_2Mg_5Si_8O_{22}(OH)_2$	$MgSi \Leftrightarrow Fe^{3+,VI}Al^{IV}$	□ $Ca_2Mg_3Fe_2^{3+}Si_6Al_2O_{22}(OH)_2$ Ferri-Tschermak
□ $Ca_2Mg_5Si_8O_{22}(OH)_2$	$CaMg \Leftrightarrow Na^{M4}Al^{VI}$	□ $Na_2Mg_3Al_2Si_8O_{22}(OH)_2$ Glaucophane
□ $Ca_2Mg_5Si_8O_{22}(OH)_2$	$MgSi_2 \Leftrightarrow Ti^{IV}Al^{IV}Al^{VI}$	□ $Ca_2Mg_5AlSi_6TiAlO_{22}(OH)_2$ Ti-Tschermak
□ $Ca_2Mg_5Si_8O_{22}(OH)_2$	$Ca \Leftrightarrow Mg^{M4}$	□ $Mg_7Si_8O_{22}(OH)_2$ Cummingtonite

ourselves to the calculation of mixing properties in the amphibole quadrilateral, there are still six binary joins for which mixing properties must be assessed (disregarding ternary and quaternary interaction terms).

To our knowledge, direct experimental data on amphibole mixtures have been obtained only for the (pseudo)binary system actinolite-cummingtonite (Camcron, 1975) at $P_{total} = P_{H_2O} = 2$ kbar and for the (pseudo)binary system tremolite-pargasite at $P_{total} = P_{H_2O} = 1$ kbar (Oba, 1980). In both cases, an extended miscibility gap (or solvus field in the second case), is evident at low T (i.e., 600 to 800 °C), which is indicative of strong positive interactions in the solid mixtures. Unmixing of other compositional terms is also evident in microprobe investigations (see Ghose, 1982 for an appropriate discussion).

Information on amphibole mixing properties comes from intracrystalline and intercrystalline Fe^{2+}-Mg distribution studies. Mueller (1960, 1961) described the coexistence of actinolite and cummingtonite in ferriferous metamorphosed bands of Bloom Lake (Quebec, Canada) based on the exchange equilibrium

$$\frac{1}{2}Ca_2^{M4}Fe_5^Y Si_8O_{22}(OH)_2 + \frac{1}{7}Mg_2^{M4}Mg_5^Y Si_8O_{22}(OH)_2$$

$$\underset{\text{Fe-actinolite}}{} \qquad \underset{\text{Mg-cummingtonite}}{}$$

$$\Leftrightarrow \frac{1}{2}Ca_2^{M4}Mg_5^Y Si_8O_{22}(OH)_2 + \frac{1}{7}Fe_2^{M4}Fe_5^Y Si_8O_{22}(OH)_2,$$

$$\underset{\text{Mg-actinolite}}{} \qquad \underset{\text{Fe-cummingtonite}}{}$$

(5.102)

where superscripts $M4$ and Y define the structural position of the element. According to Mueller (1961), the observed intracrystalline partitioning of Fe^{2+} and Mg^{2+} may be attributable to ideal mixing behavior in actinolite and to regular mixing behavior in cummingtonite. Later, in order to assess the mixing properties of cummingtonite better, Mueller (1962) proposed an intracrystalline exchange of the type

$$Fe^{M4} + Mg^{Y} \Leftrightarrow Mg^{M4} + Fe^{Y} \tag{5.103}$$

in which mixing on sites Y ($M1$, $M2$, $M3$) is ideal, but not on site $M4$. The constant of equilibrium 5.103 may be expressed as

$$K_{103} = \frac{X_{Mg,M4} X_{Fe,Y}}{X_{Fe,M4} X_{Mg,Y}} \times \frac{\gamma_{Mg,M4} \gamma_{Fe,Y}}{\gamma_{Fe,M4} \gamma_{Mg,Y}}, \tag{5.104}$$

where the γ terms are site-activity coefficients.

Adopting the model of Mueller (1962), equation 5.104 reduces to

$$K_{103} = \frac{X_{Mg,M4} X_{Fe,Y}}{X_{Fe,M4} X_{Mg,Y}} \times \frac{\gamma_{Mg,M4}}{\gamma_{Fe,M4}}. \tag{5.105}$$

In a regular mixture, the activity coefficient of component i is associated with the interaction parameter W through

$$\ln \gamma_i = \frac{W}{RT} (1 - X_i)^2 \tag{5.106}$$

thus equation 5.104 may be reexpressed in terms of interaction parameters on sites $M4$ and Y (W_{M4}, W_Y) as

$$K_{103} = \frac{X_{Fe,Y} (1 - X_{Fe,M4})}{X_{Fe,M4} (1 - X_{Fe,Y})} \times \frac{\exp\left[(1 - 2X_{Fe,Y}) \dfrac{W_Y}{RT} \right]}{\exp\left[(1 - 2X_{Fe,M4}) \dfrac{W_{M4}}{RT} \right]}. \tag{5.107}$$

Hafner and Ghose (1971) solved equation 5.107 on the basis of Mueller's (1960) data and obtained $K_{103} = 0.19$; $W_{M4}/RT = 1.58$, and $W_Y/RT = 0.54$. Their results thus imply a certain nonideality of mixing on sites Y as well.

Ghose and Weildner (1971, 1972) experimentally studied the intracrystalline distribution of Fe^{2+}-Mg^{2+} in natural cummingtonites annealed at various T conditions and $P_{total} = 2$ kbar in the presence of H_2O (3 weight %). The observed distribution of Fe^{2+} between $M4$ and Y sites may be due either to regular mixing (with $K_{103} = 0.26$, $W_{M4} = 1.09$ kcal/mole, $W_Y = 1.23$ kcal/mole at $T = 600$ °C; cf. Ghose and Weidner, 1971; and figure 5.39A) or to ideal mixing on both $M4$ and Y sites, as shown in figure 5.39B.

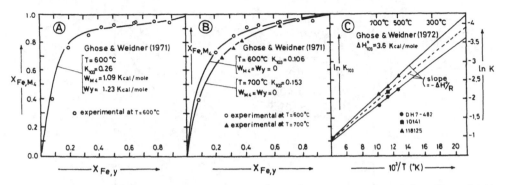

Figure 5.39 Intracrystalline distribution of Fe^{2+}-Mg^{2+} between $Y(M1, M2, M3)$ and $M4$ sites in natural cummingtonites. (A) Regular site-mixing model. (B) Ideal site-mixing model. (C) Arrhenius plot of equilibrium constant.

As we see by comparing figure 5.39A and B, the choice of the site-mixing model is, within certain limits, arbitrary: an ideal model gives hyperbolic distribution curves such as those shown in figure 5.39B. Any nonzero site interaction parameters perturb the hyperbolic distribution and induce fluctuations whose magnitude depends directly on the absolute magnitude of W. The choice of an appropriate mixing model thus rests on the sensitivity of the experimenter.

Plotting on an Arrhenius diagram (figure 5.39C) the equilibrium constants measured on several natural cummingtonites, Ghose and Weidner (1972) deduced the standard enthalpy variation associated with the intracrystalline exchange 5.103 ($\Delta H^0 = 3.6$ kcal/mole). Figure 5.39C may be utilized as a potential geothermometer to define closure temperatures for intracrystalline exchanges in natural cummingtonites.

In the example above, we did not make any distinction among $M1$, $M2$, and $M3$ site populations, because these sites were grouped in the Y notation. Studying in detail the structure of sodic amphiboles of the riebeckite-glaucophane series, Ungaretti et al. (1978) observed that the Fe^{2+}-Mg^{2+} population is not randomly distributed between $M1$ and $M3$ sites, being Fe^{2+} preferentially stabilized in $M3$. Considering the intracrystalline exchange

$$Fe^{2+}_{M3} + Mg_{M1} \Leftrightarrow Fe^{2+}_{M1} + Mg_{M3}, \tag{5.108}$$

the resulting equilibrium constant, based on the simply hyperbolic distribution of cations (figure 5.40), may be reconducted to an ideal site-mixing behavior:

$$K_{108} = \frac{a_{Mg,M3}a_{Fe,M1}}{a_{Fe,M3}a_{Mg,M1}} \equiv \frac{(1 - X_{Fe,M3})X_{Fe,M1}}{X_{Fe,M3}(1 - X_{Fe,M1})} = 0.403. \tag{5.109}$$

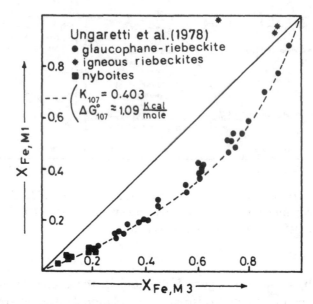

Figure 5.40 Intracrystalline distribution of Fe^{2+}-Mg^{2+} between $M1$ and $M3$ sites in sodic amphiboles. Reprinted from Ungaretti et al. (1978) with author's permission.

We recall to this purpose that "ideal site mixing" does not necessarily mean that the mixture as a whole is ideal, but that the intracrystalline site distribution may be described by a constant (without the necessity of introducing site-interaction parameters; cf. equation 5.109, for example). Generally speaking, if the adopted constant differs from 1 (as in the case of sodic amphiboles in figure 5.40 and of cummingtonites in figure 5.39), the mixture as a whole is obviously nonideal because site permutability does not reach the maximum value attained by the ideal condition (cf. section 3.8.1).

The nature of the miscibility gap between orthorhombic and monoclinic amphiboles has been recently reconsidered by Will and Powell (1992) with an extended application of Darken's Quadratic Formalism (Darken, 1967). This formalism assumes that, within a restricted compositional field ("simple regions"), excess energy terms may be described by simple mixing models (regular mixture, ideal mixture), even if the intermediate compositions necessitate more complex treatment. Under this proviso, a general binary join may be considered to be composed of two simple regions, each including one end-member, connected by a transitional zone, as depicted schematically in figure 5.41. The complex activity-composition relationship observed within the simple region is then reconducted to a regular (or even ideal) mixture between the real end-member in the simple region and a "fictive" end-member on the opposite side of the join, by appropriate selection of the standard state properties of the fictive term.

Assuming ideal mixing within the simple regions, the principle of equality of potentials between phases at equilibrium reduces to

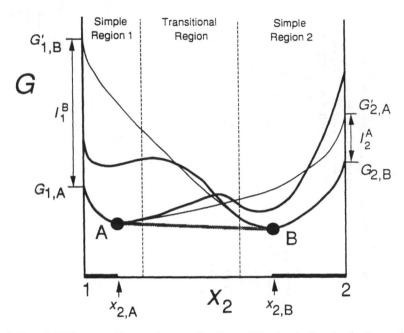

Figure 5.41 *G-X* diagram illustrating application of Darken's Quadratic Formalism to a binary join. Although mixing behavior of components 1 and 2 in phases α and β is nonideal (heavy lines), in each of the simple regions it is modeled by an ideal mixing model (light lines) by means of an appropriate choice of the fictive standard state potential μ^*. From Will and Powell (1992). Reprinted with permission of The Mineralogical Society of America.

$$\Delta\mu_i^0 = -RT \ln K_{D,i} = \mu_{i,\alpha}^* - \mu_{i,\beta}^0, \tag{5.110}$$

where $\mu_{t,\alpha}^*$ is the fictive standard state chemical potential of component i in phase α, and $K_{D,i}$ is the distribution coefficient $K_{D,i} = X_{i,\alpha}/X_{i,\beta}$.

Consideration of the actual extension of miscibility gaps in natural systems led Will and Powell (1992) to establish systematic relationships between the actual free energy of pure components and their fictive potentials in the phases of interest, as listed in table 5.51. Note that, amphiboles being multisite phases, their ideal activity in a chemically complex phase is expressed in terms of multiple product of site ionic fractions (see section 3.8.7). For anthophyllite ($\square Mg_2Mg_3Mg_2Si_4Si_4O_{22}(OH)_2$), for instance, we have

$$a_{\text{anthophyllite}} \equiv X_{\text{anthophyllite}} = X_{\square,A} X_{Mg,M4}^2 X_{Mg,M1-M3}^3 X_{Mg,M2}^2 X_{Si,T2}^4 \tag{5.111}$$

(it is assumed that Al-Si mixing takes place only on $T2$ sites and that $M1$ to $M3$ sites are energetically equivalent).

Table 5.51 Energy terms relating standard state chemical potentials of pure components ($\mu_{i,\alpha}^0$) to fictive potentials in the host phase ($\mu_{i,\beta}^*$) in the Will and Powell (1992) application of Darken's Quadratic Formalism to amphiboles.

Cummingtonite-actinolite series
$T \cong 650\ °C$

$\mu^*_{cumm,act} = \mu^0_{cumm,cumm} + 51.5 \quad (\pm 4.5)$

$\mu^*_{tr,cumm} = \mu^0_{tr,act} + 16.0 \quad (\pm 3.8)$

$\mu^*_{hb,cumm} = \mu^0_{hb,act} + 62.6 \quad (\pm 8.6)$

$\mu^*_{ed,cumm} = \mu^0_{ed,act} + 41.1 \quad (\pm 3.7)$

$\mu^*_{parg,cumm} = \mu^0_{parg,act} + 12.6 \quad (\pm 1.3)$

Orthoamphibole-actinolite series
$T \cong 650\ °C$

Calcic-sodic amphibole series
$T \cong 500\ °C$

$\mu^*_{gl,act} = \mu^0_{gl,gl} + 48.7 \quad (\pm 6.7)\ (\text{low }P)$

$\mu^*_{gl,act} = \mu^0_{gl,gl} + 31.8 \quad (\pm 2.9)\ (\text{high }P)$

$\mu^*_{rich,act} = \mu^0_{rich,gl} + 25.0 \quad (\pm 3.0)$

$\mu^*_{tr,gl} = \mu^0_{tr,act} + 46.3 \quad (\pm 3.4)$

$\mu^*_{hb,gl} = \mu^0_{hb,act} + 32.1 \quad (\pm 2.0)$

$\mu^*_{ed,gl} = \mu^0_{ed,act} + 59.8 \quad (\pm 3.8)$

$\mu^*_{parg,gl} = \mu^0_{parg,act} + 46.0 \quad (\pm 2.5)$

Orthoamphibole-actinolite series
$T \cong 650\ °C$

$\mu^*_{anth,act} = \mu^0_{anth,oam} + 35.4 \quad (\pm 4.0)$

$\mu^*_{tr,oam} = \mu^0_{tr,act} + 34.5 \quad (\pm 6.5)$

$\mu^*_{hb,oam} = \mu^0_{hb,act} + 40.1 \quad (\pm 4.8)$

$\mu^*_{ed,oam} = \mu^0_{ed,act} + 44.3 \quad (\pm 4.1)$

$\mu^*_{parg,oam} = \mu^0_{parg,act} + 54.5 \quad (\pm 4.4)$

cumm = cummingtonite ($P2_1/m$) $\square$ Mg$_2$Mg$_3$Mg$_2$Si$_4$Si$_4$O$_{22}$(OH)$_2$

tr = tremolite ($C2/m$) $\square$ Ca$_2$Mg$_3$Mg$_2$Si$_4$O$_{22}$(OH)$_2$

hb = hornblende ($C2/m$) $\square$ Ca$_2$Mg$_3$(MgAl)(Si$_3$Al)Si$_4$O$_{22}$(OH)$_2$

ed = edenite ($C2/m$) NaCa$_2$Mg$_3$Mg$_2$(Si$_3$Al)Si$_4$O$_{22}$(OH)$_2$

parg = pargasite ($C2/m$) NaCa$_2$Mg$_3$(MgAl)(Si$_2$Al$_2$)Si$_4$O$_{22}$(OH)$_2$

gl = glaucophane ($C2/m$) $\square$ Na$_2$Mg$_3$(AlAl)Si$_4$Si$_4$O$_{22}$(OH)$_2$

rich = richterite ($C2/m$) Na(CaNa)Mg$_3$Mg$_2$Si$_4$Si$_4$O$_{22}$(OH)$_2$

anth = anthophyllite ($Pnma$) $\square$ Mg$_2$Mg$_3$Mg$_2$Si$_4$Si$_4$O$_{22}$(OH)$_2$

act = actinolite; oam = orthoamphibole

5.6 Micas

Micas are *diphyllosilicates* (multiplicity = 2, periodicity = 3, dimensionality = 2). They are composed of mixed sheets in a 2:1 ratio ("talc layer") containing two tetrahedral sheets $[T_2O_5]_n$ interposed by one octahedral sheet. The tetrahedral sheets are formed from three shared corners of the tetrahedral groups $[SiO_4]$, which results in a bidimensional lattice with hexagonal meshes, as shown in figure 5.42A. At the center of each hexagon is a hydroxyl group OH or F (or S in anandite). Figure 5.42B shows the stacking that leads to the mixed 2:1 talc layer. The apexes of the tetrahedral groups are opposed to form junction planes in common with the octahedral sheet.

Micas are classified according to the charge of the mixed layer per formula unit. This depends on the diadochy level of Al^{3+} (or Fe^{3+} or, more rarely, Be^{2+}) with Si^{4+} in tetrahedral groups, and on the charge (and occupancy) of the cation in the intermediate octahedral sheet.

In each 2:1 mixed layer, the upper tetrahedral sheet is translated by $a/3$ with respect to the lower one, thus creating the octahedral oxygen coordination around the cations of the intermediate sheet. Translation may occur along any positive or negative direction defined by structural axes X_1, X_2, and X_3 of a pseudohexagonal lattice, as shown in figure 5.43.

The degree of freedom in the translational properties of the various sheets gives the *polytypism* of the phase. (*Polytypes* are structures with layers of essentially identical composition but differing in translational sequence. *Polytypes* usually have different symmetry and periodicity along axis Z). Smith and Yoder

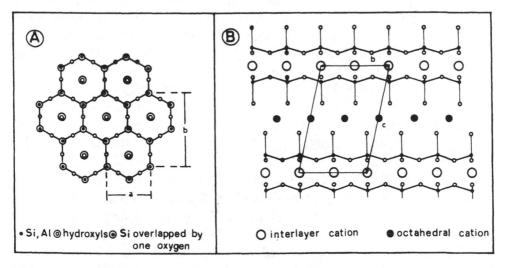

• Si, Al ⊚ hydroxyls ⊚ Si overlapped by one oxygen   ○ interlayer cation   ● octahedral cation

Figure 5.42 Structural scheme of micas. (A) Tetrahedral sheet with tetrahedral apexes directed upward. (B) Structure of mixed layers along axis *Y*; *a*, *b*, and *c* are edges of elementary cell unit.

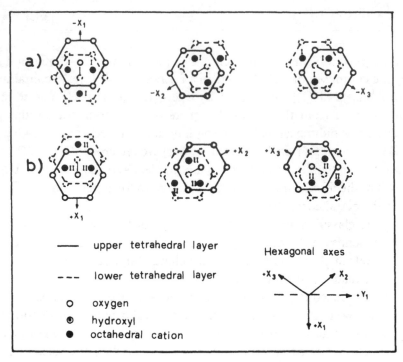

Figure 5.43 Sliding between tetrahedral sheets of mixed 2:1 talc layer. (a) Set of positions I occupied. (b) Set of positions II occupied. From Bailey (1984a). Reprinted with permission of The Mineralogical Society of America.

(1956) have shown that six standard ideal polytypes may be formed, with periodicity between one and six layers: *1M*, space group *C2/m*, $\beta = 100$; *2M*, space group *C2/c*, $\beta = 95$; *3T*, space group *P3₁12*; *2Or*, space group *Ccmm*, $\beta = 90$; *2M2*, space group *C2/c*, $\beta = 98$; *6H*, space group *P6₁22* (first number identifies periodicity—i.e., polytype *3T* ⇒ periodicity = 3; polytype *1M* ⇒ periodicity = 1, etc.). In trioctahedral micas, the most frequent polytype is *1M*, followed by *3T* and *2M1*. In dioctahedral micas, the most frequent is *2M1*, followed by *1M* and *3T* (Bailey, 1984a). The diadochy of Al (of Fe^{3+}) on Si^{4+} normally has the following ratios: 0:4, 0.5:3.5, 1:3, 2:2, and 3:1. Thus, polyanionic groups composed of two tetrahedral sheets may be of these types: $[(OH)_2Si_4O_{10}]^{6-}$; $[(OH)_2Al_{0.5}Si_{3.5}O_{10}]^{6.5-}$; $[(OH)_2AlSi_3O_{10}]^{7-}$; $[(OH)_2Al_2Si_2O_{10}]^{8-}$; $[(OH)_2Al_3SiO_{10}]^{9-}$.

The intermediate octahedral sheet is normally made up of cations of charge 2 or 3 (Mg, Al, Fe^{2+}, Fe^{3+}, or, more rarely, V, Cr, Mn, Co, Ni, Cu, Zn), but in some cases cations of charge 1 (Li) and 4 (Ti) are also found. In the infinite octahedral sheet, formed by the sharing of six corners of each octahedron, there may be full occupancy of all octahedral sites ("trioctahedral micas"); alternatively, one site out of three may be vacant ("dioctahedral micas"). Nevertheless, the primary classification of micas is based on the net charge of the mixed 2:1 layer. In "common micas" this charge is close to 1, whereas in "brittle micas" it is close to 2. In common (or "true") micas, the interlayer cations (coordination XII) counterbalancing the net charge of the mixed layers are normally K or Na (or, more rarely, Rb, Cs, NH_4, or H_3O); in brittle micas, such cations are Ca or Ba (or, more rarely, Sr). Table 5.52 lists the structural characters of some tri- and dioctahedral micas. Note the doubling of cell edge c from polytype $1M$ to polytypes $2M1$ and $2M2$, resulting in doubling of the cell volume and modification of the angle β.

Table 5.53 lists the general classification of micas with their main compositional terms. Stoichiometry obeys the general formula $XY_{2-3}Z_4O_{10}(OH,F)_2$, where X = interlayer cations, Y = octahedrally coordinated cations of the 2:1 mixed layer, and Z = tetrahedrally coordinated cations of the 2:1 mixed layer. It must be noted that several compositional terms are indeed solid mixtures of more elementary components. In particular, glauconite has a complex chemistry and an Al:Si diadochy of 0.33:3.67. (R^{2+} and R^{3+} terms in table 5.53 identify generic divalent and trivalent cations, respectively.)

5.6.1 Chemistry and Occurrence

Micas are primary phases in several types of igneous, metamorphic, and sedimentary rocks. Table 5.54 furnishes a rough scheme of main mica occurrences.

The main terms of trioctahedral common micas are phlogopite and biotite. Table 5.53 represents biotite as a mixture of Fe^{2+}- and Mg^{2+}-bearing terms, in variable proportions on site Y, and phlogopite as a pure end-member. Current nomenclature is somewhat arbitrary; trioctahedral common micas with Mg:Fe^{2+} ratios less than 2:1 are generically defined as "biotites," trioctahedral common micas with Mg:Fe^{2+} ratios higher than 2:1 are called "phlogopites." Phlogopite is a marker of thermometamorphism in magnesian limestones, where it is formed by reaction between dolomite and potash feldspar in hydrous conditions, or by reaction between muscovite and dolomite according to

$$3CaMg(CO_3)_2 + KAlSi_3O_8 + H_2O \Leftrightarrow KMg_3(Si_3Al)O_{10}(OH)_2 + 3CaCO_3 + 3CO_2 \quad (5.112.1)$$

dolomite $\quad\quad$ K-feldspar $\quad\quad$ phlogopite $\quad\quad$ calcite

and

$$4CaMg(CO_3)_2 + KAl_2(Si_3Al)O_{10}(OH)_2 \Leftrightarrow 4CaCO_3 + KMg_3(Si_3Al)O_{10}(OH)_2 + MgAl_2O_4 + 4CO_2$$

dolomite $\quad\quad$ muscovite $\quad\quad$ calcite $\quad$ phlogopite $\quad\quad$ spinel

$$(5.112.2)$$

Moreover, phlogopite is a primary mineral in leucite-bearing rocks and also sometimes occurs in ultramafic rocks. Biotite is a primary phase in several types of igneous and metamorphic rocks. In intrusive igneous rocks, it is present in granites, granodiorites, tonalites, diorites, norites, and nepheline-syenites. In effusive rocks it is less frequent, but it is still sometimes present in ryolites, trachytes, dacites, latites, andesites, and basalts, almost invariantly partially altered, with evident resorption phenomena and reaction rims. The

Table 5.52 Structural parameters for some trioctahedral and dioctahedral micas (from Smyth and Bish, 1988).

TRIOCTAHEDRAL MICAS

Name	Phlogopite	Polylithionite	Zinnwaldite
Polytype	*1M*	*1M*	*1M*
Interlayer X cations	$K_{0.77}Na_{0.16}Ba_{0.05}$	$K_{0.89}Na_{0.06}Rb_{0.05}$	$K_{0.9}Na_{0.05}$
Octahedral Y cations	Mg_3	$Al_{1.3}Li_{1.7}$	$Al_{1.05}Fe_{0.93}Li_{0.67}$
Tetrahedral Z cations	$Al_{1.05}Si_{2.95}$	$Al_{0.6}Si_{3.4}$	$Al_{0.91}Si_{3.09}$
Hydroxyls and fluorine	$(OH)_{0.7}F_{1.3}$	$(OH)_{0.5}F_{1.5}$	$(OH)_{0.8}F_{1.2}$
Molar weight (g/mole)	421.831	396.935	428.995
Molar volume (cm³/mole)	146.869	140.5	144.55
Density (g/cm³)	2.872	2.825	2.986
Formula units per unit cell	2	2	2
System	Monoclinic	Monoclinic	Monoclinic
Class	*2/m*	*2/m*	*2*
Group	*C2/m*	*C2/m*	*C2*
Cell parameters (Å):			
a	5.3078	5.20	5.296
b	9.1901	9.01	9.140
c	10.1547	10.09	10.096
β	100.08	99.28	100.83
Volume (Å³)	487.69	466.6	480.00
Reference	Hazen and Burnham (1973)	Sartori (1976)	Guggenheim and Bailey (1977)

DIOCTAHEDRAL MICAS

Name	Muscovite	Paragonite	Margarite
Polytype	*2M1*	*2M1*	*2M2*
Interlayer X cations	K	$Na_{0.92}K_{0.04}Ca_{0.02}$	$Na_{0.19}K_{0.01}Ca_{0.81}$
Octahedral Y cations	Al_2	$Al_{1.99}Fe_{0.03}Mg_{0.01}$	Al_2
Tetrahedral Z cations	Al_1Si_3	$Al_{1.06}Si_{2.94}$	$Al_{1.89}Si_{2.11}$
Hydroxyls and fluorine	$(OH)_2$	$(OH)_2$	$(OH)_2$
Molar weight (g/mole)	398.314	384.31	395.91
Molar volume (cm³/mole)	140.57	132.13	129.33
Density (g/cm³)	2.834	2.909	3.061
Formula units per unit cell	4	4	4
System	Monoclinic	Monoclinic	Monoclinic
Class	*2/m*	*2/m*	*m*
Group	*C2/c*	*C2/c*	*Cc*
Cell parameters (Å):			
a	5.1918	5.128	5.1038
b	9.1053	8.898	8.8287
c	20.0157	19.287	19.148
β	95.735	94.35	95.46
Volume (Å³)	933.56	877.51	858.89
Reference	Rothbauer (1971)	Lin and Bailey (1984)	Guggenheim and Bailey (1975, 1978)

Table 5.53 Classification and compositional terms of micas (from Bailey 1984a; modified).

Subgroup	Species	Stoichiometry
	COMMON MICAS	
Trioctahedral	Phlogopite	$KMg_3(Si_3Al)O_{10}(OH,F)_2$
	Na-phlogopite	$NaMg_3(Si_3Al)O_{10}(OH,F)_2$
	Biotite	$K(Mg,Fe)_3(Si_3Al)O_{10}(OH,F)_2$
	Annite	$KFe_3^{2+}(Si_3Al)O_{10}(OH,F)_2$
	Ferri-annite	$KFe_3^{2+}(Si_3Fe^{3+})O_{10}(OH,F)_2$
	Polylithionite (ex-lepidolite)	$K(Li_2Al)Si_4O_{10}(OH,F)_2$
	Taeniolite	$K(Mg_2Li)Si_4O_{10}(OH,F)_2$
	Siderophyllite	$K(Fe^{2+}Al)(Si_2Al_2)O_{10}(OH,F)_2$
	Zinnwaldite	$K(Li,Fe,Al)_3(Si,Al)_4O_{10}(OH,F)_2$
	Muscovite	$KAl_2(Si_3Al)O_{10}(OH,F)_2$
	Paragonite	$NaAl_2(Si_3Al)O_{10}(OH,F)_2$
	Phengite	$K[Al_{1.5}(Mg,Fe^{2+})_{0.5}](Si_{3.5}Al_{0.5})O_{10}(OH,F)_2$
Dioctahedral	Illite	$K_{0.75}(Al_{1.75}R_{0.25}^{2+})(Si_{3.5}Al_{0.5})O_{10}(OH,F)_2$
	Celadonite	$K(Mg,Fe,Al)_2Si_4O_{10}(OH,F)_2$
	Glauconite	$K(R_{1.33}^{2+}R_{0.67}^{3+})(Si_{3.67}Al_{0.33})O_{10}(OH,F)_2$
	BRITTLE MICAS	
Trioctahedral	Clintonite	$Ca(Mg_2Al)(SiAl_3)O_{10}(OH,F)_2$
	Bityite	$Ca(Al_2Li)(Si_2AlBe)O_{10}(OH,F)_2$
	Anandite	$BaFe_3^{2+}(Si_3Fe^{3+})O_{10}(OH)S$
Dioctahedral	Margarite	$CaAl_2(Si_2Al_2)O_{10}(OH,F)_2$

biotites of effusive rocks are also generally richer in Fe^{3+} and Ti (and poorer in Fe^{2+}) with respect to biotites of intrusive counterparts.

In thermal metamorphism, biotite occurs in the clorite-sericite facies as a dispersed phase in the argillitic matrix and is stable up to low-grade cornubianites. In regional metamorphism, biotite is typical of argillitic and pelitic rocks up to the staurolite-garnet facies (biotite, biotite-sericite, biotite-chlorite, and albite-biotite schists, and garnet-staurolite micaschists).

Muscovite is the main term of the common dioctahedral micas, and it is found in acidic intrusive rocks (biotite and biotite-muscovite granites), although in subordinate quantities with respect to biotite. It is a common mineral in aplitic rocks and is peculiar to fluorine metasomatism in the contact zones between granites and slates ("greisenization").

In metamorphic rocks, muscovite occurs in low-grade terrains of the regional metamorphism (albite-chlorite-sericite schists). It must be noted here that the term "sericite" identifies fine-grained white micas (muscovite, paragonite).

Table 5.55 lists analyses of natural micas taken from the collection of Deer et al. (1983).

5.6.2 Intrinsic Stability Limits

The intrinsic stability of micas as a function of P and T intensive variables and chemistry is essentially related to the capability of the tetrahedral and octahedral

Table 5.54 Occurrences of main compositional terms of mica in igneous, metamorphic and sedimentary rocks.

Term	Igneous Rocks	Metamorphic Rocks	Sedimentary Rocks
Phlogopite	Peridotites	Metamorphosed limestones, dolomites	
Biotite	Gabbros, norites, diorites, granites, pegmatites	Phyllites, chlorite-sericite schists, biotite facies of the regional metamorphism, gneisses	
Muscovite	Granites, aplites	Phyllites, schists, gneisses	Detritic and authigenic sediments
Zinnwaldite Polylithionite	Pegmatites and high-T hydrothermal veins		
Glauconite			Marine sediments (green sands)

sheets of the 2:1 mixed layer to accommodate the metal-to-oxygen interionic distances, with positional rotations of tetrahedra on the basal plane of oxygens, as discussed by Hazen and Wones (1972).

Figure 5.44 shows the conformation of the hexagonal rings on the tetrahedral sheet for limiting values of the rotational angle ($\alpha_r = 0°$ and 12°, respectively). Due to structural constraints, the value of α_r depends directly on the mean cation-to-oxygen distances in [TO$_4$] tetrahedra (d_t) and in octahedra (d_o), according to

$$\cos \alpha_r = \frac{3\sqrt{3}\, d_o \sin \psi}{4\sqrt{2}\, d_t}, \qquad (5.113)$$

where ψ is the "octahedral flattening" angle (cf. Hazen and Finger, 1982), which is also a function of d_o:

$$\sin \psi \approx 1.154 - 0.144 d_o. \qquad (5.114)$$

As already noted, the structural limits of the rotational angle are 0° and 12°. This implies that the ratio of mean octahedral and tetrahedral distances is restricted in the range

$$1.235 \leq \frac{d_o}{d_t} \leq 1.275 \qquad (5.115)$$

(Hazen and Finger, 1982). Mean d_o and d_t distances vary as a function of the ionic radii of intervening cations and, chemical composition being equal, vary

Table 5.55 Chemical analyses of natural micas (from Deer et al., 1983). Note that ionic fractions are retrieved on a 24-anion basis—i.e. double the canonical formula. (1) Muscovite from a low-grade metamorphic prasinite schist; (2) glauconite from a sandstone; (3) phlogopite from a marble; (4) biotite from a quartz-bearing latite; (5) lepidolite from a pegmatite.

Oxide	(1)	(2)	(3)	(4)	(5)
SiO_2	48.42	49.29	40.95	39.14	49.80
TiO_2	0.87	0.12	0.82	4.27	0.00
Al_2O_3	27.16	3.17	17.28	13.10	25.56
Fe_2O_3	6.57	21.72	0.43	12.94	0.08
FeO	0.81	3.19	2.38	5.05	0.00
MnO	-	Traces	Traces	0.14	0.38
MgO	Traces	3.85	22.95	12.75	0.22
CaO	Traces	0.74	0.00	1.64	0.00
Na_2O	0.35	0.12	0.16	0.70	0.40
K_2O	11.23	6.02	9.80	6.55	9.67
F	Traces	-	0.62	1.11	6.85
$H2O^{(+)}$	4.31	7.21	4.23	2.41	0.38
$H2O^{(-)}$	0.19	4.60	0.48	0.58	0.50
	99.91	100.35	100.13	100.38	102.96
$O \equiv F$	-	-	0.26	0.46	2.89
Total	99.91	100.35	99.87	99.92	100.07
IONIC FRACTIONS ON A 24-ANION BASIS (O,OH,F)					
Si	6.597	7.634	5.724	5.790	6.750
Al	1.403	0.366	2.276	2.210	1.250
Total Z	8.00	8.00	8.00	8.00	8.00
Al	2.959	0.213	0.562	0.074	2.834
Ti	0.089	0.014	0.084	0.474	-
Fe^{3+}	0.672	2.532	0.340	1.440	0.008
Fe^{2+}	0.091	0.413	0.276	0.625	-
Mn	-	-	-	0.017	0.044
Mg	-	0.889	4.776	2.811	0.044
Total Y	3.81	4.06	6.04	5.44	6.17
Ca	-	0.123	-	0.260	-
Na	0.092	0.036	0.034	0.199	0.106
K	1.952	1.190	1.746	1.236	1.674
Total X	2.04	1.35	1.78	1.70	2.02
F	-	-	0.278	0.519	2.936
OH	3.916	4.00	3.946	2.378	0.344
Total (OH,F)	3.92	4.00	4.22	2.90	3.28

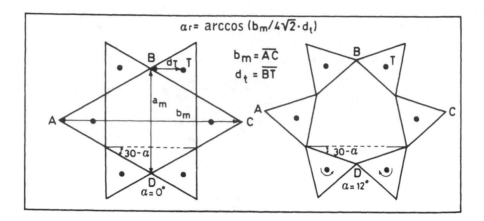

Figure 5.44 Sketch of tetrahedral rotational angle α_r for two limiting conditions, $\alpha_r = 0°$ and $\alpha_r = 12°$. From Hazen and Wones (1972). Reprinted with permission of The Mineralogical Society of America.

with T and P conditions according to the mean individual and compressibilities of the various polyhedra (see sections 1.14.2 and 1.15.1). Because the tetrahedra are relatively insensitive to P and T modifications, the effects of the two intensive variables are primarily registered by the d_o distances.

In common trioctahedral micas with Al:Si diadochy of 1:3 $[(KR_3^{2+}(Si_3Al)O_{10}(OH)_2]$, the mean T-O distance is 1.649 Å (Hazen and Burnham, 1973). Based on this value and on equations 5.113 and 5.114, we obtain

$$2.035 \leq d_o \leq 2.110. \tag{5.116}$$

Let us calculate the effects of intensive variables on the d_o distance, considering phlogopite $[KMg_3(Si_3Al)O_{10}(OH)_2]$ as an example. The octahedral cation is Mg^{2+}; the mean octahedral distance at $P = 1$ bar and $T = 25$ °C is about 2.06 Å (Hazen and Burnham, 1973). The mean polyhedral linear thermal expansion is $\alpha_l = 1.4 \times 10^{-5}(°C^{-1})$, and the mean polyhedral linear compressibility is $\beta_l = 1.7 \times 10^{-4}$ (kbar^{-1}). We know that the upper stability limit is $d_o = 2.110$ Å. We can solve the limiting P and T conditions by applying the equality

$$d_{o(T,P)} = d_{o(T_r,P_r)}\left(1 + \alpha_l T - \beta_l P\right) = 2.110, \tag{5.117}$$

from which we obtain

$$2.06\left(1 + 1.4 \times 10^{-5}T - 1.7 \times 10^{-4}P\right) = 2.110 \tag{5.118}$$

and

Table 5.56 P and T intrinsic stability limits of mica
according to Hazen and Finger (1982).

R Cation	Lower Limit $\alpha_r = 0°$	Upper Limit $\alpha_r = 12°$
General Formula $KR_3^{2+}(Si_3Al)O_{10}(OH)_2$		
Ni	$T = 11P + 200$	$T = 11P + 2800$
Mg	$T = 12P - 900$	$T = 12P + 1700$
Co	$T = 12P - 1700$	$T = 12P + 900$
Fe	$T = 13P - 2900$	$T = 13P - 300$
Mn	$T = 14P - 4600$	$T = 14P - 2000$
General Formula $K(R_{2.5}^{2+}Al_{0.5})(Si_{2.5}Al_{1.5})O_{10}(OH)_2$		
Ni	$T = 11P + 2100$	$T = 11P + 3400$
Mg	$T = 12P + 1000$	$T = 12P + 3400$
Co	$T = 12P + 300$	$T = 12P + 2800$
Fe	$T = 13P - 700$	$T = 13P + 1700$
Mn	$T = 14P - 2400$	$T = 14P$

$$T(°C) \approx 12P(\text{kbar}) + 1700. \qquad (5.119)$$

Analogously, for the lower limit,

$$2.06(1 + 1.4 \times 10^{-5}T - 1.7 \times 10^{-4}P) = 2.035 \qquad (5.120)$$

and

$$T(°C) \approx 12P(\text{kbar}) - 900. \qquad (5.121)$$

Table 5.56 lists intrinsic stability limits for common trioctahedral micas with general formulas $KR_3^{2+}(Si_3Al)O_{10}(OH)_2$ and $K(R_{2.5}^{2+}Al_{0.5})(Si_{2.5}Al_{1.5})O_{10}(OH)_2$, calculated by Hazen and Finger (1982) with the method described above. Based on these calculations, trioctahedral common micas with general formula $KR_3^{2+}(Si_3Al)O_{10}(OH)_2$ are structurally unstable for cations Fe^{2+} and Mn^{2+}, which are too large and exceed the limiting value $d_o/d_t = 1.275$. To keep the ratio within limits, it is necessary to increase the value of d_t, by means of a coupled substitution involving Al (cf. table 5.56):

$$Si^{IV} + R^{2+,VI} \Leftrightarrow Al^{VI} + Al^{IV}. \qquad (5.122)$$

Other elemental exchanges capable of reducing the d_o/d_t ratio (Hazen and Finger, 1982) are

$$R^{2+,VI} + (OH^-) \Leftrightarrow R^{3+,VI} + O^{2-} \qquad (5.123)$$

$$Al^{IV} \Leftrightarrow Fe^{3+,IV} \tag{5.124}$$

$$2R^{2+,VI} \Leftrightarrow Ti^{4+} + V_Y^{\parallel} \tag{5.125}$$

(with $V_Y^{\parallel}$ = bi-ionized cationic vacancy on site Y).

The structural model of Hazen and Wones has been confirmed experimentally. For example, the pure "annite" end-member $[KFe_3^{2+}(Si_3Al)O_{10}(OH)_2]$ appears to be geometrically unstable at all P and T conditions (table 5.56). Indeed, pure annite has never been synthesized, although slightly oxidized annites of composition $K(Fe_{2.7}^{2+}Fe_{0.3}^{3+})(Si_3Al)O_{10.3}(OH)_{1.7}$ have been prepared, with a d_o/d_t ratio of 1.275, at the limit of intrinsic stability.

5.6.3 Extrinsic Stability

Biotite-Phlogopite

Biotite is frequently found as a primary mineral in plutonic rocks, but its occurrence in effusive counterparts is quite rare and almost invariantly affected by evident signs of instability. The experimental explanation of this fact may be found in the classic paper by Yoder and Eugster (1954), which compared the P-T stability curve $KMg_3(Si_3Al)O_{10}(OH)_2$ (phlogopite) with the curves of incipient melting of granite and basalt. As figure 5.45 shows, whereas the stability curve of phlogopite crosses the curve of incipient melting of basalt at intermediate P and T conditions, it always plots to the right of the curve of incipient melting of granite at all P and T conditions (except at extremely low pressure). Phlogopite may thus form as a stable primary phase in granitic magmas, but is stable in basaltic melts only under high thermobaric conditions (i.e., above about 2.5 kbar and 1100 °C).

Further experimental studies on the stability of the magnesian phlogopite component in silica undersaturated and oversaturated systems have been carried out by various authors. Wones (1967) investigated the equilibrium

$$2KMg_3\left(Si_3Al\right)O_{10}\left(OH\right)_2 \Leftrightarrow KAlSi_2O_6 + KAlSiO_4 + 3Mg_2SiO_4 + 2H_2O. \tag{5.126}$$
$$\text{phlogopite} \qquad\qquad \text{leucite} \qquad \text{kalsilite} \qquad \text{forsterite}$$

Hewitt and Wones (1984) have shown that the equilibrium constant for the above reaction may be expressed as a polynomial function of P (bar) and T (K):

$$\log K_{126} = 2\log f_{H_2O} = -16633T^{-1} + 18.1 + 0.105(P-1)T^{-1}. \tag{5.127}$$

Wood (1976a) studied the stability of phlogopite in the presence of quartz:

$$KMg_3\left(Si_3Al\right)O_{10}\left(OH\right)_2 + 3SiO_2 \Leftrightarrow KAlSi_3O_8 + 3MgSiO_3 + H_2O. \tag{5.128}$$
$$\text{phlogopite} \qquad\qquad\qquad \text{quartz} \qquad \text{sanidine} \qquad \text{enstatite}$$

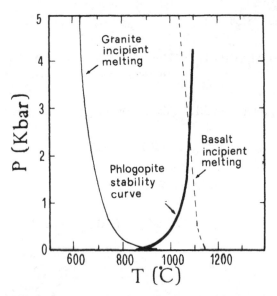

Figure 5.45 *P-T* stability curve of phlogopite compared with the incipient melting curves of granite and basalt. Reprinted from H. S. Yoder and H. P. Eugster, *Geochimica et Cosmochimica Acta,* 6, 157–185, copyright © 1954, with kind permission from Elsevier Science Ltd., The Boulevard, Langford Lane, Kidlington OX5 1GB, UK.

According to Hewitt and Wones (1984), the equilibrium constant may be expressed in the form

$$\log K_{128} = \log f_{H_2O} = -2356T^{-1} + 4.76 + 0.077(P - 1)T^{-1}. \qquad (5.129)$$

With increasing P and T, the equilibrium curve of equation 5.129 reaches an invariant condition, determined by the appearance of silicate melt ($T = 835 \, °C$, $P = 0.45$ kbar; Wones and Dodge, 1977).

Concerning the ferriferous term annite [$KFe_3^{2+}(Si_3Al)O_{10}(OH)_2$], the most important experimentally studied reactions are as follows:

$$\underset{\text{annite}}{2 \, KFe_3(Si_3Al)O_{10}(OH)_2} + H_2O \Leftrightarrow \underset{\text{sanidine}}{2 \, KAlSi_3O_8} + \underset{\text{hematite}}{3 \, Fe_2O_3} + 3 \, H_2 \qquad (5.130)$$

(Eugster and Wones, 1962);

$$\underset{\text{annite}}{KFe_3(Si_3Al)O_{10}(OH)_2} \Leftrightarrow \underset{\text{sanidine}}{KAlSi_3O_8} + \underset{\text{magnetite}}{Fe_3O_4} + H_2, \qquad (5.131)$$

with

$$\log K_{131} = \log f_{H_2} = -8113T^{-1} + 9.59 + 0.0042(P - 1)T^{-1} \qquad (5.132)$$

(Hewitt and Wones, 1981, 1984);

$$2\,KFe_3(Si_3Al)O_{10}(OH)_2 \Leftrightarrow 3\,Fe_2SiO_4 + KAlSi_2O_6 + KAlSiO_4 + 2\,H_2O, \qquad (5.133)$$
$$\underset{\text{annite}}{} \qquad \underset{\text{fayalite}}{} \quad \underset{\text{leucite}}{} \quad \underset{\text{kalsilite}}{}$$

which is analogous to reaction 5.126 valid for the magnesian component, with

$$\log K_{133} = 2\log f_{H_2O} = -17705T^{-1} + 22.44 + 0.112(P-1)T^{-1} \qquad (5.134)$$

(Eugster and Wones, 1962; Hewitt and Wones, 1984);

$$3\,KFe_3(Si_3Al)O_{10}(OH)_2 \Leftrightarrow 3\,Fe_2SiO_4 + 3\,KAlSi_2O_6 + Fe_3O_4 + 2\,H_2O + 2\,H_2, \qquad (5.135)$$
$$\underset{\text{annite}}{} \qquad \underset{\text{fayalite}}{} \quad \underset{\text{leucite}}{} \quad \underset{\text{magnetite}}{}$$

with

$$\log K_{135} = 2\log f_{H_2O} + \log f_{H_2} = -26044T^{-1} + 32.41 + 0.075(P-1)T^{-1} \qquad (5.136)$$

(Eugster and Wones, 1962; Hewitt and Wones, 1984); and

$$KFe_3(Si_3Al)O_{10}(OH)_2 + H_2 \Leftrightarrow Fe_2SiO_4 + KAlSi_2O_6 + Fe + 2\,H_2O \qquad (5.137)$$
$$\underset{\text{annite}}{} \qquad \qquad \underset{\text{fayalite}}{} \quad \underset{\text{leucite}}{} \quad \underset{\text{metal}}{}$$

$$KFe_3(Si_3Al)O_{10}(OH)_2 + 3\,H_2 \Leftrightarrow KAlSi_3O_8 + 3\,Fe + 4\,H_2O. \qquad (5.138)$$
$$\underset{\text{annite}}{} \qquad \qquad \underset{\text{sanidine}}{} \quad \underset{\text{metal}}{}$$

Figure 5.46 shows the stability field of the annite end-member at 1 kbar total pressure, based on equilibria 5.131, 5.133, and 5.135, as a function of temperature and hydrogen fugacity in the system. The figure also shows the positions of the iron-wuestite (IW) and wuestite-magnetite buffer (WM).

Muscovite-Paragonite

The P-T stability curve of muscovite intersects the incipient melting curve of granite in hydrous conditions at about 2.3 kbar total pressure and $T = 650$ °C. Thus, muscovite may crystallize as a primary phase from granitic melts above these P and T conditions (interstitial poikilitic crystals) or may form by reaction with pristine solid phases at lower P and T (muscovite as dispersed phase within feldspars, for instance). In this second type of occurrence, the following two equilibria are of particular importance:

$$KAl_2(Si_3Al)O_{10}(OH)_2 \Leftrightarrow KAlSi_3O_8 + Al_2O_3 + H_2O \qquad (5.139)$$
$$\underset{\text{muscovite}}{} \qquad \qquad \underset{\text{sanidine}}{} \quad \underset{\text{corundum}}{}$$

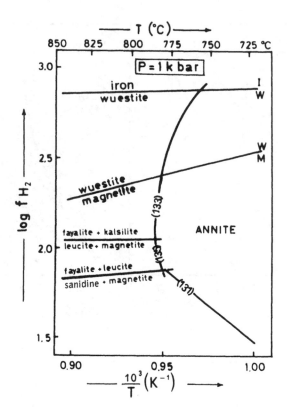

Figure 5.46 Stability field of annite at P_{total} = 1 kbar, as a function of T and f_{O_2}. From Hewitt and Wones (1984). Reprinted with permission of the Mineralogical Society of America.

$$KAl_2\left(Si_3Al\right)O_{10}\left(OH\right)_2 + SiO_2 \Leftrightarrow KAlSi_3O_8 + Al_2SiO_5 + H_2O.$$

muscovite   quartz   K-feldspar

$$(5.140)$$

Equilibrium 5.139 has been investigated by several authors. According to Hewitt and Wones (1984), the equilibrium constant may be expressed as a polynomial function of P (bar) and T (K):

$$\log K_{139} = \log f_{H_2O} = 8.5738 - 5126T^{-1} + 0.0331(P-1)T^{-1}. \qquad (5.141)$$

The univariant equilibrium 5.139 for $P_{total} = P_{H_2O}$, based on the experimental data of Chatterjee and Johannes (1974), is compared in figure 5.47A with the experimental results of Ivanov et al. (1973) for $X_{H_2O} = X_{CO_2} = 0.5$. Figure 5.47A also superimposes the melting curves of granite in the presence of a fluid phase of identical composition. Note that the composition of the fluid dramatically affects stability relations: if the amount of H_2O in the fluid is reduced by half ($X_{H_2O} = X_{CO_2} = 0.5$), the stability field of muscovite is restricted to approximately $P > 5.5$ kbar and $T > 680$ °C.

According to the experiments of Chatterjee and Johannes (1974), the equilibrium constant for equation 5.140 may be expressed within the stability field of andalusite in the form

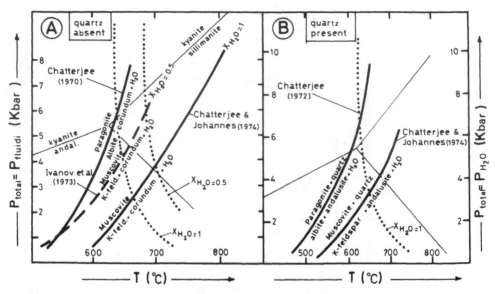

Figure 5.47 Extrinsic stability curves for muscovite and paragonite, based on (A) equilibria 5.139 and 5.144 (quartz absent) and (B) equilibria 5.140 and 5.145 (quartz present). Incipient melting curves of granite for $X_{H_2O} = 0.5$ to 1, and stability curves of Al_2SiO_5 polymorphs according to Richardson et al. (1969), are superimposed.

$$\log K_{140(andalusite)} = \log f_{H_2O}$$

$$= 8.3367 - 4682T^{-1} + 0.0163(P-1)T^{-1} \tag{5.142}$$

and, within the stability field of sillimanite, in the form

$$\log K_{140(sillimanite)} = \log f_{H_2O}$$

$$= 8.9197 - 5285T^{-1} + 0.0248(P-1)T^{-1}. \tag{5.143}$$

The validity limits for equations 5.142 and 5.143 (Al_2SiO_5 polymorphic transitions) are those of Richardson et al. (1969) (triple point at $P = 5.5$ kbar, $T = 620$ °C). In figure 5.48 we see the univariant equilibrium 5.140 plotted by Kerrick (1972) for different compositions of the fluid phase ($X_{CO_2} + X_{H_2O} = 1$). The effect of lowering the H_2O partial pressure in the fluid is analogous to previous observations for equilibrium 5.139. However, Kerrick (1972) adopted the andalusite-kyanite-sillimanite stability limits of Holdaway (1971) ($P = 3.8$ kbar, $T = 500$ °C).

The extrinsic stability relations of the sodic term paragonite [$NaAl_2(Si_3Al)O_{10}(OH)_2$] have been investigated as have those of the potassic end-member muscovite:

$$NaAl_2(Si_3Al)O_{10}(OH)_2 \Leftrightarrow NaAlSi_3O_8 + Al_2O_3 + H_2O \tag{5.144}$$
$$\text{paragonite} \qquad\qquad \text{albite} \qquad \text{corundum}$$

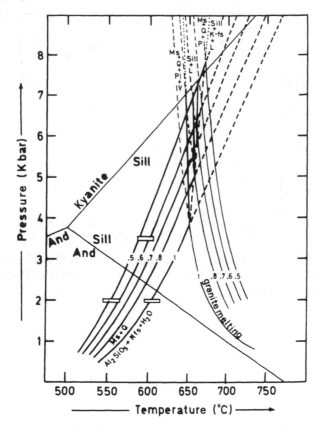

Figure 5.48 Extrinsic stability limits of muscovite, based on equilibrium 5.140 for variable amounts of H_2O component in fluid. Also shown are incipient melting curves of granite for various X_{H_2O} isopleths and stability curves of Al_2SiO_5 polymorphs, according to Holdaway (1971) From Kerrick (1972), *American Journal of Science*, 272, 946–58. Reprinted with permission of American Journal of Science.

$$
\underset{\text{paragonite}}{NaAl_2(Si_3Al)O_{10}(OH)_2} + \underset{\text{quartz}}{SiO_2} \Leftrightarrow \underset{\text{albite}}{NaAlSi_3O_8} + Al_2SiO_5 + H_2O. \tag{5.145}
$$

Evaluation of equilibria 5.144 and 5.145 is complicated by the formation of polytype *1M*, which is metastable with respect to *2M1*. The *1M* ⇔ *2M1* transformation is slow and often is not completed at low *T*. The most consistent data are those of Chatterjee (1970), who performed reversal experiments. The position of univariant equilibrium 5.144 in the *P-T* field according to this author is shown in figure 5.47A. Part B of that figure plots the position of the univariant equilibrium 5.145 according to Chatterjee (1972), compared with the stability limits for the potassic term. Note, by comparing equilibria 5.139 and 5.140 with equilibria 5.144 and 5.145, that the sodic terms are extrinsically stable, pressure being equal, at lower *T* (about 100 °C) with respect to their potassic counterparts. When studying the crystallization of dioctahedral micas from natural melts, it is therefore important to evaluate carefully the role of the Na/K ratio in the system.

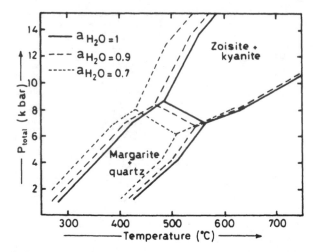

Figure 5.49 Stability field of margarite plus quartz in the $CaO-Al_2O_3-SiO_2-H_2O$ system for various values of $a_{H_2O, fluid}$. From Chatterjee (1974). Reprinted with permission of The Mineralogical Society of America.

Margarite

Chatterjee (1974) studied in detail the equilibrium

$$\underset{\text{margarite}}{CaAl_2(Si_2Al_2)O_{10}(OH)_2} \Leftrightarrow \underset{\text{anorthite}}{CaAl_2Si_2O_8} + \underset{\text{corundum}}{Al_2O_3} + H_2O \tag{5.146}$$

in the pressure range 1 to 7 kbar. According to Hewitt and Wones (1984), the equilibrium constant varies with P and T as

$$\log K_{146} = \log f_{H_2O} = 8.8978 - 4758T^{-1} - 0.0171(P-1)T^{-1}. \tag{5.147}$$

In the presence of quartz, the most significant equilibria are

$$\underset{\text{margarite}}{4CaAl_2(Si_2Al_2)O_{10}(OH)_2} + \underset{\text{quartz}}{3SiO_2} \Leftrightarrow \underset{\text{zoisite}}{2Ca_2Al_3Si_3O_{12}(OH)} + 5Al_2SiO_5 + 3H_2O$$

$$\tag{5.148}$$

and

$$\underset{\text{margarite}}{CaAl_2(Si_2Al_2)O_{10}(OH)_2} + \underset{\text{quartz}}{3SiO_2} \Leftrightarrow \underset{\text{anorthite}}{CaAl_2Si_2O_8} + Al_2SiO_5 + 3H_2O. \tag{5.149}$$

Figure 5.49 shows the P-T stability field of margarite plus quartz in the system $CaO-Al_2O_3-SiO_2-H_2O$, defined by Chatterjee (1976) for various values of H_2O activity in the fluid.

All these phase stability relations may be expressed, as we have seen, as func-

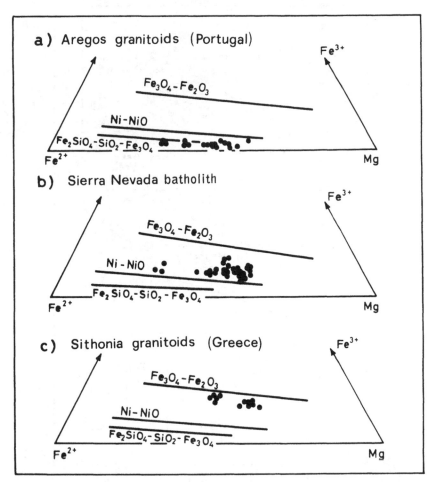

Figure 5.50 Chemistry of natural biotites as a function of the redox state of the system. Solid lines delineate the compositional trend imposed by the ruling buffer. From Speer (1984). Reprinted with permission of The Mineralogical Society of America.

tions of the fugacity of a gaseous component in the fluid phase (i.e., f_{H_2O}, f_{H_2}; see, however, section 9.1 for the significance of the term "fugacity"). It must be emphasized here that the stability and chemical composition of natural micas are also affected by the redox state of the system, which may itself be expressed as the fugacity of molecular oxygen in fluid phases. Wones and Eugster (1965) have shown that the oxidation state of biotites may be described as the result of mixing of three compositional terms: $KFe_3^{2+}AlSi_3O_{10}(OH)_2$, $KMg_3AlSi_3O_{10}(OH)_2$, and $KFe_3^{3+}AlSi_3O_{12}H_1$. In a ternary field of this type (figure 5.50), the compositional trend of micas depends on the buffer capability of the system (for the meaning of the term "buffer", see section 5.9.5).

Figure 5.51A shows in greater detail how the stability field of biotite (phlogopite-annite pseudobinary mixture) is affected by f_{O_2}-T conditions for a bulk pres-

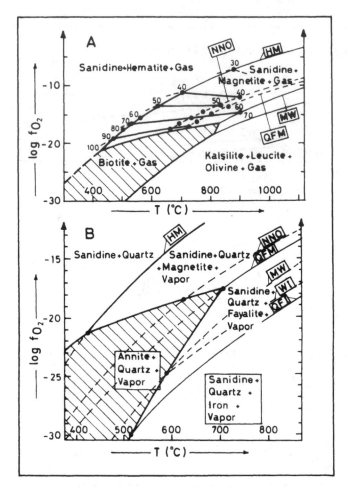

Figure 5.51 f_{O_2}-T stability field of biotite. (A) Pseudobinary phlogopite-annite mixture. Numbers at experimental points indicate observed Fe/(Fe + Mg) atom ratio. (B) Annite. HM = hematite-magnetite buffer; NNO = Ni-NiO buffer; MW = magnetite-wuestite buffer; QFM = quartz-fayalite-magnetite buffer. From Wones and Eugster (1965). Reprinted with permission of The Mineralogical Society of America.

sure of 2.07 kbar. Numbers at the various experimental points indicate the composition of the mixture. Tie lines connect points of identical composition at opposite sides of the stability region. Figure 5.51B shows the stability field of the annite end-member for the same bulk pressure. We have already noted that, for structural reasons, pure annite is intrinsically unstable at any P-T condition, and that the presence of Fe^{3+} is required to constrain the angular distortion of $[SiO_4]$ groups within structurally acceptable limits. The lower f_{O_2}-T stability limit of annite must thus be ascribed to this structural constraint (the amount of Fe^{3+} in oxidized annite obviously diminishes with decreasing f_{O_2} in the system; see also Wones and Eugster, 1965).

5.6.4 Thermodynamic Properties of End-Members and Mixing Behavior

Table 5.57 lists the thermodynamic data for mica end-members according to Helgeson et al. (1978), Berman (1988), Holland and Powell (1990) and Saxena et al. (1993). The substantial agreement among the last three sources may be ascribed to the similar optimization procedures adopted by the authors in assessing the energy properties of end-members through multivariate analysis. However, much experimental effort must still be made in order to gain proper knowledge of the energetics of micaceous components.

It is of interest here to recall the method suggested by Tardy and Garrels (1974) for the empirical estimate of the Gibbs free energy of layer silicates. This method, which is even simpler than the combinatory procedures mentioned in section 3.6, has an accuracy comparable to the discrepancy of existing experimental data.

According to Tardy and Garrels (1974), the Gibbs free energy of formation of layered silicates may be expressed as the simple sum of *fictive energies* ascribable to the constituent oxides, after suitable "decomposition" of the chemistry of the phase. Consider, for instance muscovite and paragonite: the Gibbs free energy of formation from the elements at $T = 298.15$ K and $P = 1$ bar may be expressed as the sum of the fictive silicate components ($G_{f,\text{sil}}$; see table 5.58):

$$G^0_{f,\text{KAl}_2\text{AlSi}_3\text{O}_{10}(\text{OH})_2} = \frac{1}{2}G^0_{f,\text{sil},\text{K}_2\text{O}} + \frac{3}{2}G^0_{f,\text{sil},\text{Al}_2\text{O}_3} + 3G^0_{f,\text{sil},\text{SiO}_2} + G^0_{f,\text{sil},\text{H}_2\text{O}}$$

$$= \frac{1}{2}\left(-786.6\right) + \frac{3}{2}\left(-1600.0\right) + 3\left(-856.0\right) + \left(-247.7\right) \quad (5.150)$$

$$= -5609.0$$

$$G^0_{f,\text{NaAl}_2\text{AlSi}_3\text{O}_{10}(\text{OH})_2} = \frac{1}{2}G^0_{f,\text{sil},\text{Na}_2\text{O}} + \cdots$$

$$ = -5556.3 \qquad \text{(kJ/mole)}. \quad (5.151)$$

Note that the Gibbs free energy of formation from the elements of the fictive constituent oxides does not correspond to the actual thermodynamic value. It differs from it by an empirical term ($\Delta G^0_{f,\text{silicification}}$) that accounts for the structural difference between the oxide in its stable standard form and as a formal entity present in the layered silicate (cf. table 5.58).

Table 5.59 lists Gibbs free energies of formation from the elements of mica end-members obtained with the procedure of Tardy and Garrels (1974). For comparative purposes, the same table lists Gibbs free energies of formation from the elements derived from the tabulated H^0_{f,T_r,P_r} and $S^0_{T_r,P_r}$ values (same sources as in table 5.57) by application of

$$S^0_{f,T_r,P_r} = S^0_{T_r,P_r} - \sum_i v_i S^0_{i,T_r,P_r} \quad (5.152)$$

Table 5.57 Thermodynamic properties of mica end-members in databases of Helgeson et al. (1978) (1), Saxena et al. (1993) (2), Berman (1988) (3), and Holland Powell (1990) (4). H^0_{f,T_r,P_r} = enthalpy of formation from the elements at standard state (T_r = 298.15 K, P_r = 1 bar; kJ/mole); $S^0_{T_r,P_r}$ = standard state entropy [J/(mole × K)]; $V^0_{T_r,P_r}$ = standard state volume (cm³/mole); α_0 = isobaric thermal expansion (K^{-1}); β_0 = isothermal compressibility (bar^{-1}). Heat capacity function: $C_P = K_1 + K_2 T + K_3 T^{-2} + K_4 T^2 + K_5 T^{-3} + K_6 T^{-1/2} + K_7 T^{-1}$.

End-Member	Formula	Reference	H^0_{f,T_r,P_r}	$S^0_{T_r,P_r}$	$V^0_{T_r,P_r}$
Phlogopite	KMg₃AlSi₃O₁₀(OH)₂	(1)	−6226.072	318.40	149.66
		(2)	−6203.900	334.600	149.66
		(3)	−6207.342	334.158	149.77
		(4)	−6211.76	325.0	149.64
Na-phlogopite	NaMg₃AlSi₃O₁₀(OH)₂	(4)	−6173.64	315.0	144.50
F-phlogopite	KMg₃AlSi₃O₁₀(F)₂	(1)	−6419.474	317.57	146.37
Annite	KFe₃AlSi₃O₁₀(OH)₂	(1)	−5155.504	398.317	154.32
		(2)	−5125.500	440.910	154.32
		(4)	−5149.32	414.0	154.32
Muscovite	KAl₂AlSi₃O₁₀(OH)₂	(1)	−5972.275	287.86	140.71
		(2)	−5976.300	306.40	140.82
		(3)	−5976.740	293.157	140.87
		(4)	−5981.63	289.0	140.83
Paragonite	NaAl₂AlSi₃O₁₀(OH)₂	(1)	−5928.573	277.82	132.53
		(2)	−5951.0	277.100	131.88
		(3)	−5944.208	277.70	132.16
		(4)	−5948.15	276.0	131.98
Celadonite	KMgAlSi₄O₁₀(OH)₂	(1)		313.38	157.1
		(4)	−5834.27	297.0	139.60
Fe-celadonite	KFeAlSi₄O₁₀(OH)₂	(4)	−5484.96	328.0	140.00
Margarite	CaAl₂Al₂Si₂O₁₀(OH)₂	(3)	−6236.603	265.084	129.58
		(4)	−6241.64	265.0	129.64
Eastonite	KMg₂AlAl₂Si₂O₁₀(OH)₂	(4)	−6336.56	315.0	147.51
Siderophyllite	KFe₂AlAl₂Si₂O₁₀(OH)₂	(4)	−5628.27	375.0	150.63

End-Member	Reference	K_1	$K_2 \times 10^3$	$K_3 \times 10^{-5}$	K_4	$K_5 \times 10^{-8}$
Phlogopite	(1)	420.952	120.416	−89.956	0	0
	(2)	554.42	27.835	−16.458	0	−1.0629
	(3)	610.38	0	−215.33	0	28.4104
	(4)	770.3	−36.939	−23.289	0	0
Na-phlogopite	(4)	773.5	−40.229	−25.979	0	0
F-phlogopite	(1)	402.21	111.88	−85.395	0	0
Annite	(1)	445.303	124.558	−80.793	0	0
	(2)	626.50	62.63	132.76	0	−7.947
	(4)	809.2	−70.250	−6.789	0	0
Muscovite	(1)	408.191	110.374	−106.441	0	0
	(2)	462.6	65.98	−333.74	0	47.388

***Table* 5.57** (*continued*)

End-Member	Reference	K_1	$K_2 \times 10^3$	$K_3 \times 10^{-5}$	K_4	$K_5 \times 10^{-8}$
	(3)	651.49	0	−185.232	0	27.4247
	(4)	756.4	−19.84	−21.70	0	0
Paragonite	(1)	407.647	102.508	−110.625	0	0
	(2)	502.209	32.215	−196.23	0	26.177
	(3)	577.57	0	−322.144	0	50.5008
	(4)	803.0	−31.58	2.17	0	0
Celadonite	(1)	335.77	105.855	−77.571	0	0
	(4)	741.2	−18.748	−23.688	0	0
Margarite	(3)	699.80	0	−68.077	0	7.3432
	(4)	744.4	−16.8	−20.744	0	0
Eastonite	(4)	785.5	−38.031	−21.303	0	0
Siderophyllite	(4)	815.8	−36.645	−5.645	0	0

End-Member	Reference	$K_6 \times 10^{-2}$	$K_7 \times 10^{-4}$	T Limit (K)	$\alpha_0 \times 10^5$	$\beta_0 \times 10^6$
Phlosopite	(1)	0	0	1000	-	-
	(2)	0	−5.5447	-	2.0	1.5128
	(3)	−20.838	0	1000	-	-
	(4)	−65.316	0	-	2.6063	1.6373
Na-phlogopite	(4)	−65.126	0	-	2.9758	1.6609
F-phlogopite	(1)	0	0	1000	-	-
Annite	(1)	0	0	1000	-	-
	(2)	0	−11.67	-	1.7	1.3897
	(4)	−74.03	0	-	2.7864	0.9720
Muscovite	(1)	0	0	1000	-	-
	(2)	0	1.2027	-	2.5	1.1546
	(3)	−185.232	0	967	-	-
	(4)	−69.792	0	-	2.7693	1.065
Paragonite	(1)	0	0	1000	-	-
	(2)	0	−2.1204	-	1.5	1.1620
	(3)	−14.728	0	719	-	-
	(4)	−81.51	0	-	3.1823	1.2881
Celadonite	(1)	0	0	1000	-	-
	(4)	−66.169	0	-	2.7937	1.1461
Margarite	(3)	−55.871	0	996	-	-
	(4)	−67.832	0	-	2.7769	1.0028
Eastonite	(4)	−68.937	0	-	2.5761	1.6270
Siderophyllite	(4)	−75.171	0	-	2.7219	1.1950

Table 5.58 Gibbs free energy of formation from elements at $T = 298.15$ K and $P = 1$ bar for fictive structural oxide components of layered silicates (Tardy and Garrels, 1974) compared with the actual thermodynamic values of stable oxides (data in kJ/mole).

Component	$G^0_{f,\text{oxide}}$	$G^0_{f,\text{sil}}$	$G^0_{f,\text{silicification}}$
H_2O	−237.2	−247.7	−10.5
Li_2O	−561.9	−797.5	−235.6
Na_2O	−377.4	−681.2	−303.8
K_2O	−322.2	−786.6	−464.4
CaO	−604.2	−764.8	−160.6
MgO	−569.4	−624.3	−54.9
FeO	−251.5	−268.2	−16.7
Fe_2O_3	−743.5	−743.5	0
Al_2O_3	−1582.0	−1600.0	−18.0
SiO_2	−856.0	−856.0	0

Table 5.59 Gibbs free energies of formation from the elements of micaceous components according to method of Tardy and Garrels (1974), compared with tabulated values. References as in table 5.57. Data in kJ/mole.

Component	$G^0_{f,\text{calc}}$	$\sum_i v_i S^0_i$	$G^0_{f,\text{obs}}$ (1)	(2)	(3)	(4)
$KMg_3AlSi_3O_{10}(OH)_2$	−5881.9	1571.46	−5852.5	−5835.1	−5838.4	−5840.1
$NaMg_3AlSi_3O_{10}(OH)_2$	−5818.7	1558.08	-	-	-	−5803.0
$KFe_3AlSi_3O_{10}(OH)_2$	−4813.6	1555.26	−4810.6	−4793.3	-	−4809.1
$KAl_2AlSi_3O_{10}(OH)_2$	−5609.0	1567.74	−5590.7	−5600.2	−5596.7	−5600.4
$NaAl_2AlSi_3O_{10}(OH)_2$	−5556.3	1554.36	−5547.9	−5570.2	−5563.6	−5567.0
$KMgAlSi_4O_{10}(OH)_2$	−5489.3	1562.53	-	-	-	−5456.9
$KFeAlSi_4O_{10}(OH)_2$	−5133.2	1557.13	-	-	-	−5118.5
$CaAl_2Al_2Si_2O_{10}(OH)_2$	−5924.5	1554.23	-	-	−5852.2	−5857.3
$KMg_2AlAl_2Si_2O_{10}(OH)_2$	−6001.6	1553.26	-	-	-	−5967.4
$KFe_2AlAl_2Si_2O_{10}(OH)_2$	−5928.4	1603.49	-	-	-	−5262.0
$KFe_3^{2+}Fe^{3+}Si_3O_{10}(OH)_2$	−4385.4	1591.81	-	-	-	-
$KLi_2AlSi_4O_{10}(OH)_2$	−5662.5	1588.09	-	-	-	-

$$G^0_{f,T_r,P_r} = H^0_{f,T_r,P_r} - T_r S^0_{f,T_r,P_r}, \qquad (5.153)$$

where $\sum_i v_i S^0_{i,T_r,P_r}$ is the sum of standard state entropies of the elements at stable state (Robie et al., 1978), multiplied by their stoichiometric factors.

The method of Tardy and Garrels (1974) gives approximate estimates of Gibbs free energy values for mica compounds regardless of their chemical complexity. Nevertheless, the linearity of the method cannot account for the energy of mixing contributions. Experimental data on mica mixtures are also

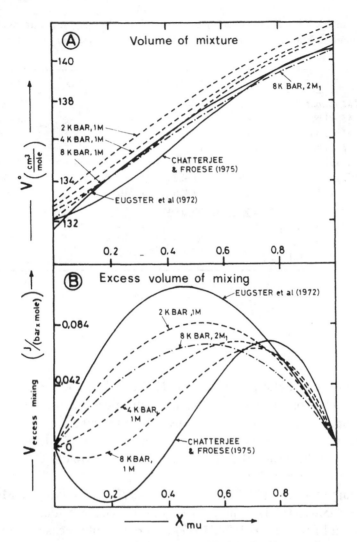

Figure 5.52 (A) Molar volumes of the $(Na,K)Al_2Si_3AlO_{10}(OH)_2$ mixture at various P conditions. (B) Corresponding excess volumes of mixing. From Blencoe (1977). Reprinted with permission of The Mineralogical Society of America.

limited to a few binary joins, and experimental observations are complicated by polymorphism (polytypism) and by metastable persistence at low T of polytypes that do not correspond to the minimum energy structure at the P, T, and X of interest. For instance, figure 5.52A shows the molar volumes of $NaAl_2Si_3AlO_{10}(OH)_2$-$KAl_2Si_3AlO_{10}(OH)_2$ (paragonite-muscovite) mixture plotted by Blencoe (1977) according to existing experimental data. Note the clear discrepancies in the molar volume of the same composition and conspicuous deviations from ideal mixing behavior. Figure 5.52B shows the corresponding excess volumes of the mixture. According to Blencoe (1977), volumetric excess terms may be represented by a subregular Margules model, with the parameters listed in

Table 5.60 Volumetric [W_V; J/(bar × mole)] enthalpic (W_H; kJ/mole), and entropic [W_S; J/(mole × K)] terms of subregular Margules model for $(Na,K)Al_2Si_3AlO_{10}(OH)_2$ binary mixture, according to various authors.

References	$W_{H,12}$	$W_{H,21}$	$W_{S,12}$	$W_{S,21}$	$W_{V,12}$	$W_{V,21}$
Blencoe (1977) *1M*, 2 kbar	-	-	-	-	0.3628	0.3151
Blencoe (1977) *1M*, 4 kbar	-	-	-	-	0.4502	0.0594
Blencoe (1977) *1M*, 8 kbar	-	-	-	-	0.5192	−0.1519
Blencoe (1977) *2M*, 8 kbar	-	-	-	-	0.3138	0.2389
Blencoe and Luth (1973)	-	-	-	-	−0.1423	−0.3222
Eugster et al. (1972)	12.895	17.422	0.711	1.653	0.344	0.527
Chatterjee and Froese (1975)	12.230	19.456	0.710	1.654	0.665	−0.456

(1) = $NaAl_2Si_3AlO_{10}(OH)_2$ and (2) = $KAl_2Si_3AlO_{10}(OH)_2$

$W_{12} = W_{H,12} - TW_{S,12} + PW_{V,12}$

$W_{21} = W_{H,21} - TW_{S,21} + PW_{V,21}$

$G_{excess\ mixing} = X_1X_2(W_{12}X_2 + W_{21}X_1)$

table 5.60. The same table also lists enthalpic and entropic subregular interaction parameters for the same mixture according to Eugster et al. (1972) and Chatterjee and Froese (1975).

Because excess volume terms are generally positive (with the exception of the values indicated by Chatterjee and Froese, 1975 and Blencoe and Luth, 1973) and because Margules parameters are also generally positive, the mixing model of table 5.60 implies the existence of a solvus field in the binary range, whose extension should widen with increasing pressure [at $T = 500$ °C and $P_{total} = P_{H_2O} = 4$ kbar, the solvus field should cover the range $0.1 < KAl_2AlSi_3O_{10}(OH)_2 < 0.85$; Chatterjee and Froese, 1975].

According to Wones (1972) and Mueller (1972), the mixture $KMg_3AlSi_3O_{10}$ $(OH)_2$-$KFe_3AlSi_3O_{10}(OH)_2$ (phlogopite-annite) is virtually ideal.

Munoz and Ludington (1974, 1977) experimentally measured the ionic exchange reaction

$$OH_{mica}^- + HF \Leftrightarrow F_{mica}^- + H_2O \qquad (5.154)$$

in muscovite, siderophyllite, annite, and phlogopite in systems F-buffered by the solid assemblage anorthite-fluorite-sillimanite-quartz at $673 < T(K) < 1000$ and $P_{fluid} = 1$ to 2 kbar. For all four phases, the *T*-dependency of the equilibrium constant was identical—i.e.:

$$KMg_3(AlSi_3)O_{10}(OH,F)_2 \rightarrow \log K_{154} = 2100T^{-1} + 1.52 \qquad (5.155)$$

$$KFe_3(AlSi_3)O_{10}(OH,F)_2 \rightarrow \log K_{154} = 2100T^{-1} + 0.41 \qquad (5.156)$$

$$K\left(Fe_2^{2+}Al\right)\left(Al_2Si_2\right)O_{10}(OH,F)_2 \rightarrow \log K_{154} = 2100T^{-1} + 0.20 \quad (5.157)$$

$$KAl_2\left(AlSi_3\right)O_{10}(OH,F)_2 \rightarrow \log K_{154} = 2100T^{-1} - 0.11. \quad (5.158)$$

Let us now consider the exchange reaction

$$KMg_3\left(AlSi_3\right)O_{10}(OH)_2 + KFe_3\left(AlSi_3\right)O_{10}(F)_2 \Leftrightarrow KMg_3\left(AlSi_3\right)O_{10}(F)_2 + KFe_3\left(AlSi_3\right)O_{10}(OH)_2.$$

OH-phlogopite $\quad$ F-annite $\qquad\qquad$ F-phlogopite $\qquad$ OH-annite

$$(5.159)$$

The constant of equilibrium 5.159 is given by the ratio of the constant relative to exchanges 5.155 and 5.156—i.e. in logarithmic notation,

$$\log K_{159} = \log K_{155} - \log K_{156} = 1.11. \quad (5.160)$$

Independently of T, reaction 5.159 thus appears to proceed spontaneously toward the right, in agreement with the "principle of Fe-F incompatibility" in silicates, postulated long ago by Ramberg (1952). This fact also implies the nonideality of the system. According to Munoz (1984), the activity coefficients of OH-phlogopite and F-phlogopite in mixture are defined by

$$\log \gamma_{OH-phlogopite} = -X_{Fe}X_F \log K_{159} \quad (5.161)$$

$$\log \gamma_{F-phlogopite} = X_{Fe}X_{OH} \log K_{159}. \quad (5.162)$$

Hinrichsen and Schurmann (1971) and Franz et al. (1977) detected the existence of a solvus field between paragonite [$NaAl_2(Si_3Al)O_{10}(OH)_2$] and margarite [$CaAl_2(Si_2Al_2)O_{10}(OH)_2$] by X-ray diffractometry and IR spectroscopy. Measurements in the pressure range $1 \leq P_{total}$ (kbar) ≤ 6 show a two-phase region that covers the compositional field $0.20 < X_{CaAl_2(Si_2Al_2)O_{10}(OH)_2} < 0.55$ at $T = 400$ °C. The mixture appears to be homogeneous at $T > 600$ °C.

5.7 Feldspars

Feldspars are *tectosilicates* (multiplicity = 4, periodicity = 4, dimensionality = 3): the tetrahedral groups [TO_4] share all their corners with neighboring [TO_4] units, thus forming a rigid tridimensional network. In feldspars, 25 to 50% of the silicon atoms are replaced by Al^{3+}. The basic structure of the network is made up of four-member rings of [TO_4] groups, two-by-two in an upward- and downward-

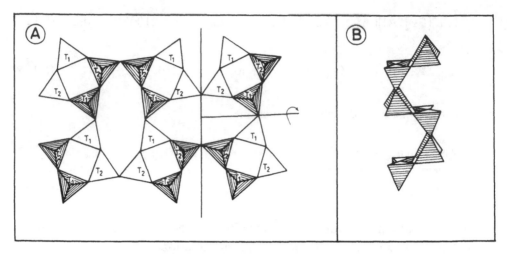

Figure 5.53 Idealized feldspar structure (topologic symmetry). (A) Projection from axis *a*, showing conformation of four-member rings composed of two nonequivalent upward-directed *T*1-*T*2 tetrahedra and two nonequivalent *T*1-*T*2 downward-directed tetrahedra. (B) Double "gooseneck" chain seen along axis *a*. Periodicity along axis *a* is 4 (about 8.4 Å).

directed (UUDD; Up-Up-Down-Down) configuration and linked to form a chain along axis *c* (figure 5.53B). The chains are interconnected, as shown in figure 5.53A. Inside the ring there is an interstitial site *M*, with VII-fold to IX-fold coordination. The structure as a whole is relatively rigid (or "inflexible" in the sense of J. V. Smith; cf. Merlino, 1976), because the [TO$_4$] groups of the same chain are alternately linked to two neighboring chains. This sort of connection prevents any cooperative rotational movement of the four-member rings.

5.7.1 Chemistry and Occurrence

Feldspars have the general formula XZ_4O_8. The *X* cation is constituted essentially of alkalis (Na^+, K^+, Rb^+) or alkaline earths (Ca^{2+}, Ba^{2+}, Sr^{2+}). The net charge of the [Z_4O_8] radical depends on the Al:Si diadochy level, and varies between 1− and 2−. The occupancy of the *M* site (univalent and/or divalent *X* cations) is thus determined by the net charge of the four-member rings. The relative amplitude of the *M* site and the need to limit the charge to 1+ or 2+ results in relatively simple chemistry because high-field-strength ions, transition elements, and Mg^{2+} are *a priori* excluded.

There are three main components of feldspar: $NaAlSi_3O_8$ (albite), $KAlSi_3O_8$ (K-feldspar) and $CaAl_2Si_2O_8$ (anorthite). These main components may also, for completeness, be associated with $BaAl_2Si_2O_8$ (celsian) and $RbAlSi_3O_8$, although these components are subordinate in natural mixtures (see table 5.61). The capability of feldspars to bear vacant sites in *M* positions is clearly demonstrated by experimental evidence. Grundy and Ito (1974), for instance, observed the existence of 13% vacant sites in M positions in an Sr-feldspar studied by single crystal X-ray diffraction. Extrinsic disorder (deviation from ideal XZ_4O_8 stoichiometry) also occurs as silica and alumina excesses: Kim and Burley (1971) observed a 5% SiO_2 excess in albite synthesized at $P = 5$ kbar and $T = 670$ °C; Goldsmith (1980) observed an Al deficiency with respect to the canonical formula in anorthite coexisting with corundum and synthesized at $P > 10$ kbar. As regards minor element and trace element levels, natural feldspars have Fe^{3+} and Ti^{4+} substituting for Si^{4+} in tetrahedral positions and Li^+, Rb^+, Cs^+, Sr^{2+}, Ba^{2+}, Fe^{2+}, and Mg^{2+} in the M sites.

Table 5.61 summarizes synthetically the observed occurrences of various trace elements in feldspars, based mainly on the indications of Smith (1983).

Feldspars are the most abundant minerals of igneous rocks, where their ubiquity and abundance of their components influence normative classifications. They are also abundant in gneisses, and may be observed in several facies of thermal and regional metamorphic regimes. Notwithstanding their alterability, they are ubiquitously present in sedimentary rocks, as authigenic and/or detritic phases. Only in carbonaceous sediments is their presence subordinate.

Table 5.62 lists the chemical compositions of natural feldspars, taken from Deer et al. (1983). Feldspar chemistry is solved on a 32-oxygen anhydrous basis, corresponding to four formula units. The X cation summation is quite close to the stoichiometric value (4); the Z cations are also close to the theoretical value (16).

5.7.2 Structural Properties and Intracrystalline Disorder

Feldspars crystallize in the monoclinic system (spatial groups $C2/m$ and $I2/m$) and the triclinic system (spatial groups $C\bar{1}$ and $P\bar{1}$. We have already noted that the topologic symmetry (i.e., the highest symmetry reached by the structure in the absence of distortion) identifies two sets of nonequivalent tetrahedra $T1$ and $T2$. In this unperturbed condition, the structure is monoclinic (e.g., high-T sanidine, $KAlSi_3O_8$). In the absence of distortion, because the tetrahedra are occupied by either Si^{4+} or Al^{3+} (in sanidine, the Al:Si proportion is 1:3), the $T1$ and $T2$ site populations are statistically identical and correspond to a stoichiometric ratio of 1:3. Whenever the Al^{3+} and Si^{4+} ions tend to order (as happens, for instance, at parity of chemical composition, with decreasing temperature), a new structural configuration is reached, with duplication of nonequivalent tetrahedral positions (i.e., from $T1$ and $T2$ to $T1_o$, $T1_m$, $T2_o$, and $T2_m$). In the condition of maximum ordering, Al^{3+} occupies only one of these sites ($T1_o$), and silicon occupies the remaining three nonequivalent positions (microline, $KAlSi_3O_8$; figure 5.54A). The site preference of aluminum for $T1_o$ may be ascribed to simple electrostatic reasons (oxygens surrounding $T1$ are more tightly linked to X cations than are oxygens surrounding $T2$, and the occupancy of $T1$ by Al^{3+} is thus energetically favored). In $KAlSi_3O_8$, the intracrystalline diffusion of Al^{3+} and Si^{4+} necessary to attain ordered distribution is difficult to achieve at a reasonable rate, because

Table 5.61 Trace elements in feldspars, according to Smith (1983).

Element	Mean Concentration	Notes
B	100 ppm	Forms pure component $NaBSiO_3$ (reedmergnerite); maximum observed amount in mixture corresponds roughly to 1 weight % B_2O_5 (Desborough, 1975).
Ga	10–100 ppm	Tendencially correlated to Na; highest concentration observed in pegmatites.
Ge	1–10 ppm	-
Ti	30–60 ppm	Occasionally high concentrations (up to 600–700 ppm) in intermediate members of plagioclase series (labradorite; bytownite).
P	20–200 ppm	Tendencially increases with increasing albite component. High concentrations observed by ion-probe analyses in perthites (Mason, 1982).
Be	1–10 ppm	Marked correlation with Na (Steele et al., 1980a). Locally high concentrations in Be-rich pegmatites.
Sn	1–10 ppm	Much higher values observed in greisens.
Fe	100–1000 ppm	Forms pure component $KFeSi_3O_8$ and is mainly present in valence state 3+.
Mg	100–3000 ppm	Positively correlated with amount of Ca^{2+}.
Mn	10–100 ppm	-
Li	1–50 ppm	Marked correlation with Na (Steele et al., 1980b, 1981).
Rb	1 ppm to %	Forms pure component $RbAlSi_3O_8$ (cf. table 5.63); marked correlation with K.
Cs	(?)	Majority of data is dubious because of presence of mica impurities (Smith, 1983).
Ta	0.01–30 ppm	Highest concentrations in alkaline feldspars.
Sr	10–100 ppm 100–5000 ppm	In alkaline feldspars of pegmatites. In other occurrences. Marked correlation with Na in comagmatic series (Steele and Smith, 1982).
Ba	100–1000 ppm	Forms pure component $BaAl_2Si_2O_8$ (celsian; cf. table 5.63); positively correlated with K. Exceptionally high values observed in Mn deposit, where celsian occurs as main component.
Pb	1–100 ppm	Exceptionally high concentrations (up to 10,000 ppm) in K-feldspars from pegmatites.
Cu	0–30 ppm	Marked correlation with amount of anorthite in mixture (Ewart et al., 1973).
REE	-	REE amounts in feldspars have been carefully studied, particularly for presence of a positive anomaly in Eu concentration, associated with presence of Eu^{2+}. This anomaly allows estimates of the redox state of the system and is discussed in chapter 10.
NH_4	1–200 ppm	Forms pure component $NH_4Al_3Si_3O_8$ (buddingtonite), sporadically found in hydrothermally altered layers.

Table 5.62 Chemical analyses of some natural feldspars (Deer et al., 1983). (1) orthoclase (Mogok, Burma); (2) Adularia (St. Gottard, Switzerland); (3) anorthoclase: inclusion in augite (Euganean Hills NE Italy); (4) sanidine from a nepheline-leucitite; (5) albite from a pegmatite; (6) anorthite from a calc-silicatic rock.

Oxide	Sample					
	(1)	(2)	(3)	(4)	(5)	(6)
SiO_2	63.66	64.28	63.70	63.62	67.84	43.88
TiO_2	-	-	-	0.08	0.00	-
Al_2O_3	19.54	19.19	21.83	19.12	19.65	36.18
Fe_2O_3	0.10	0.09	0.18	0.47	0.03	0.08
FeO	-	-	-	-	0.02	0.00
MgO	-	0.10	0.14	0.05	0.04	-
BaO	-	0.11	-	1.56	-	-
CaO	0.50	0.11	2.75	0.05	0.00	19.37
Na_2O	0.80	0.92	7.55	2.66	11.07	0.22
K_2O	15.60	15.30	3.75	12.09	0.29	0.00
$H_2O^{(+)}$	-	0.36	0.19	0.11	0.56	0.28
$H_2O^{(-)}$	-	-	-	0.00	0.30	0.08
Total	100.20	100.46	100.09	99.81	99.80	100.10
Ionic fractions on a 32 oxygens basis						
Si	11.759	11.852	11.383	11.770	11.964	8.126
Al	4.254	4.170	4.598	4.169	4.085	7.898
Fe^{3+}	0.014	0.012	0.024	0.096	0.004	0.011
Ti	-	-	-	0.011	-	-
Total Z	16.03	16.03	16.00	16.05	16.05	16.03
Mg	-	0.028	0.038	0.013	0.011	-
Na	0.286	0.329	2.616	0.954	3.785	0.079
Ca	0.099	0.022	0.526	0.010	-	3.884
K	3.676	3.599	0.855	2.855	0.066	-
Ba	-	0.008	-	0.113	-	-
Total X	4.06	3.99	4.03	3.95	3.87	3.92
$KAlSi_3O_8$	90.5	90.5	21.2	75.5	1.7	-
$NaAlSi_3O_8$	7.1	8.3	64.8	23.9	98.0	2.0
$CaAlSi_2O_2$	2.4	1.2	14.0	0.6	0.3	98.0

the energy of activation of the diffusive process (which embodies reticular distortion effects) is quite high as a result of the rigidity of the structure. Low-T persistency of metastable forms with intermediate intracrystalline disorder (adularia and orthose) is thus often observed.

In $NaAlSi_3O_8$, because the ionic size of Na^+ is much smaller than that of K^+, reticular distortion is implicit in the structure for simple topologic reasons, and migration of tetrahedral ions is energetically easier, not implying symmetry

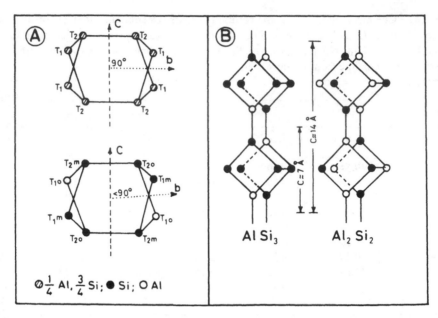

Figure 5.54 (A) Cationic occupancies in tetrahedral positions in case of complete disorder (monoclinic structure; upper drawing) and complete order (triclinic structure; lower drawing). (B) Condition of complete order in microcline and low albite with Al:Si = 1:3, compared with cationic ordering in anorthite (Al:Si = 2:2). Note doubling of edge c in anorthite.

modifications (in other words, the energy of activation of the migration process is lower). "High albite" and "low albite," both crystallizing in the triclinic system, represent the two opposite arrangements of maximum disorder and maximum order, respectively. The high-T monoclinic form "monalbite" (isostructural with sanidine) is stable only in the vicinity of the melting point (i.e., 980 to 1118 °C at $P = 1$ bar).

In tectosilicates, the Al-O-Al bonds are energetically unstable with respect to the Al-O-Si arrangement, which better conforms to Pauling's electrostatic valence principle. It is essentially this fact that leads to the alternate ordering of silicon and aluminum in Al:Si=1:1 feldspars ("aluminum avoidance principle"; Loewenstein, 1954). Thus, in anorthite ($CaAl_2Si_2O_8$), because Al:Si=1:1, at all T conditions we observe a natural invariant ordering of $[AlO_4]$ tetrahedral groups alternating with $[SiO_4]$ groups. As shown in figure 5.54, this alternation implies doubling of cell edge c.

Calorimetric data on sodium and calcium aluminosilicate crystals and glasses (Navrotsky et al., 1982, 1985) indicate that, in an exchange reaction of type

$$\left[\cdots Si - O - Si\right] + \left[\cdots Al - O - Al\right]^{2-} \Leftrightarrow 2\left[\cdots Si - O - Al\right]^{1-}, \qquad (5.163)$$

the gain in energy falls between -36 and -100 kJ per mole of substance, depending on the type of nontetrahedral coordinating cation. This value agrees quite well with two-body and three-body interionic potential lattice energy calculations for zeolites (-40 kJ/mole; Bell et al., 1992) and with more recent *ab initio* SCF estimates (Tossell, 1993; -40 kJ/mole)

Intracrystalline ordering in alkali feldspars (Al^{3+} occupancy on T sites) proceeds according to different kinetics:

1. *One-step ordering path.* Diffusion of Al^{3+} on $T1_o$ takes place at the same rate, independent of starting position ($T1_m$, $T2_o$, or $T2_m$).
2. *Ideal two-step ordering path.* Starting from maximum disorder at high temperature (high sanidine), all Al^{3+} initially moves from $T2$ to $T1$, producing a phase that may virtually be defined as "low sanidine" or "ordered orthoclase" (cf. Martin, 1974). From this initial condition, complete ordering of Al^{3+} on $T1_o$ is reached by $T1_m \rightarrow T1_o$ migration.
3. *Intermediate two-step ordering path.* The most common process in K-rich and Na-rich feldspars. It initially proceeds as in the case above, up to Al^{3+} occupancy on $T1$ sites of around 0.75; once this concentration is reached, $T1_m \rightarrow T1_o$ migration begins.

It must be emphasized that the inverse path (induced by heating of natural samples) is more complex, because of kinetic reasons related to the length of heating (see, for instance, Blasi et al., 1983).

In plagioclases ($NaAlSi_3O_8$-$CaAl_2Si_2O_8$ mixtures), ordering with decreasing T is further complicated by the Al avoidance principle: chemically homogeneous feldspars with anorthite molar amounts exceeding 2% cannot achieve perfectly ordered distribution because of excess Al^{3+}. This fact leads to exsolution phenomena, whose nature varies with the amount of anorthite (An) in the mixture. For instance, "peristerites" are unmixing products of initially homogeneous plagioclase mixtures of composition An_2-An_{16}, constituted at low T by a fully ordered albite-rich phase (An_{0-2}) and a structurally more complex phase with mean composition An_{25}.

Table 5.63 lists the structural characters of the various feldspar polymorphs, according to Smyth and Bish (1988). Note duplication of cell edge c in anorthite and celsian and the progressive increase in volume that accompanies the lowering of symmetry in $KAlSi_3O_8$.

In the absence of distortion, feldspar cell edges may be obtained directly from tetrahedral edges, as suggested by Megaw (1974). Denoting l the length of the tetrahedral edge, we have

$$a \approx \left(\frac{2}{\sqrt{3}} + \sqrt{3} \right) l \qquad (5.164)$$

Table 5.63 Structural data of main feldspar polymorphs (from Smyth and Bish, 1988).

Name	Sanidine	Orthoclase	Microcline	Rb-Feldspar
Formula unit	$KAlSi_3O_8$	$KAlSi_3O_8$	$KAlSi_3O_8$	$RbAlSi_3O_8$
Molar weight (g/mole)	278.337	278.337	278.337	324.705
Density (g/cm^3)	2.578	2.571	2.567	2.930
Molar volume (cm^3/mole)	107.958	108.283	108.425	110.825
Z	4	4	4	4
System	Monoclinic	Monoclinic	Triclinic	Monoclinic
Class	*2/m*	*2/m*	*1*	*2/m*
Group	*C2/m*	*C2/m*	*C1*	*C2/m*
Cell parameters:				
a	8.539	8.561	8.560	8.820
b	13.015	12.996	12.964	12.992
c	7.179	7.192	7.215	7.161
α			90.65	
β	115.99	116.01	115.83	116.24
γ			87.70	
Volume (Å^3)	717.15	719.13	720.07	736.01
Reference	Phillips and Ribbe (1973)	Colville and Ribbe (1968)	Brown and Bailey (1964)	Gasperin (1971)

Name	High Albite	Low Albite	Anorthite	Celsian
Formula unit	$NaAlSi_3O_8$	$NaAlSi_3O_8$	$CaAl_2Si_2O_8$	$BaAl_2Si_2O_8$
Molar weight (g/mole)	262.225	262.225	278.210	375.470
Density (g/cm^3)	2.610	2.621	2.765	3.400
Molar volume (cm^3/mole)	100.452	100.054	100.610	110.440
Z	4	4	8	8
System	Triclinic	Triclinic	Triclinic	Monoclinic
Class	*1*	*1*	*1*	*2/m*
Group	*C1*	*C1*	*P1*	*I2/m*
Cell parameters:				
a	8.161	8.142	8.173	8.627
b	12.875	12.785	12.869	13.045
c	7.110	7.159	14.165	14.408
α	93.53	94.19	93.113	90
β	116.46	116.61	115.913	115.22
γ	90.24	87.68	91.261	90
Volume (Å^3)	667.12	664.48	1336.35	1166.90
Reference	Winter et al. (1979)	Harlow and Brown (1980)	Wainwright and Starkey (1971)	Newnham and Megaw (1960)

$$b = \left(3 + 2\sqrt{\frac{2}{3}}\right)l \qquad\qquad (5.165)$$

$$c = \left(1 + \sqrt{3}\right)l \qquad\qquad (5.166)$$

$$\beta = \arcsin\left(1 + \sqrt{2}\right)\frac{l}{a} \qquad\qquad (5.167)$$

(Ribbe, 1983a), where l varies between 2.62 Å ($[SiO_4]^{4-}$ tetrahedra) and 2.88 Å ($[AlO_4]^{5-}$ tetrahedra).

In feldspars with Al:Si = 1:3 stoichiometry, the length of the tetrahedral edge is 2.8 Å. Introducing this value into the equations above gives $a = 8.1$, $b = 13.0$, $c = 7.62$ Å, and $\beta = 123°$. Comparison with experimental values shows good agreement for edge b, but the approximation is too rough for edges a and b and the angle β to propose equations 5.164 to 5.167 as a determinative method.

Concerning the effect of P and T intensive variables on feldspar structure, we must distinguish structural modifications associated with intracrystalline disorder from pure displacive effects associated with compression and thermal expansion. It may generally be stated that the triclinic structure of low-T feldspars has a sufficient degree of freedom to modify interionic distances without incurring polymorphic transitions—i.e., the (relatively inert) $[TO_4]$ groups are free to satisfy the polyhedral thermal expansion and/or compressibility of M cations without structural constraints. This fact results in the so-called "inverse relation" (i.e., structural modifications induced by cooling of the substance are similar to those induced by compression; cf. Hazen and Finger, 1982).

Table 5.64 lists isobaric thermal expansion and isothermal compressibility coefficients for feldspars. Due to the clear discrepancies existing among the various sources, values have been arbitrarily rounded off to the first decimal place.

We may envisage the almost linear variation of thermal expansion and compressibility with the amount of anorthite in the plagioclase mixture. We also see that the three polymorphs of the $KAlSi_3O_8$ component have substantially similar compressibilities, within uncertainties.

5.7.3 Thermodynamic Properties of End-Members

Polymorphic transitions render the thermodynamic description of feldspar end-members rather complex. According to Helgeson et al. (1978), the transition between the monoclinic and triclinic forms of $NaAlSi_3O_8$ may be regarded as the overlap of two high-order transitions (cf. section 2.8), one associated with displacive structural modifications and the other with ordering on tetrahedral sites. The analysis of calorimetric data by Holm and Kleppa (1968) concerning the enthalpy

Table 5.64 Isobaric thermal expansion (K^{-1}) and isothermal compressibility (bar^{-1}) of feldspars

Compound	$\alpha_0 \times 10^5$	$\beta_0 \times 10^6$	Reference
High albite	2.7	1.6	Holland and Powell (1990)
Albite	-	2.0	Birch (1966)
	2.7	1.6	Holland and Powell (1990)
	2.5	1.4	Saxena et al. (1993)
$Ab_{99}An_1$	1.9	-	Skinner (1966)
$Ab_{78}An_{22}$	-	1.7–1.8	Birch (1966)
$Ab_{77}An_{23}$	1.2	-	Skinner (1966)
$Ab_{56}An_{44}$	1.3	-	Skinner (1966)
$Ab_{52}An_{48}$	-	1.5	Birch (1966)
$Ab_{48}An_{52}$	-	1.4	Birch (1966)
Ab_5An_{95}	1.2	-	Skinner (1966)
Anorthite	-	1.1	Liebermann and Ringwood (1976)
	2.3	0.9	Saxena et al. (1993)
	1.4	1.3	Holland and Powell (1990)
Orthoclase	-	2.1	Birch (1966)
K-feldspar	1.9	1.8	Holland and Powell (1990)
Sanidine	-	1.8	Birch (1966)
	1.9	1.8	Holland and Powell (1990)
	1.4	1.4	Saxena et al. (1993)
Microcline ($Or_{91}Ab_9$)	-	1.9	Birch (1966)
Microcline ($Or_{83.5}Ab_{16.5}$)	(0.7)	-	Skinner (1966)
Adularia ($Or_{88.3}Ab_{9.3}An_{2.4}$)	1.4	-	Skinner (1966)

variation between monalbite and low albite shows that the enthalpy change associated with displacive phenomena is around 0.9 kcal/mole at T of transition ($T_{transition} = 965\ °C = 1338\ K$; Thompson et al., 1974). According to Thompson et al. (1974), the intracrystalline disorder of the $NaAlSi_3O_8$ end-member in triclinic form may be expressed by an ordering parameter Q_{od}, directly obtainable through

$$Q_{od} = \left(X_{Al,T1_o} + X_{Al,T1_m}\right) - \left(X_{Al,T2_o} + X_{Al,T2_m}\right)$$

$$= -12.523 - 3.4065b + 7.9454c, \tag{5.168}$$

where $X_{Al,T1_o}$ is the atomic fraction of Al^{3+} in $T1_o$ site and b and c are cell edges. Equation 5.168 may be reconducted to equations 5.165 and 5.166. According to Helgeson et al. (1978), displacive disorder is virtually absent at $T < 350\ °C$; moreover, the enthalpy variation connected with the displacive process (ΔH_{di}^0) may be described as a function of ordering parameter Q_{di} (it is assumed that $Q_{di} = 0.06$ at $T = 623\ K$), according to

$$\Delta H_{di}^0 = \left[2.47 - 2.63\left(Q_{di} - 0.06\right)\right]\left(T - 623\right) \qquad \left(cal/mole\right). \tag{5.169}$$

The standard molar enthalpy change connected with intracrystalline disorder (ΔH_{od}^0) is also a function of ordering parameter Q_{od}:

$$\Delta H_{od}^0 = H_{Q_{od}}^0 - H_{Q_{od}=1}^0 = 2630(1 - Q_{od}) \qquad (\text{cal}/\text{mole}). \qquad (5.170)$$

According to equation 5.170, the completely disordered form of the $NaAlSi_3O_8$ end-member (i.e., monalbite, $Q_{od} = 0$) has an enthalpy about $+2.6$ kcal/mole higher than that of the fully ordered form (low albite, $Q_{od} = 1$) as a result of simple substitutional effects.

The phase transition is energetically constituted of displacive and substitutional terms, so that we obtain

$$\Delta H_{trans}^0 = \Delta H_{di}^0 + \Delta H_{od}^0 \approx 2.6 + 0.9 \qquad (\text{kcal}/\text{mole}). \qquad (5.171)$$

Moreover, because

$$\Delta C_{P_r,od}^0 = \left(\frac{\partial \Delta H_{od}^0}{\partial T} \right)_{P_r} \qquad (5.172)$$

and

$$\Delta C_{P_r,di}^0 = \left(\frac{\partial \Delta H_{di}^0}{\partial T} \right)_{P_r}, \qquad (5.173)$$

equation 5.171 must be consistent with the similar equation for the heat capacity function:

$$\Delta C_{P_r,trans}^0 = \Delta C_{P_r,di}^0 + \Delta C_{P_r,od}^0. \qquad (5.174)$$

Nevertheless, we have already noted that the transition of $NaAlSi_3O_8$ from monoclinic to triclinic form may be regarded as the overlap of two high-order transitions, and that high-order transition implies continuity in S, H, and G at the transition point and discontinuity in the C_P function (see section 2.8). The thermodynamic parameters for the two polymorphs on the opposite sides of the transition zone must thus be constrained to fulfill the above requirements. The method followed by Helgeson et al. (1978) is as follows.

1. Extrapolation at the T_r and P_r of reference of the change in enthalpy between low albite and high albite, determined as a continuous function of ordering parameter Q.
2. Adoption of the ordered form as a stable polymorph at the T_r and P_r of reference.
3. Summation of $\Delta C_{P_r,trans}^0$, determined as a continuous function of ordering parameter Q, to the C_P function of the fictive "stable" polymorph.

We thus obtain thermodynamic parameters of a generic *albite* phase that, with T, progressively modifies its substitutional and displacive states from low-T to high-T forms.

Because for $T < 473$ K, $\Delta C^0_{P_r,\text{trans}} = 0$, Helgeson et al. (1978) adopt in the T range 298.15 to 473 K the original C_P function tabulated by Kelley (1960) valid for low albite. The albite C_P function is described, starting from 473 K, by a second set of coefficients (cf. table 5.65).

Salje (1985) interpreted overlapping (displacive plus Al-Si substitutional) phase transitions in albite in the light of Landau theory (see section 2.8.1), assigning two distinct order parameters Q_{di} and Q_{od} to displacive and substitutional disorder and expanding the excess Gibbs free energy of transition in the appropriate Landau form:

$$\Delta G_{\text{trans}} = \frac{1}{2} a_{di} \left(T - T_{c,di} \right) Q^2_{di} + \frac{1}{4} B_{di} Q^4_{di} + \frac{1}{2} a_{od} \left(T - T_{c,od} \right) Q^2_{od}$$

$$+ \frac{1}{4} B_{od} Q^4_{od} + \frac{1}{6} C_{od} Q^6_{od} + \lambda Q_{di} Q_{od} . \tag{5.175}$$

Coefficients a, B, and C in equation 5.175 have the usual meanings in the Landau expansion (see section 2.8.1) and for the (second-order) displacive transition of albite assume the values $a_{di} = 1.309$ cal/(mole $\times$ K) and $B_{di} = 1.638$ kcal/mole (Salje et al., 1985). $T_{c,di}$ is the critical temperature of transition ($T_{c,di} = B/a = 1251$ K). The corresponding coefficients of the ordering process are $a_{od} = 9.947$ cal/(mole $\times$ K), $B_{od} = -2.233$ kcal/mole, $C_{od} = 10.42$ kcal/(mole $\times$ K), and $T_{c,od} = 824.1$ K. With all three coefficients being present in the Landau expansion relative to substitutional disorder it is obvious that Salje et al. (1985) consider this transition first-order. λ is a T-dependent coupling coefficient between displacive and substitutional energy terms (Salje et al., 1985):

$$\lambda = -0.519 - 0.727T - 3.75 \times 10^{-4} T^2$$

$$+ 5.04 \times 10^{-7} T^3 \qquad \left(\text{cal/mole} \right) . \tag{5.176}$$

For a $C2/m \Leftrightarrow C\overline{1}$ transition, the relevant components of the spontaneous strain tensor are x_4 and x_6. The Gibbs free energy change associated with elastic strain is

$$\Delta G_{\text{trans},\varepsilon} = D_{di} Q_{di} \varepsilon + D_{od} Q_{od} \varepsilon + E \varepsilon^2 . \tag{5.177}$$

If $\partial \Delta G / \partial x_4 = 0$ and $\partial \Delta G / \partial x_6 = 0$, the strain components are linearly dependent on Q_{di} and Q_{od} (Salje, 1985):

$$x_4 = A' Q_{di} + B' Q_{od} \tag{5.178.1}$$

$$x_6 = C' Q_{di} + D' Q_{od} . \tag{5.178.2}$$

Replacing the generalized strain ε with strain components x_4 and x_6 and adding the elastic energy term in the Landau expansion results in equation 5.175.

As discussed by Salje et al. (1985) and also evident from equations 5.178.1 and 5.178.2, the two order parameters Q_{di} and Q_{od} are not independent of each other (although the displacive order parameter Q_{di} is largely determined by x_4 whereas the substitutional disorder Q_{od} depends more markedly on x_6). Their mutual dependence is outlined in figure 5.55.

Note that, displacive disorder being associated with a first-order transition, Q_{di} exhibits a jump at T_{trans} whose magnitude decreases from fully ordered albite (i.e., $Q_{od} = 1$) to analbite (i.e., $Q_{od} = 0$) (lower part of figure 5.5; see also section 2.8.1). Moreover, the order-disorder phase transition is stepwise for $Q_{di} = 0$ (front curve in upper part of figure 5.5) and becomes smooth for higher Q_{di} values.

The C_p change at the transition temperature resulting from displacive disorder is, according to the Landau expansion above, $\Delta C^0_{P,di} = 0.66$ cal/(mole $\times$ K). The resulting entropy change associated with Al-Si substitutional disorder is -4.97 cal/(mole $\times$ K), near the maximum configurational effect expected by the random distribution of $3Al + 1Si$ atoms on four tetrahedral sites.

According to Hovis (1974), the intracrystalline disorder in monoclinic sanidine may be derived directly from cell edges by applying

$$Q_{od} = 2\left(X_{Al,T1} - X_{Al,T2}\right) = 7.6344 - 4.3584b + 6.8615c, \qquad (5.179)$$

where $X_{Al,T1}$ and $X_{Al,T2}$ are the atomic fractions of Al^{3+} on $T1$ and $T2$ sites.

Note that the ordering parameter adopted by Hovis (1974) differs from the form adopted for the triclinic phase of the $NaAlSi_3O_8$ end-member (eq. 5.168).

Helgeson et al. (1978) have shown that the enthalpy of disordering in sanidine is similar to that observed in albite—i.e.,

$$\Delta H^0_{od} = 2650\left(1 - Q_{od}\right) \qquad \left(\text{cal}/\text{mole}\right). \qquad (5.180)$$

However, the structural transition between the triclinic and monoclinic forms (maximum microcline and high sanidine, respectively) "has no apparent effect on its thermodynamic behavior" (cf. Helgeson et al., 1978); in other words, no gain or loss of heat and no entropy variation is associated with the purely structural modification. In this respect, the $KAlSi_3O_8$ end-member differs considerably from $NaAlSi_3O_8$, for which the contribution of structural disorder to bulk transformation (displacive plus substitutional) is about one-third in terms of energy. This difference is evident if we compare the heat capacity functions of the two compounds: a cusp is seen for albite in the triclinic-monoclinic transition zone ($T = 1238$ K; cf. figure 5.56B), whereas no apparent discontinuity can be detected at the transition point ($T = 451 \pm 47$ °C at $P = 1$ bar; Hovis, 1974) for $KAlSi_3O_8$ (figure 5.56A).

Also for $KAlSi_3O_8$, it is convenient to account for phase transition effects

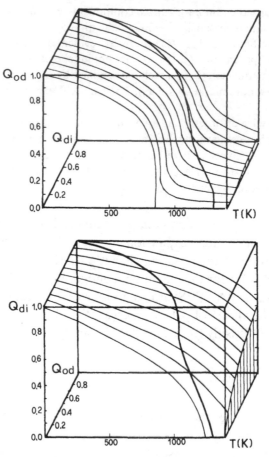

Figure 5.55 Mutual dependence of Q_{di} and Q_{od} order parameters. In the upper part of the figure is outlined the T dependence of substitutional disorder Q_{od} for different values of Q_{di} and, in the lower part, the T dependence of the displacive disorder parameter Q_{di} for different values of Q_{od}. The heavy lines on the surface of local curves represent the solution for thermal equilibrium. From E. Salje and B. Kuscholke, Thermodynamics of sodium feldspar II: experimental results and numerical calculations, *Physics and Chemistry of Minerals,* 12, 99–107, figures 5–8, copyright © 1985 by Springer Verlag. Reprinted with the permission of Springer-Verlag GmbH & Co. KG.

in the heat capacity function, as previously discussed for $NaAlSi_3O_8$. A fictive polymorph, *K-feldspar*, is thus generated, representing $KAlSi_3O_8$ in its stable state of displacive and substitutional disorder at all T conditions.

Table 5.65 lists H^0_{f,T_r,P_r}, $S^0_{T_r,P_r}$, $V^0_{T_r,P_r}$ and the heat capacity function for $NaAlSi_3O_8$, $KAlSi_3O_8$, and $CaAl_2Si_2O_8$ polymorphs according to Helgeson et al. (1978). Note that the enthalpy of formation from the elements at $T_r = 298.15$ K and $P_r = 1$ bar for *albite* and *K-feldspar* are identical to those of low-T polymorphs (low albite and maximum microcline), for the previously noted reasons. Note also that the heat capacity expressions for all $NaAlSi_3O_8$ polymorphs are initially identical in the low-T range. Moreover, as we see in figure 5.56B, the

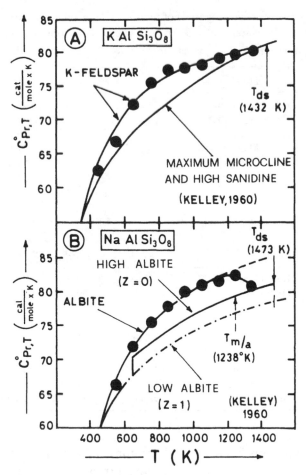

Figure 5.56 Maier-Kelley heat capacity functions for various structural forms of the $KAlSi_3O_8$ (A) and $NaAlSi_3O_8$ (B) feldspar end-members (solid curves), compared with T derivative of $H_T^0-H_{T_r}^0$ finite differences at various T (filled symbols) (from Helgeson et al., 1978; redrawn and reproduced with permission).

second heat capacity function for the fictive form *albite* (valid for $T > 473$ K) conforms quite well to experimental observations ($H_{P_r,T}^0 - H_{P_r,473}^0$ finite difference derivatives) up to the T of triclinic-monoclinic transition ($T = 1238$ K).

It is generally assumed that $CaAl_2Si_2O_8$, based on the *Al avoidance principle*, has complete ordering of Al and Si on tetrahedral sites at all temperatures. Indeed, some authors suggest that ordering is not complete at high T (see, for instance, Bruno et al., 1976). Moreover, anorthite at high T (>2000 K) in the disordered state has $I\bar{1}$ symmetry, unlike the $P\bar{1}$ symmetry at low T (Bruno et al., 1976; Chiari et al., 1978). According to Carpenter and MacConnell (1984), the $P\bar{1} \Leftrightarrow I\bar{1}$ transition is of high order (as in the previous cases). Transition temperatures between the $P\bar{1} \Leftrightarrow I\bar{1}$ forms for pure anorthite at $P = 1$ bar are estimated by Carpenter and Ferry (1984) to be around 2000 to 2250 K and (implicitly) the authors consider structural transition as coincident with substitutional order $\Leftrightarrow$

Table 5.65 Thermodynamic properties of feldspar end-members in various structural forms according to Helgeson et al. (1978). Heat capacity function: $C_P = K_1 + K_2 T + K_3 T^{-2}$; H^0_{f,T_r,P_r} in kcal/mole; $S^0_{T_r,P_r}$ in cal/(mole × K); $V^0_{T_r,P_r}$ in cm³/mole.

End-Member	Form	H^0_{f,T_r,P_r}	$S^0_{T_r,P_r}$	$V^0_{T_r,P_r}$
NaAlSi₃O₈	Low albite	−939.680	49.51	100.07
NaAlSi₃O₈	High albite	−937.050	52.30	100.43
NaAlSi₃O₈	*Albite*	−939.680	49.51	100.25
KAlSi₃O₈	Maximum microcline	−949.188	51.13	108.74
KAlSi₃O₈	High sanidine	−946.538	54.53	109.01
KAlSi₃O₈	*K-feldspar*	−949.188	51.13	108.87
CaAl₂Si₂O₃	Anorthite	−1007.772	49.10	100.79

End-Member	Form	K_1	$K_2 \times 10^3$	$K_3 \times 10^{-5}$	T_{limit}
NAlSi₃O₈	Low albite	61.70	13.90	−15.01	< 1400
NAlSi₃O₈	High albite	61.70	13.90	−15.01	> 623
		64.17	13.90	−15.01	< 1400
NAlSi₃O₈	*Albite*	61.70	13.90	−15.01	< 473
		81.880	3.554	−50.154	< 1200
KAlSi₃O₈	Maximum microcline	63.83	12.90	−17.05	< 1400
KAlSi₃O₈	High sanidine	63.83	12.90	−17.05	< 1400
KAlSi₃O₈	*K-feldspar*	76.617	4.311	−29.945	< 1400
CaAl₂Si₂O₈	Anorthite	63.311	14.794	−15.44	< 1700

disorder transition. This implicit consideration leads to an estimated transition enthalpy (ΔH_{trans}) of 3.7 ± 0.6 kcal/mole and a transition entropy (ΔS_{trans}) of 1.4 to 2.2 cal/(mole × K). The data in table 5.65 on anorthite are specific to the $P\bar{1}$ polymorph and do not comprise the transition effect that is outside the T limit of validity of the C_P function.

5.7.4 Mixing Properties

Most natural feldspars are chemically represented by the ternary system NaAlSi₃O₈-KAlSi₃O₈-CaAl₂Si₂O₈. Miscibility among alkaline terms is complete, like the miscibility between NaAlSi₃O₈ and CaAl₂Si₂O₈ ("plagioclase" series), whereas an extended miscibility gap (whose limbs vary with T and P) is observed between KAlSi₃O₈ and CaAl₂Si₂O₈. The miscibility gap field extends within the ternary system, as shown in figure 5.57, which also lists names conventionally assigned to the various compositional terms of the plagioclase series.

For better comprehension of the mixing properties in the ternary system, we must first examine in detail the binary mixtures NaAlSi₃O₈-KAlSi₃O₈, NaAlSi₃O₈-CaAl₂Si₂O₈, and KAlSi₃O₈-CaAl₂Si₂O₈.

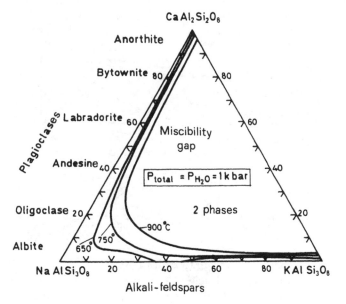

Figure 5.57 Mixing properties in ternary system $NaAlSi_3O_8$-$KAlSi_3O_8$-$CaAl_2Si_2O_8$ at $P_{total} = P_{H_2O} = 1$ kbar and various T, based on data from Seck (1971).

$NaAlSi_3O_8$-$KAlSi_3O_8$ Mixture

Figure 5.58 shows subsolidus stability relationships in the $NaAlSi_3O_8$-$KAlSi_3O_8$ system at various $P_{total} = P_{fluid}$. At low P, the system is pseudobinary, because $KAlSi_3O_8$ melts incongruently to form leucite ($KAlSi_2O_6$) plus liquid. However, melting relations are not of interest here and will be treated to some extent later. We note that, at subsolidus, an extended miscibility gap exists between Na-rich (Ab_{sm}) and K-rich (Or_{sm}) terms. This gap also expands remarkably with increasing P, due to the volume properties of $NaAlSi_3O_8$-$KAlSi_3O_8$ mixtures.

As a matter of fact, the volumes of $NaAlSi_3O_8$-$KAlSi_3O_8$ mixtures are not ideal and have positive excess terms. Deviations from simple Vegard's rule proportionality are mainly due to edge b. Kroll and Ribbe (1983) provide three polynomial expressions relating the volumes of mixtures to the molar fractions of the potassic component (X_{Or}) valid, respectively, for monoclinic symmetry (complete disorder), triclinic symmetry of perfect order (low albite and maximum microcline), and the intermediate structural state:

$$X_{Or} = -584.6683 + 2.58732V - 3.83499 \times 10^{-3}V^2$$
$$+ 1.90428 \times 10^{-6}V^3 \tag{5.181.1}$$

$$X_{Or} = -1227.8023 + 5.35958V - 7.81518 \times 10^{-3}V^2$$
$$+ 3.80771 \times 10^{-6}V^3 \tag{5.181.2}$$

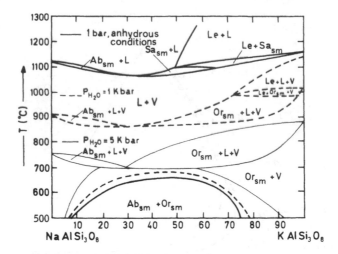

Figure 5.58 Phase stability relations in (pseudo) binary system $NaAlSi_3O_8$-$KAlSi_3O_8$ at various P and T conditions (Le = leucite; Or = orthoclase; Sa = sanidine; L = liquid; V = vapor). Here, "orthoclase" identifies simply an intermediate state of substitutional disorder.

$$X_{Or} = -929.1523 + 4.07032V - 5.96146 \times 10^{-2}V^2$$
$$+ 2.91994 \times 10^{-5}V^3. \tag{5.181.3}$$

The subsolidus diagram of figure 5.58 is extremely simplified although based on various experimental evidence. It does not refer to the structural form of the various phases at the P, T, and X of interest. Actually, in order to describe appropriately the mixing properties of $NaAlSi_3O_8$-$KAlSi_3O_8$ at subsolidus, their structural state must be taken into account. The stable form of $NaAlSi_3O_8$ immediately below the melting temperature ($T = 1118$ °C; Greig and Barth, 1938) is monoclinic ("monalbite"). If monalbite is cooled rapidly, a triclinic polymorph forms, without modification of Al and Si intracrystalline disorder, by simple collapse of the coordinating oxygens around Na as a result of attenuation of thermal vibrational motions (displacive transformation). The resulting form is called "analbite" and is unstable at all temperatures. If the cooling process is sufficiently slow, monalbite changes into high albite (triclinic) as a result of a diffusive process involving Al and Si. At $P = 1$ bar, the transition takes place at $T = 980$ °C (Kroll et al., 1980). The presence of $KAlSi_3O_8$ in the mixture lowers the temperature in an almost linear fashion (Kroll et al., 1980):

$$T_{\text{trans}}\left(°C\right) = 978 - 19.2X_{Or}. \tag{5.182}$$

According to Merkel and Blencoe (1982), the relationship among the amount of $KAlSi_3O_8$, pressure, and temperature of transition is

$$X_{Or} = 0.474 - 0.361 \frac{T_{\text{trans}}}{1000} + 25.2 P_{\text{trans}}, \qquad (5.183)$$

where T_{trans} is in K and P_{trans} is in bar.

Pure $KAlSi_3O_8$ behaves essentially like $NaAlSi_3O_8$, with high-order transformations that progressively modify the structure. With the decrease of T, high sanidine (monoclinic, perfectly disordered) is transformed into low sanidine and/or orthoclase (medium disorder) and then into microcline (triclinic, perfectly ordered). The high-T miscibility between the sodic and potassic terms is thus essentially that between a triclinic phase (high albite) and a monoclinic phase (sanidine): because miscibility is not complete, we reserve the term "miscibility gap" for the compositional extension of unmixed phases. At low T, both end-members are triclinic: if the components unmix, we have a coherent solvus (and a coherent spinodal field) in the presence of elastic strain energy, and a strain-free or "chemical" solvus (and a chemical spinodal field) in the absence of strain energy. The difference between the two unmixing phenomena is discussed in detail in sections 3.11, 3.12, and 3.13.

The petrological and mineralogical literature contains a plethora of names identifying unmixing products in the $NaAlSi_3O_8$-$KAlSi_3O_8$ system. However, the adopted terminology does not distinguish clearly between the structural states of the unmixed phases, which is necessary to refer the phenomena to a well-defined thermodynamic process.

The unmixing products of alkaline feldspars are usually defined as "perthites," which are further subdivided as follows:

1. *Macroperthites.* When the unmixed K-rich and Na-rich zones may be observed and characterized directly under the microscope;
2. *Microperthites.* When the existence of unmixing is anticipated under the microscope by widespread turbidity or the presence of spots, but is clearly detectable only by X-ray investigation;
3. *Cryptoperthites.* When unmixing is detected only by X-ray investigation.

Perthites are also subdivided according to their chemical composition into:

1. *Perthites sensu stricto.* Unmixing on an initially homogeneous K-rich phase.
2. *Antiperthites.* Unmixing on an initially homogeneous Na-rich phase.
3. *Mesoperthites.* Unmixing on an initially homogeneous phase of intermediate composition.

"The compositional relations of the sodic and potassic phases in perthites depend on whether the phases are coherent or not. If the perthitic phases are non-coherent (or if the rock consists of separate grains of sodic and potassic feldspar) their equilibrium compositions are given by the strain-free solvus. This is the solvus which has been studied extensively in the past." These sentences,

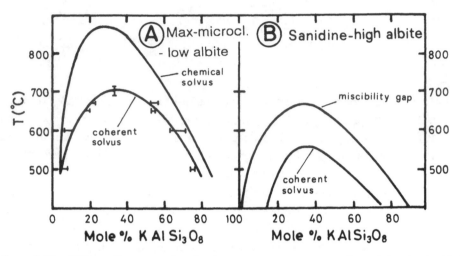

Figure 5.59 Effects of structural and coherence state on extent of unmixing in the NaAl-Si$_3$O$_8$-KAlSi$_3$O$_8$ binary join. (A) Maximum microcline–low albite, based on data from Bachinski and Müller (1971) and Yund (1974). (B) Sanidine–high albite, based on data from Thompson and Waldbaum (1969) and Sipling and Yund (1976).

taken entirely from Yund and Tullis (1983), lead us to a careful evaluation of the real nature of perthitic unmixing in rocks and of the role of coherence energy in determining feldspar phase relations. Figure 5.59A shows the approximate position of the chemical solvus between maximum microcline and low albite according to Bachinski and Müller (1971), compared with the coherent solvus determined by Yund (1974). Figure 5.59B shows the miscibility gap between sanidine and high albite and the corresponding coherent solvus, based on data from Thompson and Waldbaum (1969) and Sipling and Yund (1976). Note that the locations of the unmixing fields depend markedly on the structural state of the components (compare parts A and B) and, structural state being equal, on coherence energy. It is thus essential to determine both structure and coherence of unmixed phases in order to assess the P and T conditions of unmixing.

Table 5.66 furnishes a synthesis of Margules subregular interaction parameters according to various authors for the NaAlSi$_3$O$_8$-KAlSi$_3$O$_8$ join, collected by Ganguly and Saxena (1987). The various sets of values show important differences that lead to consistent discrepancies in the calculated Gibbs free energy values of the mixtures. This nonhomogeneity partly reflects the terminological and operative confusion existing in the literature in the treatment of feldspar thermodynamics, and partly is the result of experimental artifacts.

NaAlSi$_3$O$_8$-CaAl$_2$Si$_2$O$_8$ Mixture

Components NaAlSi$_3$O$_8$ and CaAl$_2$Si$_2$O$_8$ show complete miscibility over a wide range of P and T at subsolidus. Figure 5.60 shows the classical "lens diagram" of phase stability of plagioclases. This conformation is indicative of ideality of mixing in both liquid and solid states, and will be explained in detail in chapter 7. As

Table 5.66 Margules subregular parameters for the $NaAlSi_3O_8$-$KAlSi_3O_8$ (Ab-Or) binary join (Ganguly and Saxena, 1987). (1) Brown and Parson (1981); (2) Thompson and Waldbaum (1969); (3) Hovis and Waldbaum (1977), Thompson and Hovis (1979); (4) Haselton et al. (1983). Note that $W_{12} = W_{H,12} - TW_{S,12} + PW_{V,12}$ and $G_{\text{excess mixing}} = X_1X_2(W_{12}X_2 + W_{21}X_1)$, with $1 = NaAlSi_3O_8$, $2 = KAlSi_3O_8$, W_H cal/mole, W_S cal/(mole × K), and W_V = cal/(mole × bar).

	(1)	(2)	(3)	(4)
$W_{H,12}$	2890	6327	7474	4496
$W_{H,21}$	6100	7672	4117	6530
$W_{H,12}$ (T<1000 K)	2359	1695	2273	
$W_{H,21}$ (T<1000 K)	3810	3377	4117	
$W_{S,12}$	0.531	4.632	5.200	2.462
$W_{S,21}$	2.290	3.857	0.000	2.462
$W_{V,12}$	0.079	0.092	0.086	0.087
$W_{V,21}$	0.168	0.112	0.086	0.087

figure 5.60 shows, the structural form of mixtures at subsolidus is not unique over the entire compositional range, but can be schematically subdivided into two domains ($P\bar{1}$ and $I\bar{1}$) limited by a straight boundary between An_{59} at $T = 1000$ °C and An_{77} at $T = 1400$ °C ($P = 1$ bar; Carpenter and Ferry, 1984). If this subdivision is adopted, the two-phase field (solid plus liquid) is also necessarily subdivided into two domains: one constituted of $P\bar{1}$ mixture plus liquid, and the other of $I\bar{1}$ mixture plus liquid, as shown in the figure.

Newton et al. (1980) calorimetrically measured the enthalpy of the $NaAlSi_3O_8$-$CaAl_2Si_2O_8$ mixture at $T - 970$ K and $P = 1$ bar and found a positive excess enthalpy reproduced by a subregular Margules model:

$$H_{\text{excess mixing}} = X_1X_2\left(W_{H,12}X_2 + W_{H,21}X_1\right),$$ (5.184)

where $1 = NaAlSi_3O_8$, $2 = CaAl_2Si_2O_8$, $W_{H,12} = 6.7461$ kcal/mole, and $W_{H,21} = 2.0247$ kcal/mole. Concerning the entropy of the mixture, Kerrick and Darken (1975) proposed a simple calculation of the configurational effect of mixing based on the Al avoidance principle.

In anorthite, tetrahedral groups $[TO_4]$ lying parallel to the $(\bar{1}23)$ plane exhibit perfect alternation of Al-bearing and Si-bearing strata. Let us define N_T as the total number of tetrahedra occupied by Al and Si:

$$N_T = N_{Al} + N_{Si}.$$ (5.185)

We also define the sites occupied by Al as α and those occupied by Si as β.

If we replace half the Al with Si on a tetrahedral plane, we have random mixing over $\frac{1}{2}N_T$ sites. This sort of mixing does not contradict the Al avoidance principle. Let us now

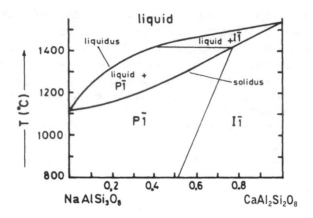

Figure 5.60 X-T phase stability diagram for the $NaAlSi_3O_8$-$CaAl_2Si_2O_8$ system at $P = 1$ bar, showing transition between forms $P\bar{1}$ and $I\bar{1}$ at subsolidus and within two-phase field. Note that form of disorder phase is usually "mediated" into crystalline class $C\bar{1}$, hence in the literature the phase transition is usually described as $C\bar{1} \Leftrightarrow I\bar{1}$. From M. A. Carpenter and J. M. Ferry, Constraints on the thermodynamic mixing properties of plagioclase feldspars, *Contributions to Mineralogy and Petrology*, 87, 138–48, figure 3, copyright © 1984 by Springer Verlag. Reprinted with the permission of Springer-Verlag GmbH & Co. KG.

consider the exchange of one Si atom in a β site with one Al atom in an α site: the exchange is possible only when the resulting configuration does not result in Al-O-Al alternation. Because the probability that an Si atom in a β site is surrounded by four other Si neighbors is $(\frac{1}{2})^4 = \frac{1}{16}$, the total number of sites available for Al-Si mixing, based on the Al avoidance principle, is

$$N_{Al \Leftrightarrow Si,\alpha} = \frac{1}{2} N_T \tag{5.186}$$

$$N_{Al \Leftrightarrow Si,\beta} = \frac{1}{2} \times \frac{1}{16} N_T. \tag{5.187}$$

Therefore, in practice, mixing takes place over α sites only. Applying this condition to the stoichiometry of the binary mixture, we obtain

$$N_{Si,\alpha} = N_{Na} \tag{5.188}$$

$$N_{Al} = 2N_{Ca} + N_{Na} \tag{5.189}$$

$$N_{Si,\alpha} + N_{Al} = 2(N_{Ca} + N_{Na}) \tag{5.190}$$

(cf. Kerrick and Darken, 1975). If we now recall the equation for permutability Q (section 3.8.1) and apply the calculation to the four atoms in the mixture (Si, Al, Na, Ca), because Na and Ca mix over one M site and Si and Al mix essentially over α sites, we obtain

$$Q = \frac{(N_{Na} + N_{Ca})!}{N_{Na}! \, N_{Ca}!} \times \frac{(N_{Si,\alpha} + N_{Al})!}{N_{Si,\alpha}! \, N_{Al}!}$$

$$= \frac{(N_{Na} + N_{Ca})!}{N_{Na}! \, N_{Ca}!} \times \frac{[2(N_{Ca} + N_{Na})]!}{N_{Na}! \, (2N_{Ca} + N_{Na})!}. \tag{5.191}$$

Application of Stirling's formula to equation 5.191 and comparison with the configurational state of pure components lead to the definition of a configurational entropy of mixing term in the form

$$S_{\text{mixing conf}} = -R \left\{ X_{Ab} \, \ln \left[X_{Ab} \left(1 - X_{An}^2 \right) \right] + X_{An} \, \ln \left[\frac{X_{An} \left(1 + X_{An} \right)^2}{4} \right] \right\}. \tag{5.192}$$

In the absence of enthalpic contributions, the activities of the components in the mixture become

$$a_{Ab} = X_{Ab}^2 \left(2 - X_{Ab} \right) \tag{5.193}$$

and

$$a_{An} = \frac{1}{4} X_{An} \left(1 + X_{An} \right)^2. \tag{5.194}$$

(In eq. 5.193 and 5.194, Ab or An simply identifies the component without any implication about its structural state.)

Figure 5.61B shows how the entropy of mixing derived by application of the Al avoidance principle compares with the one-site ideal mixing contribution. Also shown are entropy of mixing curves obtained by summing, up to the ideal one-site contribution, additional terms related to phase transition effects [variable between 1 and 2.5 cal/(mole $\times$ K)] (Carpenter and Ferry, 1984). Figure 5.61C compares the Gibbs free energy of mixing experimentally observed by Orville (1972) at $T = 700\,°C$ and $P = 1$ bar with Gibbs free energy values obtained by combining the entropy effect of the Al avoidance principle and the enthalpy of mixing experimentally observed by Newton et al. (1980) (see also figure 5.61A):

$$G_{\text{mixing}} = H_{\text{mixing}} + RT \left\{ X_1 \, \ln \left[X_1^2 \left(2 - X_1 \right) \right] + X_2 \, \ln \frac{X_2 \left(1 + X_2 \right)^2}{4} \right\}. \tag{5.195}$$

As figure 5.61C shows, the equation of Kerrick and Darken (1975), coupled with the enthalpy of mixing terms of Newton et al. (1980), satisfactorily fits the experi-

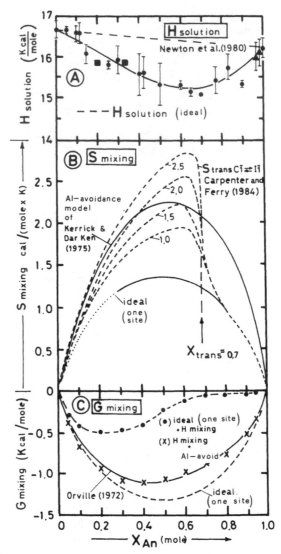

Figure 5.61 Mixing properties of plagioclases. (A) Solution enthalpy in $Pb_2B_2O_5$ at $T = 970$ K and $P = 1$ bar after Newton et al. (1980). Enthalpy of mixing of substance is obtained by subtracting value of solution enthalpy actually measured for substance from ideal solution enthalpy (dashed line). (B) Entropy of mixing of Kerrick-Darken model compared with one-site mixing configurational entropy. Dashed lines: ordering contributions that may arise from $C\bar{1} \Leftrightarrow P\bar{1}$ phase transition effects (Carpenter and Ferry, 1984). (C) Experimental Gibbs free energy of mixing curve at $T = 970$ K and $P = 1$ bar (Orville, 1972; solid line) compared with various models.

mental observations of Orville (1972). The same values cannot be reproduced either by an ideal one-site mixing model or by a regular mixture, as shown in the figure. Moreover, the phase transition contribution postulated by Carpenter and Ferry (1984) does not appear to affect the mixing properties of the substance to any significant extent.

Saxena and Ribbe (1972) have shown that the excess Gibbs free energy of mixing of the mixture, based on the data of Orville (1972), may be reproduced by a subregular Margules model:

$$G_{\text{excess mixing}} = X_1 X_2 \left(W_{12} X_2 + W_{21} X_1 \right), \qquad (5.196)$$

where $1 = NaAlSi_3O_8$, $2 = CaAl_2Si_2O_8$, $W_{12} = 252$ cal/mole, and $W_{21} = 1682$ cal/mole.

The mixing properties of plagioclases discussed above refer to high-T structural forms. The corresponding properties for low-T forms are extremely complex and only a few (and contrasting) thermochemical data exist at present. For instance, according to Kotelnikov et al. (1981), at $T = 700$ °C and $P = 1$ kbar, a miscibility gap exists between An_{67} and An_{92} (i.e., "Huttenlocher gap"; cf. Smith, 1983), although this contrasts with the analogous study by Orville (1972) at $P = 2$ kbar (same T). Moreover, we have already mentioned the existence of the "peristerite gap" at low An amounts (see Smith, 1983, for a detailed account of low-T unmixing processes in plagioclases).

$KAlSi_3O_8$-$CaAl_2Si_2O_8$ Mixtures

The subsolidus properties of the $KAlSi_3O_8$-$CaAl_2Si_2O_8$ system are essentially those of a mechanical mixture (cf. section 7.1). According to Ghiorso (1984), the miscibility gap is not influenced by pressure and may be described by a strongly asymmetric Margules model:

$$G_{\text{excess mixing}} = X_1 X_2 \left(W_{12} X_2 + W_{21} X_1 \right), \qquad (5.197)$$

where $1 = KAlSi_3O_8$, $2 = CaAl_2Si_2O_8$, and

$$W_{12} = W_{H,12} - T W_{S,12}, \qquad (5.198)$$

where $W_{H,12} = 16,125$ cal/mole, $W_{H,21} = 6688$ cal/mole, $W_{S,12} = 2.644$ cal/(mole × K), and $W_{S,21} = -4.830$ cal/(mole × K).

$NaAlSi_3O_8$-$KAlSi_3O_8$-$CaAl_2Si_2O_8$ Ternary Mixtures

The most complete set of experiments concerning mixing properties in the $NaAlSi_3O_8$-$KAlSi_3O_8$-$CaAl_2Si_2O_8$ system is given by Seck (1971) (cf. figure 5.57). Already in 1973, Saxena had rationalized the experimental results of Seck (1971) in thermodynamic form, representing ternary interactions in the framework of Wohl's model (see section 3.10.1). Later, Barron (1976) evaluated the reliability of Wohl's and Kohler's models in reproducing the data of Seck (1971) and deduced that Wohl's model better reproduced the slope of tie lines connecting the compositions of unmixed phases on opposite sides of the miscibility gap limb.

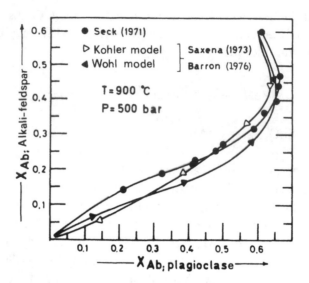

Figure 5.62 Experimentally observed distribution of $NaAlSi_3O_8$ component between plagioclases and alkaline feldspars (Seck, 1971) compared with results of Wohl's and Kohler's models. Reproduced from Barron (1976), with modifications.

Figure 5.62 shows in detail the distribution of the $NaAlSi_3O_8$ component between unmixed phases at $T = 900$ °C and $P_{H_2O} = 0.5$ kbar, as observed by Seck (1971) and compared with Wohl's and Kohler's predictions. In both models, the binary interaction parameters are those deduced by Saxena (1973) on the basis of Orville's (1972) data for the $NaAlSi_3O_8$-$CaAl_2Si_2O_8$ system, coupled with the experimental observations of Seck (1971).

A more recent model (Ghiorso, 1984) is based on the binary interaction parameters of Thompson and Hovis (1979) for the $NaAlSi_3O_8$-$KAlSi_3O_8$ join and on the experimental results of Newton et al. (1980), coupled with the Al avoidance principle of Kerrick and Darken (1975) extended to the ternary field. Ghiorso (1984) expressed the excess Gibbs free energy of mixing in the form

$$
G_{\text{excess mixing}} = W_{12}X_1X_2\left(X_2 + \frac{1}{2}X_3\right) + W_{21}X_1X_2\left(X_1 + \frac{1}{2}X_3\right)
$$

$$
+ W_{13}X_1X_3\left(X_3 + \frac{1}{2}X_2\right) + W_{31}X_1X_3\left(X_1 + \frac{1}{2}X_2\right) \tag{5.199}
$$

$$
+ W_{32}X_3X_2\left(X_2 + \frac{1}{2}X_1\right) + W_{23}X_3X_2\left(X_3 + \frac{1}{2}X_1\right) + W_{123}X_1X_2X_3,
$$

where $1 = NaAlSi_3O_8$, $2 = CaAl_2Si_2O_8$, and $3 = KAlSi_3O_8$.

Binary interaction parameters W_{23} and W_{32} and ternary interaction term W_{123} were obtained by Ghiorso (1984) by imposing isoactivity on the various compo-

Table 5.67 Interaction parameters for ternary feldspar mixtures according to Ghiorso (1984). $1 = NaAlSi_3O_8$; $2 = CaAl_2Si_2O_8$; $3 = KAlSi_3O_8$.

AB	$W_{H.AB}$, cal/mole	$W_{S.AB}$, cal/(mole × K)	$W_{V.AB}$, cal/(mole × bar)
12	6,746	0	0
21	2,025	0	0
13	7,404	5.12	0.086
31	4,078	0	0.086
23	6,688	−4.83	0
32	16,126	2.64	0
123	−3,315	−3.50	0

$W = W_{H.AB} - T\,W_{S.AB} + P\,W_{V.AB}$

nents in the unmixed phases of Seck (1971). Table 5.67 reports the resulting interaction parameters.

5.8 Silica Minerals

Silica has 22 polymorphs, although only some of them are of geochemical interest—namely, the crystalline polymorphs quartz, tridymite, cristobalite, coesite, and stishovite (in their structural modifications of low and high T, usually designated, respectively, as α and β forms) and the amorphous phases chalcedony and opal (hydrated amorphous silica). The crystalline polymorphs of silica are tectosilicates (dimensionality = 3). Table 5.68 reports their structural properties, after the synthesis of Smyth and Bish (1988). Note that the number of formula units per unit cell varies conspicuously from phase to phase. Also noteworthy is the high density of the stishovite polymorph.

Quartz is a primary phase in acidic igneous rocks (both intrusive and effusive) and in metamorphic and sedimentary rocks. Tridymite is a high-T form found in porosities of rapidly quenched effusive rocks (mainly andesites and trachites). Cristobalite is rarely found in volcanic rocks and sometimes occurs in opal gems. Coesite is a marker of high-P regimes. It has been found coexisting with garnet in the Dora-Maira Massif (Western Alps) and in various other high-P metamorphic terrains (the former Soviet Union, China). Stishovite is stable above about 80 kbar (cf. Liu and Bassett, 1986, and references therein) and is found exclusively in "impactites" (i.e., rocks formed by meteorite impact). Chalcedony and opal are amorphous phases generated by hydrothermal deposition from SiO_2-saturated solutions.

Figure 5.63A shows the relative positions of Si atoms in the α-quartz structure projected along axis z. In this simplified representation, note the hexagonal arrangement of Si atoms, which can be internally occupied by univalent cations through coupled substitution involving Al^{3+} in tetrahedral position:

Table 5.68 Structural properties of SiO_2 polymorphs (from Smyth and Bish, 1988).

Polymorph	Quartz	Coesite	Stishovite	Cristobalite	Tridymite
Formula unit	SiO_2	SiO_2	SiO_2	SiO_2	SiO_2
Molar weight (g)	60.085	60.085	60.085	60.085	60.085
Density (g/cm³)	2.648	2.909	4.287	2.318	2.269
Molar volume (cm³)	22.688	20.657	14.017	25.925	26.478
Z	3	16	2	4	48
System	Trigonal	Monoclinic	Tetragonal	Tetragonal	Monoclinic
Class	*32*	*2/m*	*4/mmm*	*422*	*m*
Group	*P3₂21*	*C2/c*	*P4₂/mnm*	*P4₁2₁2*	*Cc*
Cell parameters:					
a	4.9134	7.1464	4.1790	4.978	18.494
b	4.9134	12.3796	4.1790	4.978	4.991
c	5.4052	7.1829	2.6651	6.948	25.832
β		120.283			117.75
Volume (Å³)	113.01	548.76	46.54	172.17	2110.2
Reference	Le Page et al. (1980)	Smyth et al. (1987)	Baur and Kahn (1971)	Peacor (1973)	Kato and Nukui (1976)

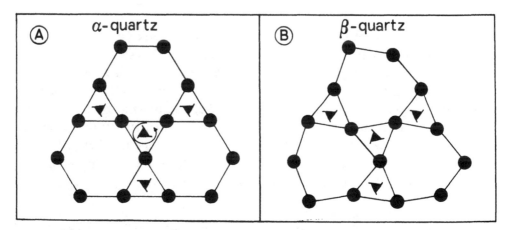

Figure 5.63 Quartz structure projected along axis z. Filled dots are Si atoms (oxygen atoms not shown). α-quartz (A) - β-quartz (B) transition takes place by simple rotation of tetrahedra on ternary helicogyre (adapted from Gottardi, 1972).

$$Si_T^x + \square \Leftrightarrow Na_\square^{\cdot} + Al_T^{|} \qquad (5.200)$$

(the hexagonal ring is actually a spiral with atoms positioned on separate planes). Figure 5.63B shows the structure of high-T β-quartz. The α-quartz–β-quartz transition, which may be described as the overlap of a first-order and a λ transition, takes place by simple rotation of tetrahedra on the ternary helicogyre.

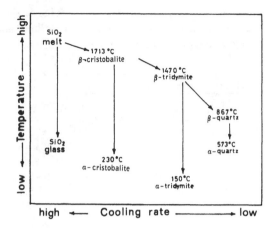

Figure 5.64 Effect of cooling rate on SiO_2 polymorphic transitions. From Putnis and McConnell (1980). Reproduced with modifications by permission of Blackwell Scientific Publications, Oxford, Great Britain.

The structural arrangements of $[SiO_4]$ groups in quartz, tridymite, and cristobalite are different, and the corresponding first-order phase transitions are difficult to complete, often giving rise to metastability. At $P = 1$ bar, for instance, β-cristobalite is stable only at $T > 1470$ °C. With decreasing T, we should observe transition to tridymite, but this transformation requires substantial rearrangement of tetrahedra and is rather sluggish. If the cooling rate is high, we observe metastable persistence of β-cristobalite down to $T = 230$ °C, where the β-cristobalite–α-cristobalite λ transition takes place. The same metastability is observed for the tridymite polymorph with respect to quartz, as shown in figure 5.64.

Table 5.69 lists thermal expansion and compressibility factors for some SiO_2 polymorphs, according to the databases of Saxena et al. (1993) and Holland and Powell (1990). Table 5.70 lists thermodynamic data for the various SiO_2 polymorphs according to various sources.

The amount of H_2O in amorphous silica (number n of H_2O molecules per unit formula) varies between 0.14 and 0.83 (Frondel, 1962). Nevertheless, the thermodynamic properties of the phase are not particularly affected by the value of n (Walther and Helgeson, 1977). The molar volume of opal is 29 cm^3/mole. The same volume of α-quartz may be adopted for chalcedony; see table 5.68 for the other polymorphs.

We have already stated that the α-β transition of quartz may be described as a λ transition overlapping a first-order transition. The heat capacity function for the two polymorphs is thus different in the two stability fields, and discontinuities are observed in the H and S values of the phase at transition temperature (T_{trans} cf. section 2.8). For instance, to calculate the thermodynamic properties of β-quartz at $T = 1000$ K and $P = 1$ bar, we

Table 5.69 Thermal expansion at $P = 1$ bar (K^{-1}) and bulk modulus at $T = 25°C$ (Mbar) for some SiO$_2$ polymorphs in the databases of Saxena et al. (1993) (1) and Holland and Powell (1990) (2). Thermal expansion at various T values of interest is obtained by applying
$\alpha_{(T)} = \alpha_0 + \alpha_1 T + \alpha_2 T^{-1} + \alpha_3 T^{-2}$

Polymorph	Thermal Expansion				Bulk Modulus		
	$\alpha_0 \times 10^5$	$\alpha_1 \times 10^8$	$\alpha_2 \times 10^4$	$\alpha_3 \times 10^2$	K_0	K'	Reference
α-quartz	2.7513	2.9868	0.05572	9.1181	0.3097	6.4	(1)
	3.5260	-	-	-	0.3846	-	(2)
β-quartz	2.0604	3.4694	1.30754	−163.76	0.5149	5.3	(1)
					0.9107	-	(2)
Coesite	0.543	0.76	0	0	0.9135	8.4	(1)
	1.066	-	-	-	0.969	-	(2)
Tridymite	2.1797	−1.53	0	0	0.25	6.0	(1)
Cristobalite	0.5	0	0	0	0.1949	6.0	(1)
Stishovite	0.23	1.2	62.0	−113.0	3.1268	6.0	(1)

Table 5.70 Thermodynamic data for silica polymorphs. Stishovite and tridymite from Saxena et al. (1993); remaining polymorphs from Helgeson et al. (1978). $H^0_{f.T_r.P_r}$ in kJ/mole; $S^0_{T_r.P_r}$ in J/(mole $\times$ K). Heat capacity function is
$C_P = K_1 + K_2 T + K_3 T^{-2} + K_4 T^2 + K_5 T^{-3} + K_6 T^{-1/2} + K_7 T^{-1}$.

Polmorph	Formula Unit	$H^0_{f.T_r.P_r}$	$S^0_{T_r.P_r}$	K_1	$K_2 \times 10^3$	$K_3 \times 10^{-5}$
Quartz (α and β)	SiO$_2$	−910.648	41.34	46.944	34.309	−11.297
				60.291	8.117	0
Cristobalite (α and β)	SiO$_2$	−906.903	43.396	58.492	13.975	−15.941
				72.760	1.297	−41.380
Tridymite	SiO$_2$	−906.913	45.116	66.6993	5.2779	−21.323
Coesite	SiO$_2$	−906.313	40.376	46.024	34.309	−11.297
				59.371	8.117	0
Stishovite	SiO$_2$	−864.000	29.50	58.12	7.002	−126.89
Chalcedony	SiO$_2$	−909.108	41.337	46.944	34.309	−11.297
Amorphous silica	SiO$_2 \times n$H$_2$O	−897.752	59.998	24.811	197.485	−95.312

Polymorph	$K_4 \times 10^7$	$K_5 \times 10^{-8}$	$K_6 \times 10^{-7}$	$K_7 \times 10^{-4}$	T_{limit} (K)
Quartz (α and β)	0	0	0	0	848
	0	0	0	0	2000
Cristobalite (α and β)	0	0	0	0	543
	0	0	0	0	2000
Tridymite	−3.5478	0	0	0	–
Coesite	0	0	0	0	848
	0	0	0	0	2000
Stishovite	0	17.928	0	1.7012	–
Chalcedony	0	0	0	0	848
Amorphous silica	0	0	0	0	622

Table 5.71 Transition energy terms for quartz and cristobalite: $\Delta \bar{S}^0_{trans}$ = molar apparent transition entropy at P = 1 bar [J/(mole × K)]; $\Delta \bar{H}^0_{trans}$ = molar apparent transition enthalpy at P = 1 bar (kJ/mole); $\Delta \bar{V}^0_{trans}$ = molar apparent transition volume at P = 1 bar (cm³/mole).

Transition	T_{trans}	$\Delta \bar{S}^0_{trans}$	$\Delta \bar{H}^0_{trans}$	$\Delta \bar{V}^0_{trans}$
α-quartz → β-quartz	848	1.431	1.213	0.372
α-cristobalite → β-cristobalite	543	2.472	1.343	(?)

proceed as follows: calculate enthalpy and entropy at T_{trans} = 848 K with the aid of the heat capacity expression valid for α-quartz (integral between T_r = 298.15 K and T = T_{trans} = 848 K); to the obtained values add the apparent standard molar α-β transition enthalpy and entropy at T_{trans}; calculate the integrals between T_{trans} and T = 1000 K with the C_P function valid for the β-phase and add the results to the previously obtained enthalpy and entropy values. Note that, at the transition point, the two polymorphs coexist and thus have the same Gibbs free energy. Table 5.71 lists α-β transition enthalpy, entropy, and volume values for quartz and cristobalite. These values are consistent with the α-phase of Helgeson et al. (1978). An extensive discussion on the energetics of SiO_2 polymorphs may be found in Helgeson et al. (1978) and Berman (1988).

5.9 Thermobarometric Properties of Silicates

Estimates of T and P conditions of equilibrium for a given mineralogical paragenesis are of fundamental importance in understanding geologic processes. From the first attempts to define metamorphic regimes (P, T, anistropic P) through index minerals, advances in physical geochemistry progressively allowed quantification of the reaction, exchange, and ordering processes in minerals that lead to consistent thermobarometric applications. Thermobarometry is now a common tool in geochemistry. We will later examine briefly the various types of processes usually considered in thermobarometric studies. As a premise, reasoning in terms of "equilibrium," the deduced P and T values correspond to the last condition at which equilibrium was complete, further exchanges being prevented by kinetic reasons (i.e., low elemental diffusivites and/or high cooling rates).

5.9.1 Intercrystalline Exchange Geothermometry

Let us consider the exchange reaction of two ions, A and B, between two phases, α and β, composed of binary mixtures (A,B)M (α) and (A,B)N (β):

$$A_\alpha + B_\beta \Leftrightarrow A_\beta + B_\alpha. \tag{5.201}$$

The Gibbs free energy change involved in reaction 5.201 is given by the sum of the chemical potentials of components in reaction (μ_i), multiplied by their respective stoichiometric factors (ν_i)—i.e.,

$$\Delta \overline{G}_{201} = \sum_i \nu_i \mu_i. \tag{5.202}$$

In the case of equation 5.201, stoichiometric factors are 1 for all components; equation 5.202 may thus be reduced to

$$\Delta \overline{G}_{201} = \sum_i \mu_i. \tag{5.203}$$

For each component,

$$\mu = \mu_i^0 + RT \ln a_i \tag{5.204}$$

is valid, where μ_i^0 is the standard state chemical potential and a_i is thermodynamic activity.

   Applying equation 5.203 to reaction 5.201 and recalling equation 5.204, we obtain

$$\Delta \overline{G}_{201} = \Delta \overline{G}_{201}^0 + RT \ln K_{201,P,T}, \tag{5.205}$$

where $\Delta \overline{G}_{201}^0$ is the Gibbs free energy change involved in reaction 5.201 at the standard state of reference condition and $K_{201,P,T}$ is the equilibrium constant at the P and T of interest. Because, at equilibrium, ΔG of the reaction is zero, we may rewrite equation 5.205 as

$$\Delta \overline{G}_{201}^0 = -RT \ln K_{201,P,T}. \tag{5.206}$$

   It is evident from equation 5.204 that the intrinsic significance of equation 5.206 is closely connected with the choice of standard state of reference. If the adopted standard state is that of the pure component at the P and T of interest, then $\Delta \overline{G}_{201}^0$ is *the Gibbs free energy of reaction between pure components at the P and T of interest.* Deriving in P the equilibrium constant, we obtain

$$\left(\frac{\partial \ln K}{\partial P} \right)_T = -\left[\frac{\partial \left(\Delta \overline{G}^0 / RT \right)}{\partial P} \right]_T = -\frac{\Delta \overline{V}^0}{RT}. \tag{5.207}$$

Nevertheless, if the standard state of reference is that of the pure component at $P = 1$ bar and T of interest, then

$$\left(\frac{\partial \ln K}{\partial P}\right)_T = 0. \tag{5.208}$$

Concerning temperature, if the standard state of reference is that of the pure component at the T and P of interest, we have

$$\left(\frac{\partial \ln K}{\partial T}\right)_P = -\left[\frac{\partial\left(\Delta\bar{G}^0/RT\right)}{\partial T}\right]_P = \frac{\Delta\bar{H}^0}{RT^2} \tag{5.209}$$

or, deriving in the inverse of T,

$$\left[\frac{\partial \ln K}{\partial(1/T)}\right]_P = -\frac{\Delta\bar{H}^0}{R}. \tag{5.210}$$

However, if the standard state is that of the pure component at $P = 1$ bar and T of interest,

$$\frac{\partial \ln K}{\partial T} = \frac{\Delta\bar{H}^0}{RT^2} \tag{5.211}$$

not dependent on P.

If we now consider the equilibrium constant expressed as activity product, then

$$K_{201,P,T} = \prod_i a_i^{v_i}. \tag{5.212}$$

In the case of reaction 5.201, stoichiometric coefficients are 1 and equation 5.212 may be reduced to

$$K_{201,P,T} = \frac{a_{A,\beta}a_{B,\alpha}}{a_{A,\alpha}a_{B,\beta}} = \left(\frac{X_{A,\beta}X_{B,\alpha}}{X_{A,\alpha}X_{B,\beta}}\right) \times \left(\frac{\gamma_{A,\beta}\gamma_{B,\alpha}}{\gamma_{A,\alpha}\gamma_{B,\beta}}\right). \tag{5.213}$$

The first term in parentheses on the right side of equation 5.213 is the distribution coefficient (K_D), and the second groups activity coefficients related to the mixing behavior of components in the two phases. The equilibrium constant is thus related to the interaction parameters of the two phases at equilibrium. For example, the equilibrium between two regular mixtures is defined as

$$\ln K_{201,P,T} = \ln K_D - \frac{W_\alpha}{RT}\left(1 - 2X_{A,\alpha}\right) + \frac{W_\beta}{RT}\left(1 - 2X_{A,\beta}\right), \qquad (5.214)$$

where W_α and W_β are interaction parameters for phases α and β, respectively, or, for two asymmetric Van Laar mixtures,

$$\ln K_{201,P,T} = \ln K_D + \frac{A_{0,\alpha}}{RT}\left(X_{A,\alpha} - X_{B,\alpha}\right) + \frac{A_{1,\alpha}}{RT}\left(6X_{B,\alpha}X_{A,\alpha} - 1\right)$$

$$+ \frac{A_{0,\beta}}{RT}\left(X_{B,\beta} - X_{A,\beta}\right) + \frac{A_{1,\beta}}{RT}\left(6X_{A,\beta}X_{B,\beta} - 1\right). \qquad (5.215)$$

where $A_{0,\alpha}$, $A_{1,\alpha}$, $A_{0,\beta}$, and $A_{1,\beta}$ are Guggenheim's parameters for phases α and β, respectively (Saxena, 1973).

Let us now examine two natural phases, α and β, coexisting at equilibrium, for which we know in detail both mixing behavior (hence W_α and W_β, or $A_{0,\alpha}$, $A_{1,\alpha}$, $A_{0,\beta}$, and $A_{1,\beta}$) and composition (molar fractions of AM, BM, AN, and BN components, or the analogous ionic fractions $X_{A,\alpha}$, $X_{A,\beta}$, $X_{B,\alpha}$, and $X_{B,\beta}$). Application of equation 5.206 allows us to define the loci of P and T conditions that satisfy equilibrium, provided that in the definitions of both $\Delta\overline{G}^0$ and the thermodynamic activity of components in the mixture we are consistent with the adopted standard state. It is obvious that the more $\Delta\overline{G}^0$ is affected by P and T intensive variables, and the more the interaction parameters describing the mixing behavior are accurate, the more efficient and precise is the thermobarometric expression. By stating "the loci of P and T conditions that satisfy equilibrium," we emphasize that we are dealing with univariant equilibria. In a P and T space, the slope of the univariant curve is defined by equations 5.207 and 5.209; if the volume of reaction is negligible, then obviously the equilibrium constant is not affected by P, and the elemental exchange furnishes only thermometric information.

Exchange geothermometry generally considers couples of elements of identical charge showing the property of diadochy—for instance, Na^+-K^+, Mg^{2+}-Fe^{2+}, or Fe^{3+}-Cr^{3+}. Intercrystalline exchange thermometers have been developed mainly on the Mg^{2+}-Fe^{2+} couple, and the investigated mineral phases comprise olivine, orthopyroxene, clinopyroxene, garnet, spinel, ilmenite, cordierite, biotite, hornblende, and cummingtonite. Table 5.72 lists bibliographic sources for these geothermometers. More detailed references may be found in Essene (1982) and Perchuk (1991).

Let us consider, for instance, the Fe^{2+}-Mg^{2+} exchange reaction between garnet and clinopyroxene:

$$\frac{1}{3}\,Mg_3Al_2Si_3O_{12} + CaFeSi_2O_6 \Leftrightarrow \frac{1}{3}\,Fe_3Al_2Si_3O_{12}\;CaMgSi_2O_6.$$

$$\underset{\text{pyrope}}{} \quad \underset{\text{hedembergite}}{} \quad \underset{\text{almandine}}{} \quad \underset{\text{diopside}}{} \qquad\qquad (5.216)$$

Table 5.72 Fe^{2+}-Mg^{2+} exchange geothermometers.

Equilibrium	References
Olivine-orthopyroxene-spinel	Ramberg and Devore (1951); Nafziger and Muan (1967); Speidel and Osborn (1967); Medaris (1969); Nishizawa and Akimoto (1973); Matsui and Nishizawa (1974); Fujii (1977); Engi (1978)
Ilmenite-orthopyroxene	Bishop (1980)
Ilmenite-clinopyroxene	Bishop (1980)
Orthopyroxene-clinopyroxene	Mori (1977); Herzberg (1978b)
Garnet-Olivine	Kawasaki and Matsui (1977); O'Neill and Wood (1979)
Garnet-Cordierite	Currie (1971); Hensen and Green (1973); Thompson (1976); Hensen (1977); Holdaway and Lee (1977); Perchuk (1991)
Garnet-clinopyroxene	Saxena (1979); Banno (1970); Oka and Matsumoto (1974); Irving (1974); Raheim and Green (1974, 1975); Raheim (1975, 1976); Mori and Green (1978); Slavinskiy (1976); Ellis and Green (1979); Ganguly (1978); Saxena (1979); Dahl (1980); Perchuk (1991); Krog (1988).
Garnet-biotite	Saxena (1969); Thompson (1976); Perchuk (1977, 1991); Goldman and Albee (1977); Ferry and Spear (1978).
Garnet-chlorite	Ghent et al. (1987); Perchuk (1991)
Garnet-staurolite	Perchuk (1991) and references therein
Garnet-amphibole	Perchuk (1991) and references therein
Garnet-chloritoid	Perchuk (1991) and references therein
Biotite-chloritoid	Perchuk (1991) and references therein
Chlorite-chloritoid	Perchuk (1991) and references therein

The constant of reaction 5.216 is plotted in a semilogarithmic Arrhenius diagram in figure 5.65. In this short of diagram (Ganguly, 1979), the equilibrium constant assumes an essentially straight path, indicating that the reaction enthalpy does not vary significantly with T at constant pressure (cf. eq. 5.210). Figure 5.65 also shows the distributive constant

$$K_D = \frac{X_{Fe,garnet} X_{Mg,cpx}}{X_{Mg,garnet} X_{Fe,cpx}},$$ (5.217)

which was experimentally measured by Wood (1976b).

By applying the ionic solution model, we may express the relationship between ionic fraction and activity in the garnet phase as

$$a_{Mg_3Al_2Si_3O_{12}} \equiv \left(X_{Mg}\gamma_{Mg}\right)^3_{garnet},$$ (5.218)

where γ_{Mg} is the activity coefficient of ion Mg^{2+} in the structural site and 3 is the stoichiometric number of sites over which substitution takes place. Analogously, for clinopyroxene,

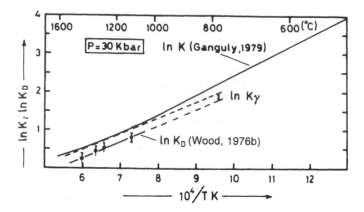

Figure 5.65 Garnet-clinopyroxene geothermometric exchange (Fe^{2+}-Mg^{2+}). Thermodynamic constant K is compared with distribution coefficient K_D: the difference between the two terms represents the effect of interactions in mixtures. Reprinted from J. Ganguly, *Geochimica et Cosmochimica Acta*, 43, 1021–1029, copyright © 1979, with kind permission from Elsevier Science Ltd., The Boulevard, Langford Lane, Kidlington 0X5 1GB, UK.

$$a_{CaMgSi_2O_6} \equiv \left(X_{Mg} \gamma_{Mg} \right)_{cpx}. \tag{5.219}$$

Applying equation 5.212 to the equilibrium constant, we have

$$K_{216} = \frac{\left[\left(X_{Fe} \gamma_{Fe} \right)^3_{garnet} \right]^{1/3}}{\left[\left(X_{Mg} \gamma_{Mg} \right)^3_{garnet} \right]^{1/3}} \times \frac{\left(X_{Mg} \gamma_{Mg} \right)_{cpx}}{\left(X_{Fe} \gamma_{Fe} \right)_{cpx}} = \frac{\left(X_{Fe} / X_{Mg} \right)_{garnet}}{\left(X_{Fe} / X_{Mg} \right)_{cpx}} \times \frac{\left(\gamma_{Fe} / \gamma_{Mg} \right)_{garnet}}{\left(\gamma_{Fe} / \gamma_{Mg} \right)_{cpx}} = K_D K_\gamma \tag{5.220}$$

and, in logarithmic notation,

$$\ln K_{216} = \ln K_D + \ln K_\gamma. \tag{5.221}$$

As figure 5.65 shows, the term $\ln K_\gamma$ definitely affects the equilibrium. This term is consistent with a subregular Margules formulation for excess Gibbs free energy terms in both phases (Ganguly, 1979; see also sections 5.3.5 and 5.4.5). Combining the various experimental evidence, Ganguly (1979) calibrated the distributive function of equation 5.217 over T and P, and proposed the following thermometric expression:

$$\ln K_D = \frac{4100 + 11.07P}{T} + c, \tag{5.222}$$

where P is in kbar, T is in K, $c = -2.4$ for $T < 1333$ K, and $c = -2.93$ for $T > 1333$ K.

As we see, equation 5.222 is ready to use for petrologic applications. Geothermometric expressions of this type are extremely useful but must be applied with caution, with careful evaluation of the P and T limits of applicability and the effect of the presence of minor components in the mixture. For instance, for the garnet-clinopyroxene geothermometer, the presence of grossular (besides pyrope and almandine) and jadeite (besides hedembergite-diopside) has been studied by Saxena (1969), Banno (1970), Oka and Matsumoto (1974), and Irving (1974), but thermometric calibration is generally based on experimental runs in compositionally simple synthetic systems. The extension to chemically complex natural phases normally involves large degrees of incertitude. Also, some geothermometric expressions operate well under certain P and T regimes but cannot be applied to P and T conditions outside the calibration range. For instance, the formulation of Ellis and Green (1979) for the geothermometer 5.216 is particularly suitable for granulitic terrains, because it expressly accounts for the Ca amount in garnet:

$$T = \frac{3104 X_{Ca,garnet} + 3030 + 10.86P}{\ln K_D + 1.9039} \qquad (5.223)$$

(T in K and P in kbar); however, its application to other thermobaric conditions is not so appropriate. The most satisfactory formulations for the exchange reaction 5.216 are those of Slavinskiy (1976), Ganguly (1979), Saxena (1979), and Dahl (1980), which are based on a large body of thermodynamic data, petrologic studies, and natural observations.

Besides caution regarding the P-T range of applicability and compositional limits for a given intracrystalline exchange expression, the role of kinetics in the attainment of equilibrium must also be emphasized. Reasoning in simple thermodynamic terms, the P and T values deduced from an ionic exchange reaction represent the final state condition. Moreover, phases at equilibrium must be compositionally homogeneous, and this corresponds (in kinetic terms) to sufficiently high ionic diffusivity for the thermometric couple in both phases.

Figure 5.66A shows the concentration profiles in an ideal intercrystalline exchange geothermometer where exchange kinetics do not play any role: elemental concentrations are constant at all distances from the interface. Figure 5.66B shows how diffusive kinetics perturb the concentration profiles of the thermometric couple: elemental diffusion is too slow with respect to the cooling rate of the system, resulting in marked compositional zonation as a function of the distance from the interface plane between the two exchanging phases. Obviously, in this case, application of a given thermometric exchange relation will result in contrasting T deductions, depending on the compositional zone of application.

The implications of diffusion kinetics for ionic exchange geothermometry have been discussed in detail by Lasaga (1983).

Ionic diffusivity varies exponentially with temperature, according to

$$D_{(T)} = A \exp\left(\frac{-E_a}{RT}\right), \qquad (5.224)$$

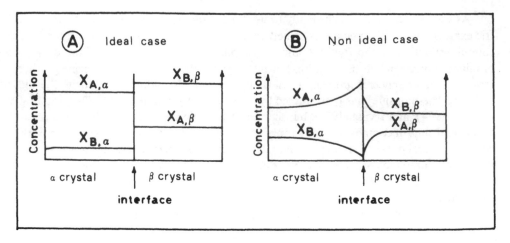

Figure 5.66 Elemental concentration profiles in a thermometric couple. (A) Ideal case; elemental concentration is constant from interface to nucleus of crystal. (B) Nonideal case: slow diffusivity generates concentration gradients from interface to nucleus of crystal.

where A is the preexponential factor and E_a is the energy of activation of the diffusive process.

Diffusivity at time t may be related to initial diffusivity at time t_0 by associating with t the corresponding temperature T:

$$D_{(t)} = D^0_{(t_0)} \exp\left\{ \frac{E_a}{R} \left[\frac{1}{T_{(t)}} - \frac{1}{T^0_{(t_0)}} \right] \right\}. \tag{5.225}$$

Assuming a linear decrease in temperature with time,

$$T_{(t)} \approx T^0_{(t_0)} - vt, \tag{5.226}$$

where v is the cooling rate of the system. Equation 5.225 is reduced to

$$D_{(t)} = D^0_{(t_0)} \exp(-\gamma t), \tag{5.227}$$

where

$$\gamma = \frac{E_a v}{R T^{0^2}_{(t_0)}}. \tag{5.228}$$

Parameter γ is extremely important, because it defines the time within which elemental exchanges proceed, beginning at the initial time t_0 when temperature was T^0. If, for instance, the energy of activation of the diffusion process is 50 kcal/mole, the cooling rate of the system is 2 °C/10^6 years, and $T^0 = 1000$ K, the resulting γ is (0.05×10^{-6}) years^{-1}. The time elapsed from the initial condition (T^0, t_0) to the (kinetic) closure of elemental exchanges is

$$t' = \frac{1}{\gamma}\left[1 - \exp(-\gamma t)\right] \le \frac{1}{\gamma} \le 2 \times 10^7 \text{ years} \tag{5.229}$$

(Lasaga, 1983).

As regards the effect of temperature on the distribution constant, if both phases are ideal, the distribution constant is equivalent to the thermodynamic constant, and we can write

$$K_D = K_{D^0} \exp\left[\frac{\Delta \overline{H}^0}{R}\left(\frac{1}{T} - \frac{1}{T^0}\right)\right], \tag{5.230}$$

where $\Delta \overline{H}^0$ is the enthalpy of the exchange reaction at standard state and K_{D^0} is the distribution constant at the initial temperature T^0. We can define K_D as a function of time, as was done previously for diffusivity (cf. equation 5.227):

$$K_{D_{(t)}} = K_{D^0} \exp(-\varepsilon' t), \tag{5.231}$$

where

$$\varepsilon' = \frac{\Delta \overline{H}^0 v}{RT^{0^2}_{(t_0)}}. \tag{5.232}$$

Lasaga (1979) has shown that the molar concentration of an exchanging element A at the interface between two phases α and β at time t $(X_{A,\alpha,0,t})$ obeys the law

$$X_{A,\alpha,0,t} = X^0_{A,\alpha} \exp(-\varepsilon t). \tag{5.233}$$

Applying Fick's law on diffusive fluxes at the interface between the two phases (cf. section 4.11):

$$-\tilde{D}_{\alpha,t} \frac{\partial X_{A,\alpha}}{\partial \chi} = -\tilde{D}_{\beta,t} \frac{\partial X_{A,\beta}}{\partial \chi} \tag{5.234}$$

and

$$-\tilde{D}_{\alpha,t} \frac{\partial X_{B,\alpha}}{\partial \chi} = -\tilde{D}_{\beta,t} \frac{\partial X_{B,\beta}}{\partial \chi}, \tag{5.235}$$

where χ is the distance from the interface, and $\tilde{D}_{\alpha,t}$ and $\tilde{D}_{\beta,t}$ are interdiffusion coefficients for elements A and B in phases α and β, Lasaga (1983) has shown that ε is related to ε' through

$$\varepsilon = \varepsilon' \frac{\sqrt{\tilde{D}_\beta^0 / \tilde{D}_\alpha^0}}{\sqrt{\tilde{D}_\beta^0 / \tilde{D}_\alpha^0}\left(1 + \frac{X_{A,\alpha}^0}{X_{B,\alpha}^0}\right)\left(\frac{X_{A,\alpha}^0}{X_{A,\beta}^0} + \frac{X_{A,\alpha}^0}{X_{B,\beta}^0}\right)}, \tag{5.236}$$

where $X_{A,\alpha}^0$ is the concentration of A in phase α, initially homogeneous at temperature T^0. If $\tilde{D}_\beta^0 \gg \tilde{D}_\alpha^0$ and $X_{B,\alpha}^0 > X_{A,\alpha}^0$, then

$$\varepsilon \approx \varepsilon' \tag{5.237}$$

and

$$\frac{\varepsilon}{\gamma} \approx \frac{\Delta\overline{H}^0}{E_\alpha}. \tag{5.238}$$

If $\tilde{D}_\beta^0 < \tilde{D}_\alpha^0$ and $X_{B,\alpha}^0 > X_{A,\alpha}^0$, then

$$\varepsilon = \varepsilon' \sqrt{\tilde{D}_\beta^0 / \tilde{D}_\alpha^0} \tag{5.239}$$

and

$$\frac{\varepsilon}{\gamma} \approx \Delta\overline{H}^0. \tag{5.240}$$

Analysis of existing data shows that ratio ε/γ in the main thermometric phases ranges between 0.01 and 0.2. The highest ε/γ values correspond to minerals with high ionic diffusivity. Knowledge of ratio ε/γ in phases used in thermometric studies allows us to establish the reliability of the obtained temperature, as a function of the grain size of the phases and the cooling rate of the system. Besides ratio ε/γ, Lasaga (1983) also defined quantity γ' as

$$\gamma' = \frac{E_\alpha v a^2}{\tilde{D}_\alpha^0 R T_{(t_0)}^{0^2}}, \tag{5.241}$$

where $\overline{D}_\alpha^0$ is the interdiffusion coefficient at temperature T^0 in phase α and a is the length of the crystal orthogonal to the interface. γ' allows us to establish whether or not the nuclei of crystals undergoing equilibration maintain the original compositions they had at crystallization (t_0, T^0). Practical calculations show that this happens for $\gamma' > 10$. Figure 5.67 shows the combined effects of parameter γ' and ratio ε/γ: for low values of γ', the

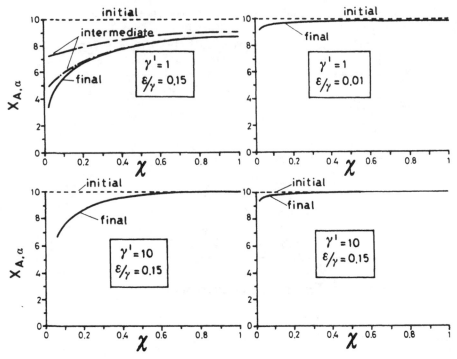

Figure 5.67 Concentration profiles in a crystal, for various values of γ' and ε/γ. Abscissa: fractional distance from crystal edge ($\chi = 0$ at edge; $\chi = 1$ at center). Reproduced with modifications from Lasaga (1983), with permission of Springer-Verlag, New York Inc.

initial composition of the nucleus is not preserved even in phases with slow diffusivity (low ε/γ). We also see that, at constant γ', the concentration profiles are more or less altered with respect to the initial condition, depending on diffusion rate. Lasaga (1983) emphasized the fact that the core of the crystal may be homogeneous at both high (i.e., > 100) and low (< 0.01) values of parameter γ'. Hence, contrary to what is commonly assumed, the homogeneity of crystal cores does not necessarily mean preservation of the initial composition acquired at crystallization.

Table 5.73 list iso-γ' temperatures preserved by some minerals at various cooling rates of the system (Lasaga, 1983). For each phase, T^0 is the initial temperature of homogeneous composition preserved at cooling rate v for $\gamma' = 10$ (crystal size = 1 mm). Consider, for instance, pyroxene: for low cooling rates (i.e., < 10 °C/10^6 y), temperatures in the range 900 to 1000 °C may be effectively considered as quenched equilibrium temperatures (i.e., either crystallization or recrystallization). However, preservation of higher T (for instance, 1100 to 1300 °C) requires very high cooling rates, to prevent alteration of concentration profiles down to the crystal nuclei. The data in table 5.73 may thus be used to evaluate the reliability of a thermometric calculation, depending on the kinetic properties of the phase.

Table 5.73 Iso-γ' temperatures for some minerals as a function of cooling rate of system. $\gamma = 10$; $a = 1$ mm (from Lasaga, 1983). (a) Based on diffusion data of Lasaga et al. (1977). (b) Based on diffusion data of Elphick et al. (1981).

| | | T (°C) | | |
| | | Garnet | | |
v (°C/10^6 y)	Pyroxene	(a)	(b)	Olivine
0.01	805	565	785	-
0.1	865	610	840	-
1	935	665	907	315
10.0	1010	720	980	370
100.0	1100	790	1050	430
1000.0	1200	860	1155	500
10000.0	1320	945	1260	585

5.9.2 Intracrystalline Exchange Geothermometry

Let us consider a solid mixture (A,B)N with two lattice positions $M1$ and $M2$ in which intracrystalline exchanges take place:

$$A_{M1} + B_{M2} \Leftrightarrow A_{M2} + B_{M1}. \tag{5.242}$$

This is the case, for instance, for the enstatite-ferrosilite solid mixture [orthopyroxene: $(Fe,Mg)SiO_3$]:

$$Mg_{M1} + Fe^{2+}_{M2} \Leftrightarrow Mg_{M2} + Fe^{2+}_{M1}. \tag{5.243}$$

If mixing on sites may be approximated by a regular model, we have

$$\ln K_{243} = \ln K_D + \frac{W_{M1}}{RT}\left(1 - 2X_{Fe,M1}\right) - \frac{W_{M2}}{RT}\left(1 - 2X_{Fe,M2}\right). \tag{5.244}$$

It is important to emphasize here that, theoretically, if a solid mixture is ideal, intracrystalline distribution is completely random (cf. section 3.8.1) and, in these conditions, the intracrystalline distribution constant is always 1 and coincides with the equilibrium constant. If the mixture is nonideal, we may observe some ordering on sites, but intracrystalline distribution may still be described without site interaction parameters. We have seen in section 5.5.4, for instance, that the distribution of Fe^{2+} and Mg^{2+} on $M1$ and $M3$ sites of riebeckite-glaucophane amphiboles may be approached by an *ideal site mixing model*—i.e.,

$$Mg_{M1} + Fe^{2+}_{M3} \Leftrightarrow Mg_{M3} + Fe^{2+}_{M1} \tag{5.245}$$

$$K_{245} = \frac{a_{Mg,M3}\, a_{Fe^{2+},M1}}{a_{Mg,M1}\, a_{Fe^{2+},M3}} = \frac{\left(1 - X_{Fe,M3}\right) X_{Fe,M1}}{\left(1 - X_{Fe,M1}\right) X_{Fe,M3}} = 0.403. \qquad (5.246)$$

In the case of reaction 5.243, Saxena and Ghose (1971) experimentally determined the distribution constant over a wide range of temperatures. According to these authors, it has an exponential dependency on $T\,(\mathrm{K})$ of the type

$$-RT \ln K_{243} = \Delta \bar{G}^0_{243} = 4479 - 1948\left(\frac{10^3}{T}\right) \qquad \left(\mathrm{cal/mole}\right). \quad (5.247)$$

In the absence of site interaction energy terms, the equality

$$K_{243} = K_D \qquad (5.248)$$

holds. We would therefore observe simple hyperbolic distribution curves similar to those for amphiboles in figure 5.39B. The presence of site interaction energy terms causes perturbations of the hyperbolic distribution curves, whose fluctuations become more marked as the values of W site parameters become higher. According to Saxena (1973), the observed distribution of Fe^{2+} and Mg^{2+} in $M1$ and $M2$ sites of orthopyroxene requires site interaction parameters that vary with $T\,(\mathrm{K})$ according to

$$W_{M1} = 3525\left(\frac{10^3}{T}\right) - 1667 \qquad \left(\mathrm{cal/mole}\right) \qquad (5.249)$$

and

$$W_{M2} = 2458\left(\frac{10^3}{T}\right) - 1261 \qquad \left(\mathrm{cal/mole}\right). \qquad (5.250)$$

The form of the distribution curves, based on equation 5.244 and application of conditions 5.247, 5.249, and 5.250, is displayed in figure 5.68.

More recently, Ganguly and Saxena (1987) reconsidered equation 5.243 in the light of all existing experimental data, proposing

$$\ln K_D = \left(0.1435 - \frac{1561.81}{T}\right) + \frac{W_{M1}}{RT}\left(1 - 2X_{Fe,M1}\right)$$
$$\qquad (5.251)$$
$$- \frac{W_{M2}}{RT}\left(1 - 2X_{Fe,M2}\right),$$

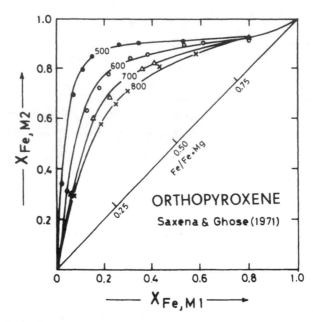

Figure 5.68 Fe^{2+}-Mg^{2+} intracrystalline distribution between $M1$ and $M2$ sites in natural orthopyroxenes of composition near a binary mixture $(Fe,Mg)SiO_3$. From Saxena and Ghose (1971). Reprinted with permission of The Mineralogical Society of America.

with W_{M1} = 1524 (cal/mole) and W_{M2} = -1080 (cal/mole). Note that the first term in parentheses in equation 5.251 corresponds to ln K_{243} in equation 5.244.

The diagrammatic form of figure 5.68 is that commonly adopted to display intracrystalline distributions (see also figures 5.39 and 5.40). However, this sort of plot has the disadvantage of losing definition as the compositional limits of the system are approached. A different representation of intracrystalline disorder is that seen for olivines (figures 5.10 and 5.12; section 5.2.5): the distribution constant is plotted against the molar fraction of one of the components in the mixture.

The effects of kinetics on intracrystalline disorder are less marked than those affecting intercrystalline exchanges, essentially because intracrystalline exchanges take place on the Å scale, whereas intercrystalline equilibration requires displacements on the mm scale. As a result, temperatures registered by intracrystalline exchange thermometers are generally lower than those derived from intercrystalline equilibria. Lastly, it must be noted that the effects of pressure on intracrystalline partitioning are not negligible, as generally assumed, because the volume properties of phases are definitely affected by the state of internal disorder.

5.9.3 Solvus Thermobarometry

Solvus thermobarometry uses the unmixing phenomena of crystalline compounds, retrieving the T and P conditions of unmixing from the compositions (and structures) of unmixed phases. The various solvi are located experimentally by physical observation of the mixing-unmixing phenomenon induced on a fixed composition by modifying the T and P. For the sake of accuracy, the experiment is usually "reversed," the miscibility gap limb being approached from opposite directions in the T-P space.

The thermodynamic significance of unmixing was stressed in chapter 3 (sections 3.11, 3.12, and 3.13), together with the intrinsic differences among solvus field, spinodal field, and miscibility gap, and the role of elastic strain on the T-X extension of the coherent solvus and coherent spinodal field with respect to chemical solvus and chemical spinodal fields, respectively. Although the significance of the various compositional fields is clear in a thermodynamic sense, very often the stability relations of unmixed phases do not seem as clear in microscopic investigation of natural assemblages. The confusion in terms used to describe unmixing phenomena in natural phases was discussed in some detail in section 5.7.4.

Feldspars

Feldspars, as we have already seen, are mixtures of albite ($NaAlSi_3O_8$), K-feldspar ($KAlSi_3O_8$), and anorthite ($CaAl_2Si_2O_8$) components (see section 5.7). Albite and anorthite are miscible to all extents (plagioclases), as are albite and K-feldspar (alkali feldspars), but the reciprocal miscibility between the two series is limited for $T < 800\ °C$. Stormer (1975) calculated the distribution of the $NaAlSi_3O_8$ component between plagioclases and alkali feldspars, accounting for the nonideality of the $NaAlSi_3O_8$-$KAlSi_3O_8$ solid mixture (Parson, 1978) and for P effects on the solvus field. Parson's (1978) model reproduces quite well the experimental observations of Seck (1971) at $T = 650\ °C$ for anorthite amounts between 5% and 45% in moles. Modifications of Stormer's (1975) model were suggested by Powell and Powell (1977a). Whitney and Stormer (1977) also proposed a low-T geothermometer based on K-feldspar amounts in mixtures. The isothermal distributions of Stormer's (1975) model at various P conditions are compared with the experimental evidence of Seck (1972) in figure 5.69. When applying this sort of thermometer, we must bear clearly in mind the significance of feldspar unmixes in the various assemblages, particularly with respect to the degree of coherence of unmixed zones. Stormer's geothermometer defines the loci of *chemical solvi* (i.e., absence of strain energy) and so cannot be applied to *coherent exsolution* or to *spinodal decomposition*. It is also worth noting that the alkali feldspar series unmixes itself partially during cooling, so that the compositions of unmixed grains must be reintegrated in order to evaluate the original composition of the solid mixture on the solvus limb. Application of Stormer's (1975) geothermometer may simply be graphical [plotting the molar amounts of $NaAlSi_3O_8$ in the two series, for a selected value of P (figure 5.69) and comparing the results with the isotherms] or analytical, through the appropriate thermometric equation.

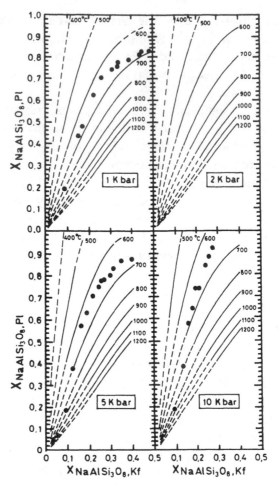

Figure 5.69 Feldspar geothermometry. From Stormer (1975). Reprinted with permission of The Mineralogical Society of America.

Stormer's (1975) model is based on the assumption that the potassic end-member has no influence on the mixing behavior of the plagioclase series and that the calcic component does not affect the mixing behavior of the K-feldspar series to any extent. With these assumptions, the problem of equilibrium between two *ternary* feldspars, normally represented by the equalities

$$\mu_{NaAlSi_3O_8, Pl} = \mu_{NaAlSi_3O_8, Kf} \tag{5.252}$$

$$\mu_{KAlSi_3O_8, Pl} = \mu_{KAlSi_3O_8, Kf} \tag{5.253}$$

$$\mu_{CaAl_2Si_2O_8, Pl} = \mu_{CaAl_2Si_2O_8, Kf}, \tag{5.254}$$

where Pl = plagioclases and Kf = alkali feldspars, is reduced to evaluation of the simple equilibrium of the albitic component (eq. 5.252). The chemical potential of the pure component in the mixture refers to the standard state potential through

$$\mu_{\mathrm{NaAlSi_3O_8},\,Pl} = \mu^0_{\mathrm{NaAlSi_3O_8},\,Pl} + RT \ln a_{\mathrm{NaAlSi_3O_8},\,Pl} \tag{5.255}$$

and

$$\mu_{\mathrm{NaAlSi_3O_8},\,Kf} = \mu^0_{\mathrm{NaAlSi_3O_8},\,Kf} + RT \ln a_{\mathrm{NaAlSi_3O_8},\,Kf}. \tag{5.256}$$

If we adopt for both phases the standard state of the pure component at the T and P of interest (and if the two phases are strictly isostructural), because

$$\mu^0_{\mathrm{NaAlSi_3O_8},\,Pl} = \mu^0_{\mathrm{NaAlSi_3O_8},\,Kf}, \tag{5.257}$$

from equations 5.255 and 5.256, and by application of equation 5.252, we obtain

$$a_{\mathrm{NaAlSi_3O_8},\,Pl} = a_{\mathrm{NaAlSi_3O_8},\,Kf}. \tag{5.258}$$

Note that the isoactivity condition of equation 5.258 is valid only for all the loci of the solvus limb (see, for instance, figure 3.10) but *not* for a spinodal limit *or* for a miscibility gap limb, because in these cases the structural state of unmixes is different, so that equation 5.257 does not hold.

If we call the distribution constant K_D, as $a = X \cdot \gamma$, from equation 5.258 we derive

$$K_D = \frac{X_{\mathrm{NaAlSi_3O_8},\,Kf}}{X_{\mathrm{NaAlSi_3O_8},\,Pl}} = \frac{\gamma_{\mathrm{NaAlSi_3O_8},\,Pl}}{\gamma_{\mathrm{NaAlSi_3O_8},\,Kf}}. \tag{5.259}$$

Moreover, if the plagioclase series is ideal (which is the case in the compositional range Ab_{100} to Ab_{45}; cf. Orville, 1972), equation 5.259 may be further reduced to

$$K_D = \frac{1}{\gamma_{\mathrm{NaAlSi_3O_8},\,Kf}}. \tag{5.260}$$

Adopting a subregular Margules model for the $\mathrm{NaAlSi_3O_8}$-$\mathrm{KAlSi_3O_8}$ (Ab-Or) binary mixture and assuming that the activity coefficient of the albite component is not affected by the presence of limited amounts of the third component in the mixture (i.e., $\mathrm{CaAl_2Si_2O_8}$), equation 5.260 may be transformed into

$$\ln K_D = -\ln \gamma_{Ab,\,Kf}$$

$$= -\frac{1}{RT}\left(1 - X_{Ab,\,Kf}\right)^2 \left[W_{Ab-Or} + 2X_{Ab,\,Kf}\left(W_{Or-Ab} - W_{Ab-Or}\right)\right]. \tag{5.261}$$

Using the interaction parameters of Thompson and Waldbaum (1969) (cf. second column of parameters in table 5.66), i.e.,

$$W_{Ab-Or} = 6326.7 + 0.0925P - 4.6321T \tag{5.262}$$

$$W_{Or-Ab} = 7671.8 + 0.1121P - 3.8565T, \qquad (5.263)$$

and combining equations 5.262 and 5.263 with equation 5.260, we obtain the thermometric equation of Stormer (1975):

$$T(K) = \cfrac{6326.7 - 9963.2 X_{Ab.Kf} + 943.3 X^2_{Ab.Kf} + 2690.2 X^3_{Ab.Kf}}{\left[-1.9872 \ln\left(\cfrac{X_{Ab.Kf}}{X_{Ab.Pl}}\right) + 4.6321 - 10.815 X_{Ab.Kf} + 7.7345 X^2_{Ab.Kf} - 1.5512 X^3_{Ab.Kf} \right]}$$
$$+ \cfrac{\left(0.0925 - 0.1458 X_{Ab.Kf} + 0.0141 X^2_{Ab.Kf} + 0.0392 X^3_{Ab.Kf}\right)P}{\left[-1.9872 \ln\left(\cfrac{X_{Ab.Kf}}{X_{Ab.Pl}}\right) + 4.6321 - 10.815 X_{Ab.Kf} + 7.7345 X^2_{Ab.Kf} - 1.5512 X^3_{Ab.Kf} \right]}. \qquad (5.264)$$

Pyroxenes

As we have already seen in section 5.4.6, unmixing between monoclinic and orthorhombic pyroxenes is conveniently interpreted in the framework of the quadrilateral field diopside-hedembergite-enstatite-ferrosilite ($CaMgSi_2O_6$-$CaFeSi_2O_6$-$Mg_2Si_2O_6$-$Fe_2Si_2O_6$). Indeed, the compositional field is ternary ($CaSiO_3$-$MgSiO_3$-$FeSiO_3$), but the molar amount of wollastonite ($CaSiO_3$) never exceeds 50% in the mixture. Initial studies on the clinopyroxene-orthopyroxene miscibility gap considered only the magnesian end-members (enstatite and diopside; see, for instance, Boyd and Schairer, 1964). Extension to the quadrilateral field was later attempted by Ross and Huebner (1975). Figure 5.70 shows the pyroxene quadrilateral at 5, 10, and 15 kbar total pressure according to Lindsley (1983).

As figure 5.70 shows, with increasing T the miscibility gap shrinks toward the iron-free join; moreover, on the Ca-poor side of the quadrilateral, miscibility relations are complicated by the formation of pigeonite over a restricted P-T-X range (cf. section 5.4.6). Once the compositions of coexisting clinopyroxene and orthopyroxene are known, figure 5.70 is a practical geothermometer.

To calculate the molar proportions of components in the unmixed phase, we proceed as follows:

Clinopyroxene

If the analysis reports the amount of ferric iron, this value is accepted and aluminum is partitioned between VI-fold and IV-fold coordinated sites, according to

$$Al_{VI} + Fe^{3+}_{VI} + Cr_{VI} + 2Ti_{VI} = Al_{IV} + Na_{M2} \qquad (5.265)$$

and

$$Al_{VI} + Al_{IV} = Al_{total}. \qquad (5.266)$$

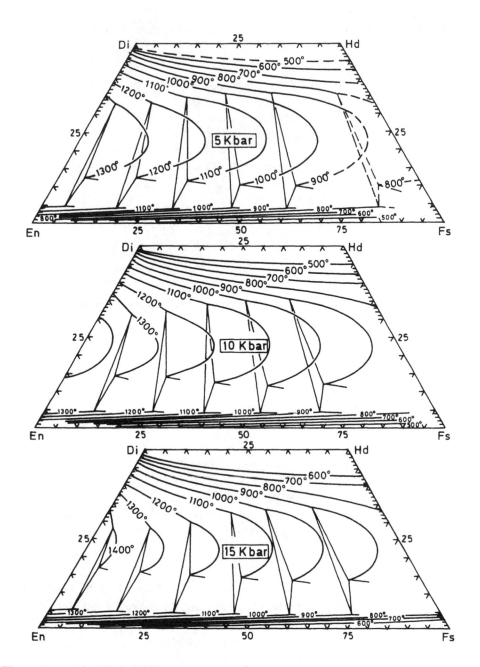

Figure 5.70 Lindsley's (1983) pyroxene geothermometer.

If the analysis reports only total iron expressed as FeO (for example, microprobe data), the amount of IV-fold coordinated aluminum is assumed to obey

$$Al_{IV} = 2 - Si. \qquad (5.267)$$

Fe^{3+}_{VI} is then defined by equations 5.265 and 5.266.

We then evaluate the molar amounts of the various clinopyroxene components as follows:

(1) Acmite = $NaFe^{3+}Si_2O_6$ = smaller amount between Na and Fe^{3+}
(2) Jadeite = $NaAlSi_2O_6$ = smaller amount between Al_{IV} and remaining Na
(3) Fe-Ca-Tschermak = $CaFeAlSiO_6$ = remaining Fe^{3+}
(4) Cr-Ca-Tschermak = $CaCrAlSiO_6$ = Cr
(5) Al-Ca-Tschermak = $CaAl_2SiO_6$ = remaining Al_{VI}
(6) If the pyroxene is an augite, then <<Ca>> = Ca; if it is a pigeonite, then <<Ca>> = 2 − (Fe^{2+} + Mg)
(7) $X = Fe^{2+}/(Fe^{2+} + Mg)$
(8) Wollastonite = $Ca_2Si_2O_6$ = (<<Ca>> + acmite − Fe-Ca-Tschermak − Cr-Ca-Tschermak − Al-Ca-Tschermak)/2
(9) Clinoenstatite = $Mg_2Si_2O_6$ = (1 − wollastonite) (1 − X)
(10) Clinoferrosilite = $Fe_2Si_2O_6$ = (1 − wollastonite)X.

The resulting molar amounts are already normalized to 1 and may be plotted in figure 5.70 for thermometric deductions.

Orthopyroxene

(1) Al_{IV} = 2 − Si
(2) $Al_{VI} = Al_{total} − Al_{IV}$
(3) Fe^{3+} calculated by applying equation 5.265
(4) $R^{3+} = Al_{VI} + Cr + Fe^{3+}$
(5) $R^{2+} = (1 − X)Mg + X Fe^{2+}$
(6) $NaR^{3+}Si_2O_6$ = smaller amount between Na and R^{3+}
(7) $NaTiAlSiO_6$ = smaller amount between Ti and Al_{IV} or remaining Na
(8) $R^{2+}TiAl_2O_6$ = smaller amount between Ti and $Al_{IV}/2$
(9) $R^{2+}R^{3+}AlSiO_6$ = smaller amount between remaining R^{3+} and Al_{IV}
(10) Remaining Ca, Fe^{2+}, and Mg normalized to form wollastonite + enstatite + ferrosilite.

On the basis of the experimental data of Lindsley and Dixon (1976) on the $CaMgSi_2O_6$-$Mg_2Si_2O_6$ binary join and accounting for the observed compositions of natural pyroxenes coexisting at equilibrium, Kretz (1982) proposed two empirical thermometric equations applicable to clinopyroxene and valid, respectively, for $T > 1080$ °C and $T < 1080$ °C:

$$T(K) = \frac{1000}{0.468 + 0.246 X_{Fe,cpx} - 0.123 \ln(1 - 2 X_{Ca,cpx})} \tag{5.268}$$

and

$$
T(\text{K}) = \frac{1000}{0.054 + 0.608\,X_{\text{Fe,cpx}} - 0.304\ln\left(1 - 2X_{\text{Ca,cpx}}\right)}, \qquad (5.269)
$$

where $X_{\text{Ca,cpx}}$ = Ca/(Ca + Mg) and $X_{\text{Fe,cpx}}$ = Fe/(Fe + Mg) in clinopyroxene.

An analogous formulation, also based on the experiments of Lindsley and Dixon (1976), came from Wells (1977):

$$
T(\text{K}) = \frac{7341}{3.355 + 2.44\,X_{\text{Fe,opx}} - \ln\left(\dfrac{a_{\text{Mg}_2\text{Si}_2\text{O}_6,\text{cpx}}}{a_{\text{Mg}_2\text{Si}_2\text{O}_6,\text{opx}}}\right)}, \qquad (5.270)
$$

where $X_{\text{Fe,opx}}$ = Fe/(Fe + Mg) in orthopyroxene.

In Wells' (1977) thermometer, the activity of the magnesian end-member in the two unmixed phases is attributable to ideal mixing of Mg and Fe on $M1$ and $M2$ sites, i.e.,

$$
a_{\text{Mg}_2\text{Si}_2\text{O}_6,\text{cpx}} = X_{\text{Mg},M1} \cdot X_{\text{Mg},M2}, \qquad (5.271)
$$

after attribution of all Ca to site $M2$ and all Al to site $M1$ (cf. Wood and Banno, 1973). This choice must be considered as purely operational, because it is inconsistent with the real nature of the mixture. It must also be emphasized that both the formulations of Kretz (1982) and Wells (1977) are based on equilibrium distribution of the magnesian component between the unmixed terms, although Kretz's (1982) formulation is explicitly derived for one of the two polymorphs. In both formulations, the effect of adding iron to the system (i.e., moving from the CaMg-Mg join into the quadrilateral) is regarded as simply linear (cf. eq. 5.268, 5.269, and 5.270), owing to the fact that, in natural clinopyroxenes, there is a linear relationship between $X_{\text{Fe,cpx}}$ and $\ln(1 - 2X_{\text{Ca,cpx}})$ (see Kretz, 1982).

5.9.4 Solid-Solid Reactions

The coexistence of various solids at equilibrium in a heterogeneous system may be opportunely described by chemical reactions involving the major components of the solid phases. If we observe, for instance, the coexistence of pyroxene, plagioclase, and quartz, we can write an equation involving sodic terms:

$$
\underset{\text{albite}}{\text{NaAlSi}_3\text{O}_8} \Leftrightarrow \underset{\text{jadeite}}{\text{NaAlSi}_2\text{O}_6} + \underset{\text{quartz}}{\text{SiO}_2} \qquad (5.272)
$$

or an equation involving calcic terms:

$$CaAl_2Si_2O_8 \Leftrightarrow CaAl_2SiO_6 + SiO_2. \qquad (5.273)$$
$$\underset{\text{anorthite}}{} \qquad \underset{\text{Ca-Tschermak}}{} \quad \underset{\text{quartz}}{}$$

If we observe the coexistence of two polymorphs of the same compound (e.g., Al_2SiO_5), we can also write

$$Al_2SiO_5 \Leftrightarrow Al_2SiO_5. \qquad (5.274)$$
$$\underset{\text{andalusite}}{} \qquad \underset{\text{kyanite}}{}$$

Coexistence of phases on opposite sides of the reaction equation implies that they are in equilibrium and that the ΔG of reaction is zero—i.e.,

$$\Delta \overline{G}_{P,T}^{0} = -RT \ln K. \qquad (5.275)$$

If the heat capacity functions of the various terms in the reaction are known and their molar enthalpy, molar entropy, and molar volume at the T_r and P_r of reference (and their isobaric thermal expansion and isothermal compressibility) are also all known, it is possible to calculate $\Delta \overline{G}_{P,T}^{0}$ at the various T and P conditions of interest, applying to each term in the reaction the procedures outlined in section 2.10, and thus defining the equilibrium constant (and hence the activity product of terms in reactions; cf. eq. 5.272 and 5.273) or the locus of the P-T points of univariant equilibrium (eq. 5.274). If the thermodynamic data are fragmentary or incomplete—as, for instance, when thermal expansion and compressibility data are missing (which is often the case)—we may assume, as a first approximation, that the molar volume of the reaction is independent of the P and T intensive variables. Adopting as standard state for all terms the state of pure component at the P and T of interest and applying

$$\Delta \overline{G}_{P,T}^{0} = \Delta \overline{H}_{P,T}^{0} - T\Delta \overline{S}_{P,T}^{0} + \left(P - P_r\right)\Delta \overline{V}_{P,T}^{0} = -RT \ln K, \qquad (5.276)$$

we may also identify the P-T locus of points where the equilibrium constant takes on the identical value. It is obvious from equation 5.276 that equilibrium among solids involving a conspicuous volume of reaction ($\Delta \overline{V}^0$) will be a good barometric function, and that one involving consistent $\Delta \overline{H}^0$ (and limited $\Delta \overline{V}^0$) will be a good thermometric function.

Table 5.74 lists the main solid-solid reactions commonly adopted in geochemistry, with their respective $\Delta \overline{H}_{P_r,T_r}^{0}$, $\Delta \overline{S}_{P_r,T_r}^{0}$, $\Delta \overline{V}_{P_r,T_r}^{0}$, and $\Delta \overline{G}_{P_r,T_r}^{0}$, calculated with the INSP and THERMO computer packages (Saxena, 1989).

Some peculiar aspects of the various reactions are briefly summarized below. The discussion is taken from Essene (1982), to whom we refer for more exhaustive treatment.

Table 5.74 Main solid-solid reaction adopted in geothermobarometry. Values of $\Delta \overline{H}^0_{P_r, T_r}$ (kJ/mole), $\Delta \overline{S}^0_{P_r, T_r}$ [J/(mole $\times$ K)], $\Delta \overline{V}^0_{P_r, T_r}$ (cm³/mole), and $\Delta \overline{G}^0_{P_r, T_r}$ (kJ/mole) calculated with INSP and THERMO computer packages (Saxena, 1989).

Paragenesis	Reaction	Reaction Number
Andalusite-kyanite	$Al_2SiO_5 \Leftrightarrow Al_2SiO_5$	1
Kyanite-sillimanite	$Al_2SiO_5 \Leftrightarrow Al_2SiO_5$	2
Andalusite-sillimanite	$Al_2SiO_5 \Leftrightarrow Al_2SiO_5$	3
Orthopyroxene-olivine-quartz	$Fe_2Si_2O_6 \Leftrightarrow Fe_2SiO_4 + SiO_2$	4
Orthopyroxene-olivine-quartz	$Mg_2Si_2O_6 \Leftrightarrow Mg_2SiO_4 + SiO_2$	5
Clinopyroxene-olivine	$CaMgSi_2O_6 + Mg_2SiO_4 \Leftrightarrow CaMgSiO_4 + Mg_2Si_2O_6$	6
Clinopyroxene-plagioclase-quartz	$NaAlSi_3O_8 \Leftrightarrow NaAlSi_2O_6 + SiO_2$	7
Clinopyroxene-plagioclase-quartz	$CaAl_2Si_2O_8 \Leftrightarrow CaAl_2SiO_6 + SiO_2$	8
Garnet-plagioclase-sillimanite-quartz	$2Al_2SiO_5 + Ca_3Al_2Si_3O_{12} + SiO_2 \Leftrightarrow 3CaAl_2Si_2O_8$	9
Garnet-plagioclase-olivine	$CaAl_2Si_2O_8 + Fe_2SiO_4 \Leftrightarrow \frac{1}{3} Ca_3Al_2Si_3O_{12} + \frac{2}{3} Fe_3Al_2Si_3O_{12}$	10
Garnet-plagioclase-olivine	$CaAl_2Si_2O_8 + Mg_2SiO_4 \Leftrightarrow \frac{1}{3} Ca_3Al_2Si_3O_{12} + \frac{2}{3} Mg_3Al_2Si_3O_{12}$	11
Garnet-plagioclase-orthopyroxene-quartz	$CaAl_2Si_2O_8 + Fe_2Si_2O_6 \Leftrightarrow \frac{1}{3} Ca_3Al_2Si_3O_{12} + \frac{2}{3} Fe_3Al_2Si_3O_{12} + SiO_2$	12
Garnet-plagioclase-orthopyroxene-quartz	$CaAl_2Si_2O_8 + Mg_2Si_2O_6 \Leftrightarrow \frac{1}{3} Ca_3Al_2Si_3O_{12} + \frac{2}{3} Mg_3Al_2Si_3O_{12} + SiO_2$	13
Garnet-Fe cordierite-sillimanite-quartz	$3Fe_2Al_4Si_5O_{18} \Leftrightarrow 2Fe_3Al_2Si_3O_{12} + 4Al_2SiO_5 + 5SiO_2$	14
Garnet-Mg cordierite-sillimanite-quartz	$3Mg_2Al_4Si_5O_{18} \Leftrightarrow 2Mg_3Al_2Si_3O_{12} + 4Al_2SiO_5 + 5SiO_2$	15
Garnet-spinel-sillimanite-quartz	$Fe_3Al_2Si_3O_{12} + 2Al_2SiO_5 \Leftrightarrow 3FeAl_2O_4 + 5SiO_2$	16
Garnet-spinel-sillimanite-quartz	$Mg_3Al_2Si_3O_{12} + 2Al_2SiO_5 \Leftrightarrow 3MgAl_2O_4 + 5SiO_2$	17
Garnet-rutile-ilmenite-sillimanite-quartz	$3FeTiO_3 + Al_2SiO_5 + 2SiO_2 \Leftrightarrow Fe_3Al_2Si_3O_{12} + 3TiO_2$	18

Reaction Number	$\Delta \overline{H}^0_{P_r, T_r}$	$\Delta \overline{S}^0_{P_r, T_r}$	$\Delta \overline{V}^0_{P_r, T_r}$	$\Delta \overline{G}^0_{P_r, T_r}$
1	−4.1	−9.1	−7.4	−1.4
2	8.2	13.5	5.7	4.1
3	4.1	4.4	−1.7	2.7
4	1.6	3.4	3.0	−2.9
5	7.5	3.1	3.7	6.5
6	41.56	7.69	4.61	39.27
7	−11.0	−44.1	−17.3	2.1
8	19.3	−18.8	−13.9	24.9

continued

Table 5.74 *(continued)*

Reaction Number	$\Delta\overline{H}^0_{P_r,T_r}$	$\Delta\overline{S}^0_{P_r,T_r}$	$\Delta\overline{V}^0_{P_r,T_r}$	$\Delta\overline{G}^0_{P_r,T_r}$
9	12.5	105.7	52.8	−18.8
10	−14.3	−39.7	−27.9	0.9
11	5.2	−29.5	−26.6	14.0
12	−12.7	−36.4	−24.8	−1.9
13	12.7	−26.4	22.9	20.5
14	−116.8	−150.2	−159.8	−72.3
15	6.8	−109.9	−160.2	39.1
16	14.6	−1.7	20.7	15.0
17	−7.9	−6.1	19.6	−6.2
18	−0.5	−17.7	−18.6	5.1

Reactions 1, 2, and 3. Al$_2$SiO$_5$ polymorphs are widely adopted as index minerals in metamorphic terranes and are easily recognizable even on the macroscopic scale. Their *P-T* stability fields have long been subjected to some uncertainty because of disagreement among authors on the exact location of the ternary invariant point ($P = 6.5$ kbar and $T = 595$ °C according to Althaus, 1967; 5.5 kbar and 620 °C according to Richardson et al., 1969; 3.8 kbar and 600 °C according to Holdaway, 1971). Essene (1982) reports the substantial preference of petrologists for Holdaway's (1971) triple point, which better conforms to thermobarometric indications arising from other phases. As already outlined in section 2.7, the recent study of Holdaway and Mukhopadhyay (1993) confirms that the triple point location of Holdaway (1971) is the most accurate.

Reactions 4 and 5. Bohlen and Boettcher (1981) studied the effect of the presence of magnesian and manganoan terms in the mixture on the distribution of ferrous component between orthopyroxene and olivine according to the reaction

$$\underset{\text{opx}}{\text{Fe}_2\text{Si}_2\text{O}_6} \Leftrightarrow \underset{\text{olivine}}{\text{Fe}_2\text{SiO}_4} + \underset{\text{quartz (α or β)}}{\text{SiO}_2}. \qquad (5.277)$$

Figure 5.71A shows univariant equilibrium curves for various molar amounts of ferrous component in the orthopyroxene mixture. The *P-T* field is split into two domains, corresponding to the structural state of the coexisting quartz (α and β polymorphs, respectively). If the temperature is known, the composition of phases furnishes a precise estimate of the *P* of equilibrium for this paragenesis. Equation 5.277 is calibrated only for the most ferriferous terms, and the geobarometer is applicable only to Fe-rich rocks such as charnockites and fayalite-bearing granitoids.

Reaction 6. This reaction was investigated in detail by Finnerty and Boyd (1978) and was more recently recalibrated by Köhler and Brey (1990). Because the Ca content in olivine is *P*-dependent, due essentially to a large solvus between monticellite and forsterite that expands with pressure (cf. section 5.2.5), this reaction should act as a sensitive barometric function. However, the enthalpy of reaction is quite high and the effect of *T* on the equilibrium is also marked. The calibrated *P-T* slope has an inflection, the origin of which is not clear at first glance.

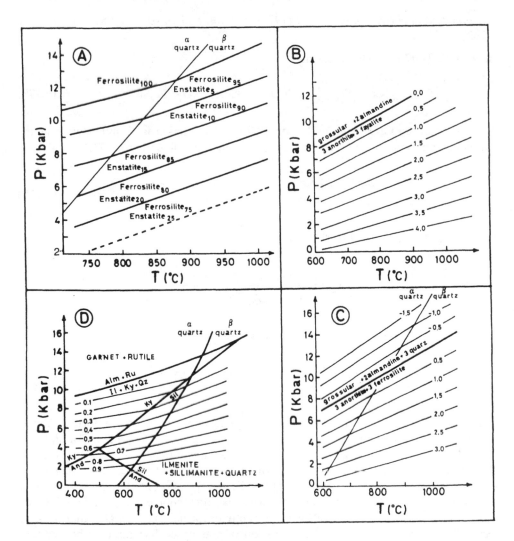

Figure 5.71 Thermobarometric equilibria plotted in *P-T* field. Numbers on univariant curves in parts B, C, and D are base-10 logarithms of respective equilibrium constants (log $K = 0$ for pure components). From Bohlen et al. (1983). Reprinted with permission of The Mineralogical Society of America.

Reactions 7 and 8. Thermobarometric estimates are complicated by order-disorder transition in albite and by the difficulty of evaluating the activity of jadeite and Ca-Tschermak in natural clinopyroxenes (cf. Essene, 1982).

Reaction 9. This reaction is the most popular geobarometer, because of the ubiquity of garnet-Al_2SiO_5-plagioclase-quartz paragenesis in medium- and high-grade metamorphic terranes. It may be affected by considerable error progression if the specimens are low in Ca (i.e., small amounts of anorthite in plagioclase and of grossular in garnet). Moreover, the slopes of the univariant curves in the *P-T* space are quite marked, and the *T* of equilibrium must be known with good approximation to avoid large errors in the derived *P* value.

For practical applications, see Ghent (1976), Newton and Haselton (1981), and Perkins (1983).

Reactions 10 and 11. The equilibrium

$$\left(Ca_{0.333}Fe_{0.666}\right)_3 Al_2Si_3O_{12} \Leftrightarrow CaAl_2Si_2O_8 + Fe_2SiO_4 \qquad (5.278)$$

was calibrated by Green and Hibberson (1970). The reaction is clearer if we split the garnet compound into the pure terms grossular and almandine (figure 5.71B):

$$Ca_3Al_2Si_3O_{12} + 2\,Fe_3Al_2Si_3O_{12} \Leftrightarrow 3\,CaAl_2Si_2O_8 + 3\,Fe_2SiO_4 \;.$$

grossular   almandine   anorthite   fayalite   (5.279)

In this case the equilibrium constant is

$$K_{279} = \frac{a^3_{CaAl_2Si_2O_8} \cdot a^3_{Fe_2SiO_4}}{a_{Ca_3Al_2Si_3O_{12}} \cdot a^2_{Fe_3Al_2Si_3O_{12}}}. \qquad (5.280)$$

The analogous reaction involving magnesian components is

$$Ca_3Al_2Si_3O_{12} + 2\,Mg_3Al_2Si_3O_{12} \Leftrightarrow 3\,CaAl_2Si_2O_8 + 3\,Mg_2SiO_4$$

grossular   pyrope   anorthite   forsterite   (5.281)

and is particularly applicable to garnet-bearing metagabbros. Consistent information may be obtained by contemporaneous application of equations 5.279 and 5.281. For all terms in the reaction, thermodynamic activity must be evaluated using appropriate mixing models. Comparison of thermobarometric information obtained with different compositional terms (e.g., eq. 5.279 and 5.281) indirectly indicates the appropriateness of the mixing models used. Incidentally, it must be noted here that interaction energies in the solid mixtures of interest are generally positive, so that activity coefficients are higher than 1 for all terms. Thus, eventual errors in the evaluation of activity coefficients may be mutually compensated in the evaluation of the equilibrium constant (eq. 5.280).

Reactions 12 and 13. The coexistence of plagioclase, orthopyroxene, garnet, and quartz may be expressed as an equilibrium involving ferrous and/or magnesian components, as seen in the preceding case:

$$Ca_3Al_2Si_3O_{12} + 2\,Fe_3Al_2Si_3O_{12} + 3\,SiO_2 \Leftrightarrow 3\,CaAl_2Si_2O_8 + 3\,Fe_2Si_2O_6$$

grossular   almandine   quartz   anorthite   ferrosilite   (5.282)

$$Ca_3Al_2Si_3O_{12} + 2Mg_3Al_2Si_3O_{12} + 3\,SiO_2 \Leftrightarrow 3\,CaAl_2Si_2O_8 + 3Mg_2Si_2O_6$$

grossular   pyrope   quartz   anorthite   enstatite   (5.283)

Reaction 5.282 was calibrated by Bohlen et al. (1982) and is particularly useful in granulites and garnet-bearing amphibolites. Figure 5.71C shows the loci of *P-T* points for which equilibrium constant K_{282} assumes identical values (base-10 logarithm). The univariant curve for pure end members ($\log K = 0$) is in boldface. The corresponding reaction involving magnesian terms was calibrated by Newton and Perkins (1982).

Reactions 14 and 15. There are still several problems regarding the calibration of these equilibria, which are discussed in detail by Essene (1982). Based on the energy terms listed in table 5.74, both T and P markedly affect reaction 14, whereas reaction 15 furnishes precise barometric indications.

Reactions 16 and 17. The reaction

$$\text{Fe}_3\text{Al}_2\text{Si}_3\text{O}_{12} + 2\,\text{Al}_2\text{SiO}_5 \Leftrightarrow 3\,\text{FeAl}_2\text{O}_4 + 5\,\text{SiO}_2$$
$$\text{almandine}\text{sillimanite}\text{hercynite}\text{quartz}$$

$$(5.284)$$

is a potentially useful thermobarometer in high-grade metamorphic terranes, where spinel, when present, is a mixture of the major components gahnite and hercynite. Experimental data concerning this reaction were produced by Richardson (1968), Holdaway and Lee (1977), and Wall and England (1979). No quantitative applications have hitherto been afforded.

Reaction 18. The reaction

$$3\,\text{FeTiO}_3 + \text{Al}_2\text{SiO}_5 + 2\,\text{SiO}_2 \Leftrightarrow \text{Fe}_3\text{Al}_2\text{Si}_3\text{O}_{12} + 3\,\text{TiO}_2$$
$$\text{ilmenite}\text{sillimanite}\text{quartz}\text{almandine}\text{rutile}$$

$$(5.285)$$

was calibrated by Bohlen et al. (1983). The loci of identical value of equilibrium constant K_{285} are plotted in the $P\text{-}T$ space of figure 5.71D (base-10 logarithm), which also shows the position of the univariant curve for pure end members ($\log K = 0$) and, superimposed, the kyanite-andalusite-sillimanite primary phase fields. As we can see, because the equilibrium is not markedly affected by temperature, it is a precise geobarometer.

An important warning regarding thermobarometric deductions based on the mineral chemistry of solid phases in heterogeneous systems concerns the correct application of the principle of equality of chemical potentials at equilibrium. Thermobarometric deductions are based essentially on the fact that in an *n*-component system the compositions of phases in an *n*-phase equilibrium are uniquely determined at any $P\text{-}T$ condition. In the simplest case, as we have seen in section 5.9.1, the actual *n*-component system may be reduced to a binary subsystem of major components, assuming more or less implicitly that the only significant chemical variations occur in this subsystem (for example, magnesian and ferroan components in garnet-clinopyroxene equilibria; cf. eq. 5.216). Also in this case, however, the principle of equality of chemical potentials at equilibrium demands that the chemical potentials of *both* components be the same in the two phases at equilibrium (cf. section 2.3). Relating the chemical potential to the standard state chemical potential (pure component at P and T of interest; cf. section 2.9.1), it may be easily seen that the equality principle applied to each component results in two equations in X (composition of the two phases) in two unknowns (P and T). The system could thus be theoretically solved (as, unfortunately, it

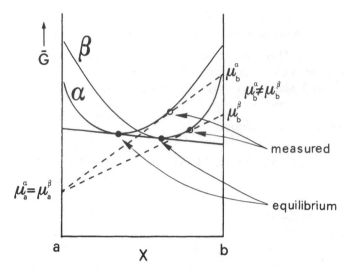

Figure 5.72 Effects of an erroneous application of the principle of equality of chemical potentials at equilibrium (from Connolly, 1992).

often is) on the basis of a single component. This, however, is a dangerous exercise.

As shown in figure 5.72, the equality of potential is satisfied for component a in mixtures α and β at the measured compositions. Neglecting the energy information arising from component b, one could deduce that the two phases, with the measured composition (open circles), coexist stably at the estimated P and T of equilibrium. However, as we can see, $\mu_{b,\alpha} \neq \mu_{b,\beta}$ and equilibrium at the deduced P and T conditions would demand different compositions (solid circles). Things may be even more dramatic with increasing chemical complexity of the system: correct solution of equilibria of type 5.282 and 5.283 in terms of chemical potentials involves, for example, five equalities at the univariant equilibrium (i.e., 10 equations in six variables, SiO_2 being a pure phase).

5.9.5 Solid-Fluid Reactions

So far we have considered only compositional terms present in solid mixtures and/or pure phases. When we wish to evaluate the chemistry of a fluid coexisting at equilibrium with a given solid paragenesis, a compositional term of the fluid phase must explicitly appear in the reaction equation. The term "fluid" here means a generic liquid or gaseous phase or a subcritical liquid-plus-gas assemblage. The state of the phase is not important when a particular standard state of reference is adopted, provided that, once the choice has been made, the adopted notation is maintained throughout the thermodynamic study. The most important molecular species appearing in solid-fluid reactions of geochemical interest are

O_2, H_2O, CO_2, H_2, and S_2. Of the reactions involving molecular oxygen (O_2), the following *buffer* equilibria are of great importance:

$$\underset{\text{wuestite}}{FeO} \Leftrightarrow \underset{\text{metal}}{Fe} + \frac{1}{2}\underset{\text{gas}}{O_2} \text{ (IW)} \qquad \log a_{O_2} = 6.80 - 27568T^{-1} \tag{5.286}$$

$$\underset{\text{bunsenite}}{NiO} \Leftrightarrow \underset{\text{metal}}{Ni} + \frac{1}{2}\underset{\text{gas}}{O_2} \text{ (NNO)} \qquad \log a_{O_2} = 9.31 - 24810T^{-1} \tag{5.287}$$

$$\underset{\text{magnetite}}{Fe_3O_4} \Leftrightarrow 3\underset{\text{wuestite}}{FeO} + \frac{1}{2}\underset{\text{gas}}{O_2} \text{ (MW)} \qquad \log a_{O_2} = 12.92 - 32638T^{-1} \tag{5.288}$$

$$\underset{\text{quartz}}{3SiO_2} + 2\underset{\text{magnetite}}{Fe_3O_4} \Leftrightarrow 3\underset{\text{fayalite}}{Fe_2SiO_4} + \underset{\text{gas}}{O_2} \text{ (QFM)} \qquad \log a_{O_2} + 10.50 - 26913T^{-1} \tag{5.289}$$

$$\underset{\text{hematite}}{3Fe_2O_3} \Leftrightarrow 2\underset{\text{magnetite}}{Fe_3O_4} + \frac{1}{2}\underset{\text{gas}}{O_2} \text{ (HM)} \qquad \log a_{O_2} = 14.41 - 24912T^{-1} \tag{5.290}$$

The significance of the term "buffer" stems from the control operated by the solid paragenesis on the activity of the O_2 component in the gaseous phase. Consider, for instance, the buffer magnetite-wuestite. Applying the usual relations among chemical potential, standard state chemical potential, and thermodynamic activity, we obtain

$$\Delta \overline{G}^0_{288} = -RT \ln \frac{a_{FeO}^3 \times a_{O_2}^{1/2}}{a_{Fe_3O_4}}. \tag{5.291}$$

If we imagine the solid phases to be composed of the pure terms FeO and Fe_3O_4, equation 5.291 may be reduced to

$$\Delta \overline{G}^0_{288} = -RT \ln a_{O_2}^{1/2} \tag{5.292}$$

and, by rearranging, to

$$a_{O_2} = \exp\left(-\frac{2\Delta \overline{G}^0_{288}}{RT}\right). \tag{5.293}$$

Because in geological literature it is customary to express the buffer effect in terms of "fugacity" of the gaseous species, it should be stressed that the adopted terminology is often misleading. As we see in equation 5.293, the buffer reaction determines the *thermodynamic activity* of the component in the gaseous phase; "activity" corresponds to "fugacity" only when the adopted standard state for the gaseous component is that of "pure perfect gas at $P = 1$ bar and $T = 298.15$ K" because, in this condition, the *standard state fugacity* is 1 (see section 9.1 for a discussion on the significance of the various terms).

The buffer reactions of equations 5.286 to 5.290 have been accurately calibrated by various authors, and the "fugacity" (actually "activity") of O_2 is usually expressed as a semilogarithmic function of absolute T. The buffer functions above are taken from Simons (1986) (eq. 5.286, 5.288, and 5.289), Huebner and Sato (1970) (eq. 5.287), and Eugster and Wones (1962) (eq. 5.290).

One peculiar oxygen buffer reaction is the well-known thermobarometric equation of Buddington and Lindsley (1964), which furnishes information on both T and the activity of molecular oxygen. The reaction uses the equilibrium among titanomagnetite (or "ulvospinel"), magnetite, hematite, and ilmenite components in the hemo-ilmenite and spinel phases:

$$Fe_2TiO_4 + Fe_3O_4 + \frac{1}{2}O_2 \Leftrightarrow FeTiO_3 + 2Fe_2O_3.$$

$$\underset{\text{spinel}}{} \quad \underset{\text{spinel}}{} \quad \underset{\text{gas}}{} \quad \underset{\text{hemo-ilmenite}}{} \quad \underset{\text{hemo-ilmenite}}{}$$

$$(5.294)$$

Equation 5.294 may be split into two partial equilibria, as shown by Powell and Powell (1977b): the exchange reaction

$$Fe_2TiO_4 + Fe_2O_3 \Leftrightarrow FeTiO_3 + Fe_3O_4,$$

$$\underset{\text{titanomagnetite}}{} \quad \underset{\text{hematite}}{} \quad \underset{\text{ilmenite}}{} \quad \underset{\text{magnetite}}{}$$

$$(5.295)$$

which is independent of the gaseous phase, and the HM buffer (eq. 5.290).

Figure 5.73 shows the loci of a_{O_2}-T points obtained by intersecting the isopleths of the titaniferous terms (X_{FeTiO_3} and $X_{Fe_2TiO_4}$) in the coexisting spinel and hemo-ilmenite phases. At $T < 600$ °C, the thermobarometer is inapplicable, because of the solvus field existing between magnetite and ulvospinel in the spinel phase. Note in the same figure the position of the HM buffer, corresponding to the absence of Ti in the system. The Buddington-Lindsley thermobarometer finds application in both igneous (see Duchesne, 1972) and metamorphic terranes (with the above low-T limits; see also Bohlen and Essene, 1977 for a discussion).

A multiple buffer of particular importance in volcanic gaseous equilibria (see also section 9.5) is

$$Fe_3O_4 + 3S_2 \Leftrightarrow 3FeS_2 + 2O_2$$

$$\underset{\text{magnetite}}{} \quad \underset{\text{gas}}{} \quad \underset{\text{pyrite}}{} \quad \underset{\text{gas}}{}$$

$$(5.296)$$

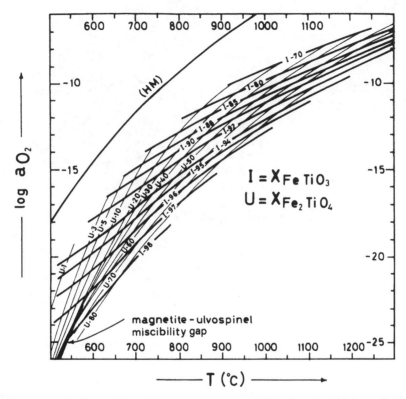

Figure 5.73 Isopleths of titaniferous components ilnenite ($FeTiO_3$) and ulvospinel (Fe_2TiO_4) in spinel and hemo-ilmenite coexisting at equilibrium in various a_{O_2}-T conditions. Position of hematite-magnetite buffer (HM) is also shown. From Spencer and Lindsley (1981). Reprinted with permission of The Mineralogical Society of America.

Assuming unitary activity of components in solid phases (pure phases and standard state = pure component at P and T of interest), we have

$$K_{296} = \frac{a_{O_2}^2}{a_{S_2}^3} = \exp\left(-\frac{\Delta \overline{G}_{296}^0}{RT}\right). \tag{5.297}$$

Let us now consider the effect of pressure on equilibrium. We assume for the sake of simplicity that fluid pressure is equal to total pressure and that gaseous components mix ideally. From

$$a_{O_2} = \frac{f_{O_2}}{f_{O_2}^0} \tag{5.298}$$

and

$$a_{S_2} = \frac{f_{S_2}}{f_{S_2}^0},\qquad(5.299)$$

where f_{O_2} is the fugacity of O_2 and $f_{O_2}^0$ is the standard state fugacity of O_2:

$$f_{O_2} = P_{O_2}\Gamma_{O_2}\qquad(5.300)$$

$$f_{S_2} = P_{S_2}\Gamma_{S_2},\qquad(5.301)$$

where P_{O_2} is the partial pressure of O_2 and Γ_{O2} is the fugacity coefficient of O_2:

$$P_{O_2} = PX_{O_2}\qquad(5.302)$$

$$P_{S_2} = PX_{S_2},\qquad(5.303)$$

where X_{O_2} is the molar fraction of O_2 in the gaseous phase, we obtain

$$PK_{296} = \frac{f_{S_2}^{0\,3}\,\Gamma_{O_2}^2\,X_{O_2}^2}{f_{O_2}^{0\,2}\,\Gamma_{S_2}^3\,X_{S_2}^3}.\qquad(5.304)$$

If we adopt as the standard state for gaseous components the state of pure perfect gas at $P = 1$ bar and $T = 298.15$ K ($f_{O_2}^0 = f_{S_2}^0 = 1$) and neglect for simplicity the fugacity coefficients, equation 5.304 combined with equation 5.297 gives

$$\frac{X_{O_2}^2}{X_{S_2}^3} = P \exp\left(-\frac{\Delta\overline{G}_{296}^0}{RT}\right).\qquad(5.305)$$

There are three variables in equation 5.305: composition (and speciation) of the fluid, P, and T. Once the chemistry of the fluid and P are known, we can deduce T; if we know the chemistry of the fluid and T, we can deduce P; and if we know P and T, we can deduce the chemistry of the fluid. This trivariance is a formidable obstacle in the application of thermobarometry in the presence of fluid phases and is often underevaluated in geology. Direct sampling of the fluid to constrain the chemical factor is in most cases impossible, or, when it is possible (e.g., for volcanic gases, geothermal gases, and fluid inclusions), the speciation state must still be evaluated and, as we will see in detail in chapter 9, speciation in the fluid phase is complex and quite variable as a function of P, T, and composition. Nevertheless, in the case of a simple oxygen buffer such as IW, NNO, WM, QFM, HM, or ilmenite-ulvospinel, the estimate of P is not as crucial, because a_{O_2} varies by several orders of magnitude as a function of T (e.g., if $a_{O_2} = 10^{-25}$ at $T = 600\ °C$ and $P = 1$ kbar, it will be 10^{-24} at $T = 600\ °C$ and $P = 10$ kbar, 10^{-23} at $P = 100$ kbar, and so on).

In the case of the ilmenite-ulvospinel thermobarometer, a_{O_2} of 10^{-25} at $T = 600\ °C$ is reached with Ti-rich compositions. If we recall that Avogadro's number is 6.022×10^{23} molecules/gfw, we may deduce that, at $P = 1$ bar and $T = 600\ °C$, less that one O_2 molecule per mole of fluid is present in the system. Things are even more dramatic at lower T (with the IW buffer, for instance, $a_{O_2} \cong 10^{-41}$ at $T = 300\ °C$ and $P = 1$ bar) or at constant T with increasing P_{total}. In light of the relative abundance of atomic oxygen (which is dominant in our systems, in most cases), these considerations force us to reflect carefully on the enormous importance of speciation in the fluid phase.

Concerning estimation of the activity of the H_2O component in the fluid, we can recall, for example, all the extrinsic stability reactions presented in section 5.6.3 for micaceous components (eq. 5.126, 5.128, 5.133, 5.139, 5.140, 5.144, 5.145, 5.146, 5.148, and 5.149) and the respective f_{H_2O} calibration functions (eq. 5.127, 5.129, 5.134, 5.141, 5.142, 5.143, and 5.147; the provisos previously outlined for O_2 are valid for the term "fugacity"). Mica may be also used to deduce the activity of H_2 in the fluid (eq. 5.131) or the combined activities of H_2O and H_2 (eq. 5.130, 5.135, 5.137, and 5.138). The application of all these functions is complicated by difficulties arising when one attempts to evaluate the activity of micaceous components in mica polymorphs, for which data on mixing behavior are fragmentary and incomplete (cf. section 5.6.4).

Concerning estimation of the activity of CO_2 in the fluid phase, besides carbonate equilibria (discussed in some detail in section 8.10), equilibria in the $CaO\text{-}SiO_2\text{-}CO_2$ system are also important. In the medium-low metamorphism of limestones, the equilibrium

$$\underset{\text{calcite}}{CaCO_3} + \underset{\text{quartz}}{SiO_2} \Leftrightarrow \underset{\text{wollastonite}}{\tfrac{1}{2}Ca_2Si_2O_6} + \underset{\text{fluid}}{CO_2}, \tag{5.306}$$

investigated in detail by Greenwood (1967), is of particular interest.

PART TWO

Geochemistry of Silicate Melts

CHAPTER SIX

Thermochemistry of Silicate Melts

6.1 Structural and Reactive Properties

The geochemical interpretation of silicate melts is essentially based on three main concepts:

1. Ionic character of melts
2. Polymeric nature of the anion matrix
3. *Lux-Flood* acid-base properties of dissolved oxides.

6.1.1 Ionic Character and Quasi-Reticular Interpretation

Electrical conductivity measurements on silicate melts indicate an essentially ionic conductivity of unipolar type (Bockris et al., 1952a,b; Bockris and Mellors, 1956; Waffe and Weill, 1975). Charge transfer is operated by cations, whereas anionic groups are essentially stationary. Transference of electronic charges (conductivity of h- and n-types) is observed only in melts enriched in transition elements, where band conduction and electron hopping phenomena are favored. We may thus state that silicate melts, like other fused salts, are ionic liquids.

In an ionic melt, coulombic forces between charges of opposite sign lead to relative short-distance ordering of ions, with anions surrounded by cations and vice versa. The probability of finding a cation replacing an anion in such ordering is effectively zero and, from a statistical point of view, the melt may be considered as a *quasi-lattice,* with two distinct reticular sites that we will define as *anion matrix* and *cation matrix.*

Let us consider a mixture of two fused salts A^+Z^- and B^+Y^-, with A^+ and B^+ cations and Y^- and Z^- anions. The bulk configurational entropy of the mixture is given by the sum of the configurational entropies of the two matrices:

$$S_{mixing} = S_{anion\,mixing} + S_{cation\,mixing}. \qquad (6.1)$$

If cations and anions are respectively of similar type, the ideal condition of maximum disorder will be reached in each matrix and the bulk entropy of mixing may be expressed by

$$
S_{\text{mixing}} = -R\left[\left(X_Z \ln X_Z + X_Y \ln X_Y\right) + \left(X_A \ln X_A + X_B \ln X_B\right)\right], \quad (6.2)
$$

where X_A is the ionic fraction of A in the appropriate matrix (i.e., $X_A + X_B = 1$).

The activity of component AZ in the ideal mixture of the two fused salts AZ and BY is expressed as

$$
a_{\text{AZ,melt}} = X_A \cdot X_Z. \quad (6.3)
$$

This "quasi-lattice" formulation of fused salts is known as *Temkin's equation*. Its application to silicate melts was provided by Richardson (1956), but it is inadequate for the compositional complexity of natural melts, mainly because, in a compositionally complex melt, the types of anions and consequently the entity of the anion matrix vary in a complicated way with composition.

Substantial further development in the theory was achieved by Toop and Samis (1962a,b) and Masson (1968), and consisted of the application of polymer theory to fused silicates.

6.1.2 Polymeric Nature of Anion Matrix: Toop-Samis and Masson Models

In polymeric models for silicate melts, it is postulated that, at each composition, for given values of P and T, the melt is characterized by an equilibrium distribution of several ionic species of oxygen, metal cations, and ionic polymers of monomeric units SiO_4^{4-}.

The charge balance of a polymerization reaction involving SiO_4^{4-} monomers may be formally described by a homogeneous reaction involving three forms of oxygen: singly bonded O^-, doubly bonded O^0 (or "bridging oxygen"), and free oxygen O^{2-} (Fincham and Richardson, 1954):

$$
\underset{\text{melt}}{2O^-} \Leftrightarrow \underset{\text{melt}}{O^0} + \underset{\text{melt}}{O^{2-}}. \quad (6.4)
$$

In fact, equation 6.4 is similar to a reaction between SiO_4^{4-} monomers:

$$
SiO_4^{4-} + SiO_4^{4-} \Leftrightarrow Si_2O_7^{6-} + O^{2-}, \quad (6.5)
$$

which in stereochemical representation may be conceived as follows:

$$O^- - \underset{\underset{O^-}{|}}{\overset{\overset{O^-}{|}}{Si}} - O^- + O^- - \underset{\underset{O^-}{|}}{\overset{\overset{O^-}{|}}{Si}} - O^- \Leftrightarrow O^- - \underset{\underset{O^-}{|}}{\overset{\overset{O^-}{|}}{Si}} - O - \underset{\underset{O^-}{|}}{\overset{\overset{O^-}{|}}{Si}} - O^- + O^{2-} . \tag{6.6}$$

Polymer chemistry shows that the larger the various polymers become, the more their reactivity becomes independent of the length of the polymer chains. This fact, known as the "principle of equal reactivity of cocondensing functional groups," has been verified in fused polyphosphate systems (which, for several properties, may be considered as analogous to silicate melts; cf. Fraser, 1977) with polymeric chains longer than 3 PO_4^{3-} units (Meadowcroft and Richardson, 1965; Cripps-Clark et al., 1974). Assuming this principle to be valid, the equilibrium constant of reaction 6.4:

$$K_4 = \frac{(O^0)(O^{2-})}{(O^-)^2} \tag{6.7}$$

(in which the terms in parentheses represent the number of moles in the melt) is always representative of the polymerization process, independent of the effective length of the polymer chains.

Toop and Samis (1962a,b) showed that, in a binary melt $MO\text{-}SiO_2$, in which MO is the oxide of a basic cation completely dissociated in the melt, the total number of bonds per mole of melt is given by

$$2(O^0) + (O^-) = 4N_{SiO_2}, \tag{6.8}$$

where N_{SiO_2} are the moles of SiO_2 in the $MO\text{-}SiO_2$ melt. The number of bridging oxygens in the melt is thus

$$(O^0) = \frac{4N_{SiO_2} - (O^-)}{2}. \tag{6.9}$$

Mass balance gives the number of moles of free oxygen per mole of melt:

$$(O^{2-}) = (1 - N_{SiO_2}) - \frac{(O^-)}{2}, \tag{6.10}$$

where $(1 - N_{SiO_2})$ are the moles of basic oxide in the melt. Equations 6.7 to 6.10 yield

Table 6.1 Terms of quadratic equation 6.12 for various values of N_{SiO_2} (Toop and Samis, 1962a).

N_{SiO_2}	$a(O^-)^2$	+	$b(O^-)$	+	c	= 0
0.10	$(4K_4 - 1)(O^-)^2$		$2.2(O^-)$		-0.72	
0.20	$(4K_4 - 1)(O^-)^2$		$2.4(O^-)$		-1.28	
0.30	$(4K_4 - 1)(O^-)^2$		$2.6(O^-)$		-1.68	
0.40	$(4K_4 - 1)(O^-)^2$		$2.8(O^-)$		-1.92	
0.50	$(4K_4 - 1)(O^-)^2$		$3.0(O^-)$		-2.00	
0.60	$(4K_4 - 1)(O^-)^2$		$3.2(O^-)$		-1.92	
0.70	$(4K_4 - 1)(O^-)^2$		$3.4(O^-)$		-1.68	
0.80	$(4K_4 - 1)(O^-)^2$		$3.6(O^-)$		-1.28	
0.90	$(4K_4 - 1)(O^-)^2$		$3.8(O^-)$		-0.72	

$$K_4 = \frac{\left[4N_{SiO_2} - (O^-)\right]\left[2 - 2N_{SiO_2} - (O^-)\right]}{4(O^-)^2}. \tag{6.11}$$

Thus, (O^-) is given by the quadratic equation

$$(O^-)^2(4K_4 - 1) + (O^-)(2 + 2N_{SiO_2}) + 8N_{SiO_2}(N_{SiO_2} - 1) = 0, \tag{6.12}$$

which may be solved for discrete values of K_4 and N_{SiO_2} (see table 6.1).

Figure 6.1 shows the solution of the system for $K_4 = 0.06$: note that the distribution of the three forms of oxygen is asymmetric over the compositional space. Moreover, bridging oxygen is the only form present in the SiO_2 monomer.

The Gibbs free energy change involved in equation 6.4 is

$$\Delta G_4^0 = -RT \ln K_4 \tag{6.13}$$

Because two moles of O^- produce one mole of O^0 and one of O^{2-}, the Gibbs free energy of mixing per mole of silicate melt is given by

$$\Delta G_{mixing} = \frac{(O^-)}{2} RT \ln K_4. \tag{6.14}$$

Figure 6.2 shows Gibbs free energy of mixing values obtained by application of equation 6.14. They are remarkably similar to the Gibbs free energy of mixing values experimentally observed on binary synthetic systems of appropriate composition. This means that the Gibbs free energy of mixing in an MO-SiO_2 melt is

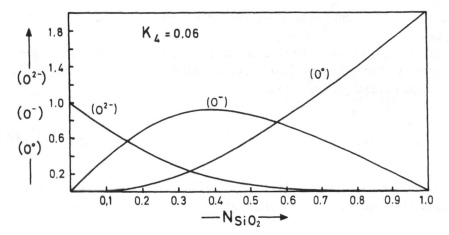

Figure 6.1 Equilibrium distribution of (O^0), (O^-), and (O^{2-}) in binary field MO-SiO$_2$ for $K_4 = 0.06$. Reprinted from G. W. Toop and C. S. Samis, *Canadian Metallurgist Quarterly*, 1, 129–152, copyright © 1962, with kind permission from Elsevier Science Ltd., The Boulevard, Langford Lane, Kidlington 0X5 1GB, UK.

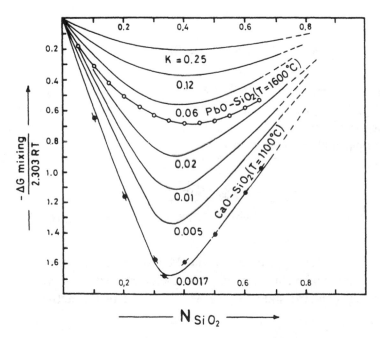

Figure 6.2 Gibbs free energy of mixing in the quasi-chemical model of Toop and Samis (1962a,b), compared with values experimentally observed in PbO-SiO$_2$ and CaO-SiO$_2$ melts. Reprinted from Toop and Samis (1962b), with kind permission of ASM International, Materials Park, Ohio.

essentially attributable to interactions among silica and the various forms of oxygen.

If we apply Gibbs-Duhem equations (cf. section 2.11) we may deduce that the activity of the dissolved basic oxide is entirely represented by the activity of free oxygen in the system, so that

$$\left(1 - N_{SiO_2}\right)\log a_{O^{2-}} + N_{SiO_2}\log a_{SiO_2} = \frac{\Delta G_{mixing}}{2.303\,RT}. \tag{6.15}$$

In Masson's model (Masson, 1965, 1968, 1972), polymerization reactions are visualized as a stepwise process of the type

$$SiO_4^{4-} + SiO_4^{4-} \Leftrightarrow Si_2O_7^{6-} + O^{2-} \tag{6.16}$$

$$Si_2O_7^{6-} + SiO_4^{4-} \Leftrightarrow Si_3O_{10}^{8-} + O^{2-}, \tag{6.17}$$

which may be generalized in the form

$$Si_nO_{3n+1}^{(2n+2)-} + SiO_4^{4-} \Leftrightarrow Si_{n+1}O_{3n+4}^{(2n+4)-} + O^{2-}. \tag{6.18}$$

The constants of equations 6.16 and 6.17 are identical, on the basis of the principle of equal reactivity of cocondensing functional groups. Adopting Temkin's model for fused salts, and thus assuming that molar fractions represent activities over the appropriate matrix, we obtain

$$\frac{X_{Si_2O_7^{6-}} \cdot X_{O^{2-}}}{X_{SiO_4^{4-}} \cdot X_{SiO_4^{4-}}} = \frac{X_{Si_3O_{10}^{8-}} \cdot X_{O^{2-}}}{X_{SiO_4^{4-}} \cdot X_{Si_2O_7^{6-}}} = \cdots \frac{X_{Si_{n+1}O_{3n+4}^{(2n+4)-}} \cdot X_{O^{2-}}}{X_{SiO_4^{4-}} \cdot X_{Si_nO_{3n+1}^{(2n+2)-}}}. \tag{6.19}$$

Simplifying this infinite set of equilibria, Masson obtained an equation relating the molar fraction of silica in an SiO_2-MO melt (with MO basic oxide completely dissociated at standard state to give $M^{2+} + O^{2-}$) to the activity of dissolved oxide a_{MO}:

$$\frac{1}{X_{SiO_2}} = 2 + \frac{1}{1 - a_{MO}} - \frac{1}{1 + a_{MO}\left(\frac{1}{K_{16}} - 1\right)}. \tag{6.20}$$

Equation 6.20 is valid only for open or "bifunctional" chains (cf. Fraser, 1977). An extension of the model to open ramified polymer chains leads to the equation

$$\frac{1}{X_{SiO_2}} = 2 + \frac{1}{1 - a_{MO}} - \frac{3}{1 + a_{MO}\left(\frac{3}{K_{16}} - 1\right)}. \tag{6.21}$$

(Whiteway et al., 1970).

Figure 6.3 shows the oxide activities of Masson's model compared with experimental values observed in $MO\text{-}SiO_2$ systems. As already described in chapter 2, the significance of activity is intimately connected with the choice of a standard state of reference, which in Masson's model is that of *completely dissociated basic oxide*. The more the actual behavior of the metal oxide in the melt differs from the reference condition (i.e., the more the metal oxide is acidic in the Lux-Flood sense of the term; see section 6.1.3), the more the thermodynamic activity of the oxide increases in the melt, silica amounts being equal (see figure 6.3). Here, as in the model of Toop and Samis (figure 6.2), we observe a direct relationship between degree of acidity of the dissolved oxide and polymerization constant K_{16}. The significance of this correlation will become clearer in the next section.

6.1.3 Acid-Base (Lux-Flood) Character of Dissolved Oxide and Its Bearing on Structure of Melt

The reactive properties of oxides have always been related to their capacity to form *salts*—i.e., "basic" oxides such as CaO, FeO, and MgO may form salts by reaction with "acidic" oxides such as SiO_2 and CO_2:

$$\text{Acid} + \text{Base} \Leftrightarrow \text{Salt}. \tag{6.22}$$

For instance,

$$SiO_2 + 2\,MgO \Leftrightarrow Mg_2\,SiO_4 \tag{6.23}$$

$$CO_2 + CaO \Leftrightarrow CaCO_3\,. \tag{6.24}$$

Moreover, in geology, a rock with a high SiO_2 content is usually defined as "acidic". Although the underlying concepts are obvious, the terminology used deserves some clarification.

The acid-base behavior of a substance in aqueous solutions and other protonated systems is conveniently described by the Brönsted-Lowry equation:

$$\text{Acid} \Leftrightarrow \text{Base} + H^+ \tag{6.25}$$

—i.e.,

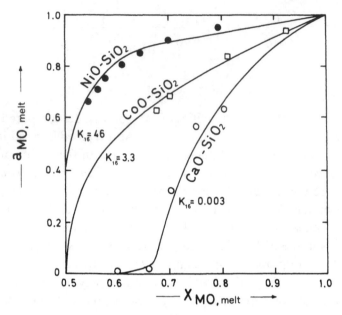

Figure 6.3 Experimental oxide activities in $MO\text{-}SiO_2$ melts compared with estimates of Masson's model. Reprinted from Masson (1968), with kind permission of American Ceramic Society Inc., Westerville, Ohio.

$$H_2SO_4 \Leftrightarrow HSO_4^- + H^+. \tag{6.26}$$

In silicate melts and other nonprotonated solvents, the Brönsted-Lowry equation is not applicable and is conveniently replaced by the Lux-Flood acid-base definition (Lux, 1939; Flood and Förland, 1947), according to which free oxygen O^{2-} replaces H^+. A "basic oxide" is one capable of furnishing oxygen ions, and an "acidic oxide" is one that associates oxygen ions:

$$\text{Base} \Leftrightarrow \text{Acid} + O^2 \tag{6.27}$$

—i.e.,

$$SiO_4^{4-} \Leftrightarrow SiO_3^{2-} + O^{2-} \tag{6.28.1}$$

$$SiO_3^{2-} \Leftrightarrow SiO_2 + O^{2-}. \tag{6.28.2}$$

A generic oxide MO will dissociate as a basic oxide when

$$MO \Leftrightarrow M^{2+} + O^{2-} \tag{6.29}$$

or as an acidic oxide when

Table 6.2 Relationship between field strength of dissolved cation and polymerization constant in MO-SiO$_2$ melts (data from Hess, 1971).

Cation	$\dfrac{Z_M{}^{z+}}{r_M{}^{z+} + r_{O^{2-}}}$	K_4
K$^+$	0.13	0.001
Na$^+$	0.18	0.001
Li$^+$	0.23	0.001
Pb^{2+}	0.30	0.002–0.01
Ca^{2+}	0.35	0.002
Mn^{2+}	0.41	0.04
Fe^{2+}	0.44	0.05
Co^{2+}	0.44	0.10
Ni^{2+}	0.46	0.35

$$MO + O^{2-} \Leftrightarrow MO_2^{2-}. \tag{6.30}$$

An "amphoteric oxide" is an oxide component with reaction properties intermediate between those of equations 6.29 and 6.30.

It is obvious that the acid-base property of a dissolved oxide may markedly affect the structure of a silicate melt. A dissolved acidic oxide associates the free oxygen, thus displacing reaction 6.4 toward the right, resulting in a marked correlation between the field strength of the dissolved cation and the polymerization constant of the melt. This correlation is shown in the values listed in table 6.2.

For instance, a quite marked effect on the structure of the silicate melt is produced by the basic oxide H$_2$O. Addition of water causes a drastic decrease in the viscosity of silicate melts: adding 6.4 weight % of H$_2$O to a granitic melt at $T = 1000\ °C$ causes a decrease in viscosity of about six orders of magnitude (i.e., from about 10^{-10} to about 10^{-4} poise^{-1}; cf. Burnham, 1975). The dissolution mechanism of H$_2$O is important in magma rheology and will be discussed more extensively in section 9.6.1.

6.2 Structure of Natural Melts

The structure of the anion matrix of silicate melts is conveniently represented in a ternary plot whose apexes are the Si^{IV}, O^0, and O$^-$ constituents of the anion groups. In this ternary representation, all the possible types of anions are aligned along a straight polymerization path, from the SiO$_4^{4-}$ monomer to the SiO$_2$ polymer (Toop and Samis, 1962a). For a given amount of silica, the value assumed by polymerization constant K_4 determines the exact position along the path (figure 6.4).

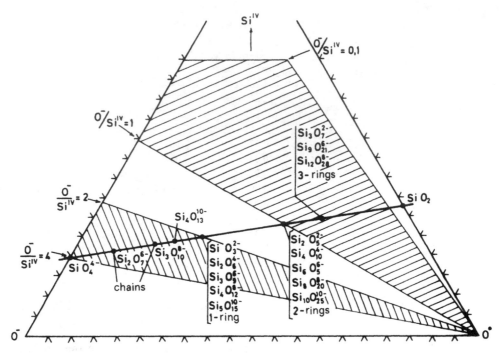

Figure 6.4 Ternary plot of singly bonded oxygen O^-, silicon Si^{IV}, and bridging oxygen O^0 in each kind of polyanion. Shaded areas: compositional ranges 6.31 and 6.33. Reprinted from Ottonello (1983), with kind permission of Theophrastus Publishing and Proprietary Co.

In natural melts, the presence of high field strength ions such as Al^{3+}, Fe^{3+}, Ti^{4+}, and P^{5+}, which, like silicon, preferentially assume a tetrahedral coordination with oxygen, complicates the structure, and the constitution of the anion matrix may not be deduced on the basis of the equations in section 6.1.2. Structural parameters valid for compositionally complex melts were proposed by Mysen et al. (1980) and Virgo et al. (1980) on the basis of the results of Raman spectroscopy. These parameters are NBO/Si and NBO/T (NBO = Non-Bridging Oxygen; T groups all tetrahedrally coordinated cations—i.e., Si^{4+}, Al^{3+}, Fe^{3+}, Ti^{4+}, P^{5+}, ...). Comparing the Raman spectra of silicate melts with various NBO/Si and NBO/T ratios with the Raman spectra of crystals of the same chemical composition, the above authors subdivided the anion structure of silicate melts into three compositional ranges:

$$4 > NBO/Si > 2 \tag{6.31}$$

$$2 > NBO/Si > 1 \tag{6.32}$$

$$1 > NBO/Si > 0.1; \quad NBO/T < 0.75 \tag{6.33}$$

The following disproportionation reactions dominate, respectively, in the three ranges:

$$2 \, Si_2O_7^{6-} \Leftrightarrow 2 \, SiO_4^{4-} + Si_2O_6^{4-} \tag{6.34}$$

$$3 \, Si_2O_6^{4-} \Leftrightarrow 2 \, SiO_4^{4-} + 2 \, Si_2O_5^{2-} \tag{6.35}$$

$$2 \, Si_2O_5^{2-} \Leftrightarrow Si_2O_6^{4-} + 2 \, SiO_2 \tag{6.36}$$

$$T_2O_5^{2-} \Leftrightarrow TO_3^{2-} + TO_2 \tag{6.37}$$

These equations do not imply any progression in the polymerization process and reduce to identities in the Fincham-Richardson formalism (cf. table 6.3). Nevertheless, the experimental evidence of Mysen et al. (1980) and Virgo et al. (1980) may be explained by the progressive polymerization steps listed in table 6.3. These equilibria are consistent with the Fincham-Richardson formalism (of which they constitute simple multiples) and obey the proportionality rules of figure 6.4.

Whiteway et al. (1970) derived an equation defining the fractional amount N_X of the various polymer units of extension X as a function of the ratio between bridging oxygen (O^0) and singly bonded oxygen in an initially totally depolymerized melt ($O^{-}*$) composed entirely of SiO_4^{4-} monomers (i.e., $\alpha = O^0/O^{-}*$):

$$N_X = \omega_X \left(\frac{2\alpha}{3} \right)^{X-1} \left(1 - \frac{2\alpha}{3} \right)^{2X+1} \tag{6.38}$$

$$\omega_X = \frac{3X}{(2X + 1)! \, X!}. \tag{6.39}$$

Table 6.4 lists the molar fractions of the various polymer units calculated for α values between 0.1 and 0.5.

Note the close connection between the values deduced by application of equation 6.38 and the ternary representation of figure 6.4. The polymerization field $0.1 < \alpha < 0.5$ corresponds roughly to the range $4 > NBO/Si > 2$. The polymer units in this range are mainly monomers, dimers, and ramified tetramers ($Si_4O_{13}^{10-}$). With increasing α, we progressively move along the polymerization path, the molar fraction of monomers decreases, and the mean length of the polymer units increases. For values of α higher than 0.5, equation 6.38 is no longer valid and is replaced by

Table 6.3 Reaction schemes for coexistence of various polymers in natural melts (from Ottonello, 1983).

Compositional Range	Disproportionation Scheme	Polymerization Reactions
$4 > NBO/Si > 2$	$2Si_2O_7^{6-} \Leftrightarrow 2SiO_4^{4-} + Si_2O_6^{4-}$	$2SiO_4^{4-} \Leftrightarrow Si_2O_7^{6-} + O^{2-}$
	$12O^- + 4O^0 \Leftrightarrow 12O^- + 4O^0$	$Si_2O_7^{6-} \Leftrightarrow Si_2O_6^{4-} + O^{2-}$
		$8O^- \Leftrightarrow 6O^- + O^0 + O^{2-}$
		$6O^- + O^0 \Leftrightarrow 4O^- + 2O^0 + O^{2-}$
$2 > NBO/Si > 1$	$3Si_2O_6^{4-} \Leftrightarrow 2SiO_4^{4-} + 2Si_2O_5^{2-}$	$2SiO_4^{4-} \Leftrightarrow Si_2O_6^{4-} + 2O^{2-}$
	$12O^- + 6O^0 \Leftrightarrow 12O^- + 6O^0$	$2Si_2O_6^{4-} \Leftrightarrow Si_2O_5^{2-} + O^{2-}$
		$8O^- \Leftrightarrow 4O^- + 2O^0 + 2O^{2-}$
		$4O^- + 2O^0 \Leftrightarrow 2O^- + 3O^0 + O^{2-}$
$1 > NBO/Si > 0.1$	$2Si_2O_5^{2-} \Leftrightarrow Si_2O_6^{4-} + 2SiO_2$	$Si_2O_6^{4-} \Leftrightarrow 2SiO_2 + 2O^{2-}$
$NBO/T < 0.75$	$4O^- + 3O^0 \Leftrightarrow 4O^- + 3O^0$	$Si_2O_6^{4-} \Leftrightarrow Si_2O_5^{2-} + O^{2-}$
		$4O^- + 2O^0 \Leftrightarrow 4O^0 + 2O^{2-}$
		$4O^- + 2O^0 \Leftrightarrow 2O^- + 3O^0 + O^{2-}$

Table 6.4 Relationship between relative amounts of bridging oxygen α and molar proportions of various polymer units (from Henderson, 1982).

α	Monomer	Dimer	Trimer	Tetramer	Other
0.1	0.813	0.142	0.033	0.009	0.003
0.2	0.651	0.196	0.078	0.036	0.039
0.3	0.512	0.197	0.101	0.059	0.131
0.4	0.394	0.170	0.123	0.064	0.249
0.5	0.296	0.132	0.078	0.053	0.441

$$N_X = \omega_X \left(\frac{2\alpha}{3}\right)^{X-1} \left(1 - \frac{2\alpha}{3}\right)^{2X+1} \left(1 - N_{O^{2-}}\right), \qquad (6.40)$$

where $N_{O^{2-}}$ is the molar fraction of free oxygen.

As regards the coordination states of the various cations in the melt, the main experimental information comes from Mössbauer spectroscopy and infrared absorption spectroscopy on silicate glasses. Because the glass-melt transition involves partial rearrangement of the structure, the interpretation of experimental evidence is not straightforward (cf. Boon and Fyfe, 1972). Treating the melt as chaotic packing of rigid spheres all of the same dimension, Bernal (1964) showed that the main coordination of the cavities between spheres is tetrahedral (73%), followed by the half-octahedron (20.3%), trigonal prism (3.2%), tetragonal dodecahedron (3.1%), and Archimedean antiprism (0.4%). Moreover, the mean co-

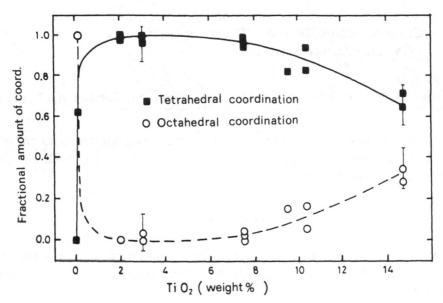

Figure 6.5 Coordination state of Ti^{4+} in glasses of TiO$_2$-SiO$_2$ system. Reprinted from R. B. Greegor, F. W. Lytle, D. R. Sanstrom, J. Wang, and P. Schultz. Investigation of TiO$_2$-SiO$_2$ glasses by X-ray absorption spectroscopy, *Journal of Non-Crystalline Solids*, 55, 27–43, copyright © 1983, with kind permission of Elsevier Science-NL, Sara Burgerhart-straat 25, 1055 KV Amsterdam, The Netherlands.

ordination number (VIII-IX) is lower than that observed in regular packing of rigid spheres (XII). Indeed, existing experimental studies on glasses show that the coordination number varies between IV and VIII-IX according to type of cation, chemistry of the system, and temperature. For instance, in alkaline glasses (Li$_2$O-SiO$_2$, Na$_2$O-SiO$_2$ and Rb$_2$O-SiO$_2$), Ni^{2+} exhibits coordination IV and VI. Moreover, the fraction of tetrahedrally coordinated sites increases with increasing T, composition being equal, and decreases with the increasing field strength of the major dissolved cation (i.e., from Rb to Li) at the same T. Experimental investigations on the structures of TiO$_2$-SiO$_2$ glasses by EXAFS (Extended X-ray Absorption Fine Structure) and XANES (X-ray Absorption Near Edge Structure) carried out by Greegor et al. (1983) show that Ti^{4+} occurs in both coordination IV and coordination VI. The proportions of the two sites vary in a complex fashion with the chemistry of the system (figure 6.5): in a pure silica glass, up to 0.05 weight % of TiO$_2$ ($X = 0.0004$) Ti^{4+} occupies exclusively tetrahedral sites. For greater amounts of TiO$_2$ ($X = 0.07$), Ti^{4+} substitutes for Si^{4+} and the VI-fold coordinated fraction remains below 5%. Above 9 weight % of dissolved TiO$_2$, the ratio in VI-fold and IV-fold coordination increases appreciably, and at 15% of dissolved TiO$_2$ a structural arrangement similar to rutile is observed (figure 6.5).

Another group of elements attentively investigated for their coordination states in silicate glasses and melts is the rare earth elements (REE). Existing studies show that Ce^{3+}, Nd^{3+}, Eu^{3+}, Er^{3+}, and Yb^{3+} occupy mainly octahedral sites, although coordinations VII and VIII may occasionally be observed (cf. Robinson, 1974, and references therein).

6.3 Effects of *P* and *T* Intensive Variables on Melting

Let us consider one mole of a pure crystalline phase at equilibrium with one mole of melt of the same composition—e.g., pure forsterite at $P = 1$ bar and $T = 2163$ K. Because the two phases are at equilibrium, we have

$$\mu_{Mg_2SiO_4,\text{olivine}} = \mu_{Mg_2SiO_4,\text{melt}}. \tag{6.41}$$

If we vary the temperature and/or pressure, equilibrium will be maintained through identical modifications of the chemical potential of the Mg_2SiO_4 component in the two phases:

$$d\mu_{Mg_2SiO_4,\text{olivine}} = d\mu_{Mg_2SiO_4,\text{melt}}. \tag{6.42}$$

Because for a pure single-component phase,

$$d\mu = d\overline{G} = -\overline{S}dT + \overline{V}dP, \tag{6.43}$$

we can write

$$-\overline{S}_{Mg_2SiO_4,\text{olivine}}dT + \overline{V}_{Mg_2SiO_4,\text{olivine}}dP = -\overline{S}_{Mg_2SiO_4,\text{melt}}dT + \overline{V}_{Mg_2SiO_4,\text{melt}}dP. \tag{6.44}$$

Collecting the terms in dP and dT, respectively, we have

$$dT\left(\overline{S}_{Mg_2SiO_4,\text{melt}} - \overline{S}_{Mg_2SiO_4,\text{olivine}}\right) = dP\left(\overline{V}_{Mg_2SiO_4,\text{melt}} - \overline{V}_{Mg_2SiO_4,\text{olivine}}\right). \tag{6.45}$$

Denoting the two terms in brackets, respectively, as $\Delta\overline{S}_{\text{fusion}}$ and $\Delta\overline{V}_{\text{fusion}}$ and rearranging, we obtain the *Clausius-Clapeyron* equation:

$$\frac{dT}{dP} = \frac{\Delta\overline{V}_{\text{fusion}}}{\Delta\overline{S}_{\text{fusion}}}. \tag{6.46}$$

Because $\Delta\overline{G} = 0$ for the two phases at equilibrium, we also have

$$\Delta\overline{S}_{\text{fusion}} = \frac{\Delta\overline{H}_{\text{fusion}}}{T} \tag{6.47}$$

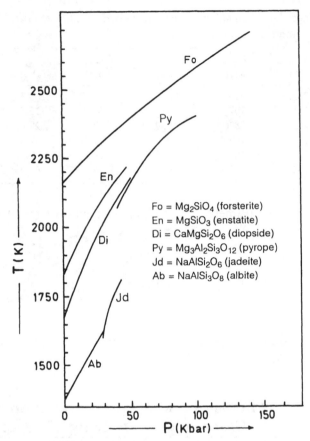

Fo

Py

En

Fo = Mg_2SiO_4 (forsterite)
En = $MgSiO_3$ (enstatite)
Di = $CaMgSi_2O_6$ (diopside)
Py = $Mg_3Al_2Si_3O_{12}$ (pyrope)
Jd = $NaAlSi_2O_6$ (jadeite)
Ab = $NaAlSi_3O_8$ (albite)

Di

Jd

Ab

Figure 6.6 Experimentally observed univariant dT/dP equilibria for several silicates of petrogenetic interest. Melting curves are satisfactorily reproduced by Simon equation with parameters listed in table 6.5.

$$\frac{dT}{dP} = \frac{T_{fusion} \cdot \Delta \overline{V}_{fusion}}{\Delta \overline{H}_{fusion}}. \tag{6.48}$$

The Clausius-Clapeyron equation describes the univariant equilibrium between crystal and melt in the P-T field. Because molar volumes and molar entropies of molten phases are generally greater than their crystalline counterparts, the two terms $\Delta \overline{S}_{fusion}$ and $\Delta \overline{V}_{fusion}$ are both positive and we almost invariably observe an increase in T_{fusion} with increasing pressure. Although extension to multicomponent systems is somewhat more complex, we can now understand why adiabatic decompression acting on a rock system generally induces partial melting.

Univariant melting curves for several pure components are shown in figure 6.6. Because silicate melts are more compressible than their crystalline counterparts, $\Delta \overline{V}_{fusion}$ decreases with increasing P, thus causing curvature in the univariant line.

Table 6.5 Parameters of Simon equation for some rock-forming silicates (after Bottinga, 1985). T_{max} and P_{max} are maximum conditions of applicability.

Compound	T_{max} (K)	T_{Simon} (K)	P_{max} (kbar)	P_{Simon} (kbar)	a (kbar)	b (adimensional)
Mg_2SiO_4	2736	2163	150	0	108.33	3.7
$Mg_3Al_2Si_3O_{12}$	2410	2073	100	40	19.79	9.25
Mg_2SiO_3	2220	1830	46.7	0	28.65	5.01
$CaMgSi_2O_6$	2172	1670	50	0	34.71	3.40
$NaAlSi_2O_6$	1817	1609	43	29	7.97	8.35
$NaAlSi_3O_8$	1625	1373	30	0	60.95	2.38

Table 6.5 lists the parameters of the Simon equation simulating the dT/dP evolution of the univariant curve. The Simon equation:

$$P = a\left[\left(\frac{T}{T_{Simon}}\right)^b - 1\right] + P_{Simon} \tag{6.49}$$

is a purely empirical fitting of experimental evidence and must be used with caution (cf. Bottinga, 1985).

6.4 Effects of Chemical Composition on Melting (Equation of Lowering of Freezing Point)

The chemical potential of the component of interest in a multicomponent phase must be referred to its potential at standard state. Equation 6.41 is thus reexpressed as follows:

$$\mu^0_{Mg_2SiO_4,\text{olivine}} + RT \ln a_{Mg_2SiO_4,\text{olivine}} = \mu^0_{Mg_2SiO_4,\text{melt}} + RT \ln a_{Mg_2SiO_4,\text{melt}}. \tag{6.50}$$

Gathering together the logarithmic terms, we obtain

$$\ln \frac{a_{Mg_2SiO_4,\text{melt}}}{a_{Mg_2SiO_4,\text{olivine}}} = \frac{\mu^0_{Mg_2SiO_4,\text{olivine}} - \mu^0_{Mg_2SiO_4\text{ melt}}}{RT}. \tag{6.51}$$

Because

$$\frac{\mu^0}{RT} = \frac{\overline{H}^0}{RT} - \frac{\overline{S}^0}{R}, \tag{6.52}$$

differentiating equation 6.51 in dT yields:

$$\frac{\partial}{\partial T}\left(\ln \frac{a_{Mg_2SiO_4,melt}}{a_{Mg_2SiO_4,olivine}}\right) = \frac{\overline{H}^0_{Mg_2SiO_4,melt} - \overline{H}^0_{Mg_2SiO_4,olivine}}{RT^2}. \qquad (6.53)$$

The term $(\overline{H}^0_{Mg_2SiO_4,melt} - \overline{H}^0_{Mg_2SiO_4,olivine})$ is the molar enthalpy of solution of pure forsterite in a pure Mg_2SiO_4 melt, and is equivalent to the molar enthalpy of fusion $\Delta\overline{H}_{fusion}$ at $T = T_{fusion}$.

Because the variation of enthalpy of solution between two temperatures may be expressed by the integral

$$\int_{T_{fusion}}^{T} d\Delta\overline{H}_{solution} = \int_{T_{fusion}}^{T} \Delta C_P \, dT, \qquad (6.54)$$

if we assume ΔC_P to be constant within the integration limits, equation 6.53 becomes

$$\ln \frac{a_{Mg_2SiO_4,melt}}{a_{Mg_2SiO_4,olivine}} = \frac{1}{R} \int_{T_{fusion}}^{T} \frac{\Delta\overline{H}_{fusion} + \Delta C_P(T - T_{fusion})}{T^2} dT, \qquad (6.55)$$

which, when integrated, gives

$$\ln \frac{a_{Mg_2SiO_4,melt}}{a_{Mg_2SiO_4,olivine}} = \frac{1}{R}\left[\left(\Delta\overline{H}_{fusion} - \Delta C_P T_{fusion}\right)\left(\frac{1}{T_{fusion}} - \frac{1}{T}\right) + \Delta C_P \ln\left(\frac{T}{T_{fusion}}\right)\right].$$

$$(6.56)$$

(see Saxena, 1973 and references therein; Wood and Fraser, 1976).

Equation 6.56 is known as the "equation of lowering of freezing point" and is valid for solid mixtures crystallizing from multicomponent melts. Like the Clausius-Clapeyron equation, it tells us how the system behaves, with changing T, to maintain equilibrium on the univariant curve. However, whereas in the Clausius-Clapeyron equation equilibrium is maintained with concomitant changes in P, here it is maintained by appropriately varying the activity of the component of interest in the melt and in the solid mixture.

Recalling the relationship between thermodynamic activity and molar concentration,

$$a_i = X_i \gamma_i, \qquad (6.57)$$

we see that, for practical purposes, the response of the system to modifications in T consists of a flow of matter between the two phases at equilibrium.

Lastly, it must be noted that, whenever a pure single-component phase crystallizes from the melt, the denominator of the term on the left side of equation 6.56 is reduced to 1 and the equation directly furnishes the activity of the dissolved component in the melt.

6.5 Mixing Properties in Melts with Constant Silica Contents

We have already seen that the degree of polymerization of the melt is controlled by the amount of silica in the system (see, for instance, figure 6.4). If we mix two fused salts with the same amount of silica and with cations of similar properties, the anion matrix is not modified by the mixing process and the Gibbs free energy of mixing arises entirely from mixing in the cation matrix—i.e.,

$$S_{mixing} = S_{cation, mixing}. \tag{6.58}$$

This hypothesis, advanced by Richardson (1956), has received some experimental confirmation (Belton et al., 1973; Rammensee and Fraser, 1982; Fraser et al., 1983, 1985). In particular, in a series of experimental studies using Knudsen cell mass spectrometry, Fraser et al. (1983, 1985) showed that the melts $NaAlSi_2O_6$-$KAlSi_2O_6$, $NaAlSi_3O_8$-$KAlSi_3O_8$, $NaAlSi_4O_{10}$-$KAlSi_4O_{10}$, and $NaAlSi_5O_{12}$-$KAlSi_5O_{12}$ mix ideally over the entire binary range. Table 6.6 lists the activity coefficients of the jadeite component $NaAlSi_2O_6$ in the molten mixture $NaAlSi_2O_6$-$KAlSi_2O_6$ obtained at various T by Fraser et al. (1983). Within experimental incertitude, these values approach 1.

The results of the experimental study of Fraser et al. (1983) on the $NaAlSi_2O_6$-$KAlSi_2O_6$ join are at first sight surprising, because the crystalline counterparts of the pure end-members (i.e., jadeite and leucite) have very differ-

Table 6.6 Activity coefficients of jadeite component $NaAlSi_2O_6$ in molten (and/or over cooled) mixture $NaAlSi_2O_6$-$KAlSi_2O_6$. Terms in parentheses indicate experimental incertitude at various T (from Fraser et al., 1983).

$X_{NaAlSi_2O_6}$	$\gamma_{NaAlSi_2O_6}$				
	1200 °C	1300 °C	1400 °C	1500 °C	1600 °C
0.4	1.032 (±0.03)	1.043 (±0.02)	1.043 (±0.02)	1.050 (±0.02)	0.998 (±0.04)
0.5	1.022	1.039	1.046	1.057	1.041
0.6	1.014	1.035	1.048	1.063	1.069
0.7	1.008	1.031	1.051	1.069	1.086
0.75	1.005	1.029	1.051	1.070	1.089 (±0.02)
0.8	1.004	1.018	1.032	1.045	1.056
0.9	1.001	1.005	1.006	1.011	1.014
1.0	1.000	1.000	1.000	1.000	1.000

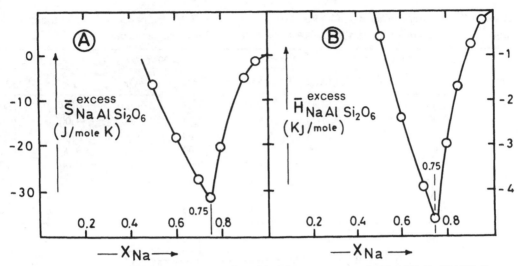

Figure 6.7 Excess partial molar properties of jadeite component in $NaAlSi_2O_6$-$KAlSi_2O_6$ molten mixture. (A) Excess partial molar entropy. (B) Excess partial molar enthalpy. Reprinted from Fraser et al. (1983), *Bulletin Mineralogique*, 106, 111–117, with permission from Masson S.A., Paris, France.

ent structural arrangements. Evidently the hypothesis of Richardson (1956) is valid and the melt does not carry any structural remnant of the crystalline aggregates after the melting event.

Nevertheless, if we analyze the excess partial molar properties of the terms in the mixture, we see well-defined minima at $X_{NaAlSi_2O_6} = 0.75$ (Fraser et al., 1983). And in figure 6.7, there are marked minima in the excess partial molar entropy and excess partial molar enthalpy of the jadeite component in the mixture at this concentration level. The compositional ratio Na:K = 3:1 corresponds to the eutectic minimum in the low-SiO_2 portion of the SiO_2-$KAlSiO_4$-$NaAlSiO_4$ system ($T = 1020$ °C at $P = 1$ bar; Schairer, 1950). Moving from the K-rich side of the join toward the sodic terms, at Na:K = 3:1, we find the field of Na-K β-nepheline, with two sites of different sizes in the ratio 3:1, occurring at liquidus.

Table 6.7 lists experimental activity coefficient values for albite and orthoclase components in the $NaAlSi_3O_8$-$KAlSi_3O_8$ molten mixture. The albite component mixes ideally at all T in the compositional range Ab_{100} to $Ab_{20}Or_{80}$, but deviates slightly from ideality for more potassic compositions.

Analyzing the partial molar properties of the terms in the mixtures, we observe a sudden change in slope for the partial molar enthalpy of the two components at $X_{Ab} = 0.2$ (figure 6.8), whereas the integral enthalpy of mixing remains near zero over the entire compositional range. Rammensee and Fraser (1982) interpret this as a result of incipient ordering in the melt structure, which is necessary to approach the structural arrangement observed in leucite, prior to its crystallization.

From the various experimental evidence available on the investigated systems,

Table 6.7 Activity coefficients of albite (Ab) and orthoclase (Or) components in molten mixtures and over cooled liquids in $NaAlSi_3O_8$-$KAlSi_3O_8$ system (after Rammensee and Fraser, 1982).

X_{Ab}	1200 °C		1300 °C		1400 °C		1500 °C	
	Ab	Or	Ab	Or	Ab	Or	Ab	Or
0	1.56	1.00	1.41	1.00	1.27	1.00	1.11	1.00
0.05	1.30	1.01	1.20	1.00	1.09	1.00	1.00	1.00
0.1	1.12	1.02	1.08	1.01	1.00	1.01	1.00	1.00
0.2	1.00	1.04	1.00	1.02	1.00	1.01	1.00	1.00
0.3	1.00	1.04	1.00	1.02	1.00	1.01	1.00	1.00
0.4	1.00	1.04	1.00	1.02	1.00	1.01	1.00	1.00
0.5	1.00	1.04	1.00	1.02	1.00	1.01	1.00	1.00
1.0	1.00	1.04	1.00	1.02	1.00	1.01	1.00	1.00

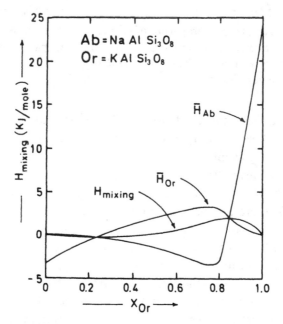

Figure 6.8 Partial molar enthalpies and integral enthalpy of mixing in pseudobinary $NaAlSi_3O_8$-$KAlSi_3O_8$ system. Reprinted from W. Rammensee and D. G. Fraser, *Geochimica et Cosmochimica Acta*, 46, 2269–2278, copyright © 1982, with kind permission from Elsevier Science Ltd., The Boulevard, Langford Lane, Kidlington OX5 1GB, UK.

Fraser et al. (1983, 1985) deduced that, in several instances, the structure of the silicate melt should mimic the structural arrangement of the solid phase at liquidus. In incongruent melting processes, the structure of the liquid differs substantially from that of the solid phase of identical stoichiometry.

The quasi-ideality of several other molten systems with constant amounts of silica may be deduced by applying the lowering of freezing point equation. A very instructive example in this sense concerns the Fe_2SiO_4-Mg_2SiO_4 system (Saxena, 1973 and references therein; Wood and Fraser, 1976). Adopting Richardson's model to describe the activity of fayalite (Fe_2SiO_4: Fa) and forsterite (Mg_2SiO_4: Fo) components in the binary melt, we obtain

$$a_{Fo,melt} = X^2_{Mg,cations} \qquad (6.59)$$

$$a_{Fa,melt} = X^2_{Fe,cations} . \qquad (6.60)$$

Because the solid Fe_2SiO_4-Mg_2SiO_4 mixture is ideal (cf. section 5.2.5), the activities of components in the solid are given by

$$a_{Fo,olivine} = X^2_{Mg,olivine} \qquad (6.61)$$

$$a_{Fa,olivine} = X^2_{Fe,olivine} . \qquad (6.62)$$

Substituting equations 6.59, 6.60, 6.61, and 6.62 into the lowering of freezing point equation, we obtain

$$2\ln\left(\frac{X_{Mg,melt}}{X_{Mg,olivine}}\right) = \frac{1}{R} \int_{T_{fusion,Fo}}^{T} \frac{\Delta\overline{H}_{fusion,Fo} + \Delta C_{P,Fo}\left(T - T_{fusion,Fo}\right)}{T^2} dT \quad (6.63)$$

$$2\ln\left(\frac{X_{Fe,melt}}{X_{Fe,olivine}}\right) = \frac{1}{R} \int_{T_{fusion,Fa}}^{T} \frac{\Delta\overline{H}_{fusion,Fa} + \Delta C_{P,Fa}\left(T - T_{fusion,Fa}\right)}{T^2} dT. \quad (6.64)$$

Wood and Fraser (1976) assigned the following molar values to the various parameters: $\Delta\overline{H}_{fusion,Fo} = 34,240$ (cal); $\Delta\overline{H}_{fusion,Fa} = 25,010$ (cal); $\Delta C_{P,Fo} = \Delta C_{P,Fa} = 7.45$ (cal/K); $T_{fusion,Fa} = 1479.5$ (K); $T_{fusion,Fo} = 2164$ (K).

Ratios $X_{Mg,melt}/X_{Mg,olivine}$ and $X_{Fe,melt}/X_{Fe,olivine}$ necessary to maintain equilibrium between the coexisting phases (solid plus melt) can be calculated by means of equations 6.63 and 6.64. The results are shown in table 6.8.

If we let a represent the $X_{Fe,melt}/X_{Fe,olivine}$ ratio and let b represent the $X_{Mg,melt}/X_{Mg,olivine}$ ratio, then, because

$$X_{Fe,melt} = a \cdot X_{Fe,olivine} \qquad (6.65)$$

and

Table 6.8 Relative amounts of Fe and Mg in olivine and melt at equilibrium, based on equations 6.63 and 6.64 (after Wood and Fraser, 1976).

Ratio		T (K)					
		1479.5	1548	1600	1683	1738	1768
$X_{Fe, melt} / X_{Fe, olivine}$	(a)	1.0	1.21	1.39	1.70	1.93	2.06
$X_{Mg, melt} / X_{Mg, olivine}$	(b)	–	0.23	0.27	0.34	0.40	0.43

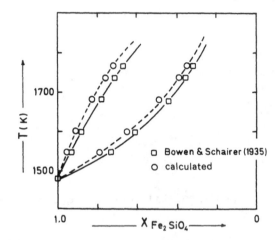

Figure 6.9 Melting relations in Mg_2SiO_4-Fe_2SiO_4 system. Experimental data confirm validity of Richardson's (1956) ideal model for liquid. Reprinted from B. J. Wood and D. G. Fraser, *Elementary Thermodynamics for Geologists*, 1976, by permission of Oxford University Press.

$$\left(1 - X_{Fe,melt}\right) = b\left(1 - X_{Fe,olivine}\right),$$ (6.66)

we obtain

$$X_{Fe,melt} = \frac{a\left(1 - b\right)}{a - b}$$ (6.67)

and

$$X_{Fe,olivine} = \frac{1 - b}{a - b}.$$ (6.68)

Figure 6.9 shows how the calculated phase boundaries compare with the experimental observations of Bowen and Schairer (1935) on the same binary join. The satisfactory reproduction of phase assemblages clearly indicates that the

Mg_2SiO_4-Fe_2SiO_4 melt is effectively ideal, in agreement with Richardson's (1956) assumptions.

6.6 Thermodynamic Properties of Complex Silicate Melts

We now examine the thermodynamic properties of natural melts, after a brief digression on the differences between vitreous and molten states, which is necessary for better understanding of the properties of interest.

6.6.1 Relationships Between Molten and Vitreous States

If we rapidly quench a silicate melt, we may obtain a glass without the formation of any crystalline aggregate. The temperature of transition between melt and glass may be operatively defined as "that temperature at which the differing thermodynamic properties of melt and glass intersect" (Berman and Brown, 1987). The supercooling phenomenon, which may easily be observed in melts with molar amounts of silica higher than 50%, involves discontinuities in the values of heat capacity, enthalpy, entropy, and Gibbs free energy. Moreover, the various thermodynamic properties for the vitreous state (and for the transition temperature) vary with the quenching rate. Figure 6.10 shows, for instance, the enthalpy and heat capacity for a $CaMgSi_2O_6$ melt compared with the corresponding properties in the crystalline and vitreous states at various T. Note in figure 6.10A that the dH/dT slope is different for vitreous and molten states. Moreover, the glass-melt transition is observed at two different temperatures, depending on quenching rate ($T = 1005$ or 905 K), and the quenching rate also affects the enthalpy of the vitreous state, which is not the same in the two cases. The enthalpy of the crystalline counterpart in all T conditions is lower than the molten and vitreous states. At the temperature of fusion T_f, the enthalpy difference between crystal and melt is the *enthalpy of fusion* of $CaMgSi_2O_6$ ($\Delta \overline{H}_{fusion}$), whereas the enthalpy differences at lower T are enthalpies of solution of pure diopside in a pure $CaMgSi_2O_6$ melt ($\Delta \overline{H}_{solution}$) and enthalpies of vitrification ($\Delta \overline{H}_{vitrification}$). Note also in figure 6.10B that the heat capacity of the vitreous state (C_{P_v}) is very similar to the heat capacity of the crystal (C_{P_c}) and that both, T being equal, are notably lower than the heat capacity of the molten state (C_{P_f}).

Lastly, we observe that the glass-melt transition involves a marked discontinuity at T of transition, reflecting the increase of vibrational freedom (rotational components appear). This discontinuity involves some complexity in the extrapolation of calorimetric data (usually obtained on glasses) from vitreous to molten states, discussed in detail by Richet and Bottinga (1983, 1986).

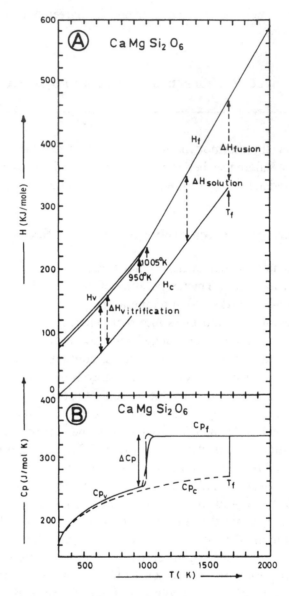

Figure 6.10 Temperature dependence of enthalpy (A) and heat capacity (B) for CaMg-Si$_2$O$_6$ component in crystalline, vitreous, and molten states at various T conditions. Reprinted with permission from Richet and Bottinga (1986), *Review of Geophysics and Space Physics,* 24, 1–25, copyright © 1986 by the American Geophysical Union.

6.6.2 Heat Capacity of Silicate Melts

We have already seen (section 6.5) that the mixing behavior of melt components with identical amounts of silica is essentially ideal. Ideal mixing implies that extensive properties of the melt, such as heat capacity at constant pressure C_P, is a linear function of the molar properties of the end-members—i.e.,

Table 6.9 Parameters for calculation of heat capacity of silicate melts: (1) after Carmichael et al. (1977) and Stebbins et al. (1984); (2) model of Richet and Bottinga (1985)

Model	Parameter	Constituent Oxide								
		SiO_2	TiO_2	Al_2O_3	Fe_2O_3	FeO	MgO	CaO	Na_2O	K_2O
(1)	K_1	80.0	111.8	157.6	229.0	78.9	99.7	99.9	102.3	97.0
		±0.9	±5.1	±3.4	±18.4	±4.9	±7.3	±7.2	±1.9	±5.1
(2)	K_1	81.37	75.21	27.21	199.7	78.94	85.78	86.05	100.6	50.13
	$K_2 \times 10^3$	0.0	0.0	94.28	0.0	0.0	0.0	0.0	0.0	15.78
	$K_3 \times 10^{-5}$	0.0	875.3	0.0	0.0	0.0	0.0	0.0	0.0	0.0

$$(1)\ \overline{C}_P \left(\frac{J}{mole \times K} \right) = \sum_i K_{1,i} X_i \ ; (2)\ \overline{C}_P \left(\frac{J}{mole \times K} \right) = \sum_i X_i (K_{1,i} + K_{2,i} T + K_{3,i} T^{-2}) + \overline{C}_{P,excess}$$

$$C_{P,melt} = \sum_i X_i \overline{C}_{P,i}^0, \qquad (6.69)$$

where X_i is the molar fraction of component i in the molten mixture and $\overline{C}_{P,i}^0$ is its molar specific heat at the standard state of pure melt. On the basis of calorimetric measurements carried out on natural and synthetic molten mixtures in the T range 1000 to 1600 K, Carmichael et al. (1977) suggested that the ideal model postulated for binary mixtures may be extended to multicomponent systems. They also observed that the heat capacity of the various components in the mixture may be considered independent of T, within the limits of experimental incertitude, through the suitable choice of standard state values (i.e., a single term in Maier-Kelley-type functions; cf. section 3.2). However, further investigations by Stebbins et al. (1984) and Richet and Bottinga (1985, 1986) showed significant deviations from the ideal mixing model. For instance, in the pseudoternary system $CaAl_2Si_2O_8$-$NaAlSi_3O_8$-$CaMgSi_2O_6$, positive deviations from ideality occur along the $NaAlSi_3O_8$-$CaMgSi_2O_6$ join and negative ones in $CaMgSi_2O_6$-$CaAl_2Si_2O_8$ molten mixtures.

Table 6.9 lists the parameters of the model of Carmichael et al. (1977) for the heat capacity of silicate melts, recalibrated by Stebbins et al. (1984) and valid for the T range 1200 to 1850 K. To obtain the heat capacity of the melt at each T condition, within the compositional limits of the system, it is sufficient to combine linearly the molar proportions of the constituent oxides multiplied by their respective $\overline{C}_P$ values (cf. equation 6.69).

Table 6.9 also lists the parameters of Richet and Bottinga's (1985) model, which, devised primarily for Al-free melts, accounts for the nonideality of molten silicate systems containing K by introducing an excess term that represents Si-K interactions:

$$\overline{C}_{P,melt} = \sum_i X_i \overline{C}_{P,i} + \overline{C}_{P,excess} \qquad (6.70)$$

$$\overline{C}_{P,\text{excess}} = W_{\text{Si-K}} X_{\text{SiO}_2} X^2_{\text{K}_2\text{O}}; \quad W_{\text{Si-K}} = 151.7\left(\frac{\text{J}}{\text{mole} \times \text{K}}\right). \tag{6.71}$$

The model of Richet and Bottinga (1985) considers the heat capacity of the melt to be variable with T and allows better reproducibility of experimental evidence in compositionally simple systems. The Stebbins-Carmichael model seems to reproduce the experimental observations on aluminosilicate melts better (Berman and Brown, 1987). Application of both models to natural melts gives substantially identical results (i.e., differences of about 1%, within the range of data uncertainty; cf. Berman and Brown, 1987).

6.6.3 Relationships Between Enthalpy-Entropy of Fusion and Enthalpy-Entropy of Formation of Melt Components at Standard State

Although in crystalline phases determination of the enthalpy of formation from the constituent elements (or from constituent oxides) may be carried out directly through calorimetric measurements, this is not possible for molten components. If we adopt as standard state the condition of "pure component at $T = 298.15$ K and $P = 1$ bar", it is obvious that this condition is purely hypothetical and not directly measurable. If we adopt the standard state of "pure component at P and T of interest", the measurement is equally difficult, because of the high melting temperature of silicates.

The molar enthalpy of molten components in the standard reference conditions of $T = 298.15$ K and $P = 1$ bar is usually obtained indirectly, by adding first the molar enthalpy of fusion to the molar enthalpy of the crystalline component at its melting point (see also figure 6.10):

$$\overline{H}_{\text{melt},T_f} = \overline{H}_{\text{crystal},T_f} + \Delta\overline{H}_{\text{fusion}}. \tag{6.72}$$

Once the enthalpy of fusion is known, because solid and liquid are at equilibrium at the melting point, the entropy of the molten component at T_f is also known through

$$\Delta\overline{G}_{\text{fusion}} = 0 \Rightarrow \Delta\overline{S}_{\text{fusion}} = \Delta\overline{H}_{\text{fusion}}/T_f. \tag{6.73}$$

If the heat capacity of the molten component is known (cf. section 6.6.2), molar enthalpy and molar entropy at the standard state of the pure melt at $T = 298.15$ and $P = 1$ bar may be readily derived by applying

$$\overline{H}^0_{\text{melt},298.15\,K,1\,\text{bar}} = \overline{H}_{\text{melt},T_f} - \int_{298.15}^{T_f} \overline{C}_{P,\text{melt}}\,dT \tag{6.74}$$

Table 6.10 Parameters for calculation of molar volume of silicate melts at various T conditions (after Stebbins et al., 1984). Resulting volume is in cm³/mole.

Component	a_i	$b_i \times 10^3$	Component	a_i	$b_i \times 10^3$
SiO_2	26.27	−0.42	MgO	12.39	0.21
TiO_2	24.22	0.63	CaO	17.68	0.15
Al_2O_3	37.89	0.38	Na_2O	30.54	0.18
Fe_2O_3	44.93	0.22	K_2O	48.92	0.41
FeO	14.50	0.19			

and

$$\overline{S}^0_{\text{melt},298.15K,1\,\text{bar}} = \overline{S}_{\text{melt},T_f} - \int_{298.15}^{T_f} \overline{C}_{P,\text{melt}} \frac{dT}{T}. \tag{6.75}$$

Because $\overline{H}_{\text{fusion}}$ is difficult to measure as a result of the high value of T_f, it may be derived indirectly through calculations involving the vitreous state (see Berman and Brown, 1987) or through the Clausius-Clapeyron equation for the crystal-melt equilibrium (cf. equation 6.48 and section 6.3).

6.6.4 Molar Volume, Thermal Expansion, and Compressibility of Silicate Melts

Like heat capacity, the molar volume of chemically complex silicate melts may also be obtained through a linear combination of the molar volumes of molten oxide components—i.e.,

$$\overline{V}_{\text{melt}} = \sum_i X_i \overline{V}_i^0. \tag{6.76}$$

Bottinga and Weill (1970) applied equation 6.76 extensively, and it was later recalibrated by Stebbins et al. (1984). In the latter model, the molar volume of the melt and its modification with T are calculated through the equation

$$\overline{V}_{\text{melt},T} = \sum_i a_i X_i + (T - 1873) \sum_i b_i X_i, \tag{6.77}$$

whose coefficients are listed in Table 6.10. Equation 6.77 also implies that thermal expansion is a simple additive function of the linear expansions of the various oxide components.

Concerning melt compressibility, existing experimental data are rather scanty. On the basis of solid/liquid univariant equilibria for six silicate components (Mg_2SiO_4, $MgSiO_3$, $CaMgSi_2O_6$, $Mg_3Al_2Si_3O_{12}$, $NaAlSi_2O_6$, and $NaAlSi_3O_8$)

Table 6.11 Isothermal bulk modulus and first derivative for silicate melt components (after Bottinga, 1985).

Component	$T_{\text{reference}}$	K_0 (Mbar)	K'
Mg_2SiO_4	2494	0.58815	3.75
$MgSiO_3$	2067	0.20555	9.79
$CaMgSi_2O_6$	1970	0.24218	5.31
$Mg_3Al_2Si_3O_{12}$	2290	0.11905	23.33
$NaAlSi_2O_6$	1735	0.04096	23.97
$NaAlSi_3O_8$	1515	0.42502	20.86

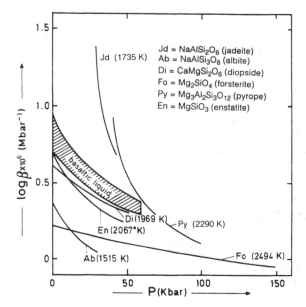

Figure 6.11 Compressibilities of silicate melt components (logarithmic scale) as a function of *P*. Values in parentheses give *T* of reference. Dashed area: compressibility of basaltic liquids according to Stolper et al. (1981). Reprinted from Bottinga (1985), with kind permission from Elsevier Science Publishers B.V, Amsterdam, The Netherlands.

and existing thermodynamic data, Bottinga (1985) calculated isothermal compressibility for the corresponding melts in the *P* range 0 to 150 kbar. This calculation, which is based on the Simon equation (eq. 6.49), gives the bulk moduli listed in table 6.11 and valid for the Birch-Murnaghan equation of state.

The calculations of Bottinga (1985) show that the compressibility of silicate melts decreases rather abruptly with increasing *P*. This fact, which has profound implications in petrogenesis, is clearly shown in figure 6.11. Note particularly the rapid decrease in compressibility for the liquid pyrope component $Mg_3Al_2Si_3O_{12}$ in the range $40 < P$ (kbar) < 60, and for the jadeite component $NaAlSi_2O_6$ in the range $29 < P < 43$. The compressibilities of the corresponding solids are lower, and the first derivative on *P* of their bulk moduli is also generally lower (compare tables 6.11 and 5.11, 5.14, 5.22, 5.35). Because at 1 bar the density of silicate melts is lower than that of the corresponding solids, the calculations of Bottinga (1985) imply that the density contrast decreases with increasing P. This fact would tendencially obstaculate the separation of molten portions from the residual solid, with increasing depth within the earth's crust and mantle.

6.7 Miscibility of Natural Melts: Ghiorso-Carmichael Model

The existence of miscibility gaps in silicate melts has been experimentally known for decades (Greig, 1927; Bowen, 1928; Bowen and Schairer, 1938; Schairer and Bowen, 1938; Roedder, 1951, 1956; Holgate, 1954) and has been directly observed both in terrestrial rocks (Anderson and Gottfried, 1971; Ferguson and Currie, 1971; De, 1974; Gelinas et al., 1976; Philpotts, 1979; Coltorti et al., 1987) and in lunar basalts (Roedder and Wieblen, 1971).

Figure 6.12 combines experimental evidence and natural observations in the form of a ternary SiO_2 - $(CaO+MgO+FeO+TiO_2)$ - $(Na_2O+K_2O+Al_2O_3)$ plot. The glassy portions of lunar basalts have compositions that conform quite well to the experimental miscibility gap of Gelinas et al. (1976), whereas terrestrial volcanites have smaller gaps.

Immiscibility phenomena in silicate melts imply positive deviations from ideality in the mixing process. Ghiorso et al. (1983) developed a mixing model applicable to natural magmas adopting the components listed in table 6.12. Because all components have the same standard state (i.e., pure melt component at the T and P of interest) and the interaction parameters used do not vary with T, we are dealing with a regular mixture of the Zeroth principle (cf. sections 2.1 and 3.8.4):

$$G_{\text{melt},P,T} = \sum_i n_i \mu_i^0 + RT \sum_i n_i \ln X_i + \frac{1}{2} n_{\text{total}} \sum_i \sum_j W_{ij} X_i X_j, \quad (6.78)$$

where μ_i^0 is the standard state chemical potential of the pure ith melt component at the P and T of interest, n_i is the number of moles of the ith component, X_i is the molar ratio of the ith component, and W_{ij} is the interaction parameter between components i and j.

In applying the model above (which is an extension of the previous version of Ghiorso and Carmichael, 1980, integrated with new experimental observations), some caution must be adopted, because of the following considerations:

1. If the heat capacity of a chemically complex melt can be obtained by a linear summation of the specific heat of the dissolved oxide constituents at all T (i.e., Stebbins-Carmichael model), the melt is by definition ideal. The addition of excess Gibbs free energy terms thus implies that the Stebbins-Carmichael model calculates only the *ideal* contribution to the Gibbs free energy of mixing.

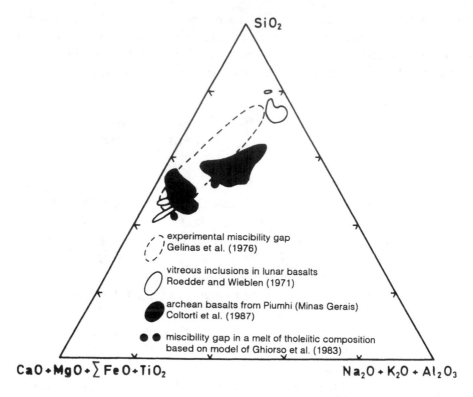

Figure 6.12 Miscibility gap in natural and synthetic silicate melts. Dots: unmixing for a tholeiitic magma, based on model of Ghiorso et al. (1983). Σ FeO: total iron as FeO.

Table 6.12 Melt components in Ghiorso-Carmichael model. Calculation first involves computation of molar fractions of various oxides, and then their combination.

Si_4O_8	$= 0.25 \times [SiO_2 - 0.5(FeO + MnO + MgO + NiO + CoO + CaO) - Na_2O - K_2O]$
Ti_4O_8	$= 0.25 \times TiO_2$
$Al_{16/3}O_8$	$= 0.375 \times Al_2O_3$
$Fe_{16/3}O_8$	$= 0.375 \times Fe_2O_3$
$Cr_{16/3}O_8$	$= 0.375 \times Cr_2O_3$
$Fe_4Si_2O_8$	$= 0.25 \times FeO$
$Mn_4Si_2O_8$	$= 0.25 \times MnO$
$Mg_4Si_2O_8$	$= 0.25 \times MgO$
$Ni_4Si_2O_8$	$= 0.25 \times CoO$
$Ca_4Si_2O_8$	$= 0.25 \times CaO$
$Na_{16/3}Si_{8/3}O_8$	$= 0.375 \times Na_2O$
$K_{16/3}Si_{8/3}O_8$	$= 0.375 \times K_2O$
$P_{16/5}O_8$	$= 0.625 \times P_2O_5$
Sr_8O_8	$= 0.125 \times SrO$
H_2O	$= 1 \times H_2O$

2. Calculation of the excess Gibbs free energy of mixing (third term on right side of eq. 6.78) involves only binary interactions. Although there is no multiple interaction model that can be reduced to the simple summation of binary interactions used here (cf. Acree, 1984; see also section 3.10), this choice is more than adequate for the state of the art, which does not allow precise location of the miscibility gap in the chemical space of interest.

3. Because the excess Gibbs free energy of mixing is independent of P, the volume, compressibility, and thermal expansion of the melt are obviously regarded as linear combinations of the values of its components (see section 6.6.4). As table 6.12 shows, the model uses a set of *ad hoc* components. The choice of components is dictated by simple convenience—i.e., the possibility of deriving realistic thermodynamic data for melt constituents through univariant equilibria involving their crystalline counterparts or through additive procedures.

The first step in the model is a normative calculation (table 6.12) to establish the molar amounts of the various melt components. For each component, the molar Gibbs free energy at the standard state of pure component at T and P of interest is given by

$$\overline{G}^0_{i,P,T,\text{melt}} = \overline{H}^0_{i,P_r,T,\text{melt}} - T\overline{S}^0_{i,P_r,T,\text{melt}} + \int_{P_r}^{P} \overline{V}^0_{i,P,T,\text{melt}} dP, \qquad (6.79)$$

where

$$\overline{H}^0_{i,P_r,T,\text{melt}} = \overline{H}^0_{i,P_r,T_r,\text{crystal}} + \int_{T_r}^{T_f} C_{P,i,\text{crystal}} \, dT$$
$$+ \Delta\overline{H}_{i,\text{fusion}} + C_{P,i,\text{melt}}\left(T - T_f\right) \qquad (6.80)$$

and

$$\overline{S}^0_{i,P_r,T,\text{melt}} = \overline{S}^0_{i,P_r,T_r,\text{crystal}} + \int_{T_r}^{T_f} C_{P,i,\text{crystal}} \, \frac{dT}{T}$$
$$+ \Delta\overline{S}_{i,\text{fusion}} + C_{P,i,\text{melt}} \, \ln\left(\frac{T}{T_f}\right). \qquad (6.81)$$

Tables 6.13 and 6.14 list sets of internally consistent thermodynamic data for crystalline and liquid components, for use in equations 6.79 to 81. Although equations 6.80 and 6.81 are simply based on the Clausius-Clapeyron approach, Ghiorso et al. (1983) use a semiempirical formulation for the volume of melt components. Its development is

Table 6.13 Thermodynamic parameters of crystalline components in Ghiorso-Carmichael model. Heat capacity function: $\overline{C}_P^0 = K_1 + K_2T + K_3T^{-2} + K_4T^{-1/2}$.

Component	$\overline{H}_{i,P_r,T_r,\text{solid}}^0$ (kcal/mole)	$\overline{S}_{i,P_r,T_r,\text{solid}}^0$ (cal/(mole × K))	$\overline{C}_P^0$			
			K_1	$K_2 \times 10^{-3}$	$K_3 \times 10^5$	K_4
Si_4O_8	−867.448	48.61	69.556	1.2432	−39.503	0
Ti_4O_8	−909.543	48.08	60.304	10.810	−9.4289	−5.3692
$Al_{16/3}O_8$	−1069.406	32.45	100.293	0.45827	−12.090	−629.73
$Fe_{16/3}O_8$	−519.023	80.99	−698.32	173.79	−65.259	21644.0
$Cr_{16/3}O_8$	−723.200	51.73	75.856	0.60525	−9.1899	−2.1698
$Fe_4Si_2O_8$	−704.730	70.90	73.020	18.720	−1.3400	0
$Mn_4Si_2O_8$	−826.038	78.02	78.510	7.4953	−18.229	0
$Mg_4Si_2O_8$	−1036.701	45.50	108.977	1.6319	−4.2733	−833.94
$Ca_4Si_2O_8$	−1109.070	36.44	98.0	0	0	0
$Na_{16/3}Si_{8/3}O_8$	−987.623	72.56	83.040	25.600	−17.253	0
$K_{16/3}Si_{8/3}O_8$	−990.064	93.15	79.280	44.933	−9.5467	0
$P_{16/5}O_8$	−575.488	44.18	13.400	86.402	0	0
Sr_8O_8	−1129.048	106.16	106.44	10.443	−5.3164	−298.14

Table 6.14 Thermodynamic parameters of melt components in Ghiorso-Carmichael model ($\Delta \overline{H}_{\text{fusion}}^0 = T_f \Delta \overline{S}_{\text{fusion}}^0$).

Component	T_f (K)	$\Delta \overline{S}_{\text{fusion}}^0$ (cal/(mole × K))	$\Delta \overline{H}_{\text{fusion}}^0$ (kcal/mole)	$\overline{C}_{P,i,\text{melt}}^0$ (cal/(mole × K))	$\overline{V}_{i,\text{melt}}^0$ (cal/(mole × bar))	
					$a_{V,i}$	$b_{V,i} \times 10^4$
Si_4O_8	1996	3.908	7.800	83.16	2.604	−0.1294
Ti_4O_8	2143	39.20	84.006	106.96	1.202	5.740
$Al_{16/3}O_8$	2327	32.08	74.650	65.71	1.767	3.397
$Fe_{16/3}O_8$	1895	34.80	65.946	122.27	2.173	3.659
$Cr_{16/3}O_8$	2603	31.76	82.671	100.00	1.854	-
$Fe_4Si_2O_8$	1490	29.57	44.059	114.60	1.920	2.834
$Mn_4Si_2O_8$	1620	26.46	42.865	116.20	2.584	-
$Mg_4Si_2O_8$	2163	37.61	81.350	128.06	2.238	0.8797
$Ca_4Si_2O_8$	2403	20.81	50.006	118.96	1.859	5.923
$Na_{16/3}Si_{8/3}O_8$	1362	24.24	33.015	113.01	2.859	4.170
$K_{16/3}Si_{8/3}O_8$	1249	25.62	31.999	114.67	3.456	7.138
$P_{16/5}O_8$	853	10.68	9.110	93.60	2.272	-
Sr_8O_8	2938	49.02	144.021	128.00	3.955	-

$$\overline{V}_{i,P_r,T,\text{melt}}^0 = \overline{V}_{i,T=0}^0 + a_{V,i}T + b_{V,i}T^2 \tag{6.82}$$

$$\int_{P_r}^{P} \overline{V}_{i,P,T,\text{melt}}^0 \, dP =$$

$$= \overline{V}_{i,P_r,T,\text{melt}}^0 \left\{ (P - P_r) - c_{V,i} \left[\frac{1}{2}\left(P^2 - P_r^2\right) - \frac{1}{2}c_{V,i}P^2 \right] \left[\frac{1}{6}\left(P^3 - P_r^3\right) \right] \right\}, \tag{6.83}$$

where $a_{V,i}$ and $b_{V,i}$ are the thermal expansion coefficients listed in table 6.14 and $c_{V,i}$ is a compressibility factor of the form

$$
c_{V,i} = \frac{1 \times 10^{-6}}{0.7551 + \dfrac{2.76}{n_i} \overline{V}^0_{i,P_r,T,\text{melt}}}.
\tag{6.84}
$$

Equation 6.83 is not valid for component Si_4O_8, for which the following equation is used:

$$
\int_{P_r}^{P} \overline{V}^0_{Si_4O_8,P,T,\text{melt}} \, dP = \overline{V}^0_{Si_4O_8,P_r,T,\text{crystal}} \left(P - 1 \right)
$$
$$
- 4.204 \times 10^{-6} \left(P^2 - P_r^2 \right).
\tag{6.85}
$$

Once the standard state potentials at the P and T of interest have been calculated ($\mu_i^0 \equiv \overline{G}_i^0$ for a pure single-component phase), the ideal and excess Gibbs free energy of mixing terms are easily obtained on the basis of the molar fractions of the various melt components and the binary interaction parameters listed in table 6.15 (cf. eq. 6.78).

It must be noted here that, because there are 16 melt components in the model, there should be 120 binary interaction parameters. Only 55 of these are listed in table 6.15, because the remaining ones are considered virtually equal to zero. Lastly, it must be noted that the Gibbs free energy of hydrated magmas cannot be obtained simply by application of equations 6.78 to 6.85, but requires additional considerations that cannot be outlined in this context (see Ghiorso and Carmichael, 1980, and Ghiorso et al., 1983, for detailed treatment of the method).

As already noted, the model can define the existence of miscibility gaps in natural magmas. Let us imagine that we have calculated the Gibbs free energy of mixing of a magmatic series with increasing amounts of component i:

$$
G_{\text{mixing,melt}} = G_{\text{ideal mixing}} + G_{\text{excess mixing}}.
\tag{6.86}
$$

We can then derive the calculated Gibbs free energy of mixing with respect to the molar amount of the component of interest, thus obtaining the difference between the chemical potential of the component in the mixture and its chemical potential at standard state:

$$
\left(\frac{\partial G_{\text{mixing,melt}}}{\partial X_i} \right)_{P,T,X_j} = RT \ln a_{i,P,T,\text{melt}} = \mu_{i,P,T,\text{melt}} - \mu^0_{i,\text{melt}}.
\tag{6.87}
$$

Table 6.15 Binary interaction parameters in Ghiorso-Carmichael model (cal/mole). $Si_4O_8 = 1$; $Ti_4O_8 = 2$; $Al_{16/3}O_8 = 3$; $Fe_{16/3}O_8 = 4$; $Fe_4Si_2O_8 = 5$; $Mn_4Si_2O_8 = 6$; $Mg_4Si_2O_8 = 7$; $Ca_4Si_2O_8 = 8$; $Na_{16/3}Si_{8/3}O_8 = 9$; $K_{16/3}Si_{8/3}O_8 = 10$; $H_2O = 11$.

	1	2	3	4	5	6	7	8	9	10
2	−29,364									
3	−78,563	−67,350								
4	2,638	−6,821	1,240							
5	−9,630	−4,595	−59,529	4,524						
6	5,525	−2,043	−1,918	212	−703					
7	−30,354	12,674	−48,675	−1,277	−57,926	−2,810				
8	−64,068	−102,442	−98,428	1,520	−59,356	699	−78,924			
9	−73,758	−101,074	−135,615	−3,717	−36,966	780	−92,611	−62,780		
10	−87,596	−40,701	−175,326	284	−84,580	−61	−45,163	−27,908	−18,130	
11	−412	−196	−71,216	−103,024	7,931	310	−20,260	−38,502	−49,213	−23,296

When the partial derivative reaches zero, the two potentials coincide and component i is present unmixed as a pure term. If unmixing takes place in solvus conditions, the thermodynamic activity of component i remains constant for the entire solvus field. However, in the case of spinodal decomposition, the activity of i within the spinodal field plots within a maximum and minimum (cf. sections 3.11 and 3.12).

For instance, figure 6.13 shows the behavior of the partial derivative of the Gibbs free energy of mixing with respect to the molar amount of Si_4O_8 component:

$$\left(\frac{\partial G_{mixing,melt}}{\partial X_{Si_4O_8,melt}} \right)_{P,T,X_j} = RT \ln a_{Si_4O_8,P,T,melt}$$

$$= 4 \left(\mu_{SiO_2,P,T,melt} - \mu^0_{SiO_2,melt} \right). \tag{6.88}$$

On the SiO_2-rich side of the diagram, spinodal decomposition (points b and d in figure 6.13) is seen at 1700 K. Ghiorso et al. (1983) showed that the topology of the Gibbs free energy of mixing surface in a multicomponent space is rather complex, with a saddle-shaped locus of spinodal decomposition. The calculations of Ghiorso et al. (1983) show that the immiscibility dome in a T-X space intersects the liquidus field of a tholeiitic magma. Table 6.16 gives the composition of a tholeiite (Rattlesnake Hill basalt) investigated by Philpotts (1979) because it was evidently an unmixed magma. Philpotts (1979) remelted the rock and was able to observe the unmixing phenomenon experimentally. During the cooling stage, two liquids were produced by unmixing at 1313 °C. Analysis of the unmixed portions indicated that one of the unmixed terms was richer in TiO_2, total iron (as FeO),

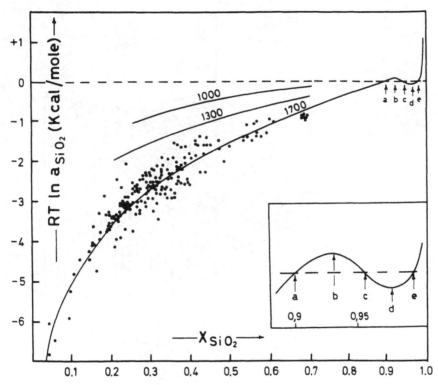

Figure 6.13 Partial derivative of Gibbs free energy of mixing of melt with respect to Si_4O_8 component plotted against molar amount of silica in melt. Enlarged area: spinodal field of critical immiscibility (spinodal decomposition). From M. S. Ghiorso and I. S. E. Carmichael, A regular solution model for meta-aluminous silicate liquids: applications to geothermometry, immiscibility, and the source regions of basic magma, *Contributions to Mineralogy and Petrology*, 71, 323–342, figure 3, copyright © 1980 by Springer Verlag. Reproduced with modifications by permission of Springer-Verlag GmbH & Co. KG.

Table 6.16 Rattlesnake Hill tholeiite (Philpotts, 1979) and its unmixes according to Ghiorso-Carmichael model (weight %; * = total iron as FeO).

Oxide	Tholeiite	Liquid 1	Liquid 2
SiO_2	52.29	45.87	53.21
TiO_2	1.17	0.00	1.26
Al_2O_3	14.75	9.06	15.28
FeO*	12.25	25.24	11.47
MnO	0.22	0.95	0.17
MgO	5.30	7.98	5.16
CaO	9.89	8.99	10.04
Na_2O	2.60	0.59	2.76
K_2O	0.33	0.00	0.36
P_2O_5	0.16	0.65	0.13

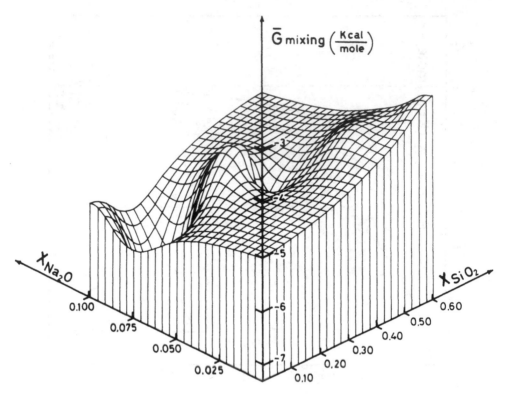

Figure 6.14 Gibbs free energy of mixing surface calculated for Archean basalts of Piumhi (Minas Gerais, Brazil) in SiO_2-Na_2O compositional space.

MnO, MgO, CaO, and P_2O_5, and the other was richer in SiO_2, Na_2O, and K_2O. Table 6.16 lists the compositions of these unmixes according to the Ghiorso-Carmichael model (also plotted in the ternary diagram of figure 6.12): the two liquids have the same compositional features described in the experiment of Philpotts (1979).

Figure 6.14 shows in stereographic form the values of the Gibbs free energy of mixing obtained for the Archean basalts of Piumhi (Minas Gerais, Brazil) on the basis of the data of Coltorti et al. (1987): note the pronounced saddle defining the spinodal decomposition field (unmix compositions also plotted in ternary diagram of figure 6.12).

The capability of evaluating with sufficient accuracy the Gibbs free energy of a silicate melt in various P-T-X conditions has obvious petrogenetic implications, besides those already outlined. For instance, the P-T loci of equilibrium of a given crystal with the melt can be determined with good approximation. Let us consider, for example, the equilibrium

$$Mg_2SiO_4 \Leftrightarrow Mg_2SiO_4 .$$
$$\text{olivine} \qquad \text{melt}$$

$$(6.89)$$

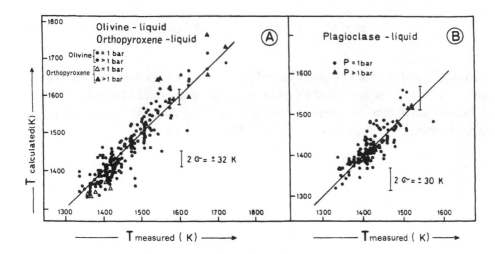

Figure 6.15 Examples of thermometric calculations based on Ghiorso-Carmichael model for silicate liquids. From M. S. Ghiorso, I. S. E. Carmichael, M. L. Rivers, and R. O. Sack, The Gibbs free energy of mixing of natural silicate liquids: an expanded regular solution approximation for the calculation of magmatic intensive variables, *Contributions to Mineralogy and Petrology,* 84, 107–145, figures 1–2, copyright © 1983 by Springer Verlag. Reprinted with the permission of Springer-Verlag GmbH & Co. KG.

For the pure component, reaction 6.89 has a constant (K_{89}) whose value under various *P-T* conditions gives the activity ratio of the magnesian component in the two phases:

$$
K_{89} = \exp\left(-\frac{\Delta \overline{G}_{89}^0}{RT}\right) = \frac{a_{Mg_2SiO_4,melt}}{a_{Mg_2SiO_4,olivine}}. \tag{6.90}
$$

For a given value of *P*, there is only one *T* condition at which equation 6.90 is satisfied. Hence, once the activity of the component in the two phases is known, the *P-T* loci of equilibrium are also known. This concept can be generalized for solid phases whose crystalline components (*M*) are stoichiometric combinations of melt components (*C*) according to

$$
M_{solid} = \sum_{i=1}^{n} v_i C_i, \tag{6.91}
$$

where the v_i terms are stoichiometric factors. The generalized calculation is

$$RT \ln K = \sum_{i=1}^{n} v_i RT \ln a_{i,\text{melt}} - RT \ln a_{i,\text{solid}}. \tag{6.92}$$

Figure 6.15 gives examples of solid-liquid geothermometry relative to the olivine-liquid (A), orthopyroxene-liquid (A), and plagioclase-liquid (B) pairs. There is satisfactory agreement between experimental and calculated values, and the thermometric procedure may thus be applied with confidence.

CHAPTER SEVEN

Introduction to Petrogenetic Systems

7.1 Phase Stability Relations in Binary Systems (Roozeboom Diagrams)

Knowledge of mineral-melt equilibria in magmas is based on experimental investigation of multicomponent synthetic systems, the simplest of which are binary joins. The interpretation of phase stability fields experimentally defined by P-T-X diagrams is straightforward, being based on equilibrium thermodynamics. The principles ruling phase stability relations are in fact only two: (1) minimization of the Gibbs free energy of the system at equilibrium, and (2) equality of chemical potentials of components in coexisting phases at equilibrium. From binary joins the extension to more complex systems is also straightforward, based on the principles listed above and on mass conservation.

Figure 7.1 shows the Gibbs free energy curves for three distinct phases that differ in mixing behavior. Phase α shows complete miscibility of components 1 and 2 in all proportions; the Gibbs free energy curve has a single downward-open concavity. The Gibbs free energy of the phase at each composition is defined by the sum of the chemical potentials of components (we recall that the chemical potential is in fact the partial derivative of the Gibbs free energy of the phase with respect to the number of moles of the component of interest):

$$\mu_{1,\alpha} = \left(\frac{\partial G_\alpha}{\partial n_{1,\alpha}} \right)_{P,T} \tag{7.1}$$

$$\mu_{2,\alpha} = \left(\frac{\partial G_\alpha}{\partial n_{2,\alpha}} \right)_{P,T} \tag{7.2}$$

$$G_\alpha = \sum_i n_{i,\alpha} \mu_{i,\alpha} \tag{7.3}$$

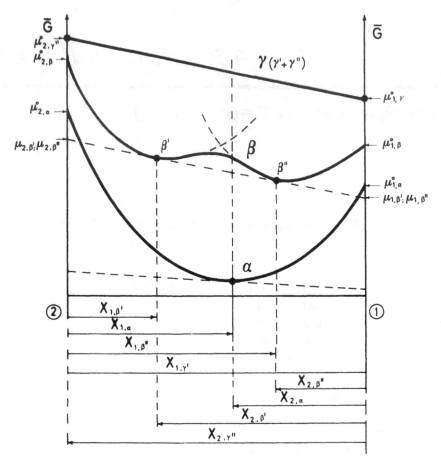

Figure 7.1 Relationships between chemical potential and composition in binary phases with different miscibility behavior: α = complete miscibility; β = partial miscibility; γ = lack of miscibility or "mechanical mixture."

$$\overline{G}_\alpha = \sum_i X_{i,\alpha}\mu_{i,\alpha} = X_{1,\alpha}\mu_{1,\alpha} + X_{2,\alpha}\mu_{2,\alpha}. \tag{7.4}$$

In equation 7.4, $\overline{G}_\alpha$ is the molar Gibbs free energy of phase α at the compositional point defined by fractional amounts $X_{1,\alpha}$ and $X_{2,\alpha}$. The Gibbs free energy curve of phase β is saddle-shaped, indicating the existence of a miscibility gap. Applying the principles of minimization of Gibbs free energy and of equality of potentials at equilibrium, we can deduce that the compositions of the two unmixed portions β' and β'' are determined by the points of tangency (and identical slope):

$$\mu_{1,\beta'} = \mu_{1,\beta''}; \mu_{2,\beta'} = \mu_{2,\beta''} \tag{7.5}$$

$$\overline{G}_{\beta'} = X_{1,\beta'}\mu_{1,\beta'} + X_{2,\beta'},\mu_{2,\beta'} \tag{7.6}$$

$$\overline{G}_{\beta''} = X_{1,\beta''}\mu_{1,\beta''} + X_{2,\beta''},\mu_{2,\beta''}. \tag{7.7}$$

Denoting $X_{\beta'}$ and $X_{\beta''}$ the molar fractions of the unmixed portions, we also have

$$\overline{G}_{\beta} = X_{\beta'}\overline{G}_{\beta'} + X_{\beta''}\overline{G}_{\beta''}. \tag{7.8}$$

The Gibbs free energy of phase γ is represented by a straight line connecting the standard state potentials of the two end-members in the mixture. Because we use the term "mixture," it is evident that the standard state of both end-members is the same and is that of "pure component." The two components are totally immiscible in any proportion and the aggregate is a "mechanical mixture" of the two components crystallized in form γ:

$$\overline{G}_{\gamma} = X_{1,\gamma}\mu_{1,\gamma}^{0} + X_{2,\gamma}\mu_{2,\gamma}^{0}. \tag{7.9}$$

Recalling the general formulation of the Gibbs free energy of a mixture:

$$G_{\text{mixture}} = \sum_{i} n_{i}\mu_{i}^{0} + G_{\text{ideal mixing}} + G_{\text{excess mixing}}, \tag{7.10}$$

it is evident that phase α has an **excess** Gibbs free energy sufficiently low (or negative) to avoid the formation of a solvus. In the case of phase β, $G_{\text{excess mixing}}$ is positive and sufficiently high to induce a zone of immiscibility. In the case of phase γ, the terms $G_{\text{ideal mixing}}$ and $G_{\text{excess mixing}}$ are mutually compensated over the whole compositional range.

7.1.1 Crystallization of Phases with Immiscible Components

Let us call the melt phase α and the solid phase with complete immiscibility of components γ. P is constant and fluids are absent. The Gibbs free energy relationships at the various T for the two phases at equilibrium are those shown in figure 7.2, with T decreasing downward from T_1 to T_6. The G-X relationships observed at the various T are then translated into a T-X stability diagram in the lower part of the figure.

At T_1, the Gibbs free energy of phase α (i.e., melt) at all compositions is lower than that of mechanical mixture $\gamma' + \gamma''$: phase α is then stable over the whole compositional range. At T_2, the chemical potential of component 1 in α is identical to the chemical potential of the same component in γ'. Moreover, the equality condition is reached at the standard state condition of the pure component: T_2 is thus the temperature of incipient crystallization of γ'. At T_3, the Gibbs free energy of α intersects mechanical mixture $\gamma' + \gamma''$ on the component 1–rich side of the diagram and touches it at the condition of pure component 2. Applying the prin-

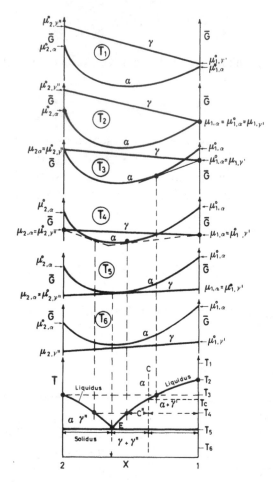

Figure 7.2 *G-X* and *T-X* plots for a binary system with a molten phase with complete miscibility of components at all *T* conditions and a solid phase in which components are totally immiscible at all proportions (mechanical mixture, $\gamma = \gamma' + \gamma''$).

ciples of equality of potentials at equilibrium and Gibbs free energy minimization, we can deduce that the zone of coexistence of α and γ' is given by the straight segment originating at $\mu_{\gamma'}^0$ and tangent to α. Moreover, T_3 is the temperature of incipient crystallization of γ''. With decreasing T, the Gibbs free energy curve of α is progressively moves upward (i.e., is less negative) and, consequently, the two-phase fields identified by the tangency zones in the potential vs. composition plot progressively expand (see condition T_4 in figure 7.2). At T_5, we reach a peculiar *T-X* condition, the *eutectic point,* where the Gibbs free energy of the mechanical mixture is identical (at the single point of tangency) to that of the melt (phase α). At this point, the three phases α, γ', and γ'' coexist stably at equilibrium. Note that, because there are two components in the system, based on the Gibbs phase rule the equilibrium is invariant: i.e., the eutectic point is defined at a single eutectic temperature. (The term "eutectic" was coined by Guthrie, 1884, to identify the lowest temperature of melting of the system.) At T_6, the Gibbs free energy of the mechanical mixture is lower than that of phase α over the entire binary range: only the mechanical mixture $\gamma' + \gamma''$ is stable at equilibrium.

Let us now consider the cooling process of a system of composition C in figure 7.2. At high T, the system is completely molten (phase α, condition T_1). As T is progressively lowered, condition T_c is reached, at which γ' begins to crystallize. The crystallization of γ' enriches the composition of the residual melt of component 2. At T_4, the melt has composition C″, while crystals γ' continue to form. The molar proportions of crystals and melts may be deduced by application of baricentric coordinates at temperature T_4 (relative lengths of the two compositional segments in the two-phase field $\alpha + \gamma'$, or "lever rule"). We reach eutectic point E at T_5: phase γ'' forms at equilibrium in stable coexistence with α and γ'; the temperature now remains constant during the crystallization of γ'', until all the residual melt is exhausted.

At T_6, the melt is absent and γ' and γ'' coexist stably in a mechanical mixture of bulk composition C. The crystallization process in a closed system is thus completely defined at all T (and P) conditions by the Gibbs free energy properties of the various phases that may form in the compositional field of interest.

7.1.2 The Diopside-Anorthite System as an Example

Figure 7.3 depicts phase stability relations in the pseudobinary system $CaMgSi_2O_6$-$CaAl_2Si_2O_8$ (diopside-anorthite). The original study of Bowen (1915) described crystallization behavior identical to the previously discussed case: a mechanical mixture (Di-An) in equilibrium with a completely miscible melt. A later investigation (Osborn, 1942) showed that the system is not strictly binary

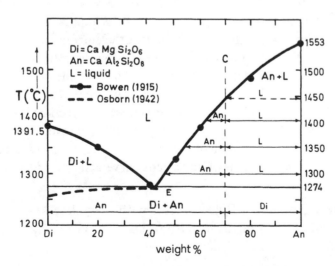

Figure 7.3 $CaMgSi_2O_6$-$CaAl_2Si_2O_8$ (diopside-anorthite) system, after Bowen (1915) and Osborn (1942). Abscissa axis: percentile weight of both components.

(i.e., cannot be defined simply on the basis of the two components $CaMgSi_2O_6$ and $CaAl_2Si_2O_8$), but is rather "pseudobinary," because a limited amount of Al_2O_3 may be dissolved in $CaMgSi_2O_6$. However, we will neglect this limited solubility here and will hereafter treat the system as purely binary.

Pure components $CaMgSi_2O_6$ and $CaAl_2Si_2O_8$ at $P = 1$ bar melt, respectively, at $T_{f,Di} = 1391.5$ °C and $T_{f,An} = 1553$ °C. The eutectic temperature is 1274 °C. The progressive decrease of incipient crystallization, from 1553 °C to 1274 °C on the An-rich side of the join and from 1391.5 °C to 1274 °C on the Di-rich side, is a result of the lowering of the freezing point expression (see chapter 6: eq. 6.56, section 6.4). If we cool a system of initial composition C, we observe incipient crystallization at $T = 1445$ °C (figure 7.3). At this temperature, the entire system is still in the liquid form. At $T = 1400$ °C, some anorthite crystals are already formed. We can determine their weight amounts by applying baricentric coordinates (or the "lever rule") to compositional segments An and L in figure 7.3—i.e.,

$$X_{An} = \frac{An}{An + L} = \frac{1}{1 + 4} = 0.20 \tag{7.11}$$

$$X_L = \frac{L}{An + L} = \frac{4}{1 + 4} = 0.80. \tag{7.12}$$

Note that the resulting fractional amounts are in weight percent, because the abscissa axis of the phase diagram reports the fractional weights of the two components (similar application of baricentric coordinates to a molar plot of type 7.2 would have resulted in molar fractions of phases in the system). Applying the lever rule at the various T, we may quantitatively follow the crystallization behavior of the system (i.e., at $T = 1350$ °C, $X_{An} \approx 0.333$ and $X_L \approx 0.666$; at $T = 1300$ °C, $X_{An} \approx 0.44$ and $X_L \approx 0.56$; and so on). It is important to recall that the baricentric coordinates lose their significance at the eutectic point, because solid/melt proportions depend on time and on the dissipation rate of the heat content of the system. Below the eutectic temperature, the lever rule can again be applied to the couple Di-An, and the resulting proportions are those of the initial system C ($X_{An} = 0.70$, $X_{Di} = 0.30$). For a *closed system* (see section 2.1), the melting behavior of an initial composition C exactly follows the reverse path of the crystallization process (i.e., beginning of melting at $T = 1274$ °C/displacement along univariant curve/abandonment of univariant curve at $T = 1450$ °C).

7.1.3 Crystallization of a Binary Phase with Complete Miscibility of Components (Roozeboom Types I, II, and III)

The intersections of the Gibbs free energy curves of the melt with those of the solid mixture of components 1 and 2 at the various T (both phases having com-

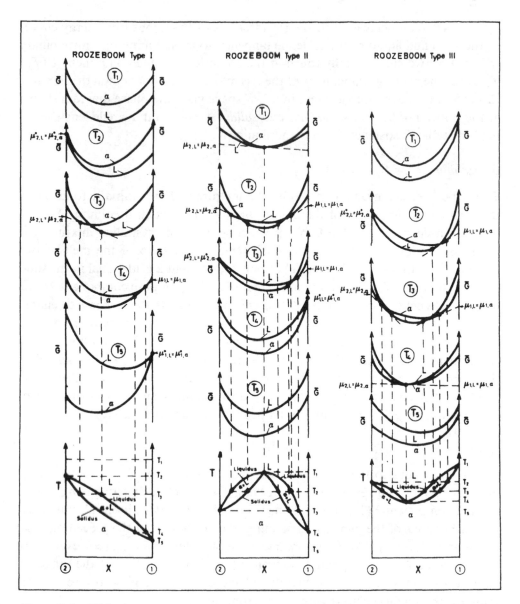

Figure 7.4 Gibbs free energy curves and phase stability relations for two binary mixtures with complete miscibility of components (types I, II, and III of Roozeboom, 1899).

plete miscibility) may give rise to three distinct configurations of the phase diagram, as outlined in figure 7.4.

Roozeboom Type I

Roozeboom type I is obtained whenever the Gibbs free energy curve of the melt initially touches that of the mixture in the condition of pure component. This takes place at the melting temperature of the more refractory component (T_2 in figure 7.4). With decreasing T, the intersections of the Gibbs free energy curves of the solid and liquid mixtures define tangency zones spanning the entire binary range. The two curves finally touch at the condition of pure component 1 (T_5): T_5 is thus the melting temperature of the less refractory component in the mixture. Projection of the two-phase zones in a T-X space gives the phase diagram shown at the bottom of figure 7.4: *liquidus* and *solidus* curves limit a single lens-shaped field of stable coexistence of solid and liquid ($\alpha + L$).

Roozeboom Type II

The Gibbs free energy curves of both solid mixture and melt initially touch at an intermediate composition of the binary join (T_1 in the central column of figure 7.4). With decreasing T, the Gibbs free energy loop of crystalline mixture α is progressively lowered, crossing the Gibbs free energy loop of the melt in two points. At T_3, the chemical potential of pure component 2 is identical in the solid and liquid phases. At this temperature (that of melting of pure component 2), the Gibbs free energy curve of the solid phase still intersects the Gibbs free energy loop of the melt in another part of the join. At T_4 (melting temperature of pure component 1), the two Gibbs free energy curves intersect at the condition of pure component 1. At lower T, the entire Gibbs free energy loop of the solid lies below the Gibbs free energy curve of the melt: the entire system is crystallized. The resulting T-X phase diagram (figure 7.4, bottom center) shows two distinct two-phase lenses ($\alpha + L$) departing from a peak on the liquidus curve.

Roozeboom Type III

This is analogous to the preceding case, but the sequence of conditions T_1, T_2, and T_3 of type II is inverted. With decreasing T, we reach the melting temperature of pure component 2 (T_2); at this T, the two Gibbs free energy loops also intersect in another zone of the join (corresponding to condition T_3 of the preceding case, see figure 7.4, center). At T_3, we observe a double intersectioning (corresponding to T_2) and at T_4 a single point of tangency (i.e., T_1). Projection of the tangency zones on a T-X space leads to two two-phase lenses converging toward a minimum on the solidus curve (figure 7.4, bottom right).

7.1.4 The Albite-Anorthite (Roozeboom Type I) and Albite-Sanidine (Roozeboom Type II) Systems as Examples

Figure 7.5 shows phase stability relations in the binary system $NaAlSi_3O_8$-$CaAl_2Si_2O_8$ (albite-anorthite), based on the experiments of Bowen (1913), Greig and Barth (1938), and Osborn (1942).

Pure anorthite (An) at $P = 1$ bar melts at $T = 1553$ °C (Osborn, 1942), and pure albite (Ab) melts at $T = 1118$ °C (Greig and Barth, 1938). The phase stability relations in the Ab-An join are Roozeboom type I. Cooling a melt of molar

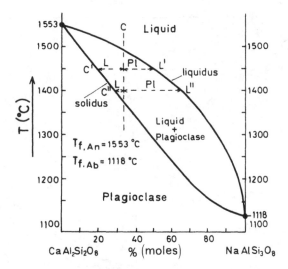

Figure 7.5 Phase stability relations in binary join $CaAl_2Si_2O_8$-$NaAlSi_3O_8$ (anorthite-albite).

composition Ab_{33} (C in figure 7.5), we observe incipient crystallization at $T = 1500\ °C$. As crystallization advances, the residual liquid becomes more and more enriched in the Ab component. The molar proportions of the solid and liquid mixtures can be derived by application of the lever rule at the T of interest (segments L and Pl in figure 7.5). That is, at $T = 1450\ °C$, the molar proportions of the two mixtures are approximately 57% plagioclase and 43% melt, and the melt has composition L′ and is in equilibrium with a solid of composition C′. At $T = 1400\ °C$, the proportion of liquid (L″ in figure 7.5) decreases to about 10% (90% of solid with composition C″) and completely disappears at about 1375 °C. Below this temperature, only the solid mixture is stable and has a molar composition corresponding to the initial system C.

Figure 7.6 shows the phase stability relations in the $NaAlSi_3O_8$-$KAlSi_3O_8$ (Ab-Sa) system. Part A of the figure has a restricted T range, to facilitate comprehension of metastability phenomena on the K-rich side of the diagram. In part B, the temperature scale is more extended, to encompass subsolidus relations.

The system is strictly binary only for $T < 1078\ °C$. In K-rich compositions at higher T, the sanidine solid mixture ($Sa_{s.m.}$) is metastable with respect to the leucite solid mixture $(K, Na)AlSi_2O_6$ ($Le_{s.m.}$). On the Na-rich side of the join ($Ab_{s.m.}$), the crystallization trend of an initial liquid mixture of composition C1 (80% Ab by weight) is similar to that of the previous example. On the K-rich side of the join, cooling of an initial composition C2 (50% Ab by weight) leads to initial formation of $Le_{s.m.}$ crystals; the residual liquid moves along the liquidus curve until it reaches the peritectic reaction point P ($T = 1078\ °C$). At P, $Le_{s.m.}$ converts into $Sa_{s.m.}$ while T remains constant. As shown in figure 7.6B, the $NaAlSi_3O_8$-$KAlSi_3O_8$ system in subsolidus conditions shows limited solubility

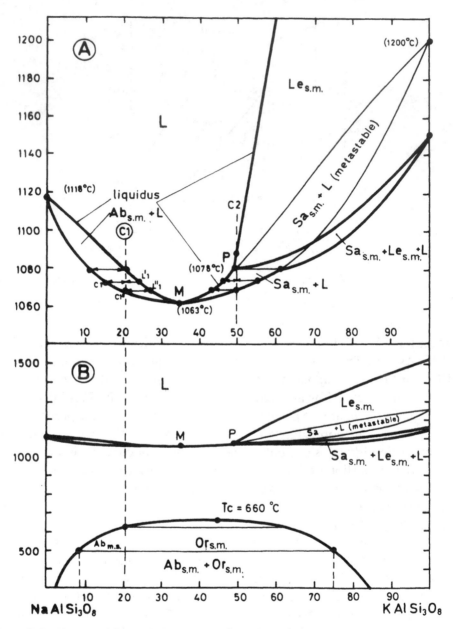

Figure 7.6 Phase stability relations in pseudobinary system $NaAlSi_3O_8$-$KAlSi_3O_8$, after Waldbaum and Thompson (1969). (A) High-T relations: loop of metastable persistency of sanidine like a Roozeboom type III. (B) Expanded T-range: with decreasing T, a solvus field opens downward.

with a consolute temperature $T_c = 660$ °C, below which the homogeneous solid mixture $(Na, K)AlSi_3O_8$ unmixes into an Na-rich $(Ab_{s.m.})$ and a K-rich $(Or_{s.m.})$ composition. This further complexity introduces us to the third type of phase relations.

7.1.5 Crystallization of Phases with Limited Miscibility (Roozeboom Types IV and V)

There are two possible geometrical configurations between a mixture with a saddle-shaped Gibbs free energy curve (indicative of unmixing) and a phase with a concave curve, indicating complete miscibility of components (figure 7.7).

Roozeboom Type IV

The Gibbs free energy curve of the solid phase with partial miscibility initially touches the curve of the liquid in the condition of pure component 1. This happens at the melting temperature of that component (i.e., $T_1 = T_{f,1}$). With decreasing T, the Gibbs free energy loop of the solid is progressively displaced downward (i.e., becomes more negative) and the equality conditions of potentials of components 1 and 2 in the liquid and in unmixing lobe β'' are reached first (T_2, T_3). As T further decreases, a particular condition (T_4) is reached: the chemical potentials of the two components are identical in the liquid and in the two lobes β' and β'': the three phases β', β'', and L coexist stably at equilibrium. At T_5, the solvus of the solid expands (tangency segment β'-β'') and there still exists a solid-liquid zone of coexistence on the component 2–rich side of the join (segment L-β'). At T_6, the chemical potential of pure component 2 is identical in the solid and liquid phases: T_6 is hence the melting temperature of the less refractory component in the join $(T_6 = T_{f,2})$. Below this temperature, only the solid phases are stable: $\beta' + \beta''$ within the solvus field, and the homogeneous mixture β outside it. Projection of the various isothermal G-X geometries onto the T-X space gives the phase diagram (figure 7.7, bottom right). The resulting phase stability relations may be conceived as a combination of a Roozeboom type I diagram with a solvus field intersecting it along the solidus curve.

Roozeboom Type V

Initially like Roozeboom type IV, Roozeboom type V differs from it at temperature T_5, where the tangency condition of the two unmixing lobes β' and β'' and liquid L finds the Gibbs free energy curve of the liquid in an intermediate position between the unmixing lobes, rather than in an external position, as in the previous case. The resulting T-X stability relations may be conceived as a Roozeboom type III diagram with a solvus field intersecting it along the solidus curve (cf. figures 7.6 and 7.7).

Lastly, let us consider the last column in figure 7.7, depicting a case in which

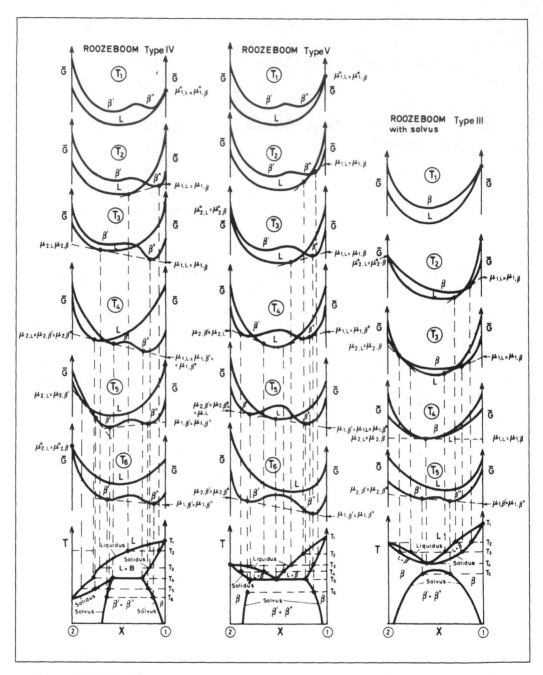

Figure 7.7 Gibbs free energy curves and T-X phase stability relations between a phase with complete miscibility of components (silicate melt L) and a binary solid mixture with partial miscibility of components (crystals β).

the compounds are completely miscible in the solid phase at high T, but become partially immiscible with decreasing T (this is the general case, from the thermochemical point of view, for the crystalline solids of our interest). Because unmixing begins at T_5, where the Gibbs free energy of the solid is entirely above that of the liquid phase, the solvus does not intersect the solidus line in the resulting T-X plot. We thus have a Roozeboom type III diagram with a solvus opening at subsolidus conditions, which was the case in figure 7.6B for the $NaAlSi_3O_8$-$KAlSi_3O_8$ system.

The unmixing fields shown in figure 7.7 are *solvi* (i.e., defined by *binodes* at the various T), but in nature we may also observe the metastable persistency of a homogeneous phase until a critical condition of unmixing is reached. The location of a *spinodal field* on a T-X diagram follows the guidelines outlined in section 3.11—i.e., the *spinodes* at the various T plot on the inflection point on the Gibbs free energy curve of the solid. Spinodal fields thus occupy an inner and more restricted portion of the T-X space with respect to the corresponding solvi.

7.1.6 Formation of Crystalline Compounds with Fixed Stoichiometry

Let us now consider a case in which components 1 and 2 form an intermediate crystalline compound (C) with precise, invariant stoichiometry (e.g., 60% of component 2 and 40% of component 1). If the chemical composition of the intermediate compound is fixed, it behaves as a mechanical mixture with respect to its pure components (i.e., zero miscibility). The presence of the intermediate compound subdivides the compositional join into two fields: mechanical mixture 2-C (γ'') and mechanical mixture C-1 (γ'). The resulting crystallization path may assume two distinct geometrical configurations, as shown in figure 7.8.

In the first column of the figure, we see the energy conditions leading to the formation of two eutectic points separated by a thermal barrier. At $T_1 = T_{f,1}$, the Gibbs free energy curve of the liquid touches the curve of the solid in the condition of pure component 1. At T_2, the chemical potential of pure crystalline component 1 is lower than that of pure liquid 1, and a zone of tangency is thus defined. As T further decreases, the Gibbs free energy curve of the liquid touches intermediate component C: at this temperature, compound C begins to crystallize (i.e., $T_3 = T_{f,C}$). At T_5, we observe the first invariant point: the potentials of the two components coincide in liquid L, intermediate compound C, and pure component 1; T_5 is thus the eutectic temperature for the 1-C side of the join. At T_6, we observe the second invariant condition, this time for the C-2 side of the join. The resulting T-X phase relations may be conceived as a simple combination of two phase diagrams relative to crystallization of mechanical mixtures γ'' and γ' (compare figure 7.8, bottom left, and figure 7.2).

Column 2 of figure 7.8 defines the Gibbs free energy relations leading to a

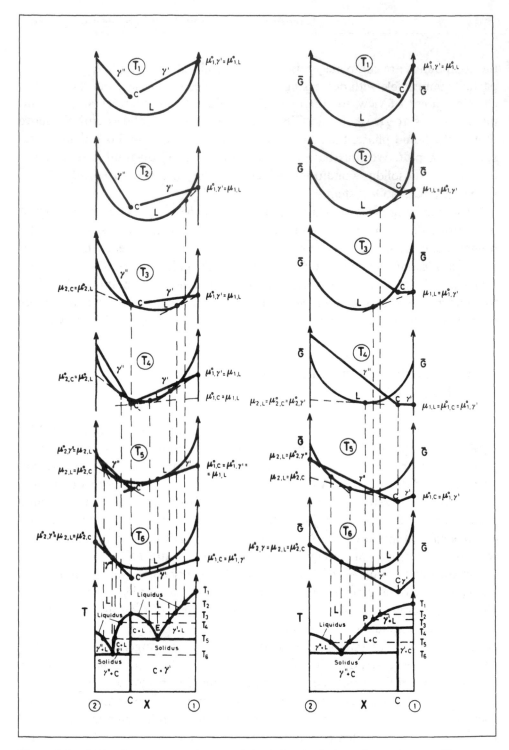

Figure 7.8 Gibbs free energy curves and T-X phase relations for an intermediate compound (C), totally immiscible with pure components. Column 1: Gibbs free energy relations leading to formation of two eutectic minima separated by a thermal barrier. Column 2: energy relations of a peritectic reaction (incongruent melting). To facilitate interpretation of phase stability fields, pure crystals of components 1 and 2 coexisting with crystals C are labeled γ' and γ'', respectively, in T-X diagrams; same notation identifies mechanical mixtures 2-C and C-1 in G-X plots.

peritectic crystallization event. At T_1, the Gibbs free energy of crystalline component 1 is identical to that of the liquid (i.e., $T_1 = T_{f,1}$). With decreasing T (T_2, T_3), zones of tangency between the Gibbs free energy curves of the liquid and component 1 are observed, but the chemical potential of intermediate compound C decreases at the same time. Lastly, at $T_4 = T_{f,C}$, we reach an invariant condition with the coexistence of three phases: compound C, liquid L, and the pure crystals of component 1. A peritectic reaction now takes place at constant T: the pure crystals of component 1 react with the liquid, forming intermediate compound C. Only when pure crystals 1 are completely exhausted does the temperature again decrease. This decrease leads to another invariant point: the eutectic point of the system, at which the liquid coexists stably with C and the pure crystals of component 2 (T_6). Below this T, the liquid is no longer stable and only the two solid phases (crystals 2 + C) are stable at equilibrium. The resulting phase diagram is shown at the bottom of figure 7.8.

7.1.7 Examples of Formation of Intermediate Compounds with Fixed Stoichiometry: Nepheline-Silica and Kalsilite-Silica Joins

Figure 7.9A shows the $NaAlSiO_4$-SiO_2 (nepheline-silica) system, after Schairer and Bowen (1956). Let us first examine the SiO_2-rich side of the join. At $P = 1$ bar, the pure component SiO_2 crystallizes in the cristobalite form (Cr) at $T = 1713$ °C (cf. figure 2.6). At $T = 1470$ °C, there is a phase transition to tridymite (Tr), which does not appreciably affect the form of the liquidus curve, which reaches the eutectic point at $T = 1062$ °C.

In the central part of the diagram, two moles of SiO_2 react with one mole of $NaAlSiO_4$ to form albite (Ab):

$$NaAlSiO_4 + 2\,SiO_2 \;\Leftrightarrow\; NaAlSi_3O_8 \;.$$

nepheline   albite

$$(7.13)$$

$NaAlSiO_4$ is immiscible with either SiO_2 or $NaAlSi_3O_8$ (in other words, its stoichiometry is fixed in the binary join of interest). When expressed in weight percent, equation 7.13 involves 54% $NaAlSiO_4$ and 46% SiO_2 (the original diagram of Schairer and Bowen, 1956, like most phase diagrams of that period, uses a weight scale rather than a molar range, simply because mixtures were dosed in weight amounts). Crystallization of any composition within $NaAlSi_3O_8$ and SiO_2 follows the classical behavior of a mechanical mixture, as discussed in section 7.1.1 (cf. figure 7.2). For example, an initial composition C1 in figure 7.9 (corresponding to 50% $NaAlSiO_4$ and 50% SiO_2 by weight) crystallizes at about 1110 °C, forming $NaAlSi_3O_8$ (Ab). With decreasing T, the progressive crystallization of Ab displaces the composition of the residual liquid toward eutectic point E′, which is

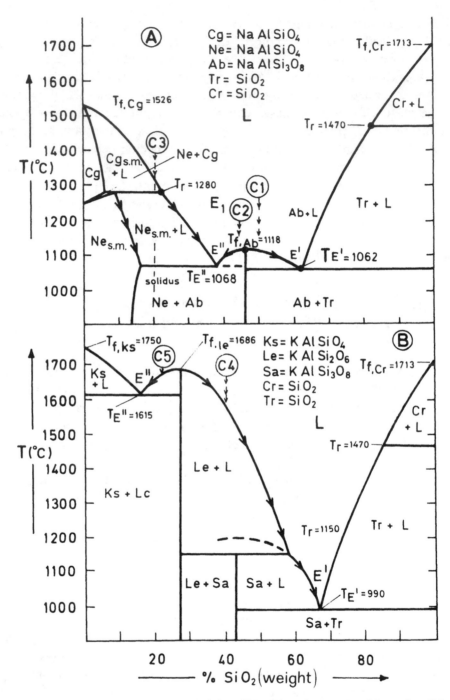

Figure 7.9 NaAlSiO$_4$-SiO$_2$ (A) and KAlSiO$_4$-SiO$_2$ (B) system at $P = 1$ bar, after Schairer and Bowen (1955, 1956).

reached at 1062 °C. The crystallization of an initial composition C2, slightly poorer in SiO_2 (i.e., 45%), displaces the composition of the residual liquid in the opposite direction: the liquid becomes more and more depleted in SiO_2 as crystallization proceeds, until eutectic point E″ is reached at $T = 1068$ °C. Although the system may be treated as binary for SiO_2 amounts exceeding 20% by weight, phase relations are more complex for less acidic compositions. Crystallizing pure $NaAlSiO_4$ in the laboratory at $T = 1526$ °C leads to the formation of the crystalline polymorph carnegieite (Cg), which is not observed in nature. Carnegieite is indeed a solid mixture ($Cg_{s.m.}$ in figure 7.9A), accepting limited amounts of SiO_2 in solution. At the peritectic temperature $T_r = 1280$ °C, the solid mixture carnegieite is transformed into solid nepheline. Nepheline, too, is a solid mixture ($Ne_{s.m.}$) whose maximum amount of SiO_2 is reached at eutectic point E″ ($T = 1068$ °C). Note also in figure 7.9A that the two-phase fields $Cg_{s.m.}$ + L and $Ne_{s.m.}$ + L are bounded on the left by solidus curves separating them from the corresponding single-phase fields Cg and $Ne_{s.m.}$. Moreover, a limited two-phase field Ne + Cg occurring between the single-phase fields Cg and $Ne_{s.m.}$ is observed experimentally at subsolidus. The $Ne_{s.m.}$ + L field is limited downward by the solidus curve at eutectic point E″ ($T = 1068$ °C), below which we find the Ne + Ab solvus.

Let us now follow the crystallization of an initial composition C3 (20% Ne and 80% SiO_2 by weight). The solid mixture $Cg_{s.m.}$ initially forms at about 1300 °C. As crystallization proceeds, the liquid becomes enriched in SiO_2, moving along the liquidus curve. At $T = 1280$ °C, crystallization stops until the congruent reaction $Cg_{s.m.} \Leftrightarrow Ne_{s.m.}$ is completed (there is invariance because there are three phases and two components). Once the congruent reaction is complete, crystallization proceeds again: both the solid mixture $Ne_{s.m.}$ and the liquid become more and more enriched in SiO_2 (moving along solidus and liquidus curves, respectively) until eutectic point E″ is reached and the remaining liquid is converted into albite plus nepheline.

Figure 7.9B shows the binary diagram for the $KAlSiO_4$-SiO_2 (kalsilite-silica) system. On the SiO_2-rich side of the join, phase boundaries are similar to those of their sodic counterpart, but two compounds with fixed stoichiometry form in the intermediate zone: sanidine ($KAlSi_3O_8$; Sa) and leucite ($KAlSi_2O_6$; Le). Crystallising an initial composition C4 (40% SiO_2 and 60% $KAlSiO_4$ by weight), we first observe the production of leucite crystals ($T = 1580$ °C). The liquid becomes more and more enriched in SiO_2 as crystallization proceeds, until the invariant peritectic point is reached at T_r and leucite is transformed into sanidine by the incongruent reaction

$$\underset{\text{leucite}}{KAlSi_2O_6} + SiO_2 \Leftrightarrow \underset{\text{sanidine}}{KAlSi_3O_8} .$$

$$(7.14)$$

However, all the SiO_2 component in the liquid is exhausted before reaction 7.14 is completed, and the resulting subsolidus assemblage is composed of Le + Sa crystals. If the initial molar amount of SiO_2 in the liquid is higher than the value dictated by the molar proportion

$$KAlSiO_4 : SiO_2 = 1 : 2, \tag{7.15}$$

peritectic reaction 7.14 achieves completion and the crystallization process continues with the production of sanidine at eutectic point E'. If, instead, the initial molar amount of SiO_2 in the liquid is lower than the value dictated by the molar proportion

$$KAlSiO_4 : SiO_2 = 1 : 1 \tag{7.16}$$

(as, for instance, C5 in figure 7.9B), crystallization proceeds with the formation of liquids increasingly impoverished in the SiO_2 component until eutectic E" is reached.

Comparing figures 7.9 and 7.8, we can fully understand the nature of phase stability relations in the examined systems: the binary $KAlSiO_4$-SiO_2 system is a simple combination of the two configurations displayed in figure 7.8, whereas the $NaAlSiO_4$-SiO_2 system shows some additional complexities on the Na-rich side of the join. However, it is important to stress that the apparent complexity of T-X phase stability diagrams always arises from the (quite simple) configuration of the Gibbs free energy curves in the various phases forming in the system. This indicates how important it is in geochemistry to establish accurately the mixing properties of the various phases involved in petrogenetic equilibria.

We have seen that the crystallization of initial compositions respectively poorer (C2) and richer (C1) in SiO_2 than $NaAlSi_3O_8$ stoichiometry follows opposite paths in the binary $NaAlSiO_4$-SiO_2 system. This occurs essentially because compound $NaAlSi_3O_8$ melts at a local T maximum (or, in thermochemical terms, with decreasing T the chemical potential of $NaAlSi_3O_8$ reaches the Gibbs free energy curve of the liquid before the two eutectic configurations; cf. figure 7.8). Analogously, in the $KAlSiO_4$-SiO_2 system, we have a local T maximum ($T = 1686$ °C) at the melting temperature of the stoichiometric compound $KAlSi_2O_6$.

These apparently insignificant facts actually have profound consequences for magma petrology, because they constitute thermal barriers below which magma differentiation proceeds in opposite chemical directions (silica undersaturation or silica oversaturation), depending on the initial composition of the system.

7.2 Extension to Ternary Systems

7.2.1 Ternary Mechanical Mixtures

In petrology, the usual representation of phase equilibria in the crystallization process of ternary systems is based on a projection from the T axis onto the compositional plane. The latter is triangular, with each side representing a binary join of the ternary system. The projection shows the liquidus surface and, at each point of the diagram, highlights the condition of incipient crystallization of the melt. To facilitate interpretation of the crystallization path followed by any initial composition in the *T-X* field, the isotherms are drawn on the liquidus surface.

Figure 7.10 shows a ternary diagram generated by the projection from the *T* axis of the phase limits for three mechanical mixtures of components 1, 2, and 3 (1–2, 1–3, and 2–3, respectively). Crystallization of an initial composition C1 will follow the path shown in figure 7.10. Composition C1 is within "primary phase field" γ''. With decreasing *T*, once the value of the isotherm passing through C1 is reached, crystallization begins with the formation of γ'' crystals of pure component 2. As a result of crystallization of γ'', the composition of the residual liquid moves away from C1 along the direction obtained by connecting corner 2 with the initial composition. During this step, the equilibria are divariant (because *P* is fixed, the variance is given by: number of components − number of phases + 1). Once the residual liquid reaches cotectic line E-E$_{III}$, determined by the intersection of primary phase field γ'' with primary phase field γ', crystals γ' (com-

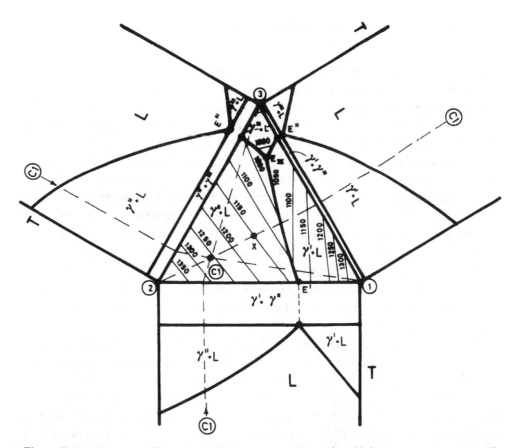

Figure 7.10 Phase stability relations in a ternary system in which components are totally immiscible at solid state, and relationships with three binary joins.

posed of pure component 1) begin to form in conjunction with γ''. The composition of the residual liquid now moves along the cotectic line in a univariant equilibrium (at each T, the chemistry of the residual liquid is now fixed, independent of initial composition). Crystallization finally ends at ternary eutectic point E_{III}, where pure crystals γ''' (component 3) are formed in conjunction with γ' and γ'' until the residual liquid is completely exhausted.

Note that the crystallization path in the T-X space does not necessarily follow the steepest descent (i.e., the trajectory is not necessarily orthogonal to the direction of the isotherms), but is determined by conjunction 2-C1 between the primary phase and the initial composition. In the case of mechanical mixtures, the trajectory in the primary phase field is thus always straight. Note also that the amount of crystals at each T is dictated by the baricentric coordinates: at $T =$ 1150 °C (point X in figure 7.10), the amount of crystals (in percent by weight or percent in moles, depending on the scale used) is given by the ratio of C1-X and 2-C1 segments. A melting process in a closed system on a ternary diagram such as that shown in figure 7.10 will follow the T-X path exactly opposite to that of the crystallization event.

7.2.2 Formation of an Intermediate Compound with Fixed Stoichiometry

Figure 7.11 shows a ternary diagram for a system with components 1–3 and 2–3 immiscible in all proportions at solid state. However, here components 1 and 2 form an intermediate compound with fixed stoichiometry (ratio 30:70).

Let us examine the crystallization process of an initial composition C1: the first crystals to form are those of the pure component 2 (γ'', $T =$ 1175 °C). Crystallization of γ'' drives the residual liquid radially away from corner 2 along direction 2-C1, until *peritectic curve* PP is reached. Here, crystals γ'' react with the liquid and are transformed into intermediate compound C_i. Displacement along the peritectic curve terminates when all phase γ'' is completely resorbed. This occurs geometrically when trajectory C_i-C1 encounters peritectic curve PP. Crystallization of C_i then proceeds, while the residual liquid moves radially away from C_i. The liquid path then reaches cotectic curve E'-E_{III} and crystals γ' begin to precipitate in conjunction with C_i. Crystallization continues along the cotectic line and, at ternary eutectic E_{III}, crystals γ''' precipitate together with C_i and γ' until the residual liquid is exhausted.

Let us now consider the crystallization process operating on a composition C2, initially poorer in component 1: this composition falls into the compositional triangle C_i-2-3. Crystallization begins in primary phase field γ'' ($T =$ 1225 °C). Precipitation of γ'' crystals drives the residual liquid radially away from C2 along direction 2-C2. At $T =$ 1100 °C, the composition of the liquid reaches cotectic line E_{III}-P, where crystals γ''' begin to form together with γ''. At peritectic point P, crystals γ'' are partially resorbed by reaction with the liquid to form intermediate compound C_i. However, the peritectic reaction is not completed, because of ex-

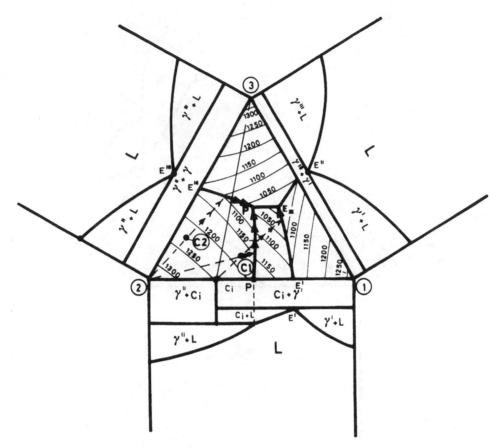

Figure 7.11 Phase stability relations in a ternary system with components 1–3 and 2–3 totally immiscible at solid state and components 1–2 forming an intermediate compound with fixed stoichiometry (C_i).

haustion of the residual melt, and at the end of crystallization the solid assemblage is composed of crystals γ'', γ''', and C_i.

The compositional paths of liquids for melting processes in closed systems of initial composition C1 or C2 are exactly the opposite of those observed in crystallization events.

7.2.3 Ternary Systems with Phases of Differing Miscibility

Figure 7.12 shows the liquidus surface of a ternary system with complete miscibility at solid state between components 1–2 and complete immiscibility at solid state between components 1–3 and 2–3. Note also that components 1 and 2 form a lens-shaped two-phase field, indicating ideality in the various aggregation states (Roozeboom type I).

The ternary system of interest closely resembles the phase relations observed in the $CaAl_2Si_2O_8$-$CaMgSi_2O_6$-$NaAlSi_3O_8$ system. We will examine this system in some detail later, when dealing with the concepts of *fractional melting* and *frac-*

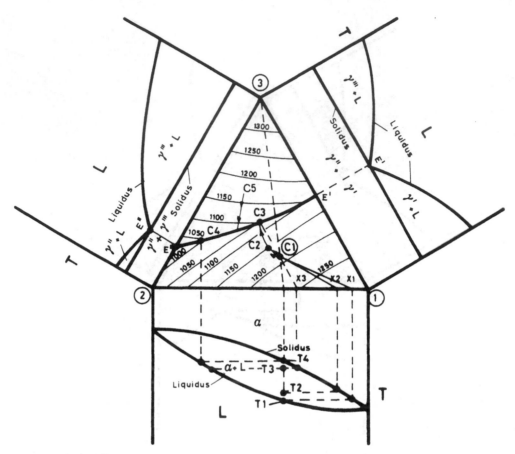

Figure 7.12 Phase stability relations in a ternary system with complete solid state miscibility of components 1–2 and complete immiscibility at solid state for components 1–3 and 2–3.

tional crystallization. For the moment, we limit ourselves to describing crystallization in a closed system (called "equilibrium crystallization" in geochemical literature) of generic composition C1 (component 1 = 50%; component 2 = 35%; component 3 = 15%). Unlike the preceding cases, in which the components were immiscible at solid state and the crystallization of pure components resulted in straight paths in the *T-X* space (see figures 7.10 and 7.11), here we have complete miscibility of components 1 and 2 in all aggregation states. The formation of compositionally continuous 1–2 crystalline mixtures results in a curved *T-X* trajectory for the residual liquid, the curvature depending on the shapes of the liquidus and solidus surfaces of the system. If we progressively cool the initial composition C1, at *T*1 = 1180 °C, we encounter the liquidus surface, and crystals X1 begin to form. The residual liquid moves radially away from X1 in the direction X1-C1. The newly formed crystals have compositions more enriched in component 2 (X2), and the liquid (C2) moves radially away from C2 along X2-C2. The entire compositional path of the liquid is the result of the various instantaneous compositional vectors at each crystallization step and has the curved shape shown in figure 7.12. This curved trajectory brings the residual liquid to cotectic line E'-E'' at point C3, and the crystals at equilibrium with the liquid have composition X3. Crystallization continues along the cotectic line with simultaneous precipitation of the 1–2 crystalline mixture and of crystals γ''' composed of pure component 3. Crystallization terminates at point C4, where the crystalline mixture 1–2 has composition X4, coinciding (in terms of relative proportions of components 1 and 2) with initial liquid C1. If the initial composition of the liquid falls in primary phase field γ''' (C5), the initial crystallization path is initially straight until it intersects the cotectic line. Crystallization then proceeds as in the preceding case.

7.3 Crystallization and Fusion in an Open System: Diopside-Albite-Anorthite as an Example

In all the preceding cases, crystallization takes place in a closed system. The loss of heat toward the exterior causes a decrease in temperature, total pressure and bulk chemistry being equal, and this modifies the Gibbs free energy relationships among the various phases of the system under the new P and T conditions. In petrology, however, it is also of interest to investigate the effects of the combined exchange of matter and energy toward the exterior. The processes taking place in this case are usually defined as "fractional crystallization" (loss of energy and matter) and "fractional melting" (gain of energy, loss of matter), whereas the corresponding terms "equilibrium crystallization" and "equilibrium melting" are applied when there is no exchange of matter with the exterior. The counterposition "fractionation-equilibrium" implicit in this terminology may give rise to confusion, and it is worth stressing here that fractional melting and fractional crystallization processes also take place at equilibrium, but that equilibrium is local. We will therefore use the corresponding terms "melting in an open system" and "crystallization in an open system" to avoid any ambiguity.

Let us now consider, as an example, crystallization in an open system in plagioclases (figure 7.13). Let us imagine cooling the initial composition C until the liquidus curve of the system is reached (C1); the crystals in equilibrium with liquid C1 have composition X1. With decreasing T, we reach condition C2, where crystals of composition X2 coexist with a liquid of composition L2. If the system is further cooled by a simple heat loss, we reach point C3. The crystals that formed during cooling at point C2 should now react with the liquid to reach composition X3. Let us now imagine that the kinetic rate of reaction is lower than the rate of heat loss toward the exterior: the inner portions of the crystals of composition X2 do not react with the liquid, and the final composition of the whole crystals is not X3 but intermediate between X3 and X2. The net result is that part of the system (inner parts of crystals) plays no part in the equilibrium:

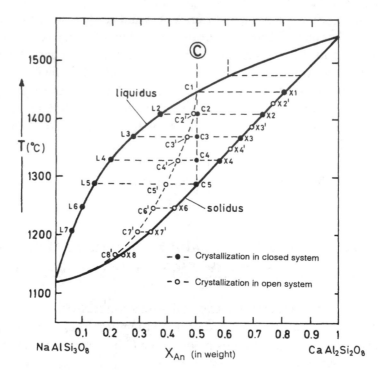

Figure 7.13 Compositional trends in plagioclase binary system during crystallization in open or closed systems.

the new bulk chemistry is no longer that of the initial system C but is displaced toward the less refractory component (i.e., more albitic). Proceeding in this way, we define a compositional path of fractionation corresponding to progressive subtraction of the inner portions of the crystals as crystallization proceeds. Note that, as a net result, crystallization is not complete at the solidus temperature of initial system C (as in the case of crystallization in a closed system), but proceeds down to the melting temperature of the low-melting component (i.e., in our case, $T_{f,Ab} = 118$ °C). The resulting crystals exhibit wide compositional zoning, independent of the initial composition of the system (in our case, crystal nuclei have composition An_{80} and the surfaces of crystals have composition An_0).

The compositional relations discussed above are explained by Rayleigh's distillation process (Rayleigh, 1896), choosing a suitable partition coefficient for the component of interest—i.e., for component An, we can write

$$\overline{X}_{An,solid} = X_{An,initial}\left(\frac{1 - F_L^{K_{An}}}{1 - F_L}\right), \tag{7.17}$$

where $\overline{X}_{An,solid}$ is the mean composition of the crystals, F_L is the fractional amount of residual liquid in the system, and K_{An} is the solid/liquid partition coefficient at interface for component An (see section 10.9.1 for derivation of eq. 7.17).

Let us now treat compositional relations in the ternary system Di-An-Ab. Figure 7.14A shows the projection of the liquidus surface of the system onto the compositional plane; the locations of cotectic line E'-E'' and of the isotherms are based on the experiments of Bowen (1913), Osborn and Tait (1952), and Schairer and Yoder (1960).

The system is not strictly ternary, because of the limited solubility of Al_2O_3 in diopside, but we will neglect this minor complication here. Because the maximum number of phases in the system is three (diopside plus plagioclase mixtures plus liquid) and the components also number three, the minimum variance of the system is 1, at constant P.

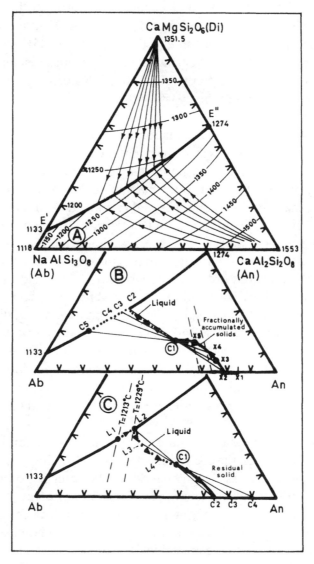

Figure 7.14 Crystallization and melting in an open system: ternary join $CaMgSi_2O_6$-$CaAl_2Si_2O_8$-$NaAlSi_3O_8$ (Di-An-Ab). Crystallization (B) and melting trends (C) are taken from Morse (1980).

In the previously discussed cases of crystallization in a closed system, at every moment of the crystallization process the residual liquid was always located on the straight line connecting the initial composition with the solid. This geometrical condition is not true for an open system, and each liquid at a given crystallization state behaves like an "initial liquid." At each point in the primary phase field it is thus essential to locate the composition of crystals in instantaneous equilibrium with the liquid and the direction of displacement of the residual liquid when an infinitesimal amount of solid is subtracted from equilibrium. These compositional vectors, or "fractionation lines," have been experimentally determined for the various systems of interest (cf. Hamilton and McKenzie, 1965). In the primary phase field of the Ab-An mixture, the fractionation lines depart from high-melting apex An and follow the trends shown in figure 7.14A. Figure 7.14B shows the fractionation line passing through initial composition C1 (85% plagioclase An_{60}, 15% diopside, by weight). During crystallization in an open system of initial composition C1, and within the primary phase field of plagioclases, the residual liquid moves along the fractionation line passing through C1. The instantaneous solid in equilibrium with initial composition C1 is An_{84}. The liquid reaches the cotectic line at C2, moving along the fractionation line. Connecting points C2 and C1, and applying the lever rule, we can deduce that the mean composition of the bulk solid precipitated up to C2 is An_{78} and that the fractional amount of residual liquid at that point (with respect to the mass of the initial system) is $F_L = 0.49$. Note that the composition of the last crystal in instantaneous equilibrium with a liquid of composition C2 is An_{65}—i.e., much richer in Ab than the mean composition of the precipitated solids. As soon as the composition of the residual liquid reaches the cotectic line, diopside begins to precipitate in conjunction with plagioclases. From this point onward, the composition of the liquid moves along the cotectic line and the bulk composition of the fractionally accumulated solids is defined at any moment by prolonging the fractionation lines of diopside (which are straight, Di being immiscible with Ab and An in any proportion) within the primary phase field of plagioclases and intersecting them with the lines connecting residual and initial liquids. Application of baricentric coordinates to segments C3-C1-X3, C4-C1-X4, and C5-C1-X5 allows us to determine the fractional amount of residual liquid in the system ($F_L = 0.43$, 0.33, and 0.17, respectively) at any point.

Melting in an open system of the same initial composition C1 begins at a temperature of 1213 °C on the cotectic line (Morse, 1980; figure 7.14C). Instantaneous subtraction of the resulting liquid drives the composition of the residual solid across the primary phase field of plagioclases until point C2 is reached on the Ab-An baseline. This happens at $T = 1229$ °C, when all the diopside in the solid paragenesis has been exhausted. The fractionally accumulated liquid (outside the system) now has bulk composition L2; the Di component (exhausted in the solid and externally accumulated) is no longer present in the system, which is thus now binary. Melting stops until the temperature reaches the solidus isotherm of composition C2 in the binary Ab-An join ($T = 1408 °C$). Once this temperature is reached, melting starts again, generating liquids that are increasingly richer in An component, up to pure An. The composition of fractionally accumulated liquids now moves along the dotted line in figure 7.14C. The amount of this liquid may be determined by applying the baricentric coordinates to segments L2-C1-C2, L3-C1-C3, and L4-C1-C4 ($F_L = 0.52$, 0.59, and 0.79, respectively). Note that the compositional paths of fractionally accumulated liquids and solids produced during open-system melting and crystallization are not opposite, as in the case of closed systems.

PART THREE

Geochemistry of Fluids

CHAPTER EIGHT

Geochemistry of Aqueous Phases

The importance of aqueous solutions in geochemistry can be appreciated if we recall that two-thirds of the surface of our planet is covered by water and that aqueous fluids of various salinities are determinant in the development of volcanic and metamorphic processes in the earth's upper mantle and crust.

Table 8.1 shows the general distribution of water masses in the hydrosphere. Although H_2O occurs mainly in oceans, 19% of the earth's H_2O mass is still trapped in lithospheric rocks. A nonnegligible H_2O mass (about 1%) is also fixed in crystalline form as ice.

8.1 General Information on Structure and Properties of Water

8.1.1 Structure of the H_2O Molecule

There are two main ways of representing the H_2O molecule: the point-charge model and molecular orbital representation.

Figure 8.1A shows the point-charge model of Verwey (1941). Point charges are located at the equilibrium distance (0.99 Å) and form bond angles of 105° between two H-O bond directions. There are six proton charges ($+6e$), located on oxygen and half a proton charge ($+\frac{1}{2}e$) on each of the two hydrogen ions. A charge of $-7e$ (resulting from the electroneutrality condition) is arbitrarily placed at 0.024 Å from the oxygen nucleus along the bisector of the bond angle. The resulting dipole moment ($\mu = 1.87 \times 10^{-18}$ esu $\times$ cm) thus agrees with experimental observations.

Figure 8.1B shows an experimental contour map of electron density for the H_2O molecule in plane y-z, after Bader and Jones (1963). The electron density is higher around the nuclei and along the bond directrix. The experimental electron density map conforms quite well to the hybrid orbital model of Duncan and Pople (1953) with the LCAO approximation.

Table 8.1 Mass distribution of H_2O in hydrosphere (g).

Oceans and seas	1.37×10^{24}
Lithosphere (hydration plus pore water)	3.2×10^{23}
Ice	1.65×10^{22}
Evaporation plus precipitation	4.5×10^{20}
Lakes and rivers	3.4×10^{19}
Atmospheric water	1.05×10^{19}

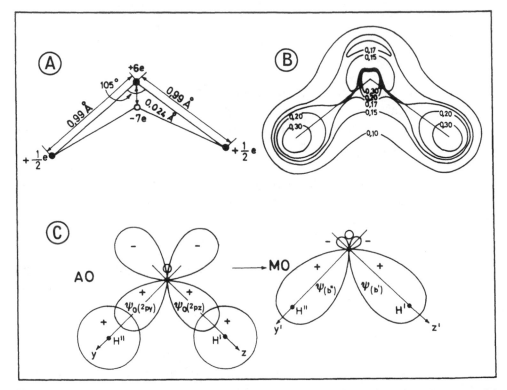

Figure 8.1 Structural models for H_2O molecule. (A) Electrostatic point-charge model (from Eisemberg and Kauzmann, 1969; redrawn). (B) Electron density map (from Bader and Jones, 1963). (C) Formation of MOs $\Psi_{(b')}$ and $\Psi_{(b'')}$ starting from AOs $2p$ and $2s$ of oxygen and $1s$ of hydrogen.

Based on the model of Duncan and Pople (1953), two molecular orbitals (b' and b'') may form as linear combinations of hybrid orbitals $2p$ and $2s$ of oxygen and orbital $1s$ of hydrogen:

$$\Psi_{(b')} = \lambda\left[\cos\varepsilon_b\psi_{O(2s)} + \sin\varepsilon_b\psi_{O(2p,b')}\right] + \xi\psi_{H'(1s)} \tag{8.1}$$

$$\Psi_{(b'')} = \lambda\left[\cos\varepsilon_b\psi_{O(2s)} + \sin\varepsilon_b\psi_{O(2p,b'')}\right] + \xi\psi_{H''(1s)} \tag{8.2}$$

and two "lone pair" orbitals (l' and l'') may form as combinations of hybrid orbitals $2s$ and $2p$ of oxygen:

$$\Psi_{(l')} = \cos \varepsilon_l \psi_{O(2s)} + \sin \varepsilon_l \psi_{O(2p,\,l')} \qquad (8.3)$$

$$\Psi_{(l'')} = \cos \varepsilon_l \psi_{O(2s)} + \sin \varepsilon_l \psi_{O(2p,\,l'')} \qquad (8.4)$$

The significance of term λ was defined in section 1.17.1. Factor ξ is analogous (ratio λ/ξ defines the polarity of the bond) and parameters ε_b and ε_l describe the hybridization state of the orbitals. Because $\cos \varepsilon_b = 0.093$ (Duncan and Pople, 1953), molecular hybrid orbitals $\Psi_{(b')}$ and $\Psi_{(b'')}$ are composed essentially of the wave functions of atomic orbitals $2p$ of oxygen and $1s$ of hydrogen. The value 0.578 obtained for $\cos \varepsilon_l$ also indicates that orbitals $\Psi_{(l')}$ and $\Psi_{(l'')}$ are essentially of type sp^3. Figure 8.1C shows the formation of hybrid MOs $\Psi_{(b')}$ and $\Psi_{(b'')}$ on plane y-z, starting from the AOs $2p$ and $1s$ of hydrogen.

8.1.2 Structure of Water: Models of Distortion of the Hydrogen Bond

An exhaustive discussion on the structure of water is beyond the scope of this textbook. The term "structure" itself, when applied to a liquid state, demands several clarifications that cannot be afforded here. We will therefore refer henceforth only very briefly to models of distortion of the hydrogen bond that, for their simplicity do not require extensive explanations.

Such models consider that transition from the crystalline to the liquid state mainly involves distortion of the hydrogen bond, but no bond rupture. The model of Pople (1951) assumes that, in the liquid state, each H_2O molecule is linked to three other molecules; distance $\overline{R}_o$ between neighboring oxygen nuclei is constant, but the distance between next neighbors varies according to the degree of distortion of the hydrogen bond. As figure 8.2A shows, in Pople's model the hydrogen bond is considered undistorted when the O-H bond direction of a molecule falls along the line connecting neighboring oxygen nuclei. As the degree of distortion increases (ϕ angles in figure 8.2A), the internal energy of the liquid also increases. This may be quantified by

$$\Delta U = K_\phi (1 - \cos \phi). \qquad (8.5)$$

In equation 8.5, K_ϕ is the *constant of distortion of the hydrogen bond.* The degree of distortion of the hydrogen bond in water varies with temperature [from 26° to 30° for $0 < T\,(°C) < 100$] and is zero in ice, where H_2O molecules are linked to form perfect tetrahedra.

The "random network model" of Bernal (1964) is a development of Pople's model. In Bernal's version, the water molecules are conceived as forming an irregular network of rings (not open chains, as in the preceding case). Most of the rings have five members, because the O-H bond in a single H_2O molecule is near the angle (108°) characteristic of a five-membered ring, but four, six, seven, or even more H_2O molecules may be present in some rings.

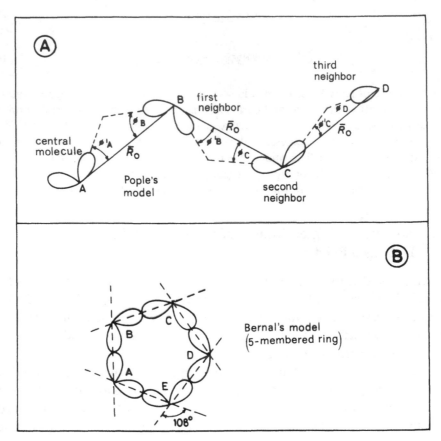

Figure 8.2 Models of distortion of hydrogen bond for water. Note that bond angle between two O-H bonds of a single molecule is 105°; formation of a five-membered ring (part B) thus implies distortion of 3°.

When we will discuss the effects of *solvent collapse* in solute-solvent interactions (section 8.11.2), we will mean local modifications of the water structure (degree of distortion of the oxygen bond; distance between neighboring oxygen nuclei) induced by the presence of electrolytes in solution. We refer to the classical text of Eisemberg and Kauzmann (1969) for a more detailed discussion on the various aggregation states of the H_2O compound.

8.1.3 Dielectric Constant of Water

The concept of *dipole moment* and its relationship to ion polarizability were discussed in section 1.8. Section 1.19 introduced the concept of *dielectric constant* of a crystalline solid and its relationship with the *polarizability* of its constituting ions (see eq. 1.168). The dielectric constant of a liquid solvent such as water represents the capacity of the solvent's molecules to shield the charges of ion

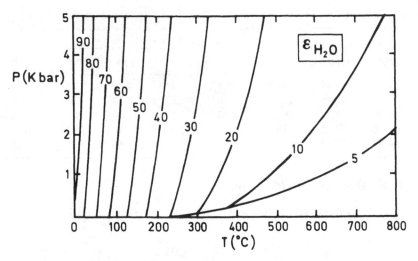

Figure 8.3 Dielectric constant of water as a function of P and T conditions. Reprinted from T. M. Seward, *Physics and Chemistry of Earth*, 13, 113–132, copyright © 1981, with kind permission from Elsevier Science Ltd., The Boulevard, Langford Lane, Kidlington 0X5 1GB, UK.

solutes by means of the solvation process. The dielectric constant of solvent ε may be related to molecular polarizability α_m and dipole moment μ of the single solvent molecules through Kirkwood's equation (Bockris and Reddy, 1970):

$$\frac{(\varepsilon - 1)(2\varepsilon + 1)}{g\varepsilon} = \left[\alpha_m + \frac{\mu^2 g}{3kT} \right] \times \frac{4\pi N_0 d}{3W}, \tag{8.6}$$

where N_0 and k are, respectively, Avogadro's number and Boltzmann's constant, W is the molecular weight of the solvent, d is the density, and g is a function of absolute temperature and density. It may be derived for water by applying Pitzer's (1983) polynomial:

$$g_{H_2O} = 1 + a_1 d + a_2 d^5 \left[\left(\frac{a_3}{T} \right)^{a_4} - 1 \right], \tag{8.7}$$

with $a_1 = 2.68$, $a_2 = 6.69$, $a_3 = 565$, and $a_4 = 0.3$.

The dielectric constant is directly proportional to the density of the solvent (hence to pressure) and inversely proportional to temperature, as shown in figure 8.3. Discrete values of the dielectric constant of water at high P and T conditions are listed in table 8.2.

Table 8.2 Dielectric constant of water at various T and P conditions, after Pitzer (1983).

P (kbar)	Temperature (°C)								
	400	450	500	550	600	650	700	750	800
1	–	12.49	9.43	7.04	5.34	4.24	3.52	3.04	2.69
2	–	16.46	13.87	11.69	9.87	8.37	7.14	6.16	5.37
3	21.78	18.72	16.20	14.10	12.31	10.80	9.51	8.42	7.49
4	23.46	20.35	17.83	15.73	13.95	12.44	11.14	10.02	9.05
5	24.82	21.63	19.63	16.94	15.17	13.65	12.35	11.21	10.22

8.1.4 Ionic Dissociation of Water

Water is partly ionized as a result of the ion dissociation process:

$$H_2O \Leftrightarrow H^+ + OH^-. \tag{8.8}$$

Dissociation constant K_{H_2O}:

$$K_{H_2O} = \frac{a_{H^+} \cdot a_{OH^-}}{a_{H_2O}} \tag{8.9}$$

varies directly with T and inversely with P, as shown in figure 8.4. Discrete values of the ionic dissociation constant for T between 0 and 60 °C and $P = 1$ bar are listed in table 8.3.

Eugster and Baumgartner (1987) suggested that the ion dissociation constant can be numerically reproduced in a satisfactory way by the polynomial

$$\log K_{H_2O} = A + \frac{B}{T} + \frac{CP}{T} + D \log d, \tag{8.10}$$

where d is water density at the T of interest and constants A, B, C, and D have the following values: $A = -4.247$; $B = -2959.4$; $C = 0.00928$; $D = 13.493$.

8.1.5 Critical Temperature and P-T Stability of Water

Figure 8.5 shows the projection on the P-V plane of the state diagram of H_2O near the critical point. This figure distinguishes four separate portions of the P-V space, limited, respectively, by the following:

1. The liquid saturation boundary curve
2. The vapor saturation boundary curve
3. The critical point
4. The critical T isotherm ($T_c = 374$ °C).

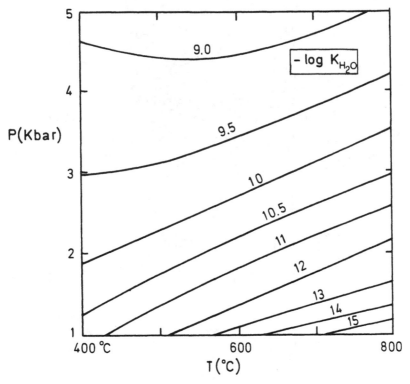

Figure 8.4 Ion dissociation constant of water as a function of P and T conditions. From Eugster and Baumgartner (1987). Reprinted with permission of The Mineralogical Society of America.

Table 8.3 Ionic dissociation constant of water for T between 0 and 60 °C at $P = 1$ bar.

T (°C)	$K_{H_2O} \times 10^{14}$	T (°C)	$K_{H_2O} \times 10^{14}$
0	0.1139	5	0.1846
10	0.2920	15	0.4505
20	0.6809	25	1.008
30	1.469	35	2.089
40	2.919	45	4.018
50	5.474	55	7.297
60	9.614		

Within the PV region delimited by the two saturation boundary curves, liquid and vapor phases coexist stably at equilibrium. To the right of the vapor saturation curve, only vapor is present: to the left of the liquid saturation curve, vapor is absent. Let us imagine inducing isothermal compression in a system composed of pure H_2O at $T = 350$ °C, starting from an initial pressure of 140 bar. The H_2O will initially be in the gaseous state up to $P < 166$ bar. At $P = 166$ bar, we reach the vapor saturation curve and the liquid phase begins to form. Any further

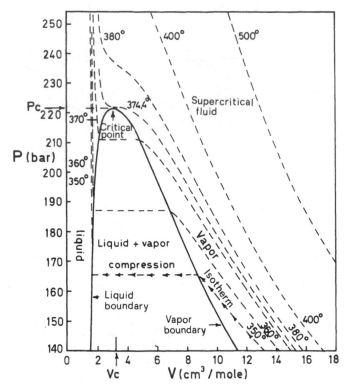

Figure 8.5 *P-V* projection of state diagram for H_2O near critical point. P_c and V_c are critical pressure and critical volume of compound.

compression on the system will not induce any increase in *P*, because the appearance of a new phase has decreased the degree of freedom of the system, which is now invariant. Compression thus simply induces an increase in the bulk density of the vapor-plus-liquid assemblage, until the liquid saturation curve is reached. At this point, the vapor phase disappears and the system is again univariant: any further compression now induces a dramatic increase in *P* (because the liquid phase is relatively incompressible with respect to the gaseous counterpart: the slopes of the isotherms on the gas and liquid sides of the saturation boundaries are very different).

If we now repeat compression along the critical isotherm T_c, we reach (at the critical point) a flexus condition, analytically defined by the partial derivatives

$$\left(\frac{\partial P}{\partial V} \right)_{T_c} = 0 \qquad (8.11.1)$$

$$\left(\frac{\partial^2 P}{\partial V^2} \right)_{T_c} = 0. \qquad (8.11.2)$$

At T higher than T_c, a liquid H_2O phase can never be formed, regardless of how much pressure is exerted on the system.

Figure 8.6A is an enlarged portion of the P-T plane for H_2O in the vicinity of the critical point ($T_c = 373.917$ °C, $P_c = 220.46$ bar).

Five P-T regions of H_2O stability are distinguished:

1. The "liquid" region, limited by the *vaporization boundary* (i.e., the liquid saturation and vapor saturation boundary curves) and the critical isobar.
2. The "supercritical liquid" region, limited to supercritical pressures and subcritical temperatures.
3. The "supercritical fluid" region, which encompasses all conditions of supercritical temperatures and pressures.
4. The "supercritical vapor" region, below the critical isobar and above the critical isotherm.

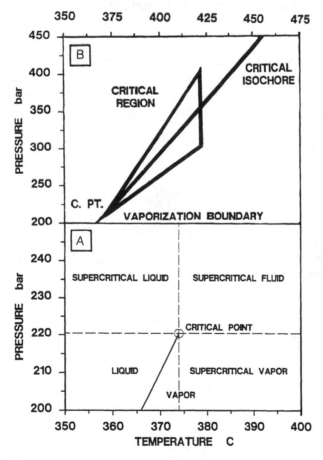

Figure 8.6 (A) P-T regions of H_2O stability. (B) Extended P-T field showing location of the critical region. The critical isochore is $\rho = 0.322778$ g/cm^3. From Johnson and Norton (1991), *American Journal of Science,* 291, 541–648. Reprinted with permission of American Journal of Science.

5. The "vapor" region, limited by the vaporization boundary and the critical isotherm.

In geochemistry, unspecified liquid-plus-gas assemblages are commonly defined as "fluids." Note, however, that a "supercritical fluid" is necessarily composed of a single gaseous phase. Another origin of confusion is use of the term "vapor" instead of "gas." The two terms in fact have *exactly the same meaning*. The term "vapor" is preferred whenever the compound of interest is *liquid* in the stable state at ambient conditions—i.e., we say "H_2O vapor" or "Hg vapor" because both H_2O and Hg are liquid in the stable state, just as we say "oxygen gas" or "hydrogen gas" because both are gaseous in the stable state. Note also that "saturated vapor" means H_2O gas stably coexistent with water (i.e., within the two-phase field) and "unsaturated vapor" or "dry vapor" means gaseous H_2O to the right of the vapor saturation boundary in the *PV* space or under supercritical conditions.

Figure 8.6B shows a wider *P-T* portion with the location of the "critical region" for H_2O, bound by the 421.85 °C isotherm and the $\rho = 0.20$ and 0.42 g/cm^3 isochores. The *PVT* properties of H_2O within the critical region are accurately described by the *nonclassical* (asymptotic scaling) equation of state of Levelt Sengers et al. (1983). Outside the critical region and up to 1000 °C and 15 kbar, *PVT* properties of H_2O are accurately reproduced by the *classical* equation of state of Haar et al. (1984). An appropriate description of the two equations of state is beyond the purposes of this textbook, and we refer readers to the excellent revision of Johnson and Norton (1991) for an appropriate treatment.

The nonclassical equation of state of Levelt Sengers et al. (1983) for the critical region of H_2O may be expressed in terms of reduced parameters (cf. section 9.2) as follows:

$$\rho_r = \left(\frac{\partial \tilde{P}}{\partial \tilde{\mu}} \right)_{\tilde{T}}, \tag{8.12}$$

where ρ_r is the reduced density ρ/ρ_c, with $\rho_c = 0.322778$ g/cm^3 (critical density). The other dimensionless parameters $\tilde{T}$, $\tilde{P}$, and $\tilde{\mu}$ are given by

$$\tilde{T} = -T_r^{-1} = -\left(\frac{T_c}{T} \right) \tag{8.13.1}$$

$$\tilde{P} = \frac{P_r}{T_r} = \frac{T_c}{P_c} \left(\frac{P}{T} \right) \tag{8.13.2}$$

$$\tilde{\mu} = \frac{\rho_c T_c}{P_c} \left(\frac{\mu}{T} \right), \tag{8.13.3}$$

where T_c and P_c are the critical temperature and critical pressure.

A classical equation of state is normally composed of a truncated Taylor series in the independent variables, normalized to the critical point conditions (e.g., van der Waals, virial expansion, etc.). All these sorts of equations yield similar (so-called "classical") asymptotic behavior in their derivative properties at the critical point.

The classical equation of state of Haar et al. (1984), valid outside the critical region, may be expressed as follows:

$$\rho = \left(PCM\right)^{1/2} \left(\frac{\partial A}{\partial \rho}\right)_T^{-1/2} \tag{8.14}$$

(Johnson and Norton, 1991), where M is the molar weight of H_2O (18.0152 g/mol), C is a conversion factor (0.02390054 cal bar^{-1}cm^{-3}) and A is the molal Helmholtz free energy function:

$$A(\rho,T) = A_{\text{base}}\left(\rho,T\right) + A_{\text{resid}}\left(\rho,T\right) + A_{\text{ideal}}\left(\rho,T\right) \tag{8.15}$$

(see Johnson and Norton, 1991, for a detailed account on the significance of the various terms in eq. 8.15). Note that the molar volume may be readily derived from equations 8.12 and 8.14 by applying $\overline{V} = M/\rho$.

Reversing equations 8.11.1 and 8.11.2, we obtain

$$\left(\frac{\partial P}{\partial V}\right)_{T_c}^{-1} = \infty \quad \text{and} \quad \left(\frac{\partial^2 P}{\partial V^2}\right)_{T_c}^{-1} = \infty. \tag{8.16}$$

Moreover, recalling what we have already seen in section 1.15 (namely, eq. 1.95, 1.96, 1.97, and 1.103), we may also derive

$$\left(\frac{\partial^2 A}{\partial V^2}\right)_{T_c}^{-1} = -\infty \quad \text{and} \quad \left(\frac{\partial^3 A}{\partial V^3}\right)_{T_c}^{-1} = -\infty, \tag{8.17}$$

where A is the Helmholtz free energy of the substance, and

$$\left(\frac{\partial^2 U}{\partial V^2}\right)_{T_c}^{-1} = -\infty \quad \text{and} \quad \left(\frac{\partial^3 U}{\partial V^3}\right)_{T_c}^{-1} = -\infty. \tag{8.18}$$

Equations 8.16 and 8.17 or 8.16 and 8.18 show that, at the critical point, "the specific first and second derivative properties of any representative equation of state will be divergent" (Johnson and Norton, 1991). This inherent divergency has profound consequences on the thermodynamic and transport properties of H_2O in the vicinity of the critical point. Figure 8.7 shows, for example, the behav-

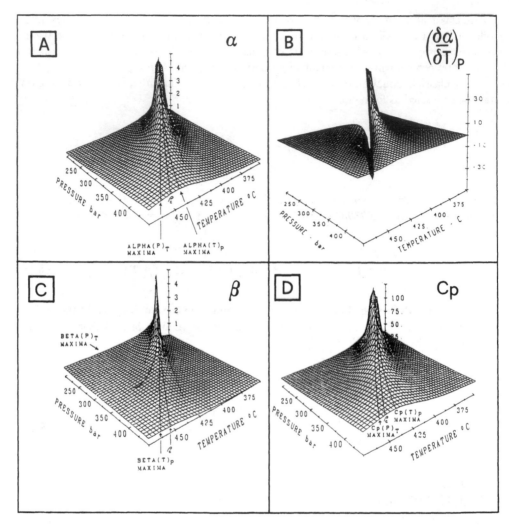

Figure 8.7   (A) Isobaric thermal expansion, (B) its first *T*-derivative, (C) isothermal compressibility, and (D) isobaric heat capacity of H_2O within the critical region, based on the equation of state of Levelt Sengers et al. (1983). From Johnson and Norton (1991), *American Journal of Science*, 291, 541–648. Reprinted with permission of American Journal of Science.

ior of the isobaric thermal expansion (α) of H_2O and of its first derivative on *T* within the critical region, computed by Johnson and Norton (1991) on the basis of the equation of state of Levelt Sengers et al. (1983). The divergent asymptotic behavior of ($\partial\alpha/\partial T$) in the vicinity of the critical point has been arbitrarily truncated within the -5×10^{-3} K^{-2} and 5×10^{-3} K^{-2} lower and upper limits for illustrative purposes. Note also in parts C and D of figure 8.7 the huge domes in the isothermal compressibility β and isobaric heat capacity C_P of H_2O within the critical region.

8.2 Electrolytic Nature of Aqueous Solutions: Some Definitions

Now that we have examined to some extent the properties of water as a "solvent," let us discuss the nature of *aqueous solutions.*

From the point of view of chemical modeling, aqueous solutions are treated as "electrolytic solutions"—i.e., solutions in which solutes are present partially or totally in ionic form. "Speciation" is the name for the characteristic distribution of ion species in a given aqueous solution in the form of simple ions, ionic couplings, and neutral molecules. Solutes in aqueous solutions are defined as "electrolytes" and may be subdivided into "nonassociated" and "associated." Nonassociated electrolytes are also defined as "strong" and mainly occur in the form of simple or simply hydrated ions. An example of a strong electrolyte is the salt NaCl, which, in aqueous solution of low ionic strength, occurs in the form of completely dissociated Na^+ and Cl^- ions.

Associated electrolytes are further subdivided into two groups: "weak electrolytes" and "ionic couplings." A weak electrolyte occurs both in the form of covalent molecules and in ionic form. Examples of weak electrolytes are various acids, some bases, and a few inorganic salts. Aqueous solutions of carbonic acid H_2CO_3, for instance, contain both undissociated neutral molecules $H_2CO_3^0$ and carbonate and bicarbonate ions CO_3^{2-} and HCO_3^-. Solutions of strong inorganic acids such as HF, when present in high molar concentration, also contain neutral molecules HF^0 besides H^+ and F^-. The term "ionic coupling" is used when the association of oppositely charged ions is due not to the formation of a strong covalent bond (as in the case of weak electrolytes) but to electrostatic interactions. For instance, in aqueous solutions of calcium sulfate $CaSO_4$, besides Ca^{2+} and SO_4^{2-} ions, neutral particles $CaSO_4^0$ are also present. In aqueous solutions of calcium hydroxide, some of the solutes are in the form of ionic couplings $Ca(OH)^+$. The distinction of weak electrolytes and ionic couplings in aqueous solutions is somewhat problematic, because in some cases both types of electrolytes have the same stoichiometry. Lastly, the term "complex ion" identifies all nonmonatomic solute ions, independent of the nature of the chemical bond. Examples of complex ions are thus $Ca(OH)^+$, CO_3^{2-}, HCO_3^-, HS^-, $ZnCl_4^{2-}$, and $B_3O_3(OH)_4^-$.

8.3 Models of "Ionic Coupling–Complexation" and "Specific Interactions"

The approach usually adopted in geochemistry to describe the speciation state of aqueous solutions is that of *ionic coupling–complexation,* based on the classical model developed by Garrels and Thompson (1962) for seawater. This model considers the various aqueous ions as species physically present in solution, in molal proportions determined by the coexistence of multiple homogeneous equilibria of the type

$$Ca^{2+} + HCO_3^- \Leftrightarrow CaHCO_3^+, \tag{8.19}$$

and related to the energy balance through application of appropriate individual ionic activity coefficients (see section 8.9).

The approach of *specific interactions,* developed primarily by Pitzer (1973) and Whitfield (1975a,b), considers all salts, from a purely formal point of view, as completely dissociated, and embodies the effects of specific interactions into particular activity coefficients, defined as "total activity coefficients" or "stoichiometric activity coefficients," with symbol γ^T. For instance, for ion i,

$$a_i = m_i^T \gamma_i^T = m_i \gamma_i, \tag{8.20}$$

where m_i^T is the total molal concentration of the species and m_i is the molal concentration of the free ion:

$$\gamma_i^T = \gamma_i \left(\frac{m_i}{m_i^T} \right). \tag{8.21}$$

The main advantage of the specific interactions model lies in the simplicity of its calculations. Also, considering dissolved salts (in their neutral salt stoichiometry—e.g., NaCl) as "components," activities and total activity coefficients are experimentally observable magnitudes.

However, the ionic coupling–complexation model is more appropriate to the actual complexity of aqueous solutions, as is evident from spectral absorption and ionic conductivity studies.

8.4 Activities and Activity Coefficients of Electrolytes in Aqueous Solutions

When speaking of "solutions," we implicitly state that the H_2O "solvent" has a different standard state of reference with respect to "solutes" (note that the term "mixture" is used when all components are treated in the same way; see section 2.1). The standard state generally used for the solvent in aqueous solutions is that of "pure solvent at P and T of interest" (or $P = 1$ bar and $T = 298.15$ K). For solutes, the "hypothetical one-molal solution referred to infinite dilution, at P and T of interest" (or $P = 1$ bar and $T = 298.15$ K) is generally used. This choice is dictated by practical considerations.

Activity a_e of a generic electrolyte $e = C_{v+}A_{v-}$, dissociating according to the scheme

$$C_{v+}A_{v-} \Leftrightarrow v^+C + v^-A , \qquad (8.22)$$

where C = cation and A = anion, is described at high dilution by Henry's law:

$$a_e = K_e m_e^v , \qquad (8.23)$$

where

$$v = v^+ + v^- . \qquad (8.24)$$

Let us now imagine plotting the activity of electrolyte e as a function of molality, elevated to stoichiometric factor v (m_e^v), as shown in figure 8.8. Graphically in this sort of plot, Henry's law constant K_e represents the slope of equation 8.23 at infinite dilution. Extending the slope of Henry's law up to value $m_e^v = 1$ (dashed line in figure 8.8) and arbitrarily fixing the ordinate scale so that activity is 1 at the point defined by the extension of the Henry's law slope at $m_e^v = 1$, we can construct the condition of "hypothetical one-molal solution referred to infinite dilution." This condition not only obeys the unitary activity implicit in the standard state definition, but also results in an activity coefficient of 1 at infinite dilution (i.e., $a_e = m_e^v$). The term "hypothetical" emphasizes the fact that the adopted reference condition does not correspond to the energy properties of the actual solution at the same concentration level (see figure 8.8).

The activity of electrolyte a_e may be related to the individual ionic activities of constituting ions a_+ and a_-:

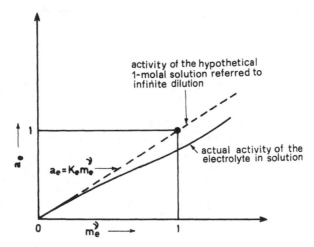

Figure 8.8 Construction of standard state of "hypothetical one-molal solution referred to infinite dilution."

$$a_e = \left(a_+\right)^{\nu+} \left(a_-\right)^{\nu-}$$ (8.25)

which, in turn, determine "mean ionic activity" $a_\pm$

$$a_\pm = \left[\left(a_+\right)^{\nu+}\left(a_-\right)^{\nu-}\right]^{1/\nu} = a_e^{1/\nu}.$$ (8.26)

Similarly, "mean ionic molality" is defined as

$$m_\pm = \left[\left(m_+\right)^{\nu+}\left(m_-\right)^{\nu-}\right]^{1/\nu} = m_e^{1/\nu},$$ (8.27)

and the "individual ionic activity coefficients" are expressed by

$$\gamma_+ = \frac{a_+}{m_+} \quad \text{and} \quad \gamma_- = \frac{a_-}{m_-},$$ (8.28)

where m_+ and m_- are moles of dissociated ions.

Lastly, "mean ionic activity coefficient" is given by

$$\gamma_\pm = \frac{a_\pm}{m_\pm} = \left[\left(\gamma_+\right)^{\nu+}\left(\gamma_-\right)^{\nu-}\right]^{1/\nu} = \gamma_e^{1/\nu}.$$ (8.29)

Among the above-defined thermodynamic entities, the individual ionic activity coefficients are particularly useful, because they allow practical calculation of the speciation state of an aqueous phase, linking individual ionic molalities to the energy balance. We will see in the following section how these coefficients may be derived.

8.5 Ionic Strength and Debye-Hückel Theory

The concept of *ionic strength,* which allows individual ionic activity coefficients to be estimated, was developed by Lewis and Randall (1921). The ionic strength of a solution is given by

$$I = \frac{1}{2} \sum_i m_i Z_i^2 ,$$ (8.30.1)

where m_i is the molality, Z_i is the charge of the ith ion, and summation is extended to all ions in solution.

The ionic strength of a one-molal solution of $CaCl_2$ is thus

$$I = \frac{1}{2}\left[m_{Ca^{2+}} \times 2^2 + m_{Cl^-} \times (-1)^2 \right] = 3. \tag{8.30.2}$$

Clearly, ionic strength corresponds to molality in the case of a uni-univalent strong electrolyte—i.e., for a one-molal solution of NaCl:

$$I = \frac{1}{2}\left[m_{Na^-} \times 1^2 + m_{Cl^-} \times (-1)^2 \right] = 1. \tag{8.30.3}$$

The concept of ionic strength is very useful in quantifying the effects of electro-static charges of all ions present in solution. Its relationship with the nature of electrostatic interactions was stressed by Lewis and Randall: "In diluted solutions the activity coefficient of a given strong electrolyte is the same in all solutions having the same ionic strength." In fact, in light of the Debye-Huckel theory, the relationship between individual ionic activity coefficient γ_i of ion i and the ionic strength of solution I is given by

$$\log \gamma_i = -\frac{AZ_i^2 \sqrt{I}}{1 + \mathring{a}_i B \sqrt{I}}. \tag{8.31}$$

A and B in equation 8.31 are characteristic parameters of the solvent (in this case, water), which can be derived in the various T and P conditions of interest by application of

$$A = \frac{1.824829 \times 10^6 \, \rho^{1/2}}{(\varepsilon T)^{3/2}} \qquad (\text{kg}^{1/2} \times \text{mole}^{-1/2}) \tag{8.32.1}$$

and

$$B = \frac{50.291586 \times 10^8 \, \rho^{1/2}}{(\varepsilon T)^{1/2}} \qquad (\text{kg}^{1/2} \times \text{mole}^{-1/2} \times \text{cm}^{-1}) \tag{8.32.2}$$

(Helgeson and Kirkham, 1974). In equations 8.32.1 and 8.32.2, ρ is solvent density, ε is the dielectric constant of the solvent, and $\mathring{a}_i$ is the "effective diameter" of ion i in solution. The latter parameter must not be confused with the ionic diameter of the ion in a condensed aggregation state, but represents the range of electrostatic interaction with the solvent molecules (see section 8.11.3).

Table 8.4 lists discrete values of solvent parameters A and B for H_2O in the

Table 8.4 Debye-Huckel constants for H_2O in T range 0 to 300 °C at saturation P (from Helgeson and Kirkham, 1974). A is in $kg^{1/2} \times mole^{1/2}$ and B is in $kg^{1/2} \times mole^{1/2} \times cm^{-1}$.

T (°C)	A	$B \times 10^{-8}$
0	0.4911	0.3244
25	0.5092	0.3283
50	0.5336	0.3325
75	0.5639	0.3371
100	0.5998	0.3422
125	0.6416	0.3476
150	0.6898	0.3533
175	0.7454	0.3592
200	0.8099	0.3655
225	0.8860	0.3721
250	0.9785	0.3792
275	1.0960	0.3871
300	1.2555	0.3965

Table 8.5 Effective ion diameters (Å) for various ions and inorganic complexes in aqueous solution (after Kielland, 1937).

$å$	Simple Ions and Aqueous Inorganic Complexes
2.5	Rb^+, Cs^+, NH_4^+, Tl^+, Ag^+
3	K^+, Cl^-, Br^-, I^-, CN^-, NO_2^-, NO_3^-,
3.5	OH^-, F^-, NCS^-, NCO^-, HS^-, ClO_3^-, ClO_4^-, BrO_3^-, IO_4^-, MnO_4^-
4	Hg_2^{2+}, SO_4^{2-}, $S_2O_3^{2-}$, $S_2O_8^{2-}$, SeO_4^{2-}, CrO_4^{2-}, HPO_4^{2-}, $S_2O_6^{2-}$, PO_4^{3-}, $Fe(CN)_6^{3-}$, $Cr(NH_3)_6^{3+}$, $Co(NH_3)_6^{3+}$, $Co(NH_3)_5H_2O^{3+}$
4–4.5	Na^+, $CdCl^+$, ClO_2^-, IO_3^-, HCO_3^-, $H_2PO_4^-$, HSO_3^-, $H_2AsO_4^-$, $Co(NH_3)_4(NO_2)_2^+$
4.5	Pb^{2+}, CO_3^{2-}, SO_3^{2-}, MoO_4^{2-}, $Co(NH_3)_5Cl^{2+}$, $Fe(CN)_5NO^{2-}$
5	Sr^{2+}, Ba^{2+}, Ra^{2+}, Cd^{2+}, Hg^{2+}, S^{2-}, $S_2O_4^{-2}$, WO_4^{2-}, $Fe(CN)_6^{4-}$
6	Li^+, Ca^{2+}, Cu^{2+}, Zn^{2+}, Sn^{2+}, Mn^{2+}, Fe^{2+}, Ni^{2+}, Co^{2+}, $Co(S_2O_3)(CN)_5^{4-}$
8	Mg^{2+}, Be^{2+}
9	H^+, Al^{3+}, Fe^{3+}, Cr^{3+}, Sc^{3+}, Y^{3+}, La^{3+}, Ce^{3+}, Pr^{3+}, Nd^{3+}, Sm^{3+}, In^{3+}, $Co(SO_3)_2(CN)_4^{5-}$
11	Th^{4+}, Zr^{4+}, Ce^{4+}, Sn^{4+}

T range 0 to 300 °C (saturation P), and table 8.5 lists effective diameter values for various ions in aqueous solution.

When the ionic strength of a solution is low, the denominator in equation 8.31 tends toward 1, and the individual ionic activity coefficient is well approximated by

$$\log \gamma_i \approx -AZ_i^2 \sqrt{I}. \tag{8.33}$$

8.6 The "Mean Salt" Method

The *mean salt method* derives implicitly from the concept of ionic strength, as expressed by Lewis and Randall (1921). It is assumed that, within the range of the ionic strength of interest, for a standard uni-univalent electrolyte such as KCl, the following assumption is valid:

$$\gamma_{K^+} = \gamma_{Cl^-} \tag{8.34}$$

From equation 8.34, with a reasonable approximation (see eq. 8.29), we can derive

$$\gamma_{\pm KCl} = \left[\left(\gamma_{K^+}\right)\left(\gamma_{Cl^-}\right)\right]^{1/2} = \gamma_{K^+} = \gamma_{Cl^-}. \tag{8.35}$$

Because the $\gamma_\pm$ values for the various salts in solution may be experimentally obtained with a satisfactory precision, equation 8.35 is used to derive the corresponding values of individual ionic activity coefficients from them. For instance, from a generic univalent chloride MCl, the $\gamma_\pm$ of which is known, we may derive the γ_+ of M^+ by applying

$$\gamma_{\pm MCl} = \left[\left(\gamma_{M^+}\right)\left(\gamma_{Cl^-}\right)\right]^{1/2} = \left[\left(\gamma_{M^+}\right)\left(\gamma_{\pm KCl}\right)\right]^{1/2} \tag{8.36.1}$$

$$\gamma_{M^+} = \gamma_{\pm MCl}^2 / \gamma_{\pm KCl}. \tag{8.36.2}$$

For a divalent chloride MCl_2,

$$\gamma_{\pm MCl_2} = \left[\left(\gamma_{M^{2+}}\right)\left(\gamma_{Cl^-}\right)^2\right]^{1/3} = \left[\left(\gamma_{M^{2+}}\right)\left(\gamma_{\pm KCl}\right)^2\right]^{1/3} \tag{8.37.1}$$

$$\gamma_{M^{2+}} = \gamma_{\pm MCl}^3 / \gamma_{\pm KCl}^2. \tag{8.37.2}$$

Similarly, the same method and the same mean salt furnish values of individual ionic activity coefficients for other ions—e.g., the sulfate group SO_4^{2-}:

$$\gamma_{\pm K_2SO_4} = \left[\left(\gamma_{K^+}\right)^2\left(\gamma_{SO_4^{2-}}\right)\right]^{1/3} = \left[\left(\gamma_{\pm KCl}\right)^2\left(\gamma_{SO_4^{2-}}\right)\right]^{1/3} \tag{8.38.1}$$

$$\gamma_{SO_4^{2-}} = \gamma_{\pm K_2SO_4}^3 / \gamma_{\pm KCl}^2. \tag{8.38.2}$$

For a salt with both cation and anion different from those of the mean salt (e.g., $CuSO_4$), double substitution is required—i.e.,

$$\gamma_{\pm CuSO_4} = \left[\left(\gamma_{Cu^{2+}} \right) \left(\gamma_{SO_4^{2-}} \right) \right]^{1/2},$$

and, stemming from equation 8.38.2,

$$\gamma_{Cu^{2+}} = \frac{\gamma_{\pm CuSO_4}^2 \gamma_{\pm KCl}^2}{\gamma_{\pm K_2SO_4}^3}. \tag{8.40}$$

Figure 8.9 shows mean activity values for various salts plotted according to the ionic strength of the solution. Note how, at low I ($I < 0.05$), the values converge for salts of the same stoichiometry (i.e., uni-univalent: NaCl, KCl: uni-divalent: Na_2SO_4, K_2SO_4).

Figure 8.10 shows the comparison between the mean salt method and the Debye-Huckel model in the estimate of individual ionic activity coefficients. The two methods give concordant results for low ionic strength values ($I < 0.1$). For ions of the same net charge, convergence toward identical values occurs at zero ionic strength. At high ionic strength, the individual ionic activity coefficients of the cations increase exponentially with I. Estimates at $I > 5$ are affected by a high degree of uncertainty.

Individual ionic activity coefficients may also be estimated more precisely

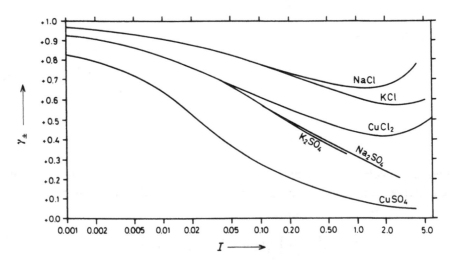

Figure 8.9 Mean activity coefficients for chlorides and sulfides, plotted following ionic strength of solution. Reprinted from Garrels and Christ (1965), with kind permission from Jones and Bartlett Publishers Inc., copyright © 1990.

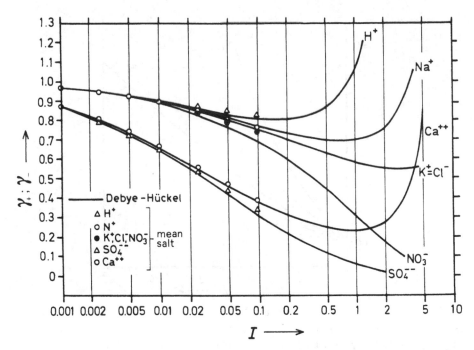

Figure 8.10 Comparison between individual ionic activity coefficients obtained with Debye-Hückel equation and with mean salt method for various ionic strength values. Reprinted from Garrels and Christ (1965), with kind permission from Jones and Bartlett Publishers Inc., copyright © 1990.

with Helgeson's (1969) equation, which introduces a corrective factor $\mathring{B}$ in the Debye-Hückel equation:

$$\log \gamma_i = -\frac{AZ_i^2 \sqrt{I}}{1 + \mathring{a}_i B \sqrt{I}} + \mathring{B}I. \tag{8.41}$$

8.7 Activity of Dissolved Neutral Species

Figure 8.11 shows activity coefficients for gaseous molecules dissolved in aqueous NaCl solutions at various molalities ($T = 15$ and $25\ °C$; $P = 1$ bar). The coefficients converge on 1 at zero ionic strength. Each species also has distinct values, at any I.

Figure 8.12 shows that, for a single neutral species (CO_2), the activity coefficient varies according to type of dissolved electrolyte and increases with increasing ionic strength of the solution. This phenomenon is known as "salting out," and generally takes place for nonpolar neutral species such as CO_2, O_2, H_2, and N_2 dissolved in aqueous solutions. For polar neutral species such as $CaSO_4^0$ and $MgSO_4^0$, the inverse phenomenon, called "salting in," is observed: in this case, the

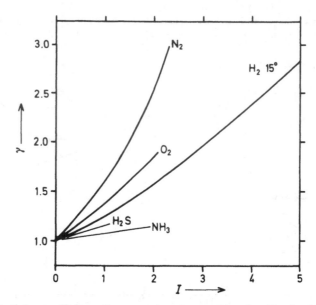

Figure 8.11 Activity coefficients for neutral gaseous molecules dissolved in aqueous solutions of various ionic strengths. All values are for $T = 25$ °C and $P = 1$ bar, except for hydrogen ($T = 15$ °C, $P = 1$ bar). Reprinted from Garrels and Christ (1965), with kind permission from Jones and Bartlett Publishers Inc., copyright © 1990.

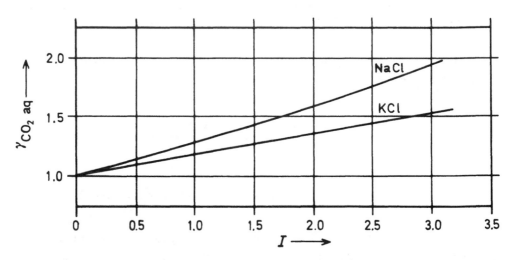

Figure 8.12 Salting-out phenomenon for aqueous CO_2. Activity coefficient of neutral species increases with increasing salinity, determining decreased solubility of aqueous CO_2 in water, T and P conditions being equal. Reprinted from Garrels and Christ (1965), with kind permission from Jones and Bartlett Publishers Inc., copyright © 1990.

activity coefficient of the neutral species in solution decreases with increasing ionic strength. In the development of speciation calculations (see section 8.9), it is often assumed that all neutral species in solution have the same activity coefficient as aqueous CO_2, calculated with a polynomial in I of the type:

$$\log \gamma_i = C_{1,CO_2} I + C_{2,CO_2} I^2 + C_{3,CO_2} I^3 + C_{4,CO_2} I^4, \qquad (8.42)$$

where C_{1,CO_2}, C_{2,CO_2}, C_{3,CO_2}, and C_{4,CO_2} are known constants.

8.8 Activity of Solvent in Aqueous Solutions

The activity of H_2O solvent a_{H_2O} in aqueous solutions decreases with increasing molality of solutes. Figure 8.13 shows the effects of progressive addition of dissolved chlorides in solution on a_{H_2O}.

In theory, once the activity of an electrolyte in solution is known, the activity of the solvent can be determined by the Gibbs-Duhem integration (see section 2.11). In practice, the calculation is prohibitive, because of the chemical complexity of most aqueous solutions of geochemical interest. Semiempirical approximations are therefore preferred, such as that proposed by Helgeson (1969), consisting of a simulation of the properties of the H_2O-NaCl system up to a solute

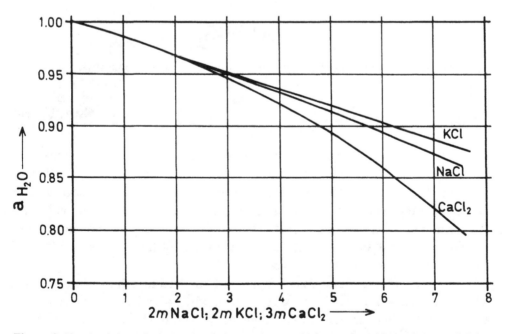

Figure 8.13 Activity of H_2O solvent in aqueous solutions of various solute molalities. Reprinted from Garrels and Christ (1965), with kind permission from Jones and Bartlett Publishers Inc., copyright © 1990.

molality of 2. The approximation adopts parameter I_E, defined as "equivalent stoichiometric ionic strength" and numerically equivalent to the molal concentration of the Na^+ or Cl^- in solution. The calculation proceeds as follows:

$$\log a_{H_2O} = \frac{-2I\varphi'}{2.303\omega_{H_2O}}, \qquad (8.43)$$

where φ' is the osmotic coefficient of water and ω_{H_2O} is the number of moles of H_2O per 10^3 g of solvent ($\omega_{H_2O} = \frac{1000}{18}$). The osmotic coefficient is then calculated with the polynomial expansion

$$\varphi' = 1 - D'J' + \frac{1}{2}W_2 I_E + \frac{1}{3}W_3 I_E^2 + \frac{1}{4}W_4 I_E^3, \qquad (8.44)$$

where

$$D' = \frac{2.303A}{W_1^3 I_E} \qquad (8.45)$$

$$J' = B' - 2\ln B' - \frac{1}{B'} \qquad (8.46)$$

$$B' = 1 + W_1 I_E^{1/2}, \qquad (8.47)$$

with W_1, W_2, W_3, and W_4 being experimentally determined constants.

8.9 Speciation

As already stated, "speciation" is the characteristic distribution of various ionic and/or neutral species in an aqueous solution. Speciation calculation, allowing practical estimation of the reactive properties of an aqueous solution, acidity, redox state, the degree of saturation of the various solids, and so on, is carried out on a thermodynamic basis starting from the chemical composition of the solution of interest and using the reaction constants of the various equilibria of the type seen in equation 8.19.

When performing the chemical analysis of an aqueous solution, we obtain a set of values representing the bulk concentration of dissolved components, but we do not discriminate the various forms of solutes in which a given species is partitioned. For instance, we can measure the total molal amount of calcium m_{Ca}^T or fluorine m_F^T, but this value is the sum of all partial molalities of ionic and neutral species containing Ca or F—i.e.,

$$m_{Ca}^{T} = m_{Ca^{2+}} + m_{CaOH^+} + m_{CaCO_3^0} + m_{CaHCO_3^+} + m_{CaHPO_4^0} + \cdots \quad (8.48)$$

$$m_{F}^{T} = m_{F^-} + m_{HF^0} + 2m_{H_2F_2^0} + 2m_{HF_2^-} + 3m_{AlF_3^0} + \cdots \quad (8.49)$$

Because we are generally able to define the chemistry of an aqueous solution containing n chemical elements by analytical procedures, n equations such as 8.48 and 8.49 exist, relating the bulk concentration of a given element m_i^T to all species actually present in solution. Associated with mass balance equations of this type may be a charge balance equation expressing the overall neutrality of the solution:

$$\sum_i Z_i m_i = 0, \quad (8.50)$$

where Z_i is the net charge of species i, and m_i is its partial molality.

Besides the above-quoted equations, a set of homogeneous reactions also exists, relating the individual ionic activities of all the complex ions in solution—e.g.,

$$CaOH^+ \Leftrightarrow Ca^{2+} + OH^- \quad (8.51)$$

$$HF^0 \Leftrightarrow H^+ + F^- \quad (8.52)$$

$$AlF_3^0 \Leftrightarrow Al^{3+} + 3F^- . \quad (8.53)$$

These equilibria may be expressed in the forms

$$\log K_{51} = \log a_{Ca^{2+}} + \log a_{OH^-} - \log a_{CaOH^+} = \log m_{Ca^{2+}} + \log m_{OH^-}$$
$$- \log m_{CaOH^+} + \log \gamma_{Ca^{2+}} + \log \gamma_{OH^-} - \log \gamma_{CaOH^+} \quad (8.54)$$

$$\log K_{52} = \log a_{H^+} + \log a_{F^-} - \log a_{HF^0} = \log m_{H^+} + \log m_{F^-}$$
$$- \log m_{HF^0} + \log \gamma_{H^+} + \log \gamma_{F^-} - \log \gamma_{HF^0} \quad (8.55)$$

$$\log K_{53} = \log a_{Al^{3+}} + 3\log a_{F^-} - \log a_{AlF_3^0} = \log m_{Al^{3+}} + 3\log m_{F^-}$$
$$- \log m_{AlF_3^0} + \log \gamma_{Al^{3+}} + 3\log \gamma_{F^-} - \log \gamma_{AlF_3^0} . \quad (8.56)$$

Because the thermodynamic constants of the various equilibria of the type exemplified by equations 8.51 to 8.53 can easily be calculated from the partial molal thermodynamic properties of the solutes, and the individual ionic activity coefficients may also be calculated, the unknowns in equations 8.54 to 8.56 are represented simply by partial molalities. When equations such as 8.48 and 8.49 are combined with equations such as 8.54 to 8.56 and with equation 8.50, the system is determined and can be solved by iterative procedures.

8.9.1 Generalities on Complexing of Metal Cations as a Function of Their Elemental Properties

The simplest approach to describing the interactions of metal cations dissolved in water with solvent molecules is the Born electrostatic model, which expresses solvation energy as a function of the dielectric constant of the solvent and, through transformation constants, of the ratio between the squared charge of the metal cation and its effective radius. This ratio, which is called the "polarizing power" of the cation (cf. Millero, 1977), defines the strength of the electrostatic interaction in a solvation-hydrolysis process of the type

$$
M^{Z+}O\begin{matrix}H\\ \diagup\\ \\ \diagdown\\ H\end{matrix} \xrightarrow{\text{solvation}} \left[M-O\begin{matrix}H\\ \diagup\\ \\ \diagdown\\ H\end{matrix} \right] \xrightarrow{\text{hydrolysis}} [M-O-H]^{(Z-1)+} + H^+.
\tag{8.57}
$$

The stronger the polarizing power of the cation, the stronger the tendency of reaction 8.57 to proceed rightward. Figure 8.14 shows the relationship between polarizing power and the degree of hydrolysis of aqueous cations in a solution with ionic strength $I = 0.65$, pH $= 8.2$, $T = 25\ °C$, and $P = 1$ bar according to Turner et al. (1981). In the logarithmic plot of figure 8.14, the ordinate axis is the coefficient of partial reaction (i.e., ratio of hydrolyzed cation with respect to total concentration). The abscissa axis does not properly indicate the polarizing power, because authors have adopted Shannon and Prewitt's (1969) ionic radii (see Turner et al., 1981, for details).

Five elemental groups can be distinguished on the basis of a diagram such as figure 8.14:

Group 1: cations with $Z^2/r < 2.5$ Å^{-1}
Group 2: cations with $2.5 < Z^2/r < 7$ Å^{-1}
Group 3: cations with $7 < Z^2/r < 11$ Å^{-1}
Group 4: cations with $11 < Z^2/r < 25$ Å^{-1}
Group 5: cations with $Z^2/r > 25$ Å^{-1}.

Observe the sharp transition from zero hydrolysis to complete hydrolysis in the region $7 < Z^2/r < 25$ Å^{-1}. The fully hydrolyzed elements belong to groups IV

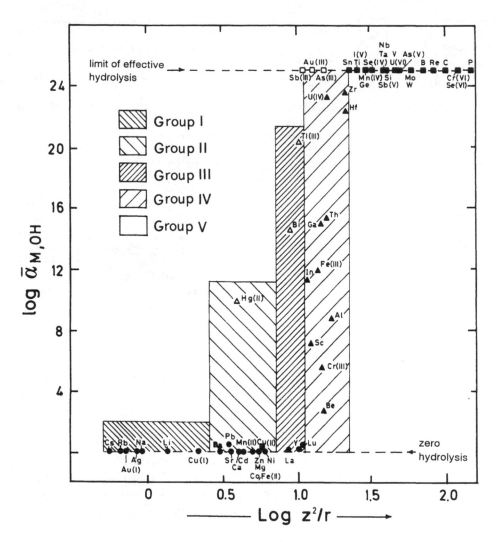

Figure 8.14 Effect of polarizing power on degree of hydrolysis of cations in water. Reprinted from D. R. Turner, M. Whitfield, and A. G. Dickson, *Geochimica et Cosmochimica Acta,* 45, 855–881, copyright © 1981, with kind permission from Elsevier Science Ltd., The Boulevard, Langford Lane, Kidlington 0X5 1GB, UK.

and V. For these elements (except for uranium, present as uranyl ion UO_2^{2+}, whose speciation is dominated by a strong interaction with carbonate ion CO_3^{2-}), complexing with any other anionic ligands that may be present in aqueous solution is negligible. Speciation of the remaining cations with the various anionic ligands is related to their tendency to form partially covalent bondings. The arrangement of the various groups on the periodic chart, according to the classification of Turner et al. (1981), is shown in figure 8.15.

Ahrland's (1973) classification distinguishes incompletely hydrolyzed ions into types *a* and *b*. Cations of type *a* form stable complexes with electronic donors of the first row of the periodic chart through electrostatic interactions; those of

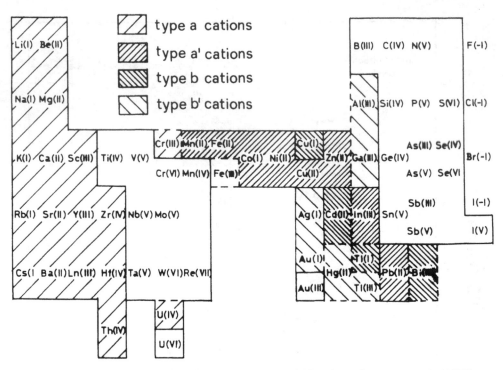

Figure 8.15 Arrangement of various ions from classification of Turner et al. (1981) on periodic chart. Reprinted from D. R. Turner, M. Whitfield, and A. G. Dickson, *Geochimica et Cosmochimica Acta*, 45, 855–881, copyright © 1981, with kind permission from Elsevier Science Ltd., The Boulevard, Langford Lane, Kidlington 0X5 1GB, UK.

type *b* form stable complexes with electronic donors of the second, third, and fourth rows (P, S, Cl: As, Se, Br; Sb, Te, I) through essentially covalent bondings. Turner et al. (1981) suggested a classification based on the different stabilities of the chloride and fluoride complexes of the cations. Denoting the stability difference as ΔK:

$$\Delta K = \log K_{MF} - \log K_{MCl}, \tag{8.58}$$

where K_{MF} and K_{MCl} are complexing constants at reference conditions $T = 25\,°C$ and $P = 1$ bar, four distinct cation groups arise:

Type *a* cations: $\Delta K > 2$
Type *a'* cations: $2 > \Delta K > 0$
Type *b* cations: $0 > \Delta K > -2$
Type *b'* cations: $\Delta K < -2.$

Figure 8.16 shows how the complexing constants of ions vary with the polarizing power of the cation, ligand being equal. Note also here that the abscissa

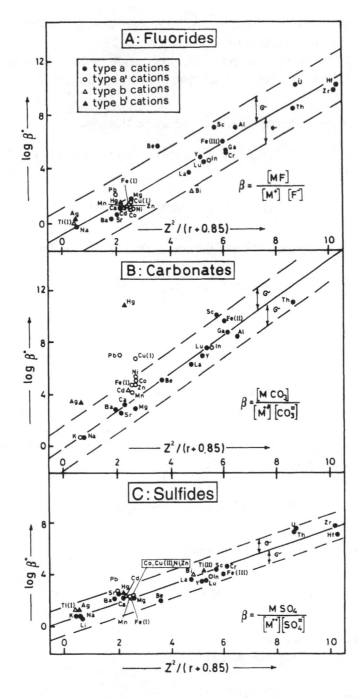

Figure 8.16 Complexing constants of various cations with different anionic ligands arranged according to ratio between squared charge and ionic radius (plus a constant term), where σ is the standard deviation on regression. Reprinted from D. R. Turner, M. Whitfield, and A. G. Dickson, *Geochimica et Cosmochimica Acta*, 45, 855–881, copyright © 1981, with kind permission from Elsevier Science Ltd., The Boulevard, Langford Lane, Kidlington 0X5 1GB, UK.

axis does not strictly represent polarizing power, because the effective radius is replaced by the ionic radius plus a constant term.

We will now examine in detail the progression of hydrolysis reactions and their influence on the stability in solution of hydrolyzed cations.

8.9.2 Energies of Hydrolysis and Solvation Processes

The initial step in the hydrolysis of a generic cation M is usually described by the equilibrium

$$M^{Z+} + H_2O \Leftrightarrow MOH^{(Z-1)+} + H^+. \tag{8.59}$$

A more realistic representation of the actual process is given by

$$M(OH_2)_n^{Z+} \Leftrightarrow M(OH)(OH_2)_{n-1}^{(Z-1)+} + H^+ \tag{8.60}$$

(Baes and Mesmer, 1981), where there is a proton loss by solvation of a water molecule. In light of equation 8.60, it is not surprising to observe that the enthalpy change associated with the process is close to the dissociation enthalpy of water (13.3 kcal/mole at $T = 25\,°C$ and $P = 1$ bar).

If the metal concentration in solution is sufficiently low to remain below the solubility product, increasing the pH, with molal concentration, T, and P remaining equal, hydrolysis proceeds by further solvation of H_2O molecules, with proton loss according to the scheme

$$M^{Z+} \rightarrow M(OH)^{(Z-1)+} \rightarrow M(OH)_2^{(Z-2)+}. \tag{8.61}$$

The process can be generalized as follows:

$$M(OH)_Y^{(Z-Y)+} + H_2O \rightarrow M(OH)_{(Y+1)}^{(Z-Y-1)+} + H^+, \tag{8.62}$$

with Y increasing from zero.

The constant of reaction 8.62 is

$$K_{62} = \frac{\left[M(OH)_{(Y+1)}^{(Z-Y-1)+} \right]\left[H^+ \right]}{\left[M(OH)_Y^{(Z-Y)+} \right]}, \tag{8.63}$$

where the terms in square brackets denote thermodynamic activity.

Baes and Mesmer (1981) observed a simple relationship between the entropy change involved in equilibrium 8.62 at reference T and P conditions ($\Delta S_{62,T_r,P_r}$) and the charge of species in reaction $(Z-Y)^+$:

$$\Delta S_{62,T_r,P_r} = -17.8 + 12.2(Z - Y) \quad \left(\frac{\text{cal}}{\text{mole} \times \text{K}}\right). \tag{8.64}$$

Because

$$\Delta G_{62,T_r,P_r} = \Delta H_{62,T_r,P_r} - T_r \Delta S_{62,T_r,P_r} = -2.303RT \log K_{62,T_r,P_r}, \tag{8.65}$$

combining equations 8.64 and 8.65 gives

$$\Delta H_{62,T_r,P_r} = -1.36 \log K_{62,T_r,P_r} - 5.3 + 3.64(Z - Y) \quad \left(\frac{\text{kcal}}{\text{mole}}\right). \tag{8.66}$$

At cation molal concentrations exceeding 10^{-3} moles per kilogram of solvent, polynuclear species may be formed, with processes of the type

$$X\,M^{Z+} + Y\,H_2O \Leftrightarrow M_X(OH)_Y^{(XZ-Y)+} + Y\,H^+, \tag{8.67}$$

whose equilibrium constant is

$$K_{67} = \left[M_X(OH)_Y^{(XZ-Y)+}\right]\left[H^+\right]^Y\left[M^{Z+}\right]^{-X}. \tag{8.68}$$

The concentrations of the various hydrolyzed species at saturation are controlled by the solubility product, with equilibria of the type

$$M(OH)_Z + Z\,H^+ \Leftrightarrow M^{Z+} + Z\,H_2O \tag{8.69}$$

$$K_{69} = \left[M^{Z+}\right]\left[H^+\right]^{-Z} \tag{8.70}$$

$$X\,M(OH)_Z + (XZ - Y)H^+ \Leftrightarrow M_X(OH)_Y^{(XZ-Y)+} + (XZ - Y)H_2O \tag{8.71}$$

$$K_{71} = \left[M_X(OH)_Y^{(XZ-Y)+}\right]\left[H^+\right]^{-(XZ-Y)}. \tag{8.72}$$

Figure 8.17 shows equilibrium relations between hydrolyzed cations and their saturation limits as a function of pH. In this semilogarithmic plot, the saturation

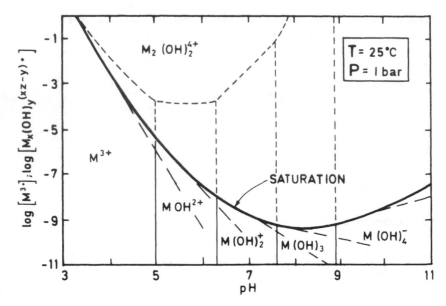

Figure 8.17 Equilibria in solution between various hydrolyzed species. From Baes and Mesmer (1981), *American Journal of Science,* 281, 935–62. Reprinted with permission of American Journal of Science.

limit defined by solubility product K_{71} is a straight line with slope XY-Z as a function of pH (disregarding minor effects of the activity coefficients). This is true for all species forming during the various hydrolysis steps with respect to their saturation limits. The overall saturation curve (bold line) is given by the envelope of the saturation limits valid for each single species. The *predominance limits* of the hydrolyzed species (isoactivity condition of neighboring species) are simply defined by pH (see eq. 8.59 and 8.62). At high metal concentrations, a field of predominance of polynuclear species opens, whose boundaries are defined by equilibria of the type exemplified by equation 8.67.

The above-discussed equilibria are of fundamental importance for an understanding of alteration processes, with special emphasis on the role of pH in water-rock interaction events.

8.10 Carbonate Equilibria

Besides controlling the redox conditions of seawater, as we will see in some detail in section 8.10.1, carbonate equilibria are important when investigating sedimentary rocks, of which carbonates are major components. The procedure normally followed in geochemistry in establishing such equilibria is the classical approach of Garrels and Christ (1965), later taken up by other authors (e.g., Helgeson, 1969; Holland, 1978; Stumm and Morgan, 1981). Five separate equilibria are taken into consideration in this model, relating aqueous (*a*), gaseous (*g*), and condensed (*c*) phases:

$$H^+_{(a)} + CO^{2-}_{3(a)} \Leftrightarrow HCO^-_{3(a)} \tag{8.73}$$

$$H^+_{(a)} + HCO^-_{3(a)} \Leftrightarrow H_2CO^*_{3(a)} \tag{8.74}$$

$$H_2O_{(a)} \Leftrightarrow H^+_{(a)} + OH^-_{(a)} \tag{8.75}$$

$$CO_{2(g)} + H_2O_{(a)} \Leftrightarrow H_2CO^*_{3a} \tag{8.76}$$

$$CaCO_{3(c)} \Leftrightarrow Ca^{2+}_{(a)} + CO^{2-}_{3(a)} . \tag{8.77}$$

The first three equilibria relate the activities of the main aqueous species, the fourth establishes the relationship between gaseous and aqueous phases, and the fifth establishes the relationship between aqueous species and the condensed phase.

The speciation state determined by the five equilibria depends on the type of system (closed or open) and imposed conditions (f_{CO_2}, pH). Two cases are particularly relevant in geochemistry:

1. Equilibrium between $CaCO_{3(c)}$ and an aqueous phase with f_{CO_2} buffered by an external reservoir (atmosphere)
2. Equilibrium between $CaCO_{3(c)}$ and an aqueous phase with a constant quantity of solutes and fixed pH.

Before we examine these two cases in detail, it is necessary to introduce some precautionary considerations.

1. Equilibria are normally treated by considering calcite as the only carbonate polymorph present in the system. Actually, as already mentioned, $CaCO_{3(c)}$ crystallizes in two forms: calcite (trigonal) and aragonite (orthorhombic). Aragonite, which at low pressure is metastable with respect to the less dense calcite (cf. table 2.1), is in fact synthesized by several marine organisms that fix $CaCO_{3(c)}$ in this form in their shells. Nevertheless, the difference in the Gibbs free energy of formation of the two polymorphs is so limited (0.2 to 0.5 kcal/mole) that the choice of type of phase does not influence calculations.
2. Carbonates in sedimentary rocks are mixtures of several components: $CaCO_3$, $CaMg(CO_3)_2$, $FeCO_3$, $SrCO_3$, and $BaCO_3$. Calculations are usually developed assuming that the condensed phase is pure calcite and assigning unitary activity to the $CaCO_3$ component. For accurate calculations, we would introduce an appropriate activity value, depending on the actual chemistry of the solid and bearing in mind that carbonates are not at all ideal mixtures but have wide miscibility gaps (cf. figure 3.11).

3. Aqueous solutions of carbon dioxide contain both $CO_{2(a)}$ and $H_2CO_{3(a)}$ neutral molecules. The $CO_{2(a)}$ molecule has a linear structure (O=C=O), whereas that of $H_2CO_{3(a)}$ is planar trigonal because of the molecular orbitals constituted of hybrid AO sp^2. Conversion from the linear form $CO_{2(a)}$ to the trigonal-planar form $H_2CO_{3(a)}$—i.e.,

$$CO_{2(a)} + H_2O_{(a)} \rightarrow H_2CO_{3(a)} \tag{8.78}$$

—is relatively slow (about 3% of $CO_{2(a)}$ is converted into $H_2CO_{3(a)}$ per second in room conditions; cf. Kern, 1960). The dehydration process

$$H_2CO_{3(a)} \rightarrow CO_{2(a)} + H_2O_{(a)} \tag{8.79}$$

is relatively faster (net rate constant $K_r = 2.0 \times 10^{-3}$ sec^{-1} at 0 °C, according to Jones et al., 1964).

The equilibrium kinetics of reactions 8.78 and 8.79 show that, at room P and T and in stationary conditions, the $CO_{2(a)}/H_2CO_{3(a)}$ ratio is near 600 (Kern, 1960). Notwithstanding this fact, in geochemistry it is customary to represent aqueous CO_2 as if it were entirely in the form of carbonic acid $H_2CO_{3(a)}$, not distinguishing planar from trigonal forms (i.e., $H_2CO_{3(a)}^*$ in eq. 8.74). This has little connection with reality and may lead to errors unless all equilibria are explicitly defined according to a single value of the Gibbs free energy of formation for aqueous neutral species.

8.10.1 Calcite in Equilibrium with an Aqueous Phase at Low Salinity and with a Gaseous Phase at Constant CO_2 Fugacity

Because the electrolyte $H_2CO_{3(a)}$ does not directly influence the acidity of the aqueous phase, it is preferable to avoid the lack of precision inherent in the definition of the form of the aqueous CO_2 by replacing equation 8.76 by

$$CO_{2(g)} + H_2O_{(a)} \Leftrightarrow HCO_{3(a)}^- + H_{(a)}^+ \tag{8.80}$$

$$CO_{2(g)} + H_2O_{(a)} \Leftrightarrow CO_{3(a)}^{2-} + 2 H_{(a)}^+. \tag{8.81}$$

The amount of $H_2CO_{3(a)}^*$ may then be determined through equation 8.74.

Table 8.6 lists equilibrium constants for reactions 8.73, 8.77, 8.80, and 8.81 along the water-vapor univariant curve. We use values of $T = 25$ °C and $P = 1$ bar in the following calculations.

Let us equilibrate some pure calcite crystals with an aqueous phase at $T = 25$ °C and $P = 1$ bar. Calcite partially dissolves to reach the solubility product (eq. 8.77).

Table 8.6 Constants of carbonate equilibria at various P and T conditions along univariant water-vapor boundary

P (bar)	T (°C)	$\log K_{77}$	$\log K_{77}^{(*)}$	$\log K_{73}$	$\log K_{80}$	$\log K_{81}$
1	25	−8.616	−8.451	10.329	−7.818	−18.147
0.006	0	−8.485	−8.317	10.553	−7.683	−18.236
0.123	50	−8.771	−8.608	10.123	−7.988	−18.111
1.013	100	−9.072	−8.912	9.759	−8.385	−18.144
4.758	150	−9.359	−9.198	9.434	−8.831	−18.265
15.537	200	−9.652	−9.487	9.115	−9.323	−18.438
39.728	250	−9.987	−9.817	8.751	−9.890	−18.641
85.805	300	−10.459	−10.282	8.235	−10.626	−18.861
165.125	350	−11.480	−11.297	7.117	−11.973	−19.090

(*)With aragonite as condensed phase at equilibrium.

Because the activity of component $CaCO_{3(c)}$ in the pure calcite phase is 1, we can write

$$\left[Ca^{2+}\right]\left[CO_3^{2-}\right] = K_{77} = 10^{-8.62}. \tag{8.82}$$

Based on equations 8.80 and 8.81 and assuming unitary fugacity for $CO_{2(g)}$ at standard state (see section 9.1), we can write

$$\left[HCO_3^-\right]\left[H^+\right] \cdot f_{CO_{2(g)}}^{-1} = K_{80} = 10^{-7.82} \tag{8.83}$$

$$\left[CO_3^{2-}\right]\left[H^+\right]^2 \cdot f_{CO_{2(g)}}^{-1} = K_{81} = 10^{-18.15} \tag{8.84}$$

and, based on equation 8.73,

$$\left[HCO_3^-\right]\left[H^+\right]^{-1}\left[CO_3^{2-}\right]^{-1} = K_{73} = 10^{10.33}. \tag{8.85}$$

We also recall that the ionic dissociation constant of pure water under room conditions is

$$\left[H^+\right]\left[OH^-\right] = 10^{-14}. \tag{8.86}$$

Let us now express the activities of the main ionic species in solution as a function of the activity of hydrogen ions and $f_{CO_{2(g)}}$. Combining equations 8.82 to 8.86, we obtain

$$\left[Ca^{2+}\right] = 10^{9.53}\left[H^+\right]^2 \cdot f_{CO_{2(g)}}^{-1} \tag{8.87}$$

$$\left[HCO_3^-\right] = 10^{-7.82}\left[H^+\right]^{-1} \cdot f_{CO_{2(g)}} \tag{8.88}$$

$$\left[CO_3^{2-}\right] = 10^{-18.15}\left[H^+\right]^{-2} \cdot f_{CO_{2(g)}} \tag{8.89}$$

$$\left[OH^-\right] = 10^{-14.00}\left[H^+\right]^{-1}. \tag{8.90}$$

System 8.87→8.90 is composed of four equations with six unknowns. If we fix the fugacity of gaseous CO_2, the unknowns are reduced to five. Note, moreover, that equation 8.85 has virtually no influence, because it can be obtained by adding equations 8.80 and 8.81. To determine the system, we introduce the electroneutrality condition

$$2m_{Ca^{2+}} + m_{H^+} = 2m_{CO_3^{2-}} + m_{HCO_3^-} + m_{OH^-} \tag{8.91}$$

(cf. section 8.9).

Let us now assume that, at first approximation, all activity coefficients of aqueous species are 1. Equation 8.91 becomes

$$2\left[Ca^{2+}\right] + \left[H^+\right] = 2\left[CO_3^{2-}\right] + \left[HCO_3^-\right] + \left[OH^-\right]. \tag{8.92}$$

Combining equation 8.92 with the system 8.87→8.90, we obtain a polynomial in $[H^+]$ that may be solved with iterative procedures for various values of $f_{CO_{2(g)}}$:

$$10^{9.83} \cdot f_{CO_{2(g)}}^{-1} \cdot \left[H^+\right]^4 + \left[H^+\right]^3$$
$$-\left(10^{-7.82} \cdot f_{CO_{2(g)}} + 10^{-14}\right)\left[H^+\right] - 10^{-17.85} \cdot f_{CO_{2(g)}} = 0. \tag{8.93}$$

Figure 8.18 shows how equation 8.93 behaves in the vicinity of zero. Equilibrium pH corresponds to the zero value on the ordinate axis.

If we adopt the present-day atmospheric $f_{CO_{2(g)}}$ value ($10^{-3.48}$), the corresponding pH, obtained through application of equation 8.93, is 8.2. It is well known that the amount of atmospheric CO_2 is steadily increasing, as a result of intensive combustion of fossil fuels. It has been deduced, from present-day exploitation rates and projections, that the amount of $CO_{2(g)}$ in the atmosphere will double in the next 40 to 70 years (i.e., $f_{CO_{2(g)}}$ will increase from $10^{-3.48}$ to $10^{-3.17}$). Correspondingly, the pH will fall by about 0.2 unit.

Equation 8.93 is asymptotic, with a flat flexus near zero (figure 8.18). Obviously, this type of equation does not allow precise pH determinations, because even slight approxi-

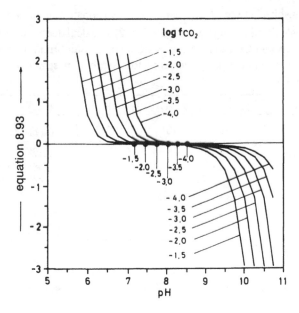

Figure 8.18 Relationship between pH of aqueous solution and f_{CO_2} of gaseous phase in equilibrium with water and calcite. Equilibrium pH is in correspondance with zero on ordinate axis.

mations in the equilibrium constants generate large variations in the solving value (the small pH differences between the procedure of Garrels and Christ, 1965, and the one developed here at quasi-parity of $f_{CO_{2(g)}}$ must be ascribed to this sort of problem). The method therefore cannot be used for high ionic strength or chemically complex solutions, in which the electroneutrality condition cannot be simplified in the form of equation 8.91. In seawater, for instance, only 88% of calcium is in the form of Ca^{2+} and 12% is in the aqueous sulfate form of $CaSO_4^0$. Similarly, the bicarbonate ion constitutes only 73% of total carbonate in solution: 13% is present as magnesium bicarbonate ($MgHCO_3^+$), 5.5% as magnesium carbonate ($MgCO_3^0$), 2.8% as calcium bicarbonate ($CaHCO_3^+$), and only 2.1% as carbonate ion. Clearly, for seawater the carbonate equilibria of equations 8.73→8.77 constitute only a particular subsystem of the n equilibria that actually determine the speciation state.

Nevertheless, it is of interest to note that precise speciation calculation conducted with an automated routine clearly indicates that, at the pH observed in seawater (8.22), the activities of Ca^{2+} and CO_3^{2-} in solution are near the solubility product of calcite and $f_{CO_{2(g)}}$ is near the atmospheric value (i.e., $f_{CO_{2(g)}} = 10^{-3.35}$). We can thus deduce that carbonate equilibria are effectively determinant in seawater and that they are conditioned by the atmospheric value of $f_{CO_{2(g)}}$. Table 8.7 shows that the increase in CO_2 in the atmosphere determines a progressive decrease of pH in seawater, in line with what was previously deduced for an aqueous solution of low salinity. In light of this marked control by atmospheric CO_2 on the acidity of aqueous solutions, during geochemical samplings of deep waters, direct contact with the atmosphere must be avoided before pH measurements are taken, because otherwise the pH turns out to be intermediate between the in situ value and the new equilibrium situation. Note here that, according to the calculation of Garrels and Christ (1965), the pH of an aqueous solution in equilibrium with calcite at $T = 25$

Table 8.7 Relationships between $f_{CO_{2(g)}}$ and pH in carbonate equilibria at $T = 25\,°$C and $P = 1$ bar. (1) = calculated through equation 8.93; (2) = calculated with EQ3NR/EQ6 routine; (3) = Garrels and Christ (1965). A = seawater in equilibrium with calcite and atmospheric CO_2; B = seawater in equilibrium with calcite; C = seawater in equilibrium with atmospheric CO_2.

	A			B	C		
Parameter	(1)	(1)	(3)	(3)	(2)	(2)	(2)
$\log f_{CO_{2(g)}}$	−3.17	−3.48	−3.50	-	−3.17	−3.35	−3.48
pH	8.00	8.20	8.40	9.90	8.06	8.22	8.32
$\log[Ca^{2+}]$	−3.30	−3.39	−3.40	−3.90	−2.65	−2.65	−2.65
$\log[HCO_3^-]$	−2.99	−3.10	−3.00	−4.05	−2.95	−2.98	−3.00
$\log[CO_3^{2-}]$	−5.32	−5.23	−4.90	−4.40	−5.22	−5.08	−5.01
$\log[OH^-]$	−6.00	−5.80	−5.60	−5.10	−5.94	−5.78	−5.68

°C and $P = 1$ bar and in the absence of atmospheric CO_2 is 9.9—i.e., much higher than the value obtained for equilibrium with atmosphere (cf. table 8.7).

8.10.2 Calcite in Equilibrium with an Aqueous Phase at Fixed pH

This is the case generally encountered in geochemical investigations: a deep water is sampled and total carbonates and pH are determined to solve the question of whether or not the sampled water was in equilibrium with calcite (or other carbonates) at the T and P of sampling. The system is clearly determined; the equilibria to be considered in this case are those of equations 8.73, 8.74, and 8.77 (eq. 8.76, 8.80, and 8.81 have no influence).

As a specific example of the problem, let us calculate the equilibria for an actual case study: a deep water from the Sarcidano region (Sardinia, Italy) in equilibrium with Mesozoic dolomites (Bertorino et al., 1981). The compositions in mEq/l of water sampled in a drilled well are listed in table 8.8. The in situ temperature is 21 °C; we assume here that the in situ T is 25 °C at 1 bar, to simplify calculations. We also assume for the sake of simplicity that the main ion species in solution are HCO_3^-, Mg^{2+}, Ca^{2+}, CO_3^{2-}, OH^-, and H^+, and that all Ca and Mg are in the ionic forms Ca^{2+} and Mg^{2+}.

The ionic strength of the solution is $I = 0.014$. Applying the Debye-Hückel equation (eq. 8.41), the appropriate solvent parameters A and B, and the effective diameters for the ions in solution listed in table 8.5, we obtain the following activity coefficients: $\gamma_{HCO_3^-} = 0.90$, $\gamma_{Mg^{2+}} = 0.68$, $\gamma_{Ca^{2+}} = 0.66$, $\gamma_{CO_3^{2-}} = 0.65$, $\gamma_{OH^-} = 0.90$, and $\gamma_{H^+} = 0.91$.

Let us admit that total inorganic carbon represents the summation of partial molalities:

$$m_{H_2CO_{3(a)}^*} + m_{HCO_{3(a)}^-} + m_{CO_{3(a)}^{2-}} = 8.1 \times 10^{-3}, \tag{8.94}$$

with

$$m_{H_2CO_{3(a)}^*} = m_{H_2CO_{3(a)}} + m_{CO_{2(a)}}. \tag{8.95}$$

Equilibrium 8.74 has constant

$$K_{74} = \frac{\gamma_{H_2CO_3^*} \cdot m_{H_2CO_3^*}}{\gamma_{H^+} \cdot m_{H^+} \cdot \gamma_{HCO_3^-} \cdot m_{HCO_3^-}} = 10^{6.4}. \tag{8.96}$$

With

$$\gamma_{H_2CO_3^*} = 1, \quad \gamma_{H^+} \cdot m_{H^+} = 10^{-7.3}, \quad \text{and} \quad \gamma_{HCO_3^-} = 0.90,$$

we obtain

$$m_{H_2CO_3^*} = 10^{-0.94} \cdot m_{HCO_3^-}. \tag{8.97}$$

Equation 8.73 yields

$$K_{73} = \frac{m_{HCO_3^-} \cdot \gamma_{HCO_3^-}}{\gamma_{H^+} \cdot m_{H^+} \cdot \gamma_{CO_3^{2-}} \cdot m_{CO_3^{2-}}} = 10^{10.33}. \tag{8.98}$$

Table 8.8 Compositions of superficial and deep waters of the Sarcidano region (Sardinia, Italy) equilibriated with Mesozoic dolomite limestones (Bertorino et al., 1981). Values in mEq/l. C_T total inorganic carbon in mmol/l.

Sample	T (°C)	Type	pH	Ca	Mg	Na	K	C_T	SO$_4$	Cl
27	14	Superficial	7.1	4.4	4.7	0.9	0.04	10	1.1	0.9
28	14	Superficial	7.1	3.5	3.5	0.6	0.02	8.1	0.7	0.7
29	16	Superficial	7.6	4.2	3.8	1.0	0.04	8.4	1.0	0.9
30	15	Superficial	7.1	3.7	4.6	1.1	0.03	9.4	0.6	1.1
36	14	Superficial	7.1	4.2	4.5	0.9	0.04	9.7	1.0	1.0
37	13	Superficial	6.9	4.8	5.2	0.9	0.05	12	1.0	1.0
38	14	Superficial	7.1	3.4	3.3	0.6	0.02	7.5	1.1	0.7
39	13	Superficial	7.1	3.9	3.9	0.6	0.02	9.1	0.2	0.7
40	15	Superficial	7.1	3.2	3.2	0.6	0.02	7.3	1.0	0.6
79	21	Well	7.3	3.5	3.5	1.5	8.1	8.1	0.9	1.4

With $\gamma_{CO_3^{2-}} = 0.65$, we obtain

$$m_{CO_3^{2-}} = 10^{-2.888} \times m_{HCO_3^-}. \tag{8.99}$$

From the equation of total inorganic carbon (eq. 8.94), combined with equations 8.97 and 8.99, we derive

$$10^{-0.94} \times m_{HCO_3^-} + m_{HCO_3^-} + 10^{-2.888} \times m_{HCO_3^-} = 10^{-2.092}. \tag{8.100}$$

Thus, $m_{HCO_3^-} = 10^{-2.139}$, $m_{H_2CO_3^*} = 10^{-3.079}$, and $m_{CO_3^{2-}} = 10^{-5.027}$.

From the total molality of Ca, assuming all calcium to be present in solution in the ionic form Ca^{2+}, we obtain

$$\gamma_{Ca^{2+}} \cdot m_{Ca^{2+}} = 0.66 \times 0.00175 = 10^{-2.937}. \tag{8.101}$$

The activity product derived for Ca^{2+} and CO_3^{2-} is thus

$$Q = \left[Ca^{2+}\right]\left[CO_3^{2-}\right] = 0.65 \times 10^{-5.027} \times 10^{-2.937} = 10^{-8.151}. \tag{8.102}$$

The solubility product of calcite at $T = 25\ °C$ and $P = 1$ bar is $K = 10^{-8.62}$.

The log $Q/K = 0.469$ value indicates slight oversaturation in calcite with an affinity to precipitation (cf. section 2.12):

$$A_P = RT \ln \frac{Q}{K} = 0.640 \qquad \left(\frac{kcal}{mole}\right). \tag{8.103}$$

Table 8.9 shows the results of detailed speciation calculations carried out with the automated routine EQ3NR/EQ6 (Wolery, 1983). The molalities of the ions in solution resulting from this calculation are tabulated in order of decreasing abundance.

In the total inorganic carbon balance of equation 8.94, we neglected the presence of some species. Table 8.9 shows that this approximation is reasonable although, for accurate calculations, we should also have considered bicarbonates $CaHCO_3^+$ and $MgHCO_3^+$ and the soluble carbonate $CaCO_3^0$. Note also that sulfates occur in negligible molar proportions and that carbonate alkalinity coincides with total alkalinity.

The precise speciation calculation indicates that four minerals are at almost perfect saturation:

Calcite: log $Q/K = 0.359$; $A_P = 0.490$ kcal/mole

Disordered dolomite: log $Q/K = 0.195$; $A_P = 0.265$ kcal/mole

Table 8.9 Speciation of a deep water from Sarcidano (sample 79 in table 8.8) according to EQ3NR/EQ6 procedure.

Species	log m	log γ	log a
HCO_3^-	−2.1548	−0.0466	−2.2014
Mg^{2+}	−2.7570	−0.1697	−2.9267
Ca^{2+}	−2.7570	−0.1796	−2.9366
Na^+	−2.8239	−0.0473	−2.8712
H_2CO_3	−3.1553	0.0013	−3.1540
$CaHCO_3^+$	−3.8423	−0.0456	−3.8879
K^+	−3.8539	−0.0488	−3.9027
$MgHCO_3^+$	−3.9945	−0.0473	−4.0418
$CaCO_3^0$	−4.9678	0.0	−4.9678
CO_3^{2-}	−5.0417	−0.1878	−5.2295
$MgCO_3$	−5.1750	0.0	−5.1750
OH^-	−6.6393	−0.0481	−6.6874
$NaCO_3^-$	−7.0938	−0.0466	−7.1404
H^+	−7.2591	−0.0409	−7.3000
$Mg(OH)^+$	−7.3537	−0.0473	−7.4010
$Cl-$	−7.4036	−0.0488	−7.4524
$Ca(OH)^+$	−8.1772	−0.0473	−8.2245
SO_4^{2-}	−8.4411	−0.1907	−8.6318
$CaSO_4^0$	−9.2570	0.0	−9.2570
$MgSO_4^0$	−9.3109	0.0	−9.3109
$NaOH$	−10.2586	0.0	−10.2586
$NaSO_4^-$	−10.7365	−0.0469	−10.7834
$NaCl$	−11.2362	0.0	−11.2362
KSO_4^-	−11.6340	−0.0481	−11.6821
HSO_4^-	−13.8794	−0.0473	−13.9267
HCl	−20.8863	0.0	−20.8863
$O_{2(a)}$	−22.9945	0.0013	−22.9932
$H_{2(a)}$	−34.6176	0.0013	−34.6163
$HS-$	−108.2457	−0.0481	−108.2937
$H_2S_{2(a)}$	−108.6132	0.0	−108.6132
S^{2-}	−117.9757	−0.1850	−118.1607

Aragonite: log Q/K = 0.194; A_P = 0.264 kcal/mole

Magnesite: log Q/K = −0.249; A_P = −0.339 kcal/mole.

Our approximation in the calculation is 150 cal/mole, which is quite satisfactory. The fact that dolomite is the mineral phase nearest to complete equilibrium is in agreement with the geology of the area and was in fact already evident at first sight from the chemical analyses, indicating similar molalities for Ca and Mg in solution. The resulting $f_{CO_{2(g)}}$ is $10^{-1.67}$, which is much higher than the atmospheric value. The previous recommendation about pH measurements is particularly obvious in this case.

In geology we often encounter conditions that are intermediate with respect to the two limiting cases mentioned above. For instance, a rainwater initially at equilibrium with atmospheric CO_2 percolates through nonreactive soils and reaches a carbonate layer: this is clearly a combination of the two cases discussed in this section and in section 8.10.1. Calculations on a semiquantitative basis in such a case have only academic interest (see Garrels and Christ, 1965, for a detailed treatment).

8.11 Thermodynamic Properties of Solutes Under High P and T Conditions: the Helgeson-Kirkham-Flowers Approach

Study of water-rock interactions occurring within the continental crust and along mid-ocean ridges, where large masses of water are involved in convective transport within the lithosphere, requires accurate estimates of the partial molal properties of solutes under high P and T conditions. In other words, in order to determine the reactivity of a given aqueous solution toward a given mineralogical assemblage with which it comes into contact, a detailed speciation calculation at the P and T of interest is necessary. As we have already seen in section 8.9, here we need knowledge of the various equilibrium constants among solutes, and these constants, to be determined, in turn require knowledge of the partial molal Gibbs free energies of the solutes at the P and T of interest. The energy model developed by Helgeson, Kirkham, and Flowers (1981) and later revised by Tanger and Helgeson (1988) allows accurate calculation of the partial molal properties of ions in aqueous solutions up to a maximum P of 5 kbar and a maximum T of 1000 °C. A brief outline of the procedure is given here (see the works of Helgeson and co-authors for more exhaustive explanations).

8.11.1 Standard State Conventions and the Additivity Principle

The standard state adopted in the Helgeson-Kirkham-Flowers (HKF) model for aqueous electrolytes and ions in solution is the classical one of a "hypothetical one-molal solution referred to infinite dilution at each P and T condition" (see section 8.4). The standard state for the solvent is that of a "pure solvent at each P and T condition." Let us recall briefly the significance of the "partial molal property" of a solute in an aqueous solution. Chapter 2 introduced the concept of the chemical potential of a component in mixture as partial derivative of the Gibbs free energy of the phase with respect to the number of moles of component (eq. 2.3). Because the Gibbs free energy of the mixture is determined by the standard state energy of the pure components in their molar proportions plus the mixing terms (ideal plus excess), the chemical potential of a given component in mixture is not a constant, but varies in a more or less complex fashion with the chemistry of the phase as a function of interactions of all the components in that mixture. We have also seen that we can consider the chemical potential of the component as made up of partial molar properties (eq. 2.59, 2.60, 2.61). These properties (i.e., partial molar enthalpy, partial molar entropy, partial molar volume) represent the discrete values assumed by the molar property at any particular composition of the phase in the system. Partial molar properties are related, in the terms outlined by equations 2.31 to 2.34 and 2.1, to the Gibbs free energy of the phase (and of the system). The significance of the partial molal properties of solutes in aqueous solutions are obviously similar, although some distinction must be made:

1. These properties are referred to the *molality* of the solute (i.e., moles of solute per 10^3 g of solvent) and not to *molarity.*

2. "Absolute" properties are distinguished from "conventional" properties; i.e., a generic standard partial molal property of an aqueous ion j ($\overline{\Xi}_j^0$) is related to the absolute property by

$$\overline{\Xi}_j^0 = \Xi_j^{0\,abs} - Z_j\,\Xi_{H^+}^{0\,abs}, \tag{8.104}$$

where Z_j is the formal charge of the jth aqueous ion and $\Xi_{H^+}^{0abs}$ is the corresponding absolute property of the hydrogen ion.

3. An "additivity principle" is adopted (Millero, 1972), according to which the conventional standard partial molal property of a generic electrolyte K is related to the absolute properties of its constituent ions through

$$\overline{\Xi}_K^0 = \sum_j v_{j,K}\,\Xi_j^{0\,abs}, \tag{8.105}$$

where $v_{j,K}$ is the stoichiometric number of moles of ion j in electrolyte K.

From equations 8.104 and 8.105, applying the electroneutrality condition, it follows that the additivity principle is also valid for conventional properties—i.e.,

$$\overline{\Xi}_K^0 = \sum_j v_{j,K}\,\overline{\Xi}_j^0. \tag{8.106}$$

Because the conventional standard partial molal properties of a hydrogen ion is zero, it follows that the conventional standard partial molal properties of a generic anion A are identical to the experimental values of the corresponding acid electrolyte. Moreover, based on equation 8.104, the standard partial molal properties of a generic cation C can be calculated, once the experimental values for aqueous electrolytes $H_{v+}A$ and $C_{v+}A_{v-}$ are known. Note here the close analogy between the additivity principle and the mean salt method, previously discussed as a method of calculating ionic activity coefficients (section 8.6).

8.11.2 Generalities of Approach

Each partial molal property of a jth aqueous ion in solution is considered to be composed of two terms: the *intrinsic property* of the solute ($\Xi_{i,j}^0$) and an *electroconstriction contribution* ($\Delta\Xi_{e,j}^0$):

$$\Xi_j^{0\,abs} = \Xi_{i,j}^{0\,abs} + \Delta\Xi_{e,j}^{0\,abs}. \tag{8.107}$$

The electroconstriction contribution derives from the structural collapse of the solvent in the immediate neighborhood of the ion ($\Delta\Xi_{c,j}^{0abs}$) and by the ion-solvation process ($\Delta\Xi_{s,j}^{0abs}$) (electrostatic solute-solvent interactions):

$$\Delta\Xi_{e,j}^{0\,abs} = \Delta\Xi_{c,j}^{0\,abs} + \Delta\Xi_{s,j}^{0\,abs}. \tag{8.108}$$

Equations 8.107 and 8.108, referring to absolute properties, are equally valid for conventional properties based on equations 8.105 and 8.106. Moreover, the additivity principle is applied to all partial contributions: thus, for a generic electrolyte K, we have

$$\Delta\Xi_{e,K}^{0} = \sum_{j} v_{j,K}\,\Delta\Xi_{e,j}^{0}, \tag{8.109}$$

$$\Delta\Xi_{c,K}^{0} = \sum_{j} v_{j,K}\,\Delta\Xi_{c,j}^{0}, \tag{8.110}$$

$$\Delta\Xi_{s,K}^{0} = \sum_{j} v_{j,K}\,\Delta\Xi_{s,j}^{0}, \tag{8.111}$$

and so on.

The HKF model groups the intrinsic properties of the solute and the effects of structural collapse of the solvent into a single term, which is defined as the contribution of "nonsolvation":

$$\Delta\Xi_{n,j}^{0\,abs} = \Xi_{i,j}^{0\,abs} + \Delta\Xi_{c,j}^{0\,abs}. \tag{8.112}$$

The partial molal property of the solute is thus composed of one nonsolvation term ($\Delta\Xi_{n,j}^{0abs}$) plus one solvation ($\Delta\Xi_{s,j}^{0abs}$) term; i.e., for a generic ion j,

$$\Xi_{j}^{0\,abs} = \Delta\Xi_{n,j}^{0\,abs} + \Delta\Xi_{s,j}^{0\,abs} \tag{8.113}$$

$$\Xi_{j}^{0} = \Delta\Xi_{n,j}^{0} + \Delta\Xi_{s,j}^{0}. \tag{8.114}$$

Similarly, for a generic electrolyte K, he have

$$\Xi_{K}^{0} = \Delta\Xi_{n,K}^{0} + \Delta\Xi_{s,K}^{0}. \tag{8.115}$$

8.11.3 Effective Electrostatic Radius, Born Coefficient, and Solvation Energy

We have already mentioned the significance of the "effective diameter" of an ion in solution, stating that it represents the distance within which a given ion in solution affects the structure of the solvent (section 8.5). We must now introduce the concept of *effective electrostatic radius*. For this purpose, we recall that the Gibbs free energy change $\Delta G_{s,j}^{0\,abs}$ involved in the solvation of a generic ion j in solution is defined by the Born equation

$$\Delta G_{s,j}^{0\,abs} = \omega_j^{abs}\left(\frac{1}{\varepsilon} - 1\right),$$

(8.116)

(Born, 1920), where ε is the dielectric constant of the solvent and ω_j^{abs} is the absolute Born coefficient for the jth ion in solution:

$$\omega_j^{abs} = \frac{N_0 e^2 Z_j^2}{2r_{e,j}},$$

(8.117)

where N_0 is Avogadro's number, e is the elementary charge, Z_j is the formal charge of ion j, and $r_{e,j}$ is its electrostatic radius. Posing

$$\eta = \frac{1}{2}N_0 e^2 = 1.66027 \times 10^5 \qquad \left(\frac{\text{Å} \cdot \text{cal}}{\text{mole}}\right),$$

(8.118)

equation 8.117 can be reexpressed as

$$\omega_j^{abs} = \frac{\eta Z_j^2}{r_{e,j}}.$$

(8.119)

Equation 8.116 is also valid for conventional magnitudes:

$$\Delta G_{s,j}^0 = \omega_j\left(\frac{1}{\varepsilon} - 1\right).$$

(8.120)

The conventional Born coefficient ω_j is related to the absolute coefficient ω_j^{abs} by

$$\omega_j = \omega_j^{abs} - Z_j\omega_{H^+}^{abs},$$

(8.121)

where $\omega_{H^+}^{abs}$ is the absolute Born coefficient of the hydrogen ion at the P and T of interest ($\omega_{H^+}^{abs} = 0.5387 \times 10^5$ cal/mole at $T = 25\ °C$ and $P = 1$ bar; cf. Helgeson and Kirkham, 1974).

Stemming from the general relations developed in the preceding section, the solvation energy of a generic electrolyte K can be related to that of its constituting ions j through the Born coefficient ω_K (Tanger and Helgeson, 1988):

$$\omega_K = \sum_j v_{j,K}\omega_j^{abs} = \sum_j \frac{v_{j,K}\eta Z_j^2}{r_{e,j}} = \sum_j v_{j,K}\omega_j$$

(8.122)

$$\Delta G^0_{s,K} = \omega_K \left(\frac{1}{\varepsilon} - 1 \right).$$

(8.123)

The solvation energy described by the Born equation is essentially electrostatic in nature. Born equations 8.116 and 8.120 are in fact similar to the Born-Landé equation (1.67) used to define the electrostatic potential in a crystal (see section 1.12.1). In light of this analogy, the effective electrostatic radius of an ion in solution $r_{e,j}$ assumes the same significance as the "equilibrium distance" in the Born-Landé equation. We may thus expect a close analogy between the crystal radius of an ion and the effective electrostatic radius of the same ion in solution.

This analogy is expressed by the empirical equation

$$r_{e,j} = r_{x,j} + |Z_j| \Gamma_{\pm},$$

(8.124)

where $r_{x,j}$ is the crystal radius of generic ion j and

$$\Gamma_{\pm} = K_Z + g.$$

(8.125)

K_Z in equation 8.125 represents a charge constant that is zero for anions and 0.94 Å for cations, and g is a solvent parameter dependent on P and T (see equations 36A, B, and C in Tanger and Helgeson, 1988).

Table 8.10 lists electrostatic radii of aqueous ions at $T = 25\,°C$ and $P = 1$ bar (Shock and Helgeson, 1988).

8.11.4 Standard Partial Molal Volume, Compressibility, and Thermal Expansion

The standard partial molal volume of a generic ion (or electrolyte) in solution can be expressed, along the lines of the preceding section, through the summation of "nonsolvation" and "solvation" contributions—i.e., omitting the subscripts (j and/or K):

$$\overline{V}^0 = \Delta\overline{V}_n^0 + \Delta\overline{V}_s^0.$$

(8.126)

Partial molal volumes can be related to the corresponding Gibbs free energy terms through the partial derivatives on P (see equations 2.28 and 2.33).

$$\Delta\overline{V}_n^0 = \left(\frac{\partial \Delta\overline{G}_n^0}{\partial P} \right)_T$$

(8.127)

$$\Delta\overline{V}_s^0 = \left(\frac{\partial \Delta\overline{G}_s^0}{\partial P} \right)_T.$$

(8.128)

The solvation contribution has the form

Table 8.10 Electrostatic radii (Å) of various aqueous ions at $T = 25\,°C$ and $P = 1$ bar, after Shock and Helgeson (1988).

Ion	r_e	Ion	r_e	Ion	r_e	Ion	r_e
Li^+	1.62	La^{3+}	3.96	Re^{4+}	4.48	BrO^-	1.77
Na^+	1.91	Ce^{3+}	3.89	Ru^{4+}	4.44	BrO_3^-	3.29
K^+	2.27	Pr^{3+}	3.88	S^{4+}	4.13	BrO_4^-	4.51
Rb^+	2.41	Nd^{3+}	3.86	Se^{4+}	4.26	Br_3^-	5.35
Cs^+	2.61	Sm^{3+}	3.82	Si^{4+}	4.18	IO^-	1.50
Ag^+	2.20	Eu^{3+}	3.80	Sn^{4+}	4.47	IO_3^-	2.51
Tl^+	2.41	Gd^{3+}	3.79	Tb^{4+}	4.57	IO_4^-	5.77
Cu^+	1.90	Tb^{3+}	3.75	Te^{4+}	4.46	I_3^-	7.41
NH_4^+	2.41	Dy^{3+}	3.74	Th^{4+}	4.78	$H_2VO_4^-$	2.55
VO_2^+	1.34	Ho^{3+}	3.73	Ti^{4+}	4.44	$HCrO_4^-$	3.29
Au^+	2.31	Er^{3+}	3.71	U^{4+}	4.73	MnO_4^-	4.17
Fr^+	2.74	Tm^{3+}	3.69	V^{4+}	4.39	ReO_4^-	4.59
Mg^{2+}	2.54	Yb^{3+}	3.68	W^{4+}	4.46	BeO_2^{2-}	2.45
Ca^{2+}	2.87	Lu^{3+}	3.67	Zr^{4+}	4.55	CO_3^{2-}	2.87
Sr^{2+}	3.00	Sc^{3+}	3.63	F^-	1.33	SiF_6^{2-}	3.92
Ba^{2+}	3.22	Am^{3+}	3.89	Cl^-	1.81	$N_2O_2^{2-}$	3.26
Ra^{2+}	3.31	As^{3+}	3.40	Br^-	1.96	HPO_4^{2-}	2.93
Mn^{2+}	2.68	Au^{3+}	3.67	I^-	2.20	$H_2P_2O_7^{2-}$	4.30
Fe^{2+}	2.62	B^{3+}	3.05	OH^-	1.40	$HAsO_4^{2-}$	3.10
Ag^{2+}	2.77	Bi^{3+}	3.78	HS^-	1.84	SO_3^{2-}	2.96
Cu^{2+}	2.60	Cr^{3+}	3.45	AlO_2^-	1.40	SO_4^{2-}	3.21
Zn^{2+}	2.62	Mn^{3+}	3.48	BO_2^-	1.36	S_2^{2-}	6.8
Cd^{2+}	2.85	Np^{3+}	3.92	BF_4^-	3.78	$S_2O_3^{2-}$	3.51
Pb^{2+}	3.08	Pa^{3+}	3.95	HCO_3^-	2.26	$S_2O_4^{2-}$	3.69
Ni^{2+}	2.57	Pm^{3+}	3.88	CN^-	2.21	$S_2O_5^{2-}$	3.78
Co^{2+}	2.60	Pu^{3+}	3.90	NO_3^-	2.97	$S_2O_6^{2-}$	3.95
Hg^{2+}	2.98	Rh^{3+}	3.50	NO_2^-	2.57	$S_2O_8^{2-}$	5.31
Sn^{2+}	2.81	Sb^{3+}	3.58	N_3^-	2.37	S_3^{2-}	3.50
Sm^{2+}	2.88	Ti^{3+}	3.58	$H_2PO_4^-$	2.18	$S_3O_6^{2-}$	4.06
Eu^{2+}	3.06	V^{3+}	3.56	$H_3P_2O_7^-$	5.22	S_4^{2-}	3.77
Yb^{2+}	2.88	Am^{4+}	4.68	$H_2AsO_4^-$	2.49	$S_4O_6^{2-}$	5.52
VO^{2+}	2.53	Ce^{4+}	4.70	$H_2AsO_3^-$	2.40	S_5^{2-}	4.08
UO_2^{2+}	2.67	Ge^{4+}	4.29	SbS_4^-	1.40	$S_5O_6^{2-}$	4.34
Hg_2^{2+}	3.50	Hf^{4+}	4.54	HO_2^-	1.65	SeO_3^{2-}	3.18
Be^{2+}	2.23	Ir^{4+}	4.44	HSO_3^-	2.84	SeO_4^{2-}	3.42
Ge^{2+}	2.61	Mn^{4+}	4.36	HSO_4^-	2.54	TeO_3^{2-}	3.18
Pd^{2+}	2.68	Mo^{4+}	4.46	HSO_5^-	5.15	HVO_4^{2-}	3.20
Pt^{2+}	2.68	Nb^{4+}	4.50	HSe^-	2.07	CrO_4^{2-}	3.40
V^{2+}	2.76	Np^{4+}	4.71	$HSeO_3^-$	2.76	$Cr_2O_7^{2-}$	5.59
Al^{3+}	3.33	Os^{4+}	4.45	$HSeO_4^-$	3.02	MnO_4^{2-}	3.45
Fe^{3+}	3.46	Pa^{4+}	4.74	$HTeO_3^-$	2.83	MoO_4^{2-}	3.26
Ga^{3+}	3.44	Pb^{4+}	4.60	HF_2^-	2.20	WO_4^{2-}	3.34
In^{3+}	3.36	Pd^{4+}	4.41	ClO^-	1.77	PO_4^{3-}	3.74
Tl^{3+}	3.77	Pr^{4+}	4.68	ClO_2^-	2.29	$HP_2O_7^{3-}$	4.93
Co^{3+}	3.45	Pt^{4+}	4.41	ClO_3^-	3.30	AsO_4^{3-}	3.95
Y^{3+}	3.74	Pu^{4+}	4.69	ClO_4^-	3.85	$P_2O_7^{4-}$	5.59

$$\Delta V_s^0 = -\frac{\omega}{\varepsilon}\left(\frac{\partial \ln \varepsilon}{\partial P}\right)_T + \left(\frac{1}{\varepsilon} - 1\right)\left(\frac{\partial \omega}{\partial P}\right)_T. \tag{8.129}$$

The nonsolvation contribution can be expressed as a function of T through

$$\Delta \overline{V}_n^0 = \sigma + \xi\left(\frac{1}{T - \theta}\right), \tag{8.130}$$

where θ is a solvent constant that, for water, assumes the value $\theta = 228$ K (singular temperature of supercooled water; Angell, 1982), and σ and ξ are coefficients not dependent on T but variable with P, based on

$$\sigma = a_1 + a_2\left(\frac{1}{\psi + P}\right) \tag{8.131}$$

$$\xi = a_3 + a_4\left(\frac{1}{\psi + P}\right). \tag{8.132}$$

In equations 8.131 and 8.132, ψ is another solvent constant that, for water, has the value $\psi = 2600$ bar, and a_1, a_2, a_3, and a_4 are characteristic parameters of ions and electrolytes in solution. These parameters, constant over P and T, are listed in table 8.11.

Combining equations 8.126 to 8.132, the standard partial molal volume can be expressed as

$$\overline{V}^0 = a_1 + a_2\left(\frac{1}{\psi + P}\right) + \left[a_3 + a_4\left(\frac{1}{\psi + P}\right)\right]\left(\frac{1}{T - \theta}\right)$$
$$- \omega Q + \left(\frac{1}{\varepsilon} - 1\right)\left(\frac{\partial \omega}{\partial P}\right)_T, \tag{8.133}$$

where Q is the Born function:

$$Q = \frac{1}{\varepsilon}\left(\frac{\partial \ln \varepsilon}{\partial P}\right)_T. \tag{8.134}$$

The compressibility of a generic ion or electrolyte in solution is given by the variation with P of the nonsolvation and solvation terms:

$$-\overline{\beta}^0 = \left(\frac{\partial \Delta \overline{V}_n^0}{\partial P}\right)_T + \left(\frac{\partial \Delta \overline{V}_s^0}{\partial P}\right)_T. \tag{8.135}$$

Adopting the same parameters previously used for calculating the standard partial molal volume, compressibility can be expressed as

Table 8.11 Volume parameters of HKF model (from Shock and Helgeson, 1988). a_1 = cal/(mole × bar); a_2 = cal/mole; a_3 = (cal × K)/(mole × bar); a_4 = (cal × K)/mole

Ion	$a_1 \times 10$	$a_2 \times 10^{-2}$	a_3	$a_4 \times 10^{-4}$
H^+	0	0	0	0
Li^+	−0.0237	−0.0690	11.5800	−2.7761
Na^+	1.8390	−2.2850	3.2560	−2.726
K^+	3.5590	−1.4730	5.4350	−2.712
Rb^+	4.2913	−0.9041	7.4070	−2.7416
Cs^+	6.1475	−0.1309	4.2094	−2.7736
NH_4^+	3.8763	2.3448	8.5605	−2.8759
Ag^+	1.7285	−3.5608	7.1496	−2.6318
Tl^+	4.3063	2.7333	4.6757	−2.8920
VO_2^+	−2.5825	−14.0871	11.2869	−2.1966
Cu^+	0.7835	−5.8682	8.0565	−2.5364
Au^+	3.3448	0.3856	5.5985	−2.7949
Mg^{2+}	−0.8217	−8.5990	8.3900	−2.390
Ca^{2+}	−0.1947	−7.2520	5.2966	−2.4792
Sr^{2+}	0.7071	−10.1508	7.0027	−2.3594
Ba^{2+}	2.7383	−10.0565	−0.0470	−2.3633
Pb^{2+}	−0.0051	−7.7939	8.8134	−2.4568
Mn^{2+}	−0.1016	−8.0295	8.9060	−2.4471
Co_2^+	−1.0748	−12.9948	16.4112	−2.2418
Ni^{2+}	−1.6942	−11.9181	10.4344	−2.2863
Cu^{2+}	−1.1021	−10.4726	9.8662	−2.3461
Zn^{2+}	−1.0677	−10.3884	9.8331	−2.3495
Cd^{2+}	0.0537	−10.7080	16.5176	−2.3363
Be^{2+}	−1.0684	−10.3901	9.8338	−2.3495
Ra^{2+}	0.6285	−6.2469	8.2053	−2.5208
Sn^{2+}	0.0786	−7.5895	8.7330	−2.4652
VO^{2+}	−1.7351	−12.0180	10.4736	−2.2822
Fe^{2+}	−0.7803	−9.6867	9.5573	−2.3786
Pd^{2+}	−0.6079	−9.2658	9.3919	−2.3960
Ag^{2+}	−0.4299	−8.8310	9.2210	−2.4139
Hg^{2+}	−0.5280	−9.0707	9.3152	−2.4040
Hg_2^{2+}	4.0126	2.0164	4.9575	−2.8624
Sm^{2+}	1.4002	−4.3625	7.4647	−2.5987
Eu^{2+}	1.8333	−3.3050	7.0490	−2.6424
Yb^{2+}	0.7707	−5.8995	8.0688	−2.5351
Al^{3+}	−3.3404	−17.1108	14.9917	−2.0716
La^{3+}	−2.7880	−14.3824	10.9602	−2.1844
Ce^{3+}	−2.9292	−14.9338	11.6196	−2.1616
Pr^{3+}	−3.2406	−14.1998	8.1257	−2.1920
Nd^{3+}	−3.3707	−14.5452	8.3211	−2.1777
Sm^{3+}	−3.2065	−15.6108	11.8857	−2.1337
Eu^{3+}	−3.1037	−15.3599	11.7871	−2.1440
Gd^{3+}	−2.9771	−15.0506	11.6656	−2.1568
Tb^{3+}	−2.9355	−14.9491	11.6257	−2.1610
Dy^{3+}	−3.0003	−15.1074	11.6879	−2.1545
Ho_3^+	−3.1198	−15.3992	11.8026	−2.1424
Er^{3+}	−3.3041	−15.8492	11.9794	−2.1238

continued

Table 8.11 (continued)

Ion	$a_1 \times 10$	$a_2 \times 10^{-2}$	a_3	$a_4 \times 10^{-4}$
Tm^{3+}	-3.2967	-15.8312	11.9724	-2.1245
Yb^{3+}	-3.4983	-16.3233	12.1658	-2.1042
Lu^{3+}	-3.5630	-16.4812	12.2279	-2.0977
Ga^{3+}	-3.4573	-16.2232	12.1264	-2.1083
In^{3+}	-3.1646	-15.5086	11.8456	-2.1379
Tl^{3+}	-2.8195	-14.6658	11.5143	-2.1727
Fe^{3+}	-3.1784	-15.5422	11.8588	-2.1365
Co^{3+}	-3.3247	-15.8994	11.9992	-2.1217
Au^{3+}	-3.0567	-15.2450	11.7419	-2.1488
Sc^{3+}	-3.1236	-15.4084	11.8062	-2.1420
Y^{3+}	-3.0140	-15.1408	11.7010	-2.1531
AlO_2^-	3.7280	3.9800	-1.5170	-2.9435
HCO_3^-	7.5621	1.1505	1.2346	-2.8266
NO_3^-	7.3161	6.7824	-4.6838	-3.0594
NO_2^-	5.5864	5.8590	3.4472	-3.0212
$H_2PO_4^-$	6.4875	8.0594	2.5823	-3.1122
OH^-	1.2527	0.0738	1.8423	-2.7821
HS^-	5.0119	4.9799	3.4765	-2.9849
HSO_3^-	6.7014	8.5816	2.3771	-3.1338
HSO_4^-	6.9788	9.2590	2.1108	-3.1618
F^-	0.6870	1.3588	7.6033	-2.8352
Cl^-	4.0320	4.8010	5.5630	-2.847
ClO_3^-	7.1665	9.7172	1.9307	-3.1807
ClO_4^-	8.1411	17.3059	-12.2254	-3.4944
Br^-	5.2690	6.5940	4.7450	-3.143
BrO_3^-	6.9617	9.2173	2.1272	-3.1600
I^-	7.7623	8.2762	1.4609	-3.1211
IO_3^-	5.7148	6.1725	3.3240	-3.0342
MnO_4^-	7.8289	11.3346	1.2950	-3.2476
ReO_4^-	8.6513	13.3428	0.5057	-3.3306
BO_2^-	-2.2428	-6.2065	-6.3216	-2.5224
BF_4^-	7.9796	11.7027	1.1504	-3.2628
CN^-	5.5125	5.6786	3.5181	-3.0138
$H_2AsO_4^-$	6.9755	9.2509	2.1140	-3.1614
$H_2AsO_3^-$	5.7934	6.3646	3.2485	-3.0421
HO_2^-	3.0671	-0.2924	5.8649	-2.7669
HSO_5^-	8.9938	14.1790	0.1771	-3.3652
HSe^-	4.8181	3.9831	4.1845	-2.9437
$HSeO_3^-$	6.5703	8.2615	2.5029	-3.1205
$HSeO_4^-$	6.3749	7.7844	2.6904	-3.1008
HF_2^-	5.2263	4.9797	3.7928	-2.9849
ClO^-	3.6325	1.0881	5.3224	-2.8240
ClO_2^-	5.5036	5.6569	3.5266	-3.0129
BrO^-	3.6325	1.0881	5.3224	-2.8240
BrO_4^-	8.5987	13.2144	0.5562	-3.3253
Br_3^-	9.0992	14.4364	0.0759	-3.3758
IO^-	2.1432	-2.5482	6.7516	-2.6737
IO_4^-	9.2968	14.9189	-0.1138	-3.3957
I_3^-	9.8500	16.2697	-0.6447	-3.4516

Table 8.11 (*continued*)

Ion	$a_1 \times 10$	$a_2 \times 10^{-2}$	a_3	$a_4 \times 10^{-4}$
$H_2VO_4^-$	6.1355	7.1997	2.9202	-3.0766
$HCrO_4^-$	8.1112	12.0239	1.0241	-3.2761
$H_3P_2O_7^-$	9.0334	14.2757	0.1391	-3.3692
CO_3^{2-}	2.8524	-3.9844	6.4142	-2.6143
HPO_4^{2-}	3.6315	1.0857	5.3233	-2.8239
SO_4^{2-}	8.3014	-1.9846	-6.2122	-2.6970
$S_2O_3^{2-}$	6.6685	12.4951	-7.7281	-3.2955
$S_2O_8^{2-}$	13.3622	24.8454	-4.0153	-3.8061
CrO_4^{2-}	5.5808	5.8454	3.4525	-3.0206
MoO_4^{2-}	6.9775	9.2559	2.1121	-3.1616
WO_4^{2-}	7.2074	8.7934	4.4896	-3.1425
BeO_2^{2-}	-0.3210	-8.5653	9.1166	-2.4249
SiF_6^{2-}	8.5311	13.0492	0.6211	-3.3185
$H_2P_2O_7^{2-}$	9.8215	16.2000	-0.6173	-3.4487
$HAsO_4^{2-}$	4.6321	3.5288	4.3630	-2.9249
SO_3^{2-}	3.6537	0.3191	7.3853	-2.7922
S_2^{2-}	5.5797	5.8426	3.4536	-3.0205
$S_2O_4^{2-}$	7.5815	10.7306	1.5324	-3.2226
$S_2O_5^{2-}$	7.9775	11.6976	1.1524	-3.2626
$S_2O_6^{2-}$	8.6225	13.2724	0.5334	-3.3277
S_3^{2-}	6.7661	8.7396	2.3150	-3.1403
$S_3O_6^{2-}$	9.0313	14.2706	0.1410	-3.3689
S_4^{2-}	7.9381	11.6012	1.1902	-3.2586
$S_4O_6^{2-}$	12.7850	23.4362	-3.4614	-3.7479
S_5^{2-}	9.1107	14.4645	0.0649	-3.3770
$S_5O_6^{2-}$	9.9535	16.5224	-0.7440	-3.4620
SeO_3^{2-}	5.0791	4.6204	3.9340	-2.9700
SeO_4^{2-}	5.6180	5.9362	3.4168	-1.2986
HVO_4^{2-}	5.2115	4.9437	3.8069	-2.9834
$Cr_2O_7^{2-}$	12.5200	22.7889	-3.2070	-3.7211
MnO_4^{2-}	6.5291	8.1609	2.5424	-3.1164
PO_4^{3-}	-0.5259	-9.0654	9.3131	-2.4042

$$\bar{\beta}^0 = \left[a_2 + a_4 \left(\frac{1}{T - \theta} \right) \right] \left(\frac{1}{\psi + P} \right)^2 + \omega \left(\frac{\partial Q}{\partial P} \right)_T$$
$$+ 2Q \left(\frac{\partial \omega}{\partial P} \right)_T - \left(\frac{1}{\varepsilon} - 1 \right) \left(\frac{\partial^2 \omega}{\partial P^2} \right)_T .$$

(8.136)

The thermal expansion of a generic ion or electrolyte in solution is given by the variation with T of the nonsolvation and solvation terms:

$$\bar{\alpha}^0 = \left(\frac{\partial \Delta \bar{V}_n^0}{\partial T} \right)_P + \left(\frac{\partial \Delta \bar{V}_s^0}{\partial T} \right)_P .$$

(8.137)

Adopting the same parameters previously used for calculating the standard partial molal volume (eq. 8.133) and compressibility (eq. 8.136), thermal expansion can be expressed as

$$
\bar{\alpha}^0 = -\left[a_3 + a_4 \left(\frac{1}{\psi + P} \right) \right] \left(\frac{1}{T - \theta} \right)^2 - \omega \left(\frac{\partial Q}{\partial T} \right)_P - Q \left(\frac{\partial \omega}{\partial T} \right)_P
$$
$$
- Y \left(\frac{\partial \omega}{\partial P} \right)_T + \left(\frac{1}{\varepsilon} - 1 \right) \left[\frac{\partial (\partial \omega / \partial P)_T}{\partial T} \right]_P ,
$$

(8.138)

where Y is the Born function:

$$
Y = \frac{1}{\varepsilon} \left(\frac{\partial \ln \varepsilon}{\partial T} \right)_P .
$$

(8.139)

8.11.5 Heat Capacity and Standard Partial Molal Entropy

Heat capacity of solutes, as partial molal volume, compressibility, and thermal expansion, is composed of the contributions of nonsolvation ($\Delta \bar{C}^0_{P,n}$) and solvation ($\Delta \bar{C}^0_{P,s}$):

$$
\bar{C}^0_P = \Delta \bar{C}^0_{P,n} + \Delta \bar{C}^0_{P,s}.
$$

The variation of $\Delta \bar{C}^0_{P,n}$ with T at reference pressure has an asymptotic form that can be described by two coefficients, constant over T and P and characteristic for each ion and electrolyte in solution (see table 8.12):

$$
\Delta \bar{C}^0_{P,n} = c_1 + c_2 \left(\frac{1}{T - \theta} \right)^2 .
$$

(8.141)

The variation of $\Delta \bar{C}^0_{P,n}$ with P and T is described by

$$
\Delta \bar{C}^0_{P,n} = c_1 + c_2 \left(\frac{1}{T - \theta} \right)^2
$$
$$
- 2T \left(\frac{1}{T - \theta} \right)^3 \left[a_3 (P - P_r) + a_4 \ln \left(\frac{\psi + P}{\psi + P_r} \right) \right].
$$

(8.142)

The solvation contribution is given by

$$
\Delta \bar{C}^0_{P,s} = \omega T X + 2 T Y \left(\frac{\partial \omega}{\partial T} \right)_P - T \left(\frac{1}{\varepsilon} - 1 \right) \left(\frac{\partial^2 \omega}{\partial T^2} \right)_P ,
$$

(8.143)

Table 8.12 Standard state heat capacity ($T_r = 25$ °C, $P_r = 1$ bar), coefficients of heat capacity, and Born coefficient in HKF model (from Shock and Helgeson, 1988). $\overline{C}_P^{\,0}$ = cal/(mole × K); c_1 = cal/(mole × K); c_2 = (cal × K)/mole; ω = cal/mole.

Ion	$\overline{C}_P^{\,0}$	c_1	$c_2 \times 10^{-4}$	$\omega \times 10^{-5}$
H^+	0	0	0	0
Li^+	14.2	19.2	−0.24	0.4862
Na^+	9.06	18.18	−2.981	0.3306
K^+	1.98	7.40	−1.791	0.1927
Rb^+	−3.0	5.7923	−3.6457	0.1502
Cs^+	−6.29	6.27	−5.736	0.0974
NH_4^+	15.74	17.45	−0.021	0.1502
Ag^+	7.9	12.7862	−1.4254	0.2160
Tl^+	−4.2	5.0890	−3.8901	0.1502
VO_2^+	31.1	30.8449	3.3005	0.7003
Cu^+	13.7	17.2831	−0.2439	0.3351
Au^+	0.6	8.1768	−2.9124	0.1800
Mg^{2+}	−5.34	20.80	−5.892	1.5372
Ca^{2+}	−7.53	9.00	−2.522	1.2366
Sr^{2+}	−10.05	10.7452	−5.0818	1.1363
Ba^{2+}	−12.30	3.80	−3.450	0.9850
Pb^{2+}	−12.70	8.6624	−5.6216	1.0788
Mn^{2+}	−4.1	16.6674	−3.8698	1.4006
Co^{2+}	−7.8	15.2014	−4.6235	1.4769
Ni^{2+}	−11.7	13.1905	−5.4179	1.5067
Cu^{2+}	−5.7	20.3	−4.39	1.4769
Zn^{2+}	−6.3	15.9009	−4.3179	1.4574
Cd^{2+}	−3.5	15.6573	−3.7476	1.2528
Be^{2+}	−1.3	22.9152	−3.2994	1.9007
Ra^{2+}	−14.4	6.2858	−5.9679	0.9290
Sn^{2+}	−11.2	11.4502	−5.3160	1.2860
VO^{2+}	−12.0	13.3910	−5.4790	1.5475
Fe^{2+}	−7.9	14.9632	−4.6438	1.4574
Pd^{2+}	−6.3	15.3780	−4.3179	1.4006
Ag^{2+}	−5.3	15.2224	−4.1142	1.3201
Hg^{2+}	2.2	18.0613	−2.5865	1.151
Hg_2^{2+}	17.1	23.7433	0.4487	0.8201
Sm^{2+}	3.7	19.6533	−2.2809	1.2285
Eu^{2+}	6.0	19.7516	−1.8124	1.0929
Yb^{2+}	0.7	17.8951	−2.8920	1.2285
Al^{3+}	−32.5	10.7	−8.06	2.8711
La^{3+}	−37.2	4.2394	−10.6122	2.1572
Ce^{3+}	−38.6	4.0445	−10.8974	2.2251
Pr^{3+}	−47.7	−1.1975	−12.7511	2.2350
Nd^{3+}	−43.2	1.6236	−11.8344	2.2550
Sm^{3+}	−43.3	1.9385	−11.8548	2.2955
Eu^{3+}	−36.6	6.0548	−10.4900	2.3161
Gd^{3+}	−35.9	6.5606	−10.3474	2.3265
Tb^{3+}	−40.5	4.2522	−11.2844	2.3685
Dy^{3+}	−31.7	9.5076	−9.4919	2.3792
Ho^{3+}	−33.3	8.6686	−9.8178	2.3899

continued

Table 8.12 (*continued*)

Ion	$\overline{C_P}^0$	c_1	$c_2 \times 10^{-4}$	$\omega \times 10^{-5}$
Er^{3+}	-34.3	8.2815	-10.0215	2.4115
Tm^{3+}	-34.3	8.4826	-10.0215	2.4333
Yb^{3+}	-36.4	7.3533	-10.4493	2.4443
Lu^{3+}	-32.0	9.5650	-9.7160	2.4554
Ga^{3+}	-30.8	13.2451	-9.3086	2.7276
In^{3+}	-34.9	8.7476	-10.1437	2.5003
Tl^{3+}	-39.3	4.7607	-11.0400	2.3474
Fe^{3+}	-34.1	11.0798	-9.9808	2.7025
Co^{3+}	-32.3	12.2500	-9.6141	2.7150
Au^{3+}	-36.2	7.5724	-10.4085	2.4554
Sc^{3+}	-35.4	8.4546	-10.2456	2.5003
Y^{3+}	-35.7	7.1634	-10.3067	2.3792
AlO_2^-	-11.9	19.1	-6.2	1.7595
HCO_3^-	-8.46	12.9395	-4.7579	1.2733
NO_3^-	-16.4	7.70	-6.725	1.0977
NO_2^-	-23.3	3.4260	-7.7808	1.1847
$H_2PO_4^-$	-7.0	14.0435	-4.4605	1.3003
OH^-	-32.79	4.15	-10.346	1.7246
HS^-	-22.17	3.42	-6.27	1.4410
HSO_3^-	-1.4	15.6949	-3.3198	1.1233
HSO_4^-	5.3	20.0961	-1.9550	1.1748
F^-	-27.23	4.46	-7.488	1.7870
Cl^-	-29.44	-4.40	-5.714	1.4560
ClO_3^-	-12.3	8.5561	-5.5401	1.0418
ClO_4^-	-5.5	22.3	-8.9	0.9699
Br^-	-30.42	-3.80	-6.811	1.3858
BrO_3^-	-20.6	3.7059	-7.2308	1.0433
I^-	-28.25	-6.27	-4.944	1.2934
IO_3^-	-16.2	7.7293	-6.3345	1.2002
MnO_4^-	-1.8	13.7427	-3.4013	0.9368
ReO_4^-	-0.2	14.3448	-3.0753	0.9004
BO_2^-	-41.0	-1.6521	-11.3863	1.7595
BF_4^-	-5.6	11.8941	-4.1753	0.9779
CN^-	-32.7	-1.1135	-9.6956	1.2900
$H_2AsO_4^-$	-0.7	16.8622	-3.1772	1.2055
$H_2AsO_3^-$	-2.9	15.8032	-3.6253	1.2305
HO_2^-	-30.2	2.7007	-9.1863	1.5449
HSO_5^-	29.2	31.2126	2.9134	0.8611
HSe^-	-12.6	11.1345	-5.6012	1.3408
$HSeO_3^-$	4.9	19.5432	-2.0365	1.1402
$HSeO_4^-$	-41.9	-8.3616	-11.5696	1.0885
HF_2^-	-33.2	-1.3751	-9.7974	1.2934
ClO^-	-49.2	-9.0630	-13.0566	1.4767
ClO_2^-	-30.5	-0.0659	-9.2474	1.2637
BrO^-	-49.2	-9.0630	-13.0566	1.4767
BrO_4^-	0.6	14.8727	-2.9124	0.9068
Br_3^-	5.6	17.2705	-1.8939	0.8490
IO^-	-64.1	-16.2398	-16.0918	1.6455
IO_4^-	7.6	18.2345	-1.4865	0.8264
I_3^-	13.1	20.8712	-0.3661	0.7628

Table 8.12 (*continued*)

Ion	$\overline{C}_P{}^0$	c_1	$c_2 \times 10^{-4}$	$\omega \times 10^{-5}$
$H_2VO_4{}^-$	0.6	17.4795	−2.9124	1.1898
$HCrO_4{}^-$	20.4	26.9872	1.1209	0.9622
$H_3P_2O_7{}^-$	29.6	31.4072	2.9949	0.8568
$CO_3{}^{2-}$	−69.5	−3.3206	−17.1917	3.3914
$HPO_4{}^{2-}$	−58.3	2.7357	−14.9103	3.3363
$SO_4{}^{2-}$	−64.38	1.64	−17.998	3.1463
$S_2O_3{}^{2-}$	−57.3	−0.0577	−14.7066	2.9694
$S_2O_8{}^{2-}$	−25.0	12.9632	−8.1271	2.3281
$CrO_4{}^{2-}$	−59.9	−1.0175	−15.2362	3.0307
$MoO_4{}^{2-}$	−47.5	7.0224	−12.7103	3.1145
$WO_4{}^{2-}$	−44.5	8.3311	−12.0992	3.0657
$BeO_2{}^{2-}$	−100.0	−18.0684	−23.5879	3.7880
$SiF_6{}^{2-}$	−47.1	4.0970	−12.6289	2.7716
$H_2P_2O_7{}^{2-}$	−20.3	18.4241	−7.1697	2.6218
$HAsO_4{}^{2-}$	−51.8	5.4710	−13.5863	3.2197
$SO_3{}^{2-}$	−76.1	−7.8368	−18.5362	3.3210
$S_2{}^{2-}$	−65.1	−3.3496	−16.2955	3.1083
$S_2O_4{}^{2-}$	−52.9	1.6707	−13.8103	2.8772
$S_2O_5{}^{2-}$	−50.5	2.6824	−13.3215	2.8343
$S_2O_6{}^{2-}$	−46.5	4.3301	−12.5066	2.7587
$S_3{}^{2-}$	−57.9	−0.3595	−14.8288	2.9749
$S_3O_6{}^{2-}$	−44.1	5.3169	−12.0178	2.7131
$S_4{}^{2-}$	−50.7	2.6081	−13.3622	2.8390
$S_4O_6{}^{2-}$	−21.3	14.6933	−7.3734	2.2805
$S_5{}^{2-}$	−43.6	5.5361	−11.9159	2.7051
$S_5O_6{}^{2-}$	−38.5	7.6266	−10.8770	2.6076
$SeO_3{}^{2-}$	−68.1	−4.5783	−16.9066	3.1658
$SeO_4{}^{2-}$	−60.2	−1.2986	−15.2973	3.0192
$HVO_4{}^{2-}$	−48.3	6.9055	−12.8733	3.1527
$Cr_2O_7{}^{2-}$	−20.4	15.0820	−7.1901	2.2654
$MnO_4{}^{2-}$	−59.3	−0.9267	−15.1140	3.0024
$PO_4{}^{3-}$	−114.9	−9.4750	−26.4397	5.6114

where X is the Born function:

$$X = \frac{1}{\varepsilon}\left[\left(\frac{\partial^2 \ln \varepsilon}{\partial T^2}\right)_P - \left(\frac{\partial \ln \varepsilon}{\partial T}\right)_P^2\right] \tag{8.144}$$

(Helgeson and Kirkham, 1974). Summing equations 8.142 and 8.143, we obtain

$$\overline{C}_P^0 = c_1 + \frac{c_2}{\left(T - \theta\right)^2} - \left[\frac{2T}{\left(T - \theta\right)^3}\right] \times \tag{8.145}$$

$$\left[a_3\left(P - P_r\right) + a_4 \ln\left(\frac{\psi + P}{\psi + P_r}\right)\right] + \omega TX + 2TY\left(\frac{\partial \omega}{\partial T}\right)_P - T\left(\frac{1}{\varepsilon} - 1\right)\left(\frac{\partial^2 \omega}{\partial T^2}\right)_P.$$

Solvation entropy contribution $\Delta \overline{S}_s^0$ is given by the integral in dT/T of $\Delta \overline{C}_{P,s}^0$ (equation 8.143):

$$\Delta \overline{S}_s^0 = \omega Y - \left(\frac{1}{\varepsilon} - 1\right)\left(\frac{\partial \omega}{\partial T}\right)_P, \quad (8.146)$$

and nonsolvation contribution $\Delta \overline{S}_n^0$ is given by the integral in dT/T of $\Delta \overline{C}_{p,n}^0$ (equation 8.142):

$$\Delta \overline{S}_n^0 = c_1 \ln \frac{T}{T_r} - \frac{c_2}{\theta}\left\{\left(\frac{1}{T-\theta}\right) - \left(\frac{1}{T_r - \theta}\right) + \frac{1}{\theta}\ln\left[\frac{T_r(T-\theta)}{T(T_r - \theta)}\right]\right\}$$
$$+ \left(\frac{1}{T-\theta}\right)^2\left[a_3(P - P_r) + a_4 \ln\left(\frac{\psi + P}{\psi + P_r}\right)\right]. \quad (8.147)$$

Combining the solvation and nonsolvation contributions, the partial molal entropy change between reference state conditions P_r and T_r and the P and T of interest is given by

$$\overline{S}_{P,T}^0 - \overline{S}_{P_r,T_r}^0 = \left(\Delta \overline{S}_n^0 - \Delta \overline{S}_{n,P_r,T_r}^0\right) + \left(\Delta \overline{S}_s^0 - \Delta \overline{S}_{s,P_r,T_r}^0\right)$$
$$= c_1 \ln\left(\frac{T}{T_r}\right) - \frac{c_2}{\theta}\left\{\left(\frac{1}{T-\theta}\right) + \left(\frac{1}{T_r - \theta}\right) + \frac{1}{\theta}\ln\left[\frac{T_r(T-\theta)}{T(T_r - \theta)}\right]\right\}$$
$$+ \left(\frac{1}{T-\theta}\right)^2\left[a_3(P - P_r) + a_4 \ln\left(\frac{\psi + P}{\psi + P_r}\right)\right]$$
$$+ \omega Y - \left(\frac{1}{\varepsilon} - 1\right)\left(\frac{\partial \omega}{\partial T}\right)_P - \omega_{P_r,T_r}Y_{P_r,T_r}. \quad (8.148)$$

8.11.6 Enthalpy and Gibbs Free Energy of Formation of Aqueous Solutes Under High P and T Conditions

We now have all the parameters and analytical forms necessary to perform enthalpy and Gibbs free energy calculations at high P and T conditions:

$$\overline{H}_{P,T}^0 = \overline{H}_{P_r,T_r}^0 + \int_{T_r}^T \overline{C}_{P_r}^0 \, dT + \int_{P_r}^P \left[\overline{V}^0 - T\left(\frac{\partial \overline{V}^0}{\partial T}\right)_P\right]_T dP \quad (8.149)$$

and

$$\overline{G}_{P,T}^0 = \overline{G}_{P_r,T_r}^0 - \overline{S}_{P_r,T_r}^0(T - T_r)$$
$$+ \int_{T_r}^T \overline{C}_{P_r}^0 \, dT - T\int_{T_r}^T \overline{C}_{P_r}^0 \, d\ln T + \int_{P_r}^P \overline{V}^0 \, dP. \quad (8.150)$$

Note that thermodynamic tabulations do not normally report the standard partial molar properties of solutes $\overline{H}^0_{T_r,P_r}$ and $\overline{G}^0_{T_r,P_r}$, but rather the enthalpy of formation $\overline{H}^0_{f,T_r,P_r}$ and the Gibbs free energy of formation from the elements at stable state under T_r and P_r reference conditions $\overline{G}^0_{f,T_r,P_r}$ (note, however, that $\overline{H}^0_{T_r,P_r}$ and $\overline{H}^0_{f,T_r,P_r}$ coincide, because the enthalpy of elements at stable state and $T_r = 25\ °C$ and $P_r = 1$ bar is conventionally set at zero). Because it is also true that

$$
\overline{H}^0_{f,P,T} - \overline{H}^0_{f,P_r,T_r} = \overline{H}^0_{P,T} - \overline{H}^0_{P_r,T_r}
\tag{8.151}
$$

and

$$
\overline{G}^0_{f,P,T} - \overline{G}^0_{f,P_r,T_r} = \overline{G}^0_{P,T} - \overline{G}^0_{P_r,T_r},
\tag{8.152}
$$

substituting the various coefficients (a_1, a_2, a_3, a_4, c_1, c_2) and the Born functions ω, Y, and X of the HKF model in equations 8.149 and 8.150 and taking into account equations 8.151 and 8.152, we obtain

$$
\overline{H}^0_{f,P,T} = \overline{H}^0_{f,P_r,T_r} + c_1\left(T - T_r\right) - c_2\left[\left(\frac{1}{T-\theta}\right) - \left(\frac{1}{T_r-\theta}\right)\right] + a_1\left(P - P_r\right)
$$
$$
+ a_2\ln\left(\frac{\psi + P}{\psi + P_r}\right) + \left[\frac{2T-\theta}{\left(T-\theta\right)^2}\right]\left[a_3\left(P-P_r\right) + a_4\ln\left(\frac{\psi + P}{\psi + P_r}\right)\right]
$$
$$
+ \omega\left(\frac{1}{\varepsilon} - 1\right) + \omega T Y - T\left(\frac{1}{\varepsilon} - 1\right)\left(\frac{\partial \omega}{\partial T}\right)_P - \omega_{P_r,T_r}\left(\frac{1}{\varepsilon_{P_r,T_r}} - 1\right)
$$
$$
- \omega_{P_r,T_r} T_r Y_{P_r,T_r}
\tag{8.153}
$$

and

$$
\overline{G}^0_{f,P,T} = \overline{G}^0_{f,P_r,T_r} - \overline{S}^0_{P_r,T_r}\left(T - T_r\right) - c_1\left[T\ln\left(\frac{T}{T_r}\right) - T + T_r\right]
$$
$$
+ a_1\left(P - P_r\right) + a_2\ln\left(\frac{\psi + P}{\psi + P_r}\right) - c_2\left\{\left[\left(\frac{1}{T-\theta}\right) - \left(\frac{1}{T_r-\theta}\right)\right]\left(\frac{\theta - T}{\theta}\right)\right.
$$
$$
\left. - \frac{T}{\theta^2}\ln\left[\frac{T_r\left(T-\theta\right)}{T\left(T_r-\theta\right)}\right]\right\} + \left(\frac{1}{T-\theta}\right)\left[a_3\left(P-P_r\right) + a_4\ln\left(\frac{\psi + P}{\psi + P_r}\right)\right]
\tag{8.154}
$$
$$
+ \omega\left(\frac{1}{\varepsilon} - 1\right) - \omega_{P_r,T_r}\left(\frac{1}{\varepsilon_{P_r,T_r}} - 1\right) + \omega_{P_r,T_r} Y_{P_r,T_r}\left(T - T_r\right).
$$

Table 8.13 lists $\overline{G}^0_{f,P_r,T_r}$, $\overline{H}^0_{f,P_r,T_r}$, $\overline{S}^0_{P_r,T_r}$, and $\overline{V}^0_{P_r,T_r}$ values for various aqueous ions, after Shock and Helgeson (1988). These values, together with the coefficients in tables

Table 8.13 Standard partial molal Gibbs free energy of formation from the elements ($\overline{G}^0_{f,P_r,T_r}$; cal/mole), standard partial molal enthalpy of formation from the elements ($\overline{H}^0_{f,P_r,T_r}$; cal/mole), standard partial molal entropy ($\overline{S}^0_{P_r,T_r}$; cal/(mole × K)), and standard partial molal volume ($\overline{V}^0_{P_r,T_r}$; cm³/mole) for aqueous ions at $T_r = 25\,°C$ and $P_r = 1$ bar (Shock and Helgeson, 1988).

Ion	$\overline{G}^0_{f,P_r,T_r}$	$\overline{H}^0_{f,P_r,T_r}$	$\overline{S}^0_{P_r,T_r}$	$\overline{V}^0_{P_r,T_r}$
H^+	0	0	0	0
Li^+	−69,933	−66,552	2.70	−0.87
Na^+	−62,591	−57,433	13.96	−1.11
K^+	−67,510	−60,270	24.15	9.06
Rb^+	−67,800	−60,020	28.8	14.26
Cs^+	−69,710	−61,670	31.75	21.42
NH_4^+	−18,990	−31,850	26.57	18.13
Ag^+	18,427	25,275	17.54	−0.8
Tl^+	−7,740	1,280	30.0	18.2
VO_2^+	−140,300	−155,300	−10.1	−33.5
Cu^+	11,950	17,132	9.7	−8.0
Au^+	39,000	47,580	24.5	11.1
Mg^{2+}	−108,505	−111,367	−33.00	−21.55
Ca^{2+}	−132,120	−129,800	−13.5	−18.06
Sr^{2+}	−134,760	−131,670	−7.53	−17.41
Ba^{2+}	−134,030	−128,500	2.3	−12.60
Pb^{2+}	−5,710	220	4.2	−15.6
Mn^{2+}	−54,500	−52,724	−17.6	−17.1
Co^{2+}	−13,000	−13,900	−27.0	−24.4
Ni^{2+}	−10,900	−12,900	−30.8	−29.0
Cu^{2+}	15,675	15,700	−23.2	−24.6
Zn^{2+}	−35,200	−36,660	−26.2	−24.3
Cd^{2+}	−18,560	−18,140	−17.4	−15.6
Be^{2+}	−83,500	−91,500	−55.7	−25.4
Ra^{2+}	−134,200	−126,100	13	−10.6
Sn^{2+}	−6,630	−2,100	−3.8	−15.5
VO^{2+}	−106,700	−116,300	−32	−29.4
Fe^{2+}	−21,870	−22,050	−25.3	−22.2
Pd^{2+}	42,200	42,080	−22.6	−20.8
Ag^{2+}	64,300	64,200	−21.0	−19.3
Hg^{2+}	39,360	40,670	−8.68	−19.6
Hg_2^{2+}	36,710	39,870	−15.66	14.4
Sm^{2+}	−123,000	−120,500	−6.2	−5.7
Eu^{2+}	−129,100	−126,100	−2.4	−2.2
Yb^{2+}	−128,500	−126,800	−11.2	−10.3
Al^{3+}	−115,377	−126,012	−75.6	−44.4
La^{3+}	−164,000	−169,600	−52.0	−38.6
Ce^{3+}	−161,600	−167,400	−49	−39.8
Pr^{3+}	−162,600	−168,800	−50	−42.1
Nd^{3+}	−160,600	−166,500	−49.5	−43.1
Sm^{3+}	−159,100	−165,200	−50.7	−42.0
Eu^{3+}	−137,300	−144,700	−53.0	−41.3

Table 8.13 (*continued*)

Ion	$\overline{G}^0_{f,P_r,T_r}$	$\overline{H}^0_{f,P_r,T_r}$	$\overline{S}^0_{P_r,T_r}$	$\overline{V}^0_{P_r,T_r}$
Gd^{3+}	−158,600	−164,200	−49.2	−40.4
Tb^{3+}	−159,500	−166,900	−54	−40.2
Dy^{3+}	−158,700	−166,500	−55.2	−40.7
Ho^{3+}	−161,400	−169,000	−54.3	−41.6
Er^{3+}	−159,900	−168,500	−58.3	−43.0
Tm^{3+}	−159,900	−168,500	−58.1	−43.0
Yb^{3+}	−153,000	−160,300	−56.9	−44.5
Lu^{3+}	−159,400	−167,900	−63.1	−32.0
Ga^{3+}	−38,000	−50,600	−79.0	−44.9
In^{3+}	−23,400	−25,000	−63.0	−42.2
Tl^{3+}	51,300	47,000	−46.0	−39.3
Fe^{3+}	−4,120	−11,850	−66.3	−42.8
Co^{3+}	32,000	22,000	−73	−43.9
Au^{3+}	103,600	96,930	−57.9	−41.3
Sc^{3+}	−140,200	−146,800	−61	−41.9
Y^{3+}	−163,800	−170,900	−60	−40.8
AlO_2^-	−198,465	−222,261	−8.4	10.0
HCO_3^-	−140,282	−164,898	23.53	24.6
NO_3^-	−26,507	−49,429	35.12	29.0
NO_2^-	−7,700	−25,000	29.4	25.0
$H_2PO_4^-$	−270,140	−309,820	21.6	31.3
OH^-	−37,595	−54,977	−2.56	−4.18
HS^-	2,860	−3,850	16.3	20.65
HSO_3^-	−126,130	−149,670	33.4	33.3
HSO_4^-	−180,630	−212,500	30.0	35.2
F^-	−67,340	−80,150	−3.15	−1.32
Cl^-	−31,379	−39,933	13.56	17.79
ClO_3^-	−1,900	−24,850	38.8	36.9
ClO_4^-	−2,040	−30,910	43.5	44.2
Br^-	−24,870	−29,040	19.40	24.85
BrO_3^-	4,450	−16,030	38.65	35.4
I^-	−12,410	−13,600	25.50	36.31
IO_3^-	−30,600	−52.900	28.3	25.9
MnO_4^-	−106,900	−129,400	45.7	42
ReO_4^-	−166,000	−188,200	48.1	48.1
BO_2^-	−162,240	−184,600	−8.9	−14.5
BF_4^-	−355,400	−376,400	43.0	43.0
CN^-	41,200	36,000	22.5	24.2
$H_2AsO_4^-$	−180,010	−217,390	28	35.1
$H_2AsO_3^-$	−140,330	−170,840	26.4	26.4
HO_2^-	−16,100	−38,320	5.7	5.7
HSO_5^-	−152,370	−185,380	50.7	50.7
HSe^-	10,500	3,800	19	−19.0
$HSeO_3^-$	−98,340	−122,980	32.3	32.3
$HSeO_4^-$	−108,100	−139,000	35.7	31.0
HF_2^-	−138,160	−155,340	22.1	22.1
ClO^-	−8,800	−25,600	10	10.0
ClO_2^-	4,100	−15,900	24.2	24.2
BrO^-	−8,000	−22,500	10	10.0

continued

Table 8.13 (continued)

Ion	$\overline{G}^0_{f,P_r,T_r}$	$\overline{H}^0_{f,P_r,T_r}$	$\overline{S}^0_{P_r,T_r}$	$\overline{V}^0_{P_r,T_r}$
BrO_4^-	28,200	3,100	47.7	47.7
Br_3^-	−25,590	−31,370	51.5	51.5
IO^-	−9,200	−25,700	−1.3	−1.3
IO_4^-	−14,000	−36,200	53	53.0
I_3^-	−12,300	−12,300	57.2	57.2
$H_2VO_4^-$	−244,000	−280,600	29	29.0
$HCrO_4^-$	−182,800	−209,900	44.0	44.0
$H_3P_2O_7^-$	−483,600	−544,100	51	51.0
CO_3^{2-}	−126,191	−161,385	−11.95	−5.02
HPO_4^{2-}	−260,310	−308,815	−8.0	5.4
SO_4^{2-}	−177,930	−217,400	4.50	13.88
$S_2O_3^{2-}$	−124,900	−155,000	16	28.5
$S_2O_8^{2-}$	−266,500	−321,400	58.4	79.0
CrO_4^{2-}	−173,940	−210,600	12.0	20.4
MoO_4^{2-}	−199,900	−238,500	6.5	30.4
WO_4^{2-}	−219,150	−257,100	9.7	32.2
BeO_2^{2-} .	−153,000	−189,000	−38	−24.6
SiF_6^{2-}	−525,700	−571,000	29.2	42.6
$H_2P_2O_7^{2-}$	−480,400	−544,600	39	52.4
$HAsO_4^{2-}$	−170,790	−216,620	−0.4	13.0
SO_3^{2-}	−116,300	−151,900	−7	5.6
S_2^{2-}	19,000	7,200	6.8	−65.1
$S_2O_4^{2-}$	−143,500	−180,100	22	35.4
$S_2O_5^{2-}$	−189,000	−232,000	25	38.4
$S_2O_6^{2-}$	−231,000	−280,400	30	43.3
S_3^{2-}	17,600	6,200	15.8	29.2
$S_3O_6^{2-}$	−229,000	−279,000	33	46.4
S_4^{2-}	16,500	5,500	24.7	38.1
$S_4O_6^{2-}$	−248,700	−292,600	61.5	74.9
S_5^{2-}	15,700	5,100	33.6	47.0
$S_5O_6^{2-}$	−229,000	−281,000	40	53.4
SeO_3^{2-}	−88,400	−121,700	3.	−16.4
SeO_4^{2-}	−105,500	−143,200	12.9	20.7
HVO_4^{2-}	−233,000	−277,000	4.	17.4
$Cr_2O_7^{2-}$	−311,000	−356,200	62.6	73.0
MnO_4^{2-}	−119,700	−156,000	14.0	27.4
PO_4^{3-}	−243,500	−305,300	−53.	−30.6

8.11 and 8.12, allow the calculation of the corresponding partial molar properties in high P and T conditions.* Practical calculations are performed by the various releases of the

---

* Actually, the various equations listed in this section are insufficient to perform the complete calculation since one would first calculate the density of H_2O through eq. 8.12 or 8.14. Equation 8.14 in its turn involves the partial derivative of the Helmholtz free energy function 8.15. Moreover, the evaluation of electrostatic properties of the solvent and of the Born functions ω, Q, Y, X involve additional equations and variables not given here for the sake of brevity (eqs. 36, 40 to 44, 49 to 52 and tables 1 to 3 in Johnson et al., 1991). In spite of this fact, the decision to outline here briefly the HKF model rests on its paramount importance in geochemistry. Moreover, most of the listed thermodynamic parameters have an intrinsic validity that transcends the model itself.

automated routine SUPCRT, distributed by the Laboratory of Theoretical Geochemistry of the University of California, Berkeley (Johnson et al. 1991).

8.12 Redox Conditions, pH, and Thermodynamic Stability of Aqueous Solutions

Redox equilibrium is defined as a process characterized by the flow of electrons from the substance being oxidized ("reducing medium") to the substance being reduced ("oxidizing medium"). For instance, ionic iron in aqueous solutions is present in two valence states, related by the redox equilibrium

$$\text{Fe}^{3+} + \text{e}^- \overset{\text{reduction}\rightarrow}{\underset{\leftarrow\text{oxidation}}{\Longleftrightarrow}} \text{Fe}^{2+}. \tag{8.155}$$

In a "reduction process" the flow of electrons is from left to right (i.e., from oxidizing to reducing media). A Gibbs free energy change ΔG^0_{155} is associated with equilibrium 8.155, so that

$$\Delta G^0_{155} = -RT \ln K_{155} = -RT \ln \frac{a_{\text{Fe}^{2+}}}{a_{\text{Fe}^{3+}} a_{\text{e}^-}}. \tag{8.156}$$

Term a_{e^-} in equation 8.156 is the "activity of the hypothetical electron in solution," which, by convention, is always assumed to be unitary.

Let us now consider a redox equilibrium involving metallic iron Fe and ferrous iron Fe^{2+} (one-molal solution at $T_r = 25\ °\text{C}$ and $P_r = 1$ bar):

$$\underset{\text{metal}}{\text{Fe}} \Longleftrightarrow \underset{\text{aqueous}}{\text{Fe}^{2+}} + 2\,\text{e}^-. \tag{8.157}$$

The Gibbs free energy change involved in equation 8.157 corresponds to the partial molal Gibbs free energy of formation of Fe^{2+} from the element at stable state (cf. table 8.13):

$$\Delta G^0_{157} = \overline{G}^0_{f,T_r,P_r,\text{Fe}^{2+}}. \tag{8.158}$$

In the case of oxidation of hydrogen, starting from the diatomic gaseous molecule, we have

$$\underset{\text{gaseous}}{\frac{1}{2}\text{H}_2} \Longleftrightarrow \underset{\text{aqueous}}{\text{H}^+ + \text{e}^-}. \tag{8.159}$$

ΔG^0_{159} is the Gibbs free energy change associated with equation 8.159, corresponding to the formation of the aqueous hydrogen ion H^+ from the gaseous molecule H_2. This amount of energy is, by convention (see, for instance, table 8.13), fixed at zero.

To the reactions of equations 8.155, 8.157, and 8.159, which imply electron transfer from the oxidized to the reduced state, the concept of electrical potential is associated through Faraday's law:

$$\Delta G = -nFE, \tag{8.160}$$

where n is the number of electrons involved in the redox equilibrium, F is Faraday's constant [$F = 96484.56$ J/(volt $\times$ mole)], and E is the electrical potential expressed in volts.

Because the Gibbs free energy change ΔG^0_{159} involved in the redox equilibrium of equation 8.159 at standard state is zero, it follows from equation 8.160 that the potential associated with this equilibrium at standard state is also zero. This fact allows us to establish a "scale of standard potentials" for the various redox couples, where zero corresponds to the oxidation of gaseous hydrogen ("normal hydrogen electrode"). The minus sign in equation 8.160 is due to the fact that, by convention, in the definition of standard potentials for metal-aqueous ion couples, the oxidizing medium is placed to the left of the reaction sign and the reducing medium to the right (i.e., a reduction process; cf. eq. 8.155), whereas the Gibbs free energy of formation of an aqueous cation from the stable metal implies the opposite (i.e., oxidation; cf. eq. 8.157 or 8.159). It must also be emphasized that absolute potentials for the various redox couples cannot be determined, and that the "standard potentials" therefore represent the potential gap measured at the terminals of a galvanic cell in which one of the electrodes is the "normal hydrogen electrode." For instance, the standard potential of zinc is assessed with a cell of this type:

$$\underset{\text{gas} \quad \text{aq}}{H_2, H^+} \Big| \Big| \underset{\text{metal} \quad \text{aq}}{Zn, \quad Z^{2+}_n}, \tag{8.161}$$

where H_2, H^+ means a one-molal solution of H^+ ions at $T = 25$ °C and $P = 1$ bar, in equilibrium with an atmosphere whose H_2 fugacity is unitary at standard state—i.e. (cf. section 9.1),

$$a_{H_2,\text{gas}} = \frac{f_{H_2\text{gas}}}{f^0_{H_2\text{gas}}} = \frac{f_{H_2,\text{gas}}}{1} = 1. \tag{8.162}$$

In equation 8.161, Zn, Zn^{2+} means metallic zinc Zn in equilibrium with a one-molal solution of Zn^{2+} under the same T and P conditions, and $||$ represents a membrane with an electron transport number of 1, in the absence of diffusive potential (the "unitary electron transport number" means that the electric conductivity is purely electronic, because ions Zn^{2+} and H^+ cannot cross the membrane).

Because, as we have already seen, the standard potential of hydrogen is zero, the electromotive force of the galvanic cell (eq. 8.161) directly gives the value of the standard potential for the Zn,Zn^{2+} redox couple. Table 8.14 lists the standard potentials for various aqueous ions. The listed values are arranged in decreasing order and are consistent with the standard partial molal Gibbs free energies of table 8.13.

Let us now imagine a redox process involving two metal-ion couples:

Table 8.14 Standard potentials for various redox couples (aqueous cation–metal; element–aqueous anion) arranged in order of decreasing E^0. Values, expressed in volts, are consistent with standard partial molal Gibbs free energy values listed in table 8.13. E^0 value for sulfur is from Nylen and Wigren (1971).

Equilibrium				E^0 (V)	Equilibrium				E^0 (V)
Li^+	+	e^-	$\rightarrow$	Li	−3.034	Al^{3+}	+	$3e^-$	$\rightarrow$	Al	−1.668
Cs^+	+	e^-	$\rightarrow$	Cs	−3.023	Mn^{2+}	+	$2e^-$	$\rightarrow$	Mn	−1.182
Rb^+	+	e^-	$\rightarrow$	Rb	−2.940	Zn^{2+}	+	$2e^-$	$\rightarrow$	Zn	−0.763
K^+	+	e^-	$\rightarrow$	K	−2.928	Ga^{2+}	+	$2e^-$	$\rightarrow$	Ga	−0.549
Sr^{2+}	+	$2e^-$	$\rightarrow$	Sr	−2.922	S	+	$2e^-$	$\rightarrow$	S^{2-}	−0.480
Ra^{2+}	+	$2e^-$	$\rightarrow$	Ra	−2.910	Fe^{2+}	+	$2e^-$	$\rightarrow$	Fe	−0.474
Ba^{2+}	+	$2e^-$	$\rightarrow$	Ba	−2.906	Cd^{2+}	+	$2e^-$	$\rightarrow$	Cd	−0.402
Ca^{2+}	+	$2e^-$	$\rightarrow$	Ca	−2.865	In^{3+}	+	$3e^-$	$\rightarrow$	In	−0.338
Eu^{2+}	+	$2e^-$	$\rightarrow$	Eu	−2.799	Tl^+	+	e^-	$\rightarrow$	Tl	−0.336
Yb^{2+}	+	$2e^-$	$\rightarrow$	Yb	−2.786	Co^{2+}	+	$2e^-$	$\rightarrow$	Co	−0.282
Na^+	+	e^-	$\rightarrow$	Na	−2.714	Ni^{2+}	+	$2e^-$	$\rightarrow$	Ni	−0.236
Sm^{2+}	+	$2e^-$	$\rightarrow$	Sm	−2.667	Sn^{2+}	+	$2e^-$	$\rightarrow$	Sn	−0.144
La^{3+}	+	$3e^-$	$\rightarrow$	La	−2.371	Pb^{2+}	+	$2e^-$	$\rightarrow$	Pb	−0.124
Y^{3+}	+	$3e^-$	$\rightarrow$	Y	−2.368	Fe^{3+}	+	$3e^-$	$\rightarrow$	Fe	−0.059
Mg^{2+}	+	$2e^-$	$\rightarrow$	Mg	−2.353	$2H^+$	+	$2e^-$	$\rightarrow$	H_2	0.000
Pr^{3+}	+	$3e^-$	$\rightarrow$	Pr	−2.350	Cu^{2+}	+	$2e^-$	$\rightarrow$	Cu	0.340
Ce^{3+}	+	$3e^-$	$\rightarrow$	Ce	−2.336	Co^{3+}	+	$3e^-$	$\rightarrow$	Co	0.462
Ho^{3+}	+	$3e^-$	$\rightarrow$	Ho	−2.333	Cu^+	+	e^-	$\rightarrow$	Cu	0.518
Nd^{3+}	+	$3e^-$	$\rightarrow$	Nd	−2.321	I_2	+	$2e^-$	$\rightarrow$	$2I-$	0.538
Er^{3+}	+	$3e^-$	$\rightarrow$	Er	−2.311	Tl^{3+}	+	$3e^-$	$\rightarrow$	Tl	0.742
Tm^{3+}	+	$3e^-$	$\rightarrow$	Tm	−2.311	Ag^+	+	e^-	$\rightarrow$	Ag	0.799
Tb^{3+}	+	$3e^-$	$\rightarrow$	Tb	−2.306	Hg^{2+}	+	$2e^-$	$\rightarrow$	Hg	0.853
Lu^{3+}	+	$3e^-$	$\rightarrow$	Lu	−2.304	Pd^{2+}	+	$2e^-$	$\rightarrow$	Pd	0.915
Dy^{3+}	+	$3e^-$	$\rightarrow$	Dy	−2.294	Br_2	+	$2e^-$	$\rightarrow$	$2Br-$	1.078
Gd^{3+}	+	$3e^-$	$\rightarrow$	Gd	−2.293	Cl_2	+	$2e^-$	$\rightarrow$	$2Cl-$	1.361
Yb^{3+}	+	$3e^-$	$\rightarrow$	Yb	−2.212	Ag^{2+}	+	$2e^-$	$\rightarrow$	Ag	1.394
Sc^{3+}	+	$3e^-$	$\rightarrow$	Sc	−2.027	Au^{3+}	+	$3e^-$	$\rightarrow$	Au	1.498
Eu^{3+}	+	$3e^-$	$\rightarrow$	Eu	−1.985	Au^+	+	e^-	$\rightarrow$	Au	1.691
Be^{2+}	+	$2e^-$	$\rightarrow$	Be	−1.810	F_2	+	$2e^-$	$\rightarrow$	$2F-$	2.920

$$\underset{\text{metal}}{\text{Fe}} + \underset{\text{aqueous}}{\text{Cu}^{2+}} \Leftrightarrow \underset{\text{aqueous}}{\text{Fe}^{2+}} + \underset{\text{metal}}{\text{Cu}} \tag{8.163}$$

or even

$$\underset{\text{metal}}{\text{Zn}} + \underset{\text{aqueous}}{\text{Cu}^{2+}} \Leftrightarrow \underset{\text{aqueous}}{\text{Zn}^{2+}} + \underset{\text{metal}}{\text{Cu}}. \tag{8.164}$$

We can visualize the reactions 8.163 and 8.164 as composed of partial equilibria—i.e., for equation 8.163,

$$\underset{\text{metal}}{\text{Fe}} \Leftrightarrow \underset{\text{aqueous}}{\text{Fe}^{2+}} + 2\,\text{e}^- \qquad \left(0.474\,\text{V}\right) \tag{8.165}$$

$$\underset{\text{aqueous}}{\text{Cu}^{2+}} + 2\,\text{e}^- \Leftrightarrow \underset{\text{metal}}{\text{Cu}} \qquad \left(0.340\,\text{V}\right). \tag{8.166}$$

Analogously, for the redox process of equation 8.164,

$$\underset{\text{metal}}{\text{Zn}} \Leftrightarrow \underset{\text{aqueous}}{\text{Zn}^{2+}} + 2\,\text{e}^- \qquad \left(0.763\,\text{V}\right) \tag{8.167}$$

$$\underset{\text{aqueous}}{\text{Cu}^{2+}} + 2\,\text{e}^- \Leftrightarrow \underset{\text{metal}}{\text{Cu}} \qquad \left(0.340\,\text{V}\right). \tag{8.168}$$

The partial equilibria of equations 8.165→8.168 reveal the usefulness of standard potentials: the Gibbs free energy change ΔG of the redox equilibrium is always given by applying Faraday's equation to the algebraic sum of the standard potentials of the redox couples in question. For equation 8.163, the bulk potential is thus

$$E^0_{163} = 0.474 + 0.340 = 0.814 \qquad \left(\text{V}\right), \tag{8.169}$$

and for equation 8.164 it is

$$E^0_{164} = 0.763 + 0.340 = 1.103 \qquad (\text{V}). \tag{8.170}$$

Based on Faraday's law (equation 8.160), the ΔG of the reaction is in both cases negative: reactions then proceed spontaneously toward the right, with the dissolution of Fe and Zn electrodes, respectively, and deposition of metallic Cu at the cathode.

Let us now consider a galvanic cell with the redox couples of equation 8.164. This cell may be composed of a Cu electrode immersed in a one-molal solution of $CuSO_4$ and a Zn electrode immersed in a one-molal solution of $ZnSO_4$ ("Daniell cell" or "Daniell element"). Equation 8.170 shows that the galvanic potential is positive: the ΔG of the reaction is negative and the reaction proceeds toward the right. If we short-circuit the cell to annul the potential, we observe dissolution of the Zn electrode and deposition of metallic Cu at the opposite electrode. The flow of electrons is from left to right: thus, the Zn electrode is the anode (metallic Zn is oxidized to Zn^{2+}; cf. eq. 8.167), and the Cu electrode is the cathode (Cu^{2+} ions are reduced to metallic Cu; eq. 8.168):

$$\text{Zn}, \text{Zn}^{2+}, \text{SO}_4^{2-} \overset{e^-}{\underset{\ominus}{\longrightarrow}} \bigg| \bigg| \underset{\oplus}{\text{Cu}, \text{Cu}^{2+}, \text{SO}_4^{2-}}. \qquad (8.171)$$

As we have already seen, the standard potentials are relative to standard reference conditions—i.e., one-molal solutions at $T_r = 25\,°C$ and $P_r = 1$ bar, in equilibrium with pure metals or pure gases. Applying the Nernst relation to a redox equilibrium such as reaction 8.163 and assuming unitary activity for the condensed phases (i.e., pure metals), we have

$$\Delta G_{163} = \Delta G_{163}^0 + RT \ln \frac{a_{\text{Fe}^{2+},\text{aqueous}}}{a_{\text{Cu}^{2+},\text{aqueous}}}. \qquad (8.172)$$

Combining equation 8.172 with the Faraday relation (eq. 8.160), we obtain

$$
\begin{aligned}
E_{163} &= E_{163}^0 - \frac{RT}{nF} \ln \frac{a_{\text{Fe}^{2+},\text{aqueous}}}{a_{\text{Cu}^{2+},\text{aqueous}}} \\
&= E_{163}^0 - \frac{RT}{nF} \ln \frac{m_{\text{Fe}^{2+}}}{m_{\text{Cu}^{2+}}} - \frac{RT}{nF} \ln \frac{\gamma_{\text{Fe}^{2+}}}{\gamma_{\text{Cu}^{2+}}}.
\end{aligned}
\qquad (8.173)
$$

Equation 8.173 establishes that in an electrolytic solution the relative proportions of oxidized and reduced species are controlled by the redox state of the system (we will see in section 8.19 that in actual fact the redox equilibrium among the various redox couples in natural, chemically complex, aqueous solutions is rarely attained). Imposing on equation 8.173 the value of the ruling redox potential (Eh)—i.e.,

$$E_{163} = \text{Eh}, \qquad (8.174)$$

the activity ratio conforms to the difference between the imposed Eh and the standard state potential:

$$\text{Eh} - E^0_{163} = -\frac{RT}{nF} \ln \frac{a_{Fe^{2+},aqueous}}{a_{Cu^{2+},aqueous}}. \tag{8.175}$$

Based on equation 8.173, the new equilibrium is attained through a change of molality of ions in solution.

Let us now consider redox limits for the thermodynamic stability of aqueous solutions. Maximum oxidation is defined by the dissociation of water molecules, with the formation of hydrogen ions and gaseous oxygen—i.e.,

$$\underset{\text{gas}}{O_2} + \underset{\text{aqueous}}{4\,H^+} + 4\,e^- \Leftrightarrow 2\,H_2O. \tag{8.176}$$

The standard potential of equation 8.176 is $E^0_{176} = 1.228$ V. At standard state, the activity of gaseous oxygen is 1 by definition, and standard potential E^0_{176} thus refers to H_2O in equilibrium with an atmosphere of pure O_2 at $T = 25$ °C and $P = 1$ bar. Applying the Nernst and Faraday relations to equation 8.176 and transforming natural logarithms into base 10 logarithms, we obtain

$$\begin{aligned}
\text{Eh} &= 1.228 + \frac{2.303\,RT}{nF}\,4 \log a_{H^+} + \frac{2.303\,RT}{nF} \log f_{O_2} \\
&= 1.228 + 0.0591 \log a_{H^+} + 0.0148 \log f_{O_2}.
\end{aligned} \tag{8.177}$$

The maximum reduction limit for the thermodynamic stability of water is defined by the reduction of hydrogen ions in solution with the formation of diatomic gaseous molecules H_2 (normal hydrogen electrode):

$$\underset{\text{aqueous}}{2\,H^+} + 2\,e^- \Leftrightarrow \underset{\text{gas}}{H_2}. \tag{8.178}$$

The redox potential dictated by equation 8.178 is related to standard potential E^0_{178} through

$$\text{Eh} = 0.000 + 0.0591 \log a_{H^+} - 0.0295 \log f_{H_2}. \tag{8.179}$$

We now recall that the ionic dissociation of water is

$$H_2O \Leftrightarrow H^+ + OH^- \tag{8.180}$$

The constant of equation 8.180 at $T = 25$ °C and $P = 1$ bar (cf. table 8.3) is

$$K_{180} = a_{H^+} a_{OH^-} = 1.008 \times 10^{-14}. \qquad (8.181)$$

Expressed in logarithmic form, equation 8.181 becomes

$$\log a_{H^+} + \log a_{OH^-} \approx -14.0. \qquad (8.182)$$

The definition of pH is

$$\mathrm{pH} = -\log a_{H^+}, \qquad (8.183)$$

hence

$$\mathrm{pH} = 14.0 + \log a_{OH^-}. \qquad (8.184)$$

Based on equation 8.184, we can subdivide the acid-base property of an aqueous solution into two fields according to the relative predominance of hydroxyls OH^- and hydrogen ions H^+—i.e.,

$$a_{H^+} > a_{OH^-} \, ; \mathrm{pH} < 7 \,(\text{acidic solutions})$$

$$a_{H^+} = a_{OH^-} \, ; \mathrm{pH} = 7 \,(\text{neutral solutions}) \qquad (8.185)$$

$$a_{H^+} < a_{OH^-} \, ; \mathrm{pH} > 7 \,(\text{basic, or } "alkaline," \text{ solutions}).$$

In analogy with acid-base neutrality at pH = 7, we can also define redox neutrality through the molecular dissociation process:

$$2\,H_2O \Leftrightarrow 2\,H_2 + O_2. \qquad (8.186)$$

Redox neutrality corresponds to the condition

$$f_{H_{2(gas)}} = 2 f_{O_{2(gas)}}. \qquad (8.187)$$

Combining equation 8.187 with equations 8.176 and 8.178 and with the acid-base neutrality concept (eq. 8.184 and 8.185), we obtain the absolute neutrality condition for a diluted aqueous solution at $T = 25$ °C and $P = 1$ bar—i.e.,

$$\mathrm{pH} = 7.0; \quad \log f_{H_2} = -27.61; \quad \log f_{O_2} = -27.91; \quad E^0 = 0.4 \,(\text{V}). \qquad (8.188)$$

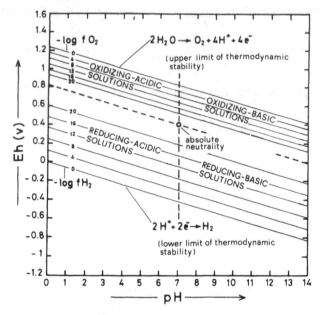

Figure 8.19 Thermodynamic stability field for aqueous solutions at $T = 25\,°C$ and $P = 1$ bar.

Figure 8.19 shows the thermodynamic stability field for diluted aqueous solutions at $T = 25\,°C$ and $P = 1$ bar. The pH-Eh range is bounded upward by the maximum oxidation limit, with a slope of -0.0591 and an intercept on the ordinate axis depending on the partial pressure of oxygen: several limiting curves are drawn for various values of f_{O_2}. We recall here that the partial pressure of oxygen in the earth's atmosphere at sea level is about 0.2 bar, so that, assuming a fugacity coefficient of 1 ($f_{O_2} = P_{O_2} = 0.2$), the intercept on the ordinate axis at pH = 0 is Eh = 1.22. The lower stability limit is defined, as we have already seen, by equation 8.179 for the various values of f_{H_2}. The absolute neutrality point subdivides the Eh-pH field of thermodynamic stability into four parts:

1. Oxidizing acidic solutions
2. Oxidizing basic solutions
3. Reducing basic solutions
4. Reducing acidic solutions.

8.13 Relative Predominance Limits of Aqueous Solutes and Eh-pH Diagrams

We have already seen that the molality ratio of species involved in redox equilibria is buffered by the redox state of the system with respect to the standard potential of the redox couple. To understand this concept better, let us examine the various speciation states of a polyvalent element such as cerium in aqueous solution. Cerium, in the absence of other anionic ligands besides OH^- groups, is mainly present in aqueous solutions in three forms, Ce^{3+}, $Ce(OH)^{3+}$, and $Ce(OH)_2^{2+}$, related by the homogeneous equilibria

$$Ce(OH)^{3+} + H^+ + e^- \Leftrightarrow Ce^{3+} + H_2O \tag{8.189}$$

$$Ce(OH)_2^{2+} + 2H^+ + e^- \Leftrightarrow Ce^{3+} + 2H_2O \tag{8.190}$$

$$Ce(OH)^{3+} + H_2O \Leftrightarrow Ce(OH)_2^{2+} + H^+. \tag{8.191}$$

The ruling equilibria between aqueous species and condensed forms are

$$\underset{\text{aqueous}}{Ce^{3+}} + 3e^- \Leftrightarrow \underset{\text{metal}}{Ce} \tag{8.192}$$

$$\underset{\text{aqueous}}{Ce^{3+}} + 3H_2O \Leftrightarrow \underset{\text{crystal}}{Ce(OH)_3} + 3H^+ \tag{8.193}$$

$$\underset{\text{aqueous}}{Ce(OH)_2^{2+}} + H_2O + e^- \Leftrightarrow \underset{\text{crystal}}{Ce(OH)_3} + H^+ \tag{8.194}$$

$$\underset{\text{crystal}}{CeO_2} + 2H^+ \Leftrightarrow \underset{\text{aqueous}}{Ce(OH)_2^{2+}}, \tag{8.195}$$

to which we can add the following equilibria among condensed forms:

$$\underset{\text{crystal}}{Ce(OH)_3} + 3H^+ + 3e^- \Leftrightarrow \underset{\text{metal}}{Ce} + 3H_2O \tag{8.196}$$

$$\underset{\text{crystal}}{CeO_2} + H_2O + H^+ + e^- \Leftrightarrow \underset{\text{crystal}}{Ce(OH)_3}. \tag{8.197}$$

If we apply the Nernst and Faraday relations to equations 8.189 and 8.190 as before (cf. eq. 8.177 and 8.179), we obtain

$$Eh = E^0_{189} - 0.0591\,pH + 0.0591\log\frac{\left[Ce(OH)^{3+}\right]}{\left[Ce^{3+}\right]} \tag{8.198}$$

and

$$Eh = E_{190}^0 - 0.1182\,pH + 0.0591 \log \frac{\left[Ce(OH)_2^{2+}\right]}{\left[Ce^{3+}\right]},\qquad(8.199)$$

where terms in square brackets denote thermodynamic activities of solutes, $E_{189}^0 = 1.715$ V, and $E_{190}^0 = 1.731$ V.

On an Eh-pH plot (figure 8.20), the isoactivity conditions for solute species:

$$\left[Ce(OH)^{3+}\right] = \left[Ce^{3+}\right]\qquad(8.200)$$

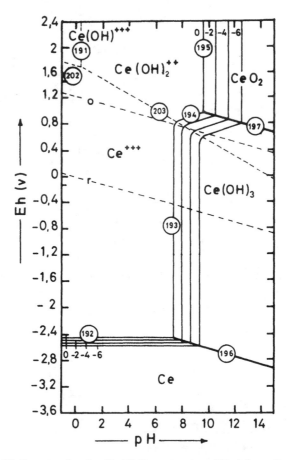

Figure 8.20 Eh-pH diagram for the Ce-H$_2$O system (modified from Pourbaix, 1966). Figures on limiting curves are base 10 logarithms of solute activity: unitary activity (i.e., one-molal solution) is identified by zero.

$$\left[Ce(OH)_2^{2+}\right] = \left[Ce^{3+}\right] \tag{8.201}$$

are defined, respectively, by the equations

$$Eh = 1.715 - 0.0591\,pH \tag{8.202}$$

$$Eh = 1.731 - 0.1182\,pH. \tag{8.203}$$

Equations 8.202 and 8.203 define the "boundaries of relative predominance" of solutes. Obviously, in the case of eq. 8.191, in which there is no electron transference, the relative predominance limits of $Ce(OH)^{3+}$ and $Ce(OH)_2^{2+}$ are determined solely by the acidity of the solution and not by its redox state—i.e.,

$$\log \frac{\left[Ce(OH)_2^{2+}\right]}{Ce(OH)^{3+}} = -0.29 + pH. \tag{8.204}$$

The limit (equation 8.204) is thus parallel to the ordinate axis (pH = 0.29; cf. figure 8.20).

As we did for equilibria between solute species, we can also define the boundaries between solute species and condensed phases. Assuming the condensed forms to be pure phases (i.e., assuming unitary activity), in the presence of metallic cerium we have

$$Eh = E_{192}^0 + 0.0197\log\left[Ce^{3+}\right], \tag{8.205}$$

with $E_{192}^0 = -2.483$ V; in the presence of $Ce(OH)_3$ we have

$$Eh = E_{194}^0 + 0.0591\,pH + 0.0591\log\left[Ce(OH)_2^{2+}\right], \tag{8.206}$$

with $E_{194}^0 = 0.422$; in the presence of $Ce(OH)_3$ we also have

$$\log\left[Ce^{3+}\right] = 22.15 + 3\,pH, \tag{8.207}$$

and in the presence of CeO_2 we have

$$\log\left[Ce(OH)_2^{2+}\right] = 19.22 - 2\,pH. \tag{8.208}$$

As figure 8.20 shows, the boundary between condensed phases and solute species is dictated by the activity of solutes, T being equal.

Lastly, the boundary between Ce and $Ce(OH)_3$ (eq. 8.196) is given by

$$Eh = E^0_{196} - 0.0591\,pH, \tag{8.209}$$

with $E^0_{196} = 2.046$, and the boundary between $Ce(OH)_3$ and CeO_2 by

$$Eh = E^0_{197} - 0.0591\,pH, \tag{8.210}$$

with $E^0_{197} = 1.559$.

Usually, in plots such as those in figure 8.20, for the sake of simplicity, the boundaries between solutes and condensed phases are not drawn for the various molal concentrations in solution, as seen for cerium, but for a selected bulk molal amount, which in geochemistry is normally 10^{-6}. This choice is dictated by the fact that, at this bulk molal concentration, the solid phase can be considered effectively inert from the point of view of reactivity (cf. Garrels and Christ, 1965). In the various heterogeneous equilibria involving aqueous solutions and condensed or gaseous phases, it is nevertheless opportune always to specify the molal concentrations to which the various boundaries refer.

The main usefulness of Eh-pH diagrams consists in the immediacy of qualitative information about the effects of redox and acid-base properties of the system on elemental solubility. Concerning, for instance, cerium, figure 8.20 immediately shows that, within the stability field of water, delimited upward by oxidation boundary curve o and downward by reduction boundary curve r, the element (in the absence of other anionic ligands besides OH^- groups) is present in solution mainly as trivalent cerium Ce^{3+} and as soluble tetravalent hydroxide $Ce(OH)_2^{2+}$. It is also evident that, with increasing pH, cerium precipitates as trivalent hydroxide $Ce(OH)_3$.

The interpretation of Eh-pH diagrams implies assumption of complete equilibrium among the various solutes and condensed forms. Although this assumption is plausible in a compositionally simple system such as that represented in figure 8.20, it cannot safely be extended to more complex natural systems, where the various redox couples are often in apparent disequilibrium. It is therefore necessary to be cautious when dealing with the concept of the "system Eh" and the various redox parameters.

8.14 "Electronic Activity," "System Eh," and Other Redox Parameters

As repeatedly noted, *standard potentials* are linked to the standard molal Gibbs free energy of formation from the elements through Faraday's equation. Let us now reconsider the equilibrium between metallic iron and aqueous ion at $T = 25\,°C$ and $P = 1$ bar:

$$\underset{\text{metal}}{Fe} \Leftrightarrow \underset{\text{aqueous}}{Fe^{2+}} + 2\,e^-.$$

(8.211)

At standard state, equation 8.211 obviously concerns standard state stable components—i.e., pure Fe metal at $T = 25\,°C$, $P = 1$ bar, and a hypothetical one-molal Fe^{2+} solution referred to infinite dilution, at the same P and T conditions (cf. section 8.4). The chemical potentials of components in reactions ($a_{Fe(metal)}$, $a_{Fe^{2+}(aqueous)}$, $a_{e^-(aqueous)}$) are those of standard state; hence, by definition, the activity of all components in reaction is 1—i.e.,

$$a_{Fe_{(metal)}} = 1; \quad a_{Fe^{2+}_{(aqueous)}} = 1; \quad a_{e^-_{(aqueous)}} = 1,$$

(8.212)

where a_{e^-} is "the activity of the hypothetical electron in solution," which is 1, as we have seen at standard state. Because, by convention, the partial molal Gibbs free energy of formation of the hypothetical aqueous electron is assumed to be zero at all P, T, and molality conditions—i.e.,

$$\overline{G}_f\,e^- = \overline{G}_f^0\,e^- = 0$$

(8.213)

—it follows that the activity of the hypothetical electron in solution a_{e^-} is always 1 under all P, T, and molality conditions. We have already seen that, based on this consideration, a_{e^-} is never explicit in Nernst equations (as in eq. 8.172 and 8.175, for example).

However, in the literature we often encounter the redox parameter "pe," which, like pH, represents the cologarithm of the activity of the hypothetical electron in solution:

$$pe = -\log a_{e^-}.$$

(8.214)

The relationship between pe and Eh is

$$pe = \frac{F}{2.303RT}\,Eh$$

(8.215)

(cf. Stumm and Morgan, 1981). As outlined by Wolery (1983), equation 8.215 implies that the Gibbs free energy of formation of the hypothetical electron in solution is not always zero, as stated in equation 8.213, but is given (cf. eq. 8.159) by

$$\overline{G}_f \, e^-_{(aqueous)} = \frac{1}{2} \overline{G}_{H_{2(gas)}} - \overline{G}_{H^+_{(aqueous)}}. \tag{8.216}$$

The incongruency between equations 8.216 and 8.213 leads us to discourage the use of parameter pe, which is formally wrong from the thermodynamic point of view. We also note that, based on equation 8.215, pe is simply a transposition in scale and magnitude of Eh (pe is adimensional, Eh is expressed in volts).

For the sake of completeness, we recall that analogies with the concept of pH do not concern solely the pe factor. In biochemistry, for instance, O_2 and H_2 fugacities in gaseous phases are often described by the rO and rH parameters, respectively:

$$rO = - \log f_{O_{2(gas)}} \tag{8.217}$$

$$rH = - \log f_{H_{2(gas)}}. \tag{8.218}$$

In the description of redox equilibria in aqueous solutions, we adopted the term "ruling Eh" and not "system Eh." By "ruling Eh" we meant the Eh value imposed on the redox couple to attain a determinate activity ratio between reduced and oxidized forms. As already noted, natural aqueous solutions often contain dissolved elements in different and contrasting redox states, clearly incompatible with a single value for the "system Eh." For instance, in almost all natural aqueous solutions, organic carbon and dissolved oxygen coexist (Wolery, 1983). In several marine sediments, methane CH_4 and bicarbonate ion HCO_3^- coexist with dissolved sulfides and sulfates (Thorstenson, 1970). Berner (1971) quotes the presence in surface seawaters of nitrogen and dissolved nitrates coexisting with free oxygen. Table 8.15 (from Jenne, 1981) lists Eh potentials for seawaters of different salinities in San Francisco Bay (California), calculated from the activities of dissolved redox species and compared with the measured Pt-electrode potential (Jenne et al., 1978).

Table 8.15 Eh values in seawaters of San Francisco Bay relative to various redox couples, compared with measured Eh values (from Jenne, 1981).

Redox Couple	Eh Potential (V)
$[NO_3]/[NO_2]$	0.44-0.43
$[NO_3]/[NH_4]$	0.33-0.32
$[Fe^{3+}]/[Fe^{2+}]$	0.12-0.05
$[SO_4]/[S^{2-}]$	−0.17-−0.20
Pt-electrode	0.39-0.36

Table 8.15 stresses the existence of marked redox disequilibria among the various couples. In this case (and also for the previously quoted cases), the concept of "system Eh" is inappropriate. Nevertheless, table 8.15 shows that the measured Eh is near the value indicated by the redox couple $[NO_3]/[NO_2]$. We can thus assume that the redox equilibrium between $[NO_3]$ and $[NO_2]$ determines the ruling Eh.

Of the various factors that cause redox disequilibria, the most effective are biologic activity (photosynthesis) and the metastable persistence of covalent complexes of light elements (C, H, O, N, S), whose bonds are particularly stable and difficult to break (Wolery, 1983). For the sake of completeness, we can also note that the apparent redox disequilibrium is sometimes actually attributable to analytical error or uncertainty (i.e., difficult determination of partial molalities of species, often extremely diluted) or even to error in speciation calculations (when using, for instance, the redox couple Fe^{3+}/Fe^{2+}, one must account for the fact that both Fe^{3+} and Fe^{2+} are partly bonded to anionic ligands so that their free ion partial molalities do not coincide with the bulk molality of the species).

8.15 Eh-pH Diagrams for Main Anionic Ligands

Knowledge of the redox states of anionic ligands is fundamental for understanding complexation processes operating in various geologic environments. The redox states of the main anionic ligands under various Eh-pH conditions within the stability range of water will be briefly described here. The predominance limits appearing in figures 8.21 to 8.24 define the isoactivity conditions of the dissolved species (cf., for instance, eq. 8.200 and 8.201). The diagrams are taken from Brookins (1988), to whom we refer readers for more exhaustive explanations. The adopted standard molal Gibbs free energy values of the various species do not differ appreciably from the tabulated values of Shock and Helgeson (1988) (table 8.13) and come mainly from Wagman et al. (1982). The two series of data are compared in table 8.16. The observed discrepancies do not greatly alter the predominance limits drawn in the figures.

Carbon is present in nature in three oxidation states (4+, 0, and 4−). The predominance limits of figure 8.21A relate to a molal concentration of total dissolved C of 10^{-3}. Carbonic acid H_2CO_3 and its ionization products occur over most of the Eh-pH stability field of water. At pH higher than approximately 6.4, the bicarbonate ion HCO_3^- predominates, and at pH higher than 10.3, the carbonate ion CO_3^{2-} predominates. Native carbon C appears to be stable at reducing conditions in a very restricted Eh range, below which aqueous methane $CH_{4(aq)}$ occurs. The development of gaseous methane is observed under reducing conditions corresponding to the lower limit of stability of water (r).

Table 8.16 Standard molal Gibbs free energies of formation from the elements for aqueous ions and complexes and condensed phases, partly adopted in constructing the Eh-pH diagrams in figure 8.21. Data in kcal/mole. Values in parentheses: Shock and Helgeson's (1988) tabulation. Sources of data: (1) Wagman et al. (1982); (2) Garrels and Christ (1965); (3) Pourbaix (1966); (4) Berner (1971)

Ion or Complex	$\overline{G}^0_{f,T_r,P_r}$		Reference	Ion or Complex	$\overline{G}^0_{f,T_r,P_r}$		Reference
BO_2	-162.26	(-162.24)	(1)	S^{2-}	20.51	(19.00)	(1)
$B_2O_{3(cr)}$	-285.29		(1)	HS^-	2.89	(2.86)	(1)
$B_4O_7^{2-}$	-622.56		(1)	H_2S	-6.65		(1)
BH_4^-	27.33		(1)	SO_3^{2-}	-116.28	(-116.30)	(1)
$B(OH)_4^{-}$	-275.61		(1)	HSO_3^-	-126.13	(-126.13)	(1)
H_3BO_3	-231.54		(1)	SO_4^{2-}	-177.95	(-177.93)	(1)
$H_2BO_3^-$	-217.60		(3)	HSO_4^-	-180.67	(-180.63)	(1)
HBO_3^{2-}	-200.29		(3)	F^-	-66.63	(-67.34)	(1)
BO_3^{3-}	-181.48		(3)	Cl^-	-31.36	(-31.38)	(1)
CH_4	-8.28		(1)	ClO_3^-	-1.90	(1.90)	(1)
CH_2O	-31.00		(2)	ClO_4^-	-2.04	(-2.04)	(1)
H_2CO_3	-149.00		(1)	Br^-	-24.85	(24.87)	(1)
HCO_3^-	-140.24	(-140.28)	(1)	BrO_3^-	4.46	(4.45)	(1)
CO_3^{2-}	-126.15	(-126.19)	(1)	BrO_4^-	28.23	(28.20)	(1)
NO_3	-25.99	(-26.51)	(1)	I^-	-12.33	(-12.41)	(1)
NH_4^+	-18.96	(-18.99)	(1)	IO_3^-	-30.59	(-30.60)	(1)
$NH_{3(gas)}$	-3.98		(4)	IO_4^-	-13.98	(-14.00)	(1)
H_3PO_4	-273.07		(1)	$H_2PO_4^-$	-270.14		(1)
HPO_4^{2-}	-260.31	(-260.41)	(1)	PO_4^{2-}	-243.48	(-243.50)	(1)

Figure 8.21B shows the Eh-pH diagram for nitrogen, calculated for $f_{N_2} = 0.8$ bar (total molality of dissolved N species $= 10^{-3.3}$). The gaseous molecule N_2 is stable over most of the Eh-pH field. The ammonium ion NH_4^+ predominates under reducing conditions, and the nitrate ion NO_3^- predominates under oxidizing conditions. Gaseous ammonia forms only under extremely reducing and alkaline conditions. The fields of ammonium and nitrate ions expand metastably within the stability field of gaseous nitrogen (the limit of metastable predominance of the two species is drawn as a dashed line within the N_2 field in figure 8.21B).

Figure 8.21C shows the Eh-pH diagram for phosphorus at a solute total molality of 10^{-4}. Within the stability field of water, phosphorus occurs as orthophosphoric acid H_3PO_4 and its ionization products. The predominance limits are dictated by the acidity of the solution and do not depend on redox conditions.

Sulfur (figure 8.21D) is present in aqueous solutions in three oxidation states ($2-$, 0, and $6+$). The field of native S, at a solute total molality of 10^{-3}, is very limited and is comparable to that of carbon (for both extension and Eh-pH range). Sulfide complexes occupy the lower part of the diagram. The sulfide-sulfate transition involves a significant amount of energy and defines the limit of predominance above which sulfates occur.

The extension of the field of native elements (S, C) depends on the adopted

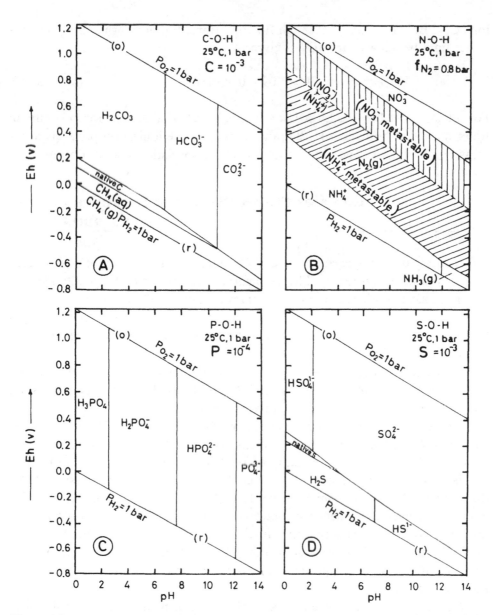

Figure 8.21 Eh-pH diagrams for main anionic ligands. From Brookins (1988). Reprinted with permission of Springer-Verlag, New York.

molality values of solutes (i.e., it expands with increasing, and shrinks with decreasing, molal concentrations). The molal concentrations used for constructing figure 8.21 are mean values for natural waters, but they may vary considerably.

The above-discussed anionic ligands are the most important ones from the viewpoint of redox properties. For the sake of completeness, we can also consider the redox behavior of halides and boron. Halides stable in water are the simple ions Cl^-, Br^-, F^-, and I^-. At high Eh conditions, IO_3^- ion forms; the predominance limit between I^- and IO_3^- is given by

$$Eh = 1.09 - 0.0591\,pH. \qquad (8.219)$$

For all the other halides, Eh-pH conditions have no influence. Boron occurs in water mainly as boric acid H_3BO_3 and its progressive ionization products at increasing pH. Redox conditions do not affect the speciation state of boron.

8.16 Complex Eh-pH Diagrams: Examples of Iron, Manganese, and Copper

Information on the speciation states of solutes and their equilibria with condensed phases furnished by Eh-pH diagrams is often simply qualitative and should be used only in the initial stages of investigations. The chemical complexity of natural aqueous solutions and the persistent metastability and redox disequilibrium induced by organic activity are often obstacles to rigorous interpretation of aqueous equilibria.

Let us now consider the effects of chemical complexity on the predominance limits of solutes at environmental P and T conditions. The elements in question are emblematic for their particular importance in ore formation.

Figure 8.22A shows the Eh-pH diagram of iron in the Fe-O-H system at $T = 25\,°C$ and $P = 1$ bar. The diagram is relatively simple: the limits of predominance are drawn for a solute total molality of 10^{-6}. Within the stability field of water, iron is present in the valence states $2+$ and $3+$. In figure 8.22A, it is assumed that the condensed forms are simply hematite Fe_2O_3 and magnetite Fe_3O_4. Actually, in the $3+$ valence state, metastable ferric hydroxide $Fe(OH)_3$ and metastable goethite $FeOOH$ may also form, and, in the $2+$ valence state, ferrous hydroxide $Fe(OH)_2$ may form. It is also assumed that the trivalent solute ion is simply Fe^{3+}, whereas, in fact, various aqueous ferric complexes may nucleate [i.e., $Fe(OH)^{2+}$, $Fe(OH)_2^+$, etc.].

Figure 8.22B shows the effect of adding carbon to the system (molality of solute carbonates $= 10^{-2}$): in reducing alkaline states, the field of siderite $FeCO_3$ opens; in the other Eh-pH conditions, the limits are unchanged.

Adding sulfur to the system (molality of solutes $= 10^{-6}$; figure 8.22C), a wide stability field opens for pyrite FeS_2 in reducing conditions and, almost at the lower stability limit of water, a limited field of pyrrhotite FeS is observed.

Figure 8.22 shows the effect of silica in the system (aqueous solution saturated with respect to amorphous silica): under reducing alkaline conditions, the metasilicate $FeSiO_3$ forms at the expense of magnetite through the equilibrium

$$Fe_3O_4 + 3\,SiO_2 + 2\,H^+ + 2e^- \Leftrightarrow 3\,FeSiO_3 + H_2O \qquad (8.220)$$

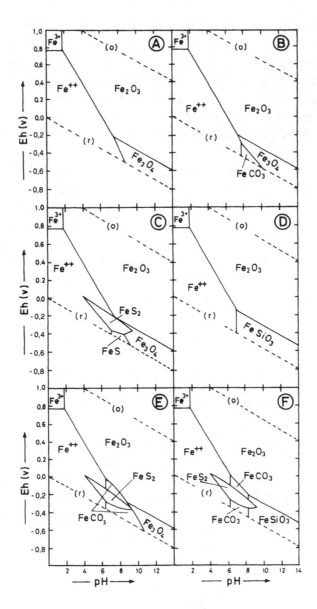

Figure 8.22 Eh-pH diagrams for iron-bearing aqueous solutions. Chemical complexity increases from A to F.

with a limit defined by

$$Eh = 0.272 - 0.0591\,pH. \tag{8.221}$$

Reaction 8.220 takes place in the silicate bands present in iron ores: iron silicates occur in such bands in coexistence with opaline amorphous silica (Klein and Bricker, 1972). A complex Eh-pH diagram for the Fe-C-S-O-H system is shown in figure 8.22E. The stability field of pyrrhotite FeS disappears and is replaced by siderite $FeCO_3$, which also limits the field of pyrite. When silicon is also present in the system (figure 8.22F), magnetite does not nucleate as a stable phase.

The predominance limits shown in figure 8.22 are analytically summarized in table 8.17. Compare figures 8.22 and 8.21 to better visualize the redox state of the anionic ligands at the various Eh-pH conditions of interest (particularly the sulfide-sulfate transition and carbonate limits). We remand to Garrels and Christ (1965) for a more detailed account on the development of complex Eh-pH diagrams.

Let us now consider the effects of Eh-pH conditions on the speciation state and solubility of manganese in aqueous solutions. Manganese complexes have been carefully studied in the last decade, owing to the discovery on ocean floors of economically important metalliferous deposits (nodules and crusts) in which Mn compounds are dominant.

Figure 8.23A shows a simplified Eh-pH diagram for the Mn-O-H system. Within the stability field of water, manganese occurs in three valence states (2+, 3+, and 4+). Figure 8.23A shows the condensed phases relative to the three valence states as the hydroxide pyrochroite $Mn(OH)_2$ (2+), multiple oxide hausmannite Mn_3O_4 (2+, 3+), sesquioxide Mn_2O_3 (3+), and oxide pyrolusite MnO_2 (4+).

A wide field is occupied by the Mn^{2+} ion. The condensed phases nucleate only in alkaline solutions, and the stability fields of the various solids expand with increasing Eh. The precipitation of Mn oxides and hydroxides is actually a complex phenomenon involving the initial nucleation of metastable compounds. Hem and Lind (1983) showed that, at f_{O_2} near the upper stability limit of water and at pH in the range 8.5 to 9.5, the type of nucleating solid depends on temperature: at $T = 25 \,°C$ metastable hausmannite Mn_3O_4 forms, whereas at T near 0 °C the mixed hydroxides feitknechtite, manganite, and groutite [all with formula unit $MnO(OH)$] precipitate. All these compounds are metastable in experimental Eh-pH conditions and transform progressively into the tetravalent compound MnO_2.

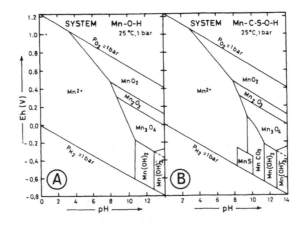

Figure 8.23 Eh-pH diagrams for manganese. From Brookins (1988). Reprinted with permission of Springer-Verlag, New York.

Table 8.17 Main predominance limits of aqueous complexes and saturation limits between solutes and condensed phases in iron-bearing aqueous solutions (see figure 8.22). Standard state Gibbs free energies of formation of species are listed in table 8.18. (c) = crystalline; (a) = aqueous; (g) = gaseous; (l) = liquid. [ ] = thermodynamic activity.

System Fe-O-H; $T = 25\ °C$; $P = 1$ bar; solutes molality $= 10^{-6}$

$Fe_2O_{3(c)} + 6H^+_{(a)} \rightarrow 2Fe^{3+}_{(a)} + 3H_2O_{(l)}$
$\log[Fe^{3+}] = -0.720 - 3pH$
$Fe^{3+}_{(a)} + e^- \rightarrow Fe^{2+}_{(a)}$
$Eh = 0.78$

$2Fe^{2+}_{(a)} + 3H_2O_{(l)} \rightarrow Fe_2O_{3(c)} + 6H^+_{(a)} + 2e^-$
$Eh = 0.728 - 0.0591 \log[Fe^{2+}] - 0.177pH$

$3Fe^{2+}_{(a)} + 4H_2O_{(l)} \rightarrow Fe_3O_{4(c)} + 8H^+_{(a)} + 2e^-$
$Eh = 0.980 - 0.0885 \log[Fe^{2+}] - 0.236pH$

$2\ Fe_3O_{4(c)} + H_2O_{(l)} \rightarrow 3\ Fe_2O_{3(c)} + 2H^+_{(a)} + 2e^-$
$Eh = 0.221 - 0.0591pH$

System Fe-C-O-H; $f{co_2}_{(g)} = 10^{-2}$; $[H_2CO_3] = 10^{-2}$

$\log[H_2CO_3] = -2.0$
$\log[HCO_3{}^-] = -6.4 + \log[H_2CO_3] + pH$

$FeCO_{3(c)} + 2H^+_{(a)} \rightarrow Fe^{2+}_{(a)} + CO_{2(g)} + H_2O_{(l)}$

$\log[Fe^{2+}] = 7.47 - \log f{co_2}_{(g)} - 2pH$

$3FeCO_{3(c)} + 4H_2O_{(l)} \rightarrow Fe_3O_{4(c)} + 3HCO^-_{3(a)} + 5H^+_{(a)} + 2e^-$
$Eh = 1.010 - 0.148pH + 0.0885 \log[HCO_3{}^-]$

System Fe-S-O-H; solutes molality $= 10^{-6}$

$FeS_{(c)} + H^+_{(a)} \rightarrow Fe^{2+}_{(a)} + HS^-_{(a)}$
$\log[Fe^{2+}] = -4.40 - pH - \log[HS^-]$

$2H_2S_{(a)} + Fe^{2+}_{(a)} \rightarrow FeS_{2(c)} + 4H^+ + 2e^-$
$Eh = 0.057 - 0.118pH - 0.0591 \log[H_2S] - 0.0295 \log[Fe^{2+}]$

$FeS_{2(c)} + 8H_2O_{(l)} \rightarrow 2HSO^-_{4(a)} + Fe^{2+}_{(a)} + 14H^+_{(a)} + 14e^-$
$Eh = 0.339 - 0.0591pH + 0.0084 \log[HSO_4{}^-] + 0.0042 \log[Fe^{2+}]$

$FeS_{(c)} + H_2S_{(a)} \rightarrow FeS_{2(c)} + 2H^+_{(a)} + 2e^-$
$Eh = 0.133 - 0.0591pH - 0.0295 \log[H_2S]$

$3FeS_{2(c)} + 28H_2O_{(l)} \rightarrow Fe_3O_{4(c)} + 6SO^{2-}_{4(a)} + 56H^+_{(a)} + 44e^-$
$Eh = 0.384 - 0.075pH + 0.0080 \log[SO_4^{2-}]$

$FeS_{(c)} + HS^-_{(a)} \rightarrow FeS_{2(c)} + H^+_{(a)} + 2e^-$
$Eh = -0.340 - 0.0295pH - 0.0295 \log[HS^-]$

System Fe-Si-O-H; unitary activity of amorphous silica

$2FeSiO_{3(c)} + H_2O_{(l)} \rightarrow Fe_2O_{3(c)} + 2SiO_{2(amorphous)} + 2H^+ + 2e^-$
$Eh = 0.258 - 0.0591pH$

$FeSiO_{3(c)} + 2H^+_{(a)} \rightarrow Fe^{2+}_{(a)} + SiO_{2(amorphous)} + H_2O_{(l)}$
$\log[Fe^{2+}] = 8.03 - 2pH$

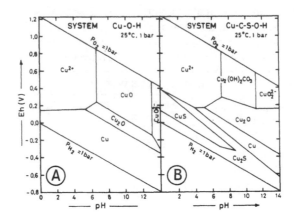

Figure 8.24 Eh-pH diagrams for copper. From Brookins (1988). Reprinted with permission of Springer-Verlag, New York.

Figure 8.23B shows the effect of adding sulfur and carbon in solution (solute molality = 10^{-3}). In reducing-alkaline conditions, both rhodochrosite $MnCO_3$ and alabandite MnS are stable. However, the field of alabandite is quite restricted, which explains why this mineral is rarely found in nature (Brookins, 1988).

Figure 8.24A shows the Eh-pH diagram for copper in the Cu-O-H system (solute molality = 10^{-6}). Within the stability field of water, copper assumes the valence states 0, +1, and +2. Native copper occupies much of the diagram, encompassing all acidity states and extending to Eh conditions higher than the sulfide-sulfate transition (cf. figure 8.21). In acidic solutions, native copper oxidizes to cupric ion Cu^{2+}. In alkaline solutions, with increasing Eh, cuprite Cu_2O and tenorite CuO form. Only in extremely alkaline solutions is the soluble compound CuO_2^{2-} stable. If carbon and sulfur are added to the system (figure 8.24B; solute molality = 10^{-3}), the field of native copper is reduced, owing to the presence of the sulfides calcocite Cu_2S and covellite CuS. At high Eh, we find the hydrated carbonate malachite $Cu_2(OH)_2CO_3$ replacing tenorite. In the presence of iron, the diagram becomes much more complex, with the formation of important ore-forming phases, such as chalcopyrite $CuFeS_2$ and bornite Cu_5FeS_4 (see, for instance, figures 7.24 and 7.25 in Garrels and Christ, 1965).

The sequence of diagrams shown in figures 8.22, 8.23, and 8.24 is only an example of an approach to the actual complexity of natural solutions. The information given by such diagrams, although qualitative, furnishes a basis for more rigorous calculations and is therefore extremely precious in the comprehension of natural phenomena. Table 8.18 lists the standard state Gibbs free energy values used in the construction of these diagrams.

8.17 Notes on Metal-Organic Complexation

Although the predominant species forming in the C-H-O system (besides H_2O) are methane CH_4, carbon dioxide CO_2 and its aqueous complexes HCO_3 and

Table 8.18 Standard state ($T = 25\ °C$; $P = 1$ bar) Gibbs free energy of formation for iron, copper, and manganese compounds, partly adopted in constructing Eh-pH diagrams of figures 8.22, 8.23, and 8.24. References: (1) Wagman et al. (1982); (2) Garrels and Christ (1965); (3) Robie et al. (1978); (4) Hem et al. (1982); (5) Shock and Helgeson (1988); (6) Bricker (1965).

$Fe = 0.0$
$Fe^{2+} = -20.30$ (2)
$Fe^{3+} = -2.52$ (2)
$Fe(OH)^{2+} = -55.91$ (2)
$Fe(OH)_2^+ = -106.2$ (2)
$FeO_2H^- = -90.6$ (2)
$Fe_{0.95}O$ (wustite) $= -58.4$ (2)
Fe_2O_3 (hematite) $= -177.1$ (2)
Fe_3O_4 (magnetite) $= -242.4$ (2)
$Fe(OH)_2 = -115.57$ (2)
$Fe(OH)_3 = -166.0$ (2)
FeS (pyrrhotite) $= -23.32$ (2)
FeS_2 (pyrite) $= -36.00$ (2)
$FeCO_3$ (siderite) $= -161.06$ (2)

$Cu = 0.00$
$Cu^+ = 11.94$ (1)
$Cu^{2+} = 15.65$ (1)
$Cu_2S = -20.60$ (chalcocite) (1)
$CuS = -12.81$ (covellite) (1)
$Cu_2O = -34.98$ (cuprite) (2)
$CuO = -31.00$ (tenorite) (1)
$CuO_2^{2-} = -43.88$ (1)
$Cu_2(CO_3)(OH)_2 = -213.58$ (malachite) (1)
$Cu_3(CO_3)_2(OH)_2 = -314.29$ (azurite) (1)

$Mn^{2+} = -54.52$ (1)
$Mn(OH)^+ = -96.80$ (1)
$Mn(OH)_3^- = -177.87$ (1)
$MnSO_4 = -232.5$ (1)
$MnO_4^- = -106.9$ (5)
$MnO_4^{2-} = -119.7$ (5)
$MnO = -86.74$ (1)
$Mn(OH)_2 = -146.99$ (1); -147.14 (6) (pyrochroite)
$MnCO_3 = -195.20$ (1); -195.04 (3) (rhodochrosite)
$MnS = -52.20$ (1)
$Mn_3O_4 = -306.69$ (1) (hausmannite)
$Mn_2O_3 = -210.59$ (1)
$Mn(OH)_3 = -181.0$ (2)
$\beta\ MnO(OH) = -129.8$ (4) (feitknechtite)
$\gamma\ MnO(OH) = -133.3$ (6) (manganite)
$\alpha\ MnO_2 = -108.3$ (6) (birnessite)
$\beta\ MnO_2 = -111.17$ (3) (pyrolusite)
$\gamma\ MnO_2 = -109.1$ (6) (γ-nsutite)

CO_3^{2-}, and graphite C, at low P and T conditions, several organic complexes form metastably, mainly as the result of bacterial activity, and persist at P and T conditions that may exceed those achieved at a burial depth on the order of 5 km, which is generally considered the limit of complete degradation.

Organic matter suspended in solution and present in sediments is an efficient complexing agent for most divalent and trivalent cations. Most organic matter in aquatic and sedimentary environments is in the form of humic and fulvic acids. About 80% of soluble organic matter in lakes and rivers is composed of fulvic acids, whereas the abundance of humic acids varies markedly from region to region (from 0.1 to about 10 mg/l of solution; cf. Reuter and Perdue, 1977). In classical terminology, "humic acids" are substances extracted from soils and sediments by alkaline solutions and precipitated through acidification, whereas "fulvic acids" refer to organic matter remaining in solution after acidification. These substances are practically composed of polyelectrolytes varying in colour from yellow to black, with molecular weights ranging from few a hundred to several hundred thousand grams per mole. The relationships between macroscopic and reactive properties for these classes of substances are outlined in figure 8.25.

The capacity of complexing of humic substances is ascribed to their oxygen-based functional group (table 8.19).

The most common apical structure complexing metal cations in solution is

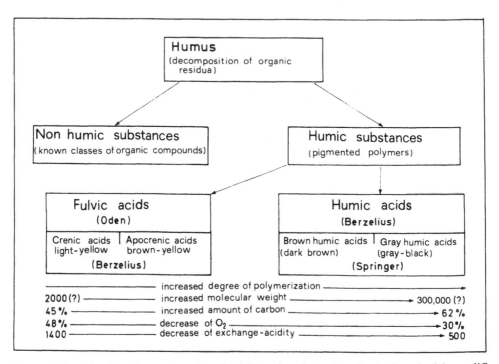

Figure 8.25 Classification and properties of humic substances. Reproduced with modifications from Stevenson (1983), with kind permission of Theophrastus Publishing and Proprietary Co.

Table 8.19 Oxygen-based functional groups in humic and fulvic acids (from Stevenson, 1982). Concentrations expressed in mEq/100 g.

Acid Type	Total Acidity	COOH	OH-Acid	OH-Alcoholic	C=O
Humic	560–890	150–570	210–570	20–496	10–560
Fulvic	640–1420	520–1120	30–570	260–950	120–420

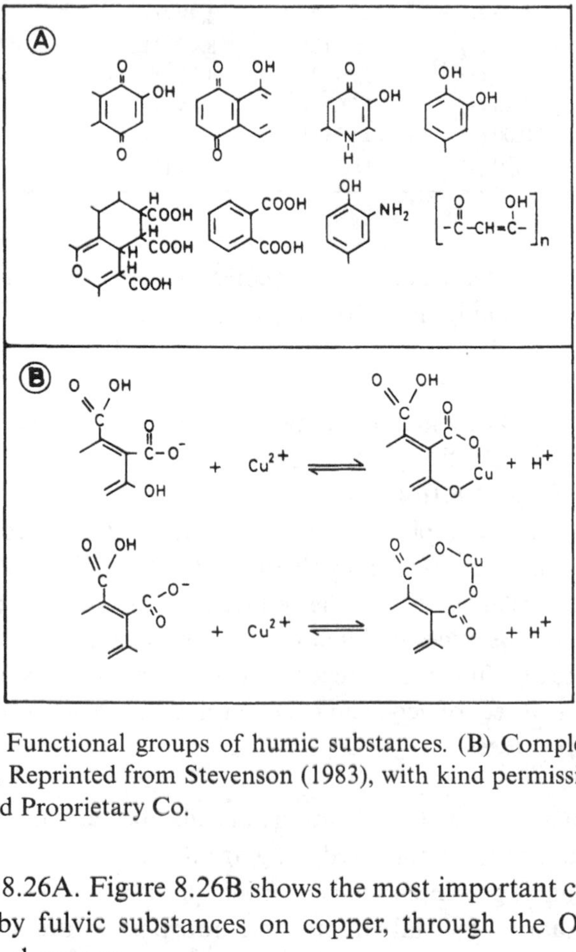

Figure 8.26 (A) Functional groups of humic substances. (B) Complexing schemes for fulvic substances. Reprinted from Stevenson (1983), with kind permission of Theophrastus Publishing and Proprietary Co.

shown in figure 8.26A. Figure 8.26B shows the most important complexing pathways followed by fulvic substances on copper, through the OH-phenolic and COOH functional groups.

As shown in figure 8.25, the acidic-exchange capacity of humic substances varies widely; a representative mean value is in the range 1.5 to 5 mEq/g. Assuming a mean carbon content of 56% in polymer units, Stevenson (1983) deduced that the ratio between complexed divalent cations and carbon atoms was around 1:20 to 1:60. Complexing capability is also greatly affected by the pH and ionic strength of the solution (Guy et al., 1975). Generally, moreover, for elements present in trace concentration levels, complexing by humic substances is more efficient in pure water than in saline solutions, where organic matter is partly bonded to the major dissolved cations (Ca^{2+}, Mg^{2+}; see Mantoura et al., 1978).

Table 8.20 Metal-organic complexing (log β^0) for some trace elements in aqueous solution at pH = 8.2, T = 25 °C. I: ionic strength of solution. References: (1) Turner et al. (1981); (2) Mantoura et al. (1978); (3) Stevenson (1976).

Ion	$I = 0$	$I = 0.02$	$I = 0.65$	% Chelate Metal
Mn^{2+}	5.00 (1)	4.53 (2)	3.61 (1)	0.01
Co^{2+}	5.20 (1)	4.73 (2)	3.81 (1)	0.01
Ni^{2+}	5.78 (1)	5.31 (2)	4.39 (1)	0.04
Cu^{2+}	9.84 (1)	9.37 (2)	8.45 (1)	47
Zn^{2+}	5.73 (1)	5.26 (2)	4.34 (1)	0.03
Cd^{2+}	5.23 (1)	4.76 (2)	3.84 (1)	0.001
Hg^{2+}	20.00 (1)	19.53 (2)	18.61 (1)	0.0
Pb^{2+}	8.70 (1)	-	7.31 (1)	2

Table 8.20 lists metal-organic complexing constants for some trace metals in waters of various ionic strengths, according to Stevenson (1976), Mantoura et al. (1978), and Turner et al. (1981). The last column in table 8.20 lists the percentage of chelate metal in seawater ($I = 0.65$), which is significant for Cu^{2+}. According to Turner et al. (1981), the metal-organic complexing in seawater is also significant for rare earths and fully hydrolyzed elements forming stable complexes with carbonate ions (cf. figure 8.16B).

In addition to their complexing capability, humic substances can also reduce oxidized forms of metal cations and polyanions ($Fe^{3+} \rightarrow Fe^{2+}$; $MoO_4^{2-} \rightarrow Mo^{5+}$; $VO_3 \rightarrow VO^{2+}$; see Szilagyi, 1971). The reduced forms are then embodied in the structure of humic substances by functional-group fixation.

Besides humic and fulvic substances in surface waters and soils, organic compounds occur in a variety of geochemical environments. Their role is determinant in the following processes.

1. Metal complexation (and fixation) in soils and sedimentary basins
2. Deposition of ore metals (reducing agents)
3. Formation of graphite in metamorphic processes
4. Complexation of metal cations and transport in ore-forming solutions derived from sedimentary basins by organic acid anionic complexes present in oil field brines
5. Generation, transport, and deposition of petroleum
6. Formation of secondary porosity in oil reservoir formations
7. Origin of life.

Impressive improvements in our knowledge of thermodynamic properties of the various organic ligands and of our capability of calculating metal-organic complexation constants at the various P and T conditions of interest come from the systematizations of Shock and Helgeson (1990) and Shock and Koretsky (1995), which evaluated equation of state parameters to be used in the revised

HKF model (see section 8.11). The main observations on which the generalizations were based are as follows.

1. The standard partial molal properties of aqueous electrolytes consisting of inorganic cations and organic anions exhibit P-T-dependent behavior similar to that of inorganic aqueous electrolytes (i.e., at pressures corresponding to liquid-vapor equilibrium, with increasing T they reach a maximum at intermediate T values and then approach $-\infty$ as T approaches the critical point of H_2O; see figure 8.27A).

2. The standard partial molal properties of neutral organic aqueous molecules behave similarly to their inorganic counterparts (i.e., at pressures corresponding to liquid-vapor equilibrium, with increasing T they reach a minimum at intermediate T values and then approach $+\infty$ as T approaches the critical point of H_2O; see figure 8.27B).

3. The Born coefficients of the HKF model for the various organic species are simply related to their standard partial molal entropies at 25 °C and 1 bar through the effective charge (Shock et al., 1989).

4. The standard partial molal volumes ($\overline{V}^0$), heat capacities ($\overline{C}_P^0$), and entropies ($\overline{S}^0$) of aqueous n-polymers, together with their standard partial molal enthalpies ($\Delta\overline{H}_f^0$) and Gibbs free energies of formation from the elements ($\Delta\overline{G}_f^0$), are linear functions of the number of moles of carbon atoms in the alkyl chains (figure 8.28).

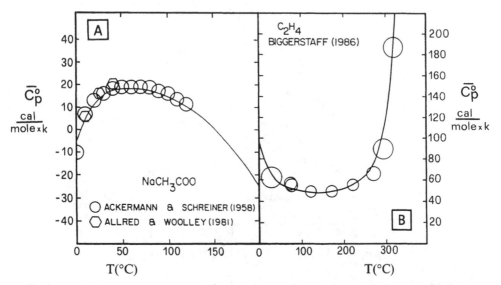

Figure 8.27 Experimental values of standard partial molal heat capacity of sodium acetate (A) and ethylene (B) in water as a function of T(°C) at pressures corresponding to water-vapor equilibrium. Interpolating curves generated by the HKF model equations. Reprinted from E. L. Shock and H. C. Helgeson, *Geochimica et Cosmochimica Acta,* 54, 915–946, copyright © 1990, with kind permission from Elsevier Science Ltd., The Boulevard, Langford Lane, Kidlington 0X5 1GB, UK.

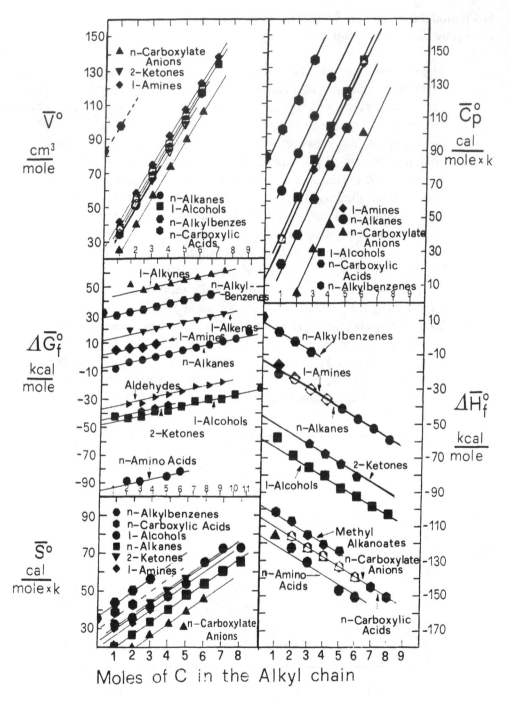

Figure 8.28 Correlation of standard partial molal properties of aqueous *n*-polymers with the number of moles of carbon atoms in the alkyl chain. Reprinted from E. L. Shock and H. C. Helgeson, *Geochimica et Cosmochimica Acta,* 54, 915–946, copyright © 1990, with kind permission from Elsevier Science Ltd., The Boulevard, Langford Lane, Kidlington 0X5 1GB, UK.

The various polymerization steps along the series of alcohols may be visualized in stereochemical representation as follows:

$$
\underset{\text{methanol}}{\text{H}-\overset{\overset{\textstyle H}{|}}{\underset{\underset{\textstyle H}{|}}{C}}-\text{OH}}+ \text{C} \Leftrightarrow \underset{\text{ethanol}}{\text{H}-\overset{\overset{\textstyle H}{|}}{\underset{\underset{\textstyle H}{|}}{C}}-\overset{\overset{\textstyle H}{|}}{\underset{\underset{\textstyle H}{|}}{C}}-\text{OH}}+ \text{C} \Leftrightarrow \underset{\text{propanol}}{\text{H}-\overset{\overset{\textstyle H}{|}}{\underset{\underset{\textstyle H}{|}}{C}}-\overset{\overset{\textstyle H}{|}}{\underset{\underset{\textstyle H}{|}}{C}}-\overset{\overset{\textstyle H}{|}}{\underset{\underset{\textstyle H}{|}}{C}}-\text{OH}}. \tag{8.222}
$$

The linearity in the standard partial molal properties of the aqueous polymers as a function of the length of the chain (in terms of C atoms) observed by Shock and Helgeson (1990) is the result of the principle of equal reactivity of co-condensing functional groups (which, as we have already seen in section 6.1.2, also holds for silica polymers). This principle is, however, strictly valid only when the length of the chain is sufficiently elevated and small departures are observed for chains with one or two carbon atoms (cf. figure 8.28).

The generalized behavior is consistent with an equation of the form

$$
\overline{\Xi}^0_j = \overline{m}n + \overline{p} \tag{8.223}
$$

(Shock and Helgeson, 1990), where $\overline{\Xi}^0_j$ is the regressed standard state partial molal property of the jth aqueous polymer, n is the number of moles of carbon atoms in the alkyl chain, and $\overline{m}$ and $\overline{p}$ are, respectively, slope and intercept of the regression lines shown in figure 8.28. Values of the regression constants for various aqueous polymers according to Shock and Helgeson (1990) are listed in table 8.21.

Table 8.21 Correlation parameters between the standard partial molal properties of aqueous n-polymers and number of moles of carbon in the alkyl chain ($T = 25\ °C$, $P = 1$ bar; Shock and Helgeson, 1990). See section 8.11 and tables 8.11, 8.12, and 8.13 for identification of symbols and units.

Property	$\overline{S}^0_{P_r,T_r}$	$\overline{C}^0_P$	$\overline{V}^0_{P_r,T_r}$	$\overline{G}^0_{f,P_r,T_r}$	$\overline{H}^0_{f,P_r,T_r}$
Slope	6.7	21.2	15.8	2.05	−5.67
Group	INTERCEPTS				
n-alkanes	12.8	47.0	19.6	−7.6	−13.5
1-alkenes	-	-	-	11.9	-
n-alkynes	-	-	-	42.2	-
n-alkylbenzenes	31.3	81.3	81.6	28.2	8.6
1-alcohols	23.2	19.0	22.6	−47.8	−58.4
Aldehydes	-	-	-	−36.7	-
2-ketones	23.2	-	18.9	−44.8	−45.2
1-amines	20.0	14.4	26.8	1.8	−12.3
n-carboxylic acid	29.3	−2.4	20.0	-	−105.0
n-amino acids	-	-	-	−95.4	−115.5
Methyl alkanoates	-	-	-	-	−97.8

Table 8.22 Partial molal thermodynamic properties of organic aqueous species at 25 °C and 1 bar and the parameters of the revised HKF model, after Shock and Helgeson (1990).

Species	$\bar{G}^0_{f,P_r,T_r}$	$\bar{H}^0_{f,P_r,T_r}$	$\bar{S}^0_{P_r,T_r}$	$\bar{C}^0_P$	$\bar{V}^0_{P_r,T_r}$	$a_1 \times 10$	$a_2 \times 10^{-2}$	a_3	$a_4 \times 10^{-4}$	c_1	$c_2 \times 10^{-4}$	$\omega \times 10^{-5}$
Methane	−8,234.	−21,010.	20.99	66.3	37.3	6.7617	8.7279	2.3212	−3.1397	42.0941	10.4707	−0.3179
Ethane	−3,886.	−24,650.	26.81	88.3	51.2	8.6340	13.3011	0.5205	−3.3288	54.1755	14.9521	−0.4060
Propane	−1,963.	−30,490.	33.37	110.6	67.0	10.7625	18.4948	−1.5133	−3.5435	66.3294	19.4946	−0.5053
n-butane	36.	−36,230.	40.02	133.9	82.8	12.8905	23.6960	−3.5683	−3.7585	79.0569	24.2408	−0.6061
n-pentane	2,130.	−41,560.	47.5	153.	98.6	15.0143	28.8771	−5.5949	−3.9727	89.2070	28.1315	−0.7193
n-hexane	4,420.	−47,400.	52.9	175.2	114.4	17.1487	34.0927	−7.6530	−4.1883	101.4642	32.6536	−0.8011
n-heptane	6,470.	−52,950.	60.0	196.4	130.2	19.2745	39.2810	−9.6879	−4.4028	112.8980	36.9721	−0.9086
n-octane	8,580.	−59,410.	63.8	217.6	146.0	21.4171	44.5126	−11.7437	−4.6191	124.7924	41.2905	−0.9662
Ethylene	19,450.	8,570.	28.7	62.5	45.5	7.8560	12.6391	−1.8737	−3.3014	39.1	9.7	−0.4
1-propene	17,910.	−290.	36.7	83.7	61.3	9.9655	16.5518	−0.7564	−3.4632	50.0997	14.0151	−0.5558
1-butene	20,310.	−5,635.	43.4	104.9	77.1	12.0932	21.7476	−2.7991	−3.6779	61.5894	18.3335	−0.6572
1-pentene	22,470.	−11,200.	50.1	126.1	92.9	14.2210	26.9431	−4.8413	−3.8927	73.0791	22.6520	−0.7587
1-hexene	24,370.	−17,025.	56.8	147.3	108.7	16.3488	32.1387	−6.8838	−4.1075	84.5689	26.9704	−0.8602
1-heptene	26,450.	−22,670.	63.5	168.5	124.5	18.4766	37.3342	−8.9260	−4.3223	96.0586	31.2889	−0.9616
1-octene	28,720.	−28,120.	70.2	189.7	140.3	20.6044	42.5297	−10.9682	−4.5371	107.5483	35.6073	−1.0631
Ethyne	51,890.	50,700.	30.0	53.0	49.3	8.3577	12.6272	0.7833	−3.3009	33.0423	7.7615	−0.4543
1-propyne	47,880.	38,970.	36.7	74.2	65.1	10.4855	17.8227	−1.2588	−3.5157	44.5321	12.0799	−0.5558
1-butyne	50,030.	33,400.	43.4	95.4	80.9	12.6132	23.0185	−3.3011	−3.7305	56.0218	16.3984	−0.6572
1-pentyne	52,160.	27,800.	50.1	116.6	96.7	14.7410	28.2140	−5.3433	−3.9453	67.5117	20.7168	−0.7587
1-hexyne	54,420.	22,340.	56.8	137.8	112.5	16.8688	33.4096	−7.3857	−4.1601	79.0012	25.0353	−0.8602
1-heptyne	56,730.	16,980.	63.7	159.0	128.3	18.9956	38.6015	−9.4245	−4.3747	90.4632	29.3537	−0.9647
1-octyne	58,860.	11,330.	70.2	180.2	144.1	21.1244	43.8006	−11.4705	−4.5896	101.9808	33.6721	−1.0631
Benzene	32,000.	12,230.	35.5	86.3	83.5	13.1244	18.1143	11.8225	−3.5277	80.8626	1.7799	−0.1976
Toluene	30,260.	3,280.	43.9	102.8	97.71	15.0259	21.9153	12.1349	−3.6849	93.9240	2.8953	−0.3248
Ethylbenzene	32,440.	−2,500.	49.8	120.5	113.8	17.1974	26.2574	12.4885	−3.8644	108.3695	4.0918	−0.4142
n-propylbenzene	34,170.	−8,630.	56.1	144.8	129.6	19.3272	30.5152	12.8372	−4.0404	128.4523	5.7345	−0.5096
n-butylbenzene	36,110.	−14,430.	62.8	166.0	145.4	21.4550	34.7658	13.1931	−4.2162	145.8054	7.1676	−0.6110
n-pentylbenzene	38,210.	−19,750.	69.5	187.2	161.2	23.5828	39.0164	13.5490	−4.3918	163.1584	8.6007	−0.7125

Table 8.22 (continued)

Species	$\bar{G}^0_{f,P_r,T_r}$	$\bar{H}^0_{f,P_r,T_r}$	$\bar{S}^0_{P_r,T_r}$	$\bar{C}^0_P$	$\bar{V}^0_{P_r,T_r}$	$a_1 \times 10$	$a_2 \times 10^{-2}$	a_3	$a_4 \times 10^{-4}$	c_1	$c_2 \times 10^{-4}$	$\omega \times 10^{-5}$
n-hexylbenzene	40,390.	−25,590.	76.2	208.4	177.	25.7106	43.2732	13.8894	−4.5678	180.5115	10.0338	−0.8140
n-heptylbenzene	42,640.	−31,090.	82.9	229.6	192.8	27.8384	47.5238	14.2453	−4.7435	197.8642	11.4670	−0.9154
n-octylbenzene	44,690.	−36,760.	89.6	250.8	208.6	29.9662	51.7744	14.6016	−4.9193	215.2173	12.9001	−1.0169
Methanol	−42,050.	−58,870.	32.2	37.8	38.17	6.9383	5.5146	11.4018	−3.0069	39.4852	−1.4986	−0.1476
Ethanol	−43,330.	−68,650.	35.9	62.2	55.08	9.2333	9.9581	12.1445	−3.1906	60.0175	0.1507	−0.2037
1-propanol	−41,910.	−75,320.	41.4	84.3	70.70	11.3426	14.3252	12.1079	−3.3711	78.3142	1.6447	−0.2869
1-butanol	−38,840.	−80,320.	46.9	104.4	86.60	13.4902	17.9400	14.1816	−3.5205	94.8858	3.0034	−0.3702
1-pentanol	−38,470.	−87,730.	53.4	125.2	102.68	15.6573	22.6230	13.6464	−3.7141	111.9216	4.4095	−0.4687
1-hexanol	−35,490.	−92,690.	64.8	144.4	118.65	17.7843	27.4289	12.5890	−3.9128	126.8936	5.7074	−0.6413
1-heptanol	−32,000.	−97,270.	72.1	174.2	133.43	19.7695	31.3977	12.9129	−4.0769	151.5815	7.7219	−0.7519
1-octanol	−30,250.	−103,060.	72.3	195.	149.2	21.9264	35.7093	13.2760	−4.2551	169.4962	9.1280	−0.7549
Phenol	−12,585.	−36,640.	45.8	75.3	86.17	13.4370	18.7373	11.8793	−3.5535	69.9366	1.0363	−0.3536
Acetone	−38,500.	−61,720.	44.4	57.7	66.92	10.8100	13.4912	11.4350	−3.3366	54.9496	−0.1534	−0.3324
2-butanone	−36,730.	−67,880.	50.3	80.4	82.52	12.9145	17.6967	11.7839	−3.5105	73.7083	1.3810	−0.4217
2-pentanone	−34,390.	−73,460.	56.3	101.6	98.0	15.0021	21.8653	12.1372	−3.6828	91.1588	2.8142	−0.5126
2-hexanone	−32,480.	−79,220.	63.4	122.8	113.7	17.1141	26.0886	12.4806	−3.8574	108.4560	4.2473	−0.6201
2-heptanone	−30,430.	−84,890.	70.1	144.0	129.5	19.2419	30.3392	12.8366	−4.0331	125.8091	5.6804	−0.7216
2-octanone	−28,380.	−90,560.	76.8	165.2	145.3	21.3697	34.5960	13.1769	−4.2091	143.1620	7.1135	−0.8231
Formic acid	−88,982.	−101,680.	38.9	19.0	34.69	6.3957	4.6630	10.7209	−2.9717	22.1924	−3.1196	−0.3442
Acetic acid	−94,760.	−116,100.	42.7	40.56	52.01	8.8031	12.4572	3.5477	−3.2939	40.8037	−0.9218	−0.2337
Propanoic acid	−93,430.	−122,470.	49.4	56.0	67.9	10.9213	13.7115	11.4582	−3.3457	54.313	−0.246	−0.4
Butanoic acid	−91,190.	−127,950.	56.1	80.5	84.61	13.1708	18.2050	11.8362	−3.5315	72.9853	1.3878	−0.5096
Pentanoic acid	−89,210.	−133,690.	62.8	103.3	100.5	15.3109	22.4858	12.1803	−3.7085	91.7185	2.9291	−0.6110
Hexanoic acid	−87,080.	−139,290.	69.5	125.2	116.55	17.4729	26.8090	12.5312	−3.8872	109.6753	4.4095	−0.7125
Heptanoic acid	−85,150.	−145,080.	76.2	146.4	132.3	19.5938	31.0465	12.8851	−4.0624	127.0284	5.8426	−0.8140
Octanoic acid	−83,400.	−151,050.	82.9	167.6	148.1	21.7216	35.2971	13.2410	−4.2381	144.3812	7.2758	−0.9154
Formate	−83,862.	−101,680.	21.7	−22.0	26.16	5.7842	6.3405	3.2606	−3.0410	17.0	−12.4	1.3003
Acetate	−88,270.	−116,180.	20.6	6.2	40.5	7.7525	8.6996	7.5825	−3.1385	26.3	−3.86	1.3182

continued

Table 8.22 (continued)

Species	$\overline{G}^0_{f,P_r,T_r}$	$\overline{H}^0_{f,P_r,T_r}$	$\overline{S}^0_{P_r,T_r}$	$\overline{C}^0_P$	$\overline{V}^0_{P_r,T_r}$	$a_1 \times 10$	$a_2 \times 10^{-2}$	a_3	$a_4 \times 10^{-4}$	c_1	$c_2 \times 10^{-4}$	$\omega \times 10^{-5}$
Propanoate	−86,770.	−122,630.	26.5	30.9	54.95	9.6992	15.9017	−0.5003	−3.4363	52.3	−4.2	1.2276
Butanoate	−84,610.	−127,930.	31.8	44.5	70.3	11.7724	20.9657	−2.4947	−3.6456	42.8123	6.0300	1.1469
Pentanoate	−82,600.	−134,410.	38.3	78.8	86.31	13.9304	26.2319	−4.5589	−3.8633	62.0175	13.0170	1.0496
Hexanoate	−80,450.	−139,890.	45.3	100.	102.21	16.0700	31.4589	−6.6186	−4.0794	73.4573	17.3354	0.9427
Heptanoate	−78,390.	−145,560.	52.0	112.2	118.6	18.2787	36.8480	−8.7285	−4.3022	79.6770	19.8205	0.8417
Octanoate	−76,340.	−151,230.	58.7	133.4	134.4	20.4065	42.0436	−10.7711	−4.5170	91.1664	24.1390	0.7402
Methanamine	5,040.	−16,320.	30.5	37.0	41.88	7.4547	6.7824	10.8884	−3.0593	39.0324	−1.5527	−0.1219
Ethanamine	6,300.	−23,830.	33.6	56.8	58.61	9.7281	11.3257	11.2651	−3.2471	55.6799	−0.2142	−0.1688
1-propanamine	7,010.	−30,680.	40.8	78.	74.15	11.8178	15.5013	11.6113	−3.4197	72.9633	1.2188	−0.2799
1-butanamine	9,410.	−36,110.	47.2	101.	89.80	13.9266	19.7205	11.9476	−3.5941	91.9109	2.7736	−0.3748
1-pentanamine	12,800.	−40,650.	53.5	123.	105.7	16.0701	24.0048	12.3010	−3.7713	110.0099	4.2608	−0.4702
1-hexanamine	14,860.	−46,320.	60.2	144.	121.6	18.2115	28.2822	12.6611	−3.9481	127.1903	5.6804	−0.5717
1-heptanamine	16,910.	−51,990.	66.9	162.8	137.6	20.3667	32.5921	13.0098	−4.1263	142.4729	6.9513	−0.6731
1-octanamine	18,960.	−57,660.	73.6	184.0	153.2	22.4671	36.7896	13.3574	−4.2998	159.8259	8.3844	−0.7746
Glycine	−88,618.	−122,846.	37.84	9.4	43.25	7.6046	7.0825	10.9119	−3.0717	14.1998	−3.4185	−0.2330
Alanine	−88,800.	−132,130.	40.0	33.8	60.45	9.9472	11.7629	11.3023	−3.2652	34.9465	−1.7690	−0.2658
α-ABA	−87,120.	−138,180.	46.7	53.1	76.5	12.1092	16.0862	11.6529	−3.4439	50.6605	−0.4643	−0.3672
Valine	−85,300.	−147,300.	42.6	72.2	90.79	14.0856	20.0343	11.9842	−3.6071	67.7090	0.8267	−0.3051
Leucine	−82,000.	−151,070.	51.5	95.1	107.57	16.3362	24.5312	12.3587	−3.7930	86.2214	2.3748	−0.4399
Isoleucine	−82,200.	−150,900.	52.8	91.6	105.45	16.0395	23.9414	12.3010	−3.7686	83.0208	2.1382	−0.4596
Serine	−122,100.	−170,800.	46.5	28.1	60.62	9.9372	11.7454	11.2943	−3.2645	29.1225	−2.1543	−0.3642
Threonine	−120,000.	−179,100.	53.2	50.2	76.86	12.1251	16.1194	11.6532	−3.4453	47.2520	−0.6604	−0.4657
Aspartic acid	−172,400.	−226,370.	54.8	30.4	71.79	11.4232	14.7124	11.5470	−3.3871	29.9488	−1.9989	−0.4899
Glutamic acid	−173,000.	−232,000.	70.5	42.3	89.36	13.7471	19.3606	11.9216	−3.5793	38.0235	−1.1944	−0.7276
Asparagine	−128,650.	−186,660.	55.2	29.9	77.18	12.1587	16.1812	11.6715	−3.4478	29.4616	−2.0327	−0.4959
Glutamine	−126,600.	−192,330.	61.9	44.7	94.36	14.4753	20.8173	12.0386	−3.6395	41.2938	−1.0322	−0.5974
Phenylalanine	−49,500.	−110,080.	52.9	91.8	121.92	18.2927	28.4440	12.6752	−3.9548	83.1794	2.1517	−0.4611
Tryptophan	−26,900.	−97,800.	36.6	100.4	144.0	21.3975	34.6481	13.1912	−4.2113	92.8723	2.7330	−0.2143
Tyrosine	−87,300.	−157,400.	45.5	71.5	123.	18.4784	28.8127	12.7114	−3.9700	66.7005	0.7794	−0.3490
Methionine	−120,200.	−177,600.	65.7	70.	105.3	15.9529	23.7688	12.2862	−3.7615	62.5883	0.6780	−0.6550

Table 8.23 Standard partial molal thermodynamic properties of aqueous metal complexes of monovalent organic acid ligands at 25°C and 1 bar and the parameters of the revised HKF model, after Shock and Koretsky (1995). See section 8.11 and tables 8.11, 8.12, and 8.13 for identification of symbols and units. For: formate; Prop: propanoate: But: *n*-butanoate; Pent: *n*-pentanoate; Glyc: glicoate; Lac: lactate; Gly: glycinate; Ala: alanate. The residual charge on the ligand refers to the metal-organic complex, i.e. Mn(For)⁺ means Mn(COOH)⁺ and Mn(For)₂ means Mn(COOH)₂.

Ligand	$\bar{G}^0_{f,P_r,T_r}$	$\bar{H}^0_{f,P_r,T_r}$	$\bar{S}^0_{P_r,T_r}$	$\bar{C}^0_P$	$\bar{V}^0_{P_r,T_r}$	$a_1 \times 10$	$a_2 \times 10^{-2}$	a_3	$a_4 \times 10^{-4}$	c_1	$c_2 \times 10^{-4}$	$\omega \times 10^{-5}$
					MANGANESE							
(For)⁺	−140,680.	−155,740.	7.4	38.4	16.1	4.1094	2.2545	4.8590	−2.8721	32.7312	4.7962	0.4379
(For)₂	−226,030.	−259,600.	26.8	77.8	53.0	9.0055	14.2060	0.1701	−3.3662	51.4842	12.8126	−0.0300
(Prop)⁺	−143,020.	−176,110.	12.2	136.6	44.8	8.0255	11.8128	1.1111	−3.2672	89.6329	24.7960	0.3686
(Prop)₂	−230,820.	−300,040.	37.9	312.3	113.9	17.3347	34.5476	−7.8345	−4.2071	188.8879	60.5707	−0.0300
(But)⁺	−140,790.	−181,340.	17.5	161.9	60.2	10.0986	16.8770	−0.8847	−3.4766	103.6776	29.9377	0.2873
(But)₂	−226,377.	−310,010.	50.9	372.5	146.3	21.7755	45.3895	−12.0924	−4.6553	224.2130	72.8487	−0.0300
(Pent)⁺	−138,600.	−187,650.	24.0	225.5	76.2	12.2563	22.1440	−2.9511	−3.6943	140.0855	42.9054	0.1895
(Pent)₂	−222,010.	−322,030.	65.1	524.5	180.2	26.4073	56.6981	−16.5352	−5.1228	313.3047	103.8147	−0.0300
(Glyc)⁺	−177,830.	−208,590.	11.9	91.3	29.7	5.9609	6.7761	3.0811	−3.0590	63.1381	15.5712	0.3735
(Glyc)₂	−300,160.	−364,740.	37.2	204.1	81.9	12.9661	23.8786	−3.6359	−3.7660	125.5108	38.5424	−0.0300
(Lac)⁺	−178,980.	−217,760.	17.6	128.8	46.1	8.1746	12.1772	0.9667	−3.2823	84.2783	23.2081	0.2832
(Lac)₂	−302,970.	−383,050.	50.3	293.6	116.6	17.7108	35.4655	−8.1945	−4.2450	177.9788	56.7789	−0.0300
(Gly)⁺	−134,770.	−165,800.	25.0	67.0	33.7	6.4299	7.9186	2.6372	−3.1063	47.0493	10.6185	0.1739
(Gly)₂	−214,100.	−278,850.	64.6	146.1	90.2	14.1002	26.6481	−4.7253	−3.8805	91.4846	26.7157	−0.0300
(Ala)⁺	−133,580.	−173,180.	29.0	112.3	51.8	8.8873	13.9167	0.2842	−3.3542	73.0237	19.8433	0.1124
(Ala)₂	−212,070.	−294,250.	71.5	254.2	128.5	19.3395	39.4414	−9.7539	−4.4094	154.8617	48.7440	−0.0300
					COBALT							
(For)⁺	−99,400.	−118,150.	−5.4	30.1	7.9	3.0614	−0.3037	5.8638	−2.7663	29.6264	3.1004	0.6305
(For)₂	−184,930.	−223,370.	10.0	61.6	43.9	7.7655	11.1773	1.3615	−3.2410	41.9649	9.5039	−0.0300
(Prop)⁺	−101,560.	−138,350.	−0.6	128.3	36.7	6.9775	9.2545	2.1152	−3.1615	86.5252	23.1002	0.5608
(Prop)₂	−189,410.	−263,490.	21.1	296.0	104.8	16.0946	31.5192	−6.6433	−4.0819	179.3685	57.2620	−0.0300
(But)⁺	−99,980.	−144,230.	4.7	153.5	52.1	9.0507	14.3185	0.1200	−3.3708	100.5727	28.2419	0.4799
(But)₂	−186,140.	−274,660.	33.3	356.3	137.2	20.5354	42.3610	−10.9009	4.5301	214.6936	69.5400	−0.0300
(Pent)⁺	−97,930.	−150,670.	11.2	217.2	68.1	11.2020	19.5901	−1.9551	−3.5888	136.9956	41.2096	0.3837
(Pent)₂	−182,030.	−286,940.	48.3	508.3	171.1	25.1672	53.6696	−15.3437	−4.9976	303.7853	100.5060	−0.0300
(Glyc)⁺	−136,870.	−171,330.	−0.9	83.0	21.6	4.9137	4.2158	4.0944	−2.9532	60.0521	13.8754	0.5681
(Glyc)₂	−260,100.	−329,560.	20.4	187.9	72.9	11.7260	20.8501	−2.4444	−3.6408	115.9914	35.2337	−0.0300

continued

Table 8.23 (continued)

Ligand	$\overline{G}^0_{f,P_r,T_r}$	$\overline{H}^0_{f,P_r,T_r}$	$\overline{S}^0_{P_r,T_r}$	$\overline{C}^0_P$	$\overline{V}^0_{P_r,T_r}$	$a_1 \times 10$	$a_2 \times 10^{-2}$	a_3	$a_4 \times 10^{-4}$	c_1	$c_2 \times 10^{-4}$	$\omega \times 10^{-5}$
(Lac)$^+$	−137,390.	−179,860.	4.8	120.5	38.0	7.1281	9.6212	1.9722	−3.1766	81.2111	21.5123	0.4799
(Lac)$_2$	−261,460.	−346,410.	33.5	277.4	107.5	16.4707	32.4370	−7.0030	−4.1198	168.4594	53.4702	−0.0300
(Gly)$^+$	−95,200.	−129,080.	15.0	58.7	25.5	5.3683	5.3273	3.6539	−2.9991	43.5722	8.9227	0.3260
(Gly)$_2$	−175,960.	−243,430.	55.0	129.8	81.2	12.8601	23.6197	−3.5341	−3.7553	81.9652	23.4070	−0.0300
(Ala)$^+$	−94,100.	−136,250.	20.0	104.0	43.6	7.8202	11.3148	1.2990	−3.2467	69.3955	18.1475	0.2482
(Ala)$_2$	−173,790.	−259,270.	60.0	238.0	119.4	18.0995	36.4127	−8.5626	−4.2842	145.3423	45.4353	−0.0300
NICKEL												
(For)$^+$	−97,310.	−117,570.	−10.6	21.3	2.8	2.3868	−1.9558	6.5236	−2.6980	25.2125	1.3130	0.7096
(For)$_2$	−182,850.	−223,290.	3.2	44.4	38.2	6.9840	9.2687	2.1133	−3.1621	31.9308	6.0164	−0.0300
(Prop)$^+$	−99,640.	−137,940.	−5.8	119.5	31.6	6.3025	7.6097	2.7546	−3.0935	82.1010	21.3127	0.6388
(Prop)$_2$	−187,630.	−263,710.	14.3	278.9	99.1	15.3131	29.6105	−5.8911	−4.0030	169.3347	53.7744	−0.0300
(But)$^+$	−97,930.	−143,690.	−0.5	144.8	46.9	8.3767	12.6711	0.7713	−3.3027	96.1758	26.4544	0.5608
(But)$_2$	−184,120.	−274,630.	26.5	339.2	131.5	19.7540	40.4521	−10.1489	4.4512	204.6596	66.0525	−0.0300
(Pent)$^+$	−95,870.	−150,130.	6.0	208.4	62.9	10.5339	17.9393	−1.3007	−3.5205	132.5696	39.4221	0.4615
(Pent)$_2$	−180,000.	−286,890.	41.5	491.2	165.4	24.3858	51.7607	−14.5917	−4.9187	293.7512	97.0185	−0.0300
(Glyc)$^+$	−135,150.	−171,130.	−6.1	74.2	16.5	4.2391	2.5699	4.7380	−2.8851	55.6378	12.0879	0.6472
(Glyc)$_2$	−258,710.	−330,150.	13.6	170.8	67.2	10.9446	18.9412	−1.6924	−3.5619	105.9573	31.7462	−0.0300
(Lac)$^+$	−135,600.	−179,580.	−0.4	111.7	32.9	6.4541	7.9800	2.6079	−3.1088	76.8142	19.7248	0.5608
(Lac)$_2$	−259,960.	−346,900.	26.8	260.3	101.8	15.6892	30.5284	−6.2512	−4.0409	158.4254	49.9827	−0.0300
(Gly)$^+$	−94,540.	−129,290.	12.0	49.9	20.4	4.6813	3.6515	4.3097	−2.9299	38.8214	7.1352	0.3686
(Gly)$_2$	−176,530.	−246,060.	48.0	112.7	75.4	12.0787	21.7107	−2.7819	−3.6764	71.9311	19.9195	−0.0300
(Ala)$^+$	−93,530.	−137,130.	15.0	95.2	38.5	7.1451	9.6640	1.9536	−3.1784	64.9703	16.3600	0.3260
(Ala)$_2$	−174,550.	−262,970.	50.0	220.8	113.8	17.3180	34.5040	−7.8105	−4.2053	135.3083	41.9478	−0.0300
COPPER												
(For)$^+$	−70,890.	−88,300.	−0.2	34.8	7.7	3.0050	−0.4430	5.9211	−2.7606	31.6869	4.0629	0.5536
(For)$_2$	−156,550.	−193,180.	16.8	70.8	43.7	7.7315	11.0979	1.3851	−3.2377	47.3676	11.3819	−0.0300
(Prop)$^+$	−74,120.	−109,578.	4.6	133.0	36.5	6.9197	9.1126	2.1719	−3.1556	88.5486	24.0627	0.4799
(Prop)$_2$	−163,000.	−235,270.	27.9	305.2	104.5	16.0606	31.4336	−6.6036	−4.0784	184.7714	59.1399	−0.0300
(But)$^+$	−71,850.	−114,770.	10.0	158.3	51.8	8.9947	14.1844	0.1682	−3.3653	102.6486	29.2044	0.4046
(But)$_2$	−158,470.	−245,180.	40.1	365.5	137.0	20.5014	42.2754	−10.8613	−4.5266	220.0963	71.4180	−0.0300
(Pent)$^+$	−69,820.	−121,220.	16.4	221.9	67.8	11.1516	19.4472	−1.8933	−3.5828	139.0314	42.1721	0.3041
(Pent)$_2$	−154,380.	−257,470.	55.1	517.5	170.8	25.1332	53.5902	−15.3200	−4.9943	309.1882	102.3839	−0.0300

Table 8.23 (continued)

Ligand	$\overline{G}^0_{f,P_r,T_r}$	$\overline{H}^0_{f,P_r,T_r}$	$\overline{S}^0_{P_r,T_r}$	$\overline{C}^0_P$	$\overline{V}^0_{P_r,T_r}$	$a_1 \times 10$	$a_2 \times 10^{-2}$	a_3	$a_4 \times 10^{-4}$	c_1	$c_2 \times 10^{-4}$	$\omega \times 10^{-5}$
(Glyc)+	−109,440.	−142,560.	4.3	87.7	21.4	4.8555	4.0750	4.1470	−2.9474	62.0660	14.8378	0.4862
(Glyc)2	−233,020.	−300,660.	27.2	197.1	72.6	11.6921	20.7705	−2.4205	−3.6376	121.3943	37.1116	−0.0300
(Lac)+	−110,350.	−151,480.	10.0	125.2	37.8	7.0703	9.4856	2.0147	−3.1710	83.2379	22.4748	0.3993
(Lac)2	−235,050.	−318,180.	40.3	286.6	107.3	16.4367	32.3514	−6.9634	−4.1163	173.8623	55.3481	−0.0300
(Gly)+	−71,300.	−102,410.	25.0	63.4	25.3	5.2864	5.1238	3.7413	−2.9907	44.9395	9.8852	0.1739
(Gly)2	−156,480.	−221,770.	63.0	139.0	80.9	12.8262	23.5338	−3.4944	−3.7518	87.3679	25.2850	−0.0300
(Ala)+	−70,600.	−109,970.	30.0	108.7	43.4	7.7387	11.1165	1.3762	−3.2385	70.7759	19.1100	0.0974
(Ala)2	−154,650.	−237,360.	70.0	247.2	119.2	18.0655	36.3271	−8.5230	−4.2807	150.7452	47.3132	−0.0300
ZINC												
(For)+	−121,480.	−140,700.	−4.3	33.5	8.0	3.0712	−0.2795	5.8532	−2.7673	31.4547	3.7879	0.6143
(For)2	−206,910.	−245,730.	11.4	68.1	44.1	7.7824	11.2204	1.3417	−3.2428	45.8240	10.8453	−0.0300
(Prop)+	−123,680.	−160,939.	0.5	131.7	36.8	6.9879	9.2834	2.0962	−3.1627	88.3708	23.7877	0.5464
(Prop)2	−211,460.	−285,920.	22.5	302.6	104.9	16.1116	31.5558	−6.6473	−4.0834	183.2279	58.6033	−0.0300
(But)+	−121,820.	−166,540.	5.8	156.9	52.2	9.0597	14.3386	0.1167	−3.3717	102.3810	28.9294	0.4615
(But)2	−207,670.	−296,560.	34.7	362.9	137.4	20.5524	42.4038	−10.9205	−4.5319	218.5528	70.8814	−0.0300
(Pent)+	−119,680.	−172,900.	12.3	220.6	68.2	11.2174	19.6055	−1.9494	−3.5894	138.7888	41.8971	0.3636
(Pent)2	−203,420.	−308,690.	49.8	514.9	171.2	25.1842	53.7125	−15.3636	−4.9994	307.6444	101.8474	−0.0300
(Glyc)+	−159,620.	−194,550.	0.2	86.4	21.7	4.9216	4.2387	4.0776	−2.9541	61.8303	14.5628	0.5464
(Glyc)2	−283,310.	−353,140.	21.8	194.5	73.0	11.7430	20.8930	−2.4643	−3.6426	119.8505	36.5751	−0.0300
(Lac)+	−160,730.	−200,060.	18.0	123.9	38.1	7.0755	9.4969	2.0126	−3.1715	81.3400	22.1998	0.2792
(Lac)2	−285,380.	−364,730.	55.0	284.0	107.7	16.4877	32.4799	−7.0229	−4.1216	172.3186	54.8116	−0.0300
(Gly)+	−117,820.	−151,610.	18.0	62.1	25.6	5.3677	5.3226	3.6628	−2.9989	45.1185	9.6102	0.2792
(Gly)2	−199,140.	−267,410.	55.0	136.4	81.3	12.8771	23.6625	−3.5537	−3.7571	85.8243	24.7484	−0.0300
(Ala)+	−116,610.	−160,450.	17.0	107.4	43.8	7.8514	11.3891	1.2746	−3.2497	71.8106	18.8350	0.2956
(Ala)2	−197,110.	−283,390.	60.0	244.5	119.6	18.1164	36.4558	−8.5823	−4.2860	149.2015	46.7767	−0.0300
CADMIUM												
(For)+	−104,930.	−121,320.	7.7	40.0	17.7	4.3381	2.8097	4.6479	−2.8951	33.5224	5.0712	0.4379
(For)2	−191,520.	−226,400.	27.1	80.4	54.9	9.2604	14.8299	−0.0792	−3.3920	53.0278	13.3492	−0.0300
(Prop)+	−107,910.	−142,340.	12.5	138.0	46.5	8.2526	12.3660	0.8957	−3.2901	90.3786	25.0710	0.3636

continued

Table 8.23 (continued)

Ligand	$\overline{G}^0_{f,P_r,T_r}$	$\overline{H}^0_{f,P_r,T_r}$	$\overline{S}^0_{P_r,T_r}$	$\overline{C}^0_{P_r,T_r}$	$\overline{V}^0_{P_r,T_r}$	$a_1 \times 10$	$a_2 \times 10^{-2}$	a_3	$a_4 \times 10^{-4}$	c_1	$c_2 \times 10^{-4}$	$\omega \times 10^{-5}$
(Prop)$_2$	−196,520.	−267,040.	38.2	314.9	115.7	17.5895	35.1655	−8.0682	−4.2326	190.4317	61.1072	−0.0300
(But)$^+$	−105,290	−147,170.	17.8	163.2	61.9	10.3259	17.4297	−1.0970	−3.4994	104.4311	30.2127	0.2832
(But)$_2$	−191,460.	−276,420.	50.4	375.2	148.2	22.0303	46.0136	−12.3417	−4.6811	225.7566	73.3853	−0.0300
(Pent)$^+$	−103,380.	−153,760.	24.3	226.9	77.9	12.4829	22.6985	−3.1719	−3.7173	140.8184	43.1804	0.1832
(Pent)$_2$	−187,390.	−288,730.	65.5	527.2	182	26.6621	57.3222	−16.7845	−5.1486	314.8482	104.3513	−0.0300
(Glyc)$^+$	−142,280.	−174,380.	12.2	92.7	31.4	6.1880	7.3293	2.8656	−3.0819	63.8836	15.8461	0.3686
(Glyc)$_2$	−265,389.	−331,280.	37.5	206.7	83.8	13.2210	24.5024	−3.8851	−3.7918	127.0546	39.0789	−0.0300
(Lac)$^+$	−143,410.	−183,520.	17.9	130.2	47.8	8.4019	12.7361	0.7390	−3.3054	85.0322	23.4831	0.2792
(Lac)$_2$	−267,700.	−349,090.	50.7	296.3	118.5	17.9656	36.0834	−8.4279	−4.2706	179.5224	57.3155	−0.0300
(Gly)$^+$	−100,240.	−132,090	27.0	68.4	35.3	6.6477	8.4525	2.4231	−3.1283	47.5446	10.8935	0.1417
(Gly)$_2$	−180,580.	−246,610.	65.0	148.7	92.1	14.3550	27.2722	−4.9746	−3.9063	93.0282	27.2523	−0.0300
(Ala)$^+$	−100,080.	−141,020.	29.2	113.7	53.4	9.1151	14.4742	0.0630	−3.3773	73.7914	20.1183	0.1098
(Ala)$_2$	−179,940.	−263,420.	71.9	256.8	130.4	19.5944	40.0653	−10,0034	−4.4352	156.4053	49.2806	−0.0300

LEAD

(For)$^+$	−92,150.	−100,690.	37.1	19.1	17.7	4.1867	2.4389	4.7962	−2.8797	17.2647	0.8546	−0.0100
(For)$_2$	−177,720.	−202,040.	65.7	40.0	54.9	9.2604	14.8299	−0.0792	−3.3920	29.3579	5.1222	−0.0300
(Prop)$^+$	−95,670.	−122,250.	41.9	117.3	46.5	8.1013	12.0011	1.0303	−3.2750	74.1249	20.8544	−0.0838
(Prop)$_2$	−184,200.	−244,160.	76.8	274.5	115.7	17.5895	35.1655	−8.0682	−4.2326	166.7618	52.8802	−0.0300
(But)$^+$	−92,820.	−126,860	47.2	142.5	61.9	10.1748	17.0643	−0.9612	−3.4843	88.1802	25.9961	−0.1639
(But)$_2$	−179,080.	−253,470.	89	334.8	148.2	22.0303	46.0136	−12.3417	−4.6811	202.0867	65.1583	−0.0300
(Pent)$^+$	−90,790.	−133,320.	53.7	206.9	77.9	12.3324	22.3315	−3.0281	−3.7021	124.5859	38.9638	−0.2620
(Pent)$_2$	−175,020.	−265,790.	104	486.8	182	26.6621	57.3222	−16.7845	−5.148	291.1785	96.1242	−0.0300
(Glyc)$^+$	−130,020.	−154,270.	41.6	72.0	47.8	8.2809	12.4416	0.8529	−3.2932	47.6306	11.6296	−0.0788
(Glyc)$_2$	−253,620.	−308,950.	76.1	166.4	118.5	17.9656	36.0834	−8.4279	−4.2706	103.3846	30.8519	−0.0300
(Lac)$^+$	−131,350.	−163,610.	47.3	109.5	47.8	8.2516	12.3686	0.8843	−3.2902	68.8031	19.2665	−0.1656
(Lac)$_2$	−256,300.	−327,120.	89.2	255.9	118.5	17.9656	36.0834	−8.4279	−4.2706	155.8527	49.0884	−0.0300
(Gly)$^+$	−88,450.	−112,310.	56.9	47.7	35.3	6.4949	8.0790	2.5712	−3.1129	31.2483	6.6769	−0.3103
(Gly)$_2$	−168,350.	−222,990.	106.3	108.3	92.1	14.3550	27.2722	−4.9746	−3.9063	69.3582	19.0253	−0.0300
(Ala)$^+$	−87,190.	−120,280.	58.7	93.0	53.4	8.9640	14.1088	0.1992	−3.3622	57.5413	15.9017	−0.3372
(Ala)$_2$	−166,270.	−239,190.	110.4	216.4	130.4	19.5944	40.0653	−10.0034	−4.4352	132.7356	41.0535	−0.0300

Discrete values of the HKF model parameters for various organic aqueous species are listed in table 8.22. Table 8.23 lists standard partial molal thermodynamic properties and HKF model parameters for aqueous metal complexes of monovalent organic acid ligands, after Shock and Koretsky (1995).

8.18   Redox State and Biologic Activity

We have already mentioned that photosynthesis and other biochemical processes are the main causes of disequilibrium in aqueous solutions. The conversion of luminous energy into chemical energy (formation of stable covalent bonds) involves local lowering of the redox state. For instance, the conversion of carbon dioxide into glucose:

$$\frac{1}{4}CO_{2(gas)} + H^+ + e^- \Leftrightarrow \frac{1}{24}C_6H_{12}O_6 + \frac{1}{4}H_2O \qquad (8.224)$$

is a reducing process whose standard potential is $E^0_{224} = -0.426$ V. This low Eh value is normally attained in photosynthetic processes. However nonphotosynthesizing organisms tend to reestablish equilibrium, with catalytic decomposition of photosynthesis products. The energy gained in the redox process is partly exploited in the synthesis of new cells. Figure 8.29 represents the main redox processes taking place through mediation of microbes in electron transfer from reduced to oxidized phase. The sequence of reactions (top to bottom) is that which may be expected on thermochemical bases, according to the Gibbs free energy changes involved in the various processes (table 8.24).

Table 8.24 lists the main redox exchanges mediated by biological activity and their equilibrium constants at pH = 7.

In a closed aqueous system containing organic matter (exemplified by formaldehyde CH_2O), oxidation of organic matter initially implies reduction of free oxygen, followed by reduction of NO_3^-. These processes involve a progressive decrease in Eh. In the presence of manganese, contemporaneously with the reduction of nitrates there is also reduction of MnO_2. At lower Eh, reduction of iron complexes ($FeOOH_{(crystal)}$ to ferrous ion Fe^{2+}) takes place. If Eh decreases sufficiently, fermentation reactions take place, accompanied by reduction of sulfates and carbonates. The sequence of reactions in table 8.24 reflects the vertical distribution of components in an aqueous system rich in nutrients (eutrophic) and can be compared with an "ecological sequence" of microorganisms that act as mediators of redox processes (*aerobic–heterotropic–denitrificant–fermentant–sulfate reducer–methaniferous bacterials;* cf. Stumm and Morgan, 1981).

Let us now briefly review the effects of biologic activity on the various systems.

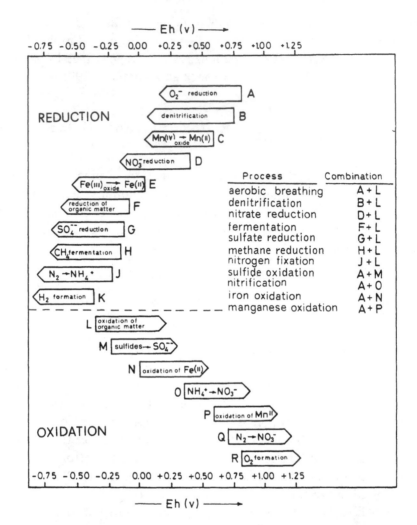

Figure 8.29 Sequence of redox equilibria mediated by biologic activity. From W. Stumm and J. J. Morgan (1981), *Aquatic Chemistry,* copyright © 1981 by John Wiley and Sons. Reprinted by permission of John Wiley & Sons. The various equilibrium constants are listed in table 8.24.

8.18.1 Effects on Carbon

It is estimated that photosynthesis annually fixes about 54×10^{14} moles of C. About 99.95% of the fixed carbon is produced by living organisms and only 0.05% (0.025×10^{14} moles) is buried in sedimentary processes in the form of native carbon. The organic complexes of carbon present in aqueous solutions are thermodynamically unstable (with the exception of methane CH_4). The decomposition of acetic acid CH_3COOH and formaldehyde CH_2O into graphite, methane, and carbon dioxide are spontaneous thermochemical processes, although their reaction kinetics are rather sluggish, leading to metastable persistence of reactants (Stumm and Morgan, 1981):

Table 8.24 Redox equilibria mediated by biological activity in eutrophic systems (from Stumm and Morgan, 1981). pH = 7; $[HCO_3^-] = 10^{-3}$; (g) = gaseous, (s) = solid.

Reduction		log K
(A)	$1/4\ O_{2(g)} + H^+ + e^- \Leftrightarrow 1/2\ H_2O$	+13.75
(B)	$1/5\ NO_3^- + 6/5\ H^+ + e^- \Leftrightarrow 1/10\ N_{2(g)} + 3/5\ H_2O$	+12.65
(C)	$1/2\ MnO_{2(s)} + 1/2\ HCO_3^- + 3/2\ H^+ + e^- \Leftrightarrow 1/2\ MnCO_{3(s)} + H_2O$	+8.9
(D)	$1/8\ NO_3^- + 5/4\ H^+ + e^- \Leftrightarrow 1/8\ NH_4^+ + 3/8\ H_2O$	+6.15
(E)	$FeOOH_{(s)} + HCO_3^- + 2H^+ + e^- \Leftrightarrow FeCO_{3(s)} + 2\ H_2O$	−0.8
(F)	$1/2\ CH_2O + H^+ + e^- \Leftrightarrow 1/2\ CH_3OH$	−3.01
(G)	$1/8\ SO_4^{2-} + 9/8\ H^+ + e^- \Leftrightarrow 1/8\ HS^- + 1/2\ H_2O$	−3.75
(H)	$1/8\ CO_{2(g)} + H^+ + e^- \Leftrightarrow 1/8\ CH_{4(g)} + 1/4\ H_2O$	−4.13
(J)	$1/6\ N_{2(g)} + 4/3\ H^+ + e^- \Leftrightarrow 1/3\ NH_4$	−4.68

Oxidation		log K
(L)	$1/4\ CH_2O + 1/4\ H_2O \Leftrightarrow 1/4\ CO_{2(g)} + H^+ + e^-$	+8.20
(L-1)	$1/2\ HCOO^- \Leftrightarrow 1/2\ CO_{2(g)} + 1/2\ H^+ + e^-$	+8.33
(L-2)	$1/2CH_2O + 1/2H_2O \Leftrightarrow 1/2\ HCOO^- + 3/2H^+ + e^-$	+7.68
(L-3)	$1/2\ CH_3OH \Leftrightarrow 1/2\ CH_2O + H^+ + e^-$	+3.01
(L-4)	$1/2\ CH_{4(g)} + 1/2\ H_2O \Leftrightarrow 1/2\ CH_3OH + H^+ + e^-$	−2.88
(M)	$1/8\ HS^- + 1/2H_2O \Leftrightarrow 1/8\ SO_4^{2-} + 9/8\ H^+ + e^-$	+3.75
(N)	$FeCO_{3(s)} + 2\ H_2O \Leftrightarrow FeOOH_{(s)} + HCO_3^- + 2H^+ + e^-$	+0.8
(O)	$1/8\ NH_4^+ + 3/8\ H_2O \Leftrightarrow 1/8\ NO_3^- + 5/4\ H^+ + e^-$	−6.15
(P)	$1/2\ MnCO_{3(s)} + H_2O \Leftrightarrow 1/2\ MnO_{2(s)} + 1/2\ HCO_3^- + 3/2\ H^+ + e^-$	−8.9

$$CH_3COOH \Leftrightarrow 2\,H_2O + 2\,C_{(solid)} \qquad \left(\Delta G = -103\,\frac{kJ}{mole} \right) \qquad (8.225)$$

$$CH_3COOH \Leftrightarrow CH_{4(gas)} + CO_{2(gas)} \qquad \left(\Delta G = -51\,\frac{kJ}{mole} \right) \qquad (8.226)$$

$$CH_2O \Leftrightarrow C_{(solid)} + H_2O \qquad \left(\Delta G = -107\,\frac{kJ}{mole} \right). \qquad (8.227)$$

A process of bacterial degradation of carbon compounds can be conceived essentially as "a conversion from a CO_2-predominant to a CH_4-predominant system" (Stumm and Morgan, 1981). The production of organic methane by microbial activity is a typical example of such conversion. Organic matter is first transformed into organic acids, which are then decomposed into acetic acid, gaseous hydrogen, and CO_2, and finally recombined as CH_4:

$$\text{Complex organic matter} \rightarrow \text{organic acids} \begin{cases} 4\,H_{2(gas)} + CO_{2(gas)} \rightarrow CH_4 + 2\,H_2O \\ CH_3COOH \rightarrow CH_4 + CO_{2(gas)} \end{cases}$$

$$(8.228)$$

8.18.2 Effects on Nitrogen

We have already seen that reduction of gaseous nitrogen $N_{2(g)}$ into ammonium ion NH_4^+ takes place at low Eh but still within the stability field of water. Blue-green algae are the organisms that most efficaciously fix free nitrogen into ammonium ion complexes, operating under the redox conditions established by photosynthesis (pH being equal, carbonate reduction takes place at lower Eh; cf. figure 8.21A and B). The fact that N_2 fixation is not a process as extensive as the reduction of atmospheric CO_2 (and of soluble complexes H_2CO_3, HCO_3^-, and CO_3^{2-}) is attributable to kinetic problems connected with the bond stability of the N_2 molecule (Stumm and Morgan, 1981). NH_4^+ and NO_3^- are also stable covalent complexes that may persist metastably under Eh-pH conditions pertinent to the N_2 molecule (see figure 8.21B).

8.18.3 Effects on Sulfur

The sulfate $\rightarrow$ sulfide reduction requires quite low Eh conditions (figure 8.21D). The process takes place through enzymatic mediation (enzymes oxidize organic matter at the Eh of interest). At low pH, the reducing process may result in the formation of native sulfur, as an intermediate step of the process

$$SO_4^{2-} + 2H^+ + 3H_2S \Leftrightarrow 4S_{(crystal)} + 4H_2O \qquad \left(\Delta G = -27.7 \frac{kJ}{mole} \right). \qquad (8.229)$$

As already mentioned, the Eh-pH extension of the field of native sulfur depends on the total sulfur concentration in the system. With solute molality lower than approximately 10^{-6} moles/kg, the field of native sulfur disappears.

8.19 Degree of Saturation of Condensed Phases and Affinity to Equilibrium in Heterogeneous Systems

Let us consider a generic hydrolytic equilibrium between a crystalline substance—such as K-feldspar, for instance—and an aqueous solution:

$$KAlSi_3O_8 + 4H^+ + 4H_2O \Leftrightarrow K^+ + Al^{3+} + 3H_4SiO_4^0. \qquad (8.230)$$

We can calculate the equilibrium constant of reaction 8.230 under any given P-T condition by computing the standard molal Gibbs free energy values of the various components at the P-T of interest—i.e.,

$$K_{230,P,T} = \exp\left(\frac{-\Delta G_{230}^0}{RT} \right). \qquad (8.231)$$

Applying the mass action law and assuming unitary activity for the water solvent and the condensed phase, we obtain

$$K_{230,P,T} = a_{K^+} \cdot a_{Al^{3+}} \cdot a^3_{H_4SiO^0_4} \cdot a^{-4}_{H^+}. \qquad (8.232)$$

The activities of solute species in an aqueous solution in equilibrium with K-feldspar at the P and T of interest will be those dictated by equation 8.232. Let us now imagine altering the chemistry of the aqueous solution in such a way that the activities of the aqueous species of interest differ from equilibrium activities. New activity product Q:

$$Q = a'_{K^+} \cdot a'_{Al^{3+}} \cdot \left(a'_{H_4SiO^0_4}\right)^3 \cdot \left(a'_{H^+}\right)^{-4} \qquad (8.233)$$

will differ from equilibrium constant K, and the Q/K ratio will define the distance from equilibrium of the new fluid with respect to K-feldspar at the P and T of interest. In a generalized form, we can write

$$Q = \prod_i a'^{v_i}_i, \qquad (8.234)$$

where i is a generic solute species and v_i is its corresponding stoichiometric coefficient in reaction (positive for products and negative for reactants).

It follows from equation 8.234 that

$$\log\left(\frac{Q}{K}\right) = \log\left(\prod_i a'^{v_i}_i\right) - \log K. \qquad (8.235)$$

Ratio Q/K is adimensional: if it is higher than 1 (i.e., $\log Q/K > 0$), the aqueous solution is oversaturated with respect to the solid phase; if it is zero, we have perfect equilibrium; and if it is lower than 1 (i.e., $\log Q/K < 0$), the aqueous phase is undersaturated with respect to the solid phase.

Recalling now the concept of *thermodynamic affinity* (section 2.12), we can, in a similar way, define an "affinity to equilibrium" as

$$A = -2.303 \, RT \log\left(\frac{Q}{K}\right). \qquad (8.236)$$

This parameter has the magnitude of an energy (i.e., kJ/mole, kcal/mole) and represents the energy driving toward equilibrium in a chemical potential field—i.e., the higher A becomes, the more the solid phase of interest is unstable in solution. In a heterogeneous system at equilibrium, based on the principle of equality of chemical potentials, all phases in the system are in mutual equilibrium. It follows that equation 8.236 is of general applicability to all condensed phases, and the value of affinity must be zero for all solids in equilibrium with the fluid. It is immediately evident that equations 8.235 and 8.236 furnish a precise measure of geochemical control of the degree of equilibrium attained by the fluid with respect to a given mineralogical paragenesis. However, because the equilibrium constant is a function of intensive variables T and P, plotting the affinity value for various minerals under different T (or P) conditions against the T (or P) of calculation provides precise thermometric (or barometric) information concerning the equilibrium state of the fluid with respect to the reservoir. Figure 8.30A shows a log $Q/K = f(T)$ diagram in the case of complete equilibrium between mineralogical paragenesis and fluid at a given T condition (in this case, $T = 250$ °C): note how the saturation curves for all solid phases converge to log $Q/K = 0$ at the temperature of equilibrium.

Detailed speciation calculations for the fluid coupled with mineralogical investigation of the solid paragenesis generally allow a sufficiently precise estimate of the T of equilibrium. As figure 8.30A shows, the chemistry of the fluid is buffered by wall-rock minerals, so that the saturation curves of other phases not pertaining to the system of interest are scattered (figure 8.28B).

The interpretation of log $Q/K = f(T)$ plots may be complicated by additional processes affecting the chemistry of the fluid. The main complications result from two processes that take place during the ascent of the fluid toward the surface: *adiabatic flashing* and *dilution by external aquifers*.

Figure 8.30C shows the boiling effect on the fluid in equilibrium with the paragenesis of figure 8.30A, calculated by Reed and Spycher (1984) assuming subtraction of a gaseous phase amounting to 0.2% by weight of the initial fluid and composed of 99.525% H_2O, 0.451% CO_2, 0.013% H_2S, and 0.011% H_2: the increased molal concentration of dissolved silica displaces the intersection of the saturation curve of quartz with the equilibrium condition log $Q/K = 0$ to a higher T. Moreover, the change in pH, altering the speciation state of aluminum, iron sulfides, and carbonates, modifies the shapes of the saturation curves of alkali feldspars, muscovite, pyrite, and calcite.

The effect of dilution by shallow aquifers can also be quantified. It invariably results in scattered saturation curves, proportional to the extent of contamination.

To the above considerations must be added the fact that equations of type 8.235 apply to pure compounds, whereas natural solids are mixtures of several components. It follows that the activity of the solid component in hydrolysis is never 1. Equation 8.235 must therefore be modified to take account of this fact:

$$\log\left(\frac{Q}{K}\right) = \log\left(\prod_i a_i'^{\nu_i}\right) - \log K - \log a_K. \qquad (8.237)$$

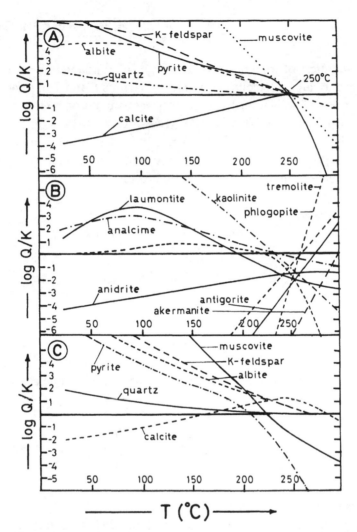

Figure 8.30 Degree of saturation of various phases as a function of T in a heterogeneous system. A = complete equilibrium; B = disequilibrium; C = dispersion of saturation curves as a result of boiling. Reprinted from M. Reed and N. Spycher, *Geochimica et Cosmochimica Acta*, 48, 1479–1492, copyright © 1984, with kind permission from Elsevier Science Ltd., The Boulevard, Langford Lane, Kidlington 0X5 1GB, UK.

a_K in eq. 8.237 is the activity of the solid component in the mixture for which hydrolytic equilibrium has been calculated. It must be emphasized that the effect of dilution in a solid mixture is often quite limited: a change from activity = 1.0 to activity = 0.5 involves a displacement of the log Q/K factor of 0.3 (T, P, and chemistry of the fluid being equal).

In the investigation of mineral-fluid equilibria, of fundamental importance are the "activity diagrams," depicting equilibria at various P and T of interest in terms of activity ratios of solutes or fugacity of gaseous components. Let us consider the system MgO-CaO-FeO-Na_2O-K_2O-Al_2O_3-SiO_2-CO_2-H_2O, as representative of the chemistry of most rock-forming minerals. Of the various oxides in the system, only H_2O, CO_2, and SiO_2 do not dissociate appreciably in aqueous solution. For the remaining oxides, dissociation is almost complete and can be represented as follows:

$$Na_2O + 2\,H^+ \Leftrightarrow 2\,Na^+ + H_2O \qquad (8.238)$$

$$CaO + 2\,H^+ \Leftrightarrow Ca^{2+} + H_2O \qquad (8.239)$$

$$Al_2O_3 + 6\,H^+ \Leftrightarrow 2\,Al^{3+} + 3\,H_2O. \qquad (8.240)$$

The dissociation of a generic oxide $M_{v+}O_{v-}$ (where v^+ and v^- are, respectively, the cationic and anionic stoichiometric factors in the formula unit) can be expressed by the general equilibrium

$$M_{v+}O_{v-} + (v+)(Z+)H^+ \Leftrightarrow \left[\frac{(v+)(Z+)}{2}\right]H_2O + (v+)M^{Z+}, \qquad (8.241)$$

where $Z+$ is the cationic charge.

Applying the mass action law to equilibrium 8.238, we can write

$$K_{244} = \frac{a_{Na^+}^2 \cdot a_{H_2O}}{a_{Na_2O} \cdot a_{H^+}^2}. \qquad (8.242)$$

The generalized form of equation 8.242 is given by

$$K = \frac{a_{M^{Z+}}^{v+} \cdot a_{H_2O}^{[(v+)(Z+)/2]}}{a_{M_{v+}O_{v-}} \cdot a_{H^+}^{(v+)(Z+)}}. \qquad (8.243)$$

Let us now rewrite equation 8.242 in logarithmic notation, assigning unitary activity to the solvent:

$$\ln a_{Na_2O} = 2 \ln\left(\frac{a_{Na^+}}{a_{H^+}}\right) - \ln K_{238}. \tag{8.244}$$

For the generic oxide $M_{\nu+}O_{\nu-}$, we have

$$\ln a_{M_{\nu+}O_{\nu-}} = (\nu +)\ln\left(\frac{a_{M^{Z+}}}{a_{H^+}^{(Z+)}}\right) - \ln K. \tag{8.245}$$

The relationship between thermodynamic activity and chemical potential is

$$\mu_{Na_2O} = \mu_{Na_2O}^0 + RT \ln a_{Na_2O}. \tag{8.246}$$

For two generic oxides i and j, the partial derivatives of equation 8.246 can be combined (Helgeson, 1967) to give

$$\frac{d\mu_i}{d\mu_j} = \frac{d \ln a_i}{d \ln a_j} = \frac{\nu_{(i)}^+}{\nu_{(j)}^+}\left[\frac{d \ln\left(\dfrac{a_{M_{(i)}^{Z+}}}{a_{H^+}^{Z+(i)}}\right)}{d \ln\left(\dfrac{a_{M_{(j)}^{Z+}}}{a_{H^+}^{Z+(j)}}\right)}\right]. \tag{8.247}$$

Equation 8.247 can be rearranged in the form

$$\frac{d \ln\left(\dfrac{a_{M_{(i)}^{Z+}}}{a_{H^+}^{Z+(i)}}\right)}{d \ln\left(\dfrac{a_{M_{(j)}^{Z+}}}{a_{H^+}^{Z+(j)}}\right)} = \frac{\nu_{(j)}^+}{\nu_{(i)}^+} \cdot \frac{d\mu_i}{d\mu_j} = \frac{\nu_{(j)}^+}{\nu_{(i)}^+} \cdot \frac{df_{(i)}}{df_{(j)}}, \tag{8.248}$$

where the last term on the right is the fugacity ratio of gaseous components. Equation 8.248 is the basis for the construction of activity diagrams.

Let us now imagine a process of hydrothermal alteration of arkose sandstones composed of Mg-chlorite, K-feldspar, K-mica, and quartz. Because precipitating SiO_2 during alteration is amorphous, we will assume the presence of amorphous silica instead of quartz, and we will consider MgO as the generic oxide i and K_2O as the generic oxide j.

The alteration of muscovite to kaolinite is described by the reaction

$$2\,KAl_3Si_3O_{10}(OH)_2 + 2\,H^+ + 3\,H_2O \Leftrightarrow 3\,Al_2Si_2O_5(OH)_4 + 2\,K^+. \quad (8.249)$$

muscovite　　　　　　　　　　　　　　　kaolinite

Assigning unitary activity to the condensed phases and to solvent H_2O, equilibrium constant K_{249} takes the form

$$K_{249} = \frac{a_{K^+}^2}{a_{H^+}^2}, \quad (8.250)$$

and, in logarithmic notation,

$$\frac{1}{2}\log K_{249} = \log a_{K^+} - \log a_{H^+} = \log a_{K^+} + pH. \quad (8.251)$$

The equilibrium between K-feldspar and muscovite is expressed as

$$3\,KAlSi_3O_8 + 2\,H^+ \Leftrightarrow KAl_3Si_3O_{10}(OH)_2 + 2\,K^+ + 6\,SiO_2, \quad (8.252)$$

K-feldspar　　　　　　　　muscovite

which yields

$$\frac{1}{2}\log K_{252} = \log a_{K^+} - \log a_{H^+} = \log a_{K^+} + pH. \quad (8.253)$$

The equilibrium between Mg-chlorite and kaolinite is given by

$$Mg_5Al_2Si_3O_{10}(OH)_8 + 10\,H^+ \Leftrightarrow Al_2Si_2O_5(OH)_4 + 5\,Mg^{2+} + SiO_2 + 7\,H_2O,$$

Mg-chlorite　　　　　　　　　　　　　kaolinite

$$(8.254)$$

from which we obtain

$$\frac{1}{5}\log K_{254} = \log a_{Mg^{2+}} + 2\,pH. \quad (8.255)$$

The equilibrium between Mg-chlorite and muscovite may be depicted as

$$3Mg_5Al_2Si_3O_{10}(OH)_8 + 2\,K^+ + 28H^+ \Leftrightarrow 2KAl_3Si_3O_{10}(OH)_2 + 15Mg^{2+} + 3SiO_2 + 24H_2O.$$

Mg-chlorite　　　　　　　　　　　　muscovite

$$(8.256)$$

The equilibrium constant of reaction 8.256 is

$$K_{256} = \frac{a_{Mg^{2+}}^{15}}{a_{K^+}^2 \cdot a_{H^+}^{28}}. \tag{8.257}$$

Multiplying and dividing by $a_{H^+}^2$ and transforming into logarithmic notation, we obtain

$$2 \log \left(\frac{a_{K^+}}{a_{H^+}} \right) + \log K_{256} = 15 \log \left(\frac{a_{Mg^{2+}}}{a_{H^+}^2} \right). \tag{8.258}$$

Lastly, the equilibrium between Mg-chlorite and K-feldspar is depicted by

$$\underset{\text{Mg-chlorite}}{Mg_5Al_2Si_3O_{10}(OH)_8} + 2\,K^+ + 8\,H^+ + 3\,SiO_2 \Leftrightarrow \underset{\text{K-feldspar}}{2\,KAlSi_3O_8} + 5\,Mg^{2+} + 8\,H_2O. \tag{8.259}$$

The equilibrium constant is

$$K_{259} = \frac{a_{Mg^{2+}}^5}{a_{K^+}^2 \cdot a_{H^+}^8}. \tag{8.260}$$

Again, multiplying and dividing by $a_{H^+}^2$ and transforming into logarithmic notation, we obtain

$$2 \log \left(\frac{a_{K^+}}{a_{H^+}} \right) + \log K_{259} = 5 \log \left(\frac{a_{Mg^{2+}}}{a_{H^+}^2} \right). \tag{8.261}$$

We can now build up the activity diagram, adopting as Cartesian axes log a_{K^+} + pH (abscissa) and log $a_{Mg^{2+}}$ + 2pH (ordinate; see figure 8.31). The kaolinite–muscovite equilibrium in such a diagram is defined by a straight line parallel to the ordinate, with an intercept at $\frac{1}{2}$ log K_{249} on the abscissa. Analogously, the equilibrium between muscovite and K-feldspar is along a straight line, parallel to the preceding one and with an intercept at $\frac{1}{2}$ log K_{252}; the kaolinite–Mg-chlorite equilibrium lies along a straight line parallel to the abscissa, with intercept at $\frac{1}{5}$ log K_{254}.

Let us now consider Mg-chlorite–muscovite and Mg-chlorite–K-feldspar equilibria. Because they depend on the activity of both Mg^{2+} and K^+ in solution, besides H^+, the slope of the univariant curve is dictated by the stoichiometric coefficients of ions in reaction: for reaction 8.256, the slope is $\frac{2}{15}$ and the intercept

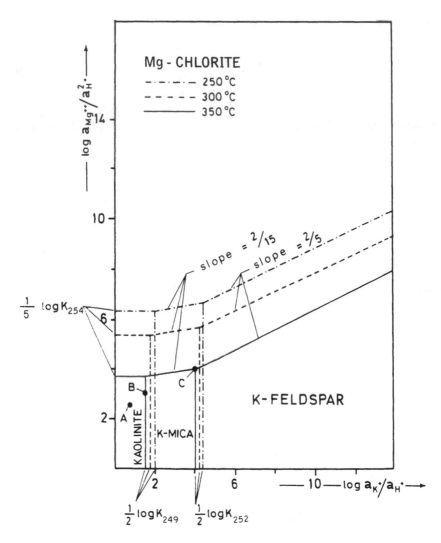

Figure 8.31 Example of activity diagram.

on the ordinate axis is $\frac{1}{15}\log K_{256}$ (cf. eq. 8.258). Analogously, for the Mg-chlorite–K-feldspar equilibrium, the univariant curve has slope $\frac{2}{5}$ and intercept at $\frac{1}{5}\log K_{259}$ (cf. eq. 8.261). As shown in figure 8.31, the intersections of univariant equilibria contour the stability fields of the various solids in reaction. A fluid of composition A in figure 8.31 is in equilibrium with kaolinite; fluid B, falling on the kaolinite–muscovite boundary at $T = 350\ °C$, is clearly buffered by the alteration of mica; and a fluid of composition C is in invariant equilibrium with the original paragenesis. Translating the above observations into $\log Q/K = f(T)$ notation (section 8.19), for a fluid of composition C we observe convergence of the saturation curves of muscovite, K-feldspar, and Mg-chlorite toward $\log Q/K = 0$ at $T = 350\ °C$ (note that, because amorphous silica has fixed unitary activity, the components are reduced to three in the computation of the degree of

Table 8.25 Equilibrium constants used for construction of figure 8.31 (activity diagram).

T (°C)	P(bar)	K_{249}	K_{252}	K_{254}	K_{256}	K_{259}
25	1	6.841	1.967	67.223	194.828	63.632
250	39.728	4.009	4.869	31.271	89.803	26.688
300	85.805	3.576	5.242	26.217	75.075	21.531
350	165.125	3.019	5.470	18.356	52.048	13.703

freedom and, because P and T are fixed, three phases determine invariance in a pseudoternary system).

To construct the diagram, we must know the various equilibrium constants at the P and T of interest. It is thus necessary to compute the standard molal Gibbs free energies of water, solid phases and aqueous ions in reaction. For solids, the methods described in section 3.7 must be adopted. For aqueous ions, the HKF model (section 8.11) or similar types of calculations are appropriate. To avoid error progression and inconsistencies, the database used in calculating the Gibbs free energies of both aqueous species and solid compounds must be internally consistent. In the case of figure 8.31, the constants were calculated at different T and P conditions along the liquid-vapor univariant curve using the database of Helgeson et al. (1978) (computer package SUPCRT). The resulting values are listed in table 8.25. It must be emphasized that extensive compilations of activity diagrams in chemically complex systems are already available in the literature. For instance, the compilation of Helgeson et al. (1976) for the system MgO-CaO-FeO-Na$_2$O-K$_2$O-Al$_2$O$_3$-SiO$_2$-CO$_2$-H$_2$O-H$_2$S-H$_2$SO$_4$-Cu$_2$S-PbS-ZnS-Ag$_2$S-HCl presents activity diagrams drawn for various T conditions (0, 25, 60, 100, 150, 200, 250, 300°C; $P = 1$ bar) and is an excellent tool for interpreting water-rock equilibria at shallow depths.

8.21 Reaction Kinetics in Hydrous Environments

The arguments treated in the two preceding sections were developed in terms of simple equilibrium thermodynamics. The "weathering" of rocks at the earth's surface by the chemical action of aqueous solutions, and the complex water-rock interaction phenomena taking place in the upper crust, are irreversible processes that must be investigated from a kinetic viewpoint. As already outlined in section 2.12, the kinetic and equilibrium approaches are mutually compatible, both being based on firm chemical-physical principles, and have a common boundary represented by the "steady state" condition (cf. eq. 2.111).

8.21.1 Compositional Dependence of Rate Constants

The fundamentals of the kinetic approach to water-rock interaction are constituted by the quantitative assessment of reaction kinetics for mineral hydrolysis and precipitation. These may be expressed in terms of rates of change in molality of one of the main ions involved in the process. Denoting as m_i the molality of solute ion i and as n_i the number of moles of i in solution, molality is related to the mass of solvent by

$$m_i = \frac{n_i}{n_W} \Omega, \qquad (8.262)$$

where n_w are the moles of water solvent and Ω are the moles of H_2O in 1 kg of water ($\Omega = 55.51$). Differentiating equation 8.262 with respect to time t gives

$$\frac{dm_i}{dt} = \left(\frac{\Omega}{n_W} \right) \left[\frac{dn_i}{dt} - \left(\frac{n_i}{n_W} \right) \frac{dn_W}{dt} \right], \qquad (8.263)$$

(Delany et al., 1986), where dn_w/dt is the consumption (or production) rate of water molecules in the system. This term is often negligible with respect to the other elemental balances.

The rate of a generic reaction j is represented by its absolute velocity v_j, which may be obtained from experiments, provided that the mass of solvent is held constant throughout the experiment and no additional homogeneous or heterogeneous reactions concur to modify the molality of the ion in solution (Delany et al., 1986):

$$v_j \cong \frac{n_W}{\Omega v_{i,j}} \times \frac{dm_i}{dt}, \qquad (8.264)$$

where $v_{i,j}$ is the stoichiometric reaction coefficient of ion i as it appears in reaction j (positive for products and negative for reactants; i.e., positive in the case of a hydrolysis reaction).

Using a molar concentration notation instead of molality ($c_i = n_i/V_s$; cf. appendix 1), equation 8.264 is translated into

$$v_j \cong \frac{V_s}{v_{i,j}} \times \frac{dc_i}{dt}. \qquad (8.265)$$

When more than one reaction affects the mass of the ion i in solution (e.g., dissolution and/or precipitation of heterogeneous solid assemblages; homogeneous reactions in solution), the overall change in mass of component i is given by the summation of individual absolute reaction velocities v_j, multiplied by their respective stoichiometric reaction coefficients:

$$\frac{dn_i}{dt} = \sum_{j=1}^{T} v_{i,j} v_j . \qquad (8.266)$$

The contribution of each reaction to the mass change is

$$
\left(\frac{dn_i}{dt} \right)_j = v_{i,j} v_j .
$$
(8.267)

Based on equation 8.266, the molality change induced by T simultaneous reactions involving ion i is

$$
\frac{dm_i}{dt} = \left(\frac{\Omega}{n_W} \right) \sum_{j=1}^{T} \left[v_{i,j} - \left(\frac{v_{W,j}}{n_W} \right) \right] v_j
$$
(8.268)

(Delany et al., 1986), where $v_{W,j}$ is the number of water molecules involved in the jth reaction.

The absolute reaction velocity of an irreversible reaction can be related to an individual progress variable ξ_j by

$$
v_j = \frac{d\xi_j}{dt} .
$$
(8.269)

In section 2.12 we recalled the significance of the "overall reaction progress variable" ξ which represents the extent of advancement of simultaneous irreversible processes in heterogeneous systems (Temkin, 1963). The overall progress variable is related to progress variables of individual reactions by

$$
\xi = \sum_{j=1}^{T} |\xi_j|
$$
(8.270)

(Delaney et al., 1986). It follows from equations 8.269 and 8.270 that we can also define an "overall reaction rate" v, which is related to individual velocities by

$$
v = \sum_{j=1}^{T} |v_j|
$$
(8.271)

(Delany et al., 1986). The "relative rate of the jth reaction" σ_j with respect to the overall reaction progress variable ξ is given by

$$
\sigma_j = \frac{d\xi_j}{d\xi}
$$
(8.272)

(Temkin, 1963). From the various equations it follows that

$$\sigma_j = \frac{v_j}{v} \tag{8.273}$$

and

$$v = \frac{d\xi}{dt}. \tag{8.274}$$

We can generally group the rate laws adopted to describe dissolution and precipitation kinetics into three main categories (Delany et al., 1986):

1. Simple rate laws in m^{eq} or c^{eq}
2. Activity term rate laws
3. Rate laws based on transition state theory.

Simple Rate Laws in m^{eq} or c^{eq}

When forward and backward reaction rates are fast enough to achieve a steady state, reaction kinetics can be expressed as a function of steady state (or "equilibrium") concentration (m_i^{eq} or c_i^{eq}). The general equation of dissolution in terms of molality change can be written as

$$\frac{dm_i^T}{dt} = \left(\frac{s}{\Omega}\right) k_+ \left(m_i^{eq} - m_i\right)^n, \tag{8.275}$$

where dm_i^T is the differential increment of total molality of ion i in solution (free ion plus complexes), k_+ is the rate constant of forward reaction (i.e., dissolution; cf. eq. 2.110), s is the active surface area, and n is an integer corresponding to the degree of reaction. The corresponding equation for precipitation is

$$-\frac{dm_i^S}{dt} = \left(\frac{s}{\Omega}\right) k_- \left(m_i - m_i^{eq}\right)^n, \tag{8.276}$$

where m_i^S is the total molality of i in the precipitating solid and k is the rate constant of the backward reaction.

The simplest case of compositional dependence is the zero-order reaction, in which the concentration gradient is not affected by concentration. Denoting the molar concentration of the ith element (or component) as c_i (and neglecting surface area and volume of solution effects), we have

$$\left(\frac{\partial c_i}{\partial t}\right)_{T,P} = -k_+. \tag{8.277}$$

Integration in dt of equation 8.277 yields

$$c_{i,t} = c_i^0 - tk_+,$$ (8.278)

where c_i^0, is the initial molar concentration of i at time t_0.

In first-order reactions, the concentration gradient depends on the local concentration at time t:

$$\left(\frac{\partial c_i}{\partial t}\right)_{T,P} = -kc_{i,t}.$$ (8.279)

Integration in dt leads to the exponential dependency of molar concentration on time:

$$c_{i,t} = c_i^0 \exp(-kt).$$ (2.280)

In this case, the rate constant has units of t^{-1} and is usually termed "half-life of the reaction" $(t_{1/2})$, the amount of time required to reduce the initial concentration c_i^0, to half:

$$t_{1/2} = \frac{\ln 2}{k}.$$ (8.281)

As we will see in detail in chapter 11, the production of radiogenic isotopes follows perfect first-order reaction kinetics.

In molar notation, and referencing to the equilibrium concentration c_i^{eq} assuming $n = 1$ (i.e., first-order reaction), equation 8.275 can be translated into

$$\frac{dc_i}{dt} = \left(\frac{s}{V_s}\right) k_+ \left(c_i^{eq} - c_i\right)^n.$$ (8.282)

Assuming that s and V_s remain constant throughout the process, the solution of a first-order equation in the form of equation 8.282 ($n = 1$) is

$$\ln\left(\frac{c_i^{eq} - c_i}{c_i^{eq} - c_i^0}\right) = -\left(\frac{s}{V_s}\right) k_+ t,$$ (8.283)

where c_i^0 is the initial concentration in solution at $t = 0$. If $c_i^0 = 0$, a semilogarithmic plot of $(c_i^{eq} - c_i)/c_i$ versus t should exhibit a constant slope corresponding to the rate constant of the forward reaction k_+ multiplied by $-s/V_s$. As we can see in figure 8.32, the rate of dissolution of biogenic opal is found to obey such a

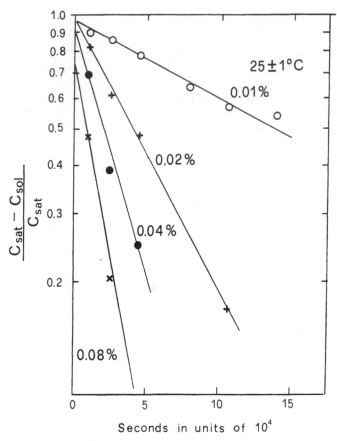

Figure 8.32 Rates of dissolution of biogenic opal in seawater at $T = 25$ °C, for various amounts of suspended opal (percent mass). Reprinted from Hurd (1972), with kind permission from Elsevier Science Publishers B.V, Amsterdam, The Netherlands.

relationship. T and V_s being equal, the negative slope depends on the percent amount of suspended opal (hence on the active surface area), but remains constant throughout the process.

By "active surface area" we meant the kinetically active part of the total surface area. According to Helgeson et al. (1984), this area is restricted to etch pits. Alternative estimates of surface areas may be obtained from measurements of specific surface area $s_\#$ whenever solid particles have a narrow size range. The specific surface area for spherical particles is given by

$$s_\# = \frac{6}{\rho D},$$ (8.284)

where ρ is grain density and D is diameter. For cubic particles, the same equation applies after substitution of edge length a for diameter:

$$s_\# = \frac{6}{\rho a}.$$
(8.285)

Experimental measurements of surface area are usually carried out by BET gas absorption techniques (see Delany et al., 1986, for an appropriate treatment).

In some case the rate laws in m^{eq} or c^{eq} are of high order—i.e., $n > 1$. If c_i is lower than the equilibrium concentration c_i^{eq}, the solution of equation 8.282 is

$$\left(c_i^{eq} - c_i\right)^{1-n} - \left(c_i^{eq} - c_i^0\right)^{1-n} = \frac{(n-1)s}{V_s} k_+ t.$$
(8.286)

To deduce the degree of reaction, one may use equation 8.286 with different integer values of n. The value of n that yields a straight line in a $(c_i^{eq} - c_i)^{1-n}$ versus t plot is the order of reaction.

The equilibrium concentration c_i^{eq} to be introduced for this purpose in equation 8.286 is the one asymptotically achieved in a c_i versus t plot, at high values of t (see Lasaga, 1981a, for a more appropriate treatment of high-order reaction kinetics).

Activity Term Rate Laws

The simplest activity term expressions of rate laws are those involving the acidity of the system. Because, generally, dissolution of a silicate can be schematically depicted as

$$\text{Mineral} + \nu H^+ \rightarrow SiO_{2(aq)}^0 + \text{cations} \pm \text{Al-hydroxides}_{(aq)}$$
(2.287)

(Lasaga, 1984), the rates of dissolution can be expected to increase with a_{H^+} in acidic solutions. The experiments, in fact, denote the pH dependency of rate constants as simple proportionalities of type:

$$k_+ \propto \left(a_{H^+}\right)^{\nu_+},$$
(8.288)

where ν_+ is a real number, depending on kinetics (not to be confused with the stoichiometric integer ν in eq. 8.287). Experimentally observed values of ν_+ are listed in table 8.26.

More complex forms of activity term rate laws involve various ions and complexes in solution. Dissolution and precipitation phenomena are in this approach regarded as a summation of individual reactions taking place at the surface of the solid. The net absolute rate is obtained as a summation of individual terms i, each with its specific rate constant k_i and activity product (Delany et al., 1986):

Table 8.26 pH dependence of mineral dissolution rates
(after Lasaga, 1984; with integrations).

Phase	ν_+	pH Range	References
K-feldspar	1.0	pH < 7	Helgeson et al. (1984)
	1.0	pH < 5	Tole and Lasaga (unpublished results)
Nepheline	−0.20	pH > 7	Tole and Lasaga (unpublished results)
Diopside	0.7	2 < pH < 6	Schott et al. (1981)
Enstatite	0.8	2 < pH < 6	Schott et al. (1981)
Forsterite	1.0	3 ≤ pH ≤ 5	Grandstaff (1980)
Chrysotile	0.24	7 ≤ pH ≤ 10	Bales and Morgan (1985)
Quartz	0.0	pH < 7	Rimstidt and Barnes (1980)
Anorthite	0.54	2 < pH < 5.6	Fleer (1982)
Sr-feldspar	1.0	pH < 4	Fleer (1982)
	−0.28	pH > 6	Fleer (1982)

$$v_j = s \sum_i k_i \prod_k a_k^{\nu_{k,i}}. \tag{8.289}$$

Plummer et al. (1978) utilized, for instance, an equation in the form of equation 8.289 to describe the dissolution of calcite. According to the authors, the simultaneous reactions concurring to define the absolute velocity of dissolution are

$$CaCO_3 + H^+ \Leftrightarrow Ca^{2+} + HCO_3^- \tag{8.290.1}$$

$$CaCO_3 + H_2CO_3^0 \Leftrightarrow Ca^{2+} + 2HCO_3^- \tag{8.290.2}$$

$$CaCO_3 + H_2O \Leftrightarrow Ca^{2+} + HCO_3^- + OH^-. \tag{8.290.3}$$

The absolute velocity of the bulk reaction is

$$v_j = s\left(k_1 a_{H^+} + k_2 a_{H_2CO_3^0} + k_3 a_{H_2O} - k_4 a_{Ca^{2+}} a_{HCO_3^-}\right). \tag{8.291}$$

Figure 8.33 shows in detail the effect of the single rate constants on the forward velocity at the various pH-P_{CO_2} conditions ($T = 25$ °C). Although the individual reactions 8.290.1–8.290.3 take place simultaneously over the entire compositional field, the bulk forward rate is dominated by reactions with single species in the field shown: away from steady state, reaction 8.290.1 is dominant, within the stippled area the effects of all three individual reactions concur to define the overall kinetic behavior, and along the lines labeled 1, 2, and 3 the forward rate corresponding to one species balances the other two.

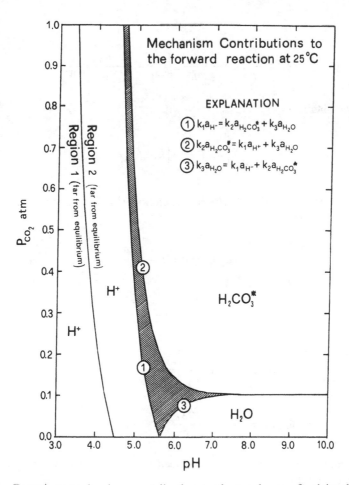

Figure 8.33 Reaction mechanism contribution to the total rate of calcite dissolution reaction as a function of pH and P_{CO_2} at 25 °C. From L. N. Plummer, T. M. L. Wigley, and D. L. Pankhurst (1978), *American Journal of Science*, 278, 179–216. Reprinted with permission of American Journal of Science.

The backward velocity k_4 is obtained by Plummer et al. (1978) by application of the principle of detailed balancing. Denoting as K_1, K_2, K_C, and K_W, respectively, the equilibrium constants of the processes

$$H_2CO_3^0 \Leftrightarrow H^+ + HCO_3^- \qquad (8.292.1)$$

$$HCO_3^- \Leftrightarrow H^+ + CO_3^{2-}$$

$$CaCO_3 \Leftrightarrow Ca^{2+} + CO_3^{2-} \qquad (8.292.3)$$

$$H_2O \Leftrightarrow H^+ + OH^-, \qquad (8.292.4)$$

we have

$$k_4 = \frac{k_1 K_2}{K_C} + \frac{k_2 K_2}{K_C K_1} a_{\text{HCO}_3^-} + \frac{k_3 K_2}{K_C K_W} a_{\text{OH}^-}, \tag{8.293}$$

from which we obtain

$$k_4 = \frac{K_2}{K_C}\left[k_1 + \frac{1}{a_{\text{H}^+}}\left(k_2 a_{\text{H}_2\text{CO}_3^0} + k_3 \right) \right]. \tag{8.294}$$

The forward rate constants (cm/sec) are semilogaritmic functions of reciprocal temperature (K) according to

$$\log k_1 = 0.198 - \frac{444}{T} \tag{8.295.1}$$

$$\log k_2 = 2.84 - \frac{2177}{T} \tag{8.295.2}$$

$$\log k_3 = -5.86 - \frac{317}{T} \quad \left(T < 298\ \text{K} \right) \tag{8.295.3}$$

$$\log k_3 = -1.10 - \frac{1737}{T} \quad \left(T > 298\ \text{K} \right). \tag{8.295.4}$$

The corresponding activation energies deduced from the Arrhenius slope (see section 8.21.2) are listed in table 8.27.

Rate Laws Based on Transition State Theory

The Transition State Theory (TST) treatment of reaction kinetics accounts for the fact that the true (microscopic) behavior of a dissolution reaction may involve the production of an intermediate complex as an essential step of the bulk process. Consider, for instance, the dissolution of silica in water. The macroscopic reaction may be written as

$$\underset{\text{solid}}{\text{SiO}_2} \rightarrow \underset{\text{aqueous}}{\text{SiO}_2^0}. \tag{8.296}$$

The true (microscopic) reaction involves, however, the formation of an activated complex in the intermediate step (Delany et al., 1986):

$$\underset{\text{solid}}{\text{SiO}_2} + v\,\text{H}_2\text{O} \rightarrow \left[\text{SiO}_2 \cdot v\,\text{H}_2\text{O} \right]^* \rightarrow \underset{\text{aqueous}}{\text{SiO}_2^0} + v\,\text{H}_2\text{O}, \tag{8.297}$$

where v molecules of water act as a catalyst of the process.

Because the energy of the activated complex is higher than the energies of reactants and of the final reaction products, it is crucial to establish how many activated complexes may form at given P-T-X conditions. This can be done by application of quantum chemistry concepts.

Limiting the system to all accessible states of energy for the molecules of interest, the probability of finding an individual molecule B in the state of energy ε_j is

$$P_j = \frac{\exp\left(-\varepsilon_j / kT\right)}{Q}, \tag{8.298}$$

where k is Boltzmann's constant and Q is the partition function

$$Q = \sum_i \exp\left(\frac{-E_i}{kT}\right) \tag{8.299}$$

(cf. section 3.1). The partition function Q is related to the partition function q_B relative to the individual molecule B by

$$q_B = \sum_j \exp\left(\frac{-\varepsilon_j}{kT}\right) \tag{8.300}$$

$$Q = q_B^{N_B} = Q_B N_B! \tag{8.301}$$

(Lasaga, 1981b), where N_B is the total number of B molecules in the system. Because the chemical potential is related to the partial derivative of the Helmholtz free energy at constant volume:

$$\mu_B = \left(\frac{\partial F}{\partial N_B}\right)_{T,V} = -kT\left[\frac{\partial \ln\left(Q/N_B!\right)}{\partial N_B}\right]_{T,V} \tag{8.302}$$

recalling the Stirling approximation (eq. 3.116) and applying equation 8.301, we obtain:

$$\mu_B = -kT \ln\left(\frac{q_B}{N_B}\right) \tag{8.303}$$

Consider now a generic reaction of type

$$\nu_A A + \nu_B B \rightarrow \nu_C C + \nu_D D. \tag{8.304}$$

Applying equation 8.303 to each term, we get, at equilibrium,

$$-kT \ln\left(\frac{q_A}{N_A}\right)^{\nu_A} - kT \ln\left(\frac{q_B}{N_B}\right)^{\nu_B} = -kT \ln\left(\frac{q_C}{N_C}\right)^{\nu_C} - kT \ln\left(\frac{q_D}{N_D}\right)^{\nu_D}. \quad (8.305)$$

Rearranging equation 8.305, we obtain

$$\frac{N_C^{\nu_C} \cdot N_D^{\nu_D}}{N_A^{\nu_A} \cdot N_B^{\nu_B}} = \frac{q_C^{\nu_C} \cdot q_D^{\nu_D}}{q_A^{\nu_A} \cdot q_B^{\nu_B}}. \quad (8.306)$$

Finally, dividing both sides of equation 8.306 by the volume of the system, we obtain

$$K = \frac{c_C^{\nu_C} \cdot c_D^{\nu_D}}{c_A^{\nu_A} \cdot c_B^{\nu_B}} = \frac{(q_C/V)^{\nu_C} \cdot (q_D/V)^{\nu_D}}{(q_A/V)^{\nu_A} \cdot (q_B/V)^{\nu_B}}. \quad (8.307)$$

Equation 8.307 is the basis of all applications founded on the TST approach.

Delany et al. (1986) give the following expressions of TST-based absolute reaction rates:

$$v_j = sk_+ q_+ \left[1 - \exp\left(\frac{A_+}{\eta RT}\right)\right] \quad (8.308)$$

$$-v_j = sk_- q_- \left[1 - \exp\left(\frac{A_+}{\eta RT}\right)\right]. \quad (8.309)$$

Equation 8.308, representing the forward reaction rate, is written in terms of forward direction parameters: A_+ is thermodynamic affinity (cf. section 2.12), q_+ is the kinetic activity product for the left side of the microscopic reaction (i.e., in the case of reaction 8.297, $q_+ = a_{SiO_2} \cdot a_{H_2O}^{\nu}$), and η is a stoichiometric factor relating the stoichiometric coefficient of the solid reactants in the macroscopic and microscopic reaction notations (in the case of reactions 8.296 and 8.297, this term is 1, because both reactions have the same stoichiometric coefficient.

8.21.2 Temperature and Pressure Dependence of Rate Constants

If equilibrium in a system is perturbed for some reason, attainment of the new equilibrium requires a certain amount of time t, depending on the rates of the single reactions, and a certain amount of energy spent in activating the reaction process. The relationship between reaction rate k and activation energy E_a has the exponential form

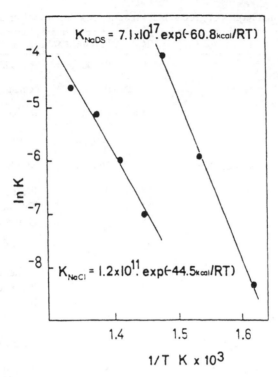

Figure 8.34 Arrhenius plot of reaction rates for equilibrium of analcite + quartz ⇔ albite in NaCl, and NaDS-bearing solutions, after Matthews (1980). Preexponential factors and activation energies can be deduced from the fitting expression. NaDS ≡ $Na_2Si_2O_5$.

$$k = k_0 \exp\left(\frac{-E_a}{RT}\right). \tag{8.310}$$

In equation 8.310, k_0 is the preexponential factor. Treating the preexponential factor as not dependent on T, with a partial derivative in $d(1/T)$ we have

$$\frac{\partial \ln k}{\partial(1/T)} = -\frac{E_a}{R}. \tag{8.311}$$

Plotting the natural logarithm of the reaction rate against the reciprocal of absolute temperature (Arrhenius plot), the activation energy of the reaction is represented by the slope of equation 8.311 at the T of interest multiplied by $-R$. If E_a is constant over T, the semilogarithmic plot gives rise to a straight line, as illustrated in figure 8.34.

Activation energies of some important geochemical reactions are listed in table 8.27.

Table 8.27 Activation energies of geochemical reactions (kJ/mole) (after Lasaga, 1981a, 1984)

Reaction	E_a	T Range (°C)	Reference
$H_4SiO_4 \rightarrow SiO_2 + 2H_2O$ quartz	50	0–300	Rimstidt and Barnes (1980)
$SiO_2 + 2H_2O \rightarrow H_4SiO_4$ quartz	67–75	0–300	Rimstidt and Barnes (1980)
$SiO_2 + 2H_2O \rightarrow H_4SiO_4$ (am)	61–65	0–300	Rimstidt and Barnes (1980)
Porcelanite $\rightarrow$ Quartz (chert) (hydrothermal)	97	300–500	Ernst and Calvert (1969)
$NaAlSi_2O_6 \cdot H_2O + SiO_2 \rightarrow NaAlSi_3O_8 + H_2O$ analcite quartz albite	186	300–500	Matthews (1980)
Anorthite dissolution	35	25–70	Fleer (1982)
Nepheline dissolution	54–71	25–80	Tole and Lassaga (unpublished)
$Mg_2SiO_4 + 4H^+ \rightarrow 2Mg^{2+} + H_4SiO_4$ forsterite	38	–	Grandstaff (1980)
$Mg_{0.77}Fe_{0.23}SiO_3$ dissolution orthopyroxene	44	–	Grandstaff (1977)
Diopside dissolution	50–150	–	Berner et al. (1980)
Enstatite dissolution	50	–	Berner et al. (1980)
Augite dissolution	79	–	Berner et al. (1980)
$CaCO_3 + H^+ \rightarrow Ca^{2+} + HCO_3^-$ calcite	8.4	5–48	Plummer et al. (1978)
$CaCO_3 + H_2CO_3^0 \rightarrow Ca^{2+} + 2HCO_3^-$ calcite	42	5–48	Plummer et al. (1978)
$CaCO_3 + H_2O \rightarrow Ca^{2+} + HCO_3^- + OH^-$ calcite	6.3 33	5–25 25–48	Plummer et al. (1978)
$CaCO_3 \rightarrow Ca^{2+} + CO_3^{2-}$ calcite	35	5–50	Sjoberg (1976)
$Ca_2C_2O_6 + Mg^{2+} \rightarrow CaMgC_2O_6 + Ca^{2+}$ calcite dolomite	205	252–295	Katz and Matthews (1977)
$FeS + S \rightarrow FeS_2$ (am) pyrite	71	–	Rickard (1975)
$^{32}SO_4 + H_2{}^{34}S \rightarrow {}^{34}SO_4 + H_2{}^{32}S$	7.5 126 201	pH < 3 4 < pH < 7 pH > 7	Ohmoto and Lasaga (1982)

Because reaction rates are also pressure-dependent (although they are much less P-dependent than T-dependent), the volume of activation of reaction V_a can be deduced in an analogous fashion from the partial derivative

$$\frac{\partial \ln k}{\partial P} = -\frac{V_a}{RT}. \tag{8.312}$$

8.22 Chemistry of Seawater: Salinity, Chlorinity, Alkalinity, and pH

The main saline constituents dissolved in seawater are Cl^-, SO_4^{2-}, Mg^{2+}, K^+, Ca^{2+}, and Na^+. Their concentration ratios are remarkably constant, as is the bulk

Table 8.28 Composition and chlorinity of seawater for a salinity of 35 (Stumm and Morgan, 1981). (1): weight concentration in g/kg; (2): weight concentration/chlorinity; (3): molarity/chlorinity.

Solute	(1)	(2)	(3)
Na^+	10.77	0.556	0.0242
Mg^{2+}	1.29	0.068	0.0027
Ca^{2+}	0.4121	0.02125	0.000530
K^+	0.399	0.0206	0.000527
Sr^{2+}	0.0079	0.00041	0.0000047
Cl^-	19.354	0.9989	0.0282
SO_4^{2-}	2.712	0.1400	0.0146
HCO_3^-	0.1424	0.00735	0.00012
Br^-	0.0673	0.00348	0.000044
F^-	0.0013	0.000067	0.0000035
B	0.0045	0.000232	0.0000213
Total	35.1605	1.8163	0.0710

amount of salts. This parameter is usually expressed as *salinity* (*S*) and represents the percent by weight of dissolved inorganic matter per 10^3 g of seawater. In the salinity calculation, the amount of Br^- and I^- is replaced by an equivalent amount of Cl^-, and carbonate ions HCO_3^- and CO_3^{2-} are converted into oxides. The *chlorinity* of seawater (*Cl*) is the amount of Cl^- and other halogens content (expressed in g per kg of seawater), equivalent to the total amount of alkalis. This parameter is practically determined by titration with $AgNO_3$, and the current definition of chlorinity is "weight in g of Ag necessary to precipitate the halogens (Cl^-, Br^-, I^-) in 328.5233 g of sea water" (Stumm and Morgan, 1981). Table 8.28 lists the mean composition of seawater at salinity 35. The first data column lists the amount of salts in g per kg of solution. In column (2), weight is related to chlorinity; column (3) shows the molarity/chlorinity ratio.

The salinity of seawater ranges from 33 to 37. Its value can be derived directly from chlorinity through the relationship

$$S = 1.805Cl + 0.30. \tag{8.313}$$

However, salinity values are easily obtained with a salinometer (which measures electrical conductivity and is appropriately calibrated with standard solutions and adjusted to account for *T* effects). The salinity of seawater increases if the loss of H_2O (evaporation, formation of ice) exceeds the atmospheric input (rain plus rivers), and diminishes near deltas and lagoons. Salinity and temperature concur antithetically to define the density of seawater. The surface temperature of the sea reflects primarily the latitude and season of sampling. The vertical thermal profile defines three zones: surface (10–100 m), where *T* is practically constant; thermoclinal (100–1000 m), where *T* diminishes regularly with depth; and abyssal

(> 1000 m), with constant low T (2–4 °C). Density at atmospheric level is usually denoted by parameter σ:

$$\sigma = \left[\text{Density} \frac{\text{kg}}{\text{dm}^3} - 1 \right] \times 1000 \qquad (8.314)$$

(thus, for seawater with density 1.025 at atmospheric level, $\sigma = 25$).

The density of seawater varies from a maximum of $\sigma = 29$, observed in deep antarctic waters, to a minimum of $\sigma = 25$ in subtropical oceanic thermoclinal waters. The high density of polar waters causes them to sink beneath subtropical waters and constitutes the driving force of deep oceanic circulation.

The HCO_3^- amount listed in table 8.28 actually represent carbonate alkalinity, calculated neglecting CO_3^{2-}. The alkalinity [Alk] of a solution represents its capacity to neutralize strong acids in solution. Alkalinity can be defined as "the sum of equivalents of all species whose concentrations depend on the concentration of H^+ ions in solution, minus the concentration of H^+ ions in solution." For a solution containing borates and carbonates, alkalinity can be expressed as

$$[\text{Alk}] = [HCO_3^-] + 2[CO_3^{2-}] + [B(OH)_4^-] + [OH^-] - [H^+]. \qquad (8.315)$$

where the terms in square brackets denote molality. If we consider only carbonate equilibria, equation 8.315 reduces to

$$[\text{Alk}] = [HCO_3^-] + 2[CO_3^{2-}] + [OH^-] - [H^+], \qquad (8.316)$$

and, in the absence of CO_3^{2-} (at the normal pH of seawater, the carbonate ion is virtually absent; see figure 8.21),

$$[\text{Alk}] = [HCO_3^-] + [OH^-] - [H^+]. \qquad (8.317)$$

from which we obtain

$$[HCO_3^-] \cong [\text{Alk}]. \qquad (8.318)$$

Figure 8.35 shows the redox state and acidity of the main types of seawaters. The redox state of normal oceanic waters is almost neutral, but they are slightly alkaline in terms of pH. The redox state increases in aerated surface waters. Seawaters of euxinic basins and those rich in nutrients (eutrophic) often exhibit Eh-pH values below the sulfide-sulfate transition and below carbonate stability limits (zone of organic carbon and methane; cf. figure 8.21). We have already seen (section 8.10.1) that the pH of normal oceanic waters is buffered by carbonate equilibria. At the normal pH of seawater (pH $\cong$ 8.2), carbonate alkalinity is 2.47 mEq per kg of solution.

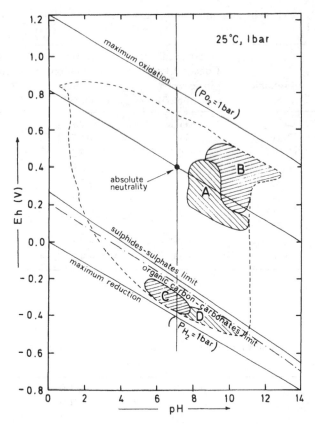

Figure 8.35 Mean Eh-pH values for various types of seawaters. A = normal oceanic waters; B = oxidized surface waters; C = euxinic basins; D = eutrophic waters. Dashed line: field of natural waters according to Baas Becking et al. (1960).

The relative constancy in the chemistry of the main saline components of seawater does not extend to minor and trace constituents, which are somewhat variable.

Interpretation of the bulk chemistry of seawater is based on five types of general considerations:

1. Chemical equilibrium between seawater and sediments
2. Exchanges with the atmosphere
3. Water-rock interactions
4. Interactions between biologic and mixing cycles
5. Kinetics of the inputs of primary constituents.

The above-listed factors concur simultaneously to define the stationary chemistry of seawater. Moreover, the control operated by the various phenomena is selective for the various types of elements: for instance, the amounts of Ca are largely controlled by precipitation of carbonates, those of Na and K by silicate hydroly-

Table 8.29 Trace element solubility control at $T = 25$ °C and $P = 1$ bar (from Turekian, 1969).

Element	Insoluble Salt	Expected Concentration in Seawater, log(mole/l)	Observed Concentration in Seawater, log(mole/l)	Observed Concentration in Reducing Muds, log(mole/l)
Lanthanum	$LaPO_4$	−11.1	−10.7	-
Thorium	$Th_3(PO_4)_4$	−11.8	−11.7	-
Cobalt	$CoCO_3$	−6.5	−8.2	−12.1
Nickel	$Ni(OH)_2$	−3.2	−6.9	−10.7
Copper	$Cu(OH)_2$	−5.8	−7.3	−26.0
Silver	$AgCl$	−4.2	−8.5	−19.8
Zinc	$ZnCO_3$	−3.7	−6.8	−14.1
Cadmium	$CdCO_3$	−5.0	−9.0	−16.2
Mercury	$Hg(OH)_2$	+1.9	−9.1	−43.7
Lead	$PbCO_3$	−5.6	−9.8	−16.6

Table 8.30 Mean composition of seawater for a salinity of 35 (Turekian, 1969). Values expressed in ppb.

H	1.10×10^8	Ti	1.0	Ba	21
He	0.0072	V	1.9	La	0.0034
Li	170	Cr	0.2	Ce	0.0012
Be	0.0006	Mn	0.4	Pr	0.00064
B	4.450	Fe	3.4	Nd	0.0028
C (inorganic)	28,000	Co	0.39	Sm	0.00045
C (organic solute)	500	Ni	6.6	Eu	0.000130
N (dissolved as N_2)	15.500	Cu	0.9	Gd	0.00070
N (as NO_2^-, NO_3^-, NH_4^+, and inorganic solute)	670	Zn	0.003	Tb	0.00014
O (dissolved as O_2)	6,000	As	2.6	Ho	0.00022
O (as H_2O)	8.33×10^8	Se	0.090	Er	0.00087
F	1,300	Br	67,300	Tm	0.00017
Ne	0.120	Kr	0.21	Yb	0.00082
Na	1.08×10^7	Rb	120	Lu	0.00015
Mg	1.29×10^6	Sr	8,100	Hf	< 0.008
Al	1	Y	0.0013	Ta	< 0.0025
Si	2,900	Zr	0.026	W	< 0.001
P	88	Nb	0.015	Rh	0.0084
S	9.04×10^5	Mo	10	Au	0.011
Cl	1.94×10^7	Ru	0.0007	Hg	0.15
Ar	450	Ag	0.28	Pb	0.03
K	3.92×10^5	Cd	0.11	Bi	0.02
Ca	4.11×10^5	Sn	0.81	Ra	1×10^{-7}
Sc	< 0.004	Sb	0.33	Th	0.0004
I	0.64	Xe	0.047	Pa	2×10^{-10}
Cs	0.30	U	3.3		

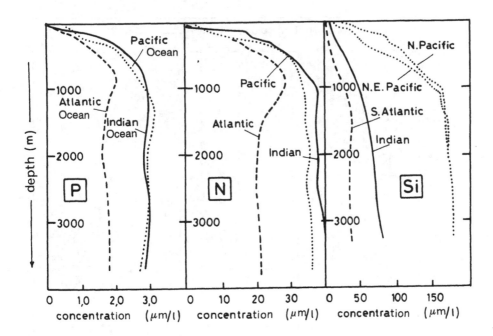

Figure 8.36 Changes in quantities of phosphorus, nitrogen, and silicon in seawater with depth.

sis, SiO_2 by silicate hydrolysis plus biogenic fixation (plankton), Mn by water-rock interactions in mid-ocean ridge zones, S by evaporite precipitation, and so on.

It is evident from the considerations above that we cannot expect the minor saline constituents of seawater to reflect the solubility product of their primary salts. For example, silver, lead, mercury, and copper form chlorides that are highly soluble in water. Nevertheless, their observed concentrations in seawater are much lower (i.e., by 3 or 4 orders of magnitude) than would be expected from the solubility products. Analogously, the solubility products of carbonates and insoluble hydroxides of transition elements suggest elemental concentrations higher than those actually observed. Table 8.29 compares hypothetical concentrations derived from the solubility products of insoluble salts with actual concentrations in oxidized surface waters and reducing muds (below the sulfide-sulfate transition). The expected concentration values were calculated by Turekian (1969), adopting the following activities of anionic ligands in solution: $\log[PO_4^{2-}] = -9,3$, $\log[CO_3^{2-}] = -5,3$; $\log[OH^-] = -6$; $\log[S^{2-}] = -9$.

Note that the observed concentrations of La and Th in seawater are in fact near the values expected from the solubility of phosphates (La and Th are effectively fixed as phosphates in almyrolithic exchanges between biogenic sediments and seawater), but the concentrations of the remaining elements are far lower than the values dictated by the solubility products.

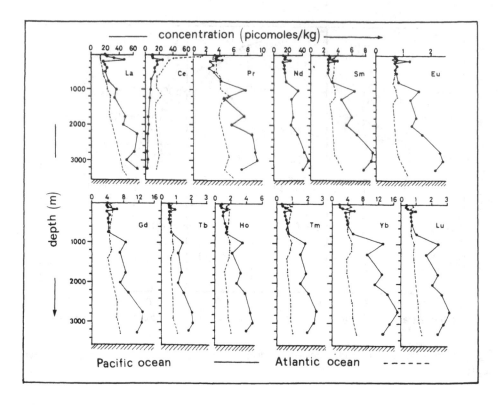

Figure 8.37 REE concentration profiles with depth in Atlantic and Pacific Oceans.

Table 8.30 shows the chemistry of seawater compiled by Turekian (1969) for major, minor, and trace constituents, expressed in parts per billion (ppb) at a mean salinity of 35. The listed values are estimates of mean amounts in solution, whereas elemental concentrations actually vary with depth. The most conspicuous variations are observed in the first 200 m from the surface, where photosynthetic processes are dominant and phosphorus and nitrogen are fixed by plankton and benthos, as well as silica and calcium, which constitute, respectively, the skeletons of planktonic algae (diatom) and the shells of foraminifera and mollusks.

Figure 8.36 shows how the elemental amounts of P (generally present as complexes of the phosphate ion PO_4^{3-}), N (present as NO_3^- complexes), and Si vary with depth in the various oceanic basins. Concentrations are expressed in micromoles per liter of solution. Note the relative constancy of concentrations below 1000 m.

Trace elements also exhibit systematic changes in concentration with depth. Figure 8.37 shows, for instance, concentration profiles of rare earth elements (REE) determined by De Baar et al. (1985) in Pacific and Atlantic waters. Note that the concentration profiles differ for the various elements in the series; in particular, the amount of Ce is quite high in surface waters of the Atlantic Ocean.

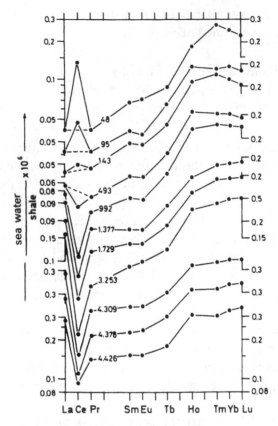

Figure 8.38 Ce anomalies in normalized REE patterns of seawater at various depths. Reproduced with modifications from H. J. W. DeBaar, M. P. Bacon, P. G. Brewer, and K. W. Bruland, *Geochimica et Cosmochimica Acta*, 49, 1943–1960, copyright © 1985, with kind permission from Elsevier Science Ltd., The Boulevard, Langford Lane, Kidlington 0X5 1GB, UK.

Figure 8.38 shows that the anomalous concentration of Ce results in a positive Ce anomaly in normalized REE patterns (relative abundances arranged in order of increasing atomic number). This anomaly, which may be ascribed to scavenging by Mn-Fe hydroxides, decreases progressively and inverts with depth.

8.23 Steady State Models and the Concept of "Time of Residence"

The kinetic interpretation of the chemistry of oceanic waters (kinetics of inputs of primary constituents; interactions between biologic and mixing cycles) leads to the development of *steady state models*, in which the relatively constant chemistry of seawater in the recent past (i.e., Phanerozoic; cf. Rubey, 1951) represents a condition of kinetic equilibrium among the dominant processes. In a system at steady state, the inputs of matter exactly balance the outputs. In the "seawater system," inputs are constituted primarily by:

1. Dissolved and particulate matter of rivers
2. Mid-ocean-ridge thermal springs and submarine eruptions
3. Atmospheric inputs.

Primary outputs are produced essentially by sedimentation and (to a much lower extent) by emissions in the atmosphere. The steady state models proposed for seawater are essentially of two types: "box models" and "tube models." In box models, oceans are visualized as neighboring interconnected boxes. Mass transfer between these boxes depends on the mean residence time in each box. The difference between mean residence times in two neighboring boxes determines the rate of flux of matter from one to the other. The box model is particularly efficient when the time of residence is derived through the chronological properties of first-order decay reactions in radiogenic isotopes. For instance, figure 8.39 shows the box model of Broecker et al. (1961), based on ^{14}C. The $^{14}C/^{12}C$ ratio, normalized to the atmospheric value, is determined for each box. According to the measured $^{14}C/^{12}C$ ratios, the maximum time of residence in each box is 1000 years, whereas the bulk time of residence in the oceans (time elapsed between precipitation plus river inputs and evaporation plus transport to continents) is 40,000 years.

The "tube models" consider the oceanic mass of water as subdivided into columns. Mass transfer between columns takes place by advection and diffusion. Examples of tube models may be found in Munk (1966) and Bieri et al. (1966), to whom we refer readers for further clarification.

The thermal motion of molecules of a given substance in a solvent medium causes dispersion and migration. If dispersion takes place by intermolecular forces acting within a gas, fluid, or solid, "molecular diffusion" takes place. In a turbulent medium, the migration of matter within it is defined as "turbulent diffusion" or "eddy diffusion." Diffusional flux J_d is the product of linear concentration gradient dC/dX multiplied by a proportionality factor generally defined as "diffusion coefficient" (D) (see section 4.11):

$$J_d = -D \times \frac{dC}{dX}.$$ (8.319)

"Advection" is generally defined as the displacement of a portion of matter with respect to the point of observation under the influence of force fields (Lerman, 1979). An advection flux [J_a; g/(sec $\times$ cm^2)] of a material with density ρ (g/cm^3) moving at velocity V (cm/sec) with respect to the point of observation is given by

$$J_a = \rho \times V.$$ (8.320)

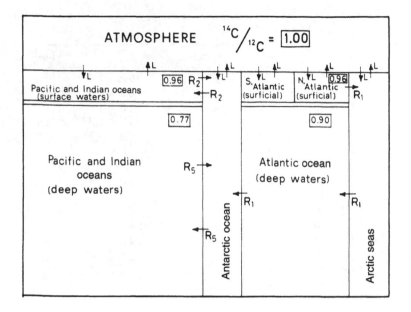

Figure 8.39 Box model for steady state chemistry of seawater. Numbers in boxes: $^{14}C/$ ^{12}C ratio normalized to atmospheric value. L: specific transfer rate between atmosphere and ocean; R: transfer rates between oceanic reservoirs. Reprinted from W. S. Broecker, R. D. Gerard, M. Ewing, and B. C. Heezen, Geochemistry and physics of ocean circulation. In *Oceanography,* M. Sears, ed., copyright © 1961 American Association for the Advancement of Science. Reprinted with permission of the American Association for the Advancement of Science, Washington.

Steady state models are based on the concept of the *mean time of residence* of the elements in the system. Mean time of residence τ for a steady state is defined as

$$\tau = \frac{\text{mass in the system}}{\text{unit time input}} = \frac{\text{mass in the system}}{\text{unit time output}}. \qquad (8.321)$$

Indexing as J_i the various input and output fluxes, we can write

$$\tau = \frac{M}{\sum_i J_i}, \qquad (8.322)$$

where mass M and fluxes have the same units (e.g., M = moles, J_i = moles/year). The "fractional time of residence" is relative to a given input:

$$\tau_i = \frac{M}{J_i}, \qquad (8.323)$$

Table 8.31 Times of residence of elements in seawater in years (data from Brewer, 1975; Stumm and Morgan, 1981).

Element(s)	t	log t	Element(s)	t	log t
Cl, Br	1×10^8	8	Ni	9.0×10^4	5.0
Na	6.8×10^7	7.8	V, Hg	8.0×10^4	4.9
B	1.3×10^7	7.1	As	5.0×10^4	4.7
Mg	1.2×10^7	7.1	Sc, Ag, Ba	4.0×10^4	4.6
K	7.0×10^6	6.8	Cu, Zn, Se	2.0×10^4	4.3
Rb, Sr	4.0×10^6	6.6	Si	1.8×10^4	4.2
U	3.0×10^6	6.5	Ti	1.3×10^4	4.1
Li, N	2.3×10^6	6.4	Ga, Mn	1.0×10^4	4.0
Ca	1.0×10^6	6.0	Sb	7.0×10^3	3.8
Cs	6.0×10^5	5.8	Cr	6.0×10^3	3.8
F	5.2×10^5	5.7	La	6.0×10^2	2.8
I	4.0×10^5	5.6	Pb	4.0×10^2	2.6
Mo, Au	2.0×10^5	5.3	Fe	2.0×10^2	2.3
P	1.8×10^5	5.2	Al	1.0×10^2	2.0
W	1.2×10^5	5.1	Th	6.0×10^1	1.8

and the "mean time of residence" is given by

$$\frac{1}{\tau} = \frac{1}{\tau_1} + \frac{1}{\tau_2} + \cdots \frac{1}{\tau_n}. \tag{8.324}$$

Indexing input and output fluxes as J_{in} and J_{out}, respectively, the mass variation in time dt is given by

$$\frac{dM}{dt} = J_{in} - J_{out}. \tag{8.325}$$

Assuming the removal rate of a given element from the system to be proportional (at first approximation) to the elemental abundance in the system, through proportionality constant K_t:

$$J_{out} = K_t \times M, \tag{8.326}$$

at steady state we have

$$\frac{dM}{dt} = 0; \quad J_{in} = K_t \times M, \tag{8.327}$$

from which we obtain

$$\frac{1}{K_t} = \frac{M}{J_{in}} = \tau. \tag{8.328}$$

The time of residence thus represents the reciprocal of the constant of removal from the system. The times of residence of elements in the "seawater system" are generally evaluated on the basis of equation 8.327. The main input flux is fluvial: the fractional time of residence based on fluvial input is thus the primary constituent of the mean time of residence (cf. eq. 8.322). An alternative way of evaluating the time of residence is based on output flux due to sedimentation, applying equation 8.326 (the two methods give concordant results). Table 8.31 lists times of residence of elements in seawater, expressed in years and arranged in decreasing series. Long times of residence (e.g., those of Cl, Br, Na, B, and Mg) are indicative of low reactivity in the system; short times of residence are connected with oversaturation and high reactivity (e.g., La, Pb, Al, Fe, and Th).

CHAPTER NINE

Geochemistry of Gaseous Phases

9.1 Perfect Gases and Gaseous Mixtures

The equation of state of perfect gases:

$$PV = nRT, \tag{9.1}$$

where n is the number of moles of component, can be applied to a gas phase composed of molecules moving freely as a result of thermal agitation, in the absence of mutual interactions. For a mixture of perfect gases the same law holds, taking into account the fact that, in this case, n represents the summation of all molecular species in mixture—i.e.,

$$n = \sum_i n_i = n\left(X_1 + X_2 + X_3 + \cdots + X_i\right), \tag{9.2}$$

where $X_1, X_2, \ldots, X_i$ are the molar fractions of the various gaseous species.

The activity of a perfect gas, as for any substance, is unitary, by definition, at standard state. Moreover, for a perfect gas, activity is (numerically) equivalent to pressure, at all pressures. Let us consider the relationship existing, with T held constant, between the chemical potential of component i in gaseous phase g at 1 bar ($\mu_{i,1,T,g}$) and at pressure P ($\mu_{i,P,T,g}$):

$$\mu_{i,P,T,g} = \mu_{i,1,T,g} + \int_1^P \overline{V}_{i,g}\, dP. \tag{9.3}$$

Because for a pure phase partial molar volume $\overline{V}_{i,g}$ is equivalent to molar volume $\overline{V}_g$, and since we have defined as standard state the condition of the pure component at $P = 1$ bar and T of interest:

$$\mu_{i,1,T,g} \equiv \mu_{i,g}^0, \tag{9.4}$$

we have

$$\mu_{i,P,T,g} = \mu_{i,g}^0 + \int_1^P \overline{V}_g \, dP. \qquad (9.5)$$

Because

$$RT \ln a_{i,g} = \mu_{i,P,T,g} - \mu_{i,g}^o, \qquad (9.6)$$

from equations 9.1, 9.5, and 9.6 for $n = 1$, we obtain

$$RT \ln a_{i,g} = \int_1^P RT \frac{dP}{P}, \qquad (9.7)$$

which, when integrated, gives

$$RT \ln a_{i,g} = RT \ln P. \qquad (9.8)$$

Hence

$$a_{i,g} = P. \qquad (9.9)$$

In the more general case of a standard state pressure of reference different from 1, we have

$$RT \ln a_{i,g} = RT \ln\left(\frac{P}{P_r}\right) \qquad (9.10)$$

and

$$a_{i,g} = \frac{P}{P_r}. \qquad (9.11)$$

A real gas does not obey the equation of state of perfect gases (eq. 9.1) because of the interactions among gaseous molecules. The higher the pressure and the lower the temperature, the greater the deviation from ideality. At very low P, intermolecular distances are great and interactions between gaseous molecules are virtually absent. For almost all gaseous species of geochemical interest, the approximation to a perfect gas is acceptable only over a very small P range.

Interaction forces among gaseous molecules are generally subdivided into

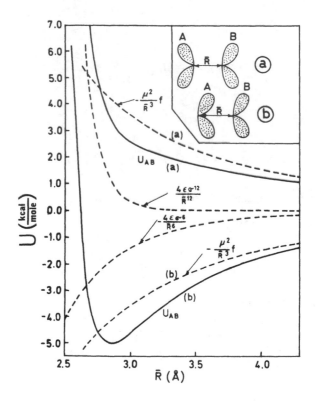

Figure 9.1 Molecular interaction potentials in Stockmayer's (1941) model for H_2O vapor. (a): antiparallel dipolar moments; (b): parallel dipolar moments. Reprinted from D. Eisemberg and W. Kauzmann, *The Structures and Properties of Water,* 1969, by permission of Oxford University Press.

long-range forces (electrostatic, inductive, and dispersive forces) and *short-range forces* (repulsive forces). Figure 9.1 shows the energy potentials of two gaseous molecules of H_2O (A-B), according to the classical model of Stockmayer (1941).

Bulk potential U_{AB} obeys the equation

$$U_{AB} = -\frac{\mu^2 f}{\overline{R}^3} - \frac{4\varepsilon\sigma^6}{\overline{R}^6} + \frac{4\varepsilon\sigma^{12}}{\overline{R}^{12}}. \qquad (9.12)$$

The first term on the right in the sum of potentials in equation 9.12 is the electrostatic term (represented here as a simple dipole moment). Factor f is the angular function of dipole orientation:

$$f = \sin\theta_A \sin\theta_B \cos(\phi_A - \phi_B) - 2\cos\phi_A \cos\phi_B. \qquad (9.13)$$

Angles θ and ϕ in equation 9.13 define the dipole orientation with respect to a coordinate system centered on the oxygen nuclei. The second term on the right in equation 9.12 represents the dispersive energy (independent of molecular orientation), and the third term represents the repulsive interactions (also independent of orientation). σ is the "collision diameter of the gaseous molecule" and represents the distance at which attractive dipole moments and repulsive energy terms mutually compensate (i.e., when $\overline{R} = \sigma$, the series of potentials of eq. 9.12 reduces to the first term), and ε represents the "characteristic energy" of the gaseous molecule. It is of interest to recall that the σ value observed by Stockmayer (1941) for gaseous H_2O is identical to the hydrogen-bond distance in ice I (Eisemberg and Kauzmann, 1969).

Note in figure 9.1 that bulk potential U_{AB} depends greatly on the mutual orientation of dipoles: molecules with parallel dipoles form more stable bonds than molecules with antiparallel dipoles. Note, moreover, as already stated, that repulsive and dispersive forces are not affected by the state of dipole orientation.

Deviations from the equation of state of perfect gases may be expressed in terms of either pressure or volume. It is obvious that specific interactions among the molecules of a real gas are translated into modifications of volume with respect to the volume occupied by n moles of an ideal perfect gas at the same temperature T, or, T and volume being held constant, into modifications of pressure exercised by the gaseous system toward the exterior. Analytically, this may be expressed through the concept of "fugacity." Fugacity f_i of a real pure gas i is obtained through

$$RT \ln f_i = RT \ln P + \int_{P_r}^{P} \left(\overline{V}_i - \overline{V}_{\text{perfect}} \right) dP, \tag{9.14}$$

where $\overline{V}_i$ is the molar volume of the real gas and $\overline{V}_{\text{perfect}}$ is the volume occupied by one mole of perfect gas under the same P and T conditions. The relationship between fugacity f_i and pressure P_i of a real gas i is given by

$$f_i = P_i \Gamma_i. \tag{9.15}$$

In equation 9.15, Γ_i is the "fugacity coefficient":

$$\Gamma_i = \exp\left[\int_{P_r}^{P} \frac{\left(\overline{V}_i - \overline{V}_{\text{perfect}} \right) dP}{RT} \right]. \tag{9.16}$$

Because Γ_i is adimensional, it is evident that fugacity has the dimension of pressure. It is important to emphasize that it is conceptually erroneous to consider the fugacity of a gaseous species as the equivalent of the thermodynamic activity of a solid or liquid component (as is often observed in literature). The relationship between activity a_i and fugacity f_i of a gaseous component is given by

$$a_i = \frac{f_i}{f_i^0}, \tag{9.17}$$

where f_i^0 is fugacity at standard state. Fugacity $f_i^0 = 1$ only when $P_i = 1$ bar and T and P are such that $\Gamma_i \equiv 1$ (cf. eq. 9.15). If the adopted standard state is, for instance, pure gas at the P and T of interest and P is high (or T is low), $f_i^0 \neq 1$, and hence $a_i \neq f_i$.

9.2 Real Gases at High P and T Conditions

Figure 9.2 shows how fugacity coefficient Γ_i of H_2, O_2, and CO_2 behaves with increasing P at various T (0 °C, 100 °C, 200 °C). Tables 9.1 and 9.2 list discrete values attained by Γ_i for H_2O and CO_2 gases under various P and T conditions.

Values of Γ_i for gases of geochemical interest for which experimental data are

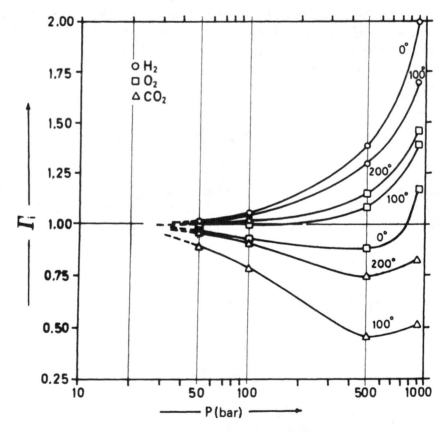

Figure 9.2 Fugacity coefficients of H_2, O_2, and CO_2 plotted as functions of P at various T. Reprinted from Garrels and Christ (1965) with kind permission from Jones and Bartlett Publishers Inc. Copyright © 1990.

Table 9.1 Fugacity coefficients of CO_2 under various P and T conditions (Mel'nik, 1972).

P (atm)	T (K)											
	400	500	600	700	800	900	1000	1100	1200	1300	1400	1500
500	0.57	0.85	1.02	1.11	1.15	1.17	1.19	1.20	1.20	1.20	1.20	1.20
1,000	0.63	0.95	1.16	1.28	1.34	1.37	1.38	1.38	1.37	1.37	1.37	1.37
1,500	0.82	1.20	1.44	1.56	1.62	1.62	1.62	1.59	1.59	1.57	1.55	1.53
2,000	1.16	1.60	1.85	1.95	2.00	1.95	1.91	1.86	1.82	1.78	1.76	1.71
3,000	2.46	3.00	3.18	3.15	2.95	2.08	2.46	2.34	2.29	2.19	2.09	2.04
4,000	5.49	5.60	5.60	5.18	4.67	3.89	3.55	3.31	3.16	2.95	2.75	2.63
5,000	12.5	11.3	9.88	8.52	6.92	5.75	5.01	4.57	4.27	3.89	3.63	3.39
6,000	28.4	22.1	17.4	13.9	10.5	8.32	7.24	6.31	5.75	5.13	4.68	4.27
7,000	64.2	42.7	30.2	22.4	16.2	12.6	10.5	8.91	7.94	6.92	6.17	5.50
8,000	143	81.4	51.8	35.7	24.6	18.6	15.1	12.6	10.7	9.12	7.94	7.08
9,000	319	154	88.1	56.3	38.0	28.8	22.4	17.8	14.8	12.3	10.5	9.33
10,000	703	289	148	80.9	56.2	41.7	31.6	24.6	20.0	16.2	13.5	11.8

Table 9.2 Fugacity coefficients of H_2O ($\Gamma_i \times 10^2$; Helgeson and Kirkham, 1974).

T (°C)	P (kbar)									
	0.5	1	2	3	4	5	6	7	8	9
200	3.66	2.36	0.19	0.20	0.24	0.30	0.38	0.50	0.66	0.89
250	8.59	5.46	0.43	0.45	0.51	0.62	0.77	0.97	1.25	1.62
300	16.59	10.50	0.82	0.83	0.92	10.9	13.2	16.3	20.5	25.9
350	27.63	17.52	13.5	13.5	14.8	17.1	20.3	24.6	30.3	37.5
400	40.83	26.22	20.2	19.9	21.6	24.5	28.7	34.3	41.4	50.4
450	54.16	35.94	27.8	27.3	29.3	32.9	38.0	44.7	53.3	64.0
500	64.81	45.89	36.1	35.3	37.6	41.8	47.8	55.6	65.4	77.6
550	72.59	55.28	44.5	43.5	46.1	51.0	57.7	66.5	77.5	90.9
600	78.48	63.58	52.7	51.7	54.6	60.0	67.5	77.1	89.0	103.4
650	82.97	70.57	60.3	59.5	62.8	68.6	76.7	87.0	99.0	115.0
700	86.46	76.33	67.3	66.8	70.4	76.7	85.3	96	109.5	125.4
800	91.37	84.87	78.9	79.5	83.9	91.0	100.4	112	126.3	142.9

not available in the literature can be obtained through the "universal chart of gases" (figure 9.3). Construction of this chart is based on the fact that the thermodynamic properties of all real gases are similar functions of P and T, when the parameters of the equation of state are reduced with respect to the values of critical temperature T_c, critical pressure P_c, and critical volume V_c of the compound of interest. We can thus introduce the concepts of *reduced pressure* P_r, *reduced temperature* T_r, and *reduced volume* V_r:

$$P_{r,i} = \frac{P_i}{P_{c,i}} \qquad (9.18)$$

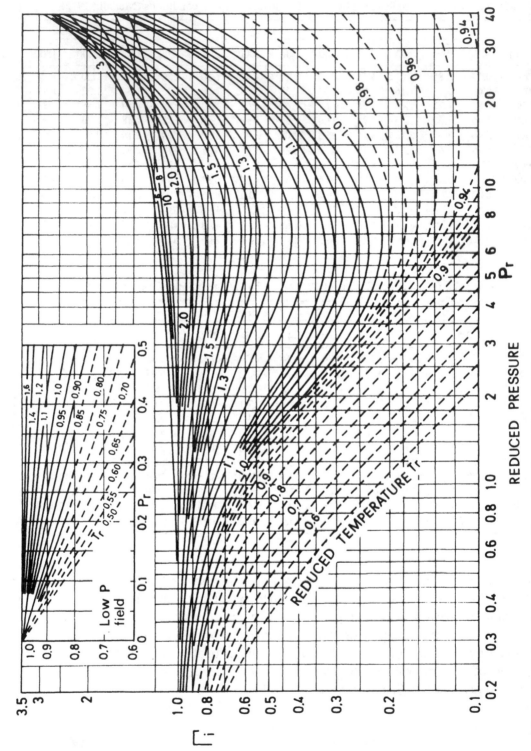

Figure 9.3 Universal chart of gases. From Garrels and Christ (1965); reprinted with kind permission from Jones and Bartlett Publishers Inc. Copyright © 1990.

Table 9.3 Critical temperatures (T_c) and critical pressures (P_c) for gaseous species of geochemical interest (from Garrels and Christ, 1965).

Compound	T_c (°C)	T_c (K)	P_c (atm)
CO_2	31.1	304.3	73
CS_2	273	546	76
CO	−139	134	35
COS	105	378	61
Cl_2	144.0	417.2	76.1
He	−267.9	5.3	2.26
H_2	−239.9	33.3	12.8
HBr	90	363	84
HCl	51.4	324.6	81.6
HI	151	424	82
H_2Se	138	411	88
H_2S	100.4	373.6	88.9
Hg	> 1550	> 1823	> 200
CH_2	−82.5	190.7	45.8
N_2	−147.1	126.1	33.5
O_2	−118.8	154.4	49.7
SiF_4	−1.5	271.7	50
SiH_4	−3.5	269.7	48
$SnCl_4$	318.7	591.9	37.0
S	1040	1313	−
SO_2	157.2	430.4	77.7
SO_3	218.3	491.5	83.6
H_2O	374.0	647.2	217.7

$$T_{r,i} = \frac{T_i}{T_{c,i}} \tag{9.19}$$

$$V_{r,i} = \frac{V_i}{V_{c,i}}. \tag{9.20}$$

Values of critical temperature and critical pressure for gaseous species of geochemical interest are listed in table 9.3.

The Γ_i values allow the calculation of fugacity and, thence, the chemical potential of the gaseous component at the P and T of interest, applying

$$\mu_{i,g} = \mu_{i,g}^0 + RT \ln\left(\frac{f_i}{f_i^0}\right). \tag{9.21}$$

9.3 The Principle of Corresponding States and Other Equations of State for Real Gases

As we have already seen, the universal chart of gases assumes that all gaseous species exhibit the same sort of deviation from ideal behavior at the same values of T_r, P_r, and V_r. This fact, known as the *principle of corresponding states,* is analytically expressed by the "deviation parameter" (or "compressibility factor") Z_g. For $n = 1$,

$$Z_g = \frac{PV}{RT} = f(P_r, T_r). \tag{9.22}$$

Confirmation of the validity of the principle of corresponding states is shown in figure 9.4A, where we see the analogies in the reduced isotherms of H_2O, CO_2, and N_2 according to reduced pressure. However, figure 9.4B shows that the principle of corresponding states is only an approximation of actual behavior in the vicinity of the critical point: in the case of H_2O, there are in fact discrepancies between predicted and actual behavior.

Deviation parameter Z_g is operationally expressed as a virial expansion on P or V:

$$Z_{g(P,T)} = 1 + BP + CP^2 + DP^3 + \cdots \tag{9.23}$$

$$Z_{g(V,T)} = 1 + B'V + C'V^{-2} + D'V^{-3} + \cdots. \tag{9.24}$$

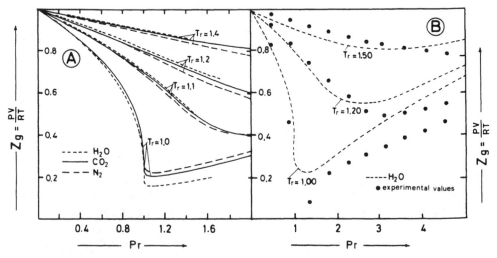

Figure 9.4 (A) Reduced isotherms of N_2, CO_2, and H_2O vapor according to reduced pressure. (B) Comparison between experimental isotherms of H_2O and T_r predicted by principle of corresponding states.

Virial expansion coefficients B, C, D and B', C', D' in equations 9.23 and 9.24 are simple functions of T and are specific to each gaseous species. Virial expansion equations such as equations 9.23 and 9.24 do not reproduce the behavior of real gases in the vicinity of the critical point and under high P and T conditions with satisfactory precision.

Deviations from ideality of real gases can also be expressed through the van der Waals equation:

$$\left(P + aV^{-2}\right)\left(V - b\right) = RT, \tag{9.25}$$

where term aV^{-2} is defined as "internal pressure" or "cohesive pressure" and term b as "co-volume." Cohesive pressure is the result of attractive forces among nonideal gaseous molecules; co-volume reflects the intermolecular distance at which repulsion between neighboring molecules becomes appreciable (see section 9.1).

If we define the distance between the centers of two gaseous molecules at the collision point as d, we have

$$b = \frac{4}{3}\pi d^{3} \tag{9.26}$$

and, for a rigid-sphere model,

$$b = \frac{32}{3}\pi r_g^3 , \tag{9.27}$$

where r_g is the radius of the gaseous molecule, conceived as a rigid sphere.

An alternative (and more accurate) form of equation of state for real gases is the Redlich-Kwong equation:

$$\left[P + \frac{a}{V\left(V + b\right)T^{1/2}}\right]\left(V - b\right) = RT. \tag{9.28}$$

Parameters a and b in equation 9.28 have the same significance as in equation 9.25. However, whereas in the van der Waals equation these parameters are constant for a given gaseous species, in the Redlich-Kwong formulation they are often assumed to be variable. For instance, Holloway (1977) adopted a coefficient a varying with T:

$$a = f(T). \tag{9.29}$$

Touret and Bottinga (1979) and Bottinga and Richet (1981) proposed

$$a = a_1\left[\left(\frac{a_3}{V}\right)^3 - \left(\frac{a_3}{V}\right)^6\right] + a_2 \tag{9.30}$$

and

$$b = \left[\ln\left(\frac{V}{a_3}\right) + b_1 \right] b_2^{-1}, \tag{9.31}$$

where a_1, a_2, a_3, b_1, and b_2 are adjustable coefficients for each gaseous species.

Kerrick and Jacobs (1982) adopted the Carnahan-Starling modification of the Redlich-Kwong equation, with polynomial expansions of coefficient a on V and T:

$$a = c + dV^{-1} + eV^{-2} \tag{9.32}$$

$$c = c_1 + c_2 T + c_3 T^2 \tag{9.33}$$

$$d = d_1 + d_2 T + d_3 T^2 \tag{9.34}$$

$$e = e_1 + e_2 T + e_3 T^2 \tag{9.35}$$

Halback and Chatterjee (1982) considered coefficient a as a function of T and coefficient b as a function of P:

$$a = a_1 + a_2 T + a_3 T^{-1} \tag{9.36}$$

$$b = \frac{\left(1 + b_1 P + b_2 P^2 + b_3 P^3\right)}{\left(b_4 + b_5 P + b_6 P^2\right)}. \tag{9.37}$$

More recently, Saxena and Fei (1987) proposed quantifying deviations from ideality of real gases through an expression they defined as "nonvirial":

$$Z_g = A + BP + CP^2 + \cdots. \tag{9.38}$$

The difference between equation 9.38 and the corresponding virial expansion of equation 9.23 lies in term A, which substitutes 1 and which is a function of T, as are the remaining coefficients. Equation 9.38 also finds application in a "corresponding states" notation:

$$Z_g = A + BP_r + CP_r^2 + \cdots. \tag{9.39}$$

Table 9.4 lists the parameters of the nonvirial expansion of equation 9.38 for H_2O and the parameters of the corresponding states notation for the remaining gaseous species.

The equation of state appropriate to a given gaseous species under high P and T conditions must be selected with care. It may generally be stated that,

Table 9.4 Parameters of nonvirial expansion of Saxena-Fei (from Ganguly and Saxena, 1987).

H_2O ($P > 1$ kbar; $T > 400$ K)	Corresponding States ($P > 1$ kbar; $T > 400$ K)	Corresponding States ($P < 1$ kbar)
A $\quad -0.7025 + 1.16 \times 10^{-3}T$ $+ 99.6799T^{-1}$	$1 - 0.5917T_r^{-2}$	1
B $\quad 0.2143T^{-1} - 3.1423 \times 10^{-14}T^3$	$0.09122T_r^{-1}$	$0.09827T_r^{-1} + 0.2709T_r^{-3}$
C $\quad -2.249 \times 10^{-6}T - 0.1459T^{-3}$ $+ 2.169 \times 10^{-15}T^2$	$-1.46 \times 10^{-4}T_r^{-2} - 2.8349$ $\times 10^{-6} \ln T_r$	$0.01472T_r^{-4} - 0.00103T_r^{-1.5}$

although the modified Redlich-Kwong equation and the nonvirial equation of Saxena-Fei are more precise than other formulations, calculations of the Redlich-Kwong equation are more difficult to perform. For instance, to evaluate the fugacity of a given gaseous phase at high P, based on

$$RT \ln f_i = \int_1^P \left(\overline{V}_i - \overline{V}_{\text{perfect}} \right) dP + RT \ln P, \tag{9.40}$$

it is necessary to perform an integration on volumes that in the case of the Redlich-Kwong formulation, is particularly complicated. As suggested by Ganguly and Saxena (1987), the Redlich-Kwong equation should be reserved for H_2O, and van der Waals-type equations should be adopted preferentially for CO_2, Ar, N_2, and CH_4 (Ross and Ree, 1980). Ganguly and Saxena (1987) also suggested adopting the model of Halback and Chatterjee (1982) for H_2O and the model of Bottinga and Richet (1981) or Kerrick and Jacobs (1982) for CO_2 at $P < 1$ kbar and $T < 400$ K. According to these authors, the nonvirial Saxena-Fei equation finds application under higher P and T conditions.

9.4 Mixtures of Real Gases: Examples in the C-O-H System

The calculation of the Gibbs free energy of a gaseous mixture can be performed, starting from the chemical potentials of the various pure gaseous components at the P and T of interest, assuming that the mixture is ideal:

$$\overline{G}_g = \sum_i \mu_{i,g}^0 X_{i,g} + \overline{G}_{\text{ideal mixing}} \tag{9.41}$$

$$\overline{G}_{\text{ideal mixing}} = RT \sum_i X_{i,g} \ln X_{i,g}. \tag{9.42}$$

However, this approximation is not valid in geochemistry, because the thermobaric regimes of interest are so wide that they prevent the assumption of ide-

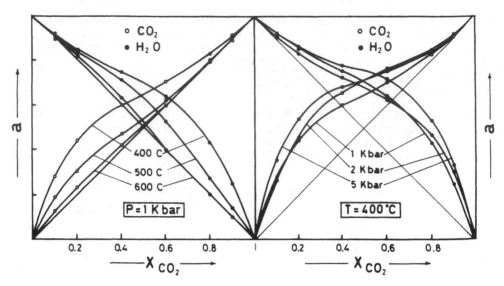

Figure 9.5 Activity-concentration relationships in binary gaseous H_2O-CO_2 mixtures. Experimental data from Shmulovich et al. (1982).

ality. Interactions among components in mixture cannot be neglected whenever the confining pressure of the system is particularly high (and the corresponding T not sufficiently high). Moreover, interactions at equal P and T conditions increase between nonpolar-nonpolar, polar-nonpolar, and polar-polar species, respectively (by "polar" we mean here a gaseous molecule with charge polarity).

Of particular interest in geochemistry are gaseous mixtures in the C-O-H system. The main gaseous species forming are H_2O, CO_2, O_2, CO, CH_4, and H_2. Among these, only H_2O exhibits charge polarity; the remaining ones are practically nonpolar (the polarity of CO is negligible for practical purposes). Figure 9.5 shows the activity diagram of the H_2O-CO_2 mixture, based on the experimental data of Shmulovich et al. (1982). The left part of the figure shows how, with P held constant, interactions decrease with increasing T, and the right part shows how, with T held constant, they increase with increasing P. The significance of this sort of deviation from ideal behavior has already been outlined in section 9.2.

As we see in figure 9.5, the mixing properties are slightly asymmetric and interactions are nonnegligible. To simulate the mixing properties of real gaseous species such as those of figure 9.5, we can combine the parameters of Stockmayer's potentials (see, for instance, Prausnitz et al., 1986) more or less empirically. For instance, if we define 1 and 2 as the interacting gaseous species, parameters $\overline{R}$, σ, ε, and μ in equation 9.12 are obtained by algebraic and geometric means:

$$\overline{R}_{12} = \frac{1}{2}\left(\overline{R}_{11} + \overline{R}_{22}\right) \tag{9.43}$$

$$\sigma_{12} = \frac{1}{2}\left(\sigma_{11} + \sigma_{22}\right) \tag{9.44}$$

$$\varepsilon_{12} = \left(\varepsilon_{11}\varepsilon_{22}\right)^{1/2} \tag{9.45}$$

$$\mu_{12} = \left(\mu_{11}\mu_{22}\right)^{1/2}. \tag{9.46}$$

Once potential U_{12}, with substitution of the various parameters in equation 9.12, has been derived, an interaction parameter W_{12} can be obtained by applying

$$W_{12} = z\left[U_{12} - \frac{1}{2}\left(U_{11} + U_{22}\right)\right] \tag{9.47}$$

(Guggenheim, 1952), where z is the number of reciprocal coordinations of species 1 and 2 in the gaseous mixture.

Saxena and Fei (1988) have shown that the observed deviations from ideality in H_2O-CO_2 mixtures can be obtained from interaction parameter W_{12} by applying

$$G_{\text{excess mixing}} = \frac{RTW_{12}X_1X_2}{\left(X_1q_1 + X_2q_2\right)}, \tag{9.48}$$

where q_1 and q_2 are the "efficacious volumes" of the gaseous molecules H_2O and CO_2. Because these parameters cannot be derived easily from first principles, the authors adopted the simplifications

$$\frac{q_1}{q_2} = \frac{V_1}{V_2} \tag{9.49}$$

and

$$q_1 + q_2 = 1, \tag{9.50}$$

where V_1 and V_2 are the volumes of molecules 1 and 2.

The activity coefficients arising from equation 9.48 are

$$\ln\gamma_1 = q_1 W_{12}\left(1 + \frac{q_1 X_1}{q_2 X_2}\right)^2 \tag{9.51}$$

and

$$\ln\gamma_2 = q_2 W_{12}\left(1 + \frac{q_2 X_2}{q_1 X_1}\right)^2. \tag{9.52}$$

Note that these coefficients are real "activity coefficients" and not "fugacity coefficients" (see section 9.1).

Table 9.5 Thermodynamic parameters for main gaseous species of geochemical interest. $\overline{H}^0_{f,P_r,T_r}$ in kJ/mole: $\overline{S}^0_{P_r,T_r}$ in J/(mole × K). $C_P = K_1 + K_2T + K_3T^2 + K_4T^{-1/2} + K_5T^{-2}$ (data from Robie et al., 1978). $T_r = 298.15$ K, $P_r = 1$ bar

Species	$\overline{H}^0_{f,P_r,T_r}$	$\overline{S}^0_{P_r,T_r}$	K_1	$K_2 \times 10^3$	$K_3 \times 10^6$	$K_4 \times 10^{-2}$	$K_5 \times 10^{-5}$
H_2O	−241.814	188.83	7.368	27.468	−4.8117	3.6174	−2.2316
CO_2	−393.510	213.79	87.82	−2.6442	0	−9.9886	7.0641
CO	−110.530	197.67	45.73	−0.097115	0	−4.1469	6.6270
O_2	0	205.15	48.318	−0.69132	0	−4.2066	4.9923
H_2	0	130.68	7.4424	11.707	−1.3899	4.1017	−5.1041
CH_4	−74.810	186.26	78.976	43.186	−10.598	−1.3202	18.836
H_2S	−20.627	205.80	26.356	26.497	−6.0244	0.43559	2.6599
N_2	0	191.61	23.941	11.068	−6.5518	0	1.9064
NH_3	−45.940	192.78	29.735	39.119	−8.2274	−1.4378	2.9243

Once the excess Gibbs free energy of mixing is known, the Gibbs free energy of the gaseous mixture is calculated as follows:

$$\overline{G}_g = \sum_i \mu^0_{i,g} X_{i,g} + \overline{G}_{\text{ideal mixing}} + \overline{G}_{\text{excess mixing}}.$$ (9.53)

The first term implies knowledge of the chemical potentials of the gaseous components at standard state. Thermodynamic parameters for the main gaseous species of geochemical interest at $T = 298.15$ K and $P = 1$ bar are listed in table 9.5. The molar volume in these T-P conditions for all species is that of an ideal gas—i.e., 24,789.2 cm^3 (or 2478.92 J/bar). The limiting T for the C_P function is 1800 K. Data are from Robie et al. (1978).

9.5 Volcanic Gases

The study of volcanic gases is of fundamental importance in understanding eruptive processes. The evolution of these processes is primarily conditioned by the solute states of the various volatile species in the magma, by the relationship between bulk pressure and partial pressure of the gaseous species above the solubility product, and by the (strongly exothermic) interactions between aquifers and rising magma (phreatomagmatic eruptions).

To understand fully the relationships existing between the dynamics of eruptive processes and the chemistry of gaseous phases, it is essential to assess the solubility relations of the various gaseous species in magmas. To do this, proper thermochemical models for silicate melts must be developed, in light of the plethora of petrologic information existing on such systems, focusing attention on the reciprocal effects of the presence of fluid phases on melt structure and of the chemistry of the melt on the solubility product of the fluid.

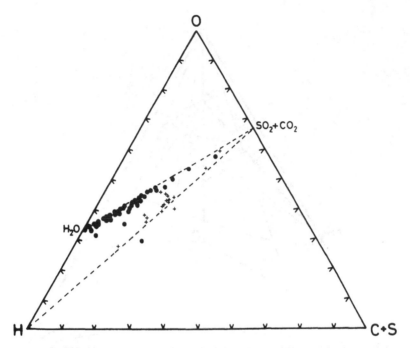

Figure 9.6 Ternary atomic composition diagram for volcanic gases. Crosses: samples from Mount Etna. From Gerlach and Nordlie (1975a), *American Journal of Science,* 275, 353–76. Reprinted with permission of American Journal of Science.

9.5.1 Atomic Compositions of Volcanic Gases

The atomic compositions of volcanic gases are conveniently described by the C-O-H-S system, because these four atoms concur to form almost all the gaseous molecules associated with magmas. Besides the above species, chlorine follows in order of abundance, mostly in the form of hydrochloric acid HCl, often conspicuous in low-T (i.e., below 600 °C) fumaroles. The amount of HCl in high-T volcanic gases does not generally exceed 1% in moles. Other atoms forming gaseous molecules in volcanic emissions are F, N, and B. Fluorine is chiefly present in the form of hydrofluoric acid HF, typically in a ratio of 1:10 with respect to HCl. Nitrogen is mainly present as N_2 and NH_3 and, although bulk amounts of N and B are extremely low, the presence of N_2 is sometimes indicative of probable atmospheric contamination. Figure 9.6 shows a ternary diagram with the atomic species O, H, and C + S at the apexes.

As figure 9.6 shows, most gaseous emissions associated with basaltic lavas have compositions lying along a tie line connecting stoichiometry

$$H : O = 2 : 1 \qquad (9.54)$$

with stoichiometry

$$(C + S) : O = 2 : 1. \qquad (9.55)$$

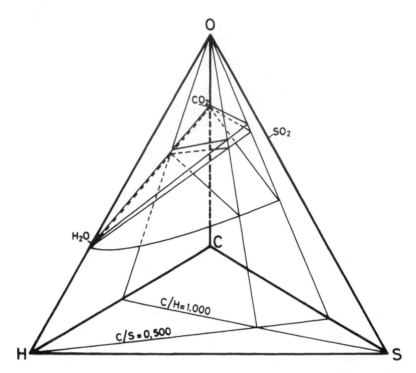

Figure 9.7 O-H-S-C compositional tetrahedron. From Gerlach and Nordlie (1975a), *American Journal of Science,* 275, 353–76. Reprinted with permission of American Journal of Science.

This indicates that the major molecular species present in volcanic gases are H_2O, CO_2, and SO_2. Some samples (mainly from Mount Etna's emissions in 1970) also contain appreciable amounts of H_2 molecules.

Figure 9.7 shows the O-H-S-C compositional tetrahedron. In this representation, the compositional limits of volcanic gases are composed of vertical planes with atomic ratios

$$C : S = 1 : 2 \tag{9.56}$$

and

$$C : H = 1 : 1, \tag{9.57}$$

respectively, the H-O-C plane, and the surfaces established by the hematite-magnetite (HM) and quartz-fayalite-magnetite (QFM) oxygen buffers:

$$3\,Fe_2O_3 \Leftrightarrow 2\,Fe_3O_4 + \frac{1}{2}O_2 \tag{9.58}$$

$$3 \, SiO_2 + 2 \, Fe_3O_4 \Leftrightarrow 3 \, Fe_2SiO_4 + O_2 \,. \tag{9.59}$$

The upper surface in the compositional tetrahedron of figure 9.7 corresponds to f_{O_2} conditions determined by buffer equation 9.58 at $T = 250 \,°C$ and represents the highest oxidation limit observable in terrestrial volcanic gases. The surface immediately below the upper surface is defined by the same buffering equilibrium at $T = 800 \,°C$, and the lower surface (buffer eq. 9.59) is attained at maximum reduction. According to Gerlach and Nordlie (1975a,b,c), this compositional space is representative of the chemistry of all terrestrial volcanic gases.

Although, as already stated, the major molecular species present in volcanic gases are H_2O, CO_2, and SO_2, several other molecules also occur, although in subordinate amounts. Figure 9.8 shows the complete speciation in gaseous phases with different C/H and C/S relative amounts at $T = 1250 \,°C$, as a function of f_{O_2}. As we can see, molecules S_2, H_2, H_2S, COS, S, SO_3, CS_2, and O_2 are also present in molar proportions dependent on f_{O_2}.

Besides buffer equations 9.58 and 9.59, figure 9.8 shows the locations of nickel–nickel oxide (NNO) and hematite-wuestite (HW) f_{O_2}-buffering equilibria:

$$Ni + \frac{1}{2} O_2 \Leftrightarrow NiO \tag{9.60}$$

$$2 \, FeO + \frac{1}{2} O_2 \Leftrightarrow Fe_2O_3 \,. \tag{9.61}$$

Another equilibrium that controls the chemistry and speciation of volcanic gases is the multiple buffer involving gaseous species S_2 and O_2:

$$Fe_3O_4 + 3 \, S_2 \Leftrightarrow 3 \, FeS_2 + 2 \, O_2 \,. \tag{9.62}$$

Application of mass-action law to equation 9.62 requires knowledge of the activity of solid components magnetite and pyrite in the respective solid phases. Moreover, in order to translate partial pressures of gaseous components into activities, the corresponding fugacity coefficients at standard state and at the state of interest, and the activity coefficients in the mixture, must be evaluated (see section 9.1). Application of the buffering equilibrium of equation 9.62 to the atomic compositions of the gaseous emissions of Kilawea (Hawaii) (Gerlach and Nordlie, 1975c) results in a speciation state rather similar to that actually observed by analysis (table 9.6). In calculating mass balance through equilibrium constant K_{62}, the authors assumed O_2 and S_2 to behave as perfect gases—i.e., $f_{O_2} = P_{O_2}$ and $f_{S_2} = P_{S_2}$. This assumption is plausible, because T is high and bulk pressure is low (see sections 9.1 and 9.2).

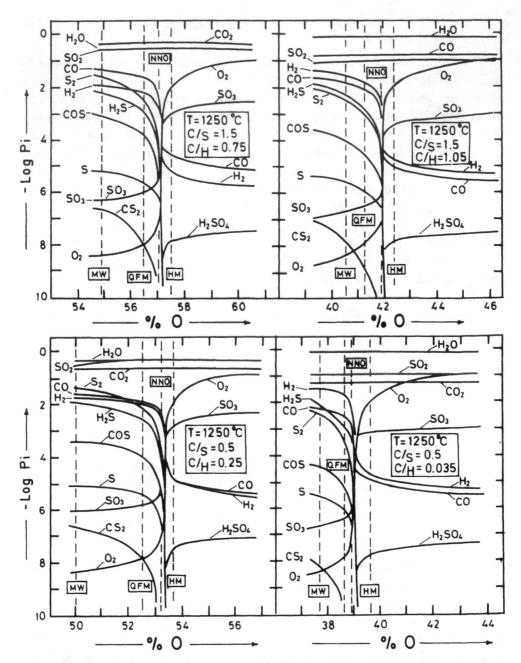

Figure 9.8 Speciation of volcanic gases with different C/H and C/S relative amounts as a function of f_{O_2} at $T = 1250\ °C$. From Gerlach and Nordlie (1975a), *American Journal of Science*, 275, 353–76. Reprinted with permission of American Journal of Science.

Table 9.6 Calculated speciation for gases of Alae Crater (Hawaii), compared with analytical results of gaseous emissions of Kilawea (Hawaii) from Gerlach and Nordlie, 1975c. $T = 800\ °C$, $P = 1$ bar, C/S = 1.5.

Gaseous Species	Alae Crater (calculated)	Kilawea Emissions (measured; Nordlie, 1971)
H_2O	4.8×10^{-1}	3.1×10^{-1}
CO_2	3.1×10^{-1}	4.1×10^{-1}
SO_2	2.0×10^{-1}	2.7×10^{-1}
H_2S	3.6×10^{-3}	2.3×10^{-3}
H_2	1.2×10^{-3}	6.9×10^{-4}
S_2	2.0×10^{-3}	2.3×10^{-3}
CO	7.2×10^{-4}	8.6×10^{-4}
COS	6.9×10^{-5}	8.9×10^{-5}
SO_3	4.3×10^{-8}	6.6×10^{-8}
CS_2	3.5×10^{-9}	4.4×10^{-9}
O_2	7.0×10^{-14}	8.6×10^{-14}
S_8	1.3×10^{-15}	2.5×10^{-15}

9.6 Solubilities of Gaseous Species in Silicate Melts

Before examining in detail the present-day knowledge of the solubilities of various gaseous species in magmas, it is necessary to describe solubility in terms of reactivity of gaseous species with melt components. A simple heterogeneous equilibrium between gas and melt involves the inert gas Ar:

$$\underset{\text{gas}}{Ar} \Leftrightarrow \underset{\text{melt}}{Ar} \quad .$$

(9.63)

This can be conceived as "mechanical solubility," which does not imply either reactivity or modifications of the molecular structure of the compound of interest. Several gaseous species do not exhibit solubility of the type exemplified by equation 9.63, but more or less complex reactions with melt components. For instance, in the case of CO_2,

$$\underset{\text{gas}}{CO_2} + \underset{\text{melt}}{O^{2-}} \Leftrightarrow \underset{\text{melt}}{CO_3^{2-}} \quad .$$

(9.64)

Equilibrium 9.64 is in fact composed of two partial reactions:

$$\underset{\text{gas}}{CO_2} \Leftrightarrow \underset{\text{melt}}{CO_2}$$

(9.65)

and

$$CO_2 + O^{2-} \Leftrightarrow CO_3^{2-}. \tag{9.66}$$
$$\text{melt} \quad \text{melt} \quad \text{melt}$$

Homogeneous equilibrium 9.66 modifies the stoichiometry, structure, and properties of the component of interest.

Generally, in the case of mechanical solubility, an extended Henry's law field is observed, whereas homogeneous equilibria of type 9.66 usually imply precocious deviations from ideality (Battino and Clever, 1966; Wilhem et al., 1977).

There are several ways of expressing the solubilities of gaseous species in melts or liquids. The Bunsen coefficient α_g defines "the volume of reduced gas at $T = 273.15\ K$, $P = 1.013$ bar adsorbed by a unit volume of solvent at T of interest and $P = 1.013$ bar." The equation used to calculate the Bunsen coefficient is

$$\alpha_g = \left[\left(V_g\ \frac{273.15}{T} \times \frac{P_g}{1.013} \right) \left(\frac{1}{V_{\text{melt}}} \right) \right] \left(\frac{1.013}{P_g} \right), \tag{9.67}$$

where V_g is the volume of reduced gas adsorbed at the T of interest, V_{melt} is the volume of solvent (melt, in our case), and P_g is the partial pressure of the gaseous species in equilibrium with the solvent. Note that equation 9.67 reduces to

$$\alpha_g = \frac{V_g}{V_{\text{melt}}} \times \frac{273.15}{T}. \tag{9.68}$$

The Ostwald coefficient L_g is "the ratio between volume of adsorbed gas and volume of solvent, measured at the same temperature":

$$L_g = \frac{V_g}{V_{\text{melt}}} = \alpha_g\ \frac{T}{273.15}. \tag{9.69}$$

Henry's constant K_h of the gas represents the ratio between the partial pressure of the species in the gaseous phase and the molar fraction of the species in solution:

$$K_h = \frac{P_g}{X_{g,\text{melt}}}. \tag{9.70}$$

Note that Henry's constant differs from equilibrium constant K, which represents the activity ratio between the two phases at equilibrium:

$$K = \frac{a_{g,\text{gas}}}{a_{g,\text{melt}}} = \frac{\Gamma_g P_g}{f^0_{g,\text{gas}}} \times \frac{1}{X_{g,\text{melt}} \gamma_{g,\text{melt}}} = K_h \times \frac{\Gamma_g}{f^0_{g,\text{gas}} \gamma_{g,\text{melt}}}, \tag{9.71}$$

where Γ_g is the fugacity coefficient in the gas phase and f_g^0 is the standard state fugacity. In the case of homogeneous equilibria such as equilibrium 9.66, the relationship between thermodynamic constant and Henry's constant is obviously more complex.

In geochemistry, the solubilities of gaseous species in melts are usually expressed through Henry's constant or as fractional solubilities in moles (molar fraction of gaseous species in the gas phase over molar fraction of gaseous species in the melt).

9.6.1 Solubility of H_2O

The H_2O molecule reacts with the bridging oxygens of the polymeric units in silicate melts to form hydroxyl groups OH on their free terminations (Fraser, 1977):

$$H_2O + O^0 \Leftrightarrow 2OH.$$
$$\text{gas} \quad\quad \text{melt} \quad\quad \text{melt}$$

(9.72)

Because H_2O is a basic oxide (i.e., it tendentially dissociates free oxygen; cf. section 6.1.3), reaction 9.72 may appear contradictory. However, reaction 9.72 clearly represents the summation of partial reactions:

$$H_2O \Leftrightarrow H_2O.$$
$$\text{gas} \quad\quad \text{melt}$$

(9.73)

$$H_2O \Leftrightarrow 2H^+ + O^{2-}$$
$$\text{melt} \quad\quad \text{melt} \quad \text{melt}$$

(9.74)

$$O^{2-} + O^0 \Leftrightarrow 2O^-$$
$$\text{melt} \quad \text{melt} \quad\quad \text{melt}$$

(9.75)

$$2O^- + 2H^+ \Leftrightarrow 2OH.$$
$$\text{melt} \quad\quad \text{melt} \quad\quad \text{melt}$$

(9.76)

If we consider a dimer, in a stereochemical representation, the bulk reaction of equation 9.72 is

$$
\begin{array}{ccccccc}
O^- & O^- & & H & O^- & & O^- \\
| & | & & / & | & & | \\
O^- - Si - O - Si - O^- & + & O & \Leftrightarrow O^- - Si - OH + & O^- - Si - OH. \\
| & | & & \backslash & | & & | \\
O^- & O^- & & H & O^- & & O^-
\end{array}
$$

(9.77)

Contrary to common assumptions, H_2O is not present in the melt completely in the form of OH groups linked to polymer chains, but is also partially present as discrete H_2O molecules. In other words, the homogeneous reaction

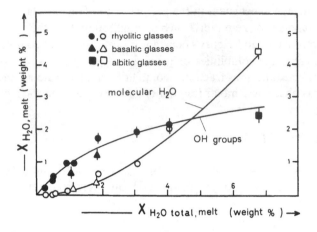

Figure 9.9 Amounts of molecular H_2O and OH hydroxyls determined in hydrated silicate glasses by IR spectroscopy. Reprinted from E. Stolper, *Geochimica et Cosmochimica Acta*, 46, 2609–2620, copyright © 1982, with kind permission from Elsevier Science Ltd., The Boulevard, Langford Lane, Kidlington 0X5 1GB, UK.

$$H_2O + O^0 \Leftrightarrow 2OH \qquad (9.78)$$
$$\text{melt} \quad \text{melt} \quad \text{melt}$$

is not completely displaced rightward. Stolper (1982) has shown that the weight proportions of H_2O and OH in melts vary nonlinearly with increasing bulk amount of H_2O. When this exceeds 5%, molecular water is dominant. Figure 9.9 shows the proportions of OH and molecular H_2O in melts with increasing amounts of dissolved H_2O, experimentally determined by infrared (IR) spectroscopy. The observed distributions are consistent with an equilibrium constant K_{78} between 0.1 and 0.3.

The value of constant K_{78} is determined in the following manner (Stolper, 1982). The constant of equilibrium 9.78 is given by the activity ratio

$$K_{78} = \frac{a_{OH,melt}^2}{a_{H_2O,melt} a_{O^0,melt}}. \qquad (9.79)$$

It is assumed, at first approximation, that the activities of the three species in homogeneous equilibrium correspond to the relative molar fractions

$$a_{H_2O,melt} \equiv X_{H_2O,melt} = \frac{n_{H_2O,melt}}{n_{H_2O,melt} + n_{OH,melt} + n_{O^0,melt}} \qquad (9.80)$$

$$a_{OH,melt} \equiv X_{OH,melt} = \frac{n_{OH,melt}}{n_{H_2O,melt} + n_{OH,melt} + n_{O^0,melt}} \qquad (9.81)$$

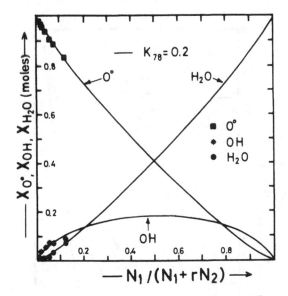

Figure 9.10 Molar fractions of components H_2O, OH and O^0 in melt calculated by Stolper (1982) for a value of $K_{78} = 0.2$. Calculated quantities are in good agreement with experimental observations. Reprinted from E. Stolper, *Geochimica et Cosmochimica Acta,* 2609–2620, copyright © 1982, with kind permission from Elsevier Science Ltd., The Boulevard, Langford Lane, Kidlington 0X5 1GB, UK.

$$a_{O^0,\text{melt}} \equiv X_{O^0,\text{melt}} = \frac{n_{O^0,\text{melt}}}{n_{H_2O,\text{melt}} + n_{OH,\text{melt}} + n_{O^0,\text{melt}}}. \tag{9.82}$$

If N_1 moles of component H_2O mix with N_2 moles of an anhydrous melt whose stoichiometric formula contains r moles of oxygen (i.e., $r = 8$ for $NaAlSi_3O_8$, $r = 2$ for SiO_2, and so on), because one mole of O^0 is necessary to form two moles of OH hydroxyls, we obtain

$$n_{O^0,\text{melt}} = rN_2 - 0.5n_{OH,\text{melt}} \tag{9.83}$$

$$n_{H_2O,\text{melt}} = N_1 - 0.5n_{OH,\text{melt}}. \tag{9.84}$$

Substituting in equation 9.79, we obtain

$$K_{78} = \frac{X_{OH,\text{melt}}^2}{X_{H_2O,\text{melt}} \cdot X_{O^0,\text{melt}}} = \frac{n_{OH,\text{melt}}^2}{\left(N_1 - 0.5n_{OH,\text{melt}}\right)\left(rN_2 - 0.5n_{OH,\text{melt}}\right)}, \tag{9.85}$$

which can be solved easily for discrete values of K_{78}.

Figure 9.10 shows molar fractions of H_2O, OH, and O^0 calculated by Stolper (1982) with $K_{78} = 0.2$ and plotted as a function of $N_1(N_1 + rN_2)$ (relative amount of oxygen in melt due to H_2O component with respect to total amount).

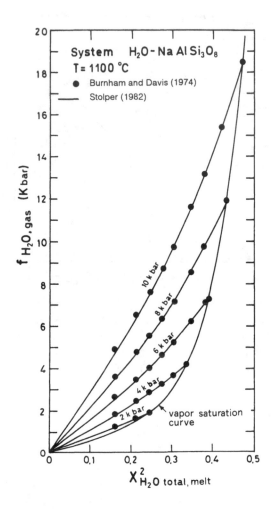

Figure 9.11 Relationship between fugacity of gaseous component H_2O and amount of same component dissolved in an albitic melt at equilibrium at $T = 1100$ °C ($K_{78} = 0.2$). Interpolants in various P conditions determine value of K_{72}, based on equation 9.87. Reprinted from E. Stolper, *Geochimica et Cosmochimica Acta,* 2609–2620, copyright © 1982, with kind permission from Elsevier Science Ltd., The Boulevard, Langford Lane, Kidlington 0X5 1GB, UK.

Figure 9.11 shows the effect of pressure on the solubility of H_2O in an albitic melt at $T = 1100$ °C, based on the experiment of Burnham and Davis (1974). At low H_2O amounts, a marked linear correlation between fugacity of H_2O in the gaseous phase and the squared molar fraction of H_2O component in the melt is observed. Note that the abscissa axis in figure 9.11 is $X^2_{H_2O total, melt}$ and thus differs from the same axis in figure 9.9. Let us now again consider equations 9.72, 9.73, and 9.78. Clearly, the constant of heterogeneous equilibrium 9.72 is given by the product of partial equilibria 9.73 and 9.78—i.e.,

$$K_{72} = K_{73} \times K_{78}. \tag{9.86}$$

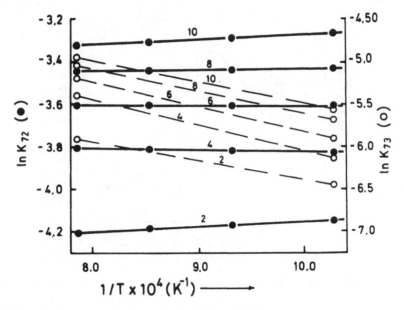

Figure 9.12 Arrhenius plot of equilibria 9.72 and 9.73.

If equilibrium 9.78 is displaced rightward (i.e., the relative amount of hydroxyls is dominant with respect to molecular H_2O), it can be shown that

$$\frac{K_{72}}{4 f^0_{H_2O,gas}} \times \left[\frac{rN_2^2 - N_1^2}{(N_1 + N_2)^2} \right] = \frac{X^2_{H_2O,melt}}{f_{H_2O,gas}} \qquad (9.87)$$

(Stolper, 1982). If the total amount of H_2O in the melt ($X_{H_2Ototal, melt}$) is sufficiently low, it is also true that $N_1 \ll N_2$ and the relationship between $X^2_{H_2Ototal, melt}$ and $f_{H_2O,gas}$ is simply linear (cf. eq. 9.87 and figure 9.11).

The interpolation of experimental points in figure 9.11 allows constant K_{72} to be determined at each pressure. This exercise, when repeated at the various T of interest, leads to the construction of the Arrhenius plot of figure 9.12 (Fraser, 1975b). The same figure also shows the Arrhenius trend relative to equation 9.73, calculated for $K_{78} = 0.2$ at $T = 1100$ °C and $K_{78} = 0.1$ at $T = 700$ °C (Stolper, 1982)—i.e.,

$$\ln K_{73} = \ln K_{72} - \ln K_{78}. \qquad (9.88)$$

Note that constant K_{73} is the reciprocal of Henry's constant when fugacity coefficient $\Gamma_g = 1$.

Lastly, figure 9.13 shows that the linear relationship between the square root of activity in the gaseous phase and the amount of H_2O component in the melt

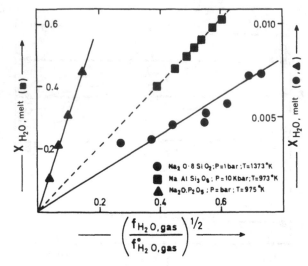

Figure 9.13 Relationship between activity of H_2O in gaseous phase and molar amount of H_2O in melt at equilibrium. Experimental data from Burnham and Davis (1974) (■); Fraser (1975b) (▲); Kurkjian and Russel (1958) (●). Reprinted from B. J. Wood and D. G. Fraser, *Elementary Thermodynamics for Geologists,* 1976, by permission of Oxford University Press.

is generally observed in all types of silicate (and also phosphate) melts (Kurkjian and Russel, 1958; Burnham and Davis, 1974; Burnham, 1975; Fraser, 1975b, 1977).

Neglecting N_1 in equation 9.87, we obtain

$$\frac{X_{H_2O,melt}}{a_{H_2O,gas}^{1/2}} = \frac{\sqrt{r}}{2} K_{72}^{1/2}.$$ (9.89)

The angular coefficient of straight-line regressions in figure 9.13 thus depends on the stoichiometry of the silicate melt component (moles of oxygen per mole of anhydrous melt) and on the value of constant K_{72}.

9.6.2 Solubility of CO_2 and SO_2

Carbon dioxide behaves as an acidic oxide and associates free oxygen according to

$$CO_2 + O^{2-} \Leftrightarrow CO_3^{2-}.$$
$$\text{melt} \quad \text{melt} \quad \text{melt}$$ (9.90)

However, not all the CO_2 is present in the melt in the form of carbonate ion: molecular CO_2 is also present, based on the heterogeneous equilibrium

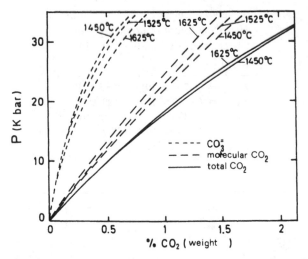

Figure 9.14 Weight concentration of molecular CO_2, carbonate ion CO_3^{2-}, and total CO_2 dissolved in albitic melts as a function of P, under various T conditions. Reproduced with modifications from E. Stolper, *Geochimica et Cosmochimica Acta*, 46, 2609–2620, copyright © 1982, with kind permission from Elsevier Science Ltd., The Boulevard, Langford Lane, Kidlington 0X5 1GB, UK.

$$\underset{\text{gas}}{CO_2} \Leftrightarrow \underset{\text{melt}}{CO_2} . \tag{9.91}$$

The complete equilibrium between gaseous phase and melt is a combination of partial equilibria 9.90 and 9.91:

$$\underset{\text{gas}}{CO_2} + \underset{\text{melt}}{(X)O^{2-}} \Leftrightarrow \underset{\text{melt}}{(X)CO_3^{2-}} + \underset{\text{melt}}{(1-X)CO_2}, \tag{9.92}$$

where X and $(1-X)$ are molar fractions of molecular CO_2 and carbonate ion in the melt, respectively, with respect to the total amount of CO_2. IR spectroscopy measurements on albitic glasses (Stolper et al., 1987) show that, at constant P, the solubility of molecular CO_2 in the melt decreases with increasing T, whereas the solubility of carbonate ion increases: the net result is that bulk solubility is practically unaffected by temperature. This fact, based on equilibrium 9.92, can be interpreted as increased reactivity of free oxygen in the melt, with increasing T. As P increases, with T held constant, we observe increased solubility of carbonate ion and, more appreciably, of molecular CO_2 (figure 9.14).

The solubility of sulfur dioxide in melts may be ascribed to two concomitant processes:

$$\underset{\text{gas}}{SO_2} + \underset{\text{melt}}{O^{2-}} \Leftrightarrow \underset{\text{melt}}{S^{2-}} + \underset{\text{gas}}{\frac{3}{2}O_2} \tag{9.93}$$

Table 9.7 Linear equations relating Henry's constant of noble gases in silicate melts to density of Liquid (ρ) (from Lux, 1987).

$$K_{h,\,He} = (-267 \pm 19)\rho + (765 \pm 49)$$
$$K_{h,\,Ne} = (-140 \pm 53)\rho + (400 \pm 140)$$
$$K_{h,\,Ar} = (-49 \pm 4)\rho + (137 \pm 10)$$
$$K_{h,\,Kr} = (-39 \pm 3)\rho + (108 \pm 8)$$
$$K_{h,\,Xe} = (-23 \pm 3)\rho + (63 \pm 7)$$

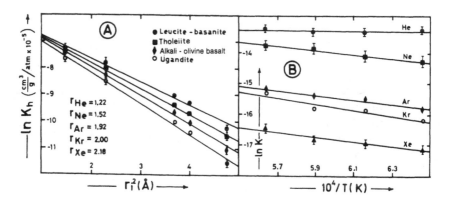

Figure 9.15 Solubility of noble gases in silicate melts as a function of kinetic radius of gaseous molecule (**A**) and temperature (**B**). Ordinate axis in part A: natural logarithm of Henry's constant; in part B, natural logarithm of equilibrium constant. Reprinted from G. Lux, *Geochimica et Cosmochimica Acta*, 51, 1549–1560, copyright © 1987, with kind permission from Elsevier Science Ltd., The Boulevard, Langford Lane, Kidlington 0X5 1GB, UK.

and

$$SO_2 + \frac{1}{2}O_2 + O^{2-} \Leftrightarrow SO_4^{2-}. \tag{9.94}$$

$$\underset{\text{gas}}{} \quad \underset{\text{gas}}{} \quad \underset{\text{melt}}{\phantom{O^{2-}}} \quad \underset{\text{melt}}{\phantom{SO_4^{2-}}}$$

Equilibria 9.93 and 9.94 are similar to those proposed by Fincham and Richardson (1954) for native sulfur. According to these equilibria, the solute states of sulfur compounds depend on oxygen fugacity; the total concentration of sulfur dissolved in the melt initially decreases with increasing f_{O_2}, based on equation 9.93, and then increases again when equilibrium 9.94 becomes dominant.

9.6.3 Solubility of Noble Gases

The solubility of noble gases in silicate melts obeys Henry's law but varies markedly with the chemistry and physical properties of the melt: it increases with the

Table 9.8 Standard molar enthalpies ($\Delta\overline{H}_i^0$) and standard entropies ($\Delta\overline{S}_i^0$) of solution of noble gases in silicate melts, after Lux (1987). (1) leucite-basanite; (2) tholeiite; (3) alkali-olivine basalt.

| Gas | ($\Delta\overline{H}_i^0$) (kcal/mole) | | | ($\Delta\overline{S}_i^0$) (cal/mole) | | |
	(1)	(2)	(3)	(1)	(2)	(3)
He	1.5 ± 4.3	4.9 ± 4.2	0.5 ± 4.2	-25.0 ± 2.5	-23.1 ± 2.5	-26.0 ± 2.5
Ne	5.1 ± 5.8	1.3 ± 5.8	12.3 ± 5.8	-23.8 ± 3.5	-26.8 ± 3.5	-20.4 ± 3.5
Ar	4.5 ± 2.2	6.8 ± 2.2	14.2 ± 2.2	-26.7 ± 1.3	-26.0 ± 1.3	-22.2 ± 1.3
Kr	3.7 ± 2.2	9.4 ± 2.2	19.0 ± 2.2	-27.7 ± 1.4	-27.7 ± 1.4	-24.9 ± 1.4
Xe	4.4 ± 4.4	2.9 ± 4.3	14.4 ± 4.4	-29.2 ± 2.7	-30.6 ± 2.7	-24.8 ± 2.7

amount of SiO_2 and decreases with the amounts of MgO and CaO in the melt, increases with melt viscosity (in agreement with the chemical effect; White et al., 1989) and decreases with increasing density. The latter fact is in agreement with essentially "*mechanical*" solubility (cf. section 9.6)—i.e., solubility is a function of the free volume available within the liquid which in turn is (approximately) inversely proportional to density. Lux (1987) furnished the parameters of a linear equation which reproduces the Henry's constant of the various gases, based on density of melt, within the field 2.1 to 2.7 g/cm^3 (Table 9.7). The resulting Henry's constant is in cm^3 of reduced volume per g of melt/atm $\times 10^{-5}$. With P, T, and composition held constant, solubility decreases with increasing atomic number and can be correlated directly to the kinetic radius of noble gas r_i through an empirical equation of the type

$$\ln K_{h,i} = a + br_i^2 \tag{9.95}$$

(Kirsten, 1968), as shown in figure 9.15A. With P and composition held constant, solubility increases slightly with T. Figure 9.15B shows an Arrhenius plot of Henry's constant for an alkaline basalt melt in the T range $1250 < T\,(°C) < 1500$ (Lux, 1987). From equality of potentials at equilibrium:

$$\mu_{i,P_r,T,\text{gas}} = \mu_{i,P_r,T,\text{melt}}, \tag{9.96}$$

we can derive

$$\ln K_i = -\frac{\Delta\overline{H}_i^0}{RT} + \frac{\Delta\overline{S}_i^0}{R}. \tag{9.97}$$

By applying equation 9.97 to plots of the type shown in figure 9.15B, Lux (1987) obtained standard molar enthalpies $\Delta\overline{H}_i^0$ and standard molar entropies $\Delta\overline{S}_i^0$ of solution of noble gases in melts, as listed in table 9.8. Enthalpies of solution are positive but close to zero, within error approximation.

Lastly, figure 9.16 shows the effects of pressure on the solubility of argon in

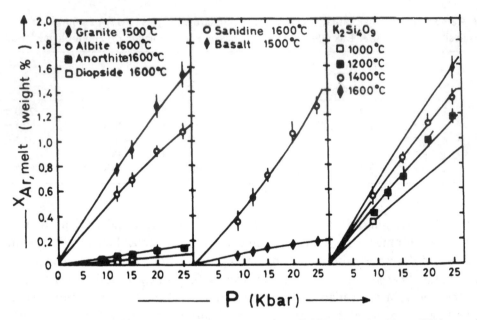

Figure 9.16 Effects of pressure on solubility of argon in silicate melts. From White et al. (1989). Reprinted with permission of The Mineralogical Society of America.

Table 9.9 Interpolation parameters giving solubility of Ar in silicate melts as a function of P. Standard state of reference for gaseous phase is pure gas Ar at $P = 15$ kbar and $T = 1600$ °C. Concentration X^0_{Ar, P_r, T_r} is adopted standard state for Ar in melt. $\overline{V}^0$ is molar volume of Ar in melt, assumed to be constant with T and P (from White et al., 1989). $\Delta \overline{H}^0_{P_r, T_r}$ is the molar enthalpy of solution at P and T of reference.

Melt	P_r (kbar)	T_r (°C)	X^0_{Ar, P_r, T_r}	$\Delta \overline{H}^0_{P_r, T_r}$ (kJ/mole)	$\overline{V}^0$ (cm^3/mole)
NaAlSi$_3$O$_8$	15	1600	0.0057 ± 0.0004	8.2 ± 0.8	22.8 ± 0.3
KAlSi$_3$O$_8$	15	1600	0.0067 ± 0.0004	7.2 ± 1.0	20.6 ± 0.3
CaAl$_2$Si$_2$O$_8$	15	1600	0.0008 ± 0.0001	14.1 ± 6.2	22.5 ± 0.5
CaMgSi$_2$O$_6$	15	1600	0.0004 ± 0.0001	14.3 ± 6.5	21.5 ± 0.6
K$_2$Si$_4$O$_9$	15	1600	0.0079 ± 0.0004	12.0 ± 0.8	22.6 ± 0.3
Granite	15	1600	0.0081 ± 0.0009	9.3 ± 0.5	23.1 ± 0.3
Basalt	15	1600	0.0012 ± 0.0002	12.9 ± 6.0	23.7 ± 0.5

silicate melts of various compositions, based on the experiments of White et al. (1989). The interpolating curves are based on

$$\ln K_{P,T} - \ln K_{P_r, T_r} = -\frac{\Delta \overline{H}^0}{R}\left(\frac{1}{T} - \frac{1}{T_r}\right) - \frac{\overline{V}^0}{RT}\left(P - P_r\right) \qquad (9.98)$$

for a standard state of pure gas Ar at $P = 15$ kbar and $T = 1600$ °C, and assuming Ar to mix ideally in the liquid over a matrix composed of Ar and O^{2-}. The interpolation parameters of equation 9.98 are listed in Table 9.9.

9.7 Fluid Equilibria in Geothermal Fields

9.7.1 Solubility of Gases in Aqueous Solutions

We have already seen that the concept of the "solubility" of a gaseous species in a liquid must be clarified by distinguishing between "purely mechanical solubility," attained whenever the gaseous molecule maintains its form in the liquid phase (as, for instance, in the case of noble gases dissolved in silicate melts), and "reactive solubility," attained whenever the gaseous molecule reacts with the host phase, forming stable complexes different from the gaseous form (see also the concept of "chemical bond" in section 1.3). Figure 9.17 shows the behavior of noble gases in liquid H_2O, within the T range between 35 °C and the critical temperature at 1 bar total pressure (373 °C). In all cases, solubility (which is purely mechanical in the case of noble gases) does not follow an Arrhenius trend but shows more or less marked upward convex patterns, reaching a maximum at intermediate T conditions and then decreasing toward zero, which is (necessarily) attained at the critical temperature of the liquid. In the case of reactive gases (such as N_2 in figure

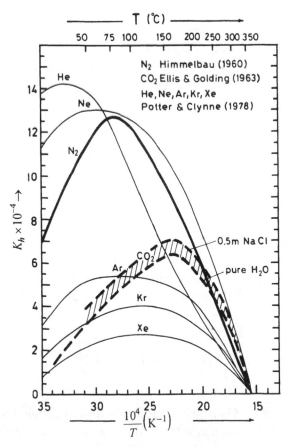

Figure 9.17 Solubility of various gas-forming molecular species in water as a function of T. Ordinate axis: $K_h \times 10^{-4}$.

Table 9.10 Constants of Potter-Clynne equation for noble gases.

Gas	A_0	A_1	A_2	A_3	A_4	A_5
He	6830.52	−187,444.82	1.974201	−2738.127	−687.751	889.593
Ne	1421.175	−55,338.362	0.185867	−516.808	−20.0201	27.2899
Ar	3145.791	−87,149.49	−516.808	−1270.795	−468.632	615.388
Kr	1098.849	−32,333.73	−20.0201	−442.515	−230.154	310.802
Xe	1074.854	−34,604.31	27.2899	−420.535	−113.399	147.401

9.17, but also O_2, H_2, CH_4, NH_3, H_2S, and CO_2), the two solubility branches of the trend, ascending and descending, are virtually straight. Because early experiments on the solubility of gaseous species in water were limited to low T (typically below 50 °C), solubility was described by means of empirical relationships accurately reproducing low-T behavior—for instance, the Benson-Krause equation:

$$\ln K_{h,i} = A(T_1/T - 1)^2 - \frac{\overline{S}_i}{R}(T_1/T - 1), \qquad (9.99)$$

where T_1 is the hypothetical temperature at which $K_{h,i} = 1$; $A = -36.855$; and $\overline{S}_i$ is the partial molal entropy of the gaseous species in aqueous solution at T_1.

More recently, extension of experiments to high T has allowed the derivation of empirical equations valid over the entire stability field of water. The equation of Potter and Clynne (1978) for noble gases, for instance, is valid in the T range $298 < T\,(\mathrm{K}) < 647$:

$$K_{h,i} = \left\{ A_0 + \frac{A_1}{T} + A_2 T + A_3 \log T + A_4 \log\left(\frac{1}{\rho}\right) + A_5 \left[\log\left(\frac{1}{\rho}\right)\right]^2 \right\}$$

$$\times 10^4 \times \Gamma_i. \qquad (9.100)$$

In equation 9.100, A_0, A_1, A_2, A_3, A_4, and A_5 are empirical constants determined for each noble gas (Table 9.10), ρ is the density of water, and Γ_i is the fugacity coefficient of gas with deviation parameter Z_g, which can be calculated as a function of the compressibility factor, as follows (see also eq. 9.16 and 9.22 in section 9.1):

$$\ln \Gamma_i = -\int_0^P (1 - Z_g)\frac{dP}{P}. \qquad (9.101)$$

Still more recently, Giggenbach (1980) proposed simple linear equations relating the molar distribution constants of the main reactive gases to T expressed in Celsius:

Table 9.11 Coefficients for Giggenbach equation (eq. 9.102) for various gaseous species in T-range $100 < T\,(°C) < 340$ (from Giggenbach, 1980).

Species	A	B
NH_3	1.4113	−0.00292
H_2S	4.0547	−0.00981
CO_2	4.7593	−0.01092
CH_4	6.0783	−0.01383
H_2	6.2283	−0.01403
N_2	6.4426	−0.01416

$$\log K_i' \equiv \log\left(\frac{X_{i,g}}{X_{i,l}}\right) = A + BT. \qquad (9.102)$$

These constants, listed in Table 9.11, were derived by Giggenbach (1980) from the respective Henry's constant by applying:

$$K_i' = K_{h,i} \times \frac{Z_{g,H_2O}}{P_{H_2O}}, \qquad (9.103)$$

where Z_{g,H_2O} is the compressibility factor of gaseous H_2O. Because the resulting values of K' vary in a simple exponential fashion with T (Figure 9.18), the maximum in Henry's constant observed for the various noble gases in Figure 9.17 can be simply ascribed, in light of equation 9.103, to the compressibility factor of H_2O vapor.

To understand fully the relationship between Henry's constant and the molar distribution constant of a given species between a gas and a liquid phase, we must recall the concept of the "compressibility factor" of a gas (see section 9.3):

$$Z_{g,i} = \frac{P_i V_i}{n_i RT}. \qquad (9.104)$$

The molar fraction of gaseous species i in gaseous mixture $(i + H_2O_{(g)})$ is given by

$$X_{i,g} = \frac{n_i}{n_i + n_{H_2O_{(g)}}}. \qquad (9.105)$$

If gaseous component i is at trace concentration level, we can also write

$$X_{i,g} = \frac{n_i}{n_i + n_{H_2O_{(g)}}} \approx \frac{n_i}{n_{H_2O_{(g)}}}. \qquad (9.106)$$

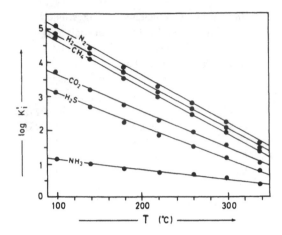

Figure 9.18 Molar distribution constants of various gaseous species as a function of *T.* Reprinted from W. F. Giggenbach, *Geochimica et Cosmochimica Acta,* 44, 2021–2032, copyright © 1980, with kind permission from Elsevier Science Ltd., The Boulevard, Langford Lane, Kidlington 0X5 1GB, UK.

Reexpressing n_i and $n_{\mathrm{H_2O_{(g)}}}$ through their respective compressibility factors, we obtain

$$X_{i,g} = \frac{P_i}{P_{\mathrm{H_2O}}} \times \frac{Z_{g,\mathrm{H_2O}}}{Z_{g,i}}. \tag{9.107}$$

Moreover, dividing by the molar fraction of species *i* in the liquid, we have

$$K_i' = \frac{X_{i,g}}{X_{i,l}} = \frac{P_i}{X_{i,l}} \times \frac{Z_{g,\mathrm{H_2O}}}{Z_{g,i}} \times \frac{1}{P_{\mathrm{H_2O}}} = K_{h,i} \times \frac{Z_{g,\mathrm{H_2O}}}{Z_{g,i}} \times \frac{1}{P_{\mathrm{H_2O}}}. \tag{9.108}$$

In the transition $K' \rightarrow K_{h,i}$, the compressibility factor of the trace component in the gaseous phase is usually neglected because, with the concentration so low, it is virtually 1 (see eq. 9.23).

9.7.2 Compositions of Geothermal Gases

The development of a gaseous phase in geothermal fields is associated with boiling of aquifers, generally induced by decompression during convective transport toward the surface or by heat transfer from reservoir rocks exceeding the *T* limit of the boiling curve at all *P.* Analysis of temperature profiles in deep geothermal fields, between 0.5 and 2 km deep, almost invariantly reveals *P-T* conditions along the boiling curve of water (i.e., at all depths, the temperature of the fluid is that of the aqueous solution at the *P* of interest; cf. Table 9.12). There are two main reasons for this.

Table 9.12 Thermal profile of water along boiling curve (from Haas, 1971).

T (°C)	P (bar)	Depth (m)	Density (g/cm^3)
100	1.0	0.0	0.958
110	1.4	4.5	0.951
120	2.0	10.4	0.943
130	2.7	18.2	0.935
140	3.6	28.2	0.926
150	4.8	40.9	0.917
160	6.2	56.8	0.907
170	7.9	76.4	0.897
180	10.0	100.5	0.887
190	12.6	129.7	0.876
200	15.6	164.7	0.865
210	19.1	206.8	0.853
220	23.2	256.4	0.840
230	28.0	314.9	0.827
240	33.5	383.3	0.814
250	39.8	462.9	0.799
260	46.9	555.2	0.784
270	55.1	661.8	0.768
280	64.2	784.6	0.751
290	74.4	925.6	0.732
300	85.9	1088	0.712
310	98.7	1273	0.691
320	122.9	1487	0.667
330	128.6	1732	0.640
340	146.1	2017	0.609
350	165.4	2350	0.573
360	186.7	2746	0.525
370	210.5	3243	0.446

1. Rocks are extremely efficient heat insulators.
2. The liquid-vapor transition for H_2O involves quite high evaporation enthalpy, which controls the thermal state of the system.

At shallower depths, the vapor phase separates from the liquid phase and migrates toward the surface, sometimes heating superficial aquifers and finally giving rise to fumarole activity. The gaseous phase is dominated by H_2O mixed with subordinate amounts of CO_2, H_2S, CH_4, H_2, N_2, and NH_3. Table 9.13 lists representative analyses of geothermal gases from the main geothermal fields of the world. It also lists the pressure of separation of the gaseous phase from the liquid, the steam fraction in the fluid ($F_{H_2O,f}$), and the "fraction of gas in steam" (X_g). The latter definition is purely operative, because H_2O vapor is a gas to all intents and purposes and the gaseous phase is not unmixed at the various P and T of interest. The need to distinguish the total amount of non-H_2O gaseous com-

Table 9.13 Gas composition in geothermal fluids (from Henley, 1984). P_{sep} = pressure of separation of gaseous phase. nd = not determined.

Locality	P_{sep} (bar)	$F_{H_2O, f}$	$X_g \left(\dfrac{\text{millimoles}}{\text{mole of steam}} \right)$	Concentration in Gas (millimoles/mole total gases)					
				CO_2	H_2S	CH_4	H_2	N_2	NH_3
(1)	2	0.3	0.2	917	44	9	8	15	6
(2)	9.8	0.2	1.2	936	64	nd	nd	nd	nd
(3)	11	0.19	10.04	956	18.4	11.8	1.01	8.89	4.65
(4)	2.87	0.19	24.8	945	11.7	28.1	3.0	2.1	10.2
(5)	7.6	0.289	5.88	822	79.1	39.8	28.6	5.1	23.1
(6)	8.6	0.414	2.95	932	55	4.1	3.6	1.2	4.3
(7)	20	0.135	0.248	962	29	1	2	6	nd
(8)	1	0.38	0.507	957	43.9	nd	nd	nd	nd
(9)	–	1.0	20.0	941	16	12	23	8	nd
(10)	–	1.0	5.9	550	48	95	150	125	nd

(1) Wairakei, New Zealand; (2) Tauhara, New Zealand; (3) Broadlands, New Zealand; (4) Ngawha, New Zealand; (5) Cerro Prieto, Mexico; (6) Mahio-Tongonan, Philippines; (7) Reykjanes, Iceland; (8) Salton Sea, California; (9) Geysers, Wyoming; (10) Larderello, Italy.

ponents from gaseous H_2O (i.e., "steam") is dictated by the fact that, as the gaseous phase flows through superficial aquifers, further enrichment in H_2O may take place by simple evaporation.

9.7.3 Thermobarometry of Gaseous Phases

The chemical composition of the gaseous phase, in terms of minor components, can be related to the kinetics of the process of separation from the liquid through simple distribution laws such as

$$\frac{C_l}{C_0} = \frac{1}{1 + F_g \left(K' - 1 \right)} \tag{9.109}$$

$$\frac{C_g}{C_0} = \frac{K'}{1 + F_g \left(K' - 1 \right)} \tag{9.110}$$

$$\frac{C_l}{C_0} = \left[1 + \Delta F_g \left(\overline{K}' - 1 \right) \right]^{-n} \tag{9.111}$$

$$\frac{C_g}{C_0} = \frac{\overline{K}'}{\left[1 + \Delta F_g \left(\overline{K}' - 1 \right) \right]^{n}} \tag{9.112}$$

$$\frac{C_l}{C_0} \approx \exp\left(-\overline{K}'F_g\right) \tag{9.113}$$

$$\frac{C_g}{C_0} \approx \overline{K}' \exp\left(-\overline{K}'F_g\right). \tag{9.114}$$

Equations 9.109 and 9.110 are valid for a boiling process in a closed system: the gaseous phase develops from a liquid with initial concentration C_0, F_g is the mass fraction of developed gas, and K' is the mass distribution constant of the component of interest between gas and liquid. Equations 9.111 and 9.112 describe a multistage separation process in which n is the number of separation stages, ΔF_g is the mass fraction of gas separated in each stage, and $\overline{K}'$ is the *mean* mass distribution constant of the process. Equations 9.113 and 9.114 refer to a boiling process in an open system: the gas is continuously removed from the system as the process advances.

The following assumptions are implicitly accepted when adopting equations 9.109 to 9.114.

1. The mass distribution constant is not affected by the concentration of the component of interest in either liquid or gas, which is a reasonable assumption when the concentration is sufficiently low (Nernst's law range).
2. The mass distribution constant does not vary appreciably with the chemistry of the liquid phase.

Assumption 2 may lead to substantial errors, because the salinity of the liquid normally increases during boiling. The consequent increase in ionic strength modifies the activity coefficients of solutes. In particular, if the gaseous species are polar and nonpolar, the activity coefficients of solutes diverge as the process advances (see section 8.7).

Let us now consider speciation in the gaseous phases. According to the literature, there are three main reactions controlling speciation in the gaseous phase:

$$\underset{\text{gas}}{CH_4} + \underset{\text{gas}}{2H_2O} \Leftrightarrow \underset{\text{gas}}{CO_2} + \underset{\text{gas}}{4H_2} \tag{9.115}$$

$$\underset{\text{gas}}{2NH_3} \Leftrightarrow \underset{\text{gas}}{N_2} + \underset{\text{gas}}{3H_2} \tag{9.116}$$

$$\underset{\text{gas}}{H_2S} \Leftrightarrow \underset{\text{solid}}{S} + \underset{\text{gas}}{H_2}. \tag{9.117}$$

Giggenbach (1980) expressed the thermodynamic constants of equilibria 9.115 and 9.116 as semilogarithmic functions of absolute T:

$$\log K_{115} = 10.76 - \frac{9323}{T} \tag{9.118}$$

$$\log K_{116} = 11.81 - \frac{5400}{T}. \tag{9.119}$$

Assuming the gaseous phase to be ideal, we can substitute partial pressures for the thermodynamic activities of the components:

$$K_{115} = \frac{P_{CO_2} \cdot P_{H_2}^4}{P_{CH_4} \cdot P_{H_2O}^2} \tag{9.120}$$

$$K_{116} = \frac{P_{N_2} \cdot P_{H_2}^3}{P_{NH_3}^2}. \tag{9.121}$$

Substituting molar fractions for partial pressures—i.e.,

$$P_i = X_{i,g} P_t, \tag{9.122}$$

where P_t is total pressure, we obtain

$$K_{115} = \frac{X_{CO_2} \cdot X_{H_2}^4}{X_{CH_4} \cdot X_{H_2O}^2} \cdot P_t^2 \tag{9.123}$$

$$K_{116} = \frac{X_{N_2} \cdot X_{H_2}^3}{X_{NH_3}^2} \cdot P_t^2. \tag{9.124}$$

For the evaluation of P_t, Giggenbach (1980) suggested an approximation based on the mean solubility of the main gaseous component (usually CO_2), because

$$P_t = P_{H_2O} + \sum_i P_i, \tag{9.125}$$

where the P_i terms are the partial pressures of gaseous components other than H_2O, assuming

$$\sum_i P_i = P_t X_g F_g \overline{K}'_{CO_2}, \tag{9.126}$$

where X_g is the fraction of gas (non-H_2O) in the system and F_g is the fraction of vapor plus non-H_2O gas in the system, one has:

$$P_t = \frac{P_{H_2O}}{\left(1 - X_g F_g \overline{K}'_{CO_2}\right)}. \tag{9.127}$$

The vapor pressure along the univariant curve of pure water can be expressed as a semilogarithmic function of the reciprocal of absolute temperature (Giggenbach, 1980):

$$\log P_{H_2O} = 5.51 - \frac{2048}{T}. \tag{9.128}$$

For aqueous solutions with mean saline amounts, a similar expression holds:

$$\log P_{H_2O} = 4.90 - \frac{1820}{T} \tag{9.129}$$

(Giggenbach, 1987). Rearranging equations 9.123 and 9.124 in light of equations 9.126 and 9.127, and accounting for equations 9.118, 9.119, and 9.128, we obtain

$$T(K) = \frac{5227}{\log X_{CH_{4,g}} - \log X_{CO_{2,g}} - 4\log X_{H_{2,g}} - 0.26 + 4\log\left(1 - X_g F_g \overline{K}'_{CO_2}\right)} \tag{9.130}$$

$$T(K) = \frac{1304}{2\log X_{NH_{3,g}} - \log X_{N_{2,g}} - 3\log X_{H_{2,g}} + 0.79 + 2\log\left(1 - X_g F_g \overline{K}'_{CO_2}\right)} \tag{9.131}$$

The precision of thermobarometric equations 9.130 and 9.131 (once T is known, P is also fixed by the water-vapor univariant curve) depends on the accuracy of the last term on the right, which becomes more precise as the fractional amount of gas in vapor X_g falls. Rearranging equations 9.130 and 9.131 with the introduction of mass distribution constants of the type defined in equation 9.102, Giggenbach (1980) transformed equations 9.130 and 9.131 into thermobarometric functions based on the chemistry of the fluid.

More recently, Chiodini and Cioni (1989) proposed a modification of equilibrium 9.115:

$$\underset{\text{gas}}{CH_4} + \underset{\text{gas}}{3CO_2} \Leftrightarrow \underset{\text{gas}}{4CO} + \underset{\text{liquid}}{2H_2O} \tag{9.132}$$

$$\underset{\text{gas}}{CO_2} + \underset{\text{gas}}{H_2} \Leftrightarrow \underset{\text{gas}}{CO} + \underset{\text{liquid}}{H_2O} \tag{9.133}$$

(substituting H_2O vapor for liquid H_2O in the two equilibria and subtracting from equilibrium 9.132 equilibrium 9.133 multiplied by 4, we return to equilibrium 9.115).

The substitution of H_2O vapor for liquid H_2O simplifies the equilibrium. Because the activity of liquid H_2O is almost 1, we obtain

$$K_{132} = \frac{X_{CO,g}^4}{X_{CH_4,g} \cdot X_{CO_2,g}^3}. \tag{9.134}$$

Based on thermodynamic data (Barin and Knacke, 1973), Chiodini and Cioni (1989) obtained

$$\log K_{132} = 8.065 - \frac{13606}{T}, \tag{9.135}$$

with T in K, which, combined with equation 9.134, gives

$$T(K) = \frac{13,606}{\log X_{CH_4,g} + 3 \log X_{CO_2,g} - 4 \log X_{CO,g} + 8.065}. \tag{9.136}$$

As we see, P_t does not appear in equation 9.136 but can be calculated, based on equilibrium 9.133 and assuming

$$P_t \approx P_{H_2O} + P_{CO_2} \tag{9.137}$$

$$\log P_t = 9.083 - \frac{2094}{T} + \log X_{CO,g} - \log X_{H_2,g} \tag{9.138}$$

(Chiodini and Cioni, 1989).

Let us now consider equilibrium 9.117. The formation of H_2S obviously implies the presence of sulfur in the system (but not necessarily native sulfur). The main crystalline sulfur compound in geothermal fields is always pyrite FeS_2. Equilibrium 9.117 can thus be rewritten as follows:

$$2H_2S + Fe^{2+} \Leftrightarrow 2H^+ + H_2 + FeS_2. \tag{9.139}$$
$$\text{gas} \quad \text{aqueous} \qquad \text{aqueous gas} \quad \text{crystal}$$

To be solved in thermometric fashion, equilibrium 9.139 requires knowledge of the pH of the solution and the amount of dissolved ferrous iron.

D'Amore and Panichi (1980) proposed an empirical geothermometer based on a modified form of the decomposition reaction of methane:

$$C + CO_2 + 6H_2 \Leftrightarrow 2CH_4 + 2H_2O \tag{9.140}$$
$$\text{crystal} \quad \text{gas} \qquad \text{gas} \qquad \text{gas} \qquad \text{liquid}$$

Table 9.14 Standard partial molal properties and coefficients of HKF model for neutral molecules in aqueous solution. $\overline{G}^0_{f,P_r,T_r}$ and $\overline{H}^0_{f,P_r,T_r}$ in cal/mole; $\overline{S}^0_{P_r,T_r}$ in cal/(mole × K)); $\overline{V}^0_{P_r,T_r}$ in cm³/mole (Shock et al., 1989).

Property	$Ne_{(a)}$	$Kr_{(a)}$	$Rn_{(a)}$	$He_{(a)}$	$Ar_{(a)}$	$Xe_{(a)}$
$\overline{G}^0_{f,P_r,T_r}$	4565	3554	2790	4658	3900	3225
$\overline{H}^0_{f,P_r,T_r}$	−870	−3650	−5000	−150	−2870	−4510
$\overline{S}^0_{P_r,T_r}$	16.74	15.06	16.0	14.02	14.30	14.62
C_p	39.7	57.6	83.0	37.3	52.8	64.8
$\overline{V}^0_{P_r,T_r}$	20.4	33.5	54.1	13.0	31.71	42.6
a_1	4.4709	6.2721	9.0862	3.4722	6.0003	7.5196
$a_2 \times 10^{-2}$	3.1352	7.5333	14.4045	0.6967	6.8698	10.5793
a_3	4.5177	2.7891	0.0884	5.4762	3.0499	1.5919
$a_4 \times 10^{-4}$	−2.9086	−3.0904	−3.3745	−2.8078	−3.0630	−3.2163
c_1	27.0977	37.8226	52.5774	26.0707	35.6471	42.1036
$c_2 \times 10^{-4}$	5.0523	8.6985	13.8725	4.5634	7.3160	10.1652
$\omega \times 10^{-5}$	−0.2535	−0.2281	−0.2423	−0.2123	−0.3073	−0.2214

Property	$N_{2(a)}$	$SO_{2(a)}$	$H_{2(a)}$	$O_{2(a)}$	$NH_{3(a)}$	$CO_{2(a)}$
$\overline{G}^0_{f,P_r,T_r}$	4347	−71,980	4236	3954	−6383	−92,250
$\overline{H}^0_{f,P_r,T_r}$	−2495	−77,194	−1000	−2900	−19,440	−98,900
$\overline{S}^0_{P_r,T_r}$	22.9	38.7	13.8	26.04	25.77	28.1
C_p	56.0	46.6	39.9	56.0	17.9	58.1
$\overline{V}^0_{P_r,T_r}$	33.3	38.5	25.2	30.38	24.43	32.8
a_1	6.2046	6.9502	5.1427	5.7889	5.0911	6.2466
$a_2 \times 10^{-2}$	7.3685	9.1890	4.7758	6.3536	2.7970	7.4711
a_3	2.8539	2.1383	3.8729	3.2528	8.6248	2.8136
$a_4 \times 10^{-4}$	−3.0836	−3.1589	−2.9764	−3.0417	−2.8946	−3.0879
c_1	35.7911	31.2101	27.6251	35.3530	20.3	40.0325
$c_2 \times 10^{-4}$	8.3726	6.4578	5.0930	8.3726	−1.17	−8.8004
$\omega \times 10^{-5}$	−0.3468	−0.2461	−0.2090	−0.3943	−0.05	−0.02

Property	$HF_{(a)}$	$H_2S_{(a)}$	$B(OH)_{3(a)}$	$HNO_{3(a)}$	$SiO_{2(a)}$	$H_3PO_{4(a)}$
$\overline{G}^0_{f,P_r,T_r}$	−71,662	−6673	−231,540	−24,730	−199,190	−273,100
$\overline{H}^0_{f,P_r,T_r}$	−76,835	−9001	−256,820	−45,410	−209,775	−307,920
$\overline{S}^0_{P_r,T_r}$	22.5	30.0	37.0	42.7	18.0	38.0
C_p	14.0	42.7	12.0	18.0	−76.1	23.6
$\overline{V}^0_{P_r,T_r}$	12.5	34.92	39.22	–	16.1	48.1
a_1	3.4753	6.5097	7.0643	–	1.9	8.2727
$a_2 \times 10^{-2}$	0.7042	6.7724	8.8547	–	1.7	12.4182
a_3	5.4732	5.9646	3.5844	–	20.0	0.8691
$a_4 \times 10^{-4}$	−2.8081	−3.0590	−3.1451	–	−2.7	−3.2924
c_1	14.3647	32.3	11.3568	15.2159	29.1	17.9708
$c_2 \times 10^{-4}$	−0.1828	4.73	−0.5902	1.2635	−51.2	1.7727
$\omega \times 10^{-5}$	−0.0007	−0.10	−0.2	−0.36	0.1291	−0.22

and on a multiple buffer involving carbonates, oxides, sulfides, and sulfates:

$$
\underset{\text{crystal}}{CaSO_4} + \underset{\text{crystal}}{FeS_2} + \underset{\text{liquid}}{3H_2O} + \underset{\text{gas}}{CO_2} \Leftrightarrow \underset{\text{crystal}}{CaCO_3} + \underset{\text{crystal}}{\frac{1}{3}Fe_3O_4} + \underset{\text{gas}}{3H_2S} + \underset{\text{gas}}{\frac{7}{3}O_2}. \quad (9.141)
$$

The thermometric equation of D'Amore and Panichi (1980) has the form

$$
T(K) = \frac{24,775}{2\log\left(\dfrac{X_{CH_4,g}}{X_{CO_2,g}}\right) - 6\log\left(\dfrac{X_{H_2,g}}{X_{CO_2,g}}\right) - 3\log\left(\dfrac{X_{H_2S,g}}{X_{CO_2,g}}\right) - 7\log P_{CO_2} + 36.05}.
$$

$$(9.142)$$

In table 9.14, for the sake of completion, we list the thermodynamic parameters of the HKF model concerning neutral molecules in solution (Shock et al., 1989). Calculation of partial molal properties of solutes (see section 8.11), combined with calculation of thermodynamic properties in gaseous phases (Table 9.5), allows rigorous estimates of the various equilibrium constants at all P and T of interest.

PART FOUR

Methods

CHAPTER TEN

Trace Element Geochemistry

10.1 Assimilation of Trace Elements in Crystals

The presence of trace elements in minerals can be ascribed to at least four main phenomena (McIntire, 1963):

1. *Surface adsorption.* Foreign ions are kept in a diffuse sheet at the surface of the crystal, as a result of electrostatic interaction with surface atoms whose bonds are not completely saturated.
2. *Occlusion.* Adsorbed impurities on the surface of the crystal are trapped by subsequent strata during crystal accretion.
3. *Substitutional solid solution with a major component.* The trace element substitutes for a major element in a regular position of the crystal lattice.
4. *interstitial solid solution.* Similar to the preceding phenomenon, but here the trace element occupies an interstitial position in the crystal lattice.

The first two processes may be operative at ultratrace concentration levels. The first one is relevant whenever the mineral has a high surface-to-mass ratio, as in the case of colloids. The last two processes are by far the most important in geochemistry and can be appropriately described in thermodynamic terms.

10.2 Raoult's and Henry's Laws

As we saw in section 3.8.1, Raoult's law describes the properties of an ideal solution, in which the thermodynamic activity of the component is numerically equivalent to its molar concentration:

$$a_i = X_i. \qquad (10.1)$$

In dilute solutions, the chemical potential of the trace component is given by

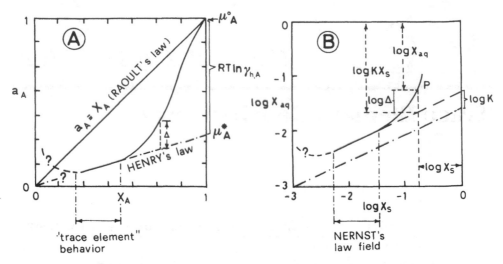

Figure 10.1 (A) Activity–molar concentration plot. Trace element concentration range is shown as a zone of constant slope where Henry's law is obeyed. Dashed lines and question marks at high dilution: in some circumstances Henry's law has a limit also toward infinite dilution. The intercept of Henry's law slope with ordinate axis defines Henry's law standard state chemical potential. (B) Deviations from Nernst's law behavior in a logarithmic plot of normalized trace/carrier distribution between solid phase s and ideal aqueous solution aq. Reproduced with modifications from Iiyama (1974), *Bullettin de la Societée Francaise de Mineralogie et Cristallographie*, 97, 143–151, by permission from Masson S.A., Paris, France. Δ in part A and log Δ in part B have the same significance, because both represent the result of deviations from Henry's law behavior in solid.

$$\mu_i = \mu_i^0 + RT \ln \gamma_i X_i, \tag{10.2}$$

where γ_i is the activity coefficient deriving from solute-solvent interactions. In the concentration field of validity of Henry's law, activity coefficient $\gamma_{h,i}$ is constant, owing to the fact that, although a solute-solvent interaction exists, it is not affected by the solute concentration, which is too low to influence the thermochemical properties of the solid. We can thus embody the interaction effects at concentration levels within the Henry's law field into a particular standard state condition called "Henry's law standard state," which gives

$$\mu_i = \mu_i^* + RT \ln X_i. \tag{10.3}$$

The relationship between the Henry's law activity coefficient of component i $\gamma_{h,i}$ and the Henry's law standard state potential μ_i^* is then

$$\gamma_{h,i} = \exp\left(\frac{\mu_i^* - \mu_i^0}{RT}\right) \equiv \exp\left(\frac{\Delta\mu_i^*}{RT}\right). \tag{10.4}$$

The excess Gibbs free energy term $\Delta\mu_i^*$ corresponds to the amount of energy required to transfer one mole of trace component i from an ideal solid solution to the solid solution of interest. As shown in figure 10.1, the more μ_i^* differs from μ_i^0, the more angular coefficient $\gamma_{h,i}$ differs from 1, which represents the condition of Raoult's law. Trace element distribution between phases of geochemical interest is governed by Henry's law only within restricted solid solution ranges (here we use the term "solution" instead of "mixture," because the Henry's law standard state of the trace element differs from the standard state adopted for the carrier, which is normally that of pure solvent at the P and T of interest: cf. section 2.1).

10.3 High Concentration Limits for Henry's Law in Crystalline Solutions

Let us consider the distribution of a trace element A and a carrier B between a crystal and an ideal aqueous solution at equilibrium. Defining

$$X_s = \frac{X_A}{X_B} \tag{10.5}$$

the molar ratio in the solid and

$$X_{aq} = \frac{m_A}{m_B} \tag{10.6}$$

as the molality ratio in aqueous solution, the equilibrium distribution for two ideal solutions is given by

$$X_{aq} = K_{(P,T)}X_s, \tag{10.7}$$

where $K_{(P,T)}$ is the thermodynamic constant valid at given P-T conditions. If solute-solvent interactions in the solid are embodied in the definition of Henry's law standard state, equation 10.7 is also valid within the concentration limits of Henry's law in the solid. In a logarithmic plot of concentrations (figure 10.1B), the validity zone of equation 10.7 is represented by a straight line with an angular coefficient of 1, or "Nernst's law range," and the intercept of its prolongation on the ordinate axis at $\log X_s = 0$ defines the value of $K_{(P,T)}$. Deviations from Henry's law behavior in solid are functions of the trace element concentration in the system and can be expressed as

$$\log \Delta = \log X_{aq} - \left(\log K + \log X_s\right). \tag{10.8}$$

If the solid solution is regular, with an interaction parameter W (cf. section 3.8.4), the equilibrium distribution curve is defined by

$$\ln X_{aq} = \ln K + \ln X_s + \left(1 - 2X_{A,s}\right)\frac{W}{RT} \qquad (10.9)$$

(Iiyama, 1974). Posing $K + W/RT = K'$ and transforming the logarithms into common logarithms, we obtain

$$\log X_{aq} = \log K' + \log X_s - \frac{2W}{RT} X_{A,s} \log e. \qquad (10.10)$$

Adopting equation 10.10, Iiyama (1974) calculated values of W necessary to justify observed deviations from Henry's law behavior for trace elements in silicates. The resulting theoretical equilibrium distributions are shown in Figure 10.2A and B. As pointed out by the author, deviations from linearity in log-log distributions observed at molar concentration levels around 10^{-3} to 10^{-2} (as frequently happens in mineral-solution hydrothermal equilibria; see, for example, Roux, 1971a,b; Iiyama, 1972, 1974; Volfinger, 1976) imply extremely high values of W—i.e., strong interactions between solute and solvent in the crystal structure.

Alternative ways of accounting for precocious deviations from ideality in solid-liquid distributions involve consideration of configurational entropy effects in the solid solution process. Two of these are the "two ideal sites" model (Matsui and Banno, 1962; Grover and Orville, 1969) and the "local lattice distortion" model (Iiyama, 1974).

10.3.1 Two Ideal Sites Model

If substitution of a trace element for a carrier takes place in two distinct lattice positions in the crystal, the distribution equation

$$X_{aq} = KX_s \qquad (10.11)$$

is replaced by

$$X_{I,s} = K_I X_{aq} \qquad (10.12)$$

$$X_{II,s} = K_{II} X_{aq}, \qquad (10.13)$$

where subscripts I and II identify magnitudes relative to sites of types I and II in the crystal. If we call m and n, respectively, the relative proportions of sites of

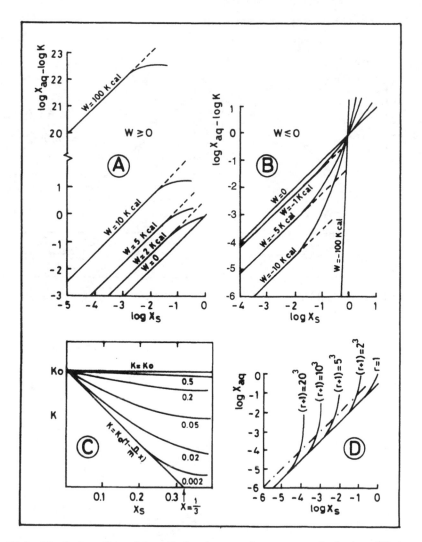

Figure 10.2 Deviations from Nernst's law in crystal–aqueous solution equilibria, as obtained from application of various thermodynamic models. (A and B) Regular solution (Iiyama, 1974). (C) Two ideal sites model (Roux, 1971a). (D) Model of local lattice distortion (Iiyama, 1974). Reprinted from Ottonello (1983), with kind permission of Theophrastus Publishing and Proprietary Co.

types I and II in the crystal ($m + n = 1$), the relationship between partial distribution constants K_I, K_{II}, and K is

$$K = \frac{mK_I + nK_{II} + X_{aq}K_I K_{II}}{1 + X_{aq}\left(nK_I + mK_{II}\right)}. \tag{10.14}$$

Eliminating X_{aq} from equations 10.11 and 10.14, we obtain an expression of K as a function of K_s:

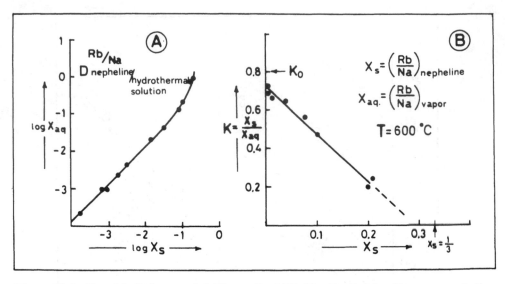

Figure 10.3 Two ideal sites model. Normalized Rb/Na distribution between nepheline (X_s) and hydrothermal solution (X_{aq}) (data from Roux, 1974).

$$K = \frac{1}{2} K_0 \left\{ 1 - aX_s + \left[\left(1 - aX_s\right)^2 + bX_s \right]^{1/2} \right\},$$ (10.15)

where

$$K_0 = mK_I + nK_{II}$$ (10.16)

$$a = \frac{mK_{II} + nK_I}{K_0}$$ (10.17)

$$b = \frac{K_I K_{II}}{K_0^2}.$$ (10.18)

Values of equation 10.15 are plotted in figure 10.2C for various K_{II}/K_I ratios between the limiting cases:

$$K = K_0 \quad \text{when} \quad \frac{K_{II}}{K_I} = 1$$ (10.19)

and

$$K = K_0 \left(1 - \frac{n}{m} X_s \right) \quad \text{when} \quad \frac{K_{II}}{K_I} = 0.$$ (10.20)

As we see, the more K_I differs from K_{II} and the more X_s approaches m/n, the more marked deviation from ideality becomes.

It was shown by Roux (1974) that the two ideal sites model applies perfectly to partitioning of Rb distribution between nepheline and hydrothermal solutions. Based on the experimental work of Roux (1974), deviation from ideality in the normalized Rb/Na distribution between nepheline (X_s) and hydrothermal solution (X_{aq}) is detectable for X_s values higher than 10^{-2} (figure 10.3A) Distribution constant K in the Nernst's law range is $K_0 = 0.82$ (Roux, 1974), and the modification of K with increasing X_s is well described by equation 10.20, within experimental approximation (figure 10.3B).

It may be noted in figure 10.3B that prolongation of the interpolant encounters the abscissa axis at $X_s = \frac{1}{3}$, which is exactly the proportion of the two structural sites where Rb/Na substitutions take place in nepheline. According to figure 10.3B and equation 10.20, it is plausible to admit that Rb is fixed almost exclusively in the larger of the two structural sites.

10.3.2 Local Lattice Distortion Model

This model (Iiyama, 1974) can be applied whenever trace-carrier substitution takes place in a single site within the crystal. According to this model, when trace element A substitutes for carrier B in a given lattice position, it causes local deformation of the structure in the neighborhood of the site where substitution takes place. This deformation, which increases as the crystal-chemical properties of trace element and carrier differ, prevents further substitutions from occurring in the vicinity of the site. A "forbidden zone" of r sites is thus created, clearly modifying the configurational properties of the phase.

In a crystal containing N sites occupied by B and A atoms, the number of possible configurations as a function of dimensions of the forbidden zone (in terms of r site positions) is given by

$$Q = \frac{N\big[N - (r+1)\big]\big[N - 2(r+1)\big]\cdots\big[N - (N_A - 1)(r+1)\big]}{N_A!} \tag{10.21}$$

(cf. section 3.8.1). The entropy of mixing is

$$S_{mixing} = k \ln Q, \tag{10.22}$$

where k is Boltzmann's constant. Applying the equality

$$N_B = N - N_A \tag{10.23}$$

and the Stirling approximation

$$\ln N! = N(\ln N - 1),$$ (10.24)

the entropy of mixing is

$$S_{mixing} = -k\left[N_A \ln \frac{N_A}{N} + \left(\frac{N_B - rN_A}{r+1}\right)\ln\left(\frac{N_B - rN_A}{N}\right)\right].$$ (10.25)

Considering an athermal solution ($H_{mixing} = 0$ and $G_{mixing} = -T \times S_{mixing}$; see section 3.8.3) composed of n_A moles of AM and n_B moles of BM, where M is the common anionic group, we obtain

$$G_{mixing} = RT\left[n_A \ln\left(\frac{n_A}{n_A + n_B}\right) + \left(\frac{n_B - rn_A}{r+1}\right)\ln\left(\frac{n_B - rn_A}{n_A + n_B}\right)\right].$$ (10.26)

The excess Gibbs free energy of mixing is

$$G_{excess\,mixing} = G_{mixing} - G_{ideal\,mixing},$$ (10.27)

where $G_{ideal\,mixing}$ is the Gibbs free energy of mixing when $r = 0$:

$$G_{excess\,mixing} = RT\left[\left(\frac{n_B - rn_A}{r+1}\right)\ln\left(\frac{n_B - rn_A}{n_A + n_B}\right) - n_B \ln\left(\frac{n_B}{n_A + n_B}\right)\right].$$ (10.28)

Differentiating equation 10.28 with respect to n_A, we obtain the activity coefficient of component AM in the solid solution (B_X, A_{1-X})M:

$$\ln \gamma_A = \left(\frac{\partial G_{excess\,mixing}}{\partial n_A}\right) = \frac{-r}{r+1}\ln(X_B - X_A),$$ (10.29)

where

$$X_A = \frac{N_A}{N} = \frac{n_A}{n_A + n_B}$$ (10.30)

and

$$X_B = \frac{N_B}{N} = \frac{n_B}{n_A + n_B}.$$ (10.31)

Figure 10.2D shows normalized distributions of elements A and B for various values of r. The more the forbidden zone (in terms of r sites) is large, the more precocious are deviations from ideality. Although the approximation to athermal solutions is not strictly valid in the case of trace elements in crystals, solid/liquid

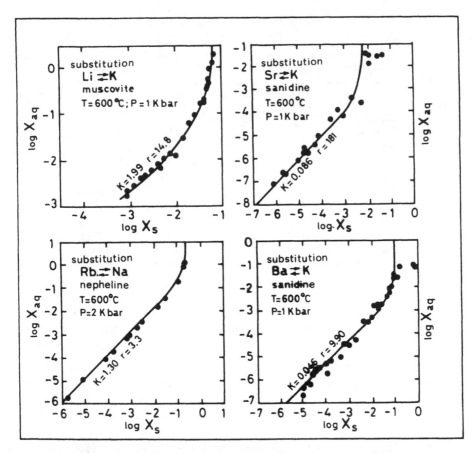

Figure 10.4 Applications of the local lattice distortion model to mineral-hydrothermal solution equilibria. Experimental data from Iiyama (1972) for Ba/K and Sr/K, Volfinger (1970) for Li/K, and Roux (1971b) for Rb/Na. Reprinted from Iiyama (1974), *Bullettin de la Societée Francaise de Mineralogie et Cristallographie,* 97, 143–151, with permission from Masson S.A., Paris, France.

distributions calculated with the local lattice distortion model reproduce experimental evidence fairly well. As described by Iiyama (1974), this occurs because any enthalpic contribution to the Gibbs free energy of mixing in the solid phase (H_{mixing}), arising from solute-solvent interactions, varies linearly with X_A in the Henry's law concentration field.

Figure 10.4 shows normalized Ba/K and Sr/K distributions between sanidine and a hydrothermal solution (Iiyama, 1972), Li/K between muscovite and a hydrothermal solution (Voltinger, 1970), and Rb/Na between nepheline and a hydrothermal solution (Roux, 1971b), interpreted through the local lattice distortion model, by an appropriate choice of the Nernst's law mass distribution constant K and the lattice distortion propagation factor r.

Table 10.1 lists r values calculated for normalized distributions of alkalis and alkaline earths between silicates and hydrothermal solutions and for normalized distributions of rare earth elements (REE) between pyroxenes and silicate melts.

Table 10.1 Linear propagation of lattice distortion (in terms of number of sites r) calculated for silicate/hydrothermal solutions (Iiyama, 1974) and silicate/melt equilibria (Ottonello et al., 1978).

Mineral	Trace Element	Carrier	r
Muscovite	Li	K	2.45
Sanidine	Sr	K	5.65
Albite	Ba	K	2.15
Albite	Sr	Na	3.22
Nepheline	Rb	Na	7.00
Clinopyroxene	Nd	Ca	10.00
Clinopyroxene	Ce	Ca	10.00
Clinopyroxene	Lu	Ca	75.00
Orthopyroxene	Lu	Mg	70.00
Orthopyroxene	La	Mg	165.00

Table 10.2 lists upper concentration limits for Henry's law behavior of various trace elements in silicate crystals for specific P and T conditions. In some cases, these limits are only approximate or conjectural. The table is not exhaustive, because it reflects the state of the art in the early 1980s and is subject to substantial modifications in response to the advance of experimental knowledge since that time.

10.4   Lower Concentration Limits for Henry's Law Behavior in Silicate Crystals

Henry's law in crystals has low concentration limits (figure 10.1A), ascribable to modifications in the solid solution process, which, at high dilution, is greatly affected by the intrinsic and extrinsic defects in the crystal (Wriedt and Darken, 1965; Morlotti and Ottonello, 1982). Figure 10.5 shows some experimental evidence concerning solid solutions of REE in silicates. Deviations from Nernst's law observed in figure 10.5 may be ascribed to defect equilibria between REE and cationic vacancies present in the solid (De Vore, 1955; Wood, 1976c; Harrison and Wood, 1980; Morlotti and Ottonello, 1982; see also chapter 4).

In logarithmic plots of simple weight concentrations, as in figure 10.5, the Nernst's law region appears as a straight line with an angular coefficient of 1. At the low concentration limits for Henry's law behavior in the solid, we observe a sudden change in slope, and, at high dilution, the new distribution has an angular coefficient lower than 1 (i.e., $\alpha < 45°$). Both the slope rupture point and the angular coefficient valid at high dilution are dictated by the extrinsic defects in the solid (Morlotti and Ottonello, 1982).

Correct evaluation of solution mechanisms of a given trace element in silicate

Table 10.2 Upper concentration limits for Henry's law behavior in silicates and oxides. Abbreviations: Ab = albite; Sa = sanidine; Pl = plagioclase; Fel = alkaline feldspar; Rb-fel = Rb-feldspar; Ne = nepheline; Mu = muscovite; Ol = olivine; Di = diopside; Cpx = clinopyroxene; Oxp = orthypoxene; Amph = amphibole; Par = pargasite; Gr = garnet; Ilm = ilmenite; Arm = armalcolite (from Ottonello, 1983).

Equilibrium	T (°C)	P (bar)	Trace	Carrier	Henry's Law Limit	Reference
Ab–hydrothermal solution	600	2000	Rb	Na	(Rb/Na)Ab = 3×10^{-4}	1
	600	1000	Rb	Na	(Rb)Ab = 33 ppm	6
	600	880	Rb	Na	(Rb)Ab = 1600 ppm	3
	600	2000	Cs	Na	(Cs/Na)Ab = 3×10^{-5}	1
	600	1000	Cs	Na	(Cs)Ab = 5 ppm	6
	600	880	Cs	Na	(Cs)Ab = 500 ppm	3
	600	1000	Sr	Na	(Sr)Ab = 1340 ppm	6
	600	1000	Ca	Na	(Ca)Ab = 914 ppm	6
	600	880	K	Na	(K)Ab = 150 ppm	3
Sa–hydrothermal solution	800	1000	Cs	K	(Cs/K)Sa = 10^{-2}	7
	700	1000	Cs	K	(Cs)Sa = 60 ppm	8
	600	1000	Cs	K	(Cs)Sa = 716 ppm	6
	500	1000	Cs	K	(Cs/K)Sa = 5×10^{-3}	7
	400	1000	Cs	K	(Cs/K)Sa = 10^{-2}	9
	600	800	Cs	K	(Cs)Sa = 480 ppm	3
	600	1000	Rb	K	(Rb)Sa = 32,300 ppm	6
	600	800	Rb	K	(Rb)Sa $\approx$ 10,000 ppm	3
	600	1000	Ba	K	(Ba)Sa = 4920 ppm	6
	600	1000	Sr	K	(Sr)Sa = 157 ppm	6
	600	1000	Na	K	(Na)Sa = 8300 ppm	6
	600	800	Na	K	(Na)Sa = 8500 ppm	3
	600	1000	Ca	K	(Ca)Sa = 72 ppm	6
Pl–melt	1190	1	Sm	Ca	(Sm)Pl = 3800 ppm	4
	1190	1	Sm	Ca	(Sm)Pl $\approx$ 5000 ppm	5
	1190	1	Ba	Ca	(Ba)Pl = 14,500 ppm	4
	1190	1	Sr	Ca	(Sr)Pl = 70,200 ppm	4
Fel–melt	780	8000	Rb	K	(Rb)Fel $\approx$ 4000 ppm	10
	780	8000	Sr	K?Ca?	(Sr)Fel $\approx$ 3000 ppm	10
	780	8000	Ba	K?Ca?	(Ba)Fel $\approx$ 40,000 ppm	10
Rb-fel–hydrothermal solution	600	800	Na	Rb	(Na)Rb-fel = 4000 ppm	3
	600	800	K	Rb	(K)Rb-fel = 12,000 ppm	3
	600	800	Cs	Rb	(Cs)Rb-fel = 400 ppm	3
Ne–hydrothermal solution	600	2000	Rb	Na	(Rb/Na)Ne = 10^{-3}	1
	600	2000	Cs	Na	(Cs/Na)Ne = 10^{-4}	1
Mu–hydrothermal solution	600	1000	Li	K	(Li/K)Mu = 7×10^{-3}	11
	400	1000	Cs	K	(Cs/K)Mu = 10^{-4}	9
Ol–melt	1184	1	Ni	Mg	(Ni)Ol $\approx$ 47,000 ppm	12
	1350	1	Ni	Mg	(Ni)Ol $\approx$ 18,000 ppm	12
	1025	20,000	Ni	Mg	(Ni)Ol $\approx$ 700 ppm*	13
	1075	10,000	Ni	Mg	(Ni)Ol $\approx$ 1000 ppm*	13
	1025	20,000	Sm	?	(Sm)Ol $\approx$ 0.7 ppm?	14

continued

Table 10.2 (continued)

Equilibrium	T (°C)	P (bar)	Trace	Carrier	Henry's Law Limit	Reference
Di–melt	1300	1	Ni	Mg	(Ni)Di ≈ 80,000 ppm	15
	1300	1	Co	Mg	(Co)Di = 30,000 ppm	15
Cpx–melt	1025	20,000	Ni	Mg	(Ni)Cpx ≈ 20 ppm	16
	950	10,000	Sm	Ca	(Sm)Cpx = 17 ppm	17
Opx–melt	1025	20,000	Ni	Mg	(Ni)Opx ≈ 100 ppm?	16
	1075	10,000	Sm	Mg	(Sm)Opx = 1 ppm	14
Amph–melt	1000	15,000	Ni	Mg	(Ni)Amph ≈ 400 ppm?	16
Par–melt	1000	15,000	Sm	unknown	(Sm)Par = 1.8 ppm	17
Gr–melt	1025	20,000	Ni	unknown	(Ni)Gr ≈ 50 ppm	13
	950	20,000	Sm	unknown	(Sm)Gr = 8 ppm	18
Ilm–melt	1128	1	Zr	unknown	(Zr)Ilm = 0.5 ppm	19
	1127	1	Nb	unknown	(Nb)Ilm = 1.5 ppm	19
Arm–melt	1127	1	Zr	unknown	(Zr)Arm = 2 ppm	19
	1127	1	Nb	unknown	(Nb)Arm = 2.5 ppm	19

(1) Roux (1971a)   (2)Iiyama (1972)   (3) Lagache and Sabatier (1973)
(4) Drake and Weill (1975) (5) Drake and Holloway (1978) (6) Iiyama (1973)
(7) Eugster (1954)   (8) Lagache (1971)   (9) Volfinger (1976)
(10) Long (1978)   (11) Volfinger (1970)   (12) Leeman and Lindstrom (1978)
(13) Mysen (1976a)   (14) Mysen (1976b)   (15) Lindstrom and Weill (1978)
(16) Mysen (1977a)   (17) Mysen (1978)   (18) Mysen (1977b)
(19) McCallum and Charette (1978)
≈ Approximated values.
? Suggested values.
* Contrasting evidence (see Drake and Holloway, 1978).

crystals thus requires knowledge of the defects in the phases. As we saw in chapter 4, defect equilibria in crystals are conveniently described by referring to the f_{O_2} condition. For instance, oxygen deficiency may generate in the crystal doubly ionized Schottky vacancies $V_O^{..}$ or interstitials $M_i^{..}$ wheras oxygen excess may result in cationic vacancies $V_M^{\|}$. Intrinsic disorder is also always present under intrinsic oxygen partial pressure conditions $P_{O_2}^*$. Table 10.3 shows the energetically plausible defect equilibria that may affect the distribution of trivalent REE between pyrope $Mg_3Al_2Si_3O_{12}$ and silicate melt (see Kröger, 1964, for exhaustive treatment of the defect state of crystals; see also chapter 4 for the significance of the adopted symbols). The proposed equilibria, which refer to the various f_{O_2} conditions and trace element concentrations of interest, can be adapted to other silicate crystals with opportune modifications in the "carrier" notations (Mg, in this case). All the equilibria ruling the REE distribution within the various concentration ranges are then translated into a bulk distribution equation whose single terms on the right each apply to a specific concentration range. For instance, constant K_4 in

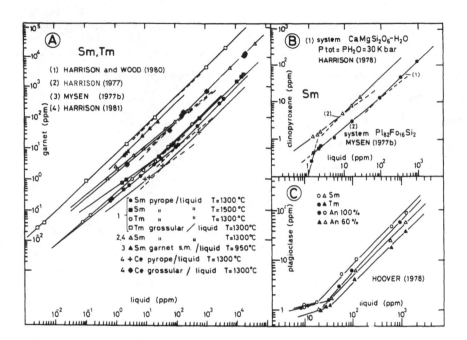

Figure 10.5 Experimentally observed deviations from Nernst's law behavior in solid/melt REE distribution. (A) Weight distribution of Sm and Tm between garnet and melt, experimental sources specified in figure. (B) Weight distribution of Sm between clinopyroxene and silicate melt. (C) Weight distribution of Sm and Tm between plagioclase and silicate melt. Systems are $An_{55}Di_{45}$ and $Ab_{60}An_{30}Di_{10}$ at $T = 1300$ °C and $P = 1$ bar. From R. Morlotti and G. Ottonello, Solution of rare earth elements in silicate solid phases: Henry's law revisited in light of defect chemistry, *Physics and Chemistry of Minerals*, 8, 87–97, figures 2–3, 1982, copyright © 1982 by Springer Verlag. Reprinted with the permission of Springer-Verlag GmbH & Co. KG.

table 10.5, relative to equilibrium (4) in the same table, describes defect process involving preexisting intrinsic defects, whereas constant K_7 describes the creation of extrinsic defects that take place when intrinsic defects are saturated (i.e., at higher concentrations).

Deviations from Nernst's law in log-distribution diagrams such as figure 10.5 are represented by the simple equation:

$$\ln \gamma_A = \ln\left(\frac{X_A^*}{X_A}\right) - \frac{1}{\tan \alpha}\left(\frac{X_A^*}{X_A}\right), \qquad (10.32)$$

where X_A is the trace element concentration in the solid, X_A^* is the lower concentration limit for Nernst's distribution, and $\tan \alpha$ is the slope of non-Nernstian distribution at high dilution (see table 10.4). Equation 10.32 can be interpreted in light of the defect equilibria listed in table 10.3, assuming that slope rupture

Table 10.3 Defect equilibria affecting REE solubility in pyrope and their effects in bulk garnet/melt REE distribution (Morlotti and Ottonello, 1982)

A: Oxygen fugacity corresponding to the intrinsic oxygen partial pressure: $f_{O_2} = P_{O_2}^*$

Defect equilibrium: $Mg_{Mg}^x \Leftrightarrow Mg_i^{\cdot\cdot} + V_{Mg}^{\parallel}$ (1)

Equilibrium constant: $K_F = [Mg_i^{\cdot\cdot}] [V_{Mg}^{\parallel}]^*$ (2)

Defect concentration: $[V_{Mg}^{\parallel}]^* = K_F^{1/2}$ (3)

Relative concentrations	Corresponding equilibria and their constants	
$[REE_{Mg}^{\cdot}] << [V_{Mg}^{\parallel}]^*$	$REE_{(l)}^{3+} + V_{Mg}^{\parallel} \Leftrightarrow REE_{Mg}^{\cdot}$	(4)
	$K_D = [REE_{Mg}^{\cdot}] / [REE_{(l)}^{3+}] = K_4 K_F^{1/2}$	(5)
$[REE_{Mg}^{\cdot}] \lesssim [V_{Mg}^{\parallel}]^*$	$K_D = K_4 K_F^{1/2} - K_4 [REE_{Mg}^{\cdot}]$	(6)
$[REE_{Mg}^{\cdot}] \gtrsim [V_{Mg}^{\parallel}]^*$	$REE_{(l)}^{3+} + \frac{3}{2} Mg_{Mg}^x \Leftrightarrow \frac{3}{2} Mg_{(l)}^{2+} + REE_{Mg}^{\cdot} + \frac{1}{2} V_{Mg}^{\parallel}$	(7)
	$K_D = K_7 [REE_{Mg}^{\cdot}]^{-1/2}$	(8)
$[REE_{Mg}^{\cdot}] \gtrsim [V_{Mg}^{\parallel}]^*$	$REE_{(l)}^{3+} \Leftrightarrow REE_{Mg}^{\cdot} + e^-$	(9)
	$K_D = K_9 [REE_{Mg}^{\cdot}]^{-1}$	(10)
$[REE_{Mg}^{\cdot}] >> [V_{Mg}^{\parallel}]^*$	$REE_{(l)}^{3+} + Al_{(l)}^{3+} + Mg_{Mg}^x + Si_{Si}^x \Leftrightarrow REE_{Mg}^{\cdot} + Al_{Si}^{\parallel} + Mg_{(l)}^{2+} + Si_{(l)}^{4+}$	(11)
	$K_D = K_{11} K_{D, Al} K_{D, Mg}^{-1} K_{D, Si}^{-1}$	(12)

Bulk distribution:

$$K_D = K_4 K_F^{1/2} - K_4 [REE_{Mg}^{\cdot}] + K_7 [REE_{Mg}^{\cdot}]^{-1/2} + K_9 [REE_{Mg}^{\cdot}]^{-1} + K_{11} K_{D, Al} K_{D, Mg}^{-1} K_{D, Si}^{-1} \quad (13)$$

B: Oxygen fugacity higher than the intrinsic oxygen partial pressure: $f_{O_2} > P_{O_2}^*$

Defect equilibrium: $\frac{1}{2} O_{2(gas)} \Leftrightarrow O_O^x + V_{Mg}^{\parallel} + 2h^{\cdot}$ (14)

Equilibrium constant: $K_{14} + 4 [V_{Mg}^{\parallel}]^3 f_{O_2}^{-1/2}$ (15)

Defect concentration: $[V_{Mg}^{\parallel}] = [V_{Mg}^{\parallel}]^* + 4^{-1/3} K_{14}^{1/3} f_{O_2}^{1/6}$ (16)

Bulk distribution:

$$K_D = K_4 K_F^{1/2} - K_4 [REE_{Mg}^{\cdot}] + 4^{-1/3} K_4 K_{14}^{1/3} f_{O_2}^{1/6} + K_7 [REE_{Mg}^{\cdot}]^{-1/2} + K_9 [REE_{Mg}^{\cdot}]^{-1} + K_{11} K_{D, Al} K_{D, Mg}^{-1} K_{D, Si}^{-1}$$
$$(17)$$

C: Oxygen fugacity lower than the intrinsic oxygen partial pressure: $f_{O_2} < P_{O_2}^*$

Defect equilibrium: $O_O^x \Leftrightarrow \frac{1}{2} O_{2(gas)} + V_O^{\cdot\cdot} + +2e^-$ (18)

Equilibrium constant: $K_{18} = 4 [V_O^{\cdot\cdot}]^3 f_{O_2}^{1/2}$ (19)

Defect concentration: $[V_{Mg}^{\parallel}] = [V_{Mg}^{\parallel}]^* - 4^{-1/3} K_{18}^{1/3} f_{O_2}^{-1/6}$ (20)

Bulk distribution:

$$K_D = K_4 K_F^{1/2} - K_4 [REE_{Mg}^{\cdot}] - 4^{-1/3} K_4 K_{18}^{1/3} f_{O_2}^{1/6} + K_7 [REE_{Mg}^{\cdot}]^{-1/2} + K_9 [REE_{Mg}^{\cdot}]^{-1} + K_{11} K4 K_{D, Al} K_{D, Mg}^{-1} K_{D, Si}^{-1}$$
$$(21)$$

Table 10.4 Least-squares fits of type $\ln X_{solid} = a \ln X_{liquid} + b$ relating REE concentrations in silicate crystals and melts, as derived from existing experimental data. Corresponding values of R2 correlation parameter are also listed (from Morlotti and Ottonello, 1982).

Element	Solid	T(°C)	P (bar)	Low Concentration				High Concentration			
				a	b	α	R2	a	b	α	R2
Ce	Grossular	1300	30,000	0.8444	0.4091	40.18	0.9998	1.0086	−0.4544	45.25	0.9998
Ce	Pyrope	1300	30,000	0.7159	−1.6725	35.60	0.99998	0.9827	−2.6263	44.50	1
Sm	Grossular	1300	30,000	0.8305	0.5522	39.71	0.9986	0.9813	−0.1343	44.46	0.9999
Sm	Pyrope	1300	30,000	0.8010	−1.3151	38.70	0.9986	1.0086	−2.2824	45.25	0.99998
Sm	Pyrope	1500	30,000	0.7079	−0.7346	35.29	0.99998	0.9712	−1.6399	44.16	0.9997
Sm	Garnet s.s.	950	20,000	–	–			0.9469	1.21	43.44	0.991
Tm	Grossular	1300	30,000	0.9387	1.8198	43.19	0.9997	0.9878	1.6158	44.65	0.99999
Tm	Pyrope	1300	30,000	0.8535	−1.1110	40.48	0.9991	0.9537	−1.4400	43.64	0.9994
Sm	Clinopyroxene	950	20,000	0.786	−0.125	38.17	0.997	0.837	−0.249	39.93	0.994
Sm	Diopside	1400	30,000	0.8385	−1.177	39.98	0.990	0.9181	−1.442	42.56	0.9997
Sm	Anorthite	1300	1	0.2745	−0.4857	15.35	0.996	1.0389	−3.046	46.09	0.997
Sm	Plagioclase	1300	1	0.1931	−0.224	10.92	0.68	0.964	−2.379	43.95	0.998
Tm	Anorthite	1300	1	0.580	−1.786	30.11	0.94	0.975	−3.284	44.27	0.999
Tm	Plagioclase	1300	1	–	–			0.976	−2.219	44.30	0.998

points X_A^* represent saturation of intrinsic and extrinsic defects. In the case of samarium, for instance,

$$X_{Sm}^* = \frac{2}{3}\left[V_{Mg}^{\parallel}\right]^*$$

(10.33)

and

$$X_{Sm} = \left\{\frac{2}{3}\left[V_{Mg}^{\parallel}\right]^* - \left[V_{Mg}^{\parallel}\right]\right\}.$$

(10.34)

Equation 10.32 can thus be rewritten in the form

$$\ln\gamma_{Sm} = \left(\frac{1-\tan\alpha}{\tan\alpha}\right)\ln\left\{\frac{3X_{Sm}}{2\left[V_{Mg}^{\parallel}\right]^*}\right\}.$$

(10.35)

10.5 Fractionation Operated by Crystalline Solids: Onuma Diagrams

As outlined in section 10.1, the presence of trace elements in crystals is attributable to several processes, the most important one being the formation of substitutional solid solutions. The "ease" of substitution depends on the magnitude of interactions between trace element and carrier. We have already seen (section 3.8.4) that macroscopic interaction parameter W can be related to microscopic interactions in a regular solution of the zeroth principle:

$$W = N_0\left(\frac{2w_{AB} - w_{AA} - w_{BB}}{2}\right).$$

(10.36)

In qualitative terms, microscopic interactions are caused by differences in crystal chemical properties of trace element and carrier, such as ionic radius, formal charge, or polarizability. This type of reasoning led Onuma et al. (1968) to construct semilogarithmic plots of conventional mass distribution coefficients K' of various trace elements in mineral/melt pairs against the ionic radius of the trace element in the appropriate coordination state with the ligands. An example of such diagrams is shown in figure 10.6.

The arrangement of elements of the same charge gives rise to more or less smooth curves with maxima located at "optimum site dimensions" for trace-

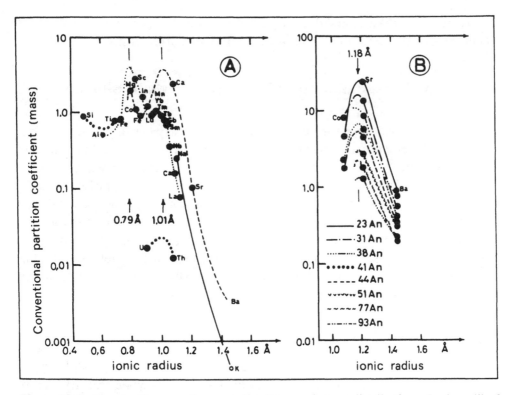

Figure 10.6 Onuma diagrams for crystal/melt trace element distributions. Ionic radii of Whittaker and Muntus (1970). (A) Augite/matrix distribution, data of Onuma et al. (1968). (B) Plagioclase/matrix distribution, data of Dudas et al. (1971), Ewart and Taylor (1969), and Philpotts and Schnetzler (1970). Reprinted from B. Jensen, *Geochimica et Cosmochimica Acta,* 37, 2227–2242, copyright © 1973, with kind permission from Elsevier Science Ltd., The Boulevard, Langford Lane, Kidlington 0X5 1GB, UK.

carrier substitutions in the crystal of interest (Onuma et al., 1968; Jensen, 1973). Figure 10.6A, for instance, shows the mass distribution of various trace elements between augitic pyroxene and the silicate matrix (assumed to represent the silicate melt), based on the data of Onuma et al. (1968). There are two maxima, at 0.79 and 1.01 Å, interpreted respectively as "optimum dimensions" of sites $M1$ and $M2$ in the crystal (Jensen, 1973). The peaks are more clearly defined for divalent ions, and the partition coefficients decrease more rapidly for elements with higher formal charges, departing from the optimum dimension. Figure 10.3B shows an Onuma diagram focused on the mass distribution of alkaline earths between plagioclase and the matrix. The distribution curves indicate the marked compositional control exerted by the solid: because the carrier is Ca, the mass distribution coefficients progressively increase with the amount of anorthite component in solid solution, whereas fractionation is not particularly affected. The main advantage of Onuma diagrams is that they can elucidate the marked fractionation induced by the crystal structure in crystal/melt equilibria. However, their interpretation is merely qualitative and does not currently allow translation of the observed phenomena into solution models for the solid phases of interest.

10.6   Assimilation of Trace Elements in Silicate Melts

As we will see in section 10.9, trace element geochemistry has important applications in the investigation of magmatic processes. However, in partial melting and partial crystallization models based on trace elements, the role ascribed to the liquid phase is often subordinate to that of the solids undergoing melting or crystallization. In such models, in fact, the fractionation properties of solids are clearly defined, but the silicate melt is often regarded as a sort of reservoir that does not exercise any structural control on trace partitioning, where trace elements are merely discharged or exhausted, depending on their relative affinities with the solids. This sort of reasoning is fundamentally wrong and may result in prejudice and vicious circles.

10.6.1   Compositional Effects of the Melt

As described in chapter 6, the main factors determining the solubility of a given element in a silicate melt are the Lux-Flood acidity of its oxide and the relative proportions of the cations of different field strengths (cation charge over squared sum of cation plus ligand radii Z/A^2) or charge densities (cation charge over ionic radius Z/r).

More properly concerning trace elements, although we may hypothesize a Henry's law solubility (i.e., molar proportions are too low to affect the extension and relative proportions of cationic and anionic matrices in the melt), it is true that their solubility is a complex function of the chemistry of the liquid phase (and of P-T conditions), as already seen for major elements.

Studying the unmixed portions of the miscibility gap between granitic melt (quartz plus feldspar normative) and ferropyroxenitic melt (pyroxene normative), Ryerson and Hess (1978) showed that high-field-strength elements such as Ti, REE, and Mn are preferentially stabilized in low polymerized liquids (e.g., ferropyroxenitic melt) with respect to highly polymerized liquids, whereas the opposite is true for low-field-strength ions such as alkaline earths. Analogous indications were obtained by Watson (1976) in the K_2O-Al_2O_3-FeO-SiO_2 system. This experimental evidence confirms that trace partitioning between silicate crystals and melts is definitely affected by the structure of the liquid. According to Ryerson and Hess (1978), for instance, the mass distribution coefficients of the various REE between clinopyroxene and ferropyroxenitic liquid are 10 times lower than the clinopyroxene/granitic liquid mass distribution coefficients. Significant effects have also been observed in other elements (Ti, Mn, K), whereas still other elements are less affected (e.g., Sr, Ba: Watson, 1976; Ryerson and Hess, 1978).

Figure 10.7 shows the effects of temperature and melt composition on the

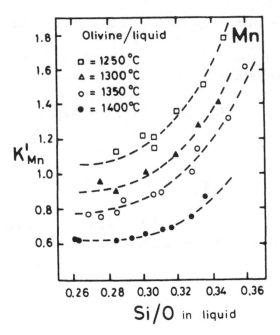

Figure 10.7 Compositional effects of melt on olivine/liquid mass distribution of Mn^{2+}. Reprinted from E. B. Watson, *Geochimica et Cosmochimica Acta,* 41, 1363–1374, copyright © 1977, with kind permission from Elsevier Science Ltd., The Boulevard, Langford Lane, Kidlington 0X5 1GB, UK.

mass distribution of manganese between olivine and silicate liquid: with decreasing melt acidity (expressed here as the Si/O atomic ratio), conventional mass distribution coefficient K' decreases for increased solubility of the element in the liquid phase, as a result of the increased relative extension of the anion matrix (see section 6.1.3).

According to Ryerson and Hess (1978), the augmented stabilization of high-field-strength elements in slightly polymerized melts is attributable to their tendency to associate with O^-, to minimize electrostatic repulsion with neighboring cations, and the preferential stabilization of low-field-strength ions in highly polymerized melts is attributable to coupled substitutions with tetrahedrally coordinated aluminum to maintain charge balance. However, we saw in chapter 6 that the preferential association with singly bonded oxygen involves low-field-strength ions (for example, preferential association of Ca^{2+} with O^- in the presence of Fe^{2+} ions). We also know that high-field-strength ions favor the polymerization reaction through acidic complexation of free oxygen O^{2-}. It is thus evident that the bulk solubility of the trace element in the liquid depends on its Lux-Flood acidity and on the relative extensions of the cation and anion matrices of the liquid phase.

10.6.2 Fractionation Effects in Melts: the Example of REE

We may now argue whether or not a silicate melt can fractionate elements of the same group. The question is rather important for REE, whose relative fractionation is usually interpreted by petrologists as a petrogenetic indicator of magmatic evolution.

According to Fraser (1975a), rare earth oxides may dissociate into the melt with both basic and acidic behavior—i.e.,

$$REE_2O_3 \Leftrightarrow 2REE^{3-} + 3O^{2-} \tag{10.37}$$

and

$$REE_2O_3 + O^{2-} \Leftrightarrow 2REEO_2^-. \tag{10.38}$$

The relative proportions of cationic (REE^{3+}) and polyanionic ($REEO_2^-$) are established according to the homogeneous equilibrium

$$REE_{(l)}^{3+} + 2O_{(l)}^{2-} \Leftrightarrow REEO_{2(l)}^-, \tag{10.39}$$

whose Gibbs free energy change is related to partial equilibria 10.37 and 10.38 by

$$\Delta G_{39} = \frac{1}{2}\left(\Delta G_{38} - \Delta G_{37}\right). \tag{10.40}$$

Rewriting equilibrium 10.39 in terms of thermodynamic activity of the involved species, we obtain

$$\Delta G_{39} = -RT \ln\left(\frac{a_{REEO_{2(l)}^-}}{a_{REE_{(l)}^{3+}} a_{O_{(l)}^{2-}}^2}\right). \tag{10.41}$$

Writing the bulk molar amount of trivalent REE in melt n in terms of partial amounts of polyanions (n_1) and cations (n_2):

$$n\,REE_{III(l)} = n_1\,REEO_{2(l)}^- + n_2\,REE_{(l)}^{3+}, \tag{10.42}$$

and adopting Temkin's model (Temkin, 1945) to define thermodynamic activity over the appropriate matrix, after some passage, one arrives at a mean activity coefficient $\bar{\gamma}_{REE}$ (relating mean thermodynamic activity to mean concentration) that is a function of the disproportionation constant K_{39} and of the activity of free oxygen in the silicate melt (Ottonello, 1983):

$$\bar{\gamma}_{REE} = \frac{2a_{O_{(l)}^{2-}} \cdot K_{39}}{1 + a_{O_{(l)}^{2-}} \cdot K_{39}}. \tag{10.43}$$

The Gibbs free energy change involved in disproportionation equilibrium 10.39 varies linearly with the field strengths of the various elements of the series:

$$\Delta G_{39} = a + b \left(\frac{Z}{A^2_{REE}} \right),$$

(10.44)

where a and b are constants. This amount of energy (hence, the various disproportionation constants K_{39}) can be evaluated from experimental observations on liquid/liquid REE partitioning (see Ottonello, 1983, for details of calculations). Table 10.5 shows a comparison between measured (m: Watson, 1976; Ryerson and Hess, 1978, 1980) and calculated (c: Ottonello, 1983) mass distributions on the basis of the amphoteric solution model. The K_{39} values obtained from the model are as follows: La = 7.5, Ce = 9, Pr = 11, Nd = 14, Sm = 21, Eu = 27, Gd = 33, Tb = 40, Dy = 50, Ho = 62, Er = 76, Tm = 95, Yb = 120, and Lu = 150. Applications of these K_{39} values to equation 10.43 leads to the mean activity coefficient trends observed in figure 10.8.

The plot in figure 10.8 clearly shows that silicate melts may exert marked fractionation on REE: acidic melts with relatively greater free oxygen activity display the highest mean activity coefficients in light REE (LREE). LREE are thus less soluble in these melts than heavy REE (HREE). The opposite is observed in basic melts, which have the highest mean activity coefficients at the HREE level. In other words, REE relative concentrations in silicate melts simply reflect their relative solubilities, as we saw for crystal structures.

10.6.3 Effects of Melt Composition on Solid/Liquid Distribution of Altervalent Elements

The Eu^{2+}/Eu^{3+} ratio in silicate melts is controlled by the redox equilibrium

$$4\,Eu^{3+}_{(l)} + 2\,O^{2-}_{(l)} \Leftrightarrow 4\,Eu^{2+}_{(l)} + O_{2(gas)} .$$

(10.45)

Morris and Haskin (1974) observed an increase in the Eu^{2+}/Eu^{3+} ratio with increasing (Al+Si)/O and high field-strength ions in the melt. This effect may be ascribed to the different acid-base properties of Eu_2O_3 and EuO (Fraser, 1975a,b). Equilibrium 10.45 may in fact be rewritten in the form

$$Eu_2O_{3(l)} \Leftrightarrow 2\,EuO_{(l)} + \frac{1}{2} O_{2(gas)} .$$

(10.46)

The activities of the two oxides in the liquid are determined by the type of dissociation. For EuO, basic dissociation is plausible:

Table 10.5 Measured (m) and calculated (c) basic liquid/acidic liquid REE mass distribution constants (Ottonello, 1983).

Reference(s)	La		Sm		Dy		Yb		Lu	
	m	c	m	c	m	c	m	c	m	c
Watson (1976)	3.70–4.31	3.68	4.19–5.15	4.15	–	–	–	–	4.18–7.13	5.69
Ryerson and Hess (1978) (system *a*)	3.08–6.90	6.86	–	–	2.20–4.64	10.9	3.03–4.18	18.2	–	–
Ryerson and Hess (1978, 1980) (system *b*)	13.8	14.5	–	–	16.57	27.37	–	–	–	–

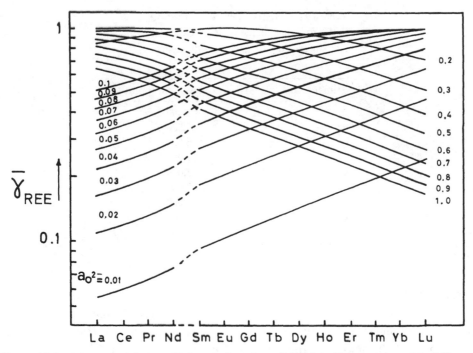

Figure 10.8 Mean activity coefficients of various REE in silicate melts with different values of activity of dissolved free oxygen $a_{O^{2-}}$, plotted as a function of increasing atomic number. Reprinted from Ottonello (1983), with kind permission of Theophrastus Publishing and Proprietary Co.

$$EuO \Leftrightarrow Eu^{2+} + O^{2-}. \tag{10.47}$$

However, Eu_2O_3 behaves as an amphoteric oxide and dissociates according to the two reaction schemes already proposed for all trivalent REE (cf. eq. 10.37 and 10.38). Denoting K_a and K_b, respectively, as the acidic and basic dissociation constants, the Eu^{3+}/Eu^{2+} oxidation ratio varies with the silica amount in the melt $(1 - X_{MO})$ according to

$$\frac{Eu^{3+}}{Eu^{2+}} = \frac{a_{O_{2(gas)}}^{1/4}}{c} \left[K_a^{1/2} \cdot a_{O_{(l)}^{2-}}^{3/2} \cdot \frac{(1 - X_{MO})}{X_{MO} + 0.25(1 - X_{MO})} \right] \tag{10.48}$$

(Wood and Fraser, 1976), where c is a constant and X_{MO} is related to $(Al+Si)/O$ in the systems investigated by Morris and Haskin (1974) through (Fraser, 1975b):

$$X_{MO} = \left(\frac{O}{Al+Si} - 2 \right) \left[\left(\frac{O}{Al+Si} - 2 \right) + 1 \right]. \tag{10.49}$$

It must be noted that the increased stabilization of the oxidized form with the basicity of the melt, experimentally observed by Morris and Haskin (1974) for

Table 10.6 Henry's law concentration
limits for trace elements in silicate melts.
? = suggested values (from Ottonello,
1983).

Element	Henry's Law Limit (ppm)
Ba	5,000
Sr	5,000
La	> 6,000
Sm	> 11,000
Lu	3,000?
Mn	> 7,000
Ti	5,000
Zr	> 10,000
Ta	> 10,000

europium, is a generalized phenomenon for all altervalent elements present at trace levels (it was observed, for instance, by Paul and Douglas, 1965a,b, and Nath and Douglas, 1965, for Ce^{3+}/Ce^{4+}, Cr^{3+}/Cr^{6+}, and Fe^{2+}/Fe^{3+}). This fact confirms that the acid-base treatment proposed by Fraser (1975a,b) can be extended to all trace elements present in melts in various oxidation states.

10.7 Limits of Validity of Henry's Law for Trace Elements in Silicate Melts

There are few data on this point. Most information comes from the experimental work of Watson (1976) on trace partitioning between immiscible liquids in the $K_2O\text{-}Al_2O_3\text{-}FeO\text{-}SiO_2$ system at 1180 °C and 1 bar. Table 10.6 shows some concentration limits based on this experimental work. At present, it is impossible to establish whether or not deviations from Henry's law take place in the acidic or basic liquids at equilibrium.

10.8 Conventional and Normalized Partition Coefficients

We have already seen that, within the range of Nernst's law, the solid/liquid partition coefficient differs from the thermodynamic constant by the ratio of the Henry's law activity coefficients in the two phases—i.e.,

$$K_i' = \frac{\gamma_{h,i,l}}{\gamma_{h,i,s}} K_i. \tag{10.50}$$

Although this equation reduces to an identity whenever solute-solvent interactions are embodied in the definition of the "Henry's law standard state" (cf. section 10.2), it must be noted that K_i' is the *molar ratio* of trace element i in the two phases and not the weight concentration ratio usually adopted in trace element geochemistry. As we will see later in this section, this double conversion (from activity ratio to molar ratio, and from molar ratio to weight concentration ratio) complicates the interpretation of natural evidence in some cases. To avoid ambiguity, we define here as "conventional partition coefficients" (with the same symbol K') all mass concentration ratios, to distinguish them from molar ratios and equilibrium constants.

In an attempt to minimize the compositional effects of phases on trace partitioning, Henderson and Kracek (1927) also introduced the concept of "normalized partition coefficient" D, which compares the relative trace/carrier (Tr/Cr) mass distributions in the two phases at equilibrium—i.e.,

$$D = \frac{(Tr/Cr)_s}{(Tr/Cr)_l}.$$

(10.51)

The formulation of this coefficient derives from the consideration that solid/liquid trace element distribution can be ascribed to the existence of simple exchange equilibria of the type

$$\underset{\text{solid}}{CrM} + \underset{\text{liquid}}{Tr} \Leftrightarrow \underset{\text{solid}}{TrM} + \underset{\text{liquid}}{Cr},$$

(10.52)

where M is the common ligand. For instance, an exchange reaction of this type may be responsible for the isomorphic substitution of Sr in plagioclase:

$$\underset{\text{solid}}{CaAl_2Si_2O_8} + \underset{\text{melt}}{Sr^{2+}} \Leftrightarrow \underset{\text{solid}}{SrAl_2Si_2O_8} + \underset{\text{melt}}{Ca^{2+}}.$$

(10.53)

In this case, trace element and carrier occupy the same structural position both in the solid phase and in the melt and are subject to the same compositional effects in both phases (i.e., extension of the cation matrix in the melt and amount of anorthite component in the solid). Figure 10.9A shows the effect of normalization: the conventional partition coefficient of Sr between plagioclase and liquid varies by about one order of magnitude under equal P-T conditions, with increasing anorthite component in solid solution, whereas normalized distribution coefficient D is virtually unaffected. Figure 10.9B shows the same effect for the Ba-Ca couple.

It is important to note that, in the example given above, trace element (Sr or Ba) and carrier (Ca) have the same formal charge. In the case of substitution of

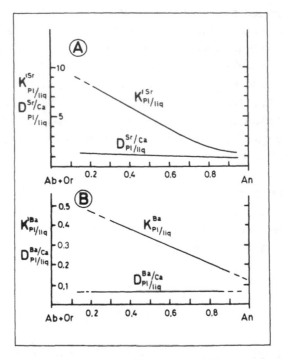

Figure 10.9 Effect of normalization on trace element distribution of Sr (A) and Ba (B) between plagioclase and silicate liquid. Distribution curves are based on various experimental evidence (Ewart et al., 1968; Ewart and Taylor, 1969; Berlin and Henderson, 1968; Carmichael and McDonald, 1961; Philpotts and Schnetzler, 1970; Drake and Weill, 1975). Ab: albite; Or: orthoclase; An: anorthite.

elements with different valences, the exchange equilibrium is more complex and the adoption of normalized partition coefficients does not eliminate the compositional effects. For instance, in the case of stabilization of a rare earth in diopside, we can write

$$\underset{\text{pyroxene}}{CaMgSi_2O_6} + \underset{\text{melt}}{REE^{3+}} + \underset{\text{melt}}{Al^{3+}} \Leftrightarrow \underset{\text{pyroxene}}{REEMgSiAlO_6} + \underset{\text{melt}}{Ca^{2+}} + \underset{\text{melt}}{Si^{4+}} . \quad (10.54)$$

In this case, trace-carrier substitution REE^{3+}-Ca^{2+} is associated with coupled substitution Si^{4+}-Al^{3+}, necessary to maintain charge balance. Simple normalization to the Ca^{2+} amount in the solid and liquid phases at equilibrium is ineffective in this case.

We have already noted that the double conversion from activity ratio to weight concentration ratio implicit in trace element geochemistry may involve complexities that must be carefully evaluated in the interpretation of natural evidence. Let us consider, for instance, the distribution of Ni between clinopyroxene and silicate liquid. Figure 10.10A shows the effect of temperature on the conven-

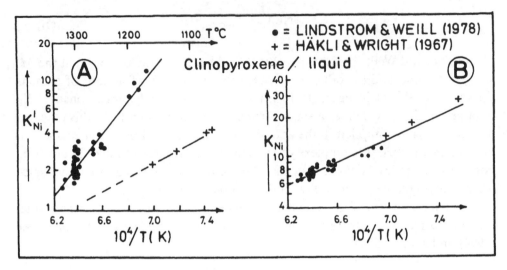

Figure 10.10 Arrhenius plot of conventional partition coefficient (A) and thermodynamic constant (B) for clinopyroxene/liquid distribution of Ni.

tional weight concentration ratio, according to the data of Lindstrom and Weill (1978) and Häkli and Wright (1967): the Arrhenius plot defines two distinct arrays, with different slopes and intercepts.

If we now consider solid/liquid Ni distribution from a thermodynamic point of view, as suggested by Lindstrom and Weill (1978):

$$\underset{\text{liquid}}{NiO} + \underset{\text{liquid}}{CaO} + \underset{\text{liquid}}{2SiO_2} \Leftrightarrow \underset{\text{pyroxene}}{CaNiSi_2O_6}, \tag{10.55}$$

the thermodynamic constant of equilibrium 10.55 is given by

$$K_{55} = \frac{\left[CaNiSi_2O_6\right]}{\left[NiO\right]\left[CaO\right]\left[SiO_2\right]^2}, \tag{10.56}$$

where the terms in square brackets denote thermodynamic activity.

We can adopt the approximations

$$\left[CaNiSi_2O_6\right] = X_{Ni,M1} \tag{10.57}$$

$$\left[SiO_2\right] = X_{SiO_2,l} \tag{10.58}$$

$$\left[CaO\right] = X_{CaO,l} \tag{10.59}$$

$$[\text{NiO}] = X_{\text{NiO},l} \tag{10.60}$$

(Lindstrom and Weill, 1978), where $X_{\text{Ni},M1}$ is the atomic fraction of Ni in the $M1$ site of pyroxene, and the other X terms represent the molar fractions of oxide in the melt. As shown in figure 10.10B, the plot of thermodynamic constant K_{55} obtained in this way defines a single Arrhenius trend. Evidently, in this case, the double conversion implicit in the adoption of conventional partition coefficients introduces an apparent complexity that is indeed ascribable to the different compositions of the investigated systems. These effects must be carefully evaluated when studying trace element distributions in petrogenetic systems.

Recent appraisals of the state of art in experimental trace partitioning studies relevant to igneous petrogenesis may be found in the review articles of Green (1994) and Jones (1995).

10.8.1 Effects of Temperature on Partition Coefficients

Let us consider the chemical potentials of a trace component (TrM) and a carrier component (CrM) in a solid mixture and in the liquid at equilibrium:

$$\mu_{\text{TrM},s} = \mu^0_{\text{TrM},s} + RT \ln a_{\text{TrM},s} \tag{10.61}$$

$$\mu_{\text{CrM},s} = \mu^0_{\text{CrM},s} + RT \ln a_{\text{CrM},s} \tag{10.62}$$

$$\mu_{\text{TrM},l} = \mu^0_{\text{TrM},l} + RT \ln a_{\text{TrM},l} \tag{10.63}$$

$$\mu_{\text{CrM},l} = \mu^0_{\text{CrM},l} + RT \ln a_{\text{CrM},l}. \tag{10.64}$$

Equations 10.61 to 10.64 yield

$$\frac{a_{\text{TrM},l}}{a_{\text{TrM},s}} = \exp\left(\frac{\mu_{\text{TrM},l} - \mu^0_{\text{TrM},l} - \mu_{\text{TrM},s} + \mu^0_{\text{TrM},s}}{RT} \right) \tag{10.65}$$

and

$$\frac{a_{\text{CrM},l}}{a_{\text{CrM},s}} = \exp\left(\frac{\mu_{\text{CrM},l} - \mu^0_{\text{CrM},l} - \mu_{\text{CrM},s} + \mu^0_{\text{CrM},s}}{RT} \right) \tag{10.66}$$

Because at equilibrium $\mu_{\text{TrM},s} = \mu_{\text{TrM},l}$ and $\mu_{\text{CrM},s} = \mu_{\text{CrM},l}$, equations 10.65 and 10.66 reduce to Vant Hoff isotherms:

$$\frac{a_{\text{TrM},l}}{a_{\text{TrM},s}} = \exp\left(\frac{\mu^0_{\text{TrM},s} - \mu^0_{\text{TrM},l}}{RT}\right) \tag{10.67}$$

and

$$\frac{a_{\text{CrM},l}}{a_{\text{CrM},s}} = \exp\left(\frac{\mu^0_{\text{CrM},s} - \mu^0_{\text{CrM},l}}{RT}\right). \tag{10.68}$$

Substituting molar fractions X and molalities m for activities in the solid and in the liquid phase, respectively, we obtain

$$K_{\text{TrM}} = \frac{a_{\text{TrM},l}}{a_{\text{TrM},s}} = \frac{m_{\text{TrM},l} \cdot \gamma_{\text{TrM},l}}{X_{\text{TrM},s} \cdot \gamma_{\text{TrM},s}} \tag{10.69}$$

and

$$K_{\text{CrM}} = \frac{a_{\text{CrM},l}}{a_{\text{CrM},s}} = \frac{m_{\text{CrM},l} \cdot \gamma_{\text{CrM},l}}{X_{\text{CrM},s} \cdot \gamma_{\text{CrM},s}}. \tag{10.70}$$

As we have already seen, within Henry's law range the activity coefficient of the trace component in the solid can be expressed as

$$\gamma_{\text{TrM},s} = \exp\left(\frac{\Delta\mu^*}{RT}\right), \tag{10.71}$$

where $\Delta\mu^*$ is the molar Gibbs free energy amount involved in solute-solvent interactions. The (molar) partition coefficient thus has the form

$$K'_{\text{TrM}} = \exp\left(\frac{\Delta\mu^*}{RT}\right) \cdot \exp\left(\frac{\Delta\mu^0_{\text{TrM}}}{RT}\right) \cdot \gamma^{-1}_{\text{TrM},l}. \tag{10.72}$$

Analogously, assuming the carrier component to obey Raoult's law in the solid, we obtain the following equation for normalized molar distribution $D_{s/l}$:

$$D_{s/l} = \exp\left(\frac{\Delta\mu^*}{RT}\right) \cdot \exp\left(\frac{\Delta\mu^0_{\text{TrM}}}{RT}\right) \cdot \exp\left(-\frac{\Delta\mu^0_{\text{CrM}}}{RT}\right) \cdot \frac{\gamma_{\text{CrM},l}}{\gamma_{\text{TrM},l}}. \tag{10.73}$$

Because

$$\Delta\mu^*_{\text{TrM}} = \Delta h^*_{\text{TrM}} - T\Delta s^*_{\text{TrM}} + P\Delta v^*_{\text{TrM}} \qquad (10.74)$$

$$\Delta\mu^0_{\text{TrM}} = \Delta h^0_{\text{TrM}} - T\Delta s^0_{\text{TrM}} + P\Delta v^0_{\text{TrM}} \qquad (10.75)$$

$$\Delta\mu^0_{\text{CrM}} = \Delta h^0_{\text{CrM}} - T\Delta s^0_{\text{CrM}} + P\Delta v^0_{\text{CrM}}, \qquad (10.76)$$

where terms in h, s, and v are, respectively, partial molar enthalpies, partial molar entropies, and partial molar volumes (cf. section 2.5), substituting equations 10.74 to 10.76 in equations 10.72 and 10.73 and deriving the logarithmic forms in dT, we obtain

$$\left(\frac{\partial \ln K'_{\text{TrM}}}{\partial T}\right)_{P,X} = \frac{1}{RT^2}\left[\left(h^0_{\text{TrM},s} - h_{\text{TrM},l}\right) - \Delta h^*_{\text{TrM}}\right] \qquad (10.77)$$

$$\left(\frac{\partial \ln D_{s/l}}{\partial T}\right)_{P,X} = \frac{1}{RT^2}\left[\left(h^0_{\text{TrM},s} - h_{\text{TrM},l}\right) - \left(h^0_{\text{CrM},s} - h_{\text{CrM},l}\right) - \Delta h^*_{\text{TrM}}\right]. \qquad (10.78)$$

The terms in parentheses in equations 10.77 and 10.78 represent the enthalpies of solution of TrM and CrM, respectively, in the melt at the T of interest (cf. section 6.4). If the enthalpy of solution does not vary appreciably with T, because the interaction term Δh^*_{TrM} cannot also be expected to vary appreciably with T, integration of equation 10.77 leads to a linear form of the type

$$\ln K'_{\text{TrM}} = A + BT^{-1}, \qquad (10.79)$$

where A and B are constants.

Linearity is more marked for normalized distributions, because possible modifications with T of the heats of solution of trace and carrier components compensate each other.

The behavior described above is exemplified in figure 10.11, where Arrhenius plots of solid/liquid conventional partition coefficients for transition elements reveal more or less linear trends, with some dispersion of points ascribable to compositional effects.

10.8.2 Effects of Pressure on Partition Coefficients

Substituting equations 10.74 to 10.76 in equations 10.72 and 10.73 and deriving in dP the logarithmic forms, we obtain

$$\left(\frac{\partial \ln K'_{\text{TrM}}}{\partial P}\right)_{T,X} = \frac{1}{RT}\left[\left(v^0_{\text{TrM},s} - v_{\text{TrM},l}\right) - \Delta v^*_{\text{TrM}}\right] \qquad (10.80)$$

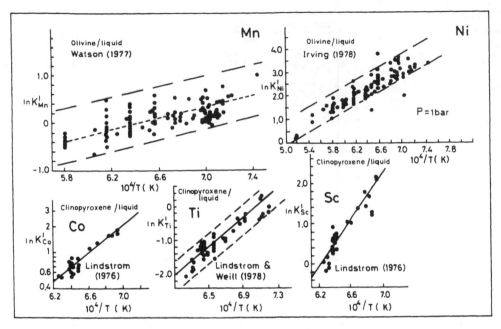

Figure 10.11 Effects of temperature on conventional solid/liquid partition coefficients of various transition elements. Sources of data are also listed.

and

$$\left(\frac{\partial \ln D_{s/l}}{\partial P}\right)_{T,X} = \frac{1}{RT}\left[\left(v^0_{\text{TrM},s} - v_{\text{TrM},l}\right) - \left(v^0_{\text{CrM},s} - v_{\text{CrM},l}\right) - \Delta v^*_{\text{TrM}}\right] \cdot (10.81)$$

The trace component in solid solution usually has partial molar volumes not far from those of the carrier, for crystal-chemical reasons, and the terms in parentheses in equation 10.81 compensate each other. The term Δv^*_{TrM}, expressing the differences in partial molar volume arising from solute-solvent interactions in the solid, is also usually negligible. As a result, the effect of P on normalized partition coefficients is usually very restricted. The same cannot be said for conventional partition coefficient K'_{TrM}, because the partial molar volumes of the trace component in the solid and liquid phases (and hence their differences) may vary appreciably with P (McIntire, 1963).

10.8.3 Effects of Oxygen Fugacity on the Solid/Liquid Distribution

Oxygen fugacity f_{O_2} directly affects the redox states of trace elements in melts (see, for instance, equation 10.45). Figure 10.12 shows the effects of f_{O_2} on the oxidation state of Ti and V in an $Na_2Si_2O_5$ melt at $T = 1085\ °C$, according to the

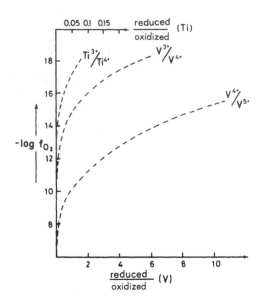

Figure 10.12 Effects of oxygen fugacity on oxidation state of Ti and V in $Na_2Si_2O_5$ melt at $T = 1085$ °C (experimental data from Johnston, 1964, 1965, and Johnston and Chelko, 1966).

experiments of Johnston (1964, 1965) and Johnston and Chelko (1966). These effects, which must not be confused with the previously described compositional effect, markedly influence solid/liquid trace element distribution, because crystal structures dramatically discriminate elements in various oxidation states. Figure 10.13 shows, for example, the effect of f_{O_2} on the conventional partition coefficient of Cr between olivine and melt (figure 10.13A) and between subcalcic pyroxene and melt (figure 10.13B). In both cases, the effect of f_{O_2} is nonlinear and varies with T in a rather complex way.

10.9 Trace Element Distribution Models for Magma Genesis and Differentiation

The evolution of an igneous system is controlled by changes in intensive variables P and T over time and by the capability of the system to exchange heat or matter with the exterior (i.e., "open," "closed," or "isolated" systems; cf. section 2.1). Note that the term "exterior" must be intended in its full thermodynamic sense: it may be the neighboring zone of a lithospheric portion undergoing melting, where liquids are conveyed, as well as the inner parts of newly formed crystals, unable to exchange matter with the residual liquid for purely kinetic reasons, during magma differentiation. We can consider here two main cases:

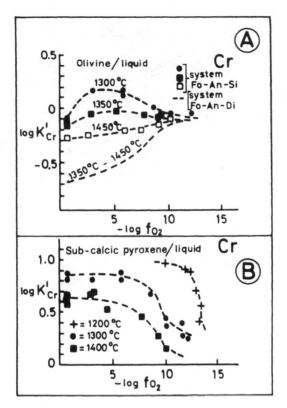

Figure 10.13 Effect of oxygen fugacity on conventional partition coefficient of Cr. (A) Olivine/liquid partitioning; experimental data of Bird (1971), Weill and McKay (1975), Huebner et al. (1976), Lindstrom (1976), and McKay and Weill (1976). (B) Subcalcic pyroxene/liquid partitioning; experimental data of Schreiber (1976). Reprinted from A.J. Irving, *Geochimica et Cosmochimica Acta,* 42, 743–770, copyright © 1978, with kind permission from Elsevier Science Ltd., The Boulevard, Langford Lane, Kidlington 0X5 1GB, UK.

1. Complete equilibrium between solid and liquid
2. Interface equilibrium between accreting (or dissolving) crystals and liquid

In the first case, the diffusion rates of the trace component in all phases of the system must be sufficiently high with respect to rates of crystal accretion or melting. Figure 10.14 shows experimental data on elemental diffusivities of alkalis in a rhyolitic melt. T conditions being equal, the diffusivities of Na and K in the melt are four to six orders of magnitude greater in the liquid than in feldspars (compare figures 10.14 and 4.8). We can thus expect the formation of chemically zoned crystals, whereas possible chemical zoning in the liquid may more reasonably be ascribed to advection.

The partition coefficient of trace element i between a given solid phase s and

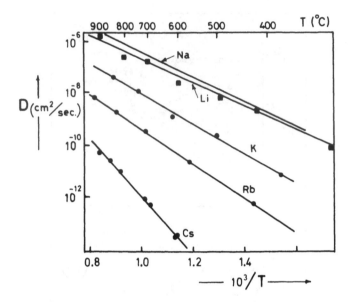

Figure 10.14 Elemental diffusivities of alkalis in a rhyolitic melt (experimental data of Gabis et al., 1979).

liquid l in a closed system is valid on each portion of the two phases at equilibrium:

$$K' = \frac{X_{i,s}}{X_{i,l}}. \tag{10.82}$$

In the case of interface equilibrium (open system conditions), the partition coefficient is valid only at the interface between solid and liquid (or at zero distance from the interface) and at time of crystallization (or melting) t:

$$K' = \frac{X_{i,s,0,t}}{X_{i,l,0,t}}. \tag{10.83}$$

It is important to stress that, when measuring the simple concentration ratios of mean abundances in the solid and liquid phases formed at interface equilibrium (as is typically done in comparative studies), the resulting (apparent) partition coefficient K_A is related to partition coefficient K', valid at the interface, by

$$K_A = \frac{\dfrac{K'}{t} \displaystyle\int_0^t X_{i,l,0,t}\, dt}{\dfrac{1}{L} \displaystyle\int_0^L X_{i,l,x,t}\, dx} \tag{10.84}$$

(Albarede and Bottinga, 1972), where t is time since the beginning of the process (at constant accretion or dissolution rates), $X_{i,l,x,t}$ is the trace element concentration measured at distance x from the interface and at time t, and L (in the case of fractional crystallization) is half the mean distance between centers of accreting crystals less half the mean thickness of the crystals. We will see later the effects of erroneous evaluation of K_A and K' when modeling fractional crystallization processes.

10.9.1 Crystallization in an Open System, or "Fractional Crystallization"

During *Rayleigh's crystallization* process (formation of crystals with equilibrium limited to the solid/liquid interface), the relative concentration of a trace element in liquid C_l with respect to the initial concentration in system C_0 is described by the Doerner-Hoskins equation (Doerner and Hoskins, 1925; we will omit hereafter subscripts in the indicization of coefficients, for the sake of simplicity):

$$C_l = C_0 \left[1 - F\right]^{K'-1},\tag{10.85}$$

where F is the mass fraction of solid in the system.

Distribution equation 10.91 has been developed independently by various authors, starting from the differential equation valid for a single component:

$$\frac{dm}{ds} = K'\left(\frac{M - m}{Q - S}\right),\tag{10.86}$$

where Q is the initial mass of the system, M is the mass of trace elements in the system, S is the mass of the solid, and m is the mass of trace element in the solid. According to Neumann et al. (1954), the integrated form of equation 10.86 is

$$m = M\left[1 - \left(1 - \frac{S}{Q}\right)^{K'}\right]\tag{10.87}$$

(when $S = 0, m = 0$).

Differentiating equation 10.87 in ds yields

$$\frac{dm}{ds} = K'\frac{M}{Q}\left(1 - \frac{S}{Q}\right)^{K'-1}.\tag{10.88}$$

Translated into our formalism, equation 10.88 becomes

$$C_s = C_0 K' \left[1 - F\right]^{K'-1} \tag{10.89}$$

and, applying equation 10.82,

$$C_l = C_0 \left[1 - F\right]^{K'-1}. \tag{10.90}$$

It must be noted that equation 10.89 is valid only at the solid/liquid interface, whereas equation 19.90 is valid over the entire crystallization range.

The mean concentration in solid $\overline{C}_s$ may be obtained by integration of equation 10.86 (Albarede and Bottinga, 1972):

$$\overline{C}_s = C_0 \frac{1 - \left[1 - F\right]^{K'}}{F}. \tag{10.91}$$

Clearly, equation 10.91 can be derived directly from equation 10.87 by substituting concentrations for masses.

Equation 10.91 also can be expressed as

$$\overline{C}_s = K_A C_0 \left[1 - F\right]^{K'-1} \tag{10.92}$$

(the identity between eq. 10.92 and eq. 10.89 at the interface where $K_A = K'$ is obvious).

The ratio between equations 10.91 and 10.90 gives the relationship between apparent partition coefficient K_A and conventional partition coefficient at interface K' (i.e., the integrated form of eq. 10.84):

$$K_A = \frac{\overline{C}_s}{C_l} = \frac{1 - \left[1 - F\right]^{K'}}{F\left[1 - F\right]^{K'-1}}. \tag{10.93}$$

Equation 10.93 clearly shows that, if apparent partition coefficient K_A is adopted instead of conventional partition coefficient K' actually valid at the solid/liquid interface to model Rayleigh's crystallization, errors arise whose magnitudes increase the more K' differs from 1 and the longer the process advances. This is clearly shown in figure 10.15, in which fractional differences $(K_A - K')/K'$ are plotted as functions of F for various values of K'.

In the case of compositional gradients in the liquid resulting from advection or slow diffusivity, the relationship between K_A and K' is more complex and depends both on the accretion rate of crystals v_A and the diffusion (or removal) rate in liquid v_D (see eq. 10.84). Values of K_A for various v_D/v_A ratios with $K' = 2$ are listed in table 10.7.

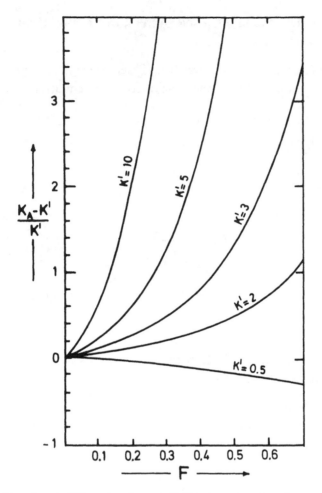

Figure 10.15 Fractional differences $(K_A - K')/K'$ plotted as functions of degree of fractional crystallization F for various values of K'. Reprinted from F. Albarede and Y. Bottinga, *Geochimica et Cosmochimica Acta*, 36, 141–156, copyright © 1972, with kind permission from Elsevier Science Ltd., The Boulevard, Langford Lane, Kidlington 0X5 1GB, UK.

Table 10.7 Relationships between apparent partition coefficient K_A and actual partition coefficient K' for various v_D/v_A ratios (from Albarede and Bottinga, 1972).

v_D/v_A	K'	K_A		
		$F = 0.1$	$F = 0.3$	$F = 0.5$
10	2	2.04	2.31	2.78
0.1	2	1.54	1.38	1.36
0.001	2	1.15	1.14	1.14

10.9.2 Crystallization in a Closed System, or "Equilibrium Crystallization"

Whenever diffusion rates within crystals and melt are high enough to ensure equilibrium in all portions of the system, the distribution equations are

$$C_s = \frac{K'C_0}{\left[1 - (1 - K')F\right]}$$

(10.94.1)

and

$$C_l = \frac{C_0}{\left[1 - (1 - K')F\right]},$$

(10.94.2)

where K' is valid in all portions of the system (and not only at the interface, as in the preceding case).

10.9.3 Particular Cases: (I) Resorption of Solid Phases During Rayleigh's Crystallization

The distribution equations for this particular case—corresponding, for instance, to a peritectic reaction taking place during cooling of a magma chamber—were evaluated by Neumann et al. (1954) starting from a modified form of general equation 10.86:

$$\frac{dm}{ds} = K'\left[\frac{M - m + \overline{C}_{Tr}PS}{Q - (1 - P)S}\right],$$

(10.95)

where P is the mass ratio between resorbed and precipitated phases and $\overline{C}_{Tr}$ is the mean concentration of the trace element in the resorbed phase (see eq. 10.86 for the remaining symbols). Neumann et al. (1954) gave the following distribution equations, which are valid at the interface:

$$C_s = \frac{K'\overline{C}_{Tr}P}{1 - P - K'}\left[\left(1 - \frac{1 - P}{Q}\right)^{[K'/(1-P)-1]} - 1\right] + \frac{K'M}{Q}\left[1 - \frac{(1 - P)S}{Q}\right]^{[K'/(1-P)-1]}$$

(10.96)

$$C_l = \frac{C_s}{K'}.$$

(10.97)

10.9.4 Particular Cases: (II) Simultaneous Crystallization of Several Phases

Conventional partition coefficient K' is replaced by bulk solid/liquid mass distribution coefficient $D_{s/l}$, corresponding to the weighted mean of the conventional partition coefficients for each solid/liquid pair:

$$D_{s/l} = \frac{X_\alpha}{K'_{l/\alpha}} + \frac{X_\beta}{K'_{l/\beta}} + \frac{X_\gamma}{K'_{l/\gamma}} + \cdots. \qquad (10.98)$$

In equation 10.98, X_α, X_β, and X_γ are the relative mass fractions of crystallizing phases ($X_\alpha + X_\beta + X_\gamma + \ldots = 1$), and $K'_{l/\alpha}$, $K'_{l/\beta}$, and $K'_{l/\gamma}$ are their respective conventional partition coefficients.

$D_{s/l}$ can be substituted for K' in equations 10.89 and 10.90 and equations 10.94.1 and 10.94.2.

10.9.5 Particular Cases: (III) Presence of Intercumulus Liquids

Like the preceding case, this case is also quite common in geology. During crystal settling at the bottom of magma chambers, portions of liquid are trapped between crystals and are not totally removed by successive filter-pressing processes. The trace element concentration in the zone of accumulus (C_{acc}) is given by the weighted summation of the relative concentrations in the solid liquid (X_{acc}) and trapped liquid ($1 - X_{acc}$). That is, for an equilibrium crystallization process,

$$C_{acc} = \frac{K'C_0}{\left[1 - \left(1 - K'\right)F\right]} \times X_{acc} + \frac{C_0}{\left[1 - \left(1 - K'\right)F\right]} \times \left(1 - X_{acc}\right) \qquad (10.99)$$

and, for a case of Rayleigh's crystallization,

$$C_{acc} = \left[C_0 K_A \left(1 - F\right)^{K_A - 1}\right] \times X_{acc} + \left[C_0 \left(1 - F\right)^{K_A - 1}\right] \times \left(1 - X_{acc}\right). \qquad (10.100)$$

The effects of the presence of intercumulus liquid on the relative distribution of trace elements during cooling of a magma chamber were investigated in detail by Albarede (1976). Figure 10.16 shows the effects of different portions of intercumulus liquid (arbitrary α values of $1 - X_{acc}$) on the relative proportions of two trace elements in the accumulus zone. Note that relative fractionation of the two elements evolves along simple logarithmic trends whenever the mass proportion of trapped liquid remains approximately constant throughout the crystallization

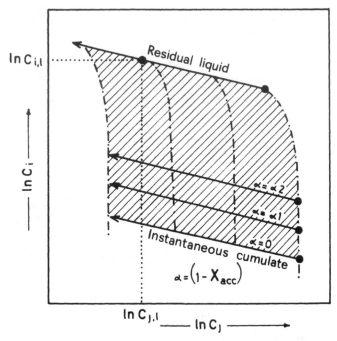

Figure 10.16 Effects of presence of intercumulus liquid on relative distribution of trace elements i and j during partial crystallization. α, α_1, and α_2 are arbitrary proportions of intercumulus liquid. Reprinted with modifications from F. Albarede, *Geochimica et Cosmochimica Acta,* 40, 667–673, copyright © 1976, with kind permission from Elsevier Science Ltd., The Boulevard, Langford Lane, Kidlington 0X5 1GB, UK.

process. Note also, however, that the field of possible trace element concentrations is quite wide (dashed area in figure 10.16), depending on the efficiency of filter-pressing.

10.9.6 Melting in a Closed System, or "Equilibrium Melting"

Following Shaw (1970), we can distinguish two types of melting of a heterogeneous solid in a closed system:

1. *Modal melting.* Solid phases melt in the same modal proportion in which they occur in the initial solid, so that the modal composition of the residual solid remains unchanged throughout the process.
2. *Nonmodal melting.* The melting proportions of solid phases differ from the modal proportions in the initial solid, so that the modal composition of the residuum is progressively modified throughout the process.

In the first case, the distribution equations are

$$C_l = \frac{C_0}{\left[D_0 + F\left(1 - D_0\right)\right]} \tag{10.101}$$

and

$$C_s = \frac{C_0 D_0}{\left[D_0 + F\left(1 - D_0\right)\right]}, \tag{10.102}$$

where F is the mass fraction of *melt* in the system (not the fraction of solid, as in the preceding cases) and D_0 is the initial solid/liquid distribution at the beginning of melting:

$$D_0 = \frac{X_{0,\alpha}}{K'_{l/\alpha}} + \frac{X_{0,\beta}}{K'_{l,\beta}} + \frac{X_{0,\gamma}}{K'_{l/\gamma}} + \cdots . \tag{10.103}$$

However, "modal melting" is highly unlikely in geological systems, because the eutectic normally differs from the modal constitution of the initial rock. Solid/liquid distribution $D_{s/l}$ thus changes progressively throughout melting (case 2), based on the melting proportions P_α, P_β, P_γ, . . . of the solid phases:

$$D_{s/l} = \frac{D_0 - PF}{1 - F}, \tag{10.104}$$

where

$$P = \frac{P_\alpha}{K'_{l/\alpha}} + \frac{P_\beta}{K'_{l/\beta}} + \frac{P_\gamma}{K'_{l/\gamma}} + \cdots . \tag{10.105}$$

In this case the distribution equations are

$$C_l = \frac{C_0}{\left[D_0 + F\left(1 - P\right)\right]} \tag{10.106}$$

and

$$C_s = \frac{C_0 D_{s/l}}{\left[D_0 + F\left(1 - D_0\right)\right]}. \tag{10.107}$$

Because $P_{\alpha,\beta,\gamma}$ and $K'_{l/\alpha,\beta,\gamma}$ represent final state conditions, equations 10.106 and 10.107 can be applied only to the entire melting range whenever melting proportions and conventional mass distribution coefficients remain constant throughout the process. In all other cases, the equations above must be associated with functions describing the evolution of $P_{\alpha,\beta,\gamma}$ and $K'_{l/\alpha,\beta\gamma}$ with the degree of partial melting F.

10.9.7 Melting in an Open System, or "Fractional Melting"

Several models have been proposed (Gast, 1968; Shaw, 1970; Hertoghen and Gijbels, 1976), but the most satisfactory are those based on iterative calculations, because they allow the evaluation of discontinuous events with sudden modifications of melting proportions, which are common in geological open systems (note that such discontinuities, or even the evolution of melting proportions, cannot be accounted for, if the distribution equations are obtained by integration at the limit). The following distribution model (Ottonello and Ranieri, 1977) is a partial modification of Gast's (1968) approach:

$$C_{s,i} = C_0 \prod_{i=1}^{n} \frac{1 - (D_i + i - 1)\Delta F}{1 - F} \tag{10.108}$$

$$D_i = \frac{D_{i-1} - P_{i-1}\Delta F}{1 - \Delta F} \tag{10.109}$$

$$C_{l,i} = D_i C_{s,i-1} \tag{10.110}$$

$$\overline{C}_{l,i} = \frac{1}{i} \sum_{i=1}^{n} C_{l,i} \tag{10.111}$$

$$P = f(F) \tag{10.112}$$

$$K'_{l/\alpha,\beta,\gamma} = f(F), \tag{10.113}$$

where $C_{s,i}$ is trace element concentration in the residual solid after removal of the ith increment of liquid, $C_{l,i}$ is trace element concentration in the ith increment of liquid, $\overline{C}_{l,i}$ is mean trace element concentration in i accumulated increments, and ΔF is the fractional mass increment of melting.

Many natural processes are closer to fractional melting than to equilibrium melting conditions. For instance, the leucosome bands in gneisses may be assimilated to fractional melting liquids, being progressively removed from the system as soon as they are produced. In such cases, the melting proportions of the solid phases may change dramatically as the process advances (cf. Presnall, 1969) and must be carefully evaluated for rigorous modeling of trace element distributions. For this purpose, it is opportune to compare the natural system to a synthetic analog, the melting behavior of which is known with sufficient precision. As we

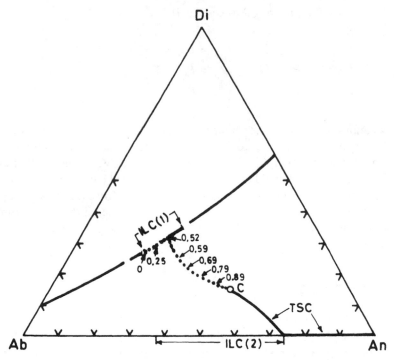

Figure 10.17 Evolution of liquid and solid during fractional melting of initial composition C in Di-Ab-An system. Curves ILC(1) and ILC(2): fractional melting liquids; TSC: residual solids; dotted line: fractionally accumulated liquids at quoted degrees of partial melting (from Ottonello and Ranieri, 1977).

saw in chapter 7, the melting proportions of synthetic systems can be properly quantified by application of petrogenetic rules.

Let us consider, for example, the system $CaMgSi_2O_6$ (Di)–$NaAlSi_3O_8$ (Ab)–$CaAl_2Si_2O_8$ (An) (figure 10.17).

Fractional melting of initial composition C produces instantaneous liquids evolving initially along curve ILC(1) and then along curve ILC(2), when all diopside in the residual solid is exhausted. The evolution of fractional melting liquids during the melting event may be described by two systems of equations $P = f(F)$ and $P = g(F)$, which are valid, respectively, in two melting ranges. For $0 < F < 0.51$ [evolution along cotectic line ILC(1)],

$$P = f(F) \equiv \begin{cases} P_\alpha = P_{Ab} = \dfrac{\left[13 - \left(9 + 40F\right)^{1/2}\right]}{20} \\[2em] P_\beta = P_{An} = \dfrac{\left(1 - P_\alpha\right)}{2.17} \\[2em] P_\gamma = P_{Di} = 1.17 P_\beta. \end{cases} \qquad (10.114)$$

For $0.52 < F < 1$ [evolution along Ab-An edge ILC(2)],

$$P = g(F) \equiv \begin{cases} P_\alpha = P_{Ab} = 1.32(1 - F) \\ P_\beta = P_{An} = 1 - P_\alpha \\ P_\gamma = P_{Di} = 0. \end{cases} \tag{10.115}$$

The equations above were determined graphically using a few points on the ILC curves of figure 10.17 and allow us to follow trace element distribution during fractional melting of initial system C ($Di_{15}An_{51}Ab_{34}$). The heavy line (TSC) in figure 10.17 describes the evolution of the residual liquid during melting; the dotted curve represents the modal evolution of fractionally accumulated liquids calculated by Morse (1976) with application of the lever rule (see chapter 7). This last curve can be obtained directly by summing the compositions of instantaneous liquids obtained through the two systems of equations 10.114 and 10.115.

10.9.8 Zone Melting

Zone melting is an industrial process used to purify slags and organic compounds, and its application to geology was initially suggested by Harris (1957). The formation of a magma by zone melting is associated with the relative displacement of a portion of lithosphere with respect to a zone of high heat flow. In such cases, the solid/liquid distribution is ascribable to two distinct processes (fractional melting and Rayleigh's crystallization) taking place, respectively, at the face and rear front of the high heat zone. During relative displacement of the molten zone along trajectory $0 \rightarrow X$, trace element concentrations in the melt are dictated by fractionation processes occurring at the two fronts:

$$C_l = C_0 \left[\frac{1}{D_{s/l}} - \left(\frac{1}{D_{s/l}} - 1 \right) \cdot \frac{X}{L} \exp(-D_{s/l}) \right]. \tag{10.116}$$

In equation 10.116, L is the length of the molten zone and X is the length of relative displacement. The application of zone melting to geological systems must be carried out with caution, because the modal proportions in the original solid, in the melt, and in the precipitating solid all generally differ markedly. Moreover, initial concentration C_0 cannot be expected to remain constant over long X trajectories. Nevertheless, zone melting probably operates in the earth's upper mantle: it has been suggested that the existence of the low-velocity layer is a result of such a process. Radiogenic elements such as U, Th, and K may be efficaciously extracted by zone melting from convecting lithospheric portions and contribute feedback to the process by heat release associated with radioactive decay.

10.10   Presence of Fluid Phases

During magma genesis and differentiation, fluid phases may develop. The fluids may exolve from the melt for two main reasons: progressive increase of concentration of gaseous species that reach the solubility product of the fluid in question, or decreased solubility in the melt during ascent to the surface (adiabatic decompression). If we call the mass fractions of solid, melt, and fluid in the system F_s, F_l, and F_f, respectively, the mass balance equation relating initial trace element concentration C_0 to the concentrations in solid, melt, and fluid is

$$C_0 = C_l F_l + C_s F_s + C_f F_f. \qquad (10.117)$$

For melting in a closed system, equation 10.117 gives

$$C_l = \frac{C_0}{\left[F_l + D_{s/l}\left(1 - F_l - F_f\right) + D_{f/l} F_f \right]} \qquad (10.118)$$

$$C_s = \frac{C_0 D_{s/l}}{\left[F_l + D_{s/l}\left(1 - F_l - F_f\right) + D_{f/l} F_f \right]} \qquad (10.119)$$

$$C_f = D_{f/l} C_l. \qquad (10.120)$$

For melting in an open system, Shaw (1978) proposed the equations

$$C_f = \frac{C_0}{D_{0,s/f} + v} \left[1 - \frac{F_l P_{s/f}}{D_{0,s/f} + v} \right]^{\left(1/P_{s/l} - 1\right)} \qquad (10.121)$$

$$C_s = \frac{C_f \left(D_{0,s/f} - P_{s/f} F_f \right)}{1 - F_l} \qquad (10.122)$$

$$C_l = \frac{C_f}{D_{f/l}} \qquad (10.123)$$

$$\overline{C}_l = \frac{C_0}{F_l} \left[1 - \left(1 - \frac{F_l P_{s/f}}{D_{0,s/f} + v} \right)^{1/P_{s/l}} \right], \qquad (10.124)$$

where ν is the constant fractional amount of fluid with respect to the initial mass of the system.

If the fluid is present during fractional crystallization, the trace element concentration in the liquid is dictated by

$$C_l = C_0 F_l^{\left(D_{s/l} + D_{f/l}G - 1\right)} \tag{10.125}$$

(Allegre et al., 1977). In equation 10.125, G is the concentration of gaseous components* dissolved in magma, which, following Holland (1972), is assumed to remain constant throughout the process.

10.11 "Hygromagmatophile" and "Incompatible" Elements

Some trace elements have a marked tendency to partition into the melt during magma genesis and differentiation. Elements showing the highest affinity for the melt (i.e., Th, Ta, Zr, Hf, La) were defined by Treuil (Treuil, 1973; Treuil and Varet, 1973; Treuil and Joron, 1975; Treuil et al., 1979) as "hygromagmatophile," and those preferentially partitioned into the melt (although not so markedly as the hygromagmatophile ones) were defined as "incompatible."

The hygromagmatophile character of elements is determined according to Treuil by the tendency to form stable complexes with anionic ligands in the melt (Treuil et al., 1979). Let us consider the formation of a complex between hygromagmatophile element M^{x+} and ligand L^{y-} present in the melt through homogeneous equilibrium of the type

$$M^{x+} + nL^{y-} \Leftrightarrow ML_n^{z-}, \tag{10.126}$$

whose constant is β. If we admit that solid/melt trace partitioning only involves cationic species M^{x+} (in agreement with the concept of "noncrystallizable groupings" of Ubbelhode, 1965, and Boon, 1971):

$$K = \frac{\left[M^{x+}\right]_s}{\left[M^{x+}\right]_l}, \tag{10.127}$$

the resulting conventional partition coefficient is

---

* The sentence "the gas phase is not considered as an additional phase" adopted by Allegre et al. (1977) for this purpose is rather cryptic. They probably meant that the fractionation laws valid for solid phases are also valid for the gaseous phase, regardless of the different state of aggregation. In our view, the gas phase is a real phase to all intents and purposes (i.e., a region of the system with peculiar and distinguishable chemical and physical properties; cf. section 2.1). Hence, the term G is operationally analogous (although not identical) to ν in equations 10.121 and 10.124.

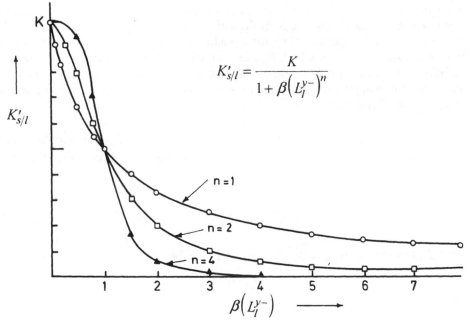

Figure 10.18 Conventional trace element partition coefficient as a function of factor $\beta(L_l^{y-})^n$. Reprinted from Treuil et al. (1979), *Bullettin Mineralogique,* 102, 402–409, with permission from Masson S.A., Paris, France.

$$K'_{s/l} = \frac{[M^{x+}]_s \gamma^{-1}_{M^{x+},s}}{[M^{x+}]_l \gamma^{-1}_{M^{x+},l} + [ML^{z-}]_l \gamma^{-1}_{ML^{z-},l}}. \qquad (10.128)$$

Introducing constant β (eq. 8 in Treuil et al., 1979):

$$K'_{s/l} = \frac{K}{\dfrac{\gamma_{M^{x+},s}}{\gamma_{M^{x+},l}} + \dfrac{\gamma_{M^{x+},l}\,\gamma^n_{L^{y-},l}}{\gamma_{ML^{z-},l}}\beta(L_l^{y-})^n}. \qquad (10.129)$$

Assuming ideal solution behavior by M^{x+}, L^{y-}, and ML^{z-} in the melt, equation 10.129 reduces to

$$K'_{s/l} = \frac{K}{1 + \beta(L_l^{y-})^n}. \qquad (10.130)$$

Based on equation 10.130 (figure 10.18), Treuil et al. (1979) suggested that hygromagmato-phile behavior is restricted to the field $\beta(L_l^{y-})^n \gg 1$, where $K'_{s/l} \ll K$, whereas for values of $(L_l^{y-})^n$ near 1 the highest variability of $K'_{s/l}$ occurs, corresponding to the field of incompatible elements.

Figure 10.18 shows that coordination number n of the complex (which depends on the relative charges of cation and ligand) has a strong influence on $K'_{s/l}$. The higher the value of n (i.e., the higher the charge of the cation for similar ligands: for example, Zr^{4+}, Hf^{4+}, Th^{4+}), the wider the field of hygromagmatophile behavior. Moreover, if the anionic ligand is a complex of type CO_3^{2-}, F^-, Cl^-, or OH^-, its concentration in the melt in saturation conditions depends on the solubility product of the gaseous phase. Modifications in fluid partial pressure may thus result in sudden changes in $K'_{s/l}$ during magma differentiation (Treuil et al., 1979).

According to Treuil, the constant concentration ratio of two hygromagmato-phile elements is indicative of fractional crystallization. From the Doerner-Hoskins equation:

$$C_l = C_0 F^{(D_{s/l}-1)},\tag{10.131}$$

where F is the fraction of melt, if $D_{s/l} \ll 1$ we obtain

$$\frac{C_l}{C_0} \approx F^{-1}\tag{10.132}$$

(Anderson and Greenland, 1969) and, for two hygromagmatophile elements i and j,

$$\frac{C_{i,l}}{C_{j,l}} = \frac{C_{i,0}}{C_{j,0}} = \text{constant.}\tag{10.133}$$

Error Δ associated with equation 10.132 can be expressed as a function of F (Allegre et al., 1977):

$$\Delta = \left[\frac{1}{F} - F^{(D_{s/l}-1)}\right]F\tag{10.134}$$

and is plotted for various values of $D_{s/l}$ in figure 10.19.

However, equation 10.133 is not sufficient to ascribe magmatic evolution to fractional crystallization if it is not supported by petrologic evidence. Adopting the formalism of Shaw (1970), it may be shown that the distribution of hygromag-matophile elements in fractionally accumulated liquids leads to the same sort of linear array. The equation

$$\overline{C}_l = C_0\left[1 - F^{1/D_{s/l}}\right]F^{-1},\tag{10.135}$$

where F is the fraction of residual solid, in fact reduces to equation 10.132 whenever $D_{s/l}$ is negligible. In the case of fractional melting, errors associated with

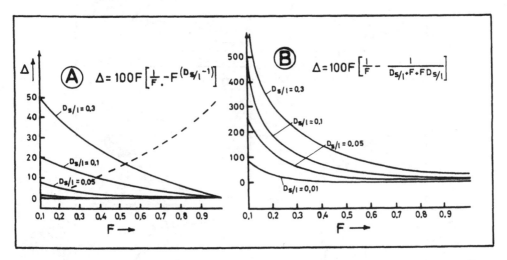

Figure 10.19 Fractional error associated with equation 10.132. (A) Rayleigh's crystallization and fractional melting. (B) Equilibrium melting.

equation 10.132 are symmetric with respect to errors associated with fractional crystallization:

$$\Delta = \left[\frac{1}{F} - F^{(1/D_{s/l})} \right] F \tag{10.136}$$

(see dashed line calculated with $D_{s/l} = 0.3$ in figure 10.19A).

But equation 10.132 cannot be applied to equilibrium melting. The introduced errors:

$$\Delta = \left[\frac{1}{F} - \frac{1}{D_{s/l} + F + FD_{s/l}} \right] F \tag{10.137}$$

are not negligible for any reasonable value of F (figure 10.19B) and for two hygromagmatophile elements i and j the relative abundances will not generally obey equation 10.133. This is especially true for low melting conditions in which $D_{s/l}$ is comparable in magnitude to F.

The differentiation index defined by equation 10.132 is a formidable tool in the petrologic investigation of volcanic suites. Because types and modal proportions of crystallizing phases change discontinuously during fractional crystallization (cf. Presnall, 1969), plotting the various chemical parameters as a function of F (defined through eq. 10.132) generally gives smooth trends with discontinuities located at particular F values corresponding to the appearance or exhaustion of a given solid. This was exemplified quite convincingly by Barberi et al. (1975) in the Boina magmatic series (Afar, Ethiopia). Figure 10.20 shows major element

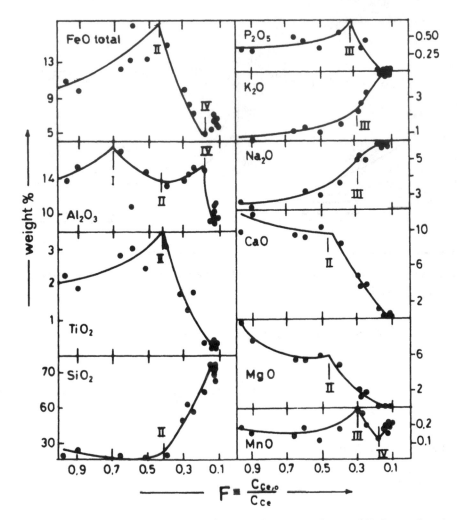

Figure 10.20 Major element concentrations in effusive products of Boina series plotted as functions of degree of fractional crystallization, based on equation 10.132. I = first discontinuity (transition from olivine-dominated to plagioclase-dominated fractionation); II = second discontinuity (appearance of Fe-Ti oxides); III = third discontinuity (field of SiO_2 oversaturated trachytes: apatite and Mn-oxides precipitate); IV = beginning of peralkalinity field. From Barberi et al. (1975), *Journal of Petrology,* 16, 22–56. Reproduced with modifications by permission of Oxford University Press.

concentrations plotted as functions of the degree of fractional crystallization F determined on the relative abundances of Ce ($C_{Ce,0}/C_{Ce}$).

The Boina emission products show a more or less continuous differentiation trend from transitional basalt to pantellerite. In the nonperalkaline field, the transition from basalt to ferro-basalt is dominated first by olivine ($F > 0.65$) and then by plagioclase ($F > 0.45$), with minor clinopyroxene. A second differentiation step with the appearance of Fe-Ti oxide crystals begins at $F = 0.45$: Fe and Ti decrease abruptly. Less marked discontinuities are also observed at $F = 0.3$ (silica-oversaturated trachytes) and $F = 0.15$ (peralkalinity field).

CHAPTER ELEVEN

Isotope Geochemistry

11.1   Isotopes in Nature

Atomic nuclei are composed mainly of *protons,* with a mass of 1.67262×10^{-24} g, or 1.007276 atomic mass units (amu, relative to $^{12}C = 12.000000$), and *neutrons,* with a mass of 1.008665 amu (see appendix 1 for a complete energy list of nuclear properties).

The number of protons Z in the nucleus establishes the identity of a chemical element. The mass number A of a given nucleus is established by the number of protons Z and neutrons N, added together. Neutrons are "neutral" in charge, and their relative numbers with respect to protons are not generally fixed. Generally, stable elements with even numbers of protons tolerate greater variability in terms of numbers of neutrons per nucleus than elements with odd numbers of protons. Ag atoms, for instance, are stable only when they have 47 protons and 60 or 62 neutrons. Each possible combination of nuclear charge Z and neutron number N is referred to as a *nuclide.* Nuclides of the same nuclear charge (same element) with the same number of protons Z and a different numbers of neutrons N are defined as *isotopes* of that element (*iso* = "same," *topos* = "place"—i.e., they occupy the same position on the periodic chart of the elements). Nuclides of the same mass number A but with different atomic numbers Z are called *isobars,* and nuclides with the same number of neutrons N but different masses A are *isotones.* The difference $N - Z$ (corresponding to $A - 2Z$) is referred to as an *isotopic number.* Finally, *isomers* are two or more nuclides having the same mass number A and atomic number Z but existing for measurable times in different quantum states, with different energies and radioactive properties.

The term "isotope" was coined by Soddy (1914) to define two or more substances of different masses occupying the same position in the periodic chart of the elements. Soddy's hypothesis was adopted to explain apparent anomalies in the relative positions of three couples of elements (Ar-K, Co-Ni, and Te-I) in the periodic chart. For instance, potassium is present in nature with three isotopes with masses of 39, 40, and 41, respectively, in the following proportions: $^{39}K = 93.26$, $^{40}K = 0.01$, and $^{41}K = 6.73$. Because the proportion of ^{39}K is dominant, the *element* K occurs in nature with mass 39.102, near the mass of *isotope* ^{39}K. Ar also has three isotopes: ^{36}Ar (0.34%), ^{38}Ar (0.06%), and ^{40}Ar (99.60%) and, obviously, the atomic weight of element Ar (39.948) is near that of isotope ^{40}Ar, which is the most abundant. Argon, which has an atomic number lower than that of potassium, has a higher mass, and this explains the (apparent) anomaly of their respective positions on Mendeleev's periodic chart.

Figure 11.1 is a chart of nuclides with N as the ordinate and Z as the abscissa. In this representation, isotones appear along horizontal lines and isotopes along the same vertical line. The opposite sort of representation is known as a "Segré chart."

All stable nuclides fall above the $N = Z$ line, with the exception of ^{1}H and ^{3}He. Up to $^{15}_7 N$, the number of neutrons does not exceed the number of protons by more than 1. Above $^{15}_7 N$, the stable nuclides diverge from the $N/Z = 1$ line upward, as a result of a progressive increase in the proportion of neutrons.

Hydrogen atoms and part of ^{4}He are believed to have been created during the "Big Bang" by proton-electron combinations. Most nuclides lighter than iron were created by nuclear fusion reactions in stellar interiors (cf. table 11.1). Nuclides heavier than the Fe-group elements (V, Cr, Mn, Fe, Co, Ni) were formed by neutron capture on Fe-group seed nuclei. Two types of neutron capture are possible: slow (*s-process*) and rapid (*r-process*).

Figure 11.2 shows two enlarged portions of the Segré chart where s-process and r-process are effective. Every box in the figure represents a possible nuclide, but only the white boxes are occupied by stable nuclides. The diagonally ruled boxes are unstable (half-lives of hours, minutes, or seconds) and the crosshatched ones are highly unstable. In the s-process (upper chain in figure), neutrons generated during He burning in stellar interiors are added to Fe-group seed nuclei. Neutron capture (n) displaces the nuclear composition rightward until it reaches an unstable box. β^- decay (electron emission, see later) then takes place and a neutron is transformed into a proton. This results in a diagonal movement in the chart of nuclides (upward and leftward).

The lower chain in figure exemplifies the r-process: several neutrons are added in a rapid sequence until a highly unstable Z-N combination is attained (crosshatched boxes), β^- decay then takes place.

The p-process takes place in supernovas by addition of protons followed by β^+ decay (positron emission). s-process nuclides are also transformed into nuclides of higher Z by photodisintegration reactions (gamma rays eject neutrons). Photodisintegration (by cosmic ray particles) is also responsible for the formation of the light nuclides ^{6}Li, ^{9}Be, ^{10}B, and ^{11}B by fragmentation of the nuclei of carbon, nitrogen, and oxygen. Table 8.1 lists in an extremely concise fashion the main nuclear reactions that lead to the formation of the light nuclides.

All elements with $Z > 83$ (Bi) are unstable and belong to chains of radioactive decay, or *decay series*. Three decay series include all radioactive elements in the $Z > 83$ part of the chart of nuclides—namely, *4n*, *4n + 2*, and *4n + 3* (because the decay takes place by α emission with mass decrease of four units, or by β emission with a negligible mass decrease, all nuclides within a series differ by

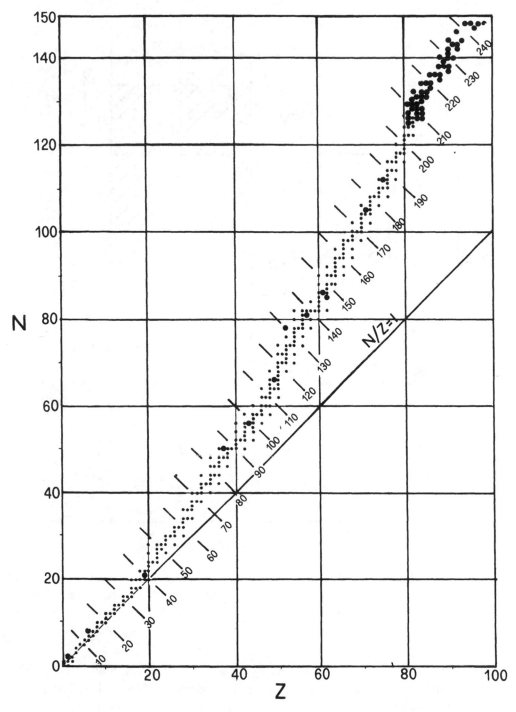

Figure 11.1 Chart of nuclides. •: stable, ○: unstable. Reproduced with modifications from Rankama (1954).

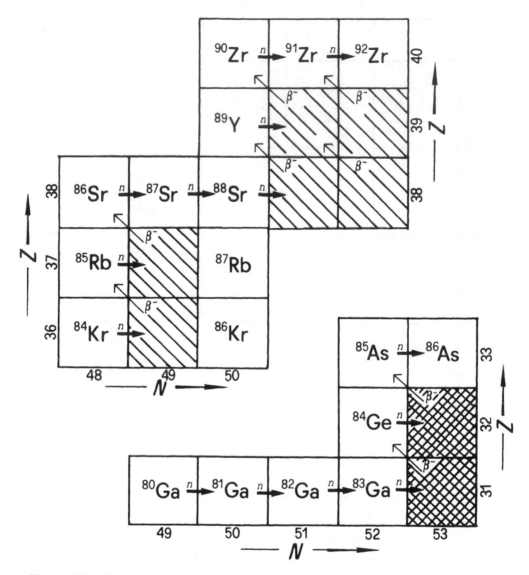

Figure 11.2 Enlarged portions of Segré chart of nuclides, showing s-process (upper chain) and r-process (lower chain). White boxes: stable nuclides; diagonally ruled boxes: unstable nuclides; crosshatched boxes: highly unstable nuclides.

multiples of four, and hence they are in a *4n*-type series, with *n* an integer). The major branches of the three decay series are shown in figure 11.3. The parent nuclide of *4n* series is ^{232}Th, which has ^{208}Pb as its stable end-product. The *4n* + *2* series has ^{238}U as parent nuclide and ^{206}Pb as stable end-product, and the *4n* + *3* series has ^{235}U as parent nuclide and ^{207}Pb as stable end-product.

All naturally occurring radioactive substances (besides those specified in the

Table 11.1 Nucleosynthesis and the main nuclear reactions taking place in stars. * = unstable; n = neutron; ν^+ = neutrino; β^+ = positron; γ = gamma radiation.

NUCLEOSYNTHESIS (3 minutes after the Big Bang → 1 hour after the Big Bang; log $T \cong 9$;
$^1H + n \rightarrow {}^2D + \gamma$ $\qquad$ $^2D + {}^1H \rightarrow {}^3He + \gamma$ $\qquad$ $^3He + {}^3He \rightarrow {}^4He + {}^1H + {}^1H$

HYDROGEN BURNING (first-generation stars; log $T \cong 7$)
$^1H + {}^1H \rightarrow {}^2D + \beta^+ + \nu^+$ $\quad$ $^2D + {}^1H \rightarrow {}^3He + \gamma$ $\qquad$ $^3He + {}^3He \rightarrow {}^4He + {}^1H + {}^1H$
$\qquad\qquad\qquad\qquad\qquad\qquad\qquad\qquad\qquad\qquad\qquad$ $^3He + {}^4He \rightarrow {}^7Be + \gamma$

$^7Be + e^- \rightarrow {}^7Li + \nu^+$ $\qquad$ $^7Li + {}^1H \rightarrow {}^8Be^* \rightarrow {}^4He + {}^4He$
$^7Be + {}^1H \rightarrow {}^8B + \gamma$ $\qquad$ $^8B \rightarrow {}^8Be^* + \beta^+ + \nu^+$ $\qquad$ $^8Be^* \rightarrow {}^4He + {}^4He$

CARBON-NITROGEN CYCLE (second- and later-generation stars)
$^{12}C + {}^1H \rightarrow {}^{13}N + \gamma$ $\qquad$ $^{13}N \rightarrow {}^{13}C + \beta^+ + \nu^+$
$^{13}C + {}^1H \rightarrow {}^{14}N + \gamma$ $\qquad$ $^{14}N + {}^1H \rightarrow {}^{15}O + \gamma$ $\qquad$ $^{15}O \rightarrow {}^{15}N + \beta^+ + \nu^+$
$\qquad\qquad\qquad\qquad\qquad\qquad\qquad\qquad\qquad\qquad\qquad$ $^{15}N + {}^1H \rightarrow {}^{12}C + {}^4He$

HELIUM-BURNING (red giants; log $T \cong 8$)
$^4He + {}^4He \rightarrow {}^8Be^*$ $\qquad\qquad$ $^8Be^* + {}^4He \rightarrow {}^{12}C + \gamma$ $\qquad$ $^{12}C + {}^4He \rightarrow {}^{16}O + \gamma$
$\qquad\qquad\qquad\qquad\qquad\qquad\qquad\qquad\qquad\qquad\qquad$ $^{16}O + {}^4He \rightarrow {}^{20}Ne + \gamma$

CARBON-BURNING reactions (log $T \cong 8.8$)
$^{12}C + {}^{12}C \rightarrow {}^{24}Mg + \gamma$
$^{12}C + {}^{12}C \rightarrow {}^{23}Mg + n$
$^{12}C + {}^{12}C \rightarrow {}^{23}Na + {}^1H$
$^{12}C + {}^{12}C \rightarrow {}^{20}Ne + \alpha$
$^{12}C + {}^{12}C \rightarrow {}^{16}O + 2\alpha$

OXYGEN-BURNING reactions (log $T \cong 9.4$)
$^{16}O + {}^{16}O \rightarrow {}^{31}S + n$
$^{16}O + {}^{16}O \rightarrow {}^{31}P + {}^1H$ $\qquad$ $^{31}P + \gamma \rightarrow {}^{30}Si + {}^1H$ $\qquad$ $^{30}Si + \gamma \rightarrow {}^{29}Si + n$
$\qquad\qquad\qquad\qquad\qquad\qquad\qquad\qquad\qquad\qquad\qquad$ $^{29}Si + \gamma \rightarrow {}^{28}Si + n$

$^{16}O + {}^{16}O \rightarrow {}^{28}Si + \alpha$

---

decay series of figure 11.3) are shown in table 11.2, together with their half-lives, isotopic abundances, and decay products. All the listed nuclides with $A \geq 40$ are the results of early cosmic events. Lighter radionuclides are continuously produced in the lower atmosphere by interaction of neutrons (produced by cosmic rays and slowed to thermal energies) with atmospheric nitrogen, oxygen, and argon nuclei. Note that β^- emission and α decay are by far the most common decay processes for natural radionuclides. Positron emission (β^+) is observed only in the decay of ^{26}Al, ^{36}Cl, ^{59}Ni and in a small branch of ^{40}K (0.001%; see section 11.7.3). Electron capture (EC) is also a quite limited process.

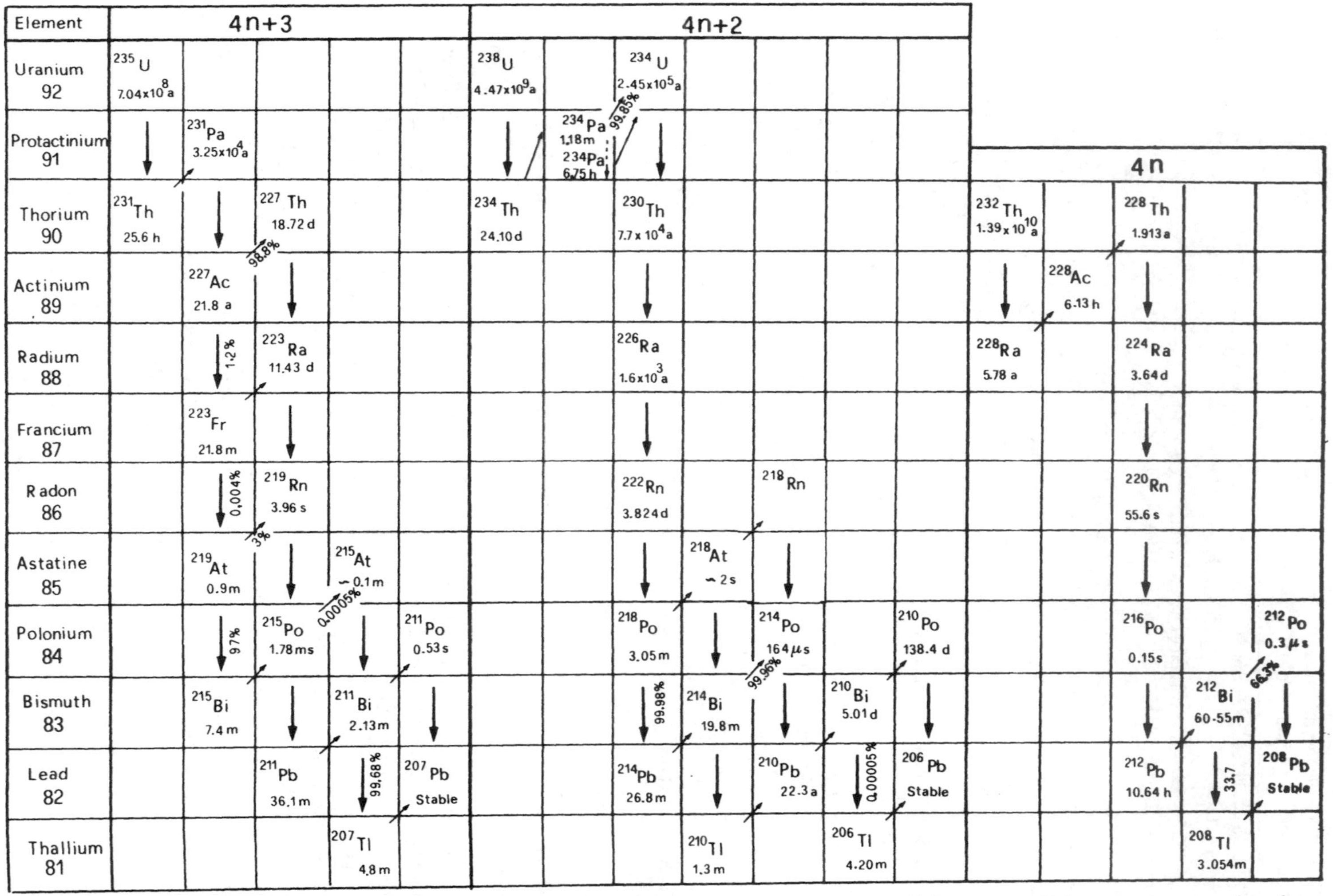

Figure 11.3 Major branches of decay chains of heavy radionuclides. light arrows: α decay: heavy arrows: β decay. a = years, d = days, h = hours, s = seconds.

Table 11.2 Naturally occurring radioactive substances. a = years. d = days.

Radionuclide	Decay Process	Half-Life	Isotopic Abundance (%)	Stable End-Product
^{3}H	β^-	12.26 a	1.38×10^{-4}	^{3}He
^{7}Be	EC	53.3 d	–	^{7}Li
^{10}Be	β^-	1.5×10^6 a	19.9	^{10}B
^{14}C	β^-	5730 a	–	^{14}N
^{26}Al	β^+, EC	7.16×10^5 a	–	^{26}Mg
^{32}Si	β^-	276 a	–	^{32}P
^{36}Cl	β^+, EC, β^-	3.08×10^5 a	–	^{36}S, ^{36}Ar
^{37}Ar	EC	34.8 d	–	^{37}Cl
^{39}Ar	β^-	269 a	–	^{39}K
^{40}K	β^-, EC, β^+	1.25×10^9 a	0.0117	^{40}Ca, ^{40}Ar
^{50}V	β^-, EC	6×10^{15} a	0.25	^{50}Cr, ^{50}Ti
^{53}Mn	EC	3.7×10^6 a	–	^{57}Cr
^{59}Ni	EC, β^+	8×10^4 a	–	^{59}Co
^{81}Kr	EC	2.13×10^5 a	–	^{81}Br
^{85}Kr	β^-	10.6 a	–	^{85}Rb
^{87}Rb	β^-	4.88×10^{10} a	27.835	^{87}Sr
^{113}Cd	β^-	9×10^{15} a	12.22	^{113}In
^{115}In	β^-	5.1×10^{14} a	95.7	^{115}Sn
^{123}Te	EC	1.2×10^{13} a	0.908	^{123}Sb
^{138}La	EC, β^-	2.69×10^{11} a	0.089	^{138}Ba, ^{138}Ce
^{142}Ce	α	$\approx 5 \times 10^{16}$ a	11.07	^{138}Ba
^{144}Nd	α	2.1×10^{15} a	23.80	^{140}Ce
^{147}Sm	α	1.06×10^{11} a	15.0	^{143}Nd
^{148}Sm	α	7×10^{15} a	11.3	^{144}Nd
^{149}Sm	α	$>1 \times 10^6$ a	13.8	^{145}Nd
^{152}Gd	α	1.1×10^{14} a	0.20	^{148}Sm
^{174}Hf	α	2.0×10^{15} a	0.162	^{170}Yb
^{176}Lu	β^-	3.57×10^{10} a	2.59	^{176}Hf
^{187}Re	β^-	4.23×10^{10} a	62.60	^{187}Os
^{190}Pt	α	6×10^{11} a	0.013	^{186}Os
^{232}Th	Decay chain	1.40×10^{10} a	100	^{208}Pb
^{234}U	Decay chain	2.44×10^5 a	0.0055	^{206}Pb
^{235}U	Decay chain	7.04×10^8 a	0.720	^{207}Pb
^{238}U	Decay chain	4.47×10^9 a	99.2745	^{206}Pb

11.2 Nuclear Energy

Nuclear mass is always lower than the sum of proton and neutron masses because of mass-energy conversion, which transforms part of the mass into binding energy. For example, ^{4}He has mass 4.002604, whereas the total mass of two protons and two neutrons is 4.032981. Based on Einstein's equation

$$E = mc^2 \tag{11.1}$$

(where c is the speed of light: 2.99792458×10^8 m · s^{-1}), the mass differential

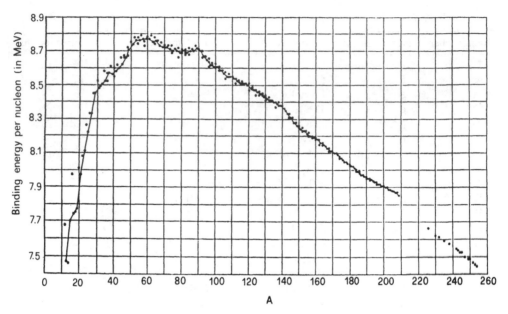

Figure 11.4 Mean binding energy per nucleon as a function of *A*. From G. Friedlander, J. W. Kennedy, E. S. Macias, and J. M. Miller, *Nuclear and Radiochemistry*, copyright © 1981 by John Wiley and Sons. Reprinted by permission of John Wiley & Sons.

$$dm = 4.032981 - 4.002604 = 0.030377 \qquad (11.2)$$

corresponds to a binding energy of 28.3 MeV (i.e., about 7.1 MeV per nuclear constituent). The average binding energy per nucleon is roughly constant (7.4 to 8.8 MeV), except in a few of the lightest nuclei. It increases slightly from light elements to $A = 60$ (iron and nickel nuclei) and then decreases more or less linearly with increasing A (figure 11.4).

Both protons and neutrons in the nucleus have an intrinsic angular momentum resulting from spinning of nucleons,

$$P = I \times \frac{h}{2\pi}, \qquad (11.3)$$

and an orbital angular momentum

$$L = m \times v \qquad (11.4)$$

that is an integral multiple of $h/2\pi$, being associated with motion along nuclear orbits with angular frequency $\omega = 2\pi v$. Because the intrinsic spin of protons and neutrons is $I = \frac{1}{2}$, the nuclear spin of nuclides with odd A is $\frac{1}{2}$; if A is even, it is zero or an integer. Because rotation of a charged particle produces a magnetic moment, nuclides with nuclear spins different from zero also have a magnetic moment of magnitude $ZeP/2mc$. The magnetic moment of a single proton is taken as the unit of nuclear magnetic moment, corresponding to $\frac{1}{658}$ of the Bohr magneton (magnetic moment of electron).

Because experimental evidence indicates that the volume and total binding energy of nuclei are proportional to the number of nucleons, it can be deduced that nuclear matter is incompressible and that each nucleon is able to interact only with a limited number of surrounding nucleons. This evidence leads to formulation of the binding energy and mass of nuclei in terms of a "charged drop" model—i.e., the nuclei are assimilated to charged liquid drops with surface tension. Total binding energy E_B is the result of various energy contributions, each with simple functional dependence on mass A and charge Z of the nucleus (Myers and Swiatecki, 1966):

$$E_B = c_1 A \left[1 - k \left(\frac{N-Z}{A} \right)^2 \right] - c_2 A^{2/3} \left[1 - k \left(\frac{N-Z}{A} \right)^2 \right]$$

$$- c_3 Z^2 A^{-1/3} + c_4 Z^2 A^{-1} + \delta.$$

(11.5)

Although equation 11.5 has only six adjustable parameters, it is able to reproduce the energy of the approximately 1200 nuclides of known mass. When binding energy is expressed in MeV, the empirically adjusted coefficients have the following values: $c_1 = 15.677$ MeV, $c_2 = 18.56$ MeV, $c_3 = 0.717$ MeV, $c_4 = 1.211$ MeV, and $k = 1.79$. δ is associated with the number of protons (if A is odd, $\delta \times A = 0$; if it is even, $\delta \times A = \pm 132$).

The first term on the right in equation 11.5 is the *volume energy*, proportional to the number of protons and neutrons in the nucleus. It contains a symmetry correction term, proportional to $(N - Z)^2 / A^2$, accounting for the fact that nuclear forces are at maximum when $N = Z$ and decrease symmetrically on both sides of $N = Z$. The second term is the *surface energy*, which accounts for the fact that nucleons at the surface of the nucleus have unsaturated forces that reduce binding energy proportionally to $A^{2/3}$ (and hence to the surface, A being proportional to volume). The third term is the electrostatic *Coulomb interaction*, accounting for Coulomb repulsion between protons, which have homologous charge. This term also reduces the binding energy of the nucleus. Because the Coulomb interaction is proportional to Z^2, it becomes increasingly important with the increase of Z and is responsible for the observed progressive relative increase of N/Z (see figure 11.1). The fourth term in equation 11.5 is a correction for Coulomb energy, resulting from the nonuniform charge distribution (or "diffuse boundary").

11.3 Nuclear Decay

All nuclides are generally subdivided into four types, depending on whether they contain even or odd numbers of protons and neutrons, as shown in table 11.3.

Table 11.3 General classification of nuclides: significance of parity is related to symmetry properties of nuclear wave functions. A nuclide is said to have odd or even parity if the sign of the wave function of the system respectively changes or not with changing sign in all spatial coordinates (see Friedlander et al., 1981 for more detailed treatment). The value assigned to $\delta x A$ is appropriate for $A > 80$. For $A < 60$, a value of ± 65 is more appropriate.

Z-N Combination	Type	Parity	Example	Number of Nuclides	$\delta x A$
Even-even	1	Even	$^{16}_{8}O$	164	-132
Even-odd	2	Odd	$^{9}_{4}Be$	55	0
Odd-even	3	Odd	$^{10}_{5}B$	50	0
Odd-odd	4	Even	$^{7}_{3}Li$	4	-132

The occurrence of nuclides in nature reflects relative stability. In general, nuclides whose mass numbers are multiples of 4 are exceptionally stable, because a group of two protons and two neutrons forms a closed shell (nuclear shells are to some extent analogous to the electron shells discussed in chapter 1).

11.3.1 Beta Decay

Beta decay is the most common decay process (either natural or artificial): a neutron is transformed into a proton by emission of a β^- particle (electron):

$$n \rightarrow p + \beta^-. \tag{11.6}$$

Alternatively, a proton is transformed into a neutron by emission of a β^+ particle (positron):

$$p \rightarrow n + \beta^+. \tag{11.7}$$

Mass number A does not change in either case.

Because the mass of the nucleus is related to the masses of protons M_p and neutrons M_n through

$$M = ZM_P + \left(A - Z\right)M_n - E_B, \tag{11.8}$$

by combining equations 11.5 and 11.8 we obtain

$$M = f_1\left(A\right)\left(Z - Z_A\right)^2 + f_2\left(A\right)\left(Z - Z_A\right) - \delta\left(A\right). \tag{11.9}$$

Because, for a constant value of A, $f_1(A)$, $f_2(A)$, and $\delta(A)$ are also constant, equation 11.9 is a parabola (for a given value of δ) with maximum at $Z = Z_A$:

$$\frac{\partial M}{\partial Z} = 0 \qquad \rightarrow \qquad Z_A = -\frac{f_2(A)}{2f_1(A)}. \qquad (11.10)$$

Figure 11.5 shows the nuclear energy parabolas for odd ($A = 125$) and even ($A = 128$) nuclides. These parabolas represent sections of the nuclear energy surface on the Z-E_B plane. Because δ is zero at any odd value of A, there is a single parabola, whereas for an even A there are two parabolas, separated along the energy axis by $2\delta/A$.

The usefulness of plots such as those in figure 11.5 stems from the fact that they give approximate values of the energy available for β decay between neighboring isobars. For odd A (figure 11.5A) there is only one stable nuclide—i.e., the one nearest to the minimum of the curve. Note here that, as is customary, energy parabolas are drawn in a reverse fashion, with the minima actually corresponding to maxima in the Z-E_B plane. Note also that in some instances the energy minimum may be achieved at a noninteger value of Z, because equation 11.5 is continuous in Z. In these cases, the actual minimum is attained at the integer value of Z nearest to the calculated one.

A nuclide of type 1 emitting β^+ or β^- particles is transformed into a type 4 nuclide, lying on a parabola translated by $2\delta/A$ (figure 11.5B). After a new β^+ or β^- emission, the type 4 nuclide is reconverted into a type 1 nuclide, and so on. In a β emission chain, nuclides of different nuclear charges alternate between the two energy parabolas in the energy–nuclear charge field. The β decay process operating on even A parent nuclides may result in two or more stable isotopes of the even-even type. For example, in figure 11.5B, both $_{52}^{128}$Te and $_{54}^{128}$Xe can be considered stable, having a binding energy lower than that of the odd parent $_{53}^{128}$I.

The energy associated with β emission during nuclear decay is not constant, but varies from a minimum value near zero to a maximum corresponding to the energy gap between the parent nuclide and the radiogenic nuclide. This apparently contrasts with the principle of energy conservation. However, as shown by Pauli (1933), a third particle participates in the process. This particle is called a "neutrino" (ν because of its very limited mass and charge. The existence of the neutrino (and of its corresponding "antineutrino," $\bar{\nu}$), postulated on purely theoretical grounds by Pauli, has only recently received experimental confirmation.

Beta particles do not transfer all the binding energy freed by the decay process; part of it resides in neutrinos. The mean β emission frequency is one-third of the energy gap involved in nuclear decay; in other words, most emitted β particles have kinetic energy corresponding to one-third of the energy gap. Only a few particles are emitted with a frequency corresponding to maximum energy. Emitted β particles may penetrate surrounding matter before losing most of their kinetic energy. Secondary electrons hit by emitted β particles also have high energy and may be expelled by their localized orbitals, causing successive ionizations.

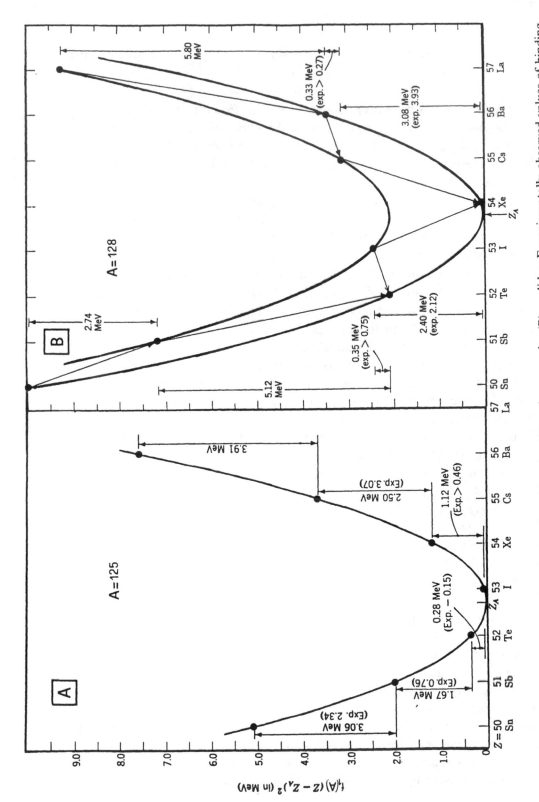

Figure 11.5 Energy versus nuclear charge plots for odd parity (A) and even parity (B) nuclides. Experimentally observed values of binding energy differences are reported in parentheses for comparative purposes. From *Nuclear and Radiochemistry,* G. Friedlander and J. W. Kennedy, Copyright © 1956 by John Wiley & Sons. Reprinted by permission of John Wiley & Sons, Ltd.

Besides the production of positive ions, secondary electrons may be captured by saturated atoms or molecules, resulting in negative ions.

If the available *decay energy* (mass difference between decaying and produced nuclides) exceeds 1.02 MeV, one positron and one electron may be annihilated reciprocally, with the emission of two γ photons of energy corresponding to the masses of the annihilated particles—i.e.,

$$E = 2mc^2 = 1.02 \text{ MeV}. \tag{11.11}$$

Let us consider, for instance, the decay process of the unstable oxygen isotope $^{14}_{8}O$, which is transformed into stable $^{14}_{7}N$ (figure 11.6). The $^{14}_{8}O$ decays, emitting two series of β^+ particles of different energies. In decay process (1) in figure 11.6, a $^{14}_{7}N$ isomer forms in an unstable excited state, and reaches stability by emission of γ rays of different energies. The first γ ray has an energy of 2.313 MeV and leads to an excited state 1.02 MeV above the ground level. The second decay step involves positron-electron interaction and leads to emission of two γ photons with energies of 1.02 MeV, which reduce the energy of the daughter nuclide at the ground state of stable $^{14}_{7}N$. The energy of the β^+ particles involved in decay process (2) is higher (4.1 MeV) and leads directly to the excited state preceding positron-electron interaction.

We can summarize the β decay process with two examples:

$$^{40}_{19}K \rightarrow {}^{40}_{20}Ca + \beta^- + \bar{\nu} + 1.312 \text{ MeV} \tag{11.12}$$

and

$$^{18}_{9}F \rightarrow {}^{18}_{8}O + \beta^+ + \nu + 1.655 \text{ MeV}. \tag{11.13}$$

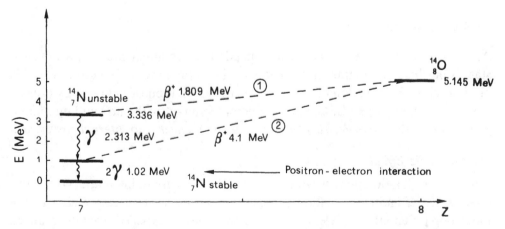

Figure 11.6 Energy scheme for β^+ decay of unstable $^{14}_{8}O$.

11.3.2 Electron Capture

Electron capture is a particular form of β decay, described for the first time by Alvarez (1938) and practically equivalent to positron emission. When the decay energy of the nuclide is not sufficient to attain the reaction

$$p \rightarrow n + \beta^+ \tag{11.14}$$

(i.e., 1.02 MeV; cf. eq. 11.11), an electron of the inner K level enters the nucleus, leaving behind a vacancy that is occupied by a more external electron. Vacancy migration proceeds toward the external orbitals and is associated with X-ray emission. An example of electron capture (EC) decay is

$$^{138}_{57}\text{La} + e^- \rightarrow {}^{138}_{56}\text{Ba} + h\nu(\gamma) - h\nu(X). \tag{11.15}$$

11.3.3 Alpha Decay

Alpha particles are composed of two protons and two neutrons. Thus they have $Z = 2$, $N = 2$, and $A = 4$ and correspond to a helium nucleus ^4_2He. The emission of α particles thus produces a decrease of 4 units in A. An unstable nuclide undergoing α decay may emit α particles of various energy and thus directly reach the ground level of the stable product. Alternatively, as in β emission, an intermediate excited state is reached, followed by γ emission. Figure 11.7 shows, for example, the decay process of $^{228}_{90}\text{Th}$, which may directly attain the ground level of $^{224}_{88}\text{Ra}$ by emission of α particles of energy 5.421 MeV or intermediate excited states by emission of α particles of lower energy, followed by γ emission.

Each γ-ray energy emission connecting two excited states corresponds to the difference between the disintegration energy associated with the two α decays, which lead to the two limiting excited conditions.

11.3.4 Gamma Transitions

As we saw above, when a radiogenic nuclide (or "radionuclide") in an excited state is produced as an intermediate step of the decay process, the ground level is often attained by emission of electromagnetic γ radiation of high energy (generally higher than 100 keV). Gamma radiations interact with surrounding matter, giving rise to the *photoelectric effect,* the *Compton effect,* and *internal conversion.*

Photoelectric Effect

Emitted γ radiation interacts with electrons of the surrounding matter. If the frequency of the emitted radiation exceeds the energy level corresponding to the ionization potential of the element, the electron may be expelled from its localized

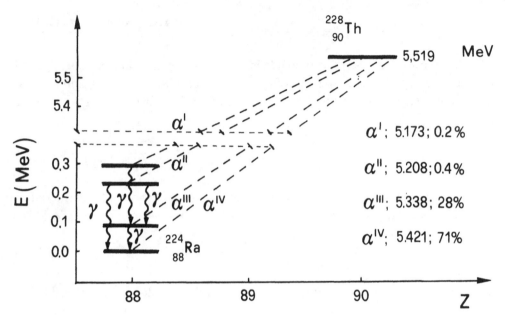

Figure 11.7 Energy scheme for α decay of $^{228}_{90}$Th.

orbital (*K*-level electrons are normally involved in the process). Because $h\nu$ is the energy of the γ radiation and W is the energy necessary to expel the electron from its orbital (or "work function of the metal," an important parameter in mass spectrometry), expulsion will take place only when $h\nu$ is greater than W. In this case, the expelled electron achieves kinetic energy corresponding to the surplus:

$$\frac{1}{2}mv^2 = h\nu - W. \tag{11.16}$$

Compton Effect

When γ radiation hits an electron, it is deviated from its original trajectory and, losing energy, changes its frequency. The increase in frequency consequent upon anelastic scattering is only a function of the angle between incident and deflected rays and does not depend on the energy of the incident radiation. The importance of this fact, known as the Compton effect, increases as the atomic number of the element decreases.

Internal Conversion

Expelled electrons belong to the *K*-level of the emitting atom. Also in this case, as in *K*-capture, the created vacancy is progressively occupied by more external electrons and is associated with X-ray emission.

The decay rate of a given radionuclide is generally constant and independent of the past history of the nuclide and of the P, T (at least below 10^8 K), and chemistry of the system. The probability $\mathcal{P}$ of the radionuclide disintegrating over time Δt is

$$\mathcal{P} = \lambda \, \Delta t, \tag{11.17}$$

where λ is the proportionality constant. Conversely, the probability of the radionuclide not disintegrating in the same lapse of time is

$$1 - \mathcal{P} = 1 - \lambda \, \Delta t. \tag{11.18}$$

This probability remains unchanged even if the radionuclide has survived the first elapsed time increment. Combining the single probabilities for n elapsed time intervals, we obtain

$$\frac{N}{N_0} = \left(1 - \lambda \frac{t}{n}\right)^n, \tag{11.19}$$

where N_0 is the number of initial radionuclides and N is the number of remaining radionuclides after disintegration at time t:

$$t = n \times \Delta t. \tag{11.20}$$

Reducing time interval Δt to an infinitesimal amount dt and applying to equation 11.19 the limit

$$\lim_{n \to \infty} \left(1 + \frac{x}{n}\right)^n = \exp(x), \tag{11.21}$$

we obtain

$$\frac{N}{N_0} = \exp(-\lambda t) \tag{11.22}$$

—i.e., the rate of a "first-order reaction" (cf. section 8.21.1).

The amount of time $t_{1/2}$ required to reduce by one-half the initial number N_0 of radionuclides is called the *half-life* of the species:

$$\ln\left(\frac{N}{N_0}\right) = \ln\left(\frac{1}{2}\right) = -\lambda t_{1/2} \qquad (11.23)$$

$$t_{1/2} = \frac{\ln 2}{\lambda} = \frac{0.69315}{\lambda}. \qquad (11.24)$$

The *average life* τ of a given radionuclide corresponds to the inverse of the decay constant—hence, to $1/\ln 2$ times the half-life $t_{1/2}$, as obtained by integration of the time of existence of all radionuclides divided by the initial number N_0:

$$\tau = -\frac{1}{N_0}\int_{t=0}^{t=\infty} t\, dN = \frac{t_{1/2}}{\ln 2} = \frac{1}{\lambda}. \qquad (11.25)$$

11.5 Growth of Radioactive Products

Because in nature the ground state of a stable nuclide is often attained by decay chains involving intermediate species decaying at different rates, it is worth evaluating the implications of the relative magnitudes of the various decay constants on the isotopic composition of the element.

If we call N_1 the number of nuclides of the first species, decaying at rate λ_1, the number of disintegrated nuclides of type 1 per increment of time dt is

$$\frac{dN_1}{dt} = -\lambda_1 N_1. \qquad (11.26)$$

The population of type 2 nuclides produced by decay of N_1 is affected both by the rate of production (corresponding to the decay rate of species 1—i.e., eq. 11.26 changed in sign) and by its own rate of disintegration:

$$\frac{dN_2}{dt} = \lambda_1 N_1 - \lambda_2 N_2. \qquad (11.27)$$

Several substitutions yield the following equation, relating the population of the second decaying species to elapsed time t:

$$N_2 = \frac{\lambda_1 N_{0,1}}{\lambda_2 - \lambda_1}\left[\exp\left(-\lambda_1 t\right) - \exp\left(-\lambda_2 t\right)\right] + N_{0,2}\exp\left(-\lambda_2 t\right), \qquad (11.28)$$

where $N_{0,1}$ and $N_{0,2}$ are the populations of species 1 and 2 at time $t = 0$.

We can now distinguish three general cases, depending on whether the first decaying species has a longer, a much longer, or a shorter half-life than that of the daughter nuclide. These three cases are *transient equilibrium, secular equilibrium,* and *nonequilibrium.*

11.5.1 Transient Equilibrium

Figure 11.8A shows the radioactivity emitted by a mixture of two independently decaying species 1 and 2, detected with detection coefficients c_1 and c_2 (independent and not necessarily identical, but identical here, for the sake of simplicity).

Total radioactivity $\mathcal{R}_T$ is the sum of separate radioactivities of the two species and depends on their half-lives ($t_{1/2,1} = 8$ hr and $t_{1/2,2} = 0.8$ hr for the cases exemplified in figure 11.8) according to

$$\mathcal{R}_T = \mathcal{R}_1 + \mathcal{R}_2 = c_1 \lambda_1 N_1 + c_2 \lambda_2 N_2. \qquad (11.29)$$

As shown in figure 11.8A, in the case of independently decaying species, composite decay curve a attains the limiting behavior imposed by the long-lived nuclide.

Let us consider now species 1 and 2 linked in a decay chain with the parent nuclide 1 being longer-lived than the daughter nuclide 2 (i.e., $\lambda_1 < \lambda_2$). After a relatively short time, the terms $\exp(-\lambda_2 t)$ and $N_{0,2} \exp(-\lambda_2 t)$ become negligible with respect to $\exp(-\lambda_1 t)$. As a result, equation 11.28 is reduced to

$$N_2 = \frac{\lambda_1 N_{0,1}}{\lambda_2 - \lambda_1} \exp\left(-\lambda_1 t\right) \qquad (11.30)$$

and, because

$$N_{0,1} \exp\left(-\lambda_1 t\right) = N_1, \qquad (11.31)$$

the final result is

$$\frac{N_1}{N_2} = \frac{\lambda_2 - \lambda_1}{\lambda_1}. \qquad (11.32)$$

As shown in figure 11.8B, because the ratio N_1/N_2 becomes constant, the slopes of the combined decay curves of the two radionuclides attain a constant value corresponding to the half-life of the longer-lived term (curves a and b in figure 11.8B). Moreover, assuming identical detection coefficients for the two species, their radioactivity ratio also attains a constant value of

$$\frac{\mathcal{R}_1}{\mathcal{R}_2} = \frac{\lambda_2 - \lambda_1}{\lambda_2}. \qquad (11.33)$$

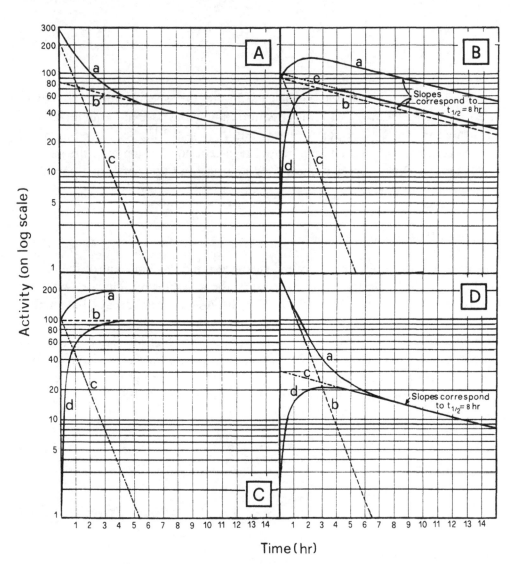

Figure 11.8 Composite decay curves for (A) mixtures of independently decaying species, (B) transient equilibrium, (C) secular equilibrium, and (D) nonequilibrium. *a:* composite decay curve; *b:* decay curve of longer-lived component (A) and parent radio nuclide (B, C, D); *c:* decay curve of short-lived radionuclide (A) and daughter radionuclide (B, C, D); *d:* daughter radioativity in a pure parent fraction (B, C, D); *e:* total daughter radioactivity in a parent-plus-daughter fraction (B). In all cases, the detection coefficients of the various species are assumed to be identical. From *Nuclear and Radiochemistry*, G. Friedlander and J. W. Kennedy, Copyright © 1956 by John Wiley and Sons. Reprinted by permission of John Wiley and Sons Ltd.

11.5.2 Secular Equilibrium

If the decay constant of the parent radionuclide is much lower than that of the daughter radionuclide (i.e., $\lambda_1 \ll \lambda_2$), equation 11.32 can be reasonably reduced to

$$\frac{N_1}{N_2} = \frac{\lambda_2}{\lambda_1}. \qquad (11.34)$$

Assuming identical detection coefficients for the two species, the radioactivity ratio obviously reduces to 1. This condition, known as "secular equilibrium," is illustrated in figure 11.8C for $t_{1/2,1} = \infty$ and $t_{1/2,2} = 0.8$ hr. Secular equilibrium can be conceived of as a limiting case of transient equilibrium with the angular coefficient of decay curves progressively approaching the zero slope condition attained in figure 11.8C.

11.5.3 Nonequilibrium

The case of "nonequilibrium" occurs whenever the decay constant of the parent nuclide is higher than the decay constant of the daughter nuclide (i.e., $\lambda_1 > \lambda_2$) and is illustrated in figure 11.8D for $\lambda_1/\lambda_2 = 10$. As the parent decays, the number of daughter nuclides progressively rises, reaching a maximum at time t_m. It then decreases in a constant slope, characteristic of its own half-life. Time t_m is found graphically by intersecting the decay curve of the parent (curve b in figure 11.8D) and the extrapolation of the final decay curve to time zero (curve c). The same parameter can be obtained analytically by applying the equation

$$t_m = \frac{\ln \lambda_2 - \ln \lambda_1}{\lambda_2 - \lambda_1}. \qquad (11.35)$$

11.6 Isotope Fractionation

11.6.1 Equilibrium Exchanges

Isotopes of the same element may fractionate during homogeneous or heterogeneous chemical reactions. The extent of fractionation increases with decreasing T and increasing mass contrast, and is favored by changes in coordination, the force constant being equal (see later in this section). The fractionation of isotopes of the same element during chemical exchanges is a quantum effect. In quantum chemistry, partition function Q (cf. section 3.1) may in a first approximation be written as a product of translational, rotational, and vibrational terms:

$$Q = Q_{\text{trans}} \times Q_{\text{rot}} \times Q_{\text{vib}}. \qquad (11.36)$$

The translational partition function Q_{trans} is equal to the classical one at all temperatures

$$Q_{\text{trans}} = \sum_i \exp\left(-\frac{\varepsilon_{k,i,x}}{kT} - \frac{\varepsilon_{k,i,y}}{kT} - \frac{\varepsilon_{k,i,z}}{kT}\right) \qquad (11.37)$$

where $\varepsilon_{k,i,x}$ is the kinetic energy of molecule i along direction x. Also the rotational partition function Q_{rot} is equal to the classical one, with the exception of hydrogen (see section 11.8.2).

In the harmonic approximation, Q_{vib} is given by

$$Q_{\text{vib}} = \prod_i \frac{\exp\left(-X_i/2\right)}{1 - \exp\left(-X_i\right)} \qquad (11.38)$$

(Bigeleisen and Mayer, 1947) where

$$X_i = \frac{h\nu_i}{kT} = \frac{\hbar\omega_i}{kT} \qquad (11.39)$$

(see section 3.3 for the significance of the symbols).

At high T (low X_i), Q_{vib} reduces to $\prod_i(1/X_i)$. Because the potential energies of molecules differing only in isotopic constituents are alike, one can define a "separative effect" based on the partition function ratio f of isotopically heavy and light molecules ($Q^\bullet$ and Q°, respectively) in such a way that the rotational and translational contributions to the partition function cancel out:

$$\left(\frac{s^\bullet}{s^\circ}\right)f = \left(\frac{s^\bullet}{s^\circ}\right)\left(\frac{Q^\bullet}{Q^\circ}\right)\left(\frac{m^\circ}{m^\bullet}\right)^{3/2}$$

$$= \prod_i \frac{X_i^\bullet}{X_i^\circ} \frac{\exp\left(-X_i^\bullet/2\right)\big/\left[1 - \exp\left(-X_i^\bullet\right)\right]}{\exp\left(-X_i^\circ/2\right)\big/\left[1 - \exp\left(-X_i^\circ\right)\right]}. \qquad (11.40)$$

The ratio of symmetry numbers $s^\bullet/s^\circ$ in equation 11.40 merely represents the relative probabilities of forming symmetrical and unsymmetrical molecules, and $m^\bullet$ and m° are the masses of exchanging molecules (the translational contribution to the partition function ratio is at all T equal to the $\frac{3}{2}$ power ratio of the inverse molecular weight). Denoting as ΔX_i the vibrational frequency shift from isotopically heavy to light molecules (i.e., $\Delta X_i = X_i^\circ - X_i^\bullet$) and assuming ΔX_i to be intrinsically positive, equation 11.40 can be transated into

$$\left(\frac{s^\bullet}{s^\circ}\right)f = \prod_i \frac{X_i^\bullet}{X_i^\bullet + \Delta X_i} \exp\left(\frac{\Delta X_i}{2}\right) \frac{1 - \exp\left[-\left(X_i^\bullet + \Delta X_i\right)\right]}{1 - \exp\left(-X_i^\bullet\right)}. \qquad (11.41)$$

The Helmholtz free energy of each component in reaction is related to the partition function Q:

$$F = -kT \ln Q \qquad (11.42)$$

Based on equation 11.41, the difference between the Helmholtz free energies of formation of two isotopic molecules with respect to their gaseous atoms depends on the shift of vibrational frequencies between heavy and light isotope-bearing compounds—i.e., according to Bigeleisen and Mayer (1947),

$$\frac{\Delta F^{\bullet} - \Delta F^{\circ}}{kT} = \sum_i \left[-\frac{1}{2}\Delta X_i + \ln\left(1 + \frac{\Delta X_i}{X_i^{\bullet}}\right) + \ln\frac{1 - \exp\left(-X_i^{\bullet}\right)}{1 - \exp\left(-X_i^{\circ}\right)} \right] \qquad (11.43)$$

$$+ \ln\left(\frac{s^{\bullet}}{s^{\circ}}\right),$$

where $\Delta F^{\bullet}$ and ΔF° are the Helmholtz free energies of formation of isotopically heavy and light molecules. If ΔX_i is small, which is the case for all isotopes except hydrogen, equation 11.43 can be reduced to

$$\frac{\Delta F^{\bullet} - \Delta F^{\circ}}{kT} = -\sum_i \left[\frac{1}{2} - \frac{1}{X_i^{\bullet}} + \frac{1}{\exp\left(X_i^{\bullet}\right) - 1} \right]\Delta X_i + \ln\left(\frac{s^{\bullet}}{s^{\circ}}\right), \qquad (11.44)$$

and the separative effect $(s^{\bullet}/s^{\circ})\, f$ becomes

$$\left(\frac{s^{\bullet}}{s^{\circ}}\right)f = 1 + \sum_i \left[\frac{1}{2} - \frac{1}{X_i^{\bullet}} + \frac{1}{\exp\left(X_i^{\bullet}\right) - 1} \right]\Delta X_i. \qquad (11.45)$$

If $X_i^{\bullet}$ is also small, then the terms in brackets in equation 11.45 approach $X_i^{\bullet}/12$ and the separative effect reduces to

$$\left(\frac{s^{\bullet}}{s^{\circ}}\right)f = 1 + \sum_i \frac{X_i^{\bullet}\,\Delta X_i}{12}. \qquad (11.46)$$

Because ΔX_i is positive, $(s^{\bullet}/s^{\circ})\, f$ is always greater than 1, and the heavy isotope preferentially stabilizes in the condensed phase whereas the light isotope favors the gaseous phase.

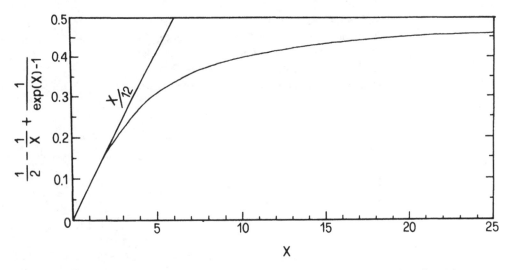

Figure 11.9 Separative effect per unit shift, plotted against nondimensionalized frequencies $X = h\nu/kt = \hbar\omega/kT$. Reprinted from Bigeleisen and Mayer (1947), with permission from the American Institute of Physics.

At low temperatures and light frequencies, the separative effect per unit shift (the terms in brackets in eq. 11.44 and 11.45) approaches $\frac{1}{2}$ (figure 11.9) and the Helmholtz free energy difference approaches the differences in zero-point energies. At high T (low frequencies), the separative effect per unit shift approaches zero and the total separative effect $(s^\bullet/s^\circ)f$ approaches 1, so that no isotopic fractionation is observed.

A useful approximation for the isotopes of heavy elements (Bigeleisen and Mayer, 1947) is

$$\left(\frac{s^\bullet}{s^\circ}\right)f = 1 + \frac{M\,\Delta m}{24m^2}\,X_s^2 n, \tag{11.47}$$

where m is the mass of the central atom (isotope) surrounded by n ligands of mass M, Δm is the difference in isotopic masses, and nondimensionalized frequency X_s is related to the totally symmetric frequency arising from the stretching of bonds with central atom n_s by

$$X_s = \frac{h\nu_s}{kT} = \frac{h}{kT}\sqrt{\frac{2\pi C}{M}} \tag{11.48}$$

where C is a force constant.

Table 11.4 lists the results of calculations based on equation 11.47 for isotopic exchange reactions at $T = 300$ K for tetrahedrally and octahedrally coordinated "heavy" isotopes:

Table 11.4 Separative effect f and isotopic fractionation constant K for heavy isotopes, computed through equation 11.47. ω_s is angular totally symmetric stretching frequency derived from Raman spectra (see Bigeleisen and Mayer, 1947 for references).

Molecule	ω_s	f	K
SiF_4	800	1.111 ⎯⎯⎯⎯	
SiF_6^{2-}	600	1.109 ⎯⎯⎯⎯	1.002
$SnCl_4$	367	1.00256 ⎯⎯⎯	
$SnCl_6^{2-}$	314	1.00281 ⎯⎯⎯	1.00025

$$^{28}SiF_4 + {}^{30}SiF_6^{2-} \Leftrightarrow {}^{30}SiF_4 + {}^{28}SiF_6^{2-} \qquad (11.49)$$

$$^{120}SnCl_4 + {}^{118}SnCl_6^{2-} \Leftrightarrow {}^{118}SnCl_4 + {}^{120}SnCl_6^{2-} . \qquad (11.50)$$

Note that the separative effect is quite significant, even for the relatively "heavy" isotopes of tin. Note also the large cancellation effect on the isotopic fractionation constant, arising from superimposed partial reactions such as

$$^{120}SnCl_4 + {}^{118}Sn \rightarrow {}^{118}SnCl_4 + {}^{120}Sn \qquad (11.51)$$

and

$$^{118}SnCl_6^{2-} + {}^{120}Sn \rightarrow {}^{120}SnCl_6^{2-} + {}^{118}Sn . \qquad (11.52)$$

This cancellation (always marked in exchanges of acid-base type) can be ascribed to the decrease in force constant C of the central atom with increasing coordination number n, and is sometimes not so significant in other types of isotopic exchange reactions.

As suggested by Urey (1947), an isotopic exchange can be expressed by a reaction such as

$$a\,A^\circ + b\,B^\bullet \Leftrightarrow a\,A^\bullet + b\,B^\circ , \qquad (11.53)$$

where A°, $A^\bullet$ and B°, $B^\bullet$ are molecules of the same type that differ only in the isotopic composition of a given constituent element (open and full symbols: light and heavy), and a and b are stoichiometric coefficients.

In light of equation 11.42, the equilibrium constant for the isotopic exchange reaction of equation 11.53 reduces to

$$K_{53} = \frac{\left(Q_A^\bullet / Q_A^\circ\right)^a}{\left(Q_B^\bullet / Q_B^\circ\right)^b}.$$ (11.54)

The equivalent formulation of equation 11.38 for a crystalline solid is

$$Q = \prod_{i=1}^{3N-6} \frac{\exp\left(-X_i/2\right)}{1 - \exp\left(-X_i\right)},$$ (11.55)

where N is the number of atoms in the crystal, and the ratio of partition functions of the isotopically heavy and light compounds (Kieffer, 1982) is

$$\frac{Q^\bullet}{Q^\circ} = \prod_{i=1}^{3N-6} \left[\frac{\exp\left(-X_i^\bullet/2\right)}{1 - \exp\left(-X_i^\bullet\right)} \right] \left[\frac{1 - \exp\left(-X_i^\circ\right)}{\exp\left(-X_i^\circ/2\right)} \right].$$ (11.56)

Introducing the masses of exchanging isotopes ($m^\bullet$ and m°, respectively), the ratio of partition functions for crystalline components can be related to that of the primitive unit cell (Kieffer, 1982), thus defining "reduced" partition function ratio f (whose formulation is equivalent to that obtained by Bigeleisen and Mayer, 1947, and Urey, 1947, for gaseous molecules):

$$f = \frac{Q^{\bullet\prime}}{Q^{\circ\prime}} = \frac{Q^\bullet}{Q^\circ} \left(\frac{m^\circ}{m^\bullet} \right)^{(3/2)r},$$ (11.57)

where r is the number of exchanging isotopes per unit formula.

Kieffer (1982) proposed detailed calculation of partition function ratio f in crystalline solids through direct evaluation of the Helmholtz free energies of isotopically light and heavy compounds:

$$F = 3nN\beta + kT \int_0^{\omega_l} \ln\left\{ 1 - \left[\exp\left(\frac{-\hbar\omega}{kT} \right) \right] \right\} g(\omega) d\omega,$$ (11.58)

where

$$3nN\beta = -U_0 - \frac{1}{2} \sum_{r=1}^{3nN} \hbar\omega.$$ (11.59)

In equation 11.58, n is the number of atoms in the chemical formula, $g(\omega)$ is the frequency distribution of normal modes, and ω_l is the highest lattice vibrational

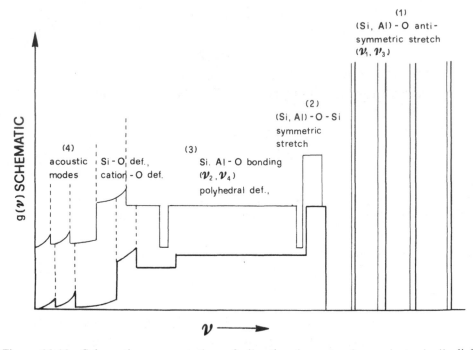

Figure 11.10 Schematic representation of vibrational spectra for an isotopically light compound (heavy bottom line) and an isotopically heavy compound (light upper line) of the same bulk stoichiometry and structure. Reprinted with permission from Kieffer (1982), *Review of Geophysics and Space Physics,* 20, 827–849, copyright © 1982 by the American Geophysical Union.

frequency. In equation 11.59, U_0 is the potential energy at the bottom of the potential well, and the summation term is zero-point energy.

As shown schematically in figure 11.10, the frequency distribution for an isotopically light compound $g°(v)$ [and the corresponding $g°(\omega)$] can be expected to differ from the frequency distribution of the isotopically heavy counterpart $g^{\bullet}(v)$ [or $g^{\bullet}(\omega)$].

Kieffer (1982) suggests a set of rules describing expected frequency shifts for different types of vibrational modes upon substitution of a heavy isotope into the mineral structure. Kieffer's (1982) rules for ^{18}O-^{16}O substitution in silicates are as follows.

Rule 1 (antisymmetric (Si,Al)-O stretching modes)
If v_1 and v_3 modes can be identified, their frequencies are assumed to decrease by 0.943 and 0.977, respectively, upon substitution of ^{18}O for ^{16}O. If v_1 and v_3 cannot be separately identified and enumerated, an average shift of 0.96 is adopted.

Rule 2 (symmetric (Si,Al)-O-Si stretching modes)
The frequency shift of symmetric (Si,Al)-O-Si stretching modes upon substitution of ^{18}O for ^{16}O is taken to be 0.99, for modes involving mainly Si motion.

Rule 3 (Si-O bending modes in orthosilicates)
If the Si-O bending modes in orthosilicates can be identified and enumerated, their frequency shift for substitution of ^{18}O for ^{16}O is taken to be 0.95.

Rule 4 (acoustic modes)
The ratio of frequencies upon substitution of ^{18}O for ^{16}O is taken to be proportional to the square root of the ratio of the mean molecular masses.

Rule 5 (remaining undifferentiated modes)
The frequency shift of the remaining modes for substitution of ^{18}O for ^{16}O must be consistent with zero fractionation at high T—i.e., at high T the frequency shifts must obey

$$\prod_{i=1}^{3s} \frac{X_i^\circ}{X_i^\bullet} = \left(\frac{m^\bullet}{m^\circ}\right)^{(3/2)r}, \tag{11.60}$$

where s is the number of atoms per unit cell.

Because

$$\ln f = \frac{F^\circ - F^\bullet}{kT} + \frac{3}{2} r \ln\left(\frac{m^\circ}{m^\bullet}\right), \tag{11.61}$$

calculations of F° and of the corresponding $F^\bullet$ for isotopically heavy compounds (through application of the above general rules for frequency shifts induced by isotopic substitution) allows determination of partition function ratio f and of the corresponding "reduced partition function" $(10^3/r) \ln f$ (see section 11.8.1).

11.6.2 Reaction Kinetics

Let us consider again an isotopic exchange reaction of the type shown in equation 11.53, such as, for instance, the already treated equilibrium between stannous chloride and an aqueous complex

$$^{120}SnCl_4 + {}^{118}SnCl_6^{2-} \Leftrightarrow {}^{118}SnCl_4 + {}^{120}SnCl_6^{2-} . \tag{11.62}$$

Reaction 11.62 can be conceived of as composed of two partial reactions—i.e.,

$$^{120}SnCl_4 + 2\,Cl^- \Leftrightarrow {}^{120}SnCl_6^{2-} \tag{11.63}$$

and

$$^{118}SnCl_4 + 2\,Cl^- \Leftrightarrow {}^{118}SnCl_6^{2-} . \tag{11.64}$$

To each of these two reactions will apply a reaction constant (K_{63} and K_{64}, respectively) that is related to the forward and backward reaction rates by

$$K_{63} = \frac{k_{+,63}}{k_{-,63}} \qquad (11.65.1)$$

and

$$K_{64} = \frac{k_{+,64}}{k_{-,64}}. \qquad (11.65.2)$$

Because at equilibrium,

$$\Delta G^0_{62} = -RT \ln K_{62} = -RT \ln\left(\frac{k_{+,63}}{k_{+,64}} \times \frac{k_{-,64}}{k_{-,63}} \right), \qquad (11.66)$$

and because we have seen that the isotopic fractionation constant at low T differs from 1 (i.e., $K_{62} = 1.0025$ at $T = 300$ K; cf. table 11.4), it follows that isotopically heavy and light molecules cannot be expected to have identical reaction rates at low T (even in the case of relatively "heavy" isotopes such as ^{118}Sn and ^{120}Sn). The effects discussed above may be particularly important when the fractional mass difference between isotopically light and heavy molecules is high (see the calculations of Sutin et al., 1961). A quantitative approach to this problem was afforded by Bigeleisen (1949) on the basis of transition state theory.

Adopting the same notation developed in the previous section, the reaction rates of isotopically heavy and light molecules are related to the differences in their respective activation energies of reaction $E^\bullet$, E°, through

$$\frac{\partial \ln(k^\bullet/k^\circ)}{\partial T} = \frac{E^\bullet - E^\circ}{\not k T^2}, \qquad (11.67)$$

where $\not k$ is Boltzmann's constant, and to shifts of vibrational spectra, through

$$\frac{\partial \ln(k^\bullet/k^\circ)}{\partial T} = \frac{\partial \ln(\tau^\bullet/\tau^\circ)}{\partial T} + \frac{X_s^{\bullet 2} - X_s^{\circ 2}}{12T}$$

$$+ \sum_{i=1}^{3N-6} \left\{ \frac{1 + (X_i^\circ - 1)\exp X_i^\circ}{\left[\exp(X_i^\circ) - 1\right]^2} - \frac{1}{2} \right\} \frac{\Delta X_i^*}{T}, \qquad (11.68)$$

where τ^* and τ° are transmission coefficients (which, for asymmetric reactions such as reactions 11.63 and 11.64, must be evaluated experimentally), and ΔX_i^* is the vibrational shift induced by formation of the activated complex.

11.6.3 Diffusion

Some aspects of thermally activated diffusion were discussed in section 5.9.1 in regard to intracrystalline exchange geothermometry. Concerning more precisely isotope diffusion, there are two main aspects that may be relevant in geochemistry:

1. Diffusion-induced isotope fractionation
2. Significance of closure temperature.

Diffusion-Induced Isotope Fractionation

Some aspects of this phenomenon are still controversial, because diffusion-induced isotopic fractionation effects may be erroneously ascribed to differential reaction kinetics (see section 11.6.2), and vice versa.

Basically, whenever isotopic exchanges occur between different phases (i.e., heterogeneous equilibria), isotopic fractionations are more appropriately described in terms of differential reaction rates. Simple diffusion laws are nevertheless appropriate in discussions of compositional gradients within a single phase—induced, for instance, by vacancy migration mechanisms, such as those treated in section 4.10—or whenever the isotopic exchange process does not affect the extrinsic stability of the phase.

Giletti et al. (1978) interpreted the results of laboratory experiments involving oxygen isotope exchanges between feldspars and hydrothermal fluids at various T in terms of self-diffusion of ^{18}O. The experiment of Giletti et al. (1978) is quite elegant: the mineral is equilibrated for a certain amount of time with isotopically altered water (40% $H_2^{18}O$) and the reacted crystals are analyzed by ion microprobe, using the primary ion beam as a drilling device and determining the $^{18}O/^{16}O$ ratio of the sputtered material (corresponding to various depths within the crystal) by ionization and mass spectrometry. Figure 11.11 shows the experimental results of Giletti et al. (1978) for adularia. The initial $^{18}O/^{16}O$ ratio of the natural material is 0.002, and the solid curve is the fitting diffusion equation. The induced isotopic fractionation is interpreted in terms of "one-dimension diffusion in a semi-infinite medium with constant surface concentration" (Crank, 1975):

$$\frac{C_x - C_1}{C_0 - C_1} = \mathrm{erf}\left[\frac{x}{2(Dt)^{1/2}}\right], \tag{11.69}$$

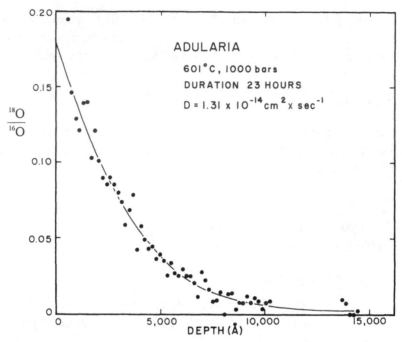

Figure 11.11 $^{18}O/^{16}O$ relative concentration profile as a function of depth in adularia hydrothermally equilibrated with an ^{18}O-enriched fluid ($H_2{}^{18}O = 40\%$). Reprinted from B. J. Giletti, M. P. Sennet, and R. A. Yund, copyright © 1978, with kind permission from Elsevier Science Ltd., The Boulevard, Langford Lane, Kidlington 0X5 1GB, UK.

where C_x is the ^{18}O concentration at distance x from the surface of the crystal, C_0 is the concentration at infinite distance, C_1 is the concentration at the surface, t is time, and D the self-diffusion coefficient.

Equation 11.69 was reexpressed by Giletti et al. (1978) in terms of $^{18}O/^{16}O$ ratios ρ_x, ρ_0, and ρ_1 as

$$\rho_x = \frac{A \times \mathrm{erf}(Y) + \rho_1}{1 - A \times \mathrm{erf}(Y)}, \tag{11.70}$$

where

$$A = \frac{\rho_0 - \rho_1}{1 + \rho_0} \tag{11.71}$$

and

$$Y = \frac{x}{2(Dt)^{1/2}}. \tag{11.72}$$

Results were then interpreted in terms of the usual form for thermally activated diffusion:

$$D = D_0 \exp\left(\frac{-E_a}{RT}\right),$$
(11.73)

where D_0 is the preexponential factor and E_a is the activation energy of diffusion (see sections 4.10 and 5.9.1). The results indicate that all feldspars have very similar oxygen exchange behavior, independent of composition and structural state, with low activation energy, averaging 24 kcal per gram formula weight of oxygen.

Unfortunately, the interpretation of Giletti et al. (1978) does not solve the problem of differential diffusivities of ^{18}O and ^{16}O. To do this, their experimental results should be interpreted in terms of interdiffusion of ^{18}O and ^{16}O. Application of Fick's first law to interdiffusion of the two species would in fact lead to the definition of an interdiffusion coefficient $\overline{D}$, so that

$$\overline{D} = \frac{D_{^{18}O} \, D_{^{16}O}}{X_{^{18}O} \, D_{^{18}O} + X_{^{16}O} \, D_{^{16}O}}$$
(11.74)

(cf. section 4.11). Because, when $X_{^{18}O} \to 0, \overline{D} \to D_{^{18}O}$, interpretation of the relative concentration profile in terms of self-diffusion is appropriate for the inner parts of the crystals, where ^{16}O is dominant, but it loses its significance near the borders of grains, where interdiffusion effects cannot be neglected (see figure 11.11).

The experiment of Giletti et al. (1978) has important implications for the isotopic reequilibration of feldspars in hydrous environments. Assuming the diffusion of ^{18}O to obey diffusion equation 11.73, with $D_0 = 2.0 \times 10^{-8}$ cm^2/s and $E_a = -24$ kcal/mole, and defining F as the fractional net ^{18}O exchange occurring with respect to total net exchange at equilibrium, Giletti et al. (1978) calculated the length of time required to achieve $F = 0.1$ and $F = 0.9$ according to temperature and particle size. The results of their calculations are shown in figure 11.12.

Any feldspar particle of radius less than 10 μm is able to exchange with a hydrothermal fluid, essentially to completion (i.e., $F \geq 0.9$), in less than 500 years, if the temperature exceeds 300 °C. However, exchanges for larger crystals may span periods of time from one million to several hundred million years, depending on size and T, before isotopic equilibrium is attained. Note also that, because of the rather low activation energy, exchanges have no single "closure temperature." This fact introduces us to the next aspect of the problem.

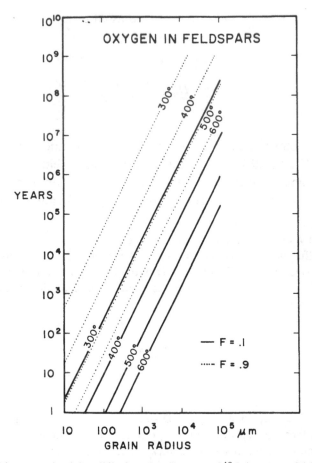

Figure 11.12 Time required for diffusional exchange of ^{18}O between feldspar and hydrothermal fluid according to particle size and temperature. $F = 0.1$ can be taken as condition of absence of exchanges and $F = 0.9$ as one of completion of exchanges. Reprinted from B. J. Giletti, M. P. Sennet, and R. A. Yund, copyright © 1978, with kind permission from Elsevier Science Ltd., The Boulevard, Langford Lane, Kidlington 0X5 1GB, UK.

Significance of Closure Temperature

The concept of *closure temperature* is particularly important in geochronology. According to Dodson (1973), closure temperature T_c can be defined as "the temperature of the system at the time corresponding to its apparent age."

The fundamental difference between the concentration gradients of stable and radiogenic isotopes can be expressed in terms of the simple equations

$$\frac{\partial C}{\partial t} = \frac{D_{t_0}}{a^2} \exp\left[\left(\frac{-t}{\tau}\right)\nabla^2 C\right] \qquad (11.75)$$

and

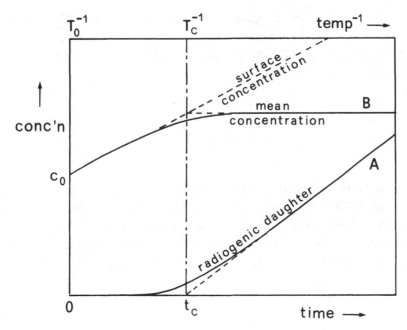

Figure 11.13 Relationship between "*geochronological closure*" (A) and "*frozen equilibrium*" (B). Diffusion parameters are identical, and T^{-1} is assumed to increase linearly with time. From M. H. Dodson, Closure temperature in cooling geochronological and petrological systems, *Contributions to Mineralogy and Petrology*, 40, 259–274, figure 2, 1973, copyright © 1973 by Springer Verlag. Reprinted with the permission of Springer-Verlag GmbH & Co. KG.

$$\frac{\partial C^*}{\partial t} = \frac{D_{t_0}^*}{a^2} \exp\left[\left(\frac{-t}{\tau^*}\right)\nabla^2 C^*\right] + L^* \qquad (11.76)$$

(Dodson, 1973). Both of these equations are developed assuming a linear decrease of temperature T with time t (see the analogy of treatment in section 5.9.1). τ is a time constant corresponding to the time taken for D to diminish by a factor of $\exp(-1)$, a is the characteristic dimension of the phase, and the asterisk denotes the radiogenic isotope.

Equations 11.75 and 11.76 both obey boundary conditions $C_x = C_0$ everywhere at t_0 and $C = C_1$ at the surface for $t \geq 0$. In equation 11.76, L^* is the rate of production of the radiogenic isotope.

As we see in figure 11.13, for a stable isotope such as ^{18}O, the mean concentration in a phase attains a constant value after the "closure" of exchanges ("closure temperature" T_c in this case is graphically determined by intersecting the prolongations of surface concentration and mean concentration; see upper dashed lines in figure 11.13). However, the radiogenic daughter isotope progressively increases its concentration, as a result of the virtual cessation of exchanges, beyond the value of T_c (in this case, T_c and the corresponding t_c are graphically determined by the prolongation of the daughter concentration curve to $C^* = 0$; see lower dashed line in figure 11.13).

The analytical representation of closure temperature for a geochronological system (Dodson, 1973) is

$$T_c = \frac{R}{E_a \ln\left(\frac{A\tau D_{t_0}}{a^2}\right)} . \qquad (11.77)$$

In equation 11.77, A is a numerical constant depending on the geometry and decay constant of the parent radionuclide. If the half-life of the parent is long in comparison with the cooling period, A takes a value of 55, 27, or 8.7 for volume diffusion from a sphere, cylinder, or plane sheet, respectively. If decay rates are faster, A progressively diminishes (see table 1 in Dodson, 1973, for numerical values).

Ganguly and Ruitz (1986) investigated the significance of closure temperature in terms of simple equilibrium thermodynamics. Assuming Rb-Sr exchanges between phases α and β to be representable in terms of the equilibrium

$$^{87}\text{Rb}_\alpha + {}^{87}\text{Sr}_\beta \Leftrightarrow {}^{87}\text{Sr}_\alpha + {}^{87}\text{Rb}_\beta, \qquad (11.78)$$

the Rb/Sr fractional amounts in the two phases can be related to the equilibrium constant through

$$K_{78} = \exp\left(\frac{-\Delta \overline{G}_{78}^0}{RT}\right) = \left[\frac{\left({}^{87}\text{Rb}/{}^{87}\text{Sr}\right)_\beta}{\left({}^{87}\text{Rb}/{}^{87}\text{Sr}\right)_\alpha}\right] \cdot K_{\gamma,78}. \qquad (11.79)$$

The term within the square brackets in equation 11.79 is the normalized (equilibrium) distribution coefficient between the minerals α and β (cf. section 10.8) at the closure condition of the mineral isochron. Ganguly and Ruitz (1986) have shown it to be essentially equal to the observed (disequilibrium) distribution coefficient between the two minerals as measured at the present time. $K_{\gamma,78}$ can be assumed to be 1, within reasonable approximation. Equation 11.79 can be calibrated by opportunely expanding $\Delta \overline{G}_{78}^0$ over P and T:

$$\Delta \overline{G}_{78}^0 = \Delta \overline{H}_{78}^0 - T\Delta \overline{S}_{78}^0 + P\Delta \overline{V}_{78}^0. \qquad (11.80)$$

Based on equations 11.79 and 11.80, the measured $D_{\text{Rb/Sr}}$ value for each mineral of a given isochron must furnish concordant indications in terms of the T and P of equilibrium. Moreover, the deduced parameters can be assumed to correspond to the condition of closure of exchanges.

Table 11.5 Some estimates of closure temperature, related to diffusion of daughter isotope. Large discrepancies and wide T_c ranges can be ascribed to differences in cooling rates of system and grain size dimensions.

Phase	Element	T_c (°C)	Reference
Biotite	Ar	373 ± 21	Berger and York (1981)
	Sr	250–311	Dodson (1979)
	Sr	280–300	Dodson (1973)
	Sr	300	Hart (1964)
	Sr	350 ± 50	Jäger et al. (1967)
	Sr	350–720	Hanson and Gast (1967)
	Sr	400	Verschure et al. (1980)
	Sr	531–833	Ganguly and Ruitz (1986)
Muscovite	Sr	308–726	Ganguly and Ruitz (1986)
	Sr	500	Jäger et al. (1967)
Hornblende	Ar	490	Harrison (1981)
	Ar	685 ± 53	Berger and York (1981)
K-feldspar	Ar	230 ± 18	Berger and York (1981)
	Sr	314–352	Dodson (1979)
	Sr	531–925	Ganguly and Ruitz (1986)
Plagioclase	Ar	176 ± 54	Berger and York (1981)
	Sr	567	Ganguly and Ruitz (1986)
Microcline	Ar	132 ± 13	Harrison and McDougall (1982)

Discrete values of closure temperatures in various minerals and mineral couples for the ^{40}Ar-^{39}Ar and Rb-Sr geochronological systems are reported in table 11.5.

11.7   Application I: Radiometric Dating

In this section we will briefly examine the various methods adopted in geochronology, all based essentially on the constancy of the decay rates of radiogenic nuclides, which, within reasonable limits, are unaffected by the physicochemical properties of the system. For more detailed treatment, specialized textbooks such as Rankama (1954), Dalrymple and Lanphere (1969), Faure and Powell (1972), Jäger and Hunziker (1979), Faure (1986), and McDougall and Harrison (1988) are recommended.

11.7.1   Rb-Sr

The element rubidium has 17 isotopes, two of which are found in nature in the following isotopic proportions: ^{85}Rb = 0.72165 and ^{87}Rb = 0.27835. Strontium has 18 isotopes, four of which occur in nature in the following average weight proportions $^{84}Sr = 0.0056$, $^{86}Sr = 0.0986$, $^{87}Sr = 0.0700$, and $^{88}Sr = 0.8258$. ^{87}Rb decays to ^{87}Sr by β^- emission, according to

$$\, ^{87}_{37}Rb \rightarrow \, ^{87}_{38}Sr + \beta^- + \bar{\nu}, \tag{11.81}$$

with a decay constant of $\lambda = 1.42 \times 10^{-11}$ a ($a = anna$, the Latin word for "years"), corresponding to a half-life $t_{1/2}$ of about 4.9×10^{10} a.

Because the quantity of unstable parent radionuclides P remaining at time t is related to the decay constant (cf. eq. 11.22) through

$$P = N_{0,P} \exp(-\lambda t), \tag{11.82}$$

and the quantity of daughter radionuclides D produced at time t is

$$D = P[\exp(\lambda t) - 1], \tag{11.83}$$

the amount of radiogenic strontium $^{87}Sr_{rad}$ produced at time t is

$$^{87}Sr_{rad} = \, ^{87}Rb[\exp(\lambda t) - 1]. \tag{11.84}$$

The bulk amount of ^{87}Sr at time t is

$$^{87}Sr_t = \, ^{87}Sr_0 + \, ^{87}Rb[\exp(\lambda t) - 1], \tag{11.85}$$

where $^{87}Sr_0$ is the amount of ^{87}Sr present at $t = 0$.

In practice, because in radiometric determinations based on mass spectrometry it is more convenient to measure mass ratios than absolute values, equation 11.85 is usually rewritten as follows, normalized to the abundance of the stable isotope ^{86}Sr:

$$\left(\frac{^{87}Sr}{^{86}Sr}\right)_t = \left(\frac{^{87}Sr}{^{86}Sr}\right)_0 + \frac{^{87}Rb}{^{86}Sr}[\exp(\lambda t) - 1]. \tag{11.86}$$

Whenever it is possible to analyze in a given rock at least two minerals that crystallized at the same initial time $t = 0$, equation 11.86 can be solved in t. On a Cartesian diagram with coordinates $^{87}Sr/^{86}Sr$ and $^{87}Rb/^{86}Sr$, equation 11.86 appears as a straight line ("isochron") with slope $\exp(\lambda t) - 1$ and intercept $(^{87}Sr/^{86}Sr)_0$. As shown in figure 11.14A, all minerals crystallized at the same t from the same initial system of composition $(^{87}Sr/^{86}Sr)_0$ rest on the same isochron, whose slope $\exp(\lambda t) - 1$ increases progressively with t.

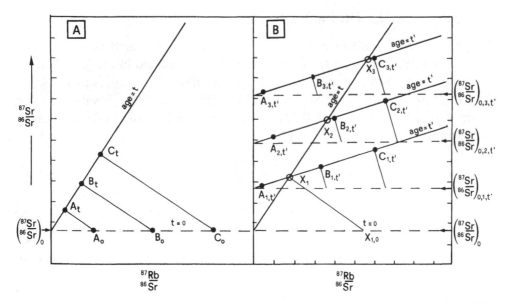

Figure 11.14 (A) Internal Rb-Sr isochron for a system composed of three crystalline phases of initial compositions A_0, B_0, and C_0 formed at time $t = 0$ and thereafter closed to isotopic exchanges up to time of measurement t, when they acquired compositions A_t, B_t, and C_t. (B) Effects of geochronological resetting resulting from metamorphism or interaction with fluids. X_1, X_2, and X_3: bulk isotopic compositions of the three rock assemblages. In cases of short-range isotopic reequilibration, the three assemblages define crystallization age and original $(^{87}Sr/^{86}Sr)_0$ of the system; the three internal isochrons (concordant in this example) define resetting age.

From the relative positions of phases A, B, and C in figure 11.14, it is obvious that the last one is richest in ^{87}Rb, so that the production of radiogenic ^{87}Sr is highest in this phase. A hypothetical phase with no initial ^{87}Rb at time $t = 0$ would lie at zero on the abscissa axis and would maintain its initial $(^{87}Sr/^{86}Sr)_0$ unaltered.

Whenever the phases in the system are reopened to isotopic exchanges, as a result of a thermal event or interaction with permeating fluids that enhance advective exchanges, time is reset to zero. To highlight isotopic remobilization events, it is useful to compare age determinations obtained from *internal isochrons* (minerals within a single rock specimen) with indications arising from *whole rock isochrons* (isotopic compositions of bulk rock specimens within a single outcrop). In the case of chronological resetting, the slope of the isochron, based on bulk rock samples from different portions of the same outcrop, may define the actual age of crystallization, if isotopic resetting was limited to short-range intercrystalline exchanges (figure 11.14B). In this case, the internal isochrons based on the isotopic compositions of the various phases within a single rock specimen define the age of resetting.

11.7.2 Sm-Nd, Lu-Hf, and La-Ce

Geochronological methods based on REE (plus Hf) have been developed only in the last two decades, because of the high precision required in analysis. Natural samarium has seven isotopes, occurring in the following proportions: $^{144}Sm = 0.0309$, $^{147}Sm = 0.1497$, $^{148}Sm = 0.1124$, $^{149}Sm = 0.1383$, $^{150}Sm = 0.0744$, $^{152}Sm = 0.2672$, and $^{154}Sm = 0.2271$. Three of these isotopes undergo α decay, being transformed into Nd isotopes:

$$^{147}_{62}Sm \rightarrow {}^{143}_{60}Nd + \alpha \qquad t_{1/2} = 1.06 \times 10^{11}\,a \qquad (11.87)$$

$$^{148}_{62}Sm \rightarrow {}^{144}_{60}Nd + \alpha \qquad t_{1/2} = 7.0 \times 10^{15}\,a \qquad (11.88)$$

$$^{149}_{62}Sm \rightarrow {}^{145}_{60}Nd + \alpha \qquad t_{1/2} = 1.0 \times 10^{16}\,a \qquad (11.89)$$

Neodymium also has seven isotopes, occurring in nature in the following proportions: $^{142}Nd = 0.2711$, $^{143}Nd = 0.1217$, $^{144}Nd = 0.2385$, $^{145}Nd = 0.0830$, $^{146}Nd = 0.1719$, $^{148}Nd = 0.0573$, and $^{150}Nd = 0.0562$. Sm-Nd dating is based on the decay of ^{147}Sm, whose half-life is the only one sufficiently short for geochronological purposes. The isochron equation is normalized to the ^{144}Nd abundance:

$$\left(\frac{^{143}Nd}{^{144}Nd}\right)_t = \left(\frac{^{143}Nd}{^{144}Nd}\right)_0 + \frac{^{147}Sm}{^{144}Nd}\left[\exp(\lambda t) - 1\right]. \qquad (11.90)$$

Lu-Hf dating is based on the β^- decay process of ^{176}Lu into ^{176}Hf:

$$^{176}_{71}Lu \rightarrow {}^{176}_{72}Hf + \beta^- + \nu. \qquad (11.91)$$

The first applications of the Lu-Hf method were attempted about 20 years ago. Lutetium has 22 isotopes, two of them occurring in nature in the following proportions: $^{175}Lu = 0.9741$ and $^{176}Lu = 0.0259$. Hafnium has 20 isotopes, six of which occur in nature in the following proportions: $^{174}Hf = 0.0016$, $^{176}Hf = 0.0520$, $^{177}Hf = 0.1860$, $^{178}Hf = 0.2710$, $^{179}Hf = 0.1370$, and $^{180}Hf = 0.3524$. The isochron equation is normalized to the ^{177}Hf abundance:

$$\left(\frac{^{176}Hf}{^{177}Hf}\right)_t = \left(\frac{^{176}Hf}{^{177}Hf}\right)_0 + \frac{^{176}Lu}{^{177}Hf}\left[\exp(\lambda t) - 1\right]. \qquad (11.92)$$

The decay constant is not known experimentally but has been deduced by comparisons with radiometric ages obtained by other methods. The obtained λ value is $(1.94 \pm 0.07) \times 10^{-11}\,a^{-1}$, corresponding to a half-life of $(3.57 \pm 0.14) \times 10^{10}\,a$.

The first application of La-Ce dating to earth sciences was proposed by Tanaka and Masuda (1982), following improvements in the technology of solid source mass spectrometry. The La-Ce method is based on the combined β^--EC decay of ^{138}La into ^{138}Ce and ^{138}Ba:

$$^{138}_{57}\text{La} \rightarrow {}^{138}_{58}\text{Ce} + \beta^- + \bar{v} \tag{11.93}$$

$$^{138}_{57}\text{La} + e^- \rightarrow {}^{138}_{56}\text{Ba} + hv(\gamma) + hv(\text{X}). \tag{11.94}$$

Lanthanum has 19 isotopes with A between 126 and 144. Only two occur in nature, in the following proportions: ^{138}La = 0.0009 and ^{139}La = 0.9991. The low relative amount of ^{138}La and its low decay rate ($\lambda = \lambda_{\text{EC}} + \lambda_{\beta-} = 6.65 \times 10^{-12}$ a^{-1}) prevented earlier application of this dating method. Cerium has 19 isotopes, with masses between 132 and 148, and two isomers at $A = 137$ and 139. Four isotopes are found in nature, in the following proportions: ^{136}Ce = 0.0019, ^{138}Ce = 0.0025, ^{140}Ce = 0.8848, and ^{142}Ce = 0.1108. The isochron equation is normalized to the ^{142}Ce abundance:

$$\left(\frac{^{138}\text{Ce}}{^{142}\text{Ce}}\right)_t = \left(\frac{^{138}\text{Ce}}{^{142}\text{Ce}}\right)_0 + \left(\frac{\lambda_{\beta-}}{\lambda}\right)\frac{^{138}\text{La}}{^{142}\text{Ce}}\left[\exp(\lambda t) - 1\right]. \tag{11.95}$$

where $\lambda_{\beta-} = 2.24 \times 10^{-12}$ a^{-1}.

Figure 11.15 shows some examples of geochronological application of REE decay processes. In figure 11.15A we see that Sm-Nd dating of gabbroic rocks of the Stillwater Complex gives consistent results when based on both mineral and total rock isochrons. This indicates that the LREE were not remobilized during the various hydrous alteration processes undergone by the host rocks. Figure 11.15B shows the first Lu-Hf isochron obtained by Patchett (1983) on eucrite-type meteorites, whose age of formation (4.55×10^9 a) was adopted to define $t_{1/2}$. Figure 11.15C shows the large uncertainty associated with ^{138}Ce/^{142}Ce determinations, which makes the application of this sort of dating problematic.

11.7.3 K-Ar

Potassium has 10 isotopes, three of which occur in nature in the following proportions: ^{39}K = 0.9326, ^{40}K = 0.0001, and ^{41}K = 0.0673. Argon (from the Greek verb *argéos*—i.e., to be inactive) is present in the earth's atmosphere (0.934% in mass). It is composed of eight isotopes, three of which are present in nature in the following proportions: ^{36}Ar = 0.0034, ^{38}Ar = 0.0006, and ^{40}Ar = 0.9960.

K-Ar geochronology is based on the natural decay of ^{40}K into ^{40}Ar (EC, $\beta^+ = 10.5\%$). However, as shown in figure 11.16, ^{40}K also decays into ^{40}Ca ($\beta^- = 89.5\%$). The ^{40}K-to-^{40}Ar decay branch is dominated by electron capture, although a limited amount of positron emission of low energy is observed (about 0.001%).

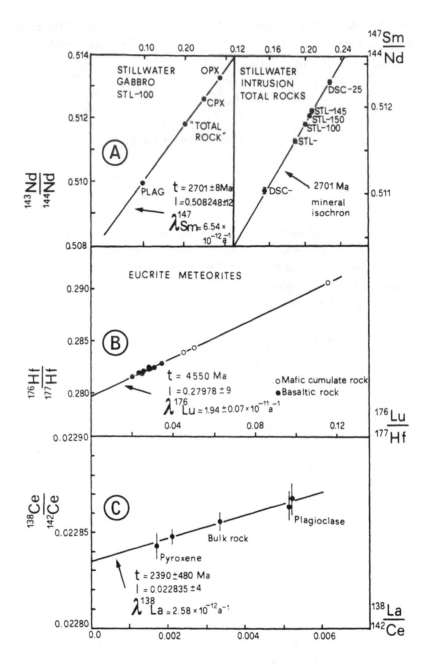

Figure 11.15 Geochronological applications of REE decay processes. (A) Total rock and mineral Sm-Nd isochrons for Stillwater Complex, Montana. (B) Lu-Hf isochron for eucrites. (C) La-Ce dating of gabbros of Bushveld Complex, South Africa. Reprinted from D. J. De Paolo and G. J. Wasserburg, *Geochimica et Cosmochimica Acta,* 43, 999–1008, copyright © 1979; J. P. Patchett, *Geochimica et Cosmochimica Acta,* 47, 81–91, copyright © 1983, with kind permission from Elsevier Science Ltd., The Boulevard, Langford Lane, Kidlington 0X5 1GB, UK; and from T. Tanaka and A. Masuda, *Nature,* 300, 515–518, copyright © 1982 Macmillan Magazines Limited.

The decay constant of the first-order reaction transforming ^{40}K into ^{40}Ar is $\lambda_{EC} = 0.581 \times 10^{-10} \ a^{-1}$ (89.3%), and the β^- decay constant producing ^{40}Ca is $\lambda_\beta = 4.962 \times 10^{-10} \ a^{-1}$ (10.7%). The resulting "branching ratio" $\lambda_{EC}/\lambda_\beta$ is 0.117 (± 0.001), and the half-life for the dual decay is $1.250(\pm 0.002) \times 10^9 \ a^{-1}$.

Because there is a double decay involved in parent P disintegration, the decay equation corresponding to the general formula of equation 11.83 reads as follows:

$$^{40}_{18}Ar_t = {}^{40}_{18}Ar_0 + \left(\frac{\lambda_{EC}}{\lambda_{EC} + \lambda_\beta}\right) \times {}^{40}_{19}K\left\{\exp\left[\left(\lambda_{EC} + \lambda_\beta\right)t\right] - 1\right\}, \quad (11.96)$$

and the corresponding age equation (obtained under the assumption that no $^{40}_{18}Ar_0$ is present in the mineral at the time of its formation) is

$$t = \frac{1}{\lambda_{EC} + \lambda_\beta} \ln\left[1 + \left(\frac{\lambda_{EC} + \lambda_\beta}{\lambda_{EC}}\right)\frac{^{40}_{18}Ar}{^{40}_{19}K}\right]$$

$$= \left(1.804 \times 10^9\right)\ln\left(1 + 9.54\frac{^{40}_{18}Ar}{^{40}_{19}K}\right). \quad (11.97)$$

K is usually determined by conventional analytical method such as atomic absorption spectroscopy, flame photometry, X-ray fluorescence, neutron activation, etc., and the corresponding amount of ^{40}K is deduced from the fractional isotopic abundance (^{40}K = 0.0118%). Analysis of ^{40}Ar is more complex, because of the high amount of ^{40}Ar present in the earth's atmosphere. It is conducted by mass spectrometry on gaseous portions extracted from minerals and rocks by melting in radiofrequency or induction furnaces in high-vacuum lines (higher than 10^{-8} torr). The released Ar is purified of reactive gaseous components such as H_2, N_2, CO_2, O_2, and H_2O by chemical traps, and a "spike" of known isotopic composition enriched in ^{38}Ar is added. The total ^{40}Ar released by a given sample is composed of radiogenic plus atmospheric amounts:

$$^{40}_{18}Ar_{total} = {}^{40}_{18}Ar_{rad} + {}^{40}_{18}Ar_{atm}. \quad (11.98)$$

Because the ^{40}Ar/^{36}Ar ratio of the earth's atmosphere is 295.5 (constant), the correction for atmospheric argon is made by measuring the amount of ^{36}Ar:

$$^{40}_{18}Ar_{rad} = {}^{40}_{18}Ar_{total} - 295.5 \times {}^{36}_{18}Ar. \quad (11.99)$$

In practice, mass spectrometric measurements are carried out on the ^{40}Ar/^{38}Ar and ^{38}Ar/^{36}Ar ratios, for the reasons previously outlined (see Dalrymple and Lanphere, 1969 for the form of the relevant chronological equation).

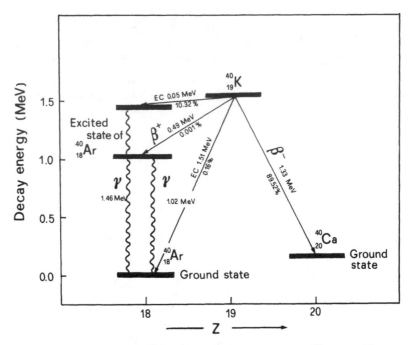

Figure 11.16 Decay scheme for ^{40}K, showing double decay to ^{40}Ca and ^{40}Ar.

In conventional K-Ar dating based on equation 11.97, it is assumed that no initial $^{40}_{18}Ar_0$ is present and that nonradiogenic ^{40}Ar results entirely from atmospheric contamination (appropriately corrected using eq. 11.99). However, an additional source of nonradiogenic argon may be the magma from which the crystal grew (^{40}Ar originated from the earth's mantle or from an old K-rich crust). Deep strata affected by thermal metamorphism may also release ^{40}Ar, which becomes an "excess component" in crystals equilibrated at shallower levels. The problem of excess ^{40}Ar may be overcome by the K-Ar isochron method of dating. Equation 11.96 is normalized to the ^{36}Ar abundance:

$$\left(\frac{^{40}Ar}{^{36}Ar}\right)_t = \left(\frac{^{40}Ar}{^{36}Ar}\right)_0$$

$$+ \left(\frac{\lambda_{EC}}{\lambda_{EC} + \lambda_\beta}\right)\left(\frac{^{40}K}{^{36}Ar}\right)\left\{\exp\left[\left(\lambda_{EC} + \lambda_\beta\right)t\right] - 1\right\}. \tag{11.100}$$

The intercept term $(^{40}Ar/^{36}Ar)_0$, which accounts for igneous, metamorphic, or atmospheric sources, is regarded as the excess contribution present at time $t = 0$, whereas the second term is the radiogenic component accumulating in the various minerals of the isochron by decay of ^{40}K. If all the minerals used to construct the isochron underwent the same geologic history and the same sort of contamination by excess ^{40}Ar, the slope of equation 11.100 would have a precise chronological

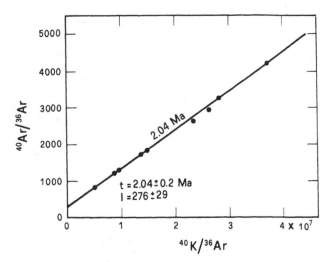

Figure 11.17 Whole rock K-Ar isochron of Tuff IB strata from Olduvai Gorge, Tanzania. From G. Faure (1986), *Principles of Isotope Geology,* 2nd edition, copyright © 1986 by John Wiley and Sons. Reprinted by permission of John Wiley & Sons.

significance. An example of a K-Ar isochron is given in figure 11.17. It is based on whole rock analyses of Tuff IB strata in the Olduvai Gorge, Tanzania (a level of paramount importance in dating the early history of humans). The slope of the isochron was recalculated by Fitch et al. (1976) on the basis of the original data of Curtis and Hay (1972) with York's (1969) algorithm (see appendix 2) and gives an age of 2.04 ± 0.02 Ma. The excess argon intercept term is 276 ± 29.

Besides the presence of excess ^{40}Ar, the main problem in K-Ar geochronology is the mobility of Ar, which may be released from the crystal lattice even under ambient P and T conditions. The Ar retention capacities of silicates seem to improve in the following sequence: feldspathoids–feldspars–micas–amphiboles–pyroxenes. Thermal events and/or alteration processes partially remove radiogenic argon from the crystal lattice, resulting in "apparent ages" younger than the actual ones. A classical study by Hart (1964) on the Precambrian gneisses of the Idaho Springs Formation, regionally metamorphosed at about 1400 Ma, reveals the Ar release effect induced by a young quartz monzonite (Eldora stock) that intruded the gneiss at about 55 Ma (figure 11.18). As figure 11.18 shows, the ages were completely reset in the contact zone and partially rejuvenated up to a distance of more than 1 km from the contact. Moreover, amphibole was less affected by resetting, at the same distance, with respect to mica and feldspar.

11.7.4 ^{40}Ar/^{39}Ar

The ^{40}Ar/^{39}Ar dating method is based on the artificial production of ^{39}Ar by neutron irradiation of K-bearing samples in nuclear reactors:

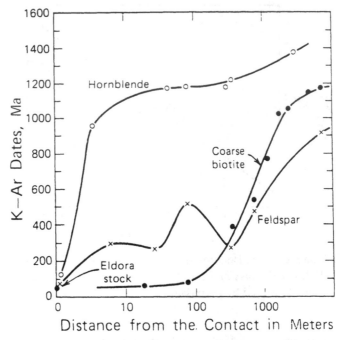

Figure 11.18 Apparent K-Ar ages of minerals from Idaho Springs Formation (Front Range, Colorado, 1350–1400 Ma) in zone subjected to contact metamorphism by intrusion of a quartz monzonite (Eldora stock, 55 Ma). Reprinted from S. R. Hart, *Journal of Geology*, (1964), 72, 493–525, copyright © 1964 by The University of Chicago, with permission of The University of Chicago Press.

$$^{39}_{19}\text{K} + n \rightarrow \,^{39}_{18}\text{Ar} + p. \tag{11.101}$$

The number of ^{39}Ar atoms formed depends on irradiation time Δt, density of neutron flux $\varphi_{(\varepsilon)}$ as a function of energy ε, and cross-section of ^{39}K atoms $\sigma_{(\varepsilon)}$:

$$^{39}_{18}\text{Ar} = \,^{39}_{19}\text{K}\left[\Delta t \int \varphi_{(\varepsilon)}\,\sigma_{(\varepsilon)}\,d\varepsilon\right]. \tag{11.102}$$

If we denote the flux production factor in square brackets in equation 11.102 as Φ, we can define a parameter J related to ^{39}Ar production in a monitor sample of known age t_m:

$$
\begin{aligned}
J &= \left(\frac{^{40}_{19}\text{K}}{^{39}_{19}\text{K}}\right)^{-1}_m \left(\frac{\lambda_{\text{EC}}}{\lambda_{\text{EC}} + \lambda_\beta}\right)^{-1} \Phi \\[2mm]
&= \left(\frac{^{39}_{18}\text{Ar}}{^{40}_{18}\text{Ar}}\right)_m \left\{\exp\left[\left(\lambda_{\text{EC}} + \lambda_\beta\right)t_m\right] - 1\right\}.
\end{aligned}
\tag{11.103}
$$

The $^{40}\text{Ar}/^{39}\text{Ar}$ ratio of a given sample after neutron irradiation is

Table 11.6 Interfering nuclear reactions during neutron irradiation of mineral samples. Boldface: principal reaction for ^{40}Ar/^{39}Ar dating procedure (from Brereton, 1970).

Produced Isotope	Calcium	Potassium	Argon	Chlorine
^{36}Ar	^{40}Ca$(n, n\alpha)$	–	–	–
^{37}Ar	^{40}Ca(n, α)	^{39}K$(n, n\,d)$	^{36}Ar(n, γ)	–
^{38}Ar	^{42}Ca$(n, n\alpha)$	^{39}K(n, d)	^{40}Ar$(n, n\,d, \beta^-)$	^{37}Cl(n, γ, β^-)
		^{41}K(n, α, β^-)		
^{39}Ar	^{42}Ca(n, α)	^{39}**K**$(\boldsymbol{n, p})$	^{38}Ar(n, γ)	–
	^{43}Ca$(n, n\alpha)$	^{40}K(n, d)	^{40}Ar(n, d, β^-)	
^{40}Ar	^{43}Ca(n, α)	^{40}K(n, p)	–	–
	^{44}Ca$(n, n\alpha)$	^{41}K(n, d)		

$$\frac{^{40}\text{Ar}}{^{39}\text{Ar}} = \frac{\exp\left[\left(\lambda_{EC} + \lambda_\beta\right)t\right] - 1}{J}, \tag{11.104}$$

and the age of the sample is

$$t = \left(\frac{1}{\lambda_{EC} + \lambda_\beta}\right)\ln\left[\left(\frac{^{40}\text{Ar}}{^{39}\text{Ar}}\right)J + 1\right]. \tag{11.105}$$

Although $^{39}_{18}$Ar is unstable, its half-life (269 a) is long enough to be considered stable with respect to the time normally involved in analyses. Ages obtained through application of equation 11.105 are normally referred to as "total argon release dates" (cf. Faure, 1986) and are subject to the same uncertainties as conventional K-Ar ages. Moreover, corrections for atmospheric Ar through equation 11.99 are complicated by the production of ^{36}Ar by ^{40}Ca$(n,n\alpha)$ reaction during irradiation (see table 11.6). More generally, neutron irradiation is responsible for a complex alteration of the pristine isotopic composition of Ar by nuclear reactions involving the neighboring isobars, as summarized in table 11.6.

According to Dalrymple and Lanphere (1971), the appropriate ^{40}Ar/^{39}Ar ratio to be introduced in age equation 11.105 can be derived from the measured ratio $(^{40}$Ar/^{39}Ar$)_{\text{meas}}$ by application of

$$\frac{^{40}\text{Ar}}{^{39}\text{Ar}} = \left[\left(\frac{^{40}\text{Ar}}{^{39}\text{Ar}}\right)_{\text{meas}} - C_1\left(\frac{^{36}\text{Ar}}{^{39}\text{Ar}}\right)_{\text{meas}} + C_1 C_2 D - C_3\right] \tag{11.106}$$

$$\times \left(1 - C_4 D\right)^{-1},$$

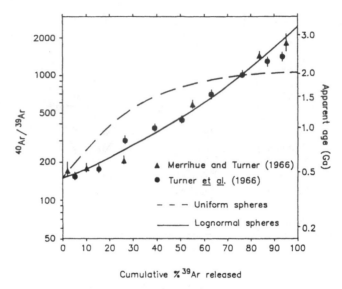

Figure 11.19 The first $^{40}Ar/^{39}Ar$ age spectrum. Reprinted from Turner et al. (1966), with kind permission from Elsevier Science Publishers B.V, Amsterdam, The Netherlands.

where $C_1 = 295.5$ is the $^{40}Ar/^{36}Ar$ ratio in the atmosphere, $C_2 = 2.72 \pm 0.014 \times 10^{-4}$ is the $^{36}Ar/^{37}Ar$ ratio induced by interfering nuclear reactions involving Ca, $C_3 = 5.9 \pm 0.42 \times 10^{-3}$ is the $^{40}Ar/^{39}Ar$ ratio induced by interfering nuclear reactions involving K, $C_4 = 6.33 \pm 0.04 \times 10^{-4}$ is the $^{39}Ar/^{37}Ar$ ratio induced by interfering nuclear reactions involving Ca, and D is the $^{37}Ar/^{39}Ar$ ratio in the sample, after correction for decay of ^{37}Ar ($t_{1/2} = 35.1$ days). We refer readers to Brereton (1970) and to Dalrymple and Lanphere (1971) for detailed accounts of the necessary corrections.

The main virtue of the $^{40}Ar/^{39}Ar$ dating method is that the isotopic composition of the various argon fractions released during stepwise heating of the sample (or "age spectrum") furnishes precious information about the nature of excess argon and the thermal history of the sample. The first interpretation in this sense was that of Turner et al. (1966), who analyzed a sample of the disturbed Bruderheim chondrite and arranged the mass spectrometric $^{40}Ar/^{39}Ar$ data in a series corresponding to the cumulative percentage of released ^{39}Ar (figure 11.19).

The data reported in figure 11.19 refer to two separate irradiations, but their distribution is consistent with the model age spectrum of a sample composed of spheres with a lognormal distribution of radii that underwent a brief episode of radiogenic $^{40}Ar*$ outgassing (90%) 0.5 Ga before present (solid line in figure 11.19). According to the single-site diffusion model of Turner (1968), if we assume homogeneous distribution of ^{39}Ar, a single (thermally activated) transport mechanism, identical diffusion rates for all Ar isotopes, and a radiogenic $^{40}Ar*$ boundary concentration of zero, a set of model age curves can be parametrically generated, corresponding to the various degrees of outgassing. As shown in figure

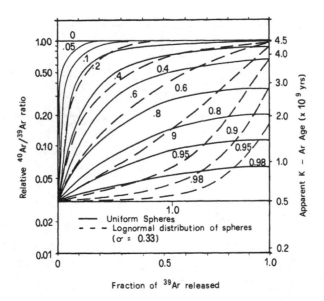

Figure 11.20 Theoretical age spectra based on Turner's (1968) single-site diffusion model. True age of sample is 4.5 Ga, and outgassing episode took place 0.5 Ga ago. Reprinted from Turner (1968).

11.20, the theoretical age spectra thus obtained differ greatly according to the size distribution of the outgassing particles (uniform radius or lognormal distribution) if the ^{40}Ar* loss exceeds 40%. The first infinitesimal increment of gas extracted represents the age of a brief thermal disturbance that affected the sample, causing a partial loss of radiogenic argon, whereas the true age is attained at maximum ^{39}Ar release.

Note that Turner's theoretical spectra for uniform spheres indicate that, if ^{40}Ar* loss exceeds 10%, the maximum age achieved at complete ^{39}Ar release does not correspond to the true age of the sample—i.e., the asymptotic rise of the curves does not reach ^{40}Ar/^{39}Ar = 1 for both lognormal and uniform distribution of radii. In these cases, assessment of ^{40}Ar* loss must be established from the form of the experimental age spectrum to correct appropriately for the lowering of maximum age. For example, in the case exemplified in figure 11.20, a ^{40}Ar* loss of 60% would lead to an apparent maximum age of 3.9 Ga for an aggregate of uniform spheres, whereas the age perturbation would be greatly reduced for an aggregate of spheres with a lognormal distribution of radii (apparent age ≈ 4.4 Ga). Partial modifications of Turner's (1968) single-site diffusion model were presented by Harrison (1983) to account for multiple ^{40}Ar* losses in samples with particularly complex thermal histories.

Further complications in ^{40}Ar/^{39}Ar age spectra arise whenever samples contain excess ^{40}Ar not produced by in situ decay of ^{40}K. In these cases, the presence of ^{40}Ar violates the zero concentration boundary condition of Turner's model, and the first infinitesimal increments of extracted argon often display extremely

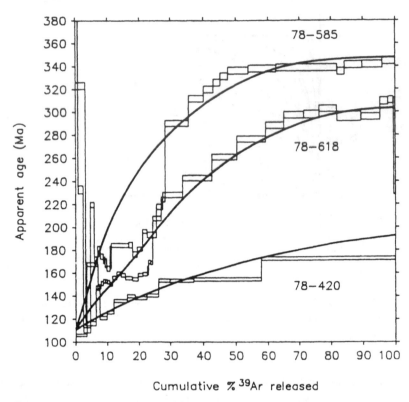

Figure 11.21 Age spectra of hornblendes of a Paleozoic gabbro (367 Ma) intruded by a granitic body during the Cretaceous (114 Ma). Samples underwent permeation of $^{40}Ar^*$ from lower crustal portions and differential losses of radiogenic argon (32, 57, and 78% respectively), proportional to distance from contact (0.3, 1, and 2.5 km). Reprinted from T. M. Harrison and I. McDougall, *Geochimica et Cosmochimica Acta,* 44, 2005–2020, copyright © 1980, with kind permission from Elsevier Science Ltd., The Boulevard, Langford Lane, Kidlington 0X5 1GB, UK.

old apparent ages, which decrease rapidly in the subsequent stepward releases. An emblematic representation of such complex age spectra is given in the study of Harrison and McDougall (1980) on hornblendes from Paleozoic gabbro intruded by a granitic body during the Cretaceous. The observed $^{40}Ar/^{39}Ar$ spectra reveal loss profiles compatible with Turner's model for an aggregate of spheres with lognormal distribution of radii (figure 11.21) with partial losses of 31, 57, and 78% of radiogenic argon and a true age of 367 Ma (coinciding with an isochron age of 366 ± 4 Ma; Harrison and McDougall, 1980). However, the presence of intergranular excess argon is evident from the initial steeply descending relative concentration profiles (which in some samples of the same outcrop generate initial apparent ages as high as 3.5 Ga; cf. Harrison and McDougall, 1980).

Additional complexities arise whenever the permeating excess argon diffuses from intergranular boundaries to lattice sites or whenever the solid phase under-

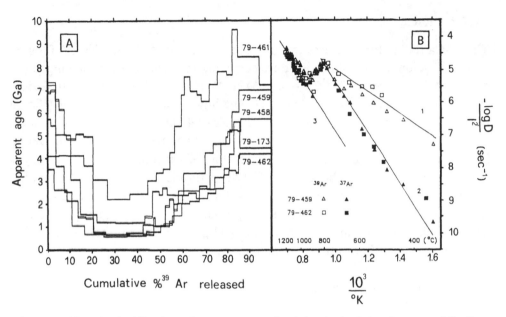

Figure 11.22 (A) Saddle-shaped age spectra of calcic plagioclases from amphibolites, Broken Hill, Australia. (B) Arrhenius plots of reactor-produced isotopes for two of the five samples, defining existence of three diffusion domains corresponding to albite-rich lamellae (domain 1) and anorthite-rich lamellae of different widths (domains 2 and 3). Reproduced with modifications from T. M. Harrison and I. McDougall (1981), with kind permission from Elsevier Science Publishers B.V, Amsterdam, The Netherlands.

goes subsolidus unmixing, generating domains with differing diffusional properties. In these cases, the age spectra have typical "saddle-shaped" conformations, like those shown for plagioclases in figure 11.22A.

In the case exemplified in figure 11.22A, the saddle-shaped conformation arises from the presence of three distinct structural domains corresponding to the products of subsolidus unmixing of plagioclase: a modulated "e" structure of alternating albite-rich and anorthite-rich lamellae (domains 1 and 2) and a coarser transitional anorthite-rich zone (domain 3). As shown in figure 11.22B, these three structural zones are reflected in three diffusion domains shown by the Arrhenius plots of the reactor-produced isotopes. The slopes for the albite-rich and anorthite-rich lamellar zones differ, indicating different activation energies of the diffusion process, whereas the anorthite-rich lamellae and coarse anorthite-rich transitional zone have the same slope (corresponding to an activation energy of about 35 kcal/mol). Their offset is attributable to a factor 10 difference in lamellar width (Harrison and McDougall, 1981). Excess ^{40}Ar and radiogenic $^{40}Ar*$ reside in different sites (anion and cation sites, respectively. At high extraction temperatures, the relatively low diffusion jump probability of excess argon in anion sites causes a retarded movement of ^{40}Ar in the crystal, resulting in extremely high apparent ages in the final extractions (see McDougall and Harrison, 1988 for an extended discussion on the diffusional mechanisms of Ar).

11.7.5 K-Ca

The K-Ca dating method uses the β^- decay branch of ^{40}K into ^{40}Ca (cf. figure 11.16). Calcium has six stable isotopes occurring in nature in the following abundances: $^{40}Ca = 0.969823$, $^{42}Ca = 0.006421$, $^{43}Ca = 0.001334$, $^{44}Ca = 0.020567$, $^{46}Ca = 0.000031$, and $^{48}Ca = 0.001824$. The high relative abundance of ^{40}Ca (a result of its mass number A, which is a multiple of 4 and thus exceptionally stable; cf. section 11.3) is one of the two main problems encountered in this sort of dating (in Ca-rich samples, the relative enrichment in ^{40}Ca resulting from ^{40}K decay is low with respect to bulk abundance). The other problem is isotopic fractionation of calcium during petrogenesis (and also during analysis; see for this purpose Russell et al., 1978). These two problems prevent extensive application of the K-Ca method, which requires extreme analytical precision. The isochron equation involves normalization to the ^{42}Ca abundance

$$\left(\frac{^{40}Ca}{^{42}Ca} \right)_t = \left(\frac{^{40}Ca}{^{42}Ca} \right)_0$$
$$+ \left(\frac{\lambda_\beta}{\lambda_{EC} + \lambda_\beta} \right) \left(\frac{^{40}K}{^{42}Ca} \right) \left\{ \exp\left[(\lambda_{EC} + \lambda_\beta)t \right] - 1 \right\}. \tag{11.107}$$

A substantial improvement in K-Ca dating was achieved by Marshall and De Paolo (1982), who combined the double-spike technique of Russell et al. (1978) with high-precision measurements on chromatographically separated sample aliquots.

The $(^{40}Ca/^{42}Ca)_t$ ratio to be introduced into equation 11.107 (hereafter, R_{corr}) was derived from the measured ratio (R_{meas}) by application of the "exponential law":

$$R_{meas,ij} = R_{corr,ij} \left(\frac{m_i}{m_j} \right)^P \tag{11.108}$$

with exponent

$$P = \frac{\ln\left(\dfrac{R_{meas,\, jk}}{R_{norm,\, jk}} \right)}{\ln\left(\dfrac{m_j}{m_k} \right)}, \tag{11.109}$$

where pedices i, j, and k identify nuclides of mass numbers 40, 42, and 44, respectively. $R_{norm,jk}$ is the normalizing ratio ($^{42}Ca/^{44}Ca$) taken to be 0.31221 (Russell et al., 1978), and m_i, m_j, m_k are, respectively, the masses of the corresponding nuclides ($^{40}Ca = 39.962591$, $^{42}Ca = 41.958618$, and $^{44}Ca = 43.955480$).

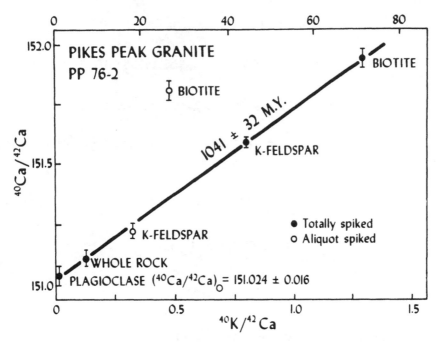

Figure 11.23 K-Ca isochron for granite batholith of Pikes Peak (Colorado). Aliquot-spiked samples not included in age calculation. Reprinted from B. D. Marshall and D. J. De Paolo, *Geochimica et Cosmochimica Acta,* 46, 2537–2545, copyright © 1982, with kind permission from Elsevier Science Ltd., The Boulevard, Langford Lane, Kidlington 0X5 1GB, UK.

Figure 11.23 shows the isochron obtained by Marshall and De Paolo (1982) for the granite batholith of Pikes Peak (Colorado). The effectiveness of the double-spike technique is evident, especially when we see that aliquot-spiked samples do not fall on the best-fit interpolant (York's algorithm; York, 1969). The obtained age (1041 ± 32 Ma) is consistent with that previously obtained with Rb-Sr whole rock analyses (1008 ± 13 Ma; see Marshall and De Paolo, 1982, for references). The initial ratio $(^{40}Ca/^{42}Ca)_0$ of 151.0 is identical, within the range of uncertainty, to upper mantle values, indicating negligible contamination by old crust components: the relative K/Ca abundance in the earth's mantle is about 0.01, a value too low to alter the "primordial" $(^{40}Ca/^{42}Ca)_0$ composition.

11.7.6 U-Th-Pb

Uranium has 15 isotopes, all unstable, with A-values from 227 to 240, and two isomers at $A = 235$. Natural uranium is composed of ^{238}U (99.2745%), ^{235}U (0.720%), and ^{234}U (0.0055%). Thorium has 12 isotopes, all unstable, with A-values from 223 to 234. Lead has 29 isotopes, four of them present in nature in the following atomic proportions: $^{204}Pb = 1.4\%$, $^{206}Pb = 24.1\%$, $^{207}Pb = 22.1\%$, and $^{208}Pb = 52.4\%$. The *4n, 4n + 2*, and *4n + 3* decay series can be schematically represented as follows:

$$^{232}_{90}\text{Th} \rightarrow {}^{208}_{82}\text{Pb} + 6\alpha + 4\beta^- \qquad \lambda_{232} = 4.9475 \times 10^{-11} a^{-1} \qquad (11,110.1)$$

$$^{238}_{92}\text{U} \rightarrow {}^{206}_{82}\text{Pb} + 8\alpha + 6\beta^- \qquad \lambda_{238} = 1.55125 \times 10^{-10} a^{-1} \qquad (11.110.2)$$

$$^{235}_{92}\text{U} \rightarrow {}^{207}_{82}\text{Pb} + 7\alpha + 4\beta^- \qquad \lambda_{235} = 9.8485 \times 10^{-10} a^{-1}, \qquad (11.110.3)$$

where λ_{232}, λ_{238}, and λ_{235} are the combined decay constants. Equation 11.83 applied to the three parent nuclides gives

$$^{208}\text{Pb} = {}^{232}\text{Th}\left[\exp\left(\lambda_{232} t\right) - 1\right] \qquad (11.111.1)$$

$$^{206}\text{Pb} = {}^{238}\text{U}\left[\exp\left(\lambda_{238} t\right) - 1\right] \qquad (11.111.2)$$

$$^{207}\text{Pb} = {}^{235}\text{U}\left[\exp\left(\lambda_{235} t\right) - 1\right]. \qquad (11.111.3)$$

Normalization to the stable isotope ^{204}Pb leads to the chronological equations

$$\left(\frac{^{208}\text{Pb}}{^{204}\text{Pb}}\right)_t = \left(\frac{^{208}\text{Pb}}{^{204}\text{Pb}}\right)_0 + \left(\frac{^{232}\text{Th}}{^{204}\text{Pb}}\right)\left[\exp\left(\lambda_{232} t\right) - 1\right] \qquad (11.112.1)$$

$$\left(\frac{^{206}\text{Pb}}{^{204}\text{Pb}}\right)_t = \left(\frac{^{206}\text{Pb}}{^{204}\text{Pb}}\right)_0 + \left(\frac{^{238}\text{U}}{^{204}\text{Pb}}\right)\left[\exp\left(\lambda_{238} t\right) - 1\right] \qquad (11.112.2)$$

$$\left(\frac{^{207}\text{Pb}}{^{204}\text{Pb}}\right)_t = \left(\frac{^{207}\text{Pb}}{^{204}\text{Pb}}\right)_0 + \left(\frac{^{235}\text{U}}{^{204}\text{Pb}}\right)\left[\exp\left(\lambda_{235} t\right) - 1\right]. \qquad (11.112.3)$$

Because of the different decay rates of ^{235}U and ^{238}U and their constant relative proportion in nature ($^{235}\text{U}/^{238}\text{U} = 0.00725$), a fourth chronological equation can be derived, based on the combination of equations 11.112.3 and 11.112.2:

$$\left(\frac{^{207}\text{Pb}}{^{206}\text{Pb}}\right)^* = \frac{\left(\dfrac{^{207}\text{Pb}}{^{204}\text{Pb}}\right)_t - \left(\dfrac{^{207}\text{Pb}}{^{204}\text{Pb}}\right)_0}{\left(\dfrac{^{206}\text{Pb}}{^{204}\text{Pb}}\right)_t - \left(\dfrac{^{206}\text{Pb}}{^{204}\text{Pb}}\right)_0} \qquad (11.113)$$

$$= 0.00725 \frac{\left[\exp\left(\lambda_{235} t\right) - 1\right]}{\left[\exp\left(\lambda_{238} t\right) - 1\right]}.$$

Table 11.7 Values of $[\exp(\lambda_{238}t) - 1]$ and $[\exp(\lambda_{235}t) - 1]$ and of $(^{207}\text{Pb}/^{206}\text{Pb})^*$ radiogenic ratio as a function of age t expressed in Ga $(1 \text{ Ga} = 1 \times 10^9 \, a)$ (from Faure, 1986).

t (Ga)	$[\exp(\lambda_{235}t) - 1]$	$[\exp(\lambda_{238}t) - 1]$	$(^{207}\text{Pb}/^{206}\text{Pb})^*$
0	0.0000	0.0000	0.04604
0.2	0.2177	0.0315	0.05012
0.4	0.4828	0.0640	0.05471
0.6	0.8056	0.0975	0.05992
0.8	1.1987	0.1321	0.06581
1.0	1.6774	0.1678	0.07250
1.2	2.2603	0.2046	0.08012
1.4	2.9701	0.2426	0.08879
1.6	3.8344	0.2817	0.09872
1.8	4.8869	0.3221	0.11000
2.0	6.1685	0.3638	0.12298
2.2	7.7292	0.4067	0.13783
2.4	9.6296	0.4511	0.15482
2.6	11.9437	0.4968	0.17436
2.8	14.7617	0.5440	0.19680
3.0	18.1931	0.5926	0.22266
3.2	22.3716	0.6428	0.25241
3.4	27.4597	0.6946	0.28672
3.6	33.6556	0.7480	0.32634
3.8	41.2004	0.8030	0.37212
4.0	50.3878	0.8599	0.42498
4.2	61.5752	0.9185	0.48623
4.4	75.1984	0.9789	0.55714
4.6	91.7873	1.0413	0.63930

Equation 11.113 is transcendental and must be solved by iterative procedures or by interpolation on tabulated values or plots representing the expected values of relative radiogenic lead abundances $(^{207}\text{Pb}/^{206}\text{Pb})^*$ with respect to t (table 11.7 and figure 11.24).

Note that, because equation 11.113 is indeterminate at the limit of $t = 0$, its solution under boundary conditions involves application of De l'Hopital's rule:

$$\lim_{t \to 0} \left[\frac{f(t)}{g(t)} \right] = \lim_{t \to 0} \left[\frac{f'(t)}{g'(t)} \right] \tag{11.114}$$

$$\lim_{t \to 0} \left\{ 0.00725 \frac{\left[\exp(\lambda_{235}t) - 1 \right]}{\left[\exp(\lambda_{238}t) - 1 \right]} \right\} = \lim_{t \to 0} \left\{ 0.00725 \frac{\left[\lambda_{235} \exp(\lambda_{235}t) \right]}{\left[\lambda_{238} \exp(\lambda_{238}t) \right]} \right\}$$

$$\tag{11.115}$$

$$= 0.00725 \frac{\lambda_{235}}{\lambda_{238}}.$$

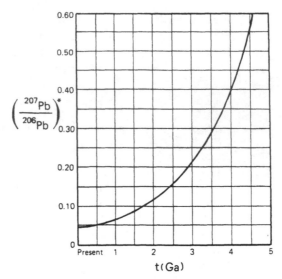

Figure 11.24 $(^{207}Pb/^{306}Pb)^*$ radiogenic ratio of a U-bearing system as a function of age t (Ga).

Ages obtained through radiometric equations 11.112.1 to 11.112.3 and equation 11.113 may be concordant, if the system remained closed to all isotopic exchanges from $t = 0$, or discordant. The second case is more commonly observed in nature and the obtained apparent ages usually have the following progression $t_{^{232}Th-^{208}Pb} < t_{^{238}U-^{206}Pb} < t_{^{235}U-^{207}Pb} < t_{^{207}Pb-^{206}Pb}$. Age determinations involving estimation of residual ^{238}U parent amounts are often subject to large error, because of the relatively easy mobilization of this element as a uranyl complex during hydrous alteration events. An emblematic representation of U-leaching effects on U-Th-Pb age systematics is given in figure 11.25. Basing their determinations on the isotopic composition of Pb (virtually unaffected by partial Pb losses), Rosholt et al. (1973) found an age of 2.79 ± 0.08 Ga for rock samples from the Granite Mountains of Wyoming. Although these rocks underwent variable gains or losses of Th and Pb, the imposed Pb-Pb age isochron is not far from the best interpolant of the $^{232}Th-^{208}Pb$ system (figure 11.25A), whereas conspicuous deviations are observed when this is superimposed on the $^{238}U-^{206}Pb$ diagram because of U losses ranging from 37 to 88% (figure 11.25B).

Limiting ourselves to the observation of isotopic abundances determined by the *4n + 2* and *4n + 3* decay series, we can construct a "concordia diagram" (Wetherill, 1956) relating $(^{206}Pb/^{238}U)^*$ and $(^{207}Pb/^{235}U)^*$ ratios developing at various t in a closed system (figure 11.26):

$$\left(\frac{^{206}Pb}{^{238}U}\right)^* = \frac{\left(\dfrac{^{206}Pb}{^{204}Pb}\right)_t - \left(\dfrac{^{206}Pb}{^{204}Pb}\right)_0}{\left(\dfrac{^{238}U}{^{204}Pb}\right)_t} = \left[\exp\left(\lambda_{238}t\right) - 1\right] \quad (11.116.1)$$

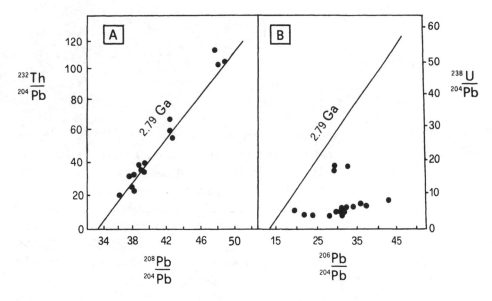

Figure 11.25 Th-Pb and U-Pb isotopic compositions of whole rocks from Granite Mountains Formation (Wyoming) based on data of Rosholt et al. (1973). 2.79 Ga isochrons based on Pb-Pb age are superimposed, to highlight effect of differential elemental remobilization.

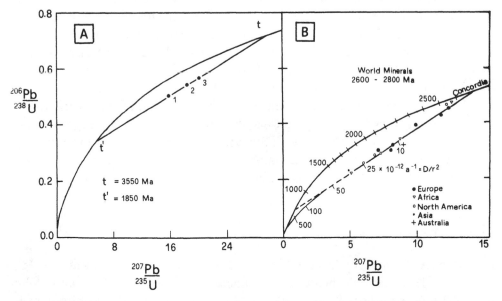

Figure 11.26 Concordia diagram. (A) Interpretation of straight discordia path as a result of a single episode of Pb loss. (B) Continuous diffusion model of Tilton (1960) applied to world minerals of a t = 2800 Ma "common" age. Reproduced with modifications from G. Faure (1986), *Principles of Isotope Geology*, 2nd edition, copyright © 1986 by John Wiley and Sons, by permission of John Wiley & Sons and from Tilton (1960) by permission of the American Geophysical Union.

$$\left(\frac{^{207}\text{Pb}}{^{235}\text{U}}\right)^* = \frac{\left(\dfrac{^{207}\text{Pb}}{^{204}\text{Pb}}\right)_t - \left(\dfrac{^{207}\text{Pb}}{^{204}\text{Pb}}\right)_0}{\left(\dfrac{^{235}\text{U}}{^{204}\text{Pb}}\right)_t} = \left[\exp\left(\lambda_{235}t\right) - 1\right]. \quad (11.116.2)$$

Each age t therefore corresponds to a certain $^{206}\text{Pb}/^{238}\text{U}$, $^{207}\text{Pb}/^{235}\text{U}$ couple, and their time evolution gives rise to a curved path or "concordia curve" in the binary field (figure 11.26A and B). The age interpretation of samples falling outside the concordia line is controversial. In the simplest acceptance, straight-line arrangements (or "discordia lines") below the concordia curve are interpreted in terms of Pb losses or U gains in the system (an eventual U loss from the system would extend the straight discordia path above the concordia curve). Figure 11.26A shows, for instance, the discordant compositions of three zircons from a gneiss of Morton and Granite Falls (Southern Minnesota; Catanzaro, 1963): because the partial loss of Pb does not significantly affect the $^{206}\text{Pb}/^{207}\text{Pb}$ ratio, the discordia line is straight and intersects the concordia curve at two points, t and t', which are interpreted, respectively, as the true age of the rock (t) and the age of remobilization (t'). However, figure 11.26B shows that U-bearing minerals from five continents with $^{206}\text{Pb}/^{207}\text{Pb}$ ages greater than 2300 Ma plot on a single discordia line, thus indicating an apparent common crystallization age of $t = 2800$ Ma, followed by a common Pb loss episode at $t' = 600$ Ma (Tilton, 1960). However, as shown by Tilton (1960), the common straight-path arrangement of the various minerals is a result of continuous diffusion of Pb from crystals. Assuming that diffusion obeys Fick's first law (cf. section 4.11) and affects only Pb, the daughter-to-parent ratios of the various minerals (assimilated to spheres of radius r with identical and constant diffusion coefficient D and uniform U distribution) evolve along a curved trajectory that may be approximated by a straight line for $D/r^2 < 50 \times 10^{-12} \, a^{-1}$. Obviously, in this case, the age of rejuvenation t' is only apparent, although initial age t maintains the same significance as in the previous model (see Faure, 1986 for alternative interpretations of the concordia diagram).

11.7.7 Re-Os

Rhenium has 12 isotopes, from $A = 185$ to $A = 196$, and two isomers at $A = 186$ and $A = 190$. Natural rhenium is composed of ^{185}Re (0.37398) and ^{187}Re (0.62602). Osmium has 28 isotopes from $A = 169$ to $A = 196$, with isomers at $A = 181, 183, 190, 191,$ and 192. Natural osmium has seven isotopes, all stable: ^{184}Os (0.00024), ^{186}Os (0.01600), ^{187}Os (0.01510), ^{188}Os (0.13286), ^{189}Os (0.16252), ^{190}Os (0.26369), and ^{192}Os (0.40958) (fractional atomic abundances after Luck and Allègre, 1983).

The ^{187}Re isotope is transformed into ^{187}Os by β^- decay:

$$\begin{smallmatrix}187\\75\end{smallmatrix}\text{Re} \rightarrow \begin{smallmatrix}187\\76\end{smallmatrix}\text{Os} + \beta^- + \bar{v} + 0.0025\,\text{MeV}.$$ (11.117)

The low decay energy prevents accurate determinations of half-life by direct counting. Reported half-lives range from 3 to 6.6×10^{10} a. The most recent direct determination (Lindner et al., 1989) assigns a half-life of $(4.23 \pm 0.13) \times 10^{10}$ a to the decay process of equation 11.117, which is fairly consistent with the indirect estimates of Hirt et al. (1963) [$(4.3 \pm 0.5) \times 10^{10}$ a]. The isochron equation is normalized to the ^{186}Os abundance:

$$\left(\frac{^{187}\text{Os}}{^{186}\text{Os}} \right)_t = \left(\frac{^{187}\text{Os}}{^{186}\text{Os}} \right)_0 + \frac{^{187}\text{Re}}{^{186}\text{Os}} \left[\exp(\lambda t) - 1 \right].$$ (11.118)

Equation 11.118 finds practical application in cosmological studies and in geology (dating of sulfide deposits and sediments). Figure 11.27A shows, for instance, the Re-Os isochron for iron meteorites and the metallic phase of chondrites, obtained by Luck and Allègre (1983). The fact that all samples fit the same isochron within analytical uncertainty has three important cosmological implications:

1. The primordial isotopic composition in the solar nebula was homogeneous ($^{187}\text{Os}/^{186}\text{Os} = 0.805 \pm 0.006$);
2. The ages of all samples are sufficiently concordant, indicating that they formed within a relatively short interval of time (about 90 Ma);
3. Because the present-day composition of the earth's mantle falls on the same isochron, the earth and the parent bodies of meteorites must have formed at about the same time from the same primordial source.

Note that the decay rate derived by Luck and Allègre (1983) from the isochron of figure 11.27, assuming a common age of 4550 Ma for all investigated samples ($\lambda_{\text{Re}} = 1.52 \pm 0.04 \times 10^{-11}$ a^{-1}), corresponds to a half-life of $4.56 \pm 0.12 \times 10^{10}$ a, which is somewhat higher than the direct determination of Lindner et al. (1989) but consistent with the direct determination of Payne and Drever (1965) [$(4.7 \pm 0.5) \times 10^{10}$ a].

Figure 11.28 shows the rhenium and osmium isotopic compositions of black shales and sulfide ores from the Yukon Territory (Horan et al., 1994). The black shale and sulfide layers are approximately isochronous. The superimposed reference isochrons bracket the depositional age of the enclosing shales. One reference line represents the minimum age (367 Ma) with an initial $(^{187}\text{Os}/^{186}\text{Os})_0$ ratio of one, consistent with the mantle isotopic composition at that age (see later). The other reference isochron is drawn for a maximum age of 380 Ma, with $(^{187}\text{Os}/^{186}\text{Os})_0 = 12$ (the maximum value measured in terrigenous sediments). Further examples of application of Re-Os dating of sediments can be found in Ravizza and Turekian (1989).

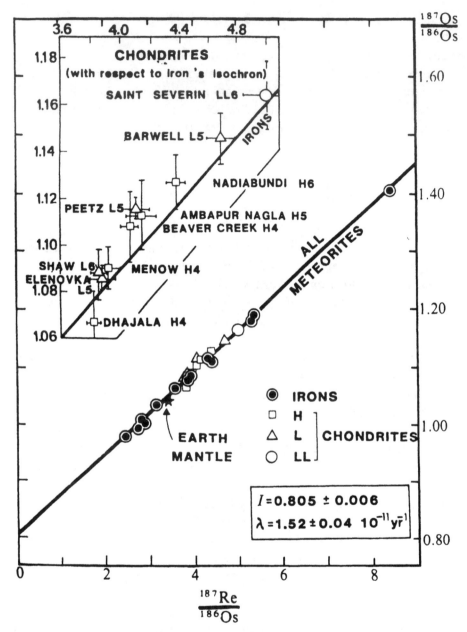

Figure 11.27 Re-Os isochron for iron meteorites and metallic phase of chondrites and earth's mantle. Reprinted with permission from J. M. Luck and C. J. Allègre, *Nature,* 302, 130–132, copyright © 1983 Macmillan Magazines Limited.

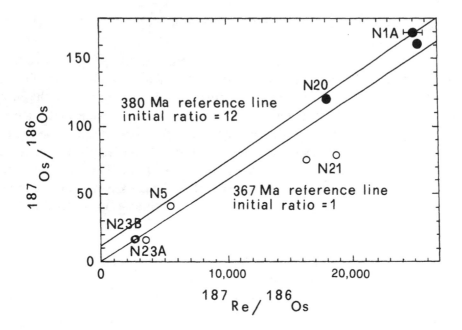

Figure 11.28 Re-Os isochron for black shales and sulfide deposits of Yukon Territory (Canada). Reprinted from M. F. Horan et al., *Geochimica et Cosmochimica Acta,* 58, 257–265, copyright © 1994, with kind permission from Elsevier Science Ltd., The Boulevard, Langford Lane, Kidlington OX5 1GB, UK.

11.7.8 ^{14}C

Carbon has eight isotopes, from $A = 9$ to $A = 16$. Natural carbon is composed of the two stable isotopes ^{12}C (0.9889) and ^{13}C (0.0111) and of radiogenic ^{14}C, continuously produced in the earth's atmosphere mainly by (n,p) reaction of slow (or "thermal") neutrons with ^{14}N:

$$^{14}_{7}N + n \rightarrow {}^{14}_{6}C + p. \tag{11.119}$$

The ^{14}C isotope reacts directly with atmospheric oxygen to produce ^{14}CO$_2$ and ^{14}CO, and undergoes isotopic exchanges with the preexisting ^{12}CO$_2$-^{12}CO and ^{13}CO$_2$-^{13}CO molecules. Mixing of the various molecules is rapidly achieved in the atmosphere (less than 2 years). Atmospheric carbon (1.4% of exchangeable carbon, mainly as CO$_2$) equilibrates readily with oceanic carbon (H$_2$CO$_3$, HCO$_3^-$, the various soluble carbonates, and very subordinate CO$_3^{2-}$, amounting in total to 95% of exchangeable carbon) and with carbon in the terrestrial biosphere and humus (3.2% of exchangeable carbon).

The decay of ^{14}C to ^{14}N takes place by β^- emission:

$$^{14}_{6}C \rightarrow {}^{14}_{7}N + \beta^- + \bar{\nu} + 0.156\,\text{MeV}. \tag{11.120}$$

The relatively rapid production rate of ^{14}C (approximately $2s^{-1}\,cm^{-2}$) coupled with high exchange kinetics (the mean time of residence in the atmosphere is relatively short—i.e. 5 to 10 years) and a comparatively long half-life of the decay process of equation 11.120 (5730 ± 40 a; Godwin, 1962) ensure that a steady state in the relative concentrations of the three carbon isotopes is quite rapidly achieved. The steady state isotopic composition of carbon results in specific radioactivity of 14.9 dpm/g in the atmosphere and 13.6 dpm/g in the biosphere (the lowering of specific radioactivity is a result of exchange kinetics among atmospheric, oceanic, and biogenic reservoirs and of minor isotope fractionation effects).

Living animals and plants acquire and maintain a constant level of ^{14}C radioactivity during their lifetimes as a result of continuous reequilibration with atmosphere and biosphere. However, when they die, these continuous exchanges cease and ^{14}C radioactivity progressively decreases according to the exponential law

$$\mathcal{R}_t = \mathcal{R}_0 \exp(-\lambda t), \tag{11.121}$$

where $\mathcal{R}_t$ is ^{14}C radioactivity measured at time t after the death of the living organism, and $\mathcal{R}_0$ is ^{14}C radioactivity in the biosphere during its lifetime. Through some substitutions, equation 11.121 can be rewritten as the chronological equation

$$t = 8.2666 \times 10^3 \ln\left(\frac{\mathcal{R}_0}{\mathcal{R}_t}\right). \tag{11.122}$$

The range of applicability of equation 11.122 depends on the limits of detection of ^{14}C in the sample. The current maximum age attained by direct radioactivity counting is about $4 \times 10^4\,a$. To measure residual ^{14}C radioactivity, the total carbon in the sample is usually converted to CO_2 and counted in the gas phase, either as purified CO_2 or after further conversion to C_2H_2 or CH_4. To enhance the amount of counted carbon, with the same detection limit (about 0.1 dpm/g), ^{14}C counters attain volumes of several liters and operate at several bars. More recent methods of direct ^{14}C detection (selective laser excitation; Van de Graaff or cyclotron acceleration) has practically doubled the range of determinable ages (Muller, 1979).

Using equation 11.122 implies knowledge of specific ^{14}C radioactivity in the biosphere at time $t = 0$. Calibration of ^{14}C radioactivity over the past 7500 years has been achieved by comparison of apparent radiocarbon ages with true ages obtained by dendrochronology (measurement of annual growth rings in trees). Discrepancies are translated into differences between the specific $\mathcal{R}_0$ radioactivities necessary to achieve concordant ages and specific present-day radioactivity, and are shown in figure 11.29 as per-mil deviations plotted against t.

Long-term sinusoidal fluctuations (period = 10,000 a, amplitude $\approx$ 10 a) are correlated with modifications of the earth's dipole moment (solid line in figure 11.29; Damon et al., 1978). Short-term fluctuations (inset in figure 11.29; Suess, 1965) can be ascribed to variations in solar activity. Both phenomena affect the intensity of the flux of cosmic rays striking the earth: both direct increases in the magnetic field of the earth and enhanced magnetic activity associated with sunspots cause enhanced deflection of the low-energy cosmic rays responsible for producing thermal neutrons. The effect of solar activity is well documented for the last three centuries. In particular, the maximum positive deviations in Δ‰ ^{14}C-radioactivity around A.D. 1700 corresponds to the "Maunder minimum," a period of almost complete absence of sunspots from 1649 to 1715. Note also in the inset of figure 11.29 the sudden present-day decrease in ^{14}C radioactivity because of massive combustion of fossil fuels (virtually ^{14}C-free), which results in world-scale dilution of ^{14}C with respect to stable isotopes ^{13}C and ^{12}C (Suess, 1965), partly offset by anthropic ^{14}C production associated with the explosion of nuclear weapons.

11.8 Application II: Stable Isotope Geothermometry

Although the scope of this book does not allow an appropriate treatment of stable isotope compositions of earth's materials (excellent monographs on this subject can be found in the literature—e.g., Hoefs, 1980; Faure, 1986), we must nevertheless introduce the significance of the various compositional parameters adopted in the literature before presenting the principles behind stable isotope geothermometry.

Let us consider again a generic isotopic exchange as represented by a reaction of the type

$$a\,A^\circ + bB^\bullet \Leftrightarrow aA^\bullet + bB^\circ \qquad (11.123)$$

(Urey, 1947), where A°, $A^\bullet$ and B°, $B^\bullet$ are, respectively, molecules of the same type differing only in the isotopic composition of a given constituting element (open and full symbols: light and heavy), and a and b are stoichiometric coefficients.

The "fractionation factor" α represents the relative distribution of heavy and light isotopes in the two phases at equilibrium (somewhat similar to the normalized distribution coefficient adopted in trace element geochemistry; cf. section 10.8):

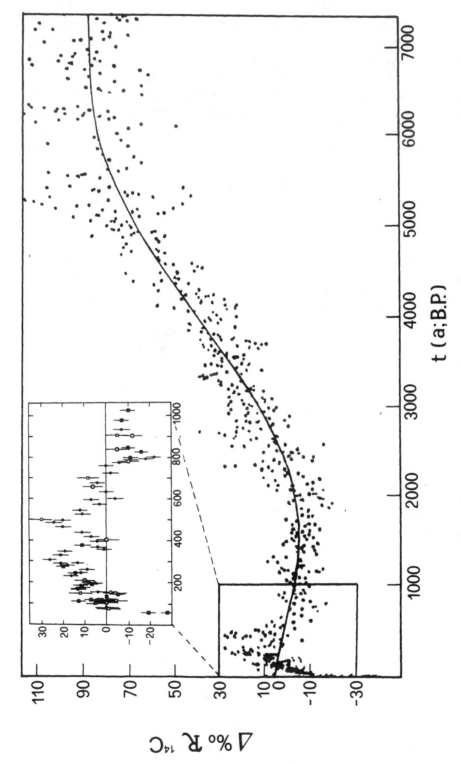

Figure 11.29 Specific radioactivity of ^{14}C expressed as per-mil deviation (Δ‰) from present-day radioactivity over the last 7500 years, derived from comparisons with dendrochronological studies. Reprinted from Damon et al. (1978a), with permission, from The Annual Review of Earth and Planetary Sciences, Volume 6, copyright © 1978 by Annual Reviews Inc. and from Suess (1965), *Journal of Geophysical Research*, 70, 5937–5952, copyright © 1965 by the American Geophysical Union.

$$\alpha = \frac{\left(\dfrac{X^{\bullet}}{X^{\circ}}\right)_A}{\left(\dfrac{X^{\bullet}}{X^{\circ}}\right)_B}, \tag{11.124}$$

where $(X^{\bullet}/X^{\circ})_A$ is the ratio of isotopic abundance in compound A. Assuming random distribution of isotopes in the exchanging compounds, the fractionation factor is related to equilibrium constant K (Epstein, 1959) by

$$\alpha = K^{1/r}, \tag{11.125}$$

where r is the number of exchanging isotopes per formula unit. In the simplest case, exchange reactions are written in such a way that only one isotope is exchanged. In this case, the fractionation factor obviously identifies with the equilibrium constant (i.e., $\alpha = K$).

Consider, for instance, the exchange

$$\frac{1}{3}\,\underset{\text{opx}}{MgSi^{16}O_3} + \frac{1}{4}\,\underset{\text{ol}}{Mg_2Si^{18}O_4} \Leftrightarrow \frac{1}{3}\,\underset{\text{opx}}{MgSi^{18}O_3} + \frac{1}{4}\,\underset{\text{ol}}{Mg_2Si^{16}O_4}. \tag{11.126}$$

The equilibrium constant of the isotopic exchange is

$$K_{126} = \frac{\left[^{18}O\right]_{opx}^{1/3}\left[^{16}O\right]_{ol}^{1/4}}{\left[^{16}O\right]_{opx}^{1/3}\left[^{18}O\right]_{ol}^{1/4}}, \tag{11.127}$$

where the terms in brackets denote activities

$$\left[^{18}O\right]_{opx} = X^3_{{}^{18}O,\,opx} \equiv X^{\bullet^n} \tag{11.128.1}$$

$$\left[^{18}O\right]_{ol} = X^4_{{}^{18}O,\,ol} \equiv X^{\bullet^m} \tag{11.128.2}$$

$$\left[^{16}O\right]_{opx} = X^3_{{}^{16}O,\,opx} \equiv X^{\circ^n} \tag{11.128.3}$$

$$\left[^{16}O\right]_{ol} = X^4_{{}^{16}O,\,ol} \equiv X^{\circ^m}. \tag{11.128.4}$$

Assuming the same notation adopted in equation 11.123, we have

$$r = a \times n = b \times m = 1, \tag{11.129}$$

and hence

$$\alpha = K_{126}. \tag{11.130}$$

Because the modifications affecting the stable isotope compositions of natural materials are rather limited (and mainly attributable to the processes discussed in sections 11.6.1 to 11.6.3), they are usually expressed in terms of Δ per-mil variations:

$$\Delta_{AB} = \left[\frac{(X^{\bullet}/X^{\circ})_A}{(X^{\bullet}/X^{\circ})_B} - 1 \right] \times 10^3 = (\alpha - 1) \times 10^3. \tag{11.131}$$

The isotopic measurements are actually performed in the laboratories quantifying the ‰ difference with respect to a standard material, expressed by the δ parameter. Denoting as st the standard material of reference, we have

$$\delta_A = \left[\frac{(X^{\bullet}/X^{\circ})_A}{(X^{\bullet}/X^{\circ})_{st}} - 1 \right] \times 10^3 \tag{11.132.1}$$

and

$$\delta_B = \left[\frac{(X^{\bullet}/X^{\circ})_B}{(X^{\bullet}/X^{\circ})_{st}} - 1 \right] \times 10^3. \tag{11.132.2}$$

Standard materials generally adopted to define the isotopic compositions of hydrogen and oxygen are the *Standard Mean Composition of Oceanic Water* (*SMOW;* Craig, 1961), the *NBS* distilled water (National Bureau of Standards), and the *Standard Light Antarctic Precipitation* (*SLAP;* International Atomic Energy Agency). For carbon (and for oxygen in carbonates), the standard materials are the *PDB* standard CO_2, produced by the University of Chicago from Cretaceous belemnites (Belemnitella Americana) of the Peedee Formation (South Carolina) and other working standards whose $\delta^{13}C$ with respect to *PDB* have been defined by Craig (1957). For sulfur, the commonly adopted reference standard is *CD troilite* (FeS) from the Canyon Diablo iron meteorite.

Based on the equations above, the relationships existing between the fractionation factor α and the δ and Δ parameters are

Table 11.8 Comparison among the various parameters adopted to define the stable isotope compositions of natural materials (after Hoefs, 1980).

δ_A	δ_B	Δ_{AB}	α_{AB}	$10^3 \ln \alpha_{AB}$
1.00	0	1	1.001	0.9995
10.00	0	10	1.01	9.95
20.00	0	20	1.02	19.80
10.00	5.00	4.98	1.00498	4.96
20.00	15.00	4.93	1.00493	4.91
30.00	20.00	9.80	1.00980	9.76
30.00	10.00	19.80	1.01980	19.61

$$\alpha_{AB} = \frac{\delta_A/10^3 + 1}{\delta_B/10^3 + 1} = \frac{\delta_A + 10^3}{\delta_B + 10^3} \tag{11.133}$$

and

$$\Delta_{AB} = \left(\frac{\delta_A + 10^3}{\delta_B + 10^3} - 1 \right) \times 10^3. \tag{11.134}$$

Because

$$10^3 \ln\left(1 + \frac{X}{1000}\right) \cong X, \tag{11.135}$$

the following approximations are often adopted in stable isotope geochemistry:

$$\delta_A - \delta_B \cong \Delta_{AB} \cong 10^3 \ln \alpha_{AB}. \tag{11.136}$$

The closeness of these approximations depend on the modulus of the first term $|\delta_A - \delta_B|$, as shown in table 11.8.

11.8.1 Oxygen

Oxygen is present in nature with three stable isotopes: ^{16}O (0.997630), ^{17}O (0.000375), and ^{18}O (0.001995). The first application of oxygen isotopic fractionation studies to geological problems was suggested by Urey (1947), who related the temperature of precipitation of calcium carbonate in seawater to the (T-dependent) isotopic fractionation induced by the exchange reaction between H_2O molecules and carbonate ion CO_3^{2-}:

$$C^{16}O_3^{2-} + 3H_2^{18}O \Leftrightarrow C^{18}O_3^{2-} + 3H_2^{16}O. \tag{11.137}$$

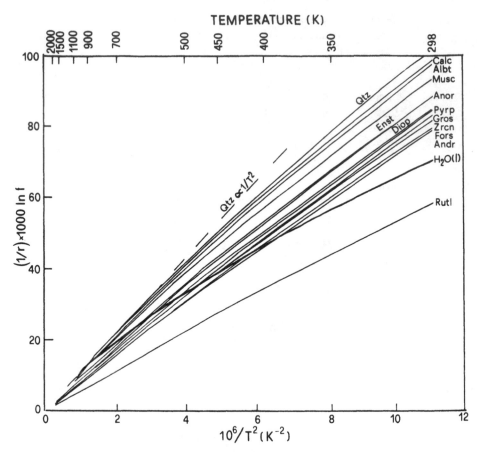

TEMPERATURE (K)

Figure 11.30 Reduced partition function for various minerals calulated by Kieffer (1982) through equation 11.61 plotted against T^{-2}. Heavy curve labeled $H_2O(l)$ is reduced partition function of water according to Becker (1971). Dashed curve is a T^{-2} extrapolation of high-T reduced partition curve for quartz. Mineral abbreviations: Qtz (quartz), Calc (calcite), Albt (albite), Musc (muscovite), Enst (clinoenstatite), Anor (anorthite), Diop (diopside), Pyrp (pyrope), Gros (grossular), Zron (zircon), Fors (forsterite), Andr (andradite), Rutl (rutile). Reprinted with permission from Kieffer (1982), *Review of Geophysics and Space Physics,* 20, 827–849, copyright © 1982 by the American Geophysical Union.

Detailed vibrational calculations (Bottinga, 1968; Becker, 1971; Kieffer, 1982) as well as accurate experiments (Clayton et al., 1989; Chiba et al., 1989, and references therein) and semiempirical generalizations (Zheng, 1991, 1993a,b) have been made since the work of Urey (1947) and ensure satisfactory knowledge of the $^{18}O/^{16}O$ fractionation properties of minerals and water and mineral-mineral couples as a function of T.

Figure 11.30 shows the reduced partition function $(1/r) \times 1000 \ln f$ calculated by Kieffer (1982) by application of equation 11.61 and plotted against T^{-2}. The adoption of this form of plot is dictated by the fact that the natural logarithm of the partition function ratio is asymptotically proportional to T^{-2} at high tem-

Table 11.9 Third-order polynomials relating reduced partition functions to inverse of squared absolute temperature $x = 10^6 \times T^{-2}$ (from Clayton and Kieffer, 1991).

Phase	a	b	c
Calcite	11.781	−0.420	0.0158
Quartz	12.116	−0.370	0.0123
Albite	11.134	−0.326	0.0104
Anorthite	9.993	−0.271	0.0082
Diopside	9.237	−0.199	0.0053
Forsterite	8.326	−0.142	0.0032
Magnetite	5.674	−0.038	0.0003

peratures (Urey, 1947; Bigeleisen and Mayer, 1947; see also section 11.6.1). Indeed, a cubic function in T^{-2} fits the calculated reduced partition functions within ± 0.02 per mil, for all temperatures above 400 K.

The Kieffer model correctly predicts the systematic change of the reduced partition functions of various minerals with structure, as indicated by Taylor and Epstein (1962). For anhydrous silicates, the decrease in the sequence framework-chain-orthosilicate reflects the decreasing frequency of antisymmetric Si-O stretching modes. The internal frequencies of the carbonate ion give a high reduced partition function at all T. The value for rutile is low because of the low frequencies of the Ti-O modes (Kieffer, 1982).

Combining the results of Kieffer's model and of laboratory experiments, Clayton and Kieffer (1991) obtained a set of third-order equations relating the reduced partition functions of various minerals to the inverse of the squared absolute temperature (table 11.9) according to

$$f_A = \left(\frac{1000}{r}\right) \ln\left(\frac{Q^\bullet}{Q^\circ}\right)_A = ax + bx^2 + cx^3 \tag{11.138}$$

$$x = 10^6 \times T^{-2}. \tag{11.139}$$

Oxygen isotopic fractionation between two minerals A and B at any given T can easily be obtained by subtracting their respective reduced partition function equations—i.e.,

$$\Delta_{AB} = f_A - f_B. \tag{11.140}$$

Equations 11.138 to 11.140 supersede the simple one-parameter equation proposed by Clayton et al. (1989) and Chiba et al. (1989):

Table 11.10 Comparison of polynomial and linear fits for mineral pairs at 1000 K. Values in italics refer to equation 11.141 (from Clayton and Kieffer, 1991).

	Qz	Ab	An	Di	Fo	Mt
Ca	−0.38	0.56	1.65	2.33	3.19	5.74
	−0.38	*0.56*	*1.61*	*2.37*	*3.29*	*5.91*
Qz		0.94	2.03	2.72	3.57	6.12
		0.94	*1.99*	*2.75*	*3.67*	*6.29*
Ab			1.09	1.78	2.63	5.18
			1.05	*1.81*	*2.73*	*5.35*
An				0.69	1.54	4.09
				0.76	*1.68*	*4.30*
Di					0.86	3.41
					0.92	*3.54*
Fo						2.55
						2.62

$$
\Delta_{AB} = a_{AB} \times \frac{10^6}{T^2}. \tag{11.141}
$$

A comparison of the two methods for $T = 1000$ K is given in table 11.10. For each entry, a positive value indicates ^{18}O enrichment in the phase on the left. The upper entry is relative to equations 11.138 to 11.140 and the lower entry to equation 11.141. Note that the table implicitly gives the parameters of the equation of Clayton et al. (1989), because at 1000 K equation 11.141 reduces to $\Delta_{AB} = a_{AB}$.

Figure 11.31 shows the results of equations 11.138 to 11.141 applied to the calcite-diopside couple compared with experimental evidence and with the indications of Kieffer's (1982) model.

Reduced partition functions for oxides were calculated by Zheng (1991) on the basis of the *"modified increment method."*

This method derives from the observation that the degree of ^{18}O enrichment in a set of cogenetic silicate minerals (expressed as oxygen isotope index I-^{18}O) can be related to bond strengths in minerals (Taylor, 1968). The oxygen isotope index for phase A is defined by

$$
\left(I-\,^{18}O\right)_A = \frac{\left(M_{16}/M_{18}\right)^{3/2}_*}{2n_O\left(M_{16}/M_{18}\right)^{3/2}_A} \sum_1^{n_{ct}} Z_{ct} \times i'_{ct-O}, \tag{11.142}
$$

where the asterisk denotes a reference phase. M_{16} and M_{18} are the formula weights of isotopically light and heavy compounds, n_O and n_{ct} are the stoichiometric numbers of oxygens and cations per formula unit. Z_{ct} is the cationic charge, and the summation extends over all cations in the formula unit, i'_{ct-O} is the "normalized ^{18}O-increment" of a cation-oxygen bond (Zheng, 1991):

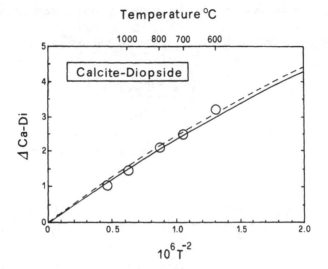

Figure 11.31 Oxygen isotopic fractionation between calcite and diopside as obtained from equations 11.131 to 11.133. Experimental data points are from Chiba et al. (1989). Dotted line: indications from vibrational calculations following Kieffer's (1982) procedure. From Clayton and Kieffer (1991), reprinted with kind permission of The Geochemical Society, Pennsylvania State University, University Park, Pennsylvania.

$$i'_{ct-O} = \left(i_{ct-O} / i_{Si-O} \right)^q . \qquad (11.143)$$

In equation 11.143, q is a parameter introduced by Zheng (1993a) to account for different types of bonds:

$$q = Z_{ct}^k \qquad k = -1, 0, +1. \qquad (11.144)$$

Integer k assumes a value of -1 in complex anions (e.g., Al^{3+} and Si^{4+} in feldspars), $+1$ in weakly bonded complex cations (e.g., Na^+ and Ca^{2+} in feldspars), and 0 for other complex cations (e.g., Fe^{2+} and Mg^{2+} in garnets).

The ^{18}O increment of a cation-oxygen bond is given by

$$i = \overline{S}_{ct-O} \ln W_{ct-O} . \qquad (11.145)$$

Zheng (1991) defined $\overline{S}_{ct-O}$ "bond-strength" as

$$\overline{S}_{ct-O} = \frac{Z_{ct}}{v_{ct} R_{ct-O}} . \qquad (11.146)$$

Although equation 11.146 agrees with the principle of electrostatic valence of Pauling (1929), normalization to the cation-to-oxygen distance R_{ct-O} to account for bond variation with the degree of ionicity is not in line with commonly adopted generalizations (see section 1.10.3).

The term $W_{\text{ct-O}}$ is related to the masses of cations and oxygen isotopes:

$$W_{\text{ct-O}} = \sqrt{\frac{\left(m_{\text{ct}} + m_{^{18}\text{O}}\right)m_{^{18}\text{O}}}{\left(m_{\text{ct}} + m_{^{18}\text{O}}\right)m_{^{16}\text{O}}}}. \tag{11.147}$$

The reduced partition function ratio of the mineral is derived from the equations above through

$$f_A = \left(\frac{1000}{r}\right)\ln\left(\frac{Q^{\bullet}}{Q^{\circ}}\right)_A = \left(\text{I}-^{18}\text{O}\right)_A \times \exp\left[\frac{\Delta E\left(1 - \left(\text{I}-^{18}\text{O}\right)_A\right)}{RT}\right]$$

$$\times \left(\frac{1000}{r}\right)\ln\left(\frac{Q^{\bullet}}{Q^{\circ}}\right)_*. \tag{11.148}$$

In equation 11.148, ΔE is regarded as an Arrhenius-slope energy term that has a value of 1 kJ/mol, and $(1000/r)\ln(Q^{\bullet}/Q^{\circ})^*$ is the reduced partition function ratio for the reference phase (quartz).

Calculated ^{18}O increments for various oxides with different coordination states are listed in table 11.11.

Calculation of reduced partition normalized fractionation factors, through the modified increment method, allows the derivation of T-dependent mineral-mineral oxygen isotopic fractionations expressed in the polynomial form

$$\Delta_{AB} = a \times 10^6\, T^{-2} + b \times 10^3\, T^{-1} + c. \tag{11.149}$$

The T-dependency selected by Zheng (1991, 1993a,b) differs from the equation of Clayton and Kieffer (1991), and the two methods do not give concordant results, especially at low T, as shown in figure 11.32.

Zheng (1993b) also extended the modified increment method to the evaluation of reduced partition function ratios in hydroxyl-bearing silicate minerals through

$$\left(\text{I}-^{18}\text{O}\right)_{\text{OH-A}} = \frac{2n_{\text{O}} - n_{\text{OH}}}{2n_{\text{O}}}\left(\text{I}-^{18}\text{O}\right)_A, \tag{11.150}$$

where n_{O} and n_{OH} represent the numbers of oxygens and hydroxyl groups in the hydrous mineral and $(\text{I}-^{18}\text{O})_A$ is the isotope index in the anhydrous counterpart.

Although largely empirical, the method of Zheng (1991, 1993a,b) has the

Table 11.11 Calculation of ^{18}O increments for silicates (from Zheng, 1993a).

Bond	Z_{ct}	n_{ct}	n_O	R_{ct-O}	m_{ct}	W_{ct-O}	$\overline{S}_{ct-O}$	i_{ct-O}
Si-O	4	4	2	1.61	28.09	1.03748	0.62112	0.02285
Si-O	4	4	3	1.62	28.09	1.03748	0.61728	0.02271
Si-O	4	4	4	1.64	28.09	1.03748	0.60976	0.02244
Si-O	4	6	3	1.76	28.09	1.03748	0.37881	0.01394
Al-O	3	4	2	1.74	26.98	1.03690	0.43103	0.01562
Al-O	3	4	3	1.75	26.98	1.03690	0.42861	0.01553
Al-O	3	4	4	1.77	26.98	1.03690	0.42373	0.01535
Al-O	3	5	3	1.84	26.98	1.03690	0.32612	0.01182
Al-O	3	6	3	1.89	26.98	1.03690	0.26455	0.00959
Al-O	3	6	4	1.91	26.98	1.03690	0.26178	0.00949
Fe-O	3	6	3	1.96	55.85	1.04631	0.25541	0.01156
Fe-O	3	6	4	1.98	55.85	1.04631	0.25253	0.01126
Fe-O	2	6	3	2.06	55.85	1.04631	0.16181	0.00733
Fe-O	2	6	4	2.08	55.85	1.04631	0.16064	0.00728
Fe-O	2	8	4	2.16	55.85	1.04631	0.11574	0.00524
Mg-O	2	6	3	2.08	24.31	1.03537	0.16026	0.00557
Mg-O	2	6	4	2.10	24.31	1.03537	0.15873	0.00552
Mg-O	2	8	4	2.25	24.31	1.03537	0.11111	0.00386
Mn-O	2	6	3	2.11	54.94	1.04613	0.15798	0.00712
Mn-O	2	6	4	2.13	54.94	1.04613	0.15649	0.00706
Mn-O	2	8	4	2.31	54.94	1.04613	0.10823	0.00488
Ca-O	2	6	3	2.36	40.08	1.04234	0.14124	0.00586
Ca-O	2	6	4	2.38	40.08	1.04234	0.14006	0.00581
Ca-O	2	7	3	2.42	40.08	1.04234	0.11806	0.00489
Ca-O	2	8	3	2.48	40.08	1.04234	0.10081	0.00418
Ca-O	2	8	4	2.50	40.08	1.04234	0.10000	0.00415
Ba-O	2	9	3	2.80	137.34	1.05381	0.07937	0.00416
K-O	1	6	3	2.74	39.10	1.04205	0.06283	0.00251
K-O	1	9	3	2.90	39.10	1.04205	0.03831	0.00158
K-O	1	10	3	2.92	39.10	1.04205	0.03425	0.00141
K-O	1	12	3	2.96	39.10	1.04205	0.02815	0.00116
Na-O	1	6	3	2.38	22.99	1.03454	0.07003	0.00238
Na-O	1	8	3	2.52	22.99	1.03454	0.04960	0.00168
Na-O	1	8	4	2.54	22.99	1.03454	0.04921	0.00167
Li-O	1	6	3	2.10	6.94	1.01729	0.07937	0.00136
Be-O	2	4	3	1.63	9.01	1.02064	0.30675	0.00627
Zn-O	2	4	3	1.96	65.37	1.04798	0.25510	0.01196
Cr-O	3	6	4	1.995	52.00	1.04551	0.25063	0.01115
Ni-O	2	6	4	2.07	58.71	1.04674	0.16103	0.00736
Ti-O	4	6	3	1.965	47.90	1.04455	0.33927	0.01479
Sn-O	4	6	3	2.05	118.69	1.05300	0.32520	0.01679
Th-O	4	8	3	2.40	232.04	1.05655	0.20833	0.01146
U-O	4	8	3	2.38	238.03	1.05665	0.21008	0.01158
Zr-O	4	8	3	2.20	91.22	1.05090	0.22727	0.01128
Hf-O	4	8	3	2.19	178.49	1.05525	0.22831	0.01228

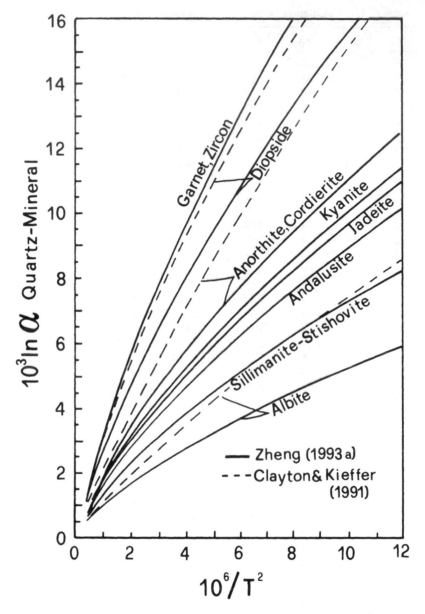

Figure 11.32 Oxygen isotopic fractionation factors between quartz and common rock-forming minerals, as obtained with modified increment method (Zheng 1993a), compared with indications based on fractionaton equations of Clayton and Kieffer (1991) (dashed lines). Reproduced with modifications from Y. F. Zheng, *Geochimica et Cosmochimica Acta,* 57, 1079–1091, copyright © 1993, with kind permission from Elsevier Science Ltd., The Boulevard, Langford Lane, Kidlington 0X5 1GB, UK.

advantage of furnishing quite extensive estimates of fractionation factors in mineral-mineral and mineral-water couples, which are valid over a wide T range (i.e., 0 to 1200 °C; see table 11.12).

11.8.2 Hydrogen

Hydrogen has two stable isotopes, $_1^1H$ (0.99985) and $_1^2H$ (0.00015), and a short-lived radioactive isotope $_1^3H$ (tritium) produced in the atmosphere by interaction of ^{14}N with cosmic ray neutrons:

$$_7^{14}N + n \rightarrow {}_1^3H + {}_6^{12}C. \tag{11.151}$$

The concentration of $_1^3H$ in the atmosphere (measured in Tritium Units, or TU; 1 TU = $_1^3H/_1^1H \times 10^{18}$) has been rapidly increasing (from 1952 to 1962) because of nuclear tests. Tritium decays to stable $_2^3He$ by β^- emission ($t_{1/2} = 12.26\ a$; see Kamensky et al., 1991 for an account of ^3H-^3He dating of groundwaters).

Isotope $_1^2H$ (deuterium), discovered by Urey et al. (1932), is usually denoted by symbol D. The large relative mass difference between H and D induces significant fractionation ascribable to equilibrium, kinetic, and diffusional effects. The main difference in the calculation of equilibrium isotopic fractionation effects in hydrogen molecules with respect to oxygen arises from the fact that the rotational partition function of hydrogen is nonclassical. Rotational contributions to the isotopic fractionation do not cancel out at high T, as in the classical approximation, and must be accounted for in the estimates of the partition function ratio f.

The classical expression for Q_{rot} is

$$Q_{rot}^* = \frac{hcB}{kT} = \frac{\hbar^2}{4IkT}, \tag{11.152}$$

where B is the rotational constant in cm^{-1}, I is the moment of inertia, and the remaining symbols have their usual meanings. Bigeleisen and Mayer (1947), to account for the nonclassical behavior of hydrogen molecules, suggested expansion of the deviation from the classical rotational partition function in the polynomial form

$$\frac{Q_{rot}}{Q_{rot}^*} = \left(1 + \frac{\sigma}{3} + \frac{\sigma^2}{15} + \frac{\sigma^3}{315} + \cdots\right), \tag{11.153}$$

where $\sigma = 1/Q_{rot}^*$.

Because the partition function ratio f is defined in such a way that the classical rotational and translational contributions are canceled, equations 11.40, 11.41 and 11.43 must be modified by introducing the ratio of the deviations from classical rotational behavior of heavy and light hydrogen molecules. For small values of σ, this contribution reduces to

Table 11.12 Oxygen isotope fractionation in quartz-mineral, mineral-water, and calcite-mineral couples, according to modified increment method of Zheng (1991, 1993a,b).
$\Delta_{AB} = a \times 10^6 T^{-2} + b \times 10^3 T^{-1} + c$ (from Zheng, 1991, 1993a,b).

Mineral	Quartz-Mineral			Mineral-Water			Calcite-Mineral		
	a	b	c	a	b	c	a	b	c
Quartz				4.48	−4.77	1.71	−0.47	0.10	0
Stishovite	0.22	1.88	−0.78	4.26	−6.64	2.08	−0.25	1.98	−0.78
K-feldspar	0.16	1.50	−0.62	4.32	−6.27	2.00	−0.30	1.60	−0.62
Albite	0.15	1.39	−0.57	4.33	−6.15	1.98	−0.32	1.49	−0.57
Anorthite	0.36	2.73	−1.14	4.12	−7.50	2.24	−0.11	2.83	−1.14
Leucite	0.22	1.91	−0.79	4.26	−6.67	2.08	−0.24	2.01	−0.79
Nepheline	0.37	2.77	−1.15	4.11	−7.53	2.24	−0.10	2.89	−1.15
Diopside	0.56	3.67	−1.53	3.92	−8.43	2.40	0.10	3.78	−1.53
Enstatite	0.51	3.45	−1.44	3.97	−8.22	2.37	0.05	3.55	−1.44
Hedenbergite	0.55	3.61	−1.51	3.93	−8.37	2.40	0.08	3.71	−1.51
Ferrosilite	0.48	3.33	−1.39	3.99	−8.09	2.35	0.02	3.43	−1.39
Jadeite	0.31	2.42	−1.01	4.17	−7.19	2.18	−0.16	2.52	−1.01
Acmite	0.30	2.38	−0.99	4.18	−7.14	2.17	−0.17	2.45	−0.99
Wollastonite	0.67	4.11	−1.72	3.81	−8.87	2.49	0.21	4.21	−1.72
Rhodonite	0.63	3.94	−1.65	3.85	−8.71	2.46	0.16	4.04	−1.65
Beryl	0.50	3.40	−1.42	3.98	−8.16	2.34	−0.03	3.50	−1.42
Cordierite	0.38	2.85	−1.19	4.10	−7.62	2.26	−0.08	2.95	−1.19
Almandine	0.72	4.26	−1.79	3.76	−9.02	2.52	0.25	4.36	−1.79
Pyrope	0.73	4.30	−1.80	3.75	−9.07	2.52	0.26	4.40	−1.80
Spessartine	0.73	4.31	−1.81	3.75	−9.07	2.52	0.26	4.41	−1.81
Grossular	0.74	4.35	−1.82	3.74	−9.11	2.52	0.27	4.45	−1.82
Andradite	0.72	4.29	−1.80	3.76	−9.05	2.52	0.26	4.38	−1.80
Uvarovite	0.72	4.28	−1.80	3.76	−9.05	2.52	0.26	4.39	−1.80
Titanite	0.67	4.11	−1.72	3.81	−8.87	2.49	0.21	4.21	−1.72
Malayaite	0.60	3.83	−1.60	3.88	−8.60	2.43	0.14	3.94	−1.60
Zircon	0.72	4.26	−1.79	3.76	−9.03	2.52	0.25	4.37	−1.79
Thorite	0.79	4.50	−1.89	3.69	−9.27	2.55	0.32	4.61	−1.89
Fayalite	0.84	4.69	−1.97	3.64	−9.46	2.59	0.38	4.79	−1.97
Forsterite	0.93	4.95	−2.09	3.55	−9.72	2.64	0.46	5.05	−2.09
Tephroite	0.86	4.75	−2.00	3.62	−9.51	2.60	0.39	4.85	−2.00
Willemite	0.69	4.18	−1.75	3.79	−8.94	2.50	0.23	4.28	−1.75
Phenacite	1.06	5.34	−2.26	3.42	−10.11	2.70	0.60	5.45	−2.26
Sillimanite	0.22	1.89	−0.78	4.26	−6.65	2.07	−0.25	1.99	−0.78
Andalusite	0.28	2.26	−0.94	4.20	−7.02	2.15	−0.19	2.36	−0.94
Kyanite	0.32	2.52	−1.05	4.16	−7.29	2.20	−0.14	2.63	−1.05
Muscovite	0.38	2.84	−1.18	4.10	−7.61	2.25	−0.08	2.97	−1.18
Paragonite	0.37	2.77	−1.15	4.11	−7.54	2.24	−0.10	2.87	−1.15
Margarite	0.38	2.84	−1.18	4.10	−7.61	2.25	−0.08	2.94	−1.18
Illite	0.34	2.60	−1.08	4.14	−7.36	2.21	−0.13	2.70	−1.08
Phengite	0.35	2.64	−1.10	4.13	−7.41	2.22	−0.12	2.74	−1.10

Table 11.12   *(continued)*

Mineral	Quartz-Mineral			Mineral-Water			Calcite-Mineral		
	a	b	c	a	b	c	a	b	c
Glauconite	0.49	3.34	−1.39	3.99	−8.11	2.34	0.02	3.44	−1.39
Phlogopite	0.62	3.92	−1.64	3.86	−8.68	2.45	0.16	4.02	−1.64
Annite	0.59	3.79	−1.59	3.89	−8.56	2.43	0.13	3.89	−1.59
Biotite	0.64	3.99	−1.67	3.84	−8.76	2.46	0.18	4.10	−1.67
Lepidolite	0.38	2.81	−1.17	4.10	−7.58	2.25	−0.09	2.91	−1.17
Hornblende	0.59	3.80	−1.59	3.89	−8.56	2.43	0.13	3.90	−1.59
Tremolite	0.53	3.52	−1.47	3.95	−8.28	2.38	0.06	3.62	−1.47
Actinolite	0.52	3.48	−1.45	3.96	−8.25	2.37	0.05	3.58	−1.45
Cummingtonite	0.52	3.48	−1.45	3.96	−8.25	2.37	0.05	3.58	−1.45
Grunerite	0.50	3.40	−1.42	3.98	−8.17	2.36	0.03	3.50	−1.42
Glaucophane	0.45	3.18	−1.32	4.03	−7.94	2.31	−0.02	3.28	−1.32
Riebeckite	0.43	3.08	−1.28	4.05	−7.85	2.30	−0.04	3.18	−1.28
Pargasite	0.71	4.22	−1.77	3.77	−8.99	2.51	0.24	4.32	−1.77
Anthophyllite	0.52	3.49	−1.46	3.96	−8.26	2.38	0.06	3.60	−1.46
Gedrite	0.63	3.96	−1.66	3.85	−8.72	2.46	0.17	4.06	−1.66
Kaolinite	0.19	1.68	−0.69	4.29	−6.44	2.03	−0.28	1.78	−0.69
Lizardite	0.51	3.42	−1.43	3.97	−8.19	2.36	0.04	3.52	−1.43
Amesite	0.53	3.53	−1.47	3.95	−8.30	2.38	0.06	3.63	−1.47
Pyrophyllite	0.08	0.86	−0.35	4.40	−5.62	1.87	−0.38	0.96	−0.35
Talc	0.28	2.27	−0.94	4.20	−7.04	2.14	−0.19	2.38	−0.94
Serpentine	0.49	3.35	−1.40	3.99	−8.12	2.35	0.02	3.45	−1.40
Pennine	0.51	3.43	−1.43	3.97	−8.19	2.36	0.04	3.53	−1.43
Clinochlore	0.51	3.43	−1.43	3.97	−8.19	2.36	0.04	3.53	−1.43
Chamosite	0.42	3.04	−1.27	4.06	−7.81	2.30	−0.04	3.14	−1.27
Thuringite	0.58	3.73	−1.56	3.90	−8.50	2.42	0.11	3.84	−1.56
Zoisite	0.43	3.07	−1.28	4.05	−7.84	2.30	−0.04	3.18	−1.28
Epidote	0.42	3.05	−1.27	4.05	−7.81	2.29	−0.04	3.15	−1.27
Vesuvianite	0.93	4.97	−2.09	3.55	−9.74	2.63	0.47	5.07	−2.09
Ilvaite	1.05	5.32	−2.24	3.43	−10.08	2.69	0.58	5.42	−2.24
Norbergite	0.86	4.75	−2.00	3.62	−9.51	2.60	0.39	4.85	−2.00
Chondrodite	0.74	4.34	−1.82	3.73	−9.11	2.52	0.27	4.45	−1.82
Humite	0.71	4.24	−1.78	3.77	−9.01	2.51	0.24	4.34	−1.78
Clinohumite	0.70	4.19	−1.75	3.78	−8.95	2.50	0.23	4.29	−1.75
Staurolite	0.39	2.90	−1.21	4.09	−7.66	2.27	−0.07	3.00	−1.21
Topaz	0.30	2.41	−1.00	4.18	−7.18	2.17	−0.16	2.52	−1.00
Datolite	0.15	1.44	−0.60	4.33	−6.21	1.99	−0.31	1.54	−0.60
Chloritoid	0.49	3.35	−1.40	3.99	−8.11	2.35	0.02	3.45	−1.40
Tourmaline	0.27	2.22	−0.92	4.21	−6.99	2.14	−0.20	2.32	−0.92
Axinite	0.28	2.26	−0.94	4.20	−7.02	2.15	−0.19	2.36	−0.94
Pectolite	0.41	2.95	−1.23	4.08	−7.71	2.28	−0.06	3.05	−1.23
Prehnite	0.30	2.40	−1.00	4.18	−7.17	2.18	0.17	2.50	−1.00
Rutile	0.79	6.82	−3.59	3.45	−10.60	2.55			
Pyrolusite	0.64	6.16	−3.23	3.61	−9.94	2.19			
Cassiterite	0.56	5.80	−3.04	3.68	−9.58	2.00			

Table 11.12 (continued)

Mineral	Quartz-Mineral			Mineral-Water			Calcite-Mineral		
	a	b	c	a	b	c	a	b	c
Plattnerite	0.56	5.80	−3.04	3.68	−9.58	2.00			
Cerianite	1.67	9.32	−4.97	2.57	−13.09	3.93			
Thorianite	1.66	9.30	−4.96	2.58	−13.07	3.92			
Uraninite	1.61	9.19	−4.90	2.63	−12.97	3.86			
Ilmaite	1.36	8.61	−4.57	2.88	−12.38	3.53			
Geikelite	1.46	8.85	−4.70	2.78	−12.63	3.66			
Hematite	1.55	9.05	−4.82	2.69	−12.82	3.78			
Corundum	2.00	9.94	−5.32	2.24	−13.71	4.28			
Magnetite	1.22	8.22	−4.35	3.02	−12.00	3.31			
Jacobsite	1.23	8.24	−4.37	3.02	−12.02	3.33			
Magnesioferrite	1.29	8.43	−4.47	2.95	−12.20	3.43			
Ulvöspinel	1.43	8.78	−4.67	2.81	−12.56	3.63			
Chromite	1.63	9.22	−4.91	2.62	−12.99	3.87			
Magnesiochromite	1.76	9.49	−5.07	2.48	−13.27	4.03			
Hercynite	1.92	9.80	−5.24	2.32	−13.58	4.20			
Spinel	2.06	10.04	−5.38	2.18	−13.81	4.34			

$$1 + \frac{\sigma^{\bullet} - \sigma^{\circ}}{3} + \frac{\left(\sigma^{\bullet} - \sigma^{\circ}\right)^2}{18} + \frac{\sigma^{\bullet 2} - \sigma^{\circ 2}}{90}. \tag{11.154}$$

In a different approach, Bottinga (1969a) evaluated the nonclassical rotational partition function for water vapor, molecular hydrogen, and methane through the asymptotic expansion of Strip and Kirkwood (1951):

$$Q_{\text{rot}} = \left[\frac{\pi}{s_x s_y s_z}\right]^{1/2} \left\{ 1 + \frac{\left[2s_x + 2s_y + 2s_z - \left(s_x s_y / s_z\right) - \left(s_y s_z / s_x\right) - \left(s_z s_x / s_y\right)\right]}{12} + \cdots \right\},$$

$$\tag{11.155}$$

where s_x is the classical rotational partition function contribution about the x-axis:

$$s_x = Q^*_{\text{rot},x} = \frac{\hbar^2}{4I_x kT}, \tag{11.156}$$

and I_x is the moment of inertia about the x-axis.

Bottinga (1969a) calculated the isotopic fractionation relative to the exchange reactions

$$\frac{1}{2}H_2O + D \Leftrightarrow \frac{1}{2}D_2O + H \tag{11.157}$$

$$\frac{1}{4}CH_4 + D \Leftrightarrow \frac{1}{4}CD_4 + H \tag{11.158}$$

$$\frac{1}{2}H_2O + \frac{1}{4}CD_4 \Leftrightarrow \frac{1}{2}D_2O + \frac{1}{4}CH_4. \tag{11.159}$$

Results of his calculations are listed in table 11.13 in terms of fractionation factors α and Δ_{AB}^*. The Δ_{AB}^* factors for methane–water vapor and hydrogen–water vapor couples are plotted in figure 11.33 as a function of T (°C).

Note that Bottinga (1969a) defines Δ_{AB}^* as

$$\Delta_{AB}^* = (\alpha - 1) \times 1000 \tag{11.160}$$

(see eq. 11.133 for the significance of α). Taking the water vapor as phase B, the parameter Δ_{AB}^* is equal to the deuterium enrichment in phase A with respect to water, and hence is equivalent to the δ notation adopted in the geochemical literature, under the assumption of equilibrium.

The hydrogen isotope fractionation between OH-bearing minerals and water is a sensitive function of T. Figure 11.34 shows the experimental results of Suzuoki and Epstein (1976). The interpolant expressions for muscovite-water, biotite-water, and hornblende-water in the T range 450 to 850 °C are

$$10^3 \ln \alpha_{\text{muscovite-water}} = -\frac{22.1 \times 10^6}{T^2} + 19.1 \tag{11.161.1}$$

$$10^3 \ln \alpha_{\text{biotite-water}} = -\frac{21.3 \times 10^6}{T^2} - 2.8 \tag{11.161.2}$$

$$10^3 \ln \alpha_{\text{hornblende-water}} = -\frac{23.9 \times 10^6}{T^2} + 7.9 \tag{11.161.3}$$

Because the slope coefficient in the equations above is practically constant, within experimental uncertainty, the T-dependency of the hydrogen fractionation factor for mica and amphibole can be generalized as

$$10^3 \ln \alpha_{\text{mineral-water}} = -\frac{22.4 \times 10^6}{T^2} \\ + 28.2 + 2X_{\text{Al}} - 4X_{\text{Mg}} - 68X_{\text{Fe}} \tag{11.162}$$

(Suzuoki and Epstein, 1976), where X_{Al}, X_{Mg}, and X_{Fe} are molar fractions of sixfold coordinated cations. Equation 1.162 implies that hydrogen isotope fractionation among coexisting mica and amphibole is temperature independent and

Table 11.13 Deuterium fractionation in the water-hydrogen-methane system, after Bottinga (1969) (wv = water vapor, me = methane, hy = hydrogen).

T (°C)	ln α		$10^3 \ln \alpha$	Δ^*_{AB}	
	wv-hy	me-hy	wv-me	me-wv	hy-wv
0	1.395	1.387	7.3	−7.3	−752.1
10	1.355	1.322	13.3	−13.3	−736.9
20	1.280	1.261	18.9	−18.7	−721.9
25	1.253	1.232	21.8	−21.2	−714.5
30	1.228	1.204	24.0	−23.7	−707.1
40	1.179	1.151	28.7	−28.2	−692.5
50	1.134	1.101	33.0	−32.4	−678.2
60	1.091	1.054	36.9	−36.3	−664.1
70	1.051	1.010	40.6	−39.8	−650.3
80	1.012	0.968	43.9	−43.0	−636.7
90	0.977	0.930	47.0	−45.9	−623.4
100	0.943	0.893	49.9	−48.6	−610.4
120	0.880	0.825	54.8	−53.4	−585.2
140	0.824	0.765	59.0	−57.3	−561.1
160	0.773	0.710	62.5	−60.6	−538.2
180	0.726	0.661	65.4	−63.3	−516.3
200	0.684	0.616	67.8	−65.5	−495.4
220	0.645	0.576	69.6	−67.2	−475.5
240	0.610	0.539	71.1	−68.6	−456.5
260	0.577	0.505	72.2	−69.7	−438.4
280	0.547	0.474	73.0	−70.4	−421.2
300	0.519	0.445	73.5	−70.9	−404.8
320	0.493	0.419	73.8	−71.2	−389.2
340	0.469	0.395	73.9	−71.3	−374.3
360	0.446	0.372	73.8	−71.2	−360.1
380	0.425	0.352	73.5	−70.9	−346.5
400	0.406	0.333	73.2	−70.5	−333.6
450	0.362	0.291	71.7	−69.2	−303.9
500	0.325	0.255	69.6	−67.3	−277.5
550	0.293	0.226	67.2	−65.0	−253.9
600	0.265	0.201	64.7	−62.3	−232.9
650	0.241	0.179	61.9	−60.0	−214.1
700	0.220	0.161	59.1	−57.4	−197.3

cannot be used as a geothermometer. Moreover, based on equation 11.162, *T* being equal, minerals in which Al-OH bonds predominate (i.e., muscovite, for example) are richer in D with respect to phases with prevalent Mg-OH bonds (biotite), and minerals with the same Mg/Fe ratio have the same D-enrichment (biotite-hornblende).

Equation 11.162 cannot be extended to other OH-bearing phases such as serpentine, chlorite, and clay minerals (illite, smectite). Concerning serpentine, the experiments conducted by Sakai and Tsutsumi (1978) on clinochrysotile at 2

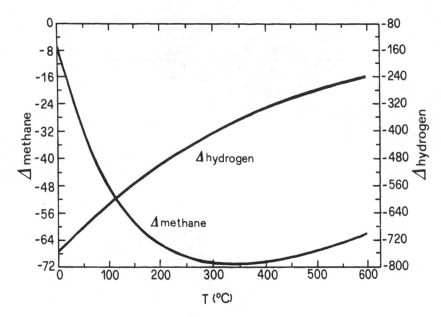

Figure 11.33 Calculated oxygen isotope fractionation factors Δ^*_{AB} for methane–water vapor and hydrogen–water vapor couples plotted against T (°C). Reprinted from Y. Bottinga, *Geochimica et Cosmochimica Acta,* 33, 49–64, copyright © 1969, with kind permission from Elsevier Science Ltd., The Boulevard, Langford Lane, Kidlington OX5 1GB, UK.

kbar water pressure in the T range 100 to 500 °C lead to the following polynomial expansion:

$$10^3 \ln \alpha_{\text{serpentine–water}} = \frac{27.5 \times 10^6}{T^2} - \frac{7.69 \times 10^4}{T} + 40.8. \qquad (11.163)$$

Note that the T-dependency in equation 11.163 is inverse with respect to equation 11.162 (see, however, Wenner and Taylor, 1971).

Concerning clay minerals (illite and smectite), Capuano (1992) proposed a simple linear dependency of the mineral-water fractionation factor over reciprocal temperature:

$$10^3 \ln \alpha_{\text{clay–water}} = -\frac{45.3 \times 10^3}{T} + 94.7. \qquad (11.164)$$

Equation 11.164 is valid in the T range 0 to 150 °C. The same sort of T-dependency (although less marked) was envisaged by Yeh (1980) (slope coefficient $= -19.6 \times 10^3$, intercept $= 25$).

One of the main applications of hydrogen and oxygen isotope thermometry in geochemistry is the estimation of the reservoir temperatures of active geothermal systems or the evaluation of the ruling T conditions during deposition or alter-

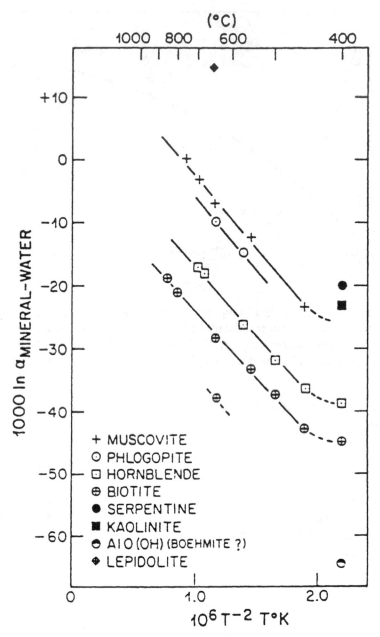

Figure 11.34 Fractionation factors versus T^{-2} for mica-water and amphibole-water couples. Reprinted from T. Suzuoki and S. Epstein, *Geochimica et Cosmochimica Acta,* 40, 1229–1240, copyright © 1976, with kind permission from Elsevier Science Ltd., The Boulevard, Langford Lane, Kidlington 0X5 1GB, UK.

ation of mineral assemblages in sedimentary basins and fossil geothermal systems. Such estimates are, however, subject to serious errors if salinity effects on the activity coefficient of HDO or $H_2^{18}O$ solvent molecules in the aqueous fluids are not attentively evaluated (the same errors could also affect generalizations of the type exemplified by eq. 11.163 and 11.164).

Because HDO or $H_2^{18}O$ molecules cannot be expected to have the same solubility in the brine of the main solvent species $H_2^{16}O$ (hereafter simply denoted H_2O), their differential behavior must be accounted for by introduction of the appropriate activity coefficient ratio (or "isotope salt effect"; cf. Horita et al., 1993a,b) Γ:

$$\Gamma_D = \frac{\gamma_{HDO}}{\gamma_{H_2O}} \qquad (11.165.1)$$

$$\Gamma_O = \frac{\gamma_{H_2^{18}O}}{\gamma_{H_2O}}. \qquad (11.165.2)$$

Denoting as $R_{activity}$ the activity ratio $[HDO]/[H_2O]$ or $[H_2^{18}O]/[H_2O]$ and as $R_{composition}$ corresponding molar ratios $(HDO)/(H_2O)$ and $(H_2^{18}O)/(H_2O)$, the isotope salt effect can also be expressed as

$$\Gamma = \frac{R_{activity}}{R_{composition}} = \frac{1 + 10^{-3}\,\delta_{activity}}{1 + 10^{-3}\,\delta_{composition}} \qquad (11.166)$$

(Horita et al., 1993a,b), where δ is the conventional delta value (‰). Combining equations 11.165 and 11.166, through some substitutions, we can derive

$$10^3 \ln \alpha_{A-brine(activity)} = 10^3 \ln \alpha_{A-brine(comp)} - 10^3 \ln \Gamma_{brine}. \qquad (11.167)$$

Both the oxygen and hydrogen isotope salt effects were determined by Horita et al. (1993a) in single-salt (NaCl, KCl, $MgCl_2$, $CaCl_2$, Na_2SO_4, and $MgSO_4$) aqueous solutions of various molalities, and their results can be represented in the simple linear form

$$10^3 \ln \Gamma = m\left(a + \frac{b}{T}\right), \qquad (11.168)$$

where m is molality, and a and b are fitting parameters whose values are listed in table 11.14.

Table 11.14 Calibration of "isotope salt effect" in single-salt solutions (eq. 11.161), after Horita et al. (1993a).

Salt	Isotope	a (mol^{-1})	b (mol$^{-1} \times$ K)	T Range (°C)
NaCl	(D)	−2.89	1503.1	10–100
	(^{18}O)	−0.015	0	25–100
KCl	(D)	−5.10	2278.4	20–100
	(^{18}O)	−0.612	230.83	25–100
CaCl$_2$	(D)	−2.34	2318.2	50–100
	(^{18}O)	−0.368	0	50–100
MgCl$_2$	(D)	+4.14	0	50–100
	(^{18}O)	+0.841	−582.73	25–100
Na$_2$SO$_4$	(D)	+0.86	0	50–100
	(^{18}O)	−0.143	0	50–100
MgSO$_4$	(D)	+8.45	−2221.8	50–100
	(^{18}O)	+0.414	−432.33	0–100

Neglecting the isotope salt effect leads to erroneous deductions concerning the T of equilibration between minerals and brines. Figure 11.35 shows, for example, the isotope salt effect in the case of kaolinite-water equilibrium. In the case illustrated, the analyzed isotopic compositions of kaolinite and aqueous solution yield a fractionation factor $10^3 \ln \alpha$ of −15 for deuterium. If the kaolinite–pure water fractionation line of Liu and Epstein (1984) is utilized, this value would suggest an equilibrium temperature of 88 °C. Imagine, however, the analyzed aqueous solution to be rich in saline components (as are the majority of geothermal fluids); in this case the actual T of equilibrium would be 81 °C (1m NaCl solution), 64 °C (3m NaCl solution), or 29 °C in the case of the Salton Sea brine (3.14m NaCl, 0.882m CaCl$_2$, 0.588m KCl). In the last case, the error induced by neglecting the isotope salt effect would thus be on the order of 300%.

The same problems encountered in mineral-water isotopic thermometry could be present in T estimates based on isotopic equilibrium between brine and dissolved species, such as SO$_4^{2-}$. Isotope geothermometers based on mineral couples or gaseous couples (e.g., CH$_4$-CO$_2$), however, are not affected by isotopic salt effect in brines (Horita et al., 1993b).

11.8.3 Carbon

Fractionation factors for the two stable carbon isotopes ^{12}C and ^{13}C have been calculated by Bottinga (1969a,b) for exchanges among graphite, calcite, carbon dioxide, and methane and for graphite-CO$_2$ and diamond-CO$_2$ equilibria. Results of vibrational calculations of Bottinga (1969a,b) in the T range 0 to 1000 °C are listed in table 11.15 and graphically displayed in figure 11.36.

The main points arising from the study of Bottinga (1969a) are as follows:

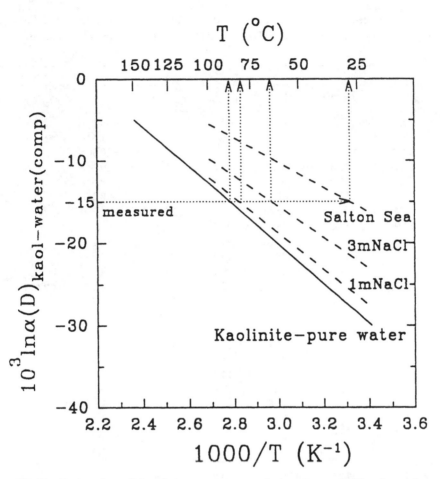

Figure 11.35 Estimation of formation temperature from measured δD values of coexisting kaolinite and aqueous solution on the isotope composition scale. The kaolinite–pure water line is from Liu and Epstein (1984). Dashed lines show the shift induced by increasing salinity of solution. Reprinted from J. Horita, D. R. Cole, and D. J. Wesolowski, *Geochimica et Cosmochimica Acta,* 57, 4703–4711, copyright © 1993, with kind permission from Elsevier Science Ltd., The Boulevard, Langford Lane, Kidlington 0X5 1GB UK.

1. The carbon isotope fractionation between CO_2 and methane and between graphite and methane is a sensitive function of temperature.
2. The isotope composition of graphite does not characterize its origin because it is not unique to any simple method of condensation.
3. The carbon fractionation factors between gaseous CO_2 and CH_4 are rather large, especially at low *T*. Temperatures deduced by application of ^{13}C fractionation geothermometers are, however, reasonable for either CO_2 or CH_4 produced by bacterial decomposition of organic matter and present in gaseous emissions associated with geothermal fields (cf. tables 9 and 10 in Bottinga, 1969a).

Table 11.15 Calculated fractionation factors (1000 ln α) for ^{13}C exchange among carbon dioxide, calcite, graphite, diamond, and methane (Bottinga 1969a,b). Cc = calcite; Gr = graphite; Dm = diamond.

T (°C)	1000 ln α_{AB}							
	Cc-CH$_4$	CO$_2$-CH$_4$	Gr-CH$_4$	Cc-Gr	CO$_2$-Gr	CO$_2$-Cc	CO$_2$-Dm	Dm-Gr
0	90.6	77.7	63.4	27.2	14.3	−13.0	2.8	11.5
10	85.5	73.7	59.3	26.2	14.4	−11.8	-	-
20	80.7	70.1	55.5	25.2	14.6	−10.7	-	-
25	78.5	68.3	53.7	24.8	14.6	−10.1	-	-
30	76.3	66.7	52.0	24.3	14.7	−9.5	-	-
40	72.2	63.6	48.8	23.4	14.8	−8.6	-	-
50	68.4	60.7	45.8	22.6	14.9	−7.7	7.0	7.8
60	64.9	58.0	43.0	21.8	15.0	−6.9	-	-
70	61.6	55.5	40.5	21.1	15.0	−6.1	-	-
80	58.5	53.2	38.1	20.4	15.1	−5.4	-	-
90	55.6	51.0	35.9	19.7	15.1	−4.7	-	-
100	52.9	48.9	33.8	19.1	15.1	−4.0	9.6	5.9
120	48.1	45.2	30.2	17.9	15.0	−2.9	-	-
140	43.8	41.9	26.9	16.9	14.9	−1.9	-	-
150	-	-	-	-	-	-	10.9	3.9
160	40.0	38.9	24.1	15.9	14.8	−1.1	-	-
180	36.6	36.3	21.6	15.0	14.6	−0.4	-	-
200	33.6	33.9	19.5	14.2	14.4	0.2	11.5	2.9
220	31.0	31.7	17.5	13.4	14.2	0.8	-	-
240	28.6	29.8	15.8	12.7	13.9	1.2	-	-
250	-	-	-	-	-	-	11.6	2.1
260	26.4	28.0	14.3	12.1	13.7	1.5	-	-
280	24.5	26.3	13.0	11.5	13.4	1.8	-	-
300	22.8	24.8	11.8	11.0	13.1	2.1	11.4	1.7
320	21.2	23.5	10.7	10.5	12.8	2.3	-	-
340	19.8	22.2	9.7	10.0	12.5	2.4	-	-
350	-	-	-	-	-	-	11.0	1.3
360	18.5	21.0	8.9	9.6	12.2	2.6	-	-
380	17.3	20.0	8.1	9.2	11.9	2.6	-	-
400	16.3	19.0	7.4	8.8	11.5	2.7	10.5	1.1
450	14.0	16.8	6.0	8.0	10.8	2.8	9.9	0.9
500	12.2	14.9	4.9	7.3	10.1	2.8	9.3	0.8
550	10.7	13.4	4.0	6.7	9.4	2.7	8.7	0.7
600	9.5	12.1	3.3	6.2	8.8	2.6	8.2	0.6
650	8.5	10.9	2.7	5.8	8.2	2.4	7.7	0.6
700	7.7	10.0	2.3	5.4	7.7	2.3	7.2	0.5
750	-	-	-	-	-	-	6.7	0.5
800	-	-	-	-	-	-	6.3	0.5
850	-	-	-	-	-	-	5.9	0.5
900	-	-	-	-	-	-	5.5	0.5
1000	-	-	-	-	-	-	4.4	0.4

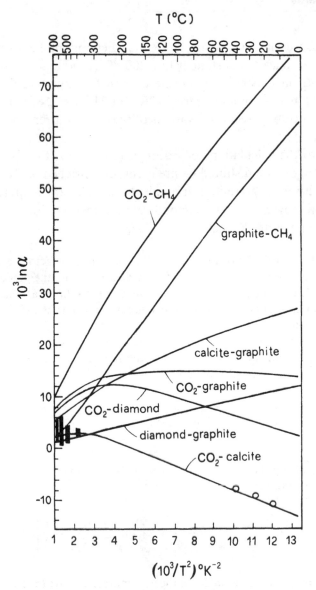

Figure 11.36 Carbon isotope fractionation among various phases in the Ca-C-O-H system, plotted as a function of T. Calcite-graphite, CO_2-graphite, CO_2-graphite, graphite-CH_4, and CO_2-CH_4 after Bottinga (1969a); diamond-graphite, CO_2-diamond, and CO_2-calcite after Bottinga (1969b). Superimposed symbols denote experimental results on CO_2-calcite fractionation after Chacko et al. (1991) (solid boxes) and Romanek et al. (1989). Reprinted from Bottinga (1969a), with kind permission from Elsevier Science Publishers B.V, Amsterdam, The Netherlands.

Moreover, as we can see in figure 11.36, the fractionation factor between CO_2 and graphite is practically constant below 200 °C (1000 ln $\alpha \approx$ 14.5) and falls progressively with increasing T. Also, the carbon fractionation factor between CO_2 and calcite has a maximum near 460 °C (1000 ln $\alpha \approx$ 2.85) and a crossover at 192 °C, confirmed by direct observations of natural occurrences in geothermal areas by Craig (1963).

According to the calculations of Bottinga (1969b) for the C-O system, the high-density form diamond should be preferentially enriched in ^{13}C with respect to the low-density form graphite (although the enrichment is quite limited; cf. figure 11.36). Moreover, the CO_2-diamond carbon fractionation curve exhibits a broad maximum near 300 °C (1000 ln $\alpha \approx$ 11).

Bottinga's (1969b) calculations of the carbon fractionation factors between carbon polymorphs and coexisting gaseous phases are carried out along the guidelines outlined in section 11.6.1. The phonon frequency distribution in the solid phases is, however, obtained by application of lattice dynamics to crystalline polymorphs. After an opportune selection of static potentials describing the interaction between neighboring carbon atoms in the structure, as a function of interatomic distances and angles, the equations of motion are obtained for all atoms in the unit cell, and the normal modes of vibration of all r particles in the crystal composed of N unit cells:

$$\int_0^{\omega_{max}} g(\omega)d\omega = 3Nr \tag{11.169}$$

are calculated by solving the $3r \times 3r$ dynamical matrix obtained by general solution of wave vectors. The vibrational partition function is then obtained by applying

$$\ln Q_{vib} = \frac{kT}{hc} \sum_n \frac{g_n}{\omega_{n+1} - \omega_n} \times \int_{X_n}^{X_{n+1}} \ln\left[\frac{\exp(-X/2)}{1 - \exp(-X)}\right] dX, \tag{11.170}$$

where the various g_n (i.e., the eigenvalues of the dynamic matrix) are the numbers of modes in the vibrational frequency interval $\omega_{n+1} - \omega_n$ (see sections 3.1, 3.3, and 11.6.1 for the significance of the remaining symbols).

Vibrational spectra generated by Bottinga (1969b) for graphite and diamond with this procedure are shown in figure 11.37. The spectrum for graphite is based on the interionic potential model of Yoshimori and Kitano (1956) and adopts the force constants of Young and Koppel (1965) (assumed to be identical for ^{12}C and ^{13}C). The histogram of figure 11.37A represents 1.79×10^6 frequencies in the irreducible segment ($\frac{1}{24}$) of the Brillouin zone, with a frequency interval $\omega_{n+1} - \omega_n$ of 1 cm^{-1}. Figure 11.37B shows the normalized phonon spectra for diamond, based on the interionic potential model of Aggarwal (1967) and the dynamic matrix equations of Herman (1959) ($\frac{1}{48}$ of the Brillouin zone; 3.46×10^6 frequencies). The corresponding Debye spectrum for $\theta_D = 1860$ K is also shown for comparative purposes. The jagged appearance of the phonon spectra is partly attributable to incomplete root sampling of phonon frequencies and to the narrow bandwidth adopted in the two histograms (Bottinga, 1969b).

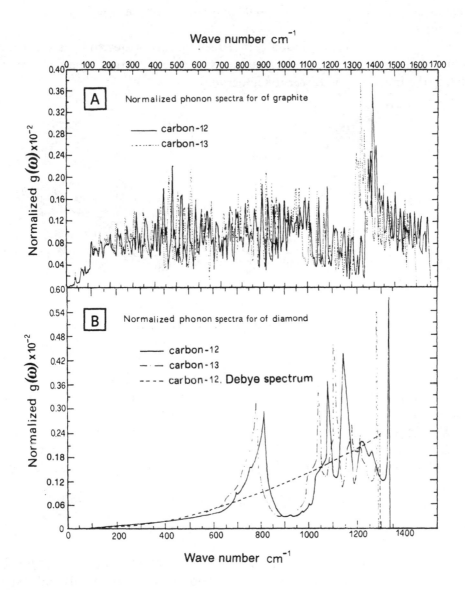

Figure 11.37 Normalized phonon spectra for ^{12}C and ^{13}C in graphite (A) and diamond (B). Reprinted from Bottinga (1969b), with kind permission from Elsevier Science Publishers B.V, Amsterdam, The Netherlands.

Concerning calcium carbonate polymorphs calcite and aragonite, the experiments of Romanek et al. (1992) at low T (10 to 40 °C) seem to indicate that the enrichment factor of ^{13}C in the solid phase for $CaCO_3$-CO_2 partitioning is reduced by about two units in the aragonite polymorph with respect to the calcite counterpart, T being equal (e.g., at $T = 10$ °C, $1000 \ln \alpha \approx 13$ for calcite-CO_2 and 11 for aragonite-CO_2; at $T = 40$ °C, $1000 \ln \alpha \approx 9$ for calcite-CO_2 and 7 for aragonite-CO_2).

Concerning isotopic equilibria between solute carbonates and gaseous CO_2, Deines et al. (1974), on the basis of all the existing data, derived linear equations relating the fractionation factor to temperature:

$$1000 \ln \alpha_{H_2CO_3^*-CO_{2(g)}} = \frac{6.3 \times 10^3}{T^2} - 0.91 \qquad (11.171.1)$$

$$1000 \ln \alpha_{HCO_3^--CO_{2(g)}} = \frac{1.099 \times 10^6}{T^2} - 4.54 \qquad (11.171.2)$$

$$1000 \ln \alpha_{CO_3^{2-}-CO_{2(g)}} = \frac{8.7 \times 10^5}{T^2} - 3.4. \qquad (11.171.3)$$

Equations 11.171.1 to 11.171.3 are, however, of limited practical application because they demand precise knowledge of the state of speciation of carbonates in aqueous solution during solid phase condensation (or late exchanges). The fact that different carbonate solute species distinctly fractionate ^{13}C is masterfully outlined by the experiments of Romanek et al. (1992), which indicate a marked control by solution pH of the fractionation between total dissolved inorganic carbon (DIC) and gaseous CO_2 (figure 11.38).

With decreasing pH, the proportion of aqueous CO_2 in solution (i.e., $H_2CO_3^*$ if one does not distinguish between simple planar O=C=O and trigonal solvated form; cf. section 8.10) increases drastically, causing a marked decrease in $^{13}\delta_{DIC}$.

11.8.4 Sulfur

Sulfur has four stable isotopes: ^{32}S (0.9502), ^{33}S (0.0075), ^{34}S (0.0421), and ^{36}S (0.0002). The $^{34}S/^{32}S$ sulfur isotope fractionation in sulfide minerals and solute sulfur species is a sensitive function of temperature (Sakai, 1968; Ohmoto, 1972). Detailed vibrational calculations of the partition function ratio and corresponding isotopic fractionation factors for solute sulfur species were first attempted by Sakai (1968). The results of his calculations are shown in figure 11.39 in terms of the $\Delta^{34}S$ enrichment factor:

$$\Delta_{AB} \cong 1000 \ln a_{AB} \cong \delta^{34}S_A - \delta^{34}S_B \qquad (11.172)$$

with H_2S as reference phase B.

Figure 11.39 makes evident the marked role played by the oxidation state of

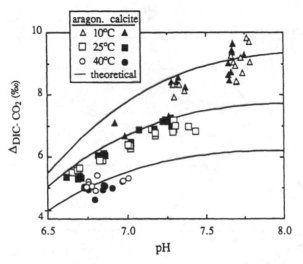

Figure 11.38 Relationship between average ^{13}C enrichment factor and pH in solute carbonates–gaseous CO_2 equilibria ($\Delta_{DIC-CO_2} = {}^{13}\delta_{DIC-CO_2}$). Reprinted from C. S. Romanek, E. L. Grossman, and J. W. Morse, *Geochimica et Cosmochimica Acta,* 56, 419–430, copyright © 1992, with kind permission from Elsevier Science Ltd., The Boulevard, Langford Lane, Kidlington 0X5 1GB, UK.

the complex (i.e., strong positive relative enrichment factors for sulfate ion SO_4^{2-} with respect to H_2S and negative enrichment factors for sulfide ion S^{2-}; see later in this section). Moreover, it is also evident that a simple inverse relation with the squared temperature holds.

Because the fractionation effect between condensed sulfates and SO_4^{2-} ions is negligible, it can be assumed that all sulfate complexes in solution (e.g., HSO_4^-, KSO_4^-, $NaSO_4^-$, . . .) have similar $\Delta^{34}S$ enrichment factors, T being equal (Ohmoto, 1972).

Sulfur isotope relative fractionation factors between various minerals and pyrite are shown in figure 11.40. As we can see, the isotopic compositions of sulfates reflect the relative fractionation effects induced by the SO_4^{2-} groups (compare figures 11.40 and 11.39).

Ohmoto and Rye (1979), through a critical examination of all the existing data, proposed a set of fractionation equations relating the fractionation factors between important ore-forming sulfides and $H_2S_{(aq)}$ to T of formation. The proposed equations obey a simple linear dependence over the inverse squared temperature (K), passing through the origin:

$$10^3 \ln \alpha_{i-H_2S_{(aq)}} = \frac{A}{T^2}. \tag{11.173}$$

The temperature of equilibrium can be readily derived by applying

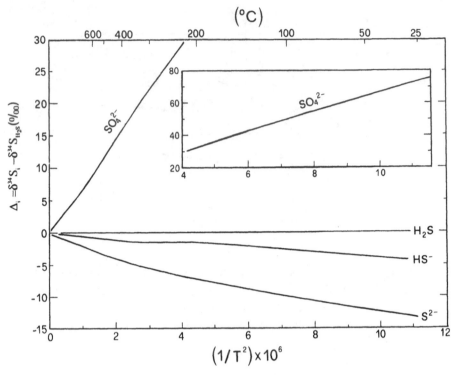

Figure 11.39 $\Delta^{34}S$ enrichment factors for solute sulfur species based on the data of Sakai (1968). Reproduced with modifications from Ohmoto, *Economic Geology,* 1972, Vol. 67, p. 533.

$$T(K) = \sqrt{\frac{A}{10^3 \ln \alpha_{i-H_2S_{(aq)}}}}. \tag{11.174}$$

The slope coefficients A for the various sulfides are listed in table 11.16.

Coupling thermometric equations such as equation 11.174 for different minerals, one obtains solid-solid fractionation expressions of the type

$$T_{AB}(K) = \sqrt{\frac{(A_A - A_B)}{10^3 \ln \alpha_{AB}}}. \tag{11.175}$$

For example, for the pyrite-galena couple, we have

$$T_{\text{pyrite-galena}}(K) = \sqrt{\frac{0.40 + 0.63}{10^3 \ln \alpha_{AB}}}. \tag{11.176}$$

Uncertainties associated with T estimates based on equation 11.176 range from ± 20 to 25 K (pyrite-galena) to ± 40 to 55 K (pyrite-pyrrhotite).

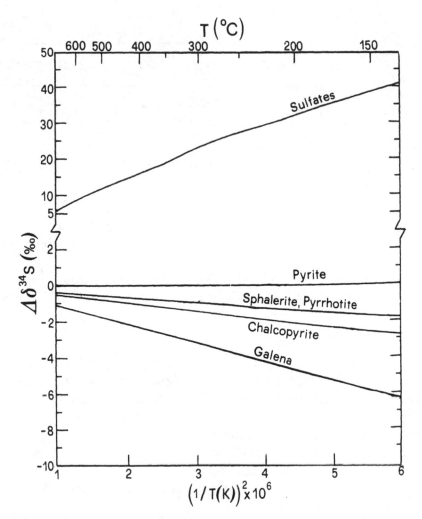

Figure 11.40 $\Delta^{34}S$ relative enrichment factors for various sulfur minerals with respect to pyrite. Reproduced with modifications from Rye and Ohmoto, *Economic Geology*, 1974, Vol. 69, p. 828.

Table 11.16 Slope coefficients relating sulfur isotopic fractionation factors between sulfides and aqueous H_2S to temperature (K) (after Ohmoto and Rye, 1979).

Phase	Composition	Slope Coefficient A
Molybdenite	MoS_2	0.45
Pyrite	FeS_2	0.40
Sphalerite	ZnS	0.10
Pyrrhotite	FeS	0.10
Chalcopyrite	$CuFeS_2$	−0.05
Covellite	CuS	−0.40
Galena	PbS	−0.63
Chalcosite	Cu_2S	−0.75
Argentite	Ag_2S	−0.80

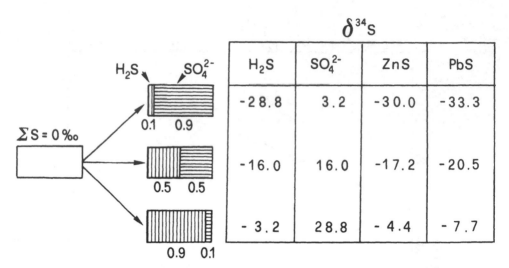

$$\delta^{34}S$$

	H$_2$S	SO$_4^{2-}$	ZnS	PbS
	-28.8	3.2	-30.0	-33.3
	-16.0	16.0	-17.2	-20.5
	-3.2	28.8	-4.4	-7.7

Figure 11.41 Variation of $\delta^{34}S$ of sulfate, $H_2S_{(aq)}$, and sulfide minerals in response to H_2S/SO_4^{2-} variations in the hydrothermal solution at $T = 200\,°C$ and $\delta^{34}S_{\Sigma S} = 0‰$. Reproduced with modifications from Rye and Ohmoto, *Economic Geology,* 1974, Vol. 69, p. 828.

The large isotopic fractionations observed between oxidized and reduced forms of sulfur compounds and aqueous complexes require accurate appraisal of the effective cogeneticity of sulfur minerals utilized as geothermometric couples, and of their equilibrium condition, to avoid erroneous deductions. In fact, besides temperature, the isotopic composition of sulfur minerals is also affected by the bulk isotopic composition of the sulfur in the system ($\delta^{34}S_{\Sigma S}$) (which is controlled by the source of sulfur) and by the proportion of oxidized and reduced sulfur species in solution (which is controlled by Eh and pH conditions and total sulfur in the system; see section 8.15).

Figure 11.41 shows, for example, the effect of oxidation state on the isotopic compositions of sulfur minerals at $T = 200\,°C$ assuming $\delta^{34}S_{\Sigma S} = 0‰$. At $T = 200\,°C$, the sulfur isotopic fractionation between SO_4^{2-} and H_2S is about 32‰. If the sulfur in the ore fluid were distributed between $H_2S_{(aq)}$ and SO_4^{2-} and the H_2S/SO_4^{2-} ratio were changed from 1:9 to 9:1, in response to a decrease in f_{O_2} sulfide minerals such as sphalerite and galena eventually precipitating from such fluids would attain the different $\delta^{34}S$ values shown in figure 11.41. At high f_{O_2}, the dominant sulfate ion concentrates ^{34}S, and H_2S is relatively enriched in ^{32}S, and hence the precipitating sulfides have low $\delta^{34}S$ values. The opposite is true at low f_{O_2}, where the relative predominance of sulfides causes the precipitating ores to be relatively richer in ^{34}S. A similar effect is induced by pH: because, with T held constant, HS^- and S^{2-} ions are depleted in ^{34}S with respect to H_2S, and because the aqueous speciation of sulfide ions is controlled by

$$H_2S_{(aq)} \Leftrightarrow H^+ + HS^- \tag{11.177}$$

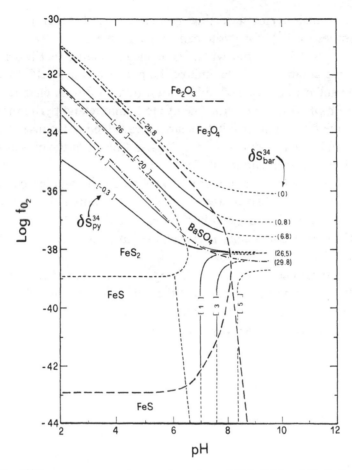

Figure 11.42 $\delta^{34}S$ contours superimposed on the stability fields of Fe-S-O minerals and barite at $T = 250\,°C$ and $\delta^{34}S_{\Sigma S} = 0‰$ Solid lines and their dashed extensions: $\delta^{34}S$ contours for pyrite (values in brackets) and barite (values in parentheses). Dashed lines: Fe-S-O mineral boundaries for a 10^{-1} molal concentration of sulfur in the system. Small dashed lines: Fe-S-O mineral boundaries for a 10^{-3} molal concentration of sulfur in the system. Dash-and-dot line: solubility product of barite. Reproduced from Ohmoto, *Economic Geology,* 1972, Vol. 72, p. 555.

$$HS^- \Leftrightarrow H^+ + S^{2-}, \qquad\qquad (11.178)$$

the increase in pH favors the formation of the relatively ^{34}S-depleted ions HS^- and S^{2-}, and hence the precipitating sulfides attain higher $\delta^{34}S$ values, T and Eh being equal.

The combined effects of f_{O_2} and pH variation on the sulfur isotopic compositions of ore-forming minerals are shown in figure 11.42 for a temperature of 250 °C and $\delta^{34}S_{\Sigma S} = 0‰$. The dashed lines delineate the stability limits of condensed phases in the Fe-S-O system for two different total sulfur concentrations (10^{-1} and 10^{-2} molal). On these lines are superimposed $\delta^{34}S$ contours (solid lines and their dashed extensions) for coexisting pyrite (values in brackets) and barite (values in parentheses). $\delta^{34}S$ for pyrite can range from $+5$ to -27 per mil and for barite from 0 to $+32$ per mil within geologically plausible Eh-pH ranges. In acidic-reducing conditions at the limit of the pyrite stability field, the $\delta^{34}S$ value of sulfide is similar to $\delta^{34}S_{\Sigma S}$ and rather insensitive to Eh-pH changes. Nevertheless, at high Eh-pH conditions, near the pyrite-hematite and pyrite-magnetite stability limits, the $\delta^{34}S$ value of sulfide is largely variable in response to limited Eh-pH changes and much dissimilar from $\delta^{34}S_{\Sigma S}$ as a result of the presence of sulfate ion in solution, which fractionates ^{34}S.

Figure 11.42 elucidates in a rather clear fashion the marked control exerted by the chemistry of the fluid on the sulfur isotopic compositions of ore-forming minerals. At the T and $\delta^{34}S_{\Sigma S}$ conditions delineated in the figure, an increase in f_{O_2} or in pH of one log unit can cause decreases in $\delta^{34}S$ values of sulfide minerals by as much as 20‰.

Although these chemical effects are important in deciphering the genesis of ore minerals, it must be emphasized that the differences in the $\delta^{34}S$ values among coexisting condensed phases (hence the fractionation factor) at equilibrium are constant in each case because they depend only on T.

APPENDIX ONE

Constants, Units, and Conversion Factors

The International Union for Pure and Applied Chemistry (IUPAC) recommends the adoption in scientific publications of the SI system notation. Tables A1.1 and A1.2 present the base units and some derived units.

In geochemical literature, one often encounters non-SI units (e.g., bar, atmosphere, calorie, angstrom, electron-volt, etc.), which can be converted to SI units by the use of simple multiplicative factors, as shown in table A1.3.

Table A1.1 SI Base Units

Magnitude	Unit	Symbol
Length	meter	m
Mass	kilogram	kg
Time	second	s
Electric current	ampere	A
Thermodynamic temperature	kelvin	K
Amount of substance	mole	mol, mole
Intensity of light	candle	cd

Table A1.2 SI Derived Units

Magnitude	Unit	Symbol	SI Derivation
Energy	joule	J	$m^2 \cdot kg \cdot s^{-2}$
Force	newton	N	$m \cdot kg \cdot s^{-2}$
Power	watt	W	$J \cdot s^{-1}$
Pressure	pascal	Pa	$m^{-1} \cdot kg \cdot s^{-2}$
Electric charge	coulomb	C	$A \cdot s$
Electric potential	volt	V	$m^2 \cdot kg \cdot s^{-3} \cdot A^{-1}$
Capacitance	farad	F	$m^{-2} \cdot kg^{-1} \cdot s^4 \cdot A^2$
Resistance	ohm	Ω	$m^2 \cdot kg \cdot s^{-3} \cdot A^{-2}$
Magnetic flux	weber	Wb	$m^2 \cdot kg \cdot s^{-2} \cdot A^{-1}$
Magnetic flux density	tesla	T	$kg \cdot s^{-2} \cdot A^{-1}$
Frequency	hertz	Hz	s^{-1}

Table A1.3 Non-SI units and their factors for conversion to the SI system

Magnitude	Unit	Symbol	SI Conversion
Length	angstrom	Å	10^{-10} m
Mass	ton	t	10^{-3} kg
Pressure	bar	bar	10^5 Pa
Pressure	atmosphere	atm	101,325 Pa
Energy	calorie	cal	4.184 J
Energy	electron-volt	eV	1.602189×10^{-19} J
Kinematic viscosity	stokes	st	10^{-4} m^{2}s^{-1}
Dynamic viscosity	poise	p	10^{-1} Pa·s

Of particular practical interest are the general energy conversion factors (Lewis and Randall, 1970) presented in table A1.4.

As we have already seen, the "mole" is the mass unit in the SI system. The molar concentrations of components in geochemical systems assume different significances as a function of the adopted reference unit:

Mole = amount of substance, expressed in weight (g), containing Avogadro's number of molecules of the substance of interest

Molality = moles of solute per kg of solvent

Molarity = moles of solute per 10^{-3} m^3 of solution

Formality = moles of solute per kg of solution

Molar fraction = moles of component of interest divided by the sum of moles of all components in the phase.

We recall that chemical analyses generally report the composition of a given phase as "weight percent (%) of constituting oxides," or as parts per mil in weight (‰) or parts per million in weight (ppm) or parts per billion in weight (ppb). To switch from weight percent to ppm, one simply multiplies the observed value by 10^4. Analogously, to switch from weight percent to ppb, the value must be multiplied by 10^7.

Finally, table A1.5 lists the principal physical constants, and table A1.6 presents some energy equivalents of atom units.

Table A1.4 Energy conversion factors

	cal·mole^{-1}	joule·mole^{-1}	cm^3·atm·mole^{-1}	cm^{-1}	eV	erg·molecule^{-1}
1 cal·mole^{-1}	1	4.1840	41.292	0.34974	4.3361×10^{-5}	6.9465×10^{-17}
1 joule·mole^{-1}	0.23901	1	9.8692	0.08359	1.0364×10^{-5}	1.6602×10^{-17}
1 cm^3·atm·mole^{-1}	0.024218	0.10133	1	8.470×10^{-3}	1.0501×10^{-6}	1.6823×10^{-18}
1 cm^{-1}	2.8593	11.963	118.07	1	1.2398×10^{-4}	1.0862×10^{-16}
1 eV	23.062	96.493	9.523×10^5	8065.7	1	1.6020×10^{-12}
1 erg·molecule^{-1}	1.4396×10^{16}	6.023×10^{16}	5.944×10^{17}	5.0348×10^{15}	6.2422×10^{11}	1

Table A1-5 Main physical constants

Constant	Symbol	SI Value	Non-SI Value
Gas constant	R	8.31451 J·K^{-1}·mole^{-1}	1.98726 cal·K^{-1}·mole^{-1}
Faraday's constant	F	96485.3 J·V^{-1}	23062.3 cal·V^{-1}
Avogadro's number	N_0	6.022137×10^{23} mole^{-1}	-
Boltzmann's constant	k	1.380658×10^{-23} J·K^{-1}	1.380658×10^{-16} erg·K^{-1}
Planck's constant	h	6.626075×10^{-34} J·s	6.626075×10^{-27} erg·s
Planck's constant ($\hbar = h/2\pi$)	$\hbar$	$1.0545727 \times 10^{-34}$ J·s	$1.0545727 \times 10^{-27}$ erg·s
Elementary charge	e	$1.6021773 \times 10^{-19}$ J	-
Atomic mass unit (amu)	u	$1.6605402 \times 10^{-27}$ kg	-
Electron rest mass	m_e	$5.4657990 \times 10^{-4}\,u$	-
Proton rest mass	m_P	$1.007276470\,u$	-
Neutron rest mass	m_n	$1.008664904\,u$	-
Bohr's radius	a_0	$0.52917725 \times 10^{-10}$ m	-
Bohr's magneton	μ_B	9.274015×10^{-24} J·kg^{-1}·s^2·A	-
Rydberg's constant	Ry	1.097373153×10^7 m^{-1}	-
Vacuum permittivity	ε_0	$8.85418782 \times 10^{-12}$ F·m^{-1}	-
Speed of light in vacuum	c	2.99792458×10^8 m·s^{-1}	-

Table A1.6 Energy conversions of atom units

Atom Unit	Energy Equivalent
Atomic mass unit (amu)	931.4943 MeV
Electron rest mass	0.5109991 MeV
Proton rest mass	938.2723 MeV
Neutron rest mass	939.5656 MeV
Temperature corresponding to 1 eV	1.160450×10^4 K
Photon wavelength associated to 1 eV	1.239852×10^{-6} m

APPENDIX TWO

Review of Mathematics

In the development of the various treated arguments, we adopted infinitesimal calculation procedures, vectorial notations (gradient, Laplacian), and some mathematical functions (Lagrange equations, Hamiltonians, error functions) that should already be known to the reader but that perhaps would be useful to review in a concise way.

A2.1   Exact Differentials and State Functions

Let F be a function of the independent variables:

$$F = F(x_1, x_2, \ldots, x_n) \tag{A2.1}$$

and let $f_1, f_2, \ldots f_n$ be functions of the single independent variables $x_1, x_2, \ldots ,x_n$, such that

$$f_1 dx_1 + f_2 dx_2 + \ldots + f_n dx_n = \sum_i f_i dx_i. \tag{A2.2}$$

Whenever the differential equation A2.2 can be equated with the differential of function F:

$$df = \left(\frac{\partial F}{\partial x_1}\right)_{x_2,\ldots x_n} dx_1 + \left(\frac{\partial F}{\partial x_2}\right)_{x_1,\ldots x_n} dx_2 + \cdots \left(\frac{\partial F}{\partial x_n}\right)_{x_1,x_2,\ldots} dx_n, \tag{A2.3}$$

then dF is an exact differential and

$$f_1 = \left(\frac{\partial F}{\partial x_1}\right)_{x_2,\ldots x_n} ; \quad f_2 = \left(\frac{\partial F}{\partial x_2}\right)_{x_1,\ldots x_n} ; \quad f_n = \left(\frac{\partial F}{\partial x_n}\right)_{x_1,x_2,\ldots} . \tag{A2.4}$$

The state functions presented in section 2.4 are examples of exact differential functions of the type exemplified by equation A2.3. Generally, reversible transformations in physical systems are described by equations of this type.

For two generic independent variables x_g and x_h, such that

$$f_g = \left(\frac{\partial F}{\partial x_g} \right)_{x_1,\ldots}$$

(A2.5)

and

$$f_h = \left(\frac{\partial F}{\partial x_h} \right)_{x_1,\ldots}$$

(A2.6)

we have

$$\left(\frac{\partial f_g}{\partial x_h} \right)_{x_1,\ldots} = \frac{\partial^2 F}{\partial x_g \partial x_h}$$

(A2.7)

and

$$\left(\frac{\partial f_h}{\partial x_g} \right)_{x_1,\ldots} = \frac{\partial^2 F}{\partial x_h \partial x_g}.$$

(A2.8)

Applying Schwarz's theorem on sequential differentiation of mixed derivatives:

$$\frac{\partial^2 F}{\partial x \partial y} = \frac{\partial}{\partial x} \left(\frac{\partial F}{\partial y} \right) = \frac{\partial}{\partial y} \left(\frac{\partial F}{\partial x} \right) = \frac{\partial^2 F}{\partial y \partial x},$$

(A2.9)

we obtain

$$\left(\frac{\partial f_g}{\partial x_h} \right)_{x_1,\ldots} = \left(\frac{\partial f_h}{\partial x_g} \right)_{x_1,\ldots}.$$

(A2.10)

Equation A2.10 represents a necessary and sufficient condition for equation A2.2 to be an exact differential and is valid for any couple of conjugate variables (f_g, x_g) and (f_h, x_h).

A2.2   Review of Implicit Functions

Let $f = f(x,y,z)$ be a function of three variables, such that

$$f(x, y, z) = 0. \tag{A2.11}$$

The function

$$z = z(x, y) \tag{A2.12}$$

is said to be an "implicitly defined function" of equation A2.11, because it is the solution in z of such an equation. Differentiating equation A2.11 produces

$$\frac{\partial f}{\partial x} dx + \frac{\partial f}{\partial y} dy + \frac{\partial f}{\partial z} dz = 0 \tag{A2.13}$$

and, from equation A2.13, keeping constant one of the three variables and assuming the partial derivatives of f to be non-null, we obtain

$$\left(\frac{\partial x}{\partial y}\right)_z = 1 \bigg/ \left(\frac{\partial y}{\partial x}\right)_z = -\left(\frac{\partial f}{\partial y}\right) \bigg/ \left(\frac{\partial f}{\partial x}\right) \tag{A2.14.1}$$

$$\left(\frac{\partial z}{\partial y}\right)_x = 1 \bigg/ \left(\frac{\partial y}{\partial z}\right)_x = -\left(\frac{\partial f}{\partial y}\right) \bigg/ \left(\frac{\partial f}{\partial z}\right) \tag{A2.14.2}$$

$$\left(\frac{\partial z}{\partial x}\right)_y = 1 \bigg/ \left(\frac{\partial x}{\partial z}\right)_y = -\left(\frac{\partial f}{\partial x}\right) \bigg/ \left(\frac{\partial f}{\partial z}\right). \tag{A2.14.3}$$

From equations A2.14.1 to A2.14.3, further relationships among partial derivatives can be obtained, such as

$$\left(\frac{\partial x}{\partial y}\right)_z \cdot \left(\frac{\partial y}{\partial z}\right)_x \cdot \left(\frac{\partial z}{\partial x}\right)_y = -1 \tag{A2.15}$$

or

$$-\left(\frac{\partial x}{\partial y}\right)_z = \left(\frac{\partial z}{\partial y}\right)_x \bigg/ \left(\frac{\partial z}{\partial x}\right)_y. \tag{A2.16}$$

Let us now consider the exact differential:

$$dx = \left(\frac{\partial x}{\partial y}\right)_z dy + \left(\frac{\partial x}{\partial z}\right)_y dz \qquad (A2.17)$$

and a further function of the two variables x and y, $\varphi(x,y)$, such that

$$d\varphi = \left(\frac{\partial \varphi}{\partial x}\right)_y dx + \left(\frac{\partial \varphi}{\partial y}\right)_x dy. \qquad (A2.18)$$

If we let x and y vary while holding z constant, we get, from the partial derivative in dx of equation A2.18,

$$\left(\frac{\partial \varphi}{\partial x}\right)_z = \left(\frac{\partial \varphi}{\partial x}\right)_y + \left(\frac{\partial \varphi}{\partial y}\right)_x \left(\frac{\partial y}{\partial x}\right)_z. \qquad (A2.19)$$

Equation A2.19 is one of the possible ways to switch from a couple of variables to another conjugate couple. Most transforms adopted in the thermodynamics of systems with two independent variables are based on equation A2.19 and its modifications (Legendre transforms, cf. section 2.4).

A2.3   Integration on Exact Differentials

For a function F:

$$F = F\left(x_1, x_2, \cdots, x_n\right) \qquad (A2.20)$$

such that dF is its exact differential:

$$dF = f_1 dx_1 + f_2 dx_2 + \cdots + f_n dx_n, \qquad (A2.21)$$

where

$$f_1 = \left(\frac{\partial F}{\partial x_1}\right)_{x_2,\dots x_n},$$

etc. (cf. eq. A2.4), we have

$$\oint \left(f_1 dx_1 + f_2 dx_2 + \cdots + f_n dx_n\right) = 0 \qquad (A2.22.2)$$

—i.e., the cyclic integral of such a function is zero. Moreover, the value of a generic contour integral between points A and B in a field of existence described by variables $x_1, x_2, \ldots, x_n$ does not depend on the path followed in integration. That is, if

$$F = F(x_1, x_2) \qquad (A\,2.23)$$

defined in a field x_1, x_2 is such that

$$dF = \left(\frac{\partial F}{\partial x_1}\right)_{x_2} dx_1 + \left(\frac{\partial F}{\partial x_2}\right)_{x_1} dx_2, \qquad (A2.24)$$

we have

$$\int_A^B dF = \int_A^B \left(\frac{\partial F}{\partial x_1}\right)_{x_2} dx_1 + \left(\frac{\partial F}{\partial x_2}\right)_{x_1} dx_2$$

$$= \int_A^B f_1 \, dx_1 + f_2 \, dx_2 = F_{(B)} - F_{(A)} \qquad (A2.25)$$

independent of the followed integration path.

A2.4   Operations on Vectors

A vector is defined by length and direction. The length V of a generic vector $\mathbf{V}$ is given by its modulus:

$$|\mathbf{V}| = V. \qquad (A\,2.26)$$

With V being the modulus of vector $\mathbf{V}$, its unit vector U is such that

$$\mathbf{V} = V\mathbf{U}. \qquad (A\,2.27)$$

The inner product $\mathbf{a} \cdot \mathbf{b}$ of vectors $\mathbf{a}$ and $\mathbf{b}$ is the inner quantity

$$\mathbf{a} \cdot \mathbf{b} = ab \cos\alpha_{ab}, \qquad (A\,2.28)$$

where a and b are the moduli of the respective vectors and α_{ab} is the minimum angle between the representatives of the unit vectors.

The vectorial product $\mathbf{a} \times \mathbf{b}$ of vectors $\mathbf{a}$ and $\mathbf{b}$ is a vector $\mathbf{c}$ whose modulus c is

$$c = ab \sin \alpha_{ab}. \tag{A 2.29}$$

The direction of $\mathbf{c}$ is orthogonal to $\mathbf{a}$ and $\mathbf{b}$ and depends on the rotation required for a to overlap on $\mathbf{b}$ (the "right-hand rule").

By decomposing the vector into its directional components, it can be shown that the modulus of the vectorial product of two vectors $\mathbf{a}$ and $\mathbf{b}$ is given by the determinant

$$\mathbf{a} \times \mathbf{b} = \begin{vmatrix} \mathbf{U}_x & \mathbf{U}_y & \mathbf{U}_z \\ a_x & a_y & a_z \\ b_x & b_y & b_z \end{vmatrix} \tag{A2.30}$$

and the mixed product of three vectors $\mathbf{a}$, $\mathbf{b}$, and $\mathbf{c}$ is given by the determinant

$$\mathbf{a} \times \mathbf{b} \cdot \mathbf{c} = \begin{vmatrix} a_x & a_y & a_z \\ b_x & b_y & b_z \\ c_x & c_y & c_z \end{vmatrix}. \tag{A2.31}$$

The vectorial operator "gradient" (with symbol nabla ∇) allows the passage from scalar to vectorial fields. For scalar Φ the vector $\nabla\Phi$ (gradient of Φ) is given by

$$\nabla\Phi = \frac{\partial\Phi}{\partial x}, \frac{\partial\Phi}{\partial y}, \frac{\partial\Phi}{\partial z}. \tag{A2.32}$$

In section 1.19, for example, $\nabla_i\Phi_{e,j}$ was the vectorial component of the electrical potential of the fth dipole $\Phi_{e,j}$ directed toward i.

The differential operator "divergence" allows the passage from vectorial to scalar fields. For a vector φ,

$$\text{div } \varphi = \nabla\varphi = \frac{\partial\varphi}{\partial x}, \frac{\partial\varphi}{\partial y}, \frac{\partial\varphi}{\partial z}. \tag{A2.33}$$

The "curl" of a vector function (or "rotation" with symbol "rot") is a vector that is formally the cross product of the operator and the vector. For a vector $\mathbf{V}$,

$$\text{rot } \mathbf{V} = \nabla \times \mathbf{V} = \begin{vmatrix} \mathbf{U}_x & \mathbf{U}_y & \mathbf{U}_z \\ \dfrac{\partial}{\partial x} & \dfrac{\partial}{\partial y} & \dfrac{\partial}{\partial z} \\ v_x & v_y & v_z \end{vmatrix}. \tag{A2.34}$$

The Laplacian operator ∇^2 performs the differential operation

$$\nabla^2 = \frac{\partial^2}{\partial x^2} + \frac{\partial^2}{\partial y^2} + \frac{\partial^2}{\partial z^2} \tag{A2.35}$$

and is applied to either scalars or vectors.

In sections 1.1.2, 1.19, and 3.1, we have seen applications of the Laplacian operator to the vectorial field.

A2.5   Lagrange Equations, Hamiltonians

The bases of classical mechanics are established by Newton's laws. In classical mechanics, the dynamic behavior of a set of particles can be determined, on the basis of Newton's laws, if one knows the initial Cartesian coordinates of each particle (X_l) at time t_0 and their velocities (dX_l/dt).

In quantum mechanics, the spatial variables are constituted by "generalized coordinates" (q_i), which replace the individual Cartesian coordinates of all single particles in the set. The Lagrangian equations of motion are the Newtonian equations transposed to the generalized coordinate system.

Let us consider a particle of mass m located at a point p and subjected to a force $\overline{F}$ given by the gradient of potential Φ in p. Denoting X_l, where $l = 1, 2$ and 3, the Cartesian coordinates in p, and denoting q_j the corresponding Lagrangian coordinates, the two systems of spatial coordinates are related by:

$$X_l = \sum_j \frac{\partial X_l}{\partial q_j} \, dq_j. \tag{A2.36}$$

For an infinitesimal displacement dq we have

$$dp = \sum_j \frac{\partial p}{\partial q_j} \, dq_j. \tag{A2.37}$$

Velocity $\mathbf{V}$, which is the derivative of displacement on time, is expressed as

$$\mathbf{V} = \frac{dp}{dt} = \sum_j \frac{\partial p}{\partial q_j} \cdot \frac{\partial q_j}{\partial t} + \frac{\partial p}{\partial t}. \tag{A2.38}$$

If constraints are fixed, the term $\partial p / \partial t$ is zero and we have

$$\frac{\partial \mathbf{V}}{\partial \dot{q}_j} = \frac{\partial p}{\partial q_j}, \quad where \ \partial \dot{q}_j = \frac{\partial q_j}{\partial t}. \tag{A2.39}$$

Because acceleration **a** is the partial derivative of velocity on time, through some substitutions, and posing

$$\mathbf{T} = \frac{1}{2} m \mathbf{V}^2, \tag{A2.40}$$

we obtain

$$m\mathbf{a} \, dp = \sum_j \left(\frac{d}{dt} \frac{\partial \mathbf{T}}{\partial \dot{q}_j} - \frac{\partial \mathbf{T}}{\partial q_j} \right) dq_j. \tag{A2.41}$$

Because to a displacement dp corresponds a potential difference $-d\Phi$ according to

$$\mathbf{F} \, dp = -d\Phi \tag{A2.42}$$

we also have

$$\mathbf{F} \, dp = -\frac{\partial \Phi}{\partial q_j} \, dq_j. \tag{A2.43}$$

Let us consider now Newton's second law for conservative systems*:

$$\sum_j \left(\mathbf{F}_j - m\mathbf{a}_j \right) dp_j = 0 \tag{A2.44}$$

Substituting equations A2.41 and A2.43 into A2.44, we get

---

* If the work exerted by a force field on a particle moving from point A to point B is independent of the path followed by the particle between the two points, the force field is said to be "conservative."

$$\sum_j \left(-\frac{\partial \Phi}{\partial q_j} - \frac{d}{dt}\frac{\partial T}{\partial \dot q_j} + \frac{\partial T}{\partial q_j} \right) dq_j = 0, \qquad \text{(A2.45)}$$

from which it follows that

$$\frac{d}{dt}\frac{\partial T}{\partial \dot q_j} - \frac{\partial T}{\partial q_j} = -\frac{\partial \Phi}{\partial q_j}. \qquad \text{(A2.46)}$$

Posing

$$L = T - \Phi, \qquad \text{(A 2.47)}$$

equation A2.46 takes the form

$$\frac{d}{dt}\frac{\partial L}{\partial \dot q_j} - \frac{\partial L}{\partial q_j} = 0. \qquad \text{(A2.48)}$$

This sort of equation, known as a "Lagrange equation of motion," is generally valid for systems with unconstrained degrees of freedom. If a system has n degrees of freedom, n second-order Lagrange equations will exist (functions of $\dot q_j$, q_j, and time: quadratic in $\dot q_j$).

We have seen in section 3.1 the application of the concept of the harmonic oscillator in the interpretation of vibrational properties of crystals. For a unidimensional harmonic oscillator, there is a single degree of freedom:

$$q = x; \quad \dot x = \frac{dx}{dt} = V; \quad \ddot x = \frac{d^2 x}{dt^2} = a \qquad \text{(A2.49.1)}$$

$$T = \frac{1}{2}m\dot x^2; \quad \Phi = \frac{K_F}{2}x^2 \qquad \text{(A2.49.2)}$$

$$L = \frac{1}{2}m\dot x^2 - \frac{K_F}{2}x^2. \qquad \text{(A2.49.3)}$$

The corresponding Lagrange equation is

$$\frac{d}{dt}\frac{\partial L}{\partial x} - \frac{\partial L}{\partial x} = 0 \qquad \text{(A2.50.1)}$$

$$\frac{d}{dt}m\dot x + K_F x = 0 \qquad \text{(A2.50.2)}$$

$$m\ddot{x} + K_F\,x = m\mathbf{a} + K_F\,x = 0. \qquad\qquad \text{(A 2.50.3)}$$

The Hamiltonian operator is

$$H = \sum_j \frac{\partial L}{\partial \dot{q}_j}\, d\dot{q}_j - L. \qquad\qquad \text{(A2.51)}$$

From the definition

$$L = T - \Phi, \qquad\qquad \text{(A 2.52)}$$

the Hamiltonian becomes

$$H = \sum_j \frac{\partial T}{\partial \dot{q}_j}\, d\dot{q}_j - T + \Phi, \qquad\qquad \text{(A2.53)}$$

but, because

$$\sum_j \frac{\partial T}{\partial \dot{q}_j}\, d\dot{q}_j = 2T, \qquad\qquad \text{(A2.54)}$$

we obtain

$$H = T + \Phi. \qquad\qquad \text{(A 2.55)}$$

As we have seen in sections 1.17 and 3.1, the quantum mechanics Hamiltonian in the Schrödinger equation has the form

$$H = -\frac{h^2}{8\pi^2 m}\,\nabla^2 + \Phi. \qquad\qquad \text{(A2.56)}$$

In the unidimensional form, the Laplacian operator ∇^2 is replaced by the simple second derivative in x— i.e. $(\partial^2\psi_{(x)}/\partial x^2)$.

A2.6   Error Function

The error function erf x and its complementary function erfc x appear in the solution of differential equations describing diffusive processes (see, for instance, section 11.6.3). They are defined by the integrals

$$\text{erf } x = \frac{2}{\sqrt{\pi}} \int_0^x \exp\!\left(-y^2\right) dy \qquad\qquad \text{(A2.57)}$$

and

$$\text{erfc } x = \frac{2}{\sqrt{\pi}} \int_x^\infty \exp\!\left(-y^2\right) dy. \tag{A2.58}$$

The limits of error function erf x are

$$\text{erf}(-\infty) = -1; \quad \text{erf } 0 = 0; \quad \text{erf } \infty = 1 \tag{A 2.59}$$

Because

$$\text{erfc } x = 1 - \text{erf } x, \tag{A 2.60}$$

the limits of erfc x are

$$\text{erfc}(-\infty) = 2; \quad \text{erfc } 0 = 1; \quad \text{erfc } \infty = 0. \tag{A 2.61}$$

Values of functions erf x and erfc x are tabulated (cf. Gautschi, 1964) but can also be rapidly obtained by the polynomials

$$\text{erfc } x = \left(1 + a_1 x + a_2 x^2 + a_3 x^3 + a_4 x^4 + a_5 x^5\right)^{-8} \tag{A 2.62}$$

(Gautschi, 1964), where $a_1 = 0.14112821$, $a_2 = 0.08864027$, $a_3 = 0.002743349$, $a_4 = -0.00039446$, $a_5 = 0.00328975$ for $0 \le x \le 3$, and

$$\text{erfc } x = \frac{\exp\!\left(-x^2\right)}{\sqrt{\pi}} \left(\frac{1}{x} - \frac{1}{2x^3} + \frac{3}{2^2 x^5} - \frac{15}{2^3 x^7} + \frac{105}{2^4 x^9} - \frac{945}{2^5 x^{11}} + \frac{10{,}365}{2^6 x^{13}} \right) \tag{A 2.63}$$

for $x \ge 3$. The function erf x is then obtained by applying equation A2.60.

A2.7 York's Algorithm for Isochrons

The least-squares treatment of the fitting of an isochron must account for the fact that the error in the y-coordinate of each point is correlated with the error in the x-coordinate.

With each point in the isochron being subjected to the requirement

$$y_i = a + bx_i \quad i = 1,\dots,n, \tag{A 2.64}$$

and denoting the observations as X_i and Y_i, the adjusted values of these observations as x_i and y_i, the weights of the various observations as $\omega(X_i)$ and $\omega(Y_i)$, and the correlations between the x and y errors as r_i, the best slope (i.e., "age") can be found by means of the equation

$$b = \frac{\sum_i Z_i^2 V_i \left[\dfrac{U_i}{\omega(Y_i)} + \dfrac{bV_i}{\omega(X_i)} - \dfrac{r_i V_i}{\alpha_i} \right]}{\sum_i Z_i^2 U_i \left[\dfrac{U_i}{\omega(Y_i)} + \dfrac{bV_i}{\omega(X_i)} - \dfrac{br_i U_i}{\alpha_i} \right]}, \tag{A2.65}$$

where

$$Z_i = \frac{\omega(X_i)\omega(Y_i)}{b^2 \omega(Y_i) + \omega(X_i) - 2br_i\alpha_i} \tag{A2.66}$$

$$U_i = X_i - \overline{X} \tag{A2.67}$$

$$V_i = Y_i - \overline{Y} \tag{A2.68}$$

$$\overline{X} = \frac{\sum_i Z_i X_i}{\sum_i Z_i} \tag{A2.69}$$

$$\overline{Y} = \frac{\sum_i Z_i Y_i}{\sum_i Z_i} \tag{A2.70}$$

$$\alpha_i = \sqrt{\omega(X_i)\omega(Y_i)}. \tag{A2.71}$$

Because U_i, V_i, and Z_i contain b, the solution of the system A2.65 $\rightarrow$ A2.71 is obtained by an iterative procedure, inserting an approximate value of b into these terms and calculating a new b from equation A2.65. The best intercept is found by use of the equation

$$a = \overline{Y} - b\overline{X}. \tag{A2.72}$$

REFERENCES

Ackermann T. and Schreiner F. (1958). Molwärmen und Entropien einiger Fettsäuren und ihrer Anionen in wässriger Lösung. *Zeits. Elektroch.*, 62:1143–1151.

Acree W. E. Jr. (1984). *Thermodynamic Properties of Nonelectrolyte Solutions.* New York: Academic Press.

Aggarwal K. G. (1967). Lattice dynamics of diamond. *Proc. Phys. Soc.*, 91:381–389.

Ahrens L. H. (1952). The use of ionization potentials, 1: Ionic radii of the elements. *Geochim. Cosmochim. Acta*, 2:155–169.

Ahrland S. (1973). Thermodynamics of the stepwise formation of metal-ion complexes in aqueous solution. *Struc. Bonding*, 15:167–188.

Ahrland S. (1975). Metal complexes present in sea water. In *The Nature of Sea Water*, E. D. Goldberg, ed. Dahlem Konferenzen.

Aikawa N., Kumazawa M., and Tokonami N. (1985). Temperature dependence of intersite distribution of Mg and Fe in olivine and the associate change of lattice parameters. *Phys. Chem. Minerals*, 12:1–8.

Akamatsu T., Kumazawa M., Aikawa N., and Takei H. (1993). Pressure effect on the divalent cation distribution in nonideal solid solution of forsterite and fayalite. *Phys. Chem. Minerals*, 19:431–444.

Akimoto S., Komada E., and Kushiro I. (1967). Effect of pressure on the melting of olivine and spinel polymorph of Fe_2SiO_4. *J. Geophys. Res.*, 72:679–689.

Akimoto S., Matsui S., and Syono S. (1976). High-pressure crystal chemistry of orthosilicates and the formation of the mantle transition zone. In *The Physics and Chemistry of Minerals and Rocks*, R. J. G. Strens, ed. New York: John Wiley.

Albarede F. (1976). Some trace element relationships among liquid and solid phases in the course of fractional crystallization of magmas. *Geochim. Cosmochim. Acta*, 40:667–673.

Albarede F. and Bottinga Y. (1972). Kinetic disequilibrium in trace element partitioning between phenocrysts and host lava. *Geochim. Cosmochim. Acta*, 36:141–156.

Allègre C. J., Treuil M., Minster J. F., Minster B., and Albarede F. (1977). Systematic use of trace element in igneous process, Part I: Fractional crystallization processes in volcanic suites. *Contrib. Mineral. Petrol.*, 60:57–75.

Allred G. C. and Woolley E. M. (1981). Heat capacities of aqueous acetic acid, sodium acetate, ammonia, and ammonium chloride at 283.15 K, 298.15 K, and 313.15 K: ΔC_P^0 for ionization of acetic acid and for dissociation of ammonium ion. *J. Soln. Chem.*, 14:549–560.

Althaus E. (1967). The triple point andalusite-sillimanite-kyanite. *Contrib. Mineral. Petrol.*, 13:31–50.

Alvarez L. W. (1938). The capture of orbital electrons by nuclei. *Phys. Rev.*, 54:486–497.

Anderson A. T. and Gottfried D. (1971). Contrasting behaviour of P, Ti and Nb in a differentiated high-alumina olivine-tholeite and a calc-alkaline suite. *Geol. Soc. Amer. Bull.*, 82:1929–1942.

Anderson A. T. and Greenland L. P. (1969). Phosphorous fractionation diagram as a quantitative indicator of crystallization differentiation of basaltic liquid. *Geochim. Cosmochim. Acta*, 33:493–505.

Anderson O. L. (1972). Patterns in elastic constants of minerals important to geophysics. In *Nature of the Solid Earth,* E. C. Robinson, ed. New York: McGraw-Hill.

Anderson O. L. (1995). *Equations of State of Solids for Geophysics and Ceramic Science.* New York: Oxford University Press.

Anderson T. F. and Kasper R. B. (1975). Oxygen self-diffusion in albite under hydrothermal conditions. *Trans. Amer. Geophys. Union.* 56:459.

Ando K., Kurokawa H., Oishi Y., and Takei H. (1981). Self-diffusion coefficient of oxygen in single-crystal forsterite. *J. Amer. Ceram. Soc.,* 64:2.

Angell C. A. (1982). Supercooled water. *Amer. Rev. Phys. Chem.,* 34:593–630.

Annersten H., Adetnuji J., and Filippidis A. (1984). Cation ordering in Fe-Mn silicate olivines. *Amer. Mineral.,* 69:1110–1115.

Anovitz L. M. and Essene E. J. (1982). Phase relations in the system $CaCO_3$-$MgCO_3$-$FeCO_3$. *Trans. Amer. Geophys. Union,* 63:464.

Anovitz L. M., Essene E. J., Metz G. W., Bohlen S. R., Westrum E. F. Jr., and Hemingway B. S. (1993). Heat capacity and phase equilibria of almandine, $Fe_3Al_2Si_3O_{12}$. *Geochim. Cosmochim. Acta,* 57: 4191–4204.

Ardell A. J., Christie J. M., and Tullis J. A. (1973). Dislocation substructures in experimentally deformed quartzites. *Cryst. Lattice Defects,* 4:275–285.

Armbruster T. and Geiger C. A. (1993). Andradite crystal chemistry: Dynamic X-site disorder and structural strain in silicate garnets. *Eur. J. Min.,* 5:59–71.

Armbruster T., Geiger C. A., and Lager G. A. (1992). Single-crystal X-ray structure study of synthetic pyrope almandine garnets at 100 and 293 K. *Amer. Mineral.,* 77:512–521.

Atkins P. W. (1978). *Physical Chemistry.* Oxford: Oxford University Press.

Baas Becking L. G. M., Kaplan I. R., and Moore O. (1960). Limits of the natural environments in terms of pH and oxidation-reduction potentials. *J. Geol.,* 68:243.

Babuska V., Fiala J., Kumazawa M., and Ohno I. (1978). Elastic Properties of Garnet Solid-Solution series. *Phys. Earth Planet. Int.,* 16:157–176.

Bachinski S. W. and Müller G. (1971). Experimental determination of microcline-low albite solvus. *J. Petrol.,* 12:329–356.

Bader R. F. W. and Jones G. A. (1963). The electron density distribution in hydride molecules, I: The water molecule. *Can. J. Chem.,* 41:586–606.

Baerlocher Ch., Hepp A., and Meier W. M. (1977). *DLS—76. A program for Simulation of Crystal Structures by Geometry Refinement.* Zurich: Institut of Crystallography and Petrography ETH.

Baes C. F. Jr. and Mesmer R. E. (1981). The thermodynamics of cation hydrolysis. *Amer. Jour. Sci.,* 281:935–962.

Bailey A. (1971). Comparison of low-temperature with high-temperature diffusion of sodium in albite. *Geochim. Cosmochim. Acta,* 35:1073–1081.

Bailey S. W. (1984a). Classification and structures of the micas. In *Reviews in Mineralogy,* vol. 13, P. H. Ribbe (series ed.), Mineralogical Society of America.

Bailey S. W. (1984b). Crystal chemistry of true micas. In *Reviews in Mineralogy,* vol. 13, P. H. Ribbe (series ed.), Mineralogical Society of America.

Bales R. C. and Morgan J. J. (1985). Dissolution kinetics of chrysotile at pH 7 to 10. *Geochim. Cosmochim. Acta,* 49:2281–2288.

Ballhausen C. J. (1954). Studies of absorption spectra, II: Theory of copper (II)—spectra. *Danske Videnskab. Selskab. Mat. Fys. Medd.,* 29:4–18.

Banno S. (1970). Classification of eclogites in terms of physical conditions of their origin. *Phys. Earth Planet Int.,* 3:405–421.

Barberi F., Ferrara G., Santacroce R., Treuil M., and Varet J. (1975). A transitional basalt-pantellerite sequence of crystallization: The Boina Centre (Afar Rift, Ethiopia). *J. Petrol.,* 16:22–56.

Barin I. and Knacke O. (1973). *Thermochemical Properties of Inorganic Substances.* Berlin-Heidelberg-New York: Springer-Verlag.

Barin I., Knacke O., and Kubaschewsky O. (1977). *Thermochemical Properties of Inorganic Substances: Supplement.* Berlin-Heidelberg-New York: Springer-Verlag.

Barnes V. E. (1930). Changes in hornblend at about 800 °C. *Amer. Mineral.,* 5:393–417.

Barr L. W. and Liliard A. B. (1971). Defects in ionic crystals. In *Physical Chemistry: An Advanced Treatise,* vol. 10, W. Jost (series ed.).

Barron L. M. (1976). A comparison of two models of ternary excess free energy. *Contrib. Mineral. Petrol.,* 57:71–81.

Bass, J. D. (1986). Elasticity of uvarovite and andradite garnets. *J. Geophys. Res.,* 91:7505–7516.

Bass J. D. (1989). Elasticity of grossular and spessartite garnets by Brillouin spectroscopy. *J. Geophys. Res.,* 94:7621–7628.

Basso R. (1985). Crystal chemical and crystallographic properties of compounds with garnet or hydrogarnet structure. *Neues Jahrb. Miner. Monat.,* 3:108–114.

Battino R. and Clever H. L. (1966). The solubilities of gases in liquids. *Chem. Rev.,* 66:395–463.

Baur W. H. (1970). Bond length variation and distorted coordination polyedra in inorganic crystals. *Trans. Amer. Cryst. Assoc.,* 6:129–155.

Baur W. H. (1978). Variation of mean Si-O bond lengths in silicon-oxygen tetrahedra. *Acta Cryst.,* B34:1754–1756.

Baur W. H. and Kahn A. A. (1971). Rutile-type compounds, IV: SiO_2, GeO_2, and a comparison with other rutile-type compounds. *Acta Cryst.,* B27:2133–2139.

Becker R. H. (1971). Carbon and oxygen isotope ratios in iron-formation and associated rocks from the Hamersley Range of Western Australia and their implications. Ph.D. diss., University of Chicago.

Bell R. G., Jackson R. A., and Catlow C. R. A. (1992). Loewenstein's rule in zeolite A: A computational study. *Zeolites,* 12:870–871.

Belton G. R., Suito H., and Gaskell D. R. (1973). Free energies of mixing in the liquid iron-cobalt orthosilicates at 1450 °C. *Met. Trans.,* 4:2541–2547.

Berger G. W. and York D. (1981). Geothermometry from $^{40}Ar/^{39}Ar$ dating experiments. *Geochim. Cosmochim. Acta,* 45:795–811.

Berlin R. and Henderson C. M. B. (1968). A re-interpretation of Sr and Ca fractionation trends in plagioclases from basic rock. *Earth Planet. Sci. Lett.,* 4:79–83.

Berman R. G. (1988). Internally consistent thermodynamic data for minerals in the system Na_2O-K_2O-CaO-MgO-FeO-Fe_2O_3-Al_2O_3-SiO_2-TiO_2-H_2O-CO_2. *J. Petrol.,* 29:445–522.

Berman R. G. (1990). Mixing properties of Ca-Mg-Fe-Mn garnets. *Amer. Mineral.,* 75:328–344.

Berman R. G. and Brown T. H. (1984). A thermodynamic model for multicomponent melts with applications to the system CaO-Al_2O_3-SiO_2. *Geochim. Cosmochim. Acta,* 48:661–678.

Berman R. G. and Brown T. H. (1985). Heat capacity of minerals in the system Na_2O-K_2O-CaO-MgO-FeO-Fe_2O_3-Al_2O_3-SiO_2-TiO_2-H_2O-CO_2: Representation, estimation, and high temperature extrapolation. *Contrib. Mineral. Petrol.,* 89:168–183.

Berman R. G. and Brown T. H. (1987). Development of models for multicomponent melts: Analysis of synthetic systems. In *Reviews in Mineralogy.* vol. 17, P. H. Ribbe (series ed.), Mineralogical Society of America.

Bernal J. D. (1964). The structure of liquids. *Proc. Roy. Soc. London,* A280:299–322.

Berner R. A. (1971). *Principles of Chemical Sedimentology.* New York: McGraw-Hill.

Berner R. A. (1980). *Early Diagenesis: A Theoretical Approach.* Princeton: Princeton University Press.

Berner R. A., Sjoberg E. L., and Schott J. (1980). Mechanisms of pyroxene and amphibole weathering, I: Experimental studies. In *Third International Symposium on Water-Rock Interaction: Proceedings,* Edmonton: Alberta Research Council.

Berthelot M. (1872). On the law which governs the distribution of a substance between two solvents. *Ann. Chim. Phys.,* 4th. ser., 26:408–417.

Bertorino G., Caboi R., Caredda A., Cidu R., Fanfani L., and Zuddas P. (1981). Caratteri idrogeochimici delle acque naturali della Sardegna meridionale, 3: Le acque della Marmilla e del Sarcidano. *Rend. S.I.M.P,* 37:951–966.

Bertrand G. L., Acree W. E. Jr., and Burchfield T. (1983). Thermodynamical excess properties of multicomponent systems: Representation and estimation from binary mixing data. *J. Solution. Chem.,* 12:327–340.

Bethe H. (1929). Splitting of terms in crystals. *Amer. Phys.,* 3:133–206.

Bieri R., Koide M., and Goldberg E. D. (1966). Noble gases contents of Pacific seawater. *J. Geophys. Res.,* 71:5243–5265.

Bigeleisen J. (1949). The relative reaction velocities of isotopic molecules. *J. Chem. Phys.,* 17:675–678.

Bigeleisen J. and Mayer M. G. (1947). Calculation of equilibrium constants for isotopic exchange reactions. *J. Chem. Phys.,* 13:261–267.

Biggerstaff D. R. (1986). The thermodynamic properties of aqueous solutions of argon, ethylene, and xenon up to 720 K and 34 MPa. Ph.D. diss., University of Delaware.

Bina C. R. and Wood B. J. (1987). The olivine-spinel transitions: Experimental and thermodynamic constraints and implications for the nature of the 400 km seismic discontinuity. *J. Geophys. Res.,* 92:4853–4859.

Birch F. (1966). Compressibility. In *Handbook of Physical Constants,* S. P. Clark Jr., ed. Memoir 97, The Geological Society of America.

Bird D. K. and Helgeson H. C. (1980). Chemical interaction of aqueous solution with epidolite-feldspar mineral assemblage in geologic systems, I: Thermodynamic analysis of phase relations in the system $CaO-FeO-Fe_2O_3-Al_2O_3-SiO_2-H_2O-CO_2$. *Amer. Jour. Sci.,* 280:907–941.

Bird M. L. (1971). Distribution of trace elements in olivines and pyroxenes: An experimental study. Ph.D. diss., University of Missouri, Rolla.

Bish D. L. (1981). Cation ordering in synthetic and natural Ni-Mg olivine. *Amer. Mineral.,* 66:770–776.

Bishop F. C. (1980). The distribution of Fe^{++} and Mg between coexisting ilmenite and pyroxene with application to geochemistry. *Amer. Jour. Sci.,* 280:46–47.

Bishop F. C., Smith J. V., and Dawson J. B. (1976). Na, P, Ti, and coordination of Si in garnet from peridotite and eclogite xenoliths. *Nature,* 260:696–697.

Blasi A., Brajkovic A., and De Pol Blasi (1983). Dry-heating conversions of low microcline to high sanidine via a one-step disordering process. *Abstr. NATO Advanced Study Inst. on Feldspars and Feldspatoids,* Rennes, France.

Blencoe J. G. (1977). Molal volumes of synthetic paragonite-muscovite micas. *Amer. Mineral.,* 62:1200–1215.

Blencoe J. G. and Luth W. C. (1973). Muscovite-paragonite solvi at 2, 4, and 8 kbar pressure. *Geol. Soc. Amer. Abstr. with Programs,* 5:553–554.

Bocchio R., Bondi M., and Domeneghetti M. C. (1979). Reexamination of shefferite, urbanite, and lindesite from central Sweden, Part I: Chemistry and crystal structure. *Per. Mineral.,* 48:181–193.

Bockris J. O'M. and Mellors G. W. (1956). Electric conductance in liquid lead silicates and borates. *J. Phys. Chem.,* 60:1321–1328.

Bockris J. O'M. and Reddy A. K. N. (1970). *Modern Electrochemistry.* New York: Plenum Press.

Bockris J. O'M., Kitchener J. A., and Davies A. E. (1952a). Electric transport in liquid silicates. *Trans. Faraday Soc.,* 48:536–548.

Bockris J. O'M., Kitchener J. A., Ignatowicz S., and Tomlinson J. W. (1952b). The electrical conductivity of silicate melts: Systems containing Ca, Mn, Al. *Discuss. Faraday Soc.,* 4:281–286.

Boeglin J. L. (1981). Mineralogie et geochimie des gisements de manganese de Conselheiro Lafaiete au Bresil et de Moanda au Gabon. These 3eme Cycle, Toulouse, 1608.

Bohlen S. R. and Boettcher A. L. (1981). Experimental investigations and geological applications of orthopyroxene geobarometry. *Amer. Mineral.,* 66: 951–964.

Bohlen S. R. and Essene E. J. (1977). Feldspar and oxide thermometry of granulites in the Adirondack Highlands. *Contrib. Mineral. Petrol.*, 62:153–169.

Bohlen S. R., Wall V. J., and Boettcher A. L. (1982). The system albite-water-carbon dioxide: A model for melting and activities of water at high pressures. *Amer. Mineral.*, 67:451–462.

Bohlen S. R., Wall V. J., and Boettcher A. L. (1983). Experimental investigation and geological application of equilibria in the system $FeO-TiO_2-Al_2O_3-SiO_2-H_2O$. *Amer. Mineral.*, 68:1049–1058.

Bokreta M. (1992). Energetics of garnets: Computational model of the thermochemical and thermophysical properties. Ph.D. diss., University of Philadelphia.

Bokreta M. and Ottonello G. (1987). Enthalpy of formation of end-member garnets. *EOS*, 68:448.

Bonatti E. (1965). Palagonite, hyaloclastites, and alteration of volcanic glass in the ocean. *Bull. Volcanol.*, 28:251–269.

Bonatti E., Ottonello G., and Hamlyn P. R. (1986). Peridotites from the Island of Zabargad (St. John)., Red Sea: Petrology and geochemistry. *J. Geophys. Res.*, 91:599–631.

Boon J. A. (1971). Mössbauer investigations in the system $Na_2O-FeO-SiO_2$. *Chem. Geol.*, 7:153–169.

Boon J. A. and Fyfe W. S. (1972). The coordination number of ferrous ions in silicate systems. *Chem. Geol.*, 10:287–298.

Borg I. Y. (1967). Optical properties and cell parameter in the glaucophane-riebeckite series. *Contrib. Mineral. Petrol.*, 15:67–92.

Born M. (1920). Volumen und Hydratationswarme der ionen. *Zeitschr. Physic*, 1:45–48.

Born M. and Von Karman T. (1913). Hüber die Verteilung der Eigenschwingungen von Punktgittern. *Physik. Zeitschr.*, 14:65–71.

Boström D. (1987). Single-crystal X-ray diffraction studies of synthetic Ni-Mg olivine solid solutions. *Amer. Mineral.*, 66:770–776.

Boström D. (1988). Experimental studies of (Ni,Mg)—and (Co,Mg)—olivine solid solutions by single crystal X-ray diffraction and solid state emf method. Ph.D. thesis, Department of Inorganic Chemistry, University of Umeå, Umeå, Sweden.

Boström D. and Rosen E. (1988). Determination of activity-composition in $(Ni,Mg)_2SiO_4$ solid solution at 1200–1600 K by solid state emf measurements. *Acta Chim. Scand.*, A42:149–155.

Boswara I. M. and Franklyn A. D. (1968). Theory of the energetics of simple defects in oxides. In *Mass Transport in Oxides*, J. B. Wachtman and A. D. Franklyn, eds. Natl. Bur. Stand. (U.S.). Spec. Publ., 296.

Bottinga Y. (1968). Calculation of fractionation factors for carbon and oxygen isotopic exchanges in the system calcite-carbon dioxide-water. *J. Phys. Chem.*, 72:800–807.

Bottinga Y. (1969a). Calculated fractionation factors for carbon and hydrogen isotope exchange in the system calcite-carbon dioxide-graphite-methane-hydrogen-water vapor. *Geochim. Cosmochim. Acta*, 33:49–64.

Bottinga Y. (1969b). Carbon isotope fractionation between graphite, diamond and carbon dioxide. *Earth Planet. Sci. Letters*, 5:301–307.

Bottinga Y. (1985). On the isothermal compressibility of silicate liquids at high pressure. *Earth Planet. Sci. Letters.*, 74:350–360.

Bottinga Y. E. and Richet P. (1981). High pressure and temperature equation of state and calculation of the therodynamic properties of gaseous carbon dioxide. *Amer. Jour. Sci.*, 281:615–660.

Bottinga Y. and Weill D. F. (1970). Densities of liquid silicates systems calculated from partial molar volumes of oxide components. *Amer. Jour. Sci.*, 269:169–182.

Bowen N. L. (1913). Melting phenomena in the plagioclase feldspar. *Amer. Jour. Sci.*, 35:577–599.

Bowen N. L. (1915). The crystallization of haplobasaltic, haplorhyolitic and related magmas. *Amer. Jour. Sci.*, 4th ser., 33:551–573.

Bowen N. L. (1928). *The Evolution of Igneous Rocks*. Princeton University Press, Princeton, N.J.

Bowen N. L. and Schairer J. F. (1935). The system $MgO-FeO-SiO_2$. *Amer. Jour. Sci.*, 29:151–217.

Bowen N. L. and Schairer J. F. (1938). Crystallization equilibrium in nepheline-albite-silica mixtures with fayalite. *J. Geol.*, 46:397–411.

Bowen N. L., Shairer J. F., and Posniak E. (1933). The system $Ca_2SiO_4-Fe_2SiO_4$. *Amer. Jour. Sci.*, 26:273–297.

Boyd F. R. (1973). The pyroxene geothermometry. *Geochim. Cosmochim. Acta*, 37:2533–2546.

Boyd F. R. and Schairer J. F. (1964). The system $MgSiO_3-CaMgSi_2O_6$. *J. Petrol.*, 5:275–309.

Bragg W. L. (1920). The arrangement of atoms in crystals. *Phil. Mag.*, 40:169–172.

Brereton N. R. (1970). Corrections for interfering isotopes in the $^{40}Ar/^{39}Ar$ dating method. *Earth Planet. Sci. Letters*, 8:427–433.

Brewer P. G. (1975). Minor elements in seawater. In *Chemical Oceanography.* vol. 2, J. P. Riley and G. Skirrow, eds. New York: Academic Press.

Bricker O. (1965). Some stability relations in the system $Mn-O_2-H_2O$ at 25° and one atmosphere total pressure. *Amer. Mineral.*, 57:284–287.

Brillouin L. (1953). *Wave Propagation in Periodic Structures.* New York: Dover.

Brinkmann U. and Laqua W. (1985). Decomposition of fayalite (Fe_2SiO_4). in an oxygen potential gradient at 1418 K. *Phys. Chem. Minerals*, 12:283–290.

Broecker W. S., Gerard R. D., Ewing M., and Heezen B. C. (1961). Geochemistry and physics of ocean circulation. In *Oceanography*, M. Sears, ed. Amer. Assn. Adv. Sci., Washington.

Brookins D. G. (1988). *Eh-pH Diagrams for Geochemistry.* Berlin-Heidelberg-New York: Springer-Verlag.

Brown B. E. and Bailey S. W. (1964). The structure of maximum microcline and the sanidine-microcline series. *Norsk Geol. Tidsskr.*, 42/2:25–36.

Brown G. E. (1970). *The crystal chemistry of the olivines.* Ph.D. Dissertation, Virginia Polytechnic Inst. and State University, Blacksburg, Virginia.

Brown G. E. (1982). Olivines and silicate spinels. In *Reviews in Mineralogy*, vol. 5, P. H. Ribbe (series ed.), Mineralogical Society of America.

Brown G. E. and Prewitt C. T. (1973). High temperature crystal chemistry of hortonolite. *Amer. Mineral.*, 58:577–587.

Brown G. E., Prewitt C. T., Papike J. J., and Sueno S. (1972). A composition of the structures of low and high pigeonite. *J. Geophys. Res.*, 77:5778–5789.

Brown G. M. (1967). Mineralogy of basaltic rocks. In *Basalts*, H. H. Hess and A. Poldervaart, eds., New York: John Wiley.

Brown I. D. and Shannon R. D. (1973). Empirical bond-strength—bond-length curves for oxides. *Acta Cryst.*, A29:266–282.

Brown W. L. and Parson I. (1981). Towards a more practical two-feldspar geothermometer. *Contrib. Mineral. Petrol.*, 76:369–377.

Bruno E., Chiari G., and Facchinelli A. (1976). Anorthite quenched from 1530 °C. I. Structure refinement. *Acta Cryst.*, B32:3270–3280.

Buddington A. F. and Lindsley D. H. (1964). Iron-titanium oxide minerals and synthetic equivalents. *J. Petrol.*, 5:310–357.

Buening D. K. and Buseck P. K. (1973). Fe-Mg lattice diffusion in olivine. *J. Geophys. Res.*, 78:6852–6862.

Buiskool Toxopeus J. M. A. and Boland J. M. (1976). Several types of natural deformation in olivine: An electron microscope study. *Tectonophysics*, 32:209–233.

Burnham C. W. (1967). Ferrosilite. *Carnegie Inst. Wash. Yb.*, 65:285–290.

Burnham C. W. (1975). Water and magmas: A mixing model. *Geochim. Cosmochim. Acta*, 39:1077–1084.

Burnham C. W. and Davis N. F. (1974). The role of H_2O in silicate melts, II: Thermodynamic and phase relations in the system $NaAlSi_3O_8-H_2O$ to 10 kbar, 700 °C to 1000 °C. *Amer. Jour. Sci.*, 274:902–940.

Burns R. G. (1970). *Mineralogical Application of Crystal Field Theory* Cambridge University Press, Cambridge.

Burns R. G. (1975). On the occurrence and stability of divalent chromium in olivines included in diamonds. *Contrib. Mineral. Petrol.*, 51:213–221.

Burns R. G. and Fyfe W. S. (1967). Crystal field theory and geochemistry of transition elements. In *Researches in Geochemistry*, vol. 2, P. H. Abelson (series ed.). New York: John Wiley.

Burns R. G. and Vaughan D. J. (1970). Interpretation of the reflectivity behavior of ore minerals. *Amer. Mineral.*, 55:1576–1586.

Buseck P. R., Nord G. L., and Veblen D. R. (1982). Subsolidus phenomena in pyroxenes. In *Reviews in Mineralogy*, vol. 7, 2d ed., P. H. Ribbe (series ed.), Mineralogical Society of America.

Cahn J. W. (1959). Free energy of a nonuniform system, II: Thermodynamic basis. *J. Chem. Phys.*, 30:1121–1124.

Cahn J. W. (1968). Spinodal decomposition. *Trans. Met. Soc. AIME*, 242:166–180.

Cahn J. W. and Hilliard J. E. (1958). Free energy of a nonuniform system, I: Interfacial free energy. *J. Chem. Phys.*, 28:258–267.

Cameron K. L. (1975). An experimental study of actinolite-cummingtonite phase relations with notes on the synthesis of Fe-rich anthophyllite. *Amer. Mineral.*, 60:375–390.

Cameron M. and Papike J. J. (1981). Structural and chemical variations in pyroxenes. *Amer. Mineral.*, 66:1–50.

Cameron M. and Papike J. J. (1982). Crystal chemistry of silicate pyroxenes. In *Reviews in Mineralogy.* vol. 7, 2d ed., P. H. Ribbe (series ed.), Mineralogical Society of America.

Cameron M., Sueno S., Prewitt C. T., and Papike J. J. (1973a). High-temperature crystal chemistry of K-fluor-richterite (abstract). *Trans. Amer. Geophys. Union*, 54:457–498.

Cameron M., Sueno S., Prewitt C. T., and Papike J. J. (1973b). High temperature crystal chemistry of acmite, diopside, hedembergite, jadeite, spodumene and ureyte. *Amer. Mineral.*, 58:596–618.

Campbell F. E. and Roeder P. (1968). The stability of olivine and pyroxene in the Ni-Mg-Si-O system. *Amer. Mineral.*, 53:257–268.

Capuano R. M. (1992). The temperature dependence of hydrogen isotope fractionation between clay minerals and water: Evidence from a geopressured system. *Geochim. Cosmochim. Acta*, 56:2547–2554.

Carmichael I. S. E. and McDonald A. (1961). The geochemistry of some natural acid glasses from the North Atlantic tertiary volcanic province. *Geochim. Cosmochim. Acta*, 25:189–222.

Carmichael I. S. E., Nicholls J., Spera F. J., Wood B. J., and Nelson S. A. (1977). High temperature properties of silicate liquids: applications to the equilibration and ascent of basic magma. *Phil. Trans. Roy. Soc. London*, A286:373–431.

Carpenter M. A. (1985). Order-disorder transformations in mineral solid solutions. In *Reviews in Mineralogy*, P. H. Ribbe (series ed.), Mineralogical Society of America, 14:187–223.

Carpenter M. A. (1988). Thermochemistry of aluminium/silicon ordering in feldspar minerals. In *Physical Properties and Thermodynamic Behaviour of Minerals*, E. K. H. Salje, ed. 265–323, D. Reidel Publishing Company.

Carpenter M. A. and Ferry J. M. (1984). Constraints on the thermodynamic mixing properties of plagioclase feldspars. *Contrib. Mineral. Petrol.*, 87:138–148.

Carpenter M. A. and McConnell J. D. (1984). Experimental delineation of the transformation in intermediate plagioclase feldspars. *Amer. Mineral.*, 69:112–121.

Carpenter M. A. and Salje E. K. H. (1994a). Thermodynamics of nonconvergent cation ordering in minerals, II: Spinels and the orthopyroxene solid solution. *Amer. Mineral.*, 79:1068–1083.

Carpenter M. A. and Salje E. K. H. (1994b). Thermodynamics of nonconvergent cation ordering in minerals, III: Order parameter coupling in potassium feldspar. *Amer. Mineral.*, 79:1084–1093.

Carpenter M. A., Powell R., and Salje E. K. H. (1994). Thermodynamics of nonconvergent cation ordering in minerals, I: An alternative approach. *Amer. Mineral.*, 79:1053–1067.

Catanzaro E. J. (1963). Zircon ages in southwestern Minnesota. *J. Geophys. Res.*, 68:2045–2048.

Catlow C. R. A. and Stoneham A. M. (1983). Ionicity in solids. *J. Phys.*, C16:4321–4338.

Catti M. (1981). A generalized Born-Mayer parametrization of the lattice energy in orthorhombic ionic crystals. *Acta Cryst.*, A37:72–76.

Chacko T., Mayeda T. K., Clayton R. N., and Goldsmith J. R. (1991). Oxygen and carbon isotope fractionations between CO_2 and calcite. *Geochim. Cosmochim. Acta*, 55:2867–2882.

Chakraborty S. and Ganguly J. (1991). Compositional zoning and cation diffusion in garnets. In *Advances in Physical Geochemistry*, vol. 8, J. Ganguly, ed. New York-Heidelberg-Berlin: Springer-Verlag.

Chakraborty S. and Ganguly J. (1992). Cation diffusion in aluminosilicate garnets: Experimental determination in spessartine-almandine diffusion couples, evaluation of effective binary diffusion coefficients, and applications. *Contrib. Mineral. Petrol.*, 111:74–86.

Charles R. W. (1974). The physical properties of the Mg-Fe richerites. *Amer. Mineral.*, 59:518–528.

Charles R. W. (1980). Amphiboles on the join pargasite-ferropargasite. *Amer. Mineral.*, 65:996–1001.

Charlu T. V., Newton R. C., and Kleppa O. J. (1975). Enthalpies of formation at 970 K of compounds in the system MgO-Al_2O_3-SiO_2 from high temperature solution calorimetry. *Geochim. Cosmochim. Acta*, 39:1487–1497.

Charlu T. V., Newton R. C., and Kleppa O. J. (1978). Enthalpy of formation of lime silicates by high temperature calorimetry, with discussion of high pressure phase equilibrium. *Geochim. Cosmochim. Acta*, 42:367–375.

Chatillon-Colinet C., Newton R. C., Perkins D., and Kleppa O. J. (1983). Thermochemistry of $(Fe^{++},Mg)SiO_3$ orthopyroxene. *Geochim. Cosmochim. Acta*, 47:1597–1603.

Chatterjee N. (1987). Evaluation of thermochemical data on Fe-Mg olivine, orthopyroxene, spinel and Ca-Fe-Mg-Al garnet. *Geochim. Cosmochim. Acta*, 51:2515–2525.

Chatterjee N. (1989). An internally consistent thermodynamic data base on minerals: Applications to the earth's crust and upper mantle. Ph.D. diss., City University of New York.

Chatterjee N. D. (1970). Synthesis and upper stability of pargasite. *Contrib. Mineral. Petrol.*, 27:244–257.

Chatterjee N. D. (1972). The upper stability limit of the assemblage paragonite + quartz and its natural occurrences. *Contrib. Mineral. Petrol.*, 34:288–303.

Chatterjee N. D. (1974). Synthesis and upper thermal stability limit of 2M-margarite, $CaAl_2[Al_2Si_2O_{10}(OH)_2]$. *Schweiz. Mineral. Petrogr. Mitt.*, 54:753–767.

Chatterjee N. D. (1976). Margarite stability and compatibility relations in the system CaO-Al_2O_3-SiO_2-H_2O as a pressure-temperature indicator. *Amer. Mineral.*, 61:699–709.

Chatterjee N. D. and Froese E. (1975). A thermodynamic study of the pseudobinary join muscovite-paragonite in the system $KAlSi_3O_8$-$NaAlSi_3O_8$-Al_2O_3-SiO_2-H_2O. *Amer. Mineral.*, 61:699–709.

Chatterjee N. D. and Johannes W. (1974). Thermal stability and standard thermodynamic properties of synthetic 2M, muscovite $KAl[AlSi_3O_{10}(OH)_2]$. *Contrib. Mineral. Petrol.*, 48:89–114.

Cheng W. and Ganguly J. (1994). Some aspects of multicomponent excess free energy models with subregular binaries. *Geochim. Cosmochim. Acta*, 58:3763–3767.

Chiari G., Facchinelli A. and Bruno E. (1978). Anorthite quenched from 1630 °C. Discussion. *Acta Cryst.*, B34:1757–1764.

Chiba H., Chacko T., Clayton R. N., and Goldsmith J. R. (1989). Oxygen isotope fractionations involving diopside, forsterite, magnetite, and calcite: Application to geothermometry. *Geochim. Cosmochim. Acta*, 53:2985–2995.

Chiodini G. and Cioni R. (1989). Gas geobarometry for hydrothermal systems and its applications to some Italian geothermal areas. *Appl. Geochem.*, 4:465–472.

Clark A. M. and Long J. V. P. (1971). Anisotropic diffusion of nickel in olivine. In *Graham Memorial Symposium on Diffusion Processes.* New York: Gordon & Breach.

Clark J. R., Appleman D. E., and Papike J. J. (1969). Crystal-chemical characterization of clinopyroxenes based on eight new structure refinements. *Min. Soc. Amer. Spec. Paper,* 2:31–50.

Clayton R. N. and Kieffer S. W. (1991). Oxygen isotopic thermometer calibrations. In *Stable Isotope Geochemistry: A Tribute to Samuel Epstein,* H. P. Taylor Jr., J. R. O'Neil, and I. R. Kaplan, eds., The Geochemical Society, Special Publication n. 3.

Clayton R. N., Goldsmith J. R., and Mayeda T. K. (1989). Oxygen isotope fractionations in quartz, albite, anorthite and calcite. *Geochim. Cosmochim. Acta,* 53:725–733.

Clugston M. J. (1978). The calculation of intermolecular forces: A critical examination of the Gordon-Kim model. *Adv. Phys.,* 27:893–912.

Cohen R. E. (1986). Thermodynamic solution properties of aluminous clinopyroxenes: Nonlinear least squares refinements. *Geochim. Cosmochim. Acta,* 50:563–575.

Coltorti M., Girardi V., and Schoerster J. (1987). Liquid immiscibility in the Archean Greenstone Belt of Piumhi (Minas Gerais, Brazil). *Lithos,* 20:77–91.

Colville A. A. and Ribbe P. H. (1968). The crystal structure of an adularia and a refinement of a structure of orthoclase. *Amer. Mineral.,* 53:25–37.

Connolly J. A. D. (1992). Phase diagram principles and computations: A review. In *Proceedings of the V Summer School of Earth and Planetary Sciences,* University of Siena Press, Siena.

Connolly J. A. D. (1994). Computer calculation of multidimensional phase diagrams: Why and how? In *Proceedings of the VII Summer School of Earth and Planetary Sciences,* University of Siena Press, Siena.

Connolly J. A. D. and Kerrick D. M. (1987). An algorithm and computer program for calculating composition phase diagrams. *CALPHAD,* 11:1–55.

Coulson C. A. (1961). *Valence.* (2d ed.) Oxford University Press, London.

Craig H. (1957). Isotopic standards for carbon and oxygen and correction factors for mass-spectrometric analysis of carbon dioxide. *Geochim. Cosmochim. Acta,* 12:133–149.

Craig H. (1961). Standards for reporting concentrations of deuterium and oxygen-18 in natural waters. *Science,* 133:1833–1934.

Craig H. (1963). The isotopic geochemistry of water and carbon in geothermal areas. In *Nuclear Geology of Geothermal Areas,* E. Tongiorgi, ed. Spoleto.

Crank J. (1975). *The Mathematics of Diffusion* Oxford University Press, London.

Cressey G. (1978). Exsolution in almandine-pyrope-grossular garnet. *Nature,* 271:533–534.

Cressey G., Schmid R., and Wood B. J. (1978). Thermodynamic properties of almandine-grossular garnet solid solutions. *Contrib. Mineral. Petrol.,* 67:397–404.

Cripps-Clark C. J., Sridhar R., Jeffes J. H. E., and Richardson F. D. (1974). Chain distribution and transition temperatures for phosphate glasses. In *Physical Chemistry of Process Metallurgy.* J. H. E. Jeffes and R. J. Tait, eds. Inst. Mining Met.

Currie K. L. (1971). The reaction 3 cordierite = 2 garnet + 4 sillimanite + 5 quartz as a geological thermometer in the Ipinicon Lake Region, Ontario. *Contrib. Mineral. Petrol.,* 33:215–226.

Currie K. L. and Curtis L. W. (1976). An application of multicomponent solution theory to jadeitic pyroxenes. *J. Geol.,* 84:179–194.

Curtis G. H. and Hay R. L. (1972). Further geological studies and potassium-argon dating at Olduvai Gorge and Ngorongoro Crater. In *Calibration of Hominoid Evolution,* W. W. Bishop and J. A. Miller, eds. Edinburgh: Scottish Academic Press.

Czaya R., (1971). Refinement of the structure of $\alpha-Ca_2SiO_4$. *Acta Cryst.,* B27:848–849.

Dahl P. S. (1980). The thermal compositional dependence of Fe^{++}–Mg distributions between co-existing garnet and pyroxene: Applications to geothermometry. *Amer. Mineral.,* 65:852–866.

Dal Negro A., Carbonin S., Molin G. M., Cundari A., and Piccirillo E. M. (1982). Intracrystalline cation distribution in natural clinopyroxenes of tholeiitic, transitional and alkaline basaltic rocks. In *Advances in Physical Geochemistry.* vol. 1, S. K. Saxena (series ed.), New York: Springer-Verlag.

Dal Negro A., Carbonin S., Domeneghetti C., Molin G. M., Cundari A., and Piccirillo E. M. (1984). Crystal chemistry and evolution of the clinopyroxene in a suite of high pressure ultramafic nodules from the newer volcanism of Victoria, Australia. *Contrib. Mineral. Petrol.*, 86:221–229.

Dalrymple G. B. and Lanphere M. A. (1969). *Potassium-Argon Dating* W. H. Freeman, San Francisco.

Dalrymple G. B. and Lanphere M. A. (1971). $^{40}Ar/^{39}Ar$ dating technique of K-Ar dating: A comparison with the conventional technique. *Earth Planet. Sci. Letters*, 12:300–308.

Damon P. E., Lerman J. C., and Long A. (1978). Temporal fluctuations of ^{14}C: Casual factors and implications. *Ann. Rev. Earth Planet. Sci.*, 6:457–464.

D'Amore F. and Panichi C. (1980). Evaluation of deep temperatures of hydrothermal systems by a new gas geothermometer. *Geochim. Cosmochim. Acta*, 44:549–556.

Darken L. S. (1948). Diffusion mobility and their interrelations through free energy in binary metallic systems. *Trans. Met. Soc. AIME*, 175:184–201.

Darken L. S. (1967). Thermodynamics of binary metallic solutions. *Trans. Met. Soc. AIME*, 239:80–89.

Das C. D., Keer H. V., and Rao R. V. G. (1963). Lattice energy and other properties of some ionic crystals. *Z. Physic. Chem.*, 224:377–383.

Davidson P. M. and Mukhopadhyay D. K. (1984). Ca-Fe-Mg olivines: Phase relations and a solution model. *Contrib. Mineral. Petrol.*, 86:256–263.

Davis L. L. and Smith D. (1993). Ni-rich olivine in minettes from Two Buttes, Colorado: A connection between potassic melts from the mantle and low Ni partition coefficients. *Geochim. Cosmochim. Acta*, 57:123–129.

Davoli P. (1987). A crystal chemical study of aegirin-augites and some evaluations on the oxidation state of Mn. *Neues Jahrb. Miner. Abh.*, 158:67–87.

De A. (1974). Silicate liquid immiscibility in the Deccan Traps and its petrogenetic significance. *Geol. Soc. Amer. Bull.*, 85:471–474.

De Baar H. J. W., Bacon M. P., Brewer P. G., and Bruland K. W. (1985). Rare earth elements in the Pacific and Atlantic Oceans. *Geochim. Cosmochim. Acta*, 49:1943–1960.

Debye P. (1912). Zur Theorie der spezifischen warmer. *Ann. Physik*, 39(4).: 789–839.

Deer W. A., Howie R. A., and Zussman J. (1978). *Rock Forming Minerals* vol. 2A, (2d ed.), New York: John Wiley.

Deer W. A., Howie R. A., and Zussman J. (1983). *An Introduction to the Rock-Forming Minerals.* Longman, Harlow Essex, England.

Deganello S. (1978). Thermal expansion from 25°C to 500°C of a few ionic radii. *Zeit. Krist.*, 147:217–227.

Deines P., Langmuir D., and Harmon R. S. (1974). Stable carbon isotope ratios and the existence of a gas phase in the evolution of carbonate ground waters. *Geochim. Cosmochim. Acta*, 38:1147–1164.

Delany J. M., Puigdomenech I., and Wolery T. J. (1986). *Precipitation Kinetics Option for the EQ6 Geochemical Reaction Path Code* Lawrence Livermore National Laboratory, Livermore, Cal., UCRL-53642.

Della Giusta A. and Ottonello G. (1993). Energy and long-range disorder in simple spinels. *Phys. Chem. Minerals*, 20:228–241.

Della Giusta A., Ottonello G., and Secco L. (1990). Precision estimates of interatomic distances using site occupancies, ionization potentials and polarizability in Pbnm silicate olivines. *Acta Cryst.*, B46:160–165.

Dempsey M. J. (1980). Evidence for structural changes in garnet caused by calcium substitution. *Contrib. Mineral. Petrol.*, 7:281–282.

Denbigh K. G. (1971). *The Principles of Chemical Equilibrium*. 3d ed., Cambridge University Press, Cambridge.

De Paolo D. J. and Wasserburg G. J. (1979). Sm-Nd age of the Stillwater complex and the mantle curve for neodymium. *Geochim. Cosmochim. Acta*, 43:999–1008.

Desborough G. A. (1975). Authigenic albite and potassium feldspar in the Green River formation, Colorado and Wyoming. *Amer. Mineral.*, 60:235–239.

De Vore G. W. (1955). The role of absorption in the fractionation and distribution of elements. *J. Geol.*, 63:159–190.

Dodson M. H. (1973). Closure temperature in cooling geochronological and petrological systems. *Contrib. Mineral. Petrol.*, 40:259–274.

Dodson M. H. (1979). Theory of cooling ages. In *Lectures in Isotope Geology*, E. Jäger and J. C. Hunziker, eds., Berlin: Springer-Verlag.

Doerner H. A. and Hoskins W. M. (1925). Coprecipitation of radium and barium sulfates. *J. Amer. Chem. Soc.*, 47:662–675.

Donnay G. (1969). Further use for the Pauling-bond concept. *Carnegie Inst. Wash. Yb.*, 68:292–295.

Donnay G. and Allmann R. (1970). How to recognize O^{2-}, OH, and H_2O in crystal structures determined by X-rays. *Amer. Mineral.*, 55:1003–1015.

Downs J. W. (1991). Electrostatic properties of minerals from X-ray diffraction data: A guide for accurate atomistic models. In *Advances in Physical Geochemistry*, vol. 8, J. Ganguly, ed. Berlin-Heidelberg-New York: Springer-Verlag.

Drake M. J. and Holloway J. R. (1978). Henry's law behavior of Sm in a natural plagioclase/melt system: Importance of experimental procedure. *Geochim. Cosmochim. Acta*, 42:679–683.

Drake M. J. and Weill D. F. (1975). The partioning of Sr, Ba, Ca, Y, Eu^{2+}, Eu^{3+}. and other REE between plagioclase feldspar and magmatic silicate liquid: An experimental study. *Geochim. Cosmochim. Acta*, 39:689–712.

Driessens F. C. M. (1968). Thermodynamics and defect chemistry of some oxide solid solutions. Parts I and II. *Ber. Buns. Phys. Chem.*, 72:754–772.

Duchesne J. C. (1972). Iron-titanum oxide minerals in the Bjerkrem-Sogndal massif, southwestern Norway. *J. Petrol.*, 13:57–81.

Dudas M. J., Schmitt R. A., and Harward M. E. (1971). Trace element partitioning between volcanic plagioclases and dacitic pyroclastic matrix. *Earth Planet. Sci. Letters*, 11:440–446.

Duncan A. B. F. and Pople J. A. (1953). The structure of some simple molecules with lone pair electrons. *Trans. Faraday Soc.*, 49:217–227.

Dunitz J. D. and Orgel L. E. (1957). Electronic properties of transition metal oxides, II: Cation distribution amongst octahedral and tetrahedral sites. *J. Phys. Chem. Solids*, 3:318–323.

Einstein A. (1907). Die Plancksche Theorie der Strahlung und die Theorie der spezifischen Wärme. *Ann. Physik*, 22:180–190.

Eisemberg D. and Kauzmann W. (1969). *The Structures and Properties of Water* Oxford University Press, London.

Ellis A. J. and Golding R. M. (1963). The solubility of carbon dioxide above 100°C in water and sodium chloride solutions. *Amer. Jour. Sci.*, 261:47–60.

Ellis D. J. and Green D. H. (1979). An experimental study of the effect of Ca upon garnet-clinopyroxene exchange equilibria. *Contrib. Mineral. Petrol.*, 71:13, 22.

Elphick S. C., Ganguly J., and Leonis T. P. (1981). Experimental study of Fe-Mg interdiffusion in aluminumsilicate garnet (abstract). *EOS Trans. Amer. Geophys. Union*, 62:411.

Engi M. (1978). Mg-Fe Exchange equilibria among spinel, olivine, orthopyroxene and cordierite. Ph.D. diss., n. 6256, Eidgenossische Technische Hochschule, Zurich, Switzerland.

Epstein S. (1959). The variation of the $^{18}O/^{16}O$ ratio in nature and some geologic applications. In *Researches in Geochemistry*, P. H. Abelson, ed. New York: John Wiley.

Ernst W. G. and Calvert S. E. (1969). An experimental study of the recrystallization of porcelanite and its bearing on the origin of some bedded cherts. *Amer. Jour. Sci.*, 267A:114–133.

Ernst W. G. and Wai C. M. (1970). Infrared, X-ray and optical study of cation ordering and dehydrogenation in natural and heat-treated sodic amphiboles. *Amer. Mineral.*, 55:1226–1258.

Essene E. J. (1982). Geologic thermometry and barometry. In *Reviews in Mineralogy*, vol. 10, P. H. Ribbe (series ed.), Mineralogical Society of America.

Eugster H. P. (1954). Distribution of cesium between sanidine and a hydrous fluid. *Carnegie Inst. Wash. Yb.*, 53:102–104.

Eugster H. P. and Baumgartner L. (1987). Mineral solubilities and speciation in supercritical metamorphic fluids. In *Reviews in Mineralogy*. vol. 17, P. H. Ribbe (series ed.), Mineralogical Society of America.

Eugster H. P. and Wones D. R. (1962). Stability relations of the ferruginos biotite, annite. *J. Petrol.*, 3:82–125.

Eugster H. P., Albee A. L., Bence A. E., Thompson J. B. Jr., and Walbaum D. R. (1972). The two phase region and excess mixing properties of paragonite-muscovite crystalline solutions. *J. Petrol.*, 13:147–179.

Ewart A. and Taylor S. R. (1969). Trace element geochemistry of rhyolitic volcanic rocks, Central North Island, New Zealand. Phenocryst data. *Contrib. Mineral. Petrol.*, 22:127–138.

Ewart A., Bryan W. B., and Gill J. B. (1973). Mineralogy and geochemistry of younger volcanic islands of Tonga S. W. Pacific. *J. Petrol.*, 14:429–465.

Ewart A., Taylor S. R., and Capp A. C. (1968). Trace and minor element geochemistry of the rhyolitic volcanic rocks, Central North Island, New Zealand. *Contrib. Mineral. Petrol.*, 18:76–85.

Faure G. (1986). *Principles of Isotope Geology.* 2d ed. New York: John Wiley.

Faure G. and Powell J. L. (1972). *Strontium Isotope Geology.* New York: Springer-Verlag.

Fei Y. and Saxena S. K. (1986). A thermochemical data base for phase equilibria in the system Fe-Mg-Si-O at high pressure and temperature. *Phys. Chem. Minerals.*, 13:311–324.

Fei Y. and Saxena S. K. (1987). An equation for the heat capacity of solids. *Geochim. Cosmochim. Acta*, 51:251–254.

Fei Y., Saxena S. K., and Eriksson G. (1986). Some binary and ternary silicate solution models. *Contrib. Mineral. Petrol.*, 94:221–229.

Ferguson J. and Currie K. L. (1971). Evidence of liquid immiscibility in alkaline ultrabasic dikes at Callender Bay, Ontario. *J. Petrol.*, 12:561–585.

Ferry J. M. and Spear F. S. (1978). Experimental calibration of the partitioning of Fe and Mg between biotite and garnet. *Contrib. Mineral. Petrol.*, 66:113–117.

Fincham C. J. B. and Richardson R. F. (1954). The behaviour of sulfur in silicate and aluminate melts. *Proc. Roy. Soc. London*, A223:40–61.

Finger L. W. (1970). Refinement of the crystal structure of an anthophyllite. *Carnegie Inst. Wash. Yb.*, 68:283–288.

Finger L. W. and Ohashi Y. (1976). The thermal expansion of diopside to 800 °C and a refinement of the crystal structure at 700 °C. *Amer. Mineral.*, 61:303–310.

Finnerty T. A. (1977). Exchange of Mn, Ca, Mg and Al between synthetic garnet, orthopyroxene, clinopyroxene and olivine. *Carnegie Inst. Wash. Yb.*, 68:290–292.

Finnerty T. A. and Boyd F. R. (1978). Pressure dependent solubility of calcium in forsterite coexisting with diopside and enstatite. *Carnegie Inst. Wash. Yb.*, 77:713–717.

Fisher K. (1966). A further refinement of the crystal structure of cummingtonite, $(Mg, Fe)_7 (Si_4O_{11})_2(OH)_2$. *Amer. Mineral.*, 51:814–818.

Fitch F. J., Miller J. A., and Hooker P. J. (1976). Single whole rock K-Ar isochron. *Geol. Mag.*, 113:1–10.

Fleer V. N. (1982). The dissolution kinetics of anorthite $(CaAl_2Si_2O_8)$ and synthetic strontium feldspar $(SrAl_2Si_2O_8)$ in aqueous solutions at temperatures below 100°C: With applications to the geological disposal of radioactive nuclear wastes. Ph.D. diss., Pennsylvania State University, University Park.

Flood H. and Förland T. (1947). The acidic and basic properties of oxides. *Acta Chem. Scand.*, 1:952–1005.

Flood H., Förland T., and Gzotheim K. (1954). Über den Zusammenhang zwischen Konzentrazion und Aktivitäten in geschmolzenen Salzmischungen. *Zeit. Anorg. Allg. Chem.*, 276:290–315.

Foland K. A. (1974). Alkali diffusion in orthoclase. In *Geochemical Transport and Kinetics*, Hoffman, Giletti, Yoder, and Yund, eds. Carnegie Institution of Washington.

Francis C. A. and Ribbe P. H. (1980). The forsterite-tephroite series, I: Crystal structure refinement. *Amer. Mineral.*, 65:1263–1269.

Franz G., Hinrichsen T., and Wannermacher E. (1977). Determination of the miscibility gap on the mean of the infrared spetroscopy. *Contrib. Mineral. Petrol.*, 59:207–236.

Fraser D. G. (1975a). Activities of trace elements in silicate melts. *Geochim. Cosmochim. Acta*, 39:1525–1530.

Fraser D. G. (1975b). An investigation of some long-chain oxi-acid systems. D.Phil. diss., University of Oxford.

Fraser D. G. (1977). Thermodynamic properties of silicate melts. In *Thermodynamics in Geology*. D. G. Fraser, ed. Reidel, Dortrecht-Holland.

Fraser D. G., Rammensee W., and Hardwick A. (1985). Determination of the mixing properties of molten silicates by Knudsen cell mass spectrometry, II: The system $(Na-K)AlSi_4O_{10}$ and $(Na-K)AlSi_5O_{12}$. *Geochim. Cosmochim. Acta*, 49:349–359.

Fraser D. G., Rammensee W., and Jones R. H. (1983). The mixing properties of melts in the system $NaAlSi_3O_6$-$KAlSi_2O_6$ determined by Knudsen-Cell Mass Spectrometry. *Bull. Mineral.*, 106:111–117.

Frey F. A. (1982). Rare earth element abundances in upper mantle rocks. In *Rare Earth Element Geochemistry*, P. Henderson, ed. Elesevier, Amsterdam.

Friedel J. (1964). *Dislocations*. Addison-Wesley, Reading, Massachusetts.

Friedlander G. and Kennedy J. W. (1956). *Nuclear and Radiochemistry*. New York: John Wiley.

Friedlander G., Kennedy J. W., Macias E. S., and Miller J. M. (1981). *Nuclear and Radiochemistry*. New York: John Wiley.

Frondel C. F. (1962). *The System of Mineralogy of James Dwight Dana and Edward Salysbury Dana*. 7th ed., vol. 8, Silica Minerals, New York: John Wiley.

Fujii T. (1977). Fe-Mg partitioning between olivine and spinel. *Carnegie Inst. Wash. Yb.*, 76:563–569.

Fujino K., Sasaki S., Takeuchi Y., and Sadanaga R. (1981). X-ray determination of electron distribution in forsterite, faialite and tephroite. *Acta Cryst.*, B37:513–518.

Fumi F. G. and Tosi M. P. (1957). Naor relations between Madelung constants for cubic ionic lattices. *Phil. Mag.*, 2:284–285.

Fumi F. G. and Tosi M. P. (1964). Ionic sizes and Born repulsive parameters in the NaCl-type alkali halides, I: The Huggins-Mayer and Pauling forms. *J. Phys. Chem. Solids*, 25:31–43.

Fung P. C. and Shaw D. M. (1978). Na, Rb and Tl distributions between phlogopite and sanidine by direct synthesis in a common vapour phase. *Geochim. Cosmochim. Acta*, 42:703–708.

Fursenko B. A. (1981). High pressure synthesis of the chromium bearing garnets $Mn_3Cr_2Si_3O_{12}$. *Doklady Acad. Sci. USSR*, 250:176–179.

Fyfe W. S., Turner F. J., and Verhoogen J. (1958). Metamorphic reactions and metamorphic facies. *Geol. Soc. Amer. Mem.*, vol. 75, 253 pp.

Gabis V., Abelard P., Ildefonse J. P., Jambon A., and Touray J. C. (1979). La diffusion a haute temperature dans les systems d'interet geologique. In *Haute Temperatures et Science de la Terre*. Centre Regional de Publications de Toulouse, Editions du C.N.R.S., 163–174.

Ganguly J. (1973). Activity-composition relation of jadeite in omphacite pyroxenes: Theoretical deductions. *Earth Planet. Sci. Letters*, 19:145–153.

Ganguly J. (1976). The energetics of natural garnet solid solution, II: Mixing of the calcium silicate end members. *Contrib. Mineral. Petrol.*, 55:81–90.

Ganguly J. (1977). Crystal chemical aspects of olivine structures. *N. Jahrb Miner. Abh.*, 130:303–318.

Ganguly J. (1979). Garnet and clinopyroxene solid solutions and geothermometry based on Fe-Mg distribution coefficient. *Geochim. Cosmochim. Acta*, 43:1021–1029.

Ganguly J. and Cheng W. (1994). Thermodynamics of (Ca,Mg,Fe,Mn)-garnet solid solution: New experiments, optimized data set, and applications to thermo-barometry. *I.M.A., 16th General Meeting* (Abstracts)., Pisa, Italy.

Ganguly J. and Kennedy G. C. (1974). The energetics of natural garnet solid solutions, I: Mixing of the aluminosilicate end members. *Contrib. Mineral. Petrol.*, 48:137–148.

Ganguly J. and Ruitz J. (1986). Time-temperature relation of mineral isochrons: A thermodynamic model, and illustrative examples for the Rb-Sr system. *Earth Planet. Sci. Letters*, 81:338–348.

Ganguly J. and Saxena S. K. (1984). Mixing properties of aluminosilicate garnets: Constraints from natural and experimental data and applications to geothermo-barometry. *Amer. Mineral.*, 69:79–87.

Ganguly J. and Saxena S. K. (1987). *Mixtures and Mineral Reactions*. Berlin-Heidelberg-New York: Springer-Verlag.

Ganguly J., Cheng W., and O'Neill H. St. C. (1993). Syntheses, volume, and structural changes of garnets in the pyrope-grossular join: Implications for stability and mixing properties. *Amer. Mineral.*, 78:583–593.

Garrels R. M. and Christ C. L. (1965). *Solutions, Minerals, and Equilibria*. New York: Harper and Row.

Garrels R. M. and Thompson M. E. (1962). A chemical model for seawater at 25°C and one atmosphere total pressure. *Amer. Jour. Sci.*, 260:57–66.

Gartner L. (1979). *Relations entre enthalpies en enthalpies libres de formation des ions, des oxydes de formule $M_nN_mO_2$. Utilization des frequences de vibration dans l'infra-rouge* Doct. Ing. Université de Strasbourg, 193 pp.

Gasparik T. (1985). Experimentally determined compositions of diopside-jadeite pyroxene in equilibrium with albite and quartz at 1200–1350 °C and 15–34 kbar. *Geochim. Cosmochim. Acta*, 89:865–870.

Gasparik T. and Lindsley D. H. (1980). Phase equilibria at high pressure of pyroxenes containing monovalent and trivalent ions. In *Reviews in Mineralogy*, vol. 7, P. H. Ribbe (series ed.), Mineralogical Society of America.

Gasperin M. (1971). Crystal structure of rubidium feldspar, $RbAlSi_3O_8$. *Acta Cryst.*, B27:854–855.

Gasperin J. and McConnel J. D. (1984). Experimental delineations of the $C\bar{1}$–$I\bar{1}$ trasformation in intermediate plagioclase feldspar. *Amer. Mineral.*, 69:112–121.

Gast P. W. (1968). Trace element fractionation and the origin of tholeiitic and alkaline magma types. *Geochim. Cosmochim. Acta*, 32:1057–1086.

Gautschi W. (1964). Error function and Fresnel integrals. In *Handbook of Mathematical Functions*, M. Abramowitz and I. A. Stegun, eds. National Bureau of Standards, Washington.

Geiger C. A., Newton R. C. and Kleppa O. J. (1987). Enthalpy of mixing of synthetic almandine-grossular and almandine-pyrope garnets from high-temperature solution calorimetry. *Geochim. Cosmochim. Acta*, 51:1755–1763.

Gelinas L., Brooks C., and Trzcienski W. E. Jr. (1976). Archean variolites-quenched immiscible liquids reexamined: A reply to criticisms. *Can. Jour. Earth Sci.*, 14:2945–2958.

Gerlach T. M. and Nordlie B. E. (1975a). The C-O-H-S- gaseous systems, Part I: Composition limits and trends in basaltic cases. *Amer. Jour. Sci.*, 275:353–376.

Gerlach T. M. and Nordlie B. E. (1975b). The C-O-H-S-gaseous systems, Part II: Temperature, atomic composition, and molecular equilibria in volcanic gases. *Amer. Jour. Sci.*, 275:377–394.

Gerlach T. M. and Nordlie B. E. (1975c). The C-O-H-S-gaseous systems, Part III: Magmatic gases compatible with oxides and sulfides in basaltic magmas. *Amer. Jour. Sci.,* 275:395–410.

Ghent E. D. (1976). Plagioclase-garnet-Al$_2$SiO$_5$-quartz: A potential geobarometer-geothermometer. *Amer. Mineral.,* 61:710–714.

Ghent E. D., Stout M. Z., Black P. M., and Brothers R. N. (1987). Chloritoid bearing rocks associated with blueschists and eclogites, northern New California. *J. Met. Geol.,* 5:239–254.

Ghiorso M. S. (1984). Activity/composition relation in the ternary feldspars. *Contrib. Mineral. Petrol.,* 87:282–296.

Ghiorso M. S. and Carmichael I. S. E. (1980). A regular solution model for meta-aluminous silicate liquids: Applications to geothermometry, immiscibility, and the source regions of basic magma. *Contrib. Mineral. Petrol.,* 71:323–342.

Ghiorso M. S., Carmichael I. S. E., Rivers M. L., and Sack R. O. (1983). The Gibbs free energy of mixing of natural silicate liquids; an expanded regular solution approximation for the calculation of magmatic intensive variables. *Contrib. Mineral. Petrol.,* 84:107–145.

Ghose S. (1982). Subsolidus reactions and microstructures in amphiboles. In *Reviews in Mineralogy,* vol. 9A, P. H. Ribbe (series ed.), Mineralogical Society of America.

Ghose S. and Wan C. (1974). Strong site preference of Co^{2+} in olivine Co$_{1.10}$Mg$_{0.90}$SiO$_4$. *Contrib. Mineral. Petrol.,* 47:131–140.

Ghose S. and Weidner J. R. (1971). Mg^{2+}-Fe^{2+} isotherms in cummingtonites at 600 and 700 °C (abstract). *Trans. Amer. Geophys. Union,* 52:381.

Ghose S. and Weidner J. R. (1972). Mg^{2+}-Fe^{2+} order-disorder in cummingtonite (Mg,Fe)$_7$Si$_8$O$_{22}$(OH)$_2$. A new geothermometer. *Earth Planet. Sci. Letters,* 16:346–354.

Ghose S., Choudhury N., Chaplot S. L., and Rao K. R. (1992). Phonon density of states and thermodynamic properties of minerals. In *Advances in Physical geochemistry,* vol. 10, S. K. Saxena (series ed.). Berlin-Heidelberg-New York: Springer-Verlag.

Ghose S., Kersten M., Langer K., Rossi G. and Ungaretti L. (1986). Crystal field spectra and Jahn-Teller effect of Mn^{3+} in clinopyroxene and clinoamphiboles from India. *Phys. Chem. Minerals,* 13:291–305.

Gibbs J. W. (1906). On the equilibrium of heterogeneous substances. In *The Scientific Papers of J. Willard Gibbs.* Vol. 1: Thermodynamics. Longmans Green (reprinted by Dover, 1961).

Giggenbach W. F. (1980). Geothermal gas equilibria. *Geochim. Cosmochim. Acta,* 44:2021–2032.

Giggenbach W. F. (1987). Redox processes governing the chemistry of fumarolic gas discharges from White Islands, New Zealand. *Appl. Geochem.,* 2:143–161.

Gilbert M. C., Helz R. T., Popp R. K., and Spear F. S. (1982). Experimental studies of amphibole stability. In *Reviews in Mineralogy,* vol. 9B, P. H. Ribbe (series ed.), Mineralogical Society of America.

Giletti B. J., Sennet M. P., and Yund R. A. (1978). Studies in diffusion, III: Oxygen in feldspars—an ion microprobe determination. *Geochim. Cosmochim. Acta,* 42:45–57.

Glasser F. P. (1960). Einige Ergebnisse von Phasengleichgewwichtsuntersuchungen in der Systemen MgO-MnO-SiO$_2$ und CaO-MnO-SiO$_2$. *Silikattechnik,* 11:362–363.

Godwin H. (1962). Half-life of radiocarbon. *Nature,* 195:984.

Goldman D. S. and Albee A. L. (1977). Correlation of Mg/Fe partitioning between garnet and biotite with ^{18}O/^{16}O partitioning between quartz and magnetite. *Amer. Jour. Sci.,* 277:750–767.

Goldschmidt V. M. (1923). Geochemical laws of the distribution of the elements. *Videnskaps. Skrift. Mat.-Natl. Kl.,* 3:1–17.

Goldschmidt V. M., Barth T., Lunde G., and Zachariasen W. (1926). Geochemische Verteilungsgesetze der Elemente, VII: Die Gesetze die Kristallchemie. *Det. Norske. Vid. Akad. Oslo I, Mat. Natl. Kl.,* 2:1–117.

Goldsmith J. R. (1980). The melting and breakdown reactions of anorthite at high pressures and temperatures. *Amer. Mineral.,* 66:1183–1188.

Goldsmith J. R. and Heard H. C. (1961). Subsolidus phase relations in the system $CaCO_3$-$MgCO_3$. *J. Geol.*, 69:45–74.

Goldsmith J. R. and Newton R. C. (1969). P-T-X relations in the system $CaCO_3$-$MgCO_3$ at high temperatures and pressures. *Amer. Jour. Sci.*, 267A:160–190.

Gordon R. G. and Kim Y. S. (1971). Theory of the forces between closed-shell atoms and molecules. *J. Chem. Phys.*, 56:3122–3133.

Gordy W. (1946). A new method of determining electronegativity from other atomic properties. *Phys. Rev.*, 69:604–607.

Gordy W. (1950). Interpretation of nuclear quadrupole couplings in molecules. *J. Chem. Phys.*, 19:792–793.

Gottardi G. (1972). *I Minerali.* Serie Geologia, Boringhieri, Bologna.

Gramaccioli C. M. and Filippini G. (1983). Lattice dynamical calculations for orthorhombic sulfur: A non-rigid molecular model. *Chem. Phys. Letters*, 108:585–588.

Grandstaff D. E. (1977). Some kinetics of bronzite orthopyroxene dissolution. *Geochim. Cosmochim. Acta*, 41:1097–1103.

Grandstaff D. E. (1980). The dissolution rate of forsterite olivine from Hawaiian beach sand. In *Third International Symposium on Water-Rock Interaction: Proceedings*, Alberta Research Council, Edmonton.

Graves J. (1977). Chemical mixing in multicomponent solutions. In *Thermodynamics in Geology*, D. E. Fraser, ed. Reidel, Dordrecht-Holland.

Grebenshchikov R. G., Romanov D. P., Sipovskii D. P., and Kosulina G. I. (1974). Study of a germanate hydroxyamphibole. *Zhur. Prikl. Khimii*, 47:1905–1910.

Greegor R. B., Lytle F. W., Sanstrom D. R., Wang J., and Schultz P. (1983). Investigation of TiO_2-SiO_2 glasses by X-ray absorption spectroscopy. *J. Non. Cryst. Solids*, 55:27–43.

Green D. H. and Hibberson W. (1970). The instability of plagioclase on peridotite at high pressure. *Lithos*, 3:209–221.

Green E. J. (1970). Predictive thermodynamic models for mineral systems, I: Quasi-chemical analysis of the halite-sylvite subsolidus. *Amer. Mineral.*, 55:1692–1713.

Green S. H. and Gordon R. G. (1974). *POTLSURF: A program to compute the interaction potential energy surface between a closed-shell molecule and an atom.* Quantum Chemistry Program Exchange No. 251, Indiana University.

Green T. H. (1994). Experimental studies of trace-element partitioning applicable to igneous petrogenesis—Sedona 16 years later. *Chem. Geol.*, 117:1–36.

Greenwood H. J. (1967). Wollastonite: Stability in H_2O-CO_2 mixtures and occurrence in a contact metamorphic aureole near Salmo, British Columbia, Canada. *Amer. Mineral.*, 52:1669–1680.

Greenwood N. N. (1970). *Ionic Crystals, Lattice Defects and Non*-stoichiometry. New York: Chemical Pub. Co.

Greig J. W. (1927). Immiscibility in silicate melts. *Amer. Jour. Sci.*, 5th Ser., 13(73).: 1–44 and 13(74).: 133–154.

Greig J. W. and Barth T. F. W. (1938). The system Na_2O-Al_2O_3-$2SiO_2$ (nepheline, carnegieite).-Na_2O-Al_2O_3-$6SiO_2$. (albite). *Amer. Jour. Sci.*, 5th. ser., 35A:93–112.

Griffin W. L. and Mottana A. (1982). Crystal chemistry of clinopyroxenes form St. Marcel manganese deposit, Val d'Aosta, Italy. *Amer. Mineral.*, 67:568–586.

Grover J. (1977). Chemical mixing in multicomponent solutions: An introduction to the use of Margules and other thermodynamic excess functions to represent non-ideal behaviour. In *Thermodynamics in Geology*, D. G. Fraser, ed. D. Reidel, Dordrecht-Holland.

Grover J. E. and Orville T. M. (1969). The partitioning of cations between coexisting single and multisite phases with application to the assemblage: Orthopyroxene-clinopyroxene and orthopyroxene-olivine. *Geochim. Cosmochim. Acta*, 33:205–226.

Grundy H. D. and Ito J. (1974). The refinement of the crystal structure of a synthetic non-stoichiometric Sr-feldspar. *Amer. Mineral.*, 68:1319–1326.

Guggenheim E. A. (1937). Theoretical basis of Raoult's Law. *Trans. Faraday Soc.*, 33:205–226.

Guggenheim E. A. (1952). *Mixtures.* Clarendon Press, Oxford.

Guggenheim S. and Bailey S. W. (1975). Refinement of the margarite structure in subgroup symmetry. *Amer. Mineral.,* 60:1023–1029.

Guggenheim S. and Bailey S. W. (1977). The refinement of zinnwaldite-IM in subgroup symmetry. *Amer. Mineral.,* 62:1158–1167.

Guggenheim S. and Bailey S. W. (1978). Refinement of the margarite structure in subgroup symmetry: Correction, further refinement, and comments. *Amer. Mineral.,* 63:186–187.

Guidotti C. V., Herd H. H., and Tuttle C. L. (1973). Composition and structural state of K-feldspars from K-feldspars + sillimanite grade rocks in northwestern Maine. *Amer. Mineral.,* 58:705–716.

Guthrie F. (1884). On eutexia. *Philos. Mag.,* 17–5:462–482.

Guy R. D., Chakrabarti C. L., and Schraumm L. L. (1975). The application of a simple chemical model of natural waters to metal fixation in particulate matter. *Canadian Jour. Chem.,* 53:661–669.

Haar L., Gallager J. G., and Kell G. S. (1979). Thermodynamic properties of fluid water. In *Contributions to the 9th Int. Conf. on Properties of Steam,* Munich, Western Germany.

Haar L., Gallager J. G., and Kell G. S. (1984). *NBS/NRC Steam Tables: Thermodynamic and transport properties and computer programs for vapor and liquid states of water in SI units.* New York: Hemisphere Pub. Co., McGraw-Hill.

Haas I. L. Jr., Robinson G. R. Jr., and Hemingway B. S. (1981). Thermodynamic tabulations for selected phases in the system $CaO-Al_2O_3-SiO_2-H_2O$ at 101.325 kpa (1 atm). between 273.15 and 1800 K. *J. Phys. Chem. Ref. Data,* 10:576–669.

Haas J. L. (1971). Effect of salinity on the maximum thermal gradient of a hydrothermal system at hydrostatic pressure. *Econ. Geol.,* 66:940–946.

Haas J. L. Jr. and Fisher J. R. (1976). Simultaneous evaluation and correlation of thermodynamic data. *Amer. Jour. Sci.,* 276:525–545.

Haasen P. (1978). *Physical Metallurgy.* Cambridge University Press, London.

Hackler R. T. and Wood B. J. (1989). Experimental determination of Fe and Mg exchange between garnet and olivine and estimation of Fe-Mg mixing properties in garnet. *Amer. Mineral.,* 74:994–999.

Hafner S. S. and Ghose S. (1971). Iron and magnesium distribution in cummingtonites. *Zeit. Kristallog.,* 133:301–326.

Häkli T. and Wright T. L. (1967). The fractionation of nickel between olivine and liquid as a geothermometer. *Geochim. Cosmochim. Acta,* 31:877–884.

Halback H. E. and Chatterjee N. D. (1982). An empirical Redlich-Kwong-type equation state for water to 1000°C and 200 kbar. *Contrib. Mineral. Petrol.,* 79:337–345.

Hamilton D. L. and MacKenzie W. S. (1965). Phase equilibrium studies in the system $NaAlSiO_4$ (nepheline)-$KAlSiO_4$ (kalsilite)-SiO_2-H_2O. *Min. Mag.,* 34:214–231.

Hanson G. N. and Gast P. W. (1967). Kinetic studies in contact metamorphic zones. *Geochim. Cosmochim. Acta,* 31:1119–1153.

Harlow, G. E. and Brown G. E. Jr. (1980). Low albite: An X-ray and neutron diffraction study. *Amer. Mineral.,* 65:986–995.

Harris P. G. (1957). Zone refining and the origin of potassic basalts. *Geochim. Cosmochim. Acta,* 12:195–208.

Harrison T. M. (1981). Diffusion of ^{40}Ar in hornblende. *Contrib. Mineral. Petrol.,* 78:324–331.

Harrison T. M. (1983). Some observations on the interpretation of $^{40}Ar/^{39}Ar$ age spectra. *Isot. Geosci.,* 1:319–338.

Harrison T. M. and McDougall I. (1980). Investigations of an intrusive contact, northwest Nelson, New Zealand, II: Diffusion of radiogenic and excess ^{40}Ar in hornblende revealed by $^{40}Ar/^{39}Ar$ age spectrum analysis. *Geochim. Cosmochim. Acta,* 44:2005–2020.

Harrison T. M. and McDougall I. (1981). Excess ^{40}Ar in metamorphic rocks from Broken Hill, New South Wales: Implications for $^{40}Ar/^{39}Ar$ age spectra and the thermal history of the region. *Earth Planet. Sci. Letters,* 55:123–149.

Harrison T. M. and McDougall I. (1982). The thermal significance of potassium feldspar K-Ar ages inferred from $^{40}Ar/^{39}Ar$ age spectrum results. *Geochim. Cosmochim. Acta*, 46:1811–1820.

Harrison W. J. (1977). An experimental study of the partitioning of samarium between garnet and liquid at high pressures. In *Papers Presented to the International Conference on Experimental Trace Elements Geochemistry*, Sedona, Arizona.

Harrison W. J. (1978). Rare earth element partitionings between garnets, pyroxenes and melts at low trace element concentration. *Carnegie Inst. Wash. Yb.*, 77:682–689.

Harrison W. J. (1981). Partition coefficients for REE between garnets and liquids: Implications of non-Henry's law behavior for models of basalt origin and evolution. *Geochim. Cosmochim. Acta*, 45:1529–1544.

Harrison W. J. and Wood B. J. (1980). An experimental investigation of the partitioning of REE between garnet and liquid with reference to the role of defect equilibria. *Contrib. Mineral. Petrol.*, 72:145–155.

Hart R. (1970). Chemical exchange between seawater and deep ocean basalts. *Earth. Planet. Sci. Letters.*, 9:269–279.

Hart S. R. (1964). The petrology and isotopic-mineral age relations of a contact zone in the Front Range, Colorado. *J. Geol.*, 72:493–525.

Harvey K. B. and Porter G. B. (1976). *Introduzione alla Chimica Fisica Inorganica*. Piccin Editore, Padova.

Haselton H. T. and Newton R. C. (1980). Thermodynamics of pyrope-grossular garnets and their stabilities at high temperatures and pressures. *J. Geophys. Res.*, 85:6973–6982.

Haselton H. T. and Westrum E. F. Jr. (1980). Low-temperature heat capacities of synthetic pyrope, grossular, and $pyrope_{60}grossular_{40}$. *Geochim. Cosmochim. Acta*, 44:701–709.

Haselton H. T., Hovis G. L., Hemingway B. S., and Robie R. A. (1983). Calorimetric investigation of the excess entropy of mixing in analbite-sanidine solid solutions: Lack of evidence for Na, K short range order and implications for two feldspar thermometry. *Amer. Mineral.*, 68:398–413.

Hawthorne F. C. (1976). The crystal chemistry of the amphiboles, V: The structure and chemistry of arfvedsonite. *Canadian Mineral.*, 346–356.

Hawthorne F. C. (1981a). Crystal chemistry of amphiboles. In *Reviews in Mineralogy*, vol. 9A, P. H. Ribbe (series ed.), Mineralogical Society of America.

Hawthorne F. C. (1981b). Some systematics of the garnet structure. *J. Solid State Chem.*, 37:157–164.

Hawthorne F. C. and Grundy H. D. (1976). The crystal chemistry of the amphiboles, IV: X-ray and neutron refinement of the crystal structure of tremolite. *Canadian Mineral.*, 14:334–345.

Hazen R. M. (1977). Effects of temperature and pressure on the crystal structure of ferromagnesian olivine. *Amer. Mineral.*, 62:286–295.

Hazen R. M. and Burnham C. W. (1973). The crystal structure of 1 layer phlogopite and annite. *Amer. Mineral.*, 58:889–900.

Hazen R. M. and Finger L. W. (1978). Crystal structure and compressibilities of pyrope and grossular to 60 kbar. *Amer. Mineral.*, 63:297–303.

Hazen R. M. and Finger L. W. (1979). Bulk-modulus volume relationship for cation-anion polyhedra. *J. Geophys. Res.*, 84:6723–6728.

Hazen R. M. and Finger L. W. (1982). *Comparative Crystal Chemistry*. New York: John Wiley.

Hazen R. M. and Prewitt C. T. (1977). Effects of temperature and pressure on interatomic distances in oxygen-based minerals. *Amer. Mineral.*, 62:309–315.

Hazen R. M. and Wones D. R. (1972). The effect of cation substitutions on the physical properties of trioctahedral micas. *Amer. Mineral.*, 57:103–129.

Helgeson H. C. (1967). Solution chemistry and metamorphism. In *Research in Geochemistry*, vol. 2, P. H. Abelson (series ed.). New York: John Wiley.

Helgeson H. C. (1969). Thermodynamics of hydrothermal system at elevated temperatures and pressures. *Amer. Jour. Sci.*, 267:729–804.

Helgeson H. C. and Kirkham D. H. (1974). Theoretical prediction of thermodynamic behavior of aqueous electrolytes at high pressures and temperatures, I: Summary of the thermodynamic/electrostatic properties of the solvent. *Amer. Jour. Sci.*, 274:1089–1198.

Helgeson H. C., Brown T. H., and Leeper R. H. (1976). *Handbook of Theoretical Activity Diagrams Depicting Chemical Equilibria in Geologic Systems Involving an Aqueous Phase at 1 atm and 0 to 300 °C.* Freeman, Cooper, San Francisco.

Helgeson H. C., Delany J., and Bird D. K. (1978). Summary and critique of the thermodynamic properties of rock-forming minerals. *Amer. Jour. Sci.*, 278A:1–229.

Helgeson H. C., Kirkham D. H., and Flowers G. C. (1981). Theoretical prediction of the thermodynamic behavior of aqueous electrolytes at high pressures and temperatures, IV: Calculation of activity coefficients, osmotic coefficients and apparent molal and standard and relative partial molal properties to 600 °C and 5 kbar. *Amer. Jour. Sci.*, 281:1249–1516.

Helgeson H. C., Murphy W. M., and Aagaard P. (1984). Thermodynamic and kinetic constraints on reaction rates among minerals and aqueous solutions, II: Rate constants, effective surface area, and the hydrolysis of feldspar. *Geochim. Cosmochim. Acta.*, 48:2405–2432.

Hem J. D. and Lind C. J. (1983). Nonequilibrium models for predicting forms of precipitated manganese oxides. *Geochim. Cosmochim. Acta*, 47:2037–2046.

Hem J. D., Roberson C. E., and Fournier R. B. (1982). Stability of -$MnOOH$ and manganese oxide deposition from spring water. *Water. Resour. Res.*, 18:563–570.

Hemingway B. S., Krupka K. M., and Robie R. A. (1981). Heat capacities of the alkali feldspars between 350 and 1000 K from differential scanning calorimetry, the thermodynamic functions of alkali feldspars from 298.15 to 1400 K, and the reaction quartz + jadeite = albite. *Amer. Mineral.*, 66:1202–1215.

Henderson L. M. and Kracek F. C. (1927). The fractional precipitation of barium and chromates. *J. Amer. Chem. Soc.*, 49:739–749.

Henderson P. (1982). *Inorganic Geochemistry.* Pergamon Press, Oxford.

Henley R. W. (1984). Chemical structure of geothermal systems. In *Reviews in Economic Geology*, vol. 1, J. M. Robertson (series ed.), Society of Economic Geologists.

Hensen B. J. (1977). Cordierite garnet bearing assemblages as geothermometers and barometers in granulite facies terranes. *Tectonophysics*, 43:73–88.

Hensen B. J. and Green D. H. (1973). Experimental study of the stability of cordierite and garnet in pelitic compositions at high pressures and temperatures. *Contrib. Mineral. Petrol.*, 38:151–166.

Hensen B. J., Schmid R., and Wood B. J. (1975). Activity-composition relationship for pyrope-grossular garnet. *Contrib. Mineral. Petrol.*, 54:161–166.

Herman F. (1959). Lattice vibrational spectrum of germanium. *J. Phys. Chem. Solids*, 8:405–418.

Hertoghen J. and Gijbels K. (1976). Calculation of trace element fractionation during partial melting. *Geochim. Cosmochim. Acta*, 40:313–322.

Herzberg C. T. (1978a). The bearing of phase equilibria in simple and complex systems on the origin and evolution of some garnet websterites. *Contrib. Mineral. Petrol.*, 66:375–382.

Herzberg C. T. (1978b). Pyroxene geothermometry and geobarometry: Experimental and thermodynamic evaluation of some subsolidus phase relations involving pyroxenes in the system $CaO\text{-}MgO\text{-}Al_2O_3\text{-}SiO_2$. *Geochim. Cosmochim. Acta*, 42:945–957.

Herzberg C. T. and Lee S. M. (1977). Fe-Mg cordierite stability in high grade pelitic rocks based on experimental, theoretical, and natural observations. *Contrib. Mineral. Petrol.*, 63:175–193.

Hess P. C. (1971). Polymer models of silicate melts. *Geochim. Cosmochim. Acta*, 35:289–306.

Hess P. C. (1977). Structure of silicate melts. *Canadian Mineral.*, 15:162–178.

Hewitt D. A. and Wones D. R. (1981). The annite-sanidine-magnetite equilibrium GAC-MAG Joint Annual Meeting, Calgary, Abstracts, 6:A-66.

Hewitt D. A. and Wones D. R. (1984). Experimental phase relations of the micas. In *Reviews in Mineralogy*. vol. 13, P. H. Ribbe (series ed.), Mineralogical Society of America.

Hillert M. (1980). Empirical methods of predicting and representing thermodynamic properties of ternary solution phases. *CALPHAD,* 4:1–12.

Himmelbau D. M. (1960). Solubilities of inert gases in water. *J. Chem. Eng. Data,* 5:10–15.

Hinrichsen T. and Schurmann K. (1971). Synthese und stabilität von glimmern im System CaO-Na_2O-Al_2O_3-SiO_2-H_2O. *Forschr. Mineral.,* 49:21.

Hinshelwood C. N. (1951). *The Structure of Physical Chemistry* Oxford University Press, London.

Hinze J. A. and Jaffe H. H. (1962). Electronegativity, I: Orbital electronegativity of neutral atoms. *J. Amer. Chem. Soc.,* 84:540–546.

Hinze J. A., Whitehead M. A., and Jaffe H. H. (1963). Electronegativity, II: Bond and orbital electronegativity. *J. Inorg. Chem.,* 38:983–984.

Hirschfelder J. O., Curtiss C. F., and Bird R. B. (1954). *The Molecular Theory of Gases and Liquids.* New York: John Wiley.

Hirschmann M. (1991). Thermodynamics of multicomponent olivines and the solution properties of $(Ni,Mg,Fe)_2SiO_4$ and $(Ca,Mg,Fe)_2SiO_4$ olivines. *Amer. Mineral.,* 76:1232–1248

Hirt B., Herr W., and Hoffmeister W. (1963). Age determinations by the rhenium-osmium method. In *Radioactive Dating,* International Atomic Energy Agency, Vienna.

Hoefs J. (1980). *Stable Isotope Geochemistry.* Berlin-Heidelberg-New York: Springer-Verlag.

Hofman A. W. and Giletti B. J. (1970). Diffusion of geochronologically important nuclides in minerals under hydrothermal conditions. *Eclogae Geol. Helv.,* 63:141–150.

Holdaway M. J. (1971). Stability of andalusite and the aluminium silicate phase diagram. *Amer. Jour. Sci.,* 271:97–131.

Holdaway M. J. (1972). Thermal stability of Al-Fe epidote as a function of fO_2 and Fe content. *Contrib. Mineral. Petrol.,* 37:307–340.

Holdaway M. J. and Lee S. M. (1977). Fe-Mg cordierite stability in high grade pelitic rocks based on experimental, theoretical and natural observations. *Contrib. Mineral. Petrol.,* 63:175–198.

Holdaway M. J. and Mukhopadhyay B. (1993). A reevaluation of the stability relations of andalusite: Thermochemical data and phase diagram for the aluminum silicates. *Amer. Mineral.,* 78:298–315.

Holder J. and Granato A. V. (1969). Thermodynamic properties of solid containing defects. *Phys. Rev.,* 182:729–741.

Holgate N. (1954). The role of liquid immiscibility in igneous petrogenesis. *J. Geol.,* 62:439–306.

Holland H. D. (1972). Granites-solutions and base metal deposits. *Econ. Geol.,* 67:281–301.

Holland H. D. (1978). *The Chemistry of Atmosphere and Oceans.* New York: Wiley Interscience.

Holland T. J. B. (1989). Dependence of entropy on volume for silicate and oxide minerals: A review and predictive model. *Amer. Mineral.,* 74:5–13.

Holland T. J. B. and Powell R. (1990). An enlarged and updated internally consistent thermodynamic dataset with uncertainties and correlations: The system K_2O-Na_2O-CaO-MgO-MnO-FeO-Fe_2O_3-Al_2O_3-TiO_2-SiO_2-C-H_2-O_2. *J. Metamorphic Geol.,* 8:89–124.

Holloway J. R. (1977). Fugacity and activity of molecular species in supercritical fluids. In *Thermodynamics in Geology,* D. G. Fraser, ed. Reidel, Dordrecht.

Holloway J. R. (1987). Igneous fluids. In *Reviews in Mineralogy,* vol. 17, P. H. Ribbe (series ed.), Mineralogical Society of America.

Holm J. L. and Kleppa O. J. (1968). Thermodynamics of the disordering process in albite. *Amer. Mineral.,* 53:123–133.

Hoover J. D. (1978). The distribution of samarium and tulium between plagioclase and liquid in the system An-Di and Ab-An-Di at T = 1300°C. *Carnegie Inst. Wash. Yb.,* 77:703–706.

Horan M. F., Morgan J. W., Grauch R. I., Coveney R. M. Jr., Murowchick J. B., and Hulbert L. J. (1994). Rhenium and osmium isotopes in black shales and Ni-Mo-PGE-rich sulfide layers, Yukon Territory, Canada, and Hunan and Guizhou provinces, China. *Geochim. Cosmochim. Acta,* 58:257–265.

Horita J., Wesolowski D. J., and Cole D. R. (1993a). The activity-composition relationship of oxygen and hydrogen isotopes in aqueous salt solutions, I: Vapor-liquid water equilibration of single salt solutions from 50 to 100°C. *Geochim. Cosmochim. Acta,* 57:2797–2817.

Horita J., Cole D. R., and Wesolowski D. J. (1993b). The activity-composition relationship of oxygen and hydrogen isotopes in aqueous salt solutions, II: Vapor-liquid water equilibration of mixed salt solutions from 50 to 100°C and geochemical implications. *Geochim. Cosmochim. Acta,* 57:4703–4711.

Hovis G. L. (1974). A solution calorimetric and X-ray investigation of Al-Si distribution in monoclinic potassium feldspars. In *The feldspars,* W. S. MacKenzie and J. Zussman, eds. Manchester: Manchester University Press.

Hovis G. L. and Waldbaum D. R. (1977). A solution calorimetric investigation of K-Na mixing in a sanidine-analbite join-exchange series. *Amer. Mineral.,* 53:1965–1979.

Hsu L. C. (1968). Selected phase relationships in the system Al-Mn-Fe-Si-O-H: A model for garnet equilibria. *J. Petrol.,* 9:40–83.

Huckenholtz H. G. and Knittel D. (1975). Uvarovite: Stability of uvarovite-grossularite solid solutions at low pressure. *Contrib. Mineral. Petrol.,* 49:211–232.

Huckenholtz H. G., Schairer J. F., and Yoder H. S. Jr. (1969). Synthesis and stability of ferri-diopside. *Mineral. Soc. Amer. Spec. Paper,* 2:163–177.

Huebner J. S. (1982). Phyroxene phase equilibria at low pressure. In *Reviews in Mineralogy,* vol. 7 (2d ed.), P. H. Ribbe (series ed.), Mineralogical Society of America.

Huebner J. S. and Papike J. J. (1970). Synthesis and crystal chemistry of sodium-potassium richterite, $(Na, K)NaCaMg_5Si_8O_{22}(OH,F)_2$. *Amer. Mineral.,* 55:1973–1993.

Huebner J. S. and Sato M. (1970). The oxygen fugacity-temperature relationships of manganese oxide and nickel oxide buffers. *Amer. Mineral.,* 55:934–952.

Huebner J. S., Lipin B. R., and Wiggins L. B. (1976). Partitioning of chromium between silicate crystal and melts. *Proc. Seventh Lunar Sci. Conf.,* 1195–1220.

Huheey J. E. (1975). *Inorganic Chemistry: Principles of Structure and Reactivity.* New York: Harper and Row.

Hurd D. C. (1972). Factors affecting solution rate of biogenic opal in sea water. *Earth Planet. Sci. Letters,* 15:411–417.

Hutner R. A., Rittner E. S., and Du Pré F. K. (1949). Concerning the work of polarization in ionic crystals of the NaCl type, II: Polarization around two adjacent charges in the rigid lattice. *J. Chem. Phys.,* 17:204–208.

Iczkowski R. P. and Margrave J. L. (1961). Electronegativity. *J. Amer. Chem. Soc.,* 83: 3547–3551.

Iiyama J. T. (1972). Fixation des elements alcalinoterreux, Ba, Sr et Ca dans les feldspaths; etude experimentale. *Proc. 24 Congres Geol. Internat.,* Section 10, 122–130.

Iiyama J. T. (1973). Behavior of trace elements in feldspars under hydrothermal conditions. In *The Feldspars,* W. S. MacKenzie and J. Zussmann, eds. Manchester: Manchester University Press.

Iiyama J. T. (1974). Substitution, deformation locale de la maille et equilibre de distribution des elements en tracs entre silicates et solution hydrothermale. *Bul. Soc. Fr. Mineral. Cristallog.,* 97:143–151.

Irefune T., Ohtani E., and Kumazawa I. (1982). Stability field of knorringite at high pressure and its application to the occurrence of Cr-rich pyrope in the upper mantle. *Phys. Earth Planet. Interiors,* 27:263–272.

Irving A. J. (1974). Geochemical and high-pressure experimental studies of garnet pyroxenite and granulite xenoliths from the Delegate basaltic pipes, Australia. *J. Petrol.,* 15:1–40.

Irving A. J. (1978). A review of experimental studies of crystal/liquid trace element partitioning. *Geochim. Cosmochim. Acta.* 42:743–770.

Isaak D. G. and Graham E. K. (1976). The elastic properties of an almandine-spessartite garnet and elasticity in the garnet solid solutions series. *J. Geophys. Res.,* 81:2483–2489.

Isaak D. G., Anderson L., and Oda H. (1992). High-temperature thermal expansion and elasticity of calcium-rich garnets. *Phys. Chem. Minerals,* 19:106–120.

Ivanov I. P., Potekhin V. Y., Dmitriyenko L. T., and Beloborodov S. M. (1973). An experimental study of T and P conditions of equilibrium of reaction: Muscovite-Kfeldspar + corundum + H_2O at $P(H_2O)<P(total)$. *Geokhimiya,* 9:1300–1310.

Jäger E. and Hunziker J. C. (1979). *Lectures in Isotope Geology* Berlin-Heidelberg-New York: Springer-Verlag.

Jäger E., Niggli E., and Wenk E. (1967). Alterbestimmungen an Glimmern der Zentralalpen. *Beitr. Geol. Karte Schweitz,* NF134, Bern.

Jahn H. A. and Teller E. (1937). Stability of polyatomic molecules in degenerate electronic states. *Proc. Roy. Soc. London,* A161:220–235.

JANAF (1974–1975). *Thermochemical Tables* U.S. Department of Commerce, National Bureau of Standards, Institute for Applied Technology, Supplements.

Jaoul O., Froidevaux C., Durham W. B., and Michaud M. (1980). Oxygen self-diffusion in forsterite: Implications for the high temperature creep mechanism. *Earth. Planet. Sci. Letters,* 47:391–397.

Jaoul O., Poumellec M., Froidevaux C., and Havette A. (1981). Silicon diffusion in forsterite: A new constraint for understanding mantle deformation. In *Anelasticity in the Earth,* F. D. Stacey et al., eds. Godyn Ser., vol. 4, AGU, Washington, D.C.

Jenne E. A. (1981). Geochemical modeling: A review. In *Waste/Rock Interactions Technology Program,* Pacific Northwest Laboratory, Publication 3574.

Jenne E. A., Girovin D. C., Ball J. W., and Burchard J. M. (1978). Inorganic speciation of silver in natural waters-fresh to marine. In *Environmental Impacts of Nucleating Agents Used in Weather Modification Programs,* D. A. Klein, ed. Dowden-Hutchison and Ross, Strasburg, Pennsylvania.

Jensen B. (1973). Patterns of trace elements partitioning. *Geochim. Cosmochim. Acta,* 37:2227–2242.

Johnson J. W. and Norton D. (1991). Critical phenomena in hydrothermal systems: State, thermodynamic, electrostatic, and transport properties of H_2O in the critical region. *Amer. Jour. Sci.,* 291:541–648.

Johnson J. W., Oelskers E. H., and Helgeson H. C. (1991). *SUPCRT92: A Software Package for Calculating the Standard Molal Thermodynamic Properties of Minerals, Gases, Aqueous Species, and Reactions from 1 to 5000 bars and 0° to 1000°C* Earth Sciences Department, L-219, Lawrence Livermore National Laboratory, Livermore, California.

Johnston W. D. (1964). Oxidation-reduction equilibria in iron-containing glass. *J. Amer. Ceram. Soc.,* 47:198–201.

Johnston W. D. (1965). Oxidation-reduction equilibria in molten Na_2O-$2SiO_2$ glass. *J. Amer. Ceram Soc.,* 48:184–190.

Johnston W. D., and Chelko A. (1966). Oxidation-reduction equilibria in molten Na_2O-$2SiO_2$ glass in contact with metallic copper and silver. *J. Amer. Ceram. Soc.,* 49:562–564.

Jones J. H. (1995). Experimental trace element partitioning. In *Rock Physics and Phase Relations. A Handbook of Physical Constants,* T. J. Ahrens, ed., AGU, Washington, D.C.

Jost W. (1960). *Diffusion in Solid State Physics.* New York: Academic Press.

Kamensky I. L., Tokarev I. V., and Tolstikhin I. N. (1991). ^3H-^3He dating: A case for mixing of young and old groundwaters. *Geochim. Cosmochim. Acta,* 55:2895–2899.

Karpinskaya T. B., Ostrovsky I. A., and Yevstigneyeva O. (1983). Synthetic pure iron garnet skiagite. *Internat. Geol. Rev.,* 25:1129–1130.

Kato K. and Nukui A. (1976). Die Kristallstructur des monoclinen Tief-trydymits. *Acta Cryst.,* B32:2486–2491.

Katz A. and Matthews A. (1977). The dolomitization of $CaCO_3$: An experimental study at 252–295°C. *Geochim. Cosmochim. Acta,* 41:297–308.

Kawasaki T. and Matsui Y. (1977). Partitioning of Fe^{2+} and Mg^{2+} between olivine and garnet. *Earth. Planet. Sci. Letters,* 37:159–166.

Kelley K. K. (1960). Contributions to the data on theoretical metallurgy, XIII: High temperature heat content, heat capacity and entropy data for the elements and inorganic compounds. *U.S. Bur. Mines Bull.,* 584, 232 pp.

Kern D. M. (1960). The hydration of carbon dioxide. *J. Chem. Educ.*, 37:14–23.

Kerrick D. M. (1972). Experimental determination of muscovite + quartz stability with $P(H_2O)$ < P (total). *Amer. Jour. Sci.*, 272:946–958.

Kerrick D. M. and Darken L. S. (1975). Statistical thermodynamic models for ideal oxide and silicate solid solutions, with applications to plagioclase. *Geochim. Cosmochim. Acta*, 39:1431–1442.

Kerrick D. M. and Jacobs G. K. (1982). A modified Redlich-Kwong equation for H_2O, CO_2 and H_2O-CO_2 mixtures at elevated pressures and temperatures. *Amer. Jour. Sci.*, 281:735–767.

Kieffer S. W. (1979a). Thermodynamics and lattice vibrations of minerals, 1: Mineral heat capacities and their relationships to simple lattice vibrational models. *Rev. Geophys. Space Phys.*, 17:1–19.

Kieffer S. W. (1979b). Thermodynamics and lattice vibrations of minerals, 2: Vibrational characteristics of silicates. *Rev. Geophys. Space Phys.*, 17:20–34.

Kieffer S. W. (1979c). Thermodynamics and lattice vibrations of minerals, 3: Lattice dynamics and an approximation for minerals with application to simple substances and framework silicates. *Rev. Geophys. Space Phys.*, 17:35–59.

Kieffer S. W. (1980). Thermodynamics and lattice vibrations of minerals, 4: Application to chain and sheet silicates and orthosilicates. *Rev. Geophys. Space Phys.*, 18:862–886.

Kieffer S. W. (1982). Thermodynamics and lattice vibrations of minerals, 5: Applications to phase equilibria, isotopic fractionation, and high-pressure thermodynamic properties. *Rev. Geophys. Space Phys.*, 20:827–849.

Kieffer S. W. (1985). Heat capacity and entropy: Systematic relations to lattice vibrations. In *Reviews in Mineralogy*, vol. 14, P. H. Ribbe (series ed.), Mineralogical Society of America.

Kieffer S. W. and Navrotsky A. (1985). Scientific perspective. In *Reviews in Mineralogy*, vol. 14, P. H. Ribbe (series ed.)., Mineralogical Society of America.

Kielland J. (1937). Individual activity of ions in aqueous solutions. *J. Amer. Chem. Soc.*, 59:1675–1678.

Kim K. and Burley B. J. (1971). Phase equilibria in the system $NaAlSi_3O_8$-$NaAlSiO_4$-H_2O with special emphasis on the stability of analcite. *Canadian Jour. Earth Sci.*, 8:311–337.

Kim K. Y., Chabildas L. C., and Ruoff A. L. (1976). Isothermal equations of state for lithium fluoride. *J. Appl. Phys.*, 47:2862–2866.

Kirby S. H. and Wegner M. W. (1978). Dislocation substructure of mantle-derived olivine as revealed by selective chemical etching and transmission electron microscopy. *Phys. Chem. Minerals*, 3:309–330.

Kirsten T. (1968). Incorporation of rare gases in solidifying enstatite melts. *J. Geophys. Res.*, 73:2807–2810.

Kiseleva I. A. (1977). Gibbs free energy of formation of calcium garnets. *Geokhimiya*, 5:705–715.

Kitayama K. and Katsura T. (1968). Activity measurements in orthosilicate and metasilicate solid solutions, I: Mg_2SiO_4-Fe_2SiO_4 and $MgSiO_3$-$FeSiO_3$ at 1204 °C. *Bull. Chem. Soc. Jpn.*, 41:1146–1151.

Kittel C. (1971). *Introduction to Solid State Physics.* New York: John Wiley.

Kittel C. (1989). How to define entropy. *Nature*, 339:170.

Klein C. and Bricker O. P. (1977). Some aspects of the sedimentary and diagenetic environment of proterozoic banded iron formation. *Econ. Geol.*, 72:1457–1470.

Klein C. Jr. (1964). Cummingtonite-grunerite series: A chemical optical and X-ray study. *Amer. Mineral.*, 49:963–982.

Klein C. Jr. and Ito J. (1968). Zincian and manganoan amphiboles from Franklin, New Jersey. *Amer. Mineral.*, 53:1264–1275.

Kohler F. (1960). Zur Berechnung der thermodynamischen Daten eines ternären Systems aus den zugehörigen binären Systemen. *Monatsh. Chem.*, 96:1228–1251.

Köhler T. P. and Brey G. P. (1990). Calcium exchange between olivine and clinopyroxene calibrated as a geothermobarometer for natural peridotites from 2 to 60 kb with applications. *Geochim. Cosmochim. Acta,* 54:2375–2388.

Kohlstedt D. L. and Goetze C. (1974). Low-stress high-temperature creep in olivine single crystals. *J. Geophys. Res.,* 79:2045–2051.

Kohlstedt D. L. and Vander Sande J. B. (1973). Transmission electron microscopy investigation of defects microstructure of four natural orthopyroxenes. *Contrib. Mineral. Petrol.,* 42:169–180.

Kostov I. (1975). Crystal chemistry and classification of silicate minerals. *Geochem. Mineral. Petrol.,* 1:5–14.

Kotelnikov A. R., Bychkov A. M., and Chernavina N. I. (1981). Experimental study of calcium distribution in granitoid plagioclase and water-salt fluid at 700°C and $P_{fl} = 1000$ atm. *Geokhimiya,* 21:707–721.

Koziol A. M. (1990). Activity-composition relationship of binary Ca-Fe and Ca-Mn garnets determined by reversed, displaced equilibrium experiments. *Amer. Mineral.,* 75:319–327.

Kretz R. (1982). Transfer and exchange equilibria in a portion of the pyroxene quadrilateral as deduced from natural and experimental data. *Geochim. Cosmochim. Acta,* 46:411–422.

Kröger F. A. (1964). *The Chemistry of Imperfect Crystals.* North Holland Pub. Co., Amsterdam.

Krogh E. J. (1988). The garnet-clinopyroxene Fe-Mg geothermometer—a reinterpretation of existing experimental data. *Contrib. Mineral. Petrol.,* 99:44–48.

Kroll H. and Ribbe P. H. (1983). Lattice parameters, composition and Al, Si order in alkali feldspars. In *Reviews in Mineralogy,* vol. 2 (2d ed.), P. H. Ribbe (series ed.), Mineralogical Society of America.

Kroll H., Bambauer H. U., and Schirmer U. (1980). The high albite-monalbite and analbite-monalbite transitions. *Amer. Mineral.,* 65:1192–1211.

Kumar M. D. (1987). Cation hydrolysis and the regulation of trace metal composition in seawater. *Geochim. Cosmochim. Acta,* 51:2137–2145.

Kurkjian C. R. and Russel L. E. (1958). Solubility of water in molten alkali silicates. *Jour. Soc. Glass Technol.,* 42:130–144.

Ladd M. F. C. (1979). *Structure and Bonding in Solid State Chemistry.* New York: Ellis Harwood Limited.

Ladd M. F. C. and Lee W. H. (1958). Solubility of some inorganic halides. *Trans. Faraday Soc.,* 54:34–39.

Ladd M. F. C. and Lee W. H. (1959). Calculation of lattice energies. *J. Inorg. Nucl. Chem.,* 11:264–271.

Ladd M. F. C. and Lee W. H. (1964). Lattice energy and related topics. In *Progress in Solid State Chemistry,* vol. 1, H. Reiss, ed. New York: MacMillan.

Lagache M. (1971). Etude experimental de la rertition du cesium entre les felspaths sodi-potassiques et des solutions hydrothermales a 700°C, 1 kb. *C. R. Acad. Sci. Ser. D.* 272:1328–1330.

Lagache M. and Sabatier G. (1973). Distribution des elements Na, K, Rb et Cs a l'etat de trace entre feldspaths alcalins et solutions hydrothermales a 650°C, 1 kbar: Données experimentales et interpretation thermodynamique. *Geochim. Cosmochim. Acta,* 37: 2617–2640.

Lager G. A. and Meagher D. M. (1978). High temperature structural study of six olivines. *Amer. Mineral.,* 63:365–377.

Landau L. D. and Lifshitz E. M. (1980). *Statistical Physics* Pergamon Press, Oxford.

Landé A. (1920). The magnitude of the atom. *Z. Physik,* 1:191–197.

Lasaga A. C. (1979). Multicomponent exchange and diffusion in silicates. *Geochim. Cosmochim. Acta,* 43:455–469.

Lasaga A. C. (1980). Defects calculations in silicates: Olivine. *Amer. Mineral.,* 65:1237–1248.

Lasaga A. C. (1981a). Rate laws of chemical reactions. In *Reviews in Mineralogy,* vol. 8, P. H. Ribbe (series ed.), Mineralogical Society of America.

Lasaga A. C. (1981b). Transition state theory. In *Reviews in Mineralogy,* vol. 8, P. H. Ribbe (series ed.), Mineralogical Society of America.

Lasaga A. C. (1981c). The atomistic basis of kinetics: Defects in minerals. In *Reviews in Mineralogy,* vol. 8, P. H. Ribbe (series ed.), Mineralogical Society of America.

Lasaga A. C. (1983). Geospeedometry: An extension of geothermometry. In *Advances in Physical Geochemistry,* vol. 3, S. K. Saxena (series ed.). Berlin-Heidelberg-New York: Springer-Verlag.

Lasaga A. C. (1984). Chemical kinetics of water-rock interactions. *J. Geophys. Res.,* 89:4009–4025.

Lasaga A. C. and Cygan R. T. (1982). Electronic and ionic polarizabilities of silicate minerals. *Amer. Mineral.,* 67:328–334.

Lasaga A. C. and Kirkpatrick R. J. (1981). Kinetics of Geochemical processes. In *Reviews in Mineralogy,* vol. 8, P. H. Ribbe (series ed.), Mineralogical Society of America.

Lasaga A. C., Richardson S. M., and Holland H. D. (1977). The mathematics of cation diffusion and exchange between silicate minerals during retrograde metamorphism. In *Energetics of Geological Processes,* S. K. Saxena and S. Bhattacharji, eds. New York: Springer-Verlag.

Lattard D. and Schreyer W. (1983). Synthesis and stability of the garnet calderite in the system Fe-Mn-Si-O. *Contrib. Mineral. Petrol.,* 84:199–214.

Leeman W. P. and Lindstrom D. J. (1978). Patritioning of Ni^{2+} between basaltic and synthetic melts and olivines—an experimental study. *Geochim. Cosmochim. Acta,* 42:801–816.

Leitner B. J., Weidner D. J., and Liebermann R. C. (1980). Elasticity of single crystal pyrope and implications for garnet solid solution series. *Phys. Earth Planet. Interiors.,* 22:111.

Le Page Y. and Donnay G. (1976). Refinement of the crystal structure of low-quartz. *Acta Cryst.,* B32:2456–2459.

Le Page Y., Calvert L. D., and Gabe E. J. (1980). Parameter variation in low-quartz between 94 and 298 K. *Phys. Chem. Solids,* 41:721–725.

Lerman A. (1979). *Geochemical Processes: Water and Sediment Environments.* New York: John Wiley.

Levelt Sengers J. M. H., Kamgar-Parsi B., Balfour F. W., and Sengers J. V. (1983). Thermodynamic properties of steam in the critical region. *J. Phys. Chem. Ref. Data,* 12:1–28.

Levitskii V. A., Golovanova Yu. G., Popov S. G., and Chentsov V. N. (1975). Thermodynamics of binary oxide system: Thermodynamic properties of nickel orthosilicate. *Russ. J. Phys. Chem.,* 49:971–974.

Lewis G. N. (1916). The atom and the molecule. *J. Amer. Chem. Soc.,* 38:762–785.

Lewis G. N. and Randall M. (1921). The activity coefficient of strong electrolytes. *J. Amer. Chem. Soc.,* 43:1112–1154.

Lewis G. N. and Randall M. (1970). *Termodinamica.* Leonardo Edizioni Scientifiche, Roma.

Liebau F. (1982). Classification of silicates. In *Reviews in Mineralogy.* vol. 5, P. H. Ribbe (series ed.), Mineralogical Society of America.

Liebermann R. C. and Ringwood A. E. (1976). Elastic properties of anorthite and the nature of the lunar crust. *Earth Planet. Sci. Letters,* 31:69–74.

Lin T. H. and Yund R. A. (1972). Potassium and sodium self-diffusion in alkali feldspar. *Contrib. Mineral. Petrol.,* 34:177–184.

Lin Y. C. and Bailey S. W. (1984). The crystal structure of paragonite—$2M_1$. *Amer. Mineral.,* 19:122–127.

Lindner M., Leich D. A., Russ G. P., Bazan J. M., and Borg R. J. (1989). Direct determination of the half-life of ^{187}Re. *Geochim. Cosmochim. Acta,* 53:1597–1606.

Lindsley D. H. (1981). The formation of pigeonite on the join hedembergite-ferrosilite at 11, 5 and 15 kbar: Experiments and a solution model. *Amer. Mineral.,* 66:1175–1182.

Lindsley D. H. (1982). Phase equilibria of pyroxenes at pressure > 1 atmosphere. In *Reviews in Mineralogy,* vol. 7 (2d ed.), P. H. Ribbe (series ed.), Mineralogical Society of America.

Lindsley D. H. (1983). Pyroxene thermometry. *Amer. Mineral.,* 68:477–493.

Lindsley D. H. and Anderson D. J. (1982). A two pyroxene thermometer. *Proc. 13th Lunar Planet. Sci. Conf.*, 887–906.

Lindsley D. H. and Dixon S. A. (1976). Diopside-enstatite equilibria at 850 to 1400°C, 5 to 35 kbar. *Amer. Jour. Sci.*, 276:1285–1301.

Lindsley D. H., Grover J. E., and Davidson P. M. (1981). The thermodynamics of the $Mg_2Si_2O_6$-$CaMgSi_2O_6$ join: A review and a new model. In *Advances in Physical Geochemistry*, vol. 1, S. K. Saxena (series ed.), New York: Springer-Verlag.

Lindstrom D. J. (1976). Experimental study of the partitioning of the transition metals between clinopyroxene and coexisting silicate liquids. Ph.D. diss., University of Oregon.

Lindstrom D. J. and Weill D. F. (1978). Partitioning of transition metals between diopside and coexisting silicate liquids, I: Nickel, cobalt, and manganese. *Geochim. Cosmochim. Acta*, 42:801–816.

Liu K. K. and Epstein S. (1984). The hydrogen isotope fractionation between kaolinite and water. *Chem. Geol. (Isot. Geosci. Sec.).*, 2:335–350.

Liu L. and Bassett A. (1986). Elements oxides and silicates. In *Oxford Monographs on Geology and Geophysics*, vol. 4, P. Allen, H. Charnock, E. R. Oxburg, and B. J. Skinner, eds. New York: Oxford University Press.

Loewenstein W. (1954). The distribution of aluminum in the tetrahedra of silicates and aluminates. *Amer. Mineral.*, 39:92–96.

London F. (1930). Properties and application of molecular forces. *Z. Phys. Chem.*, B11:222–251.

Long P. E. (1978). Experimental determination of partition coefficients for Rb, Sr and Ba between alkali feldspar and silicate liquid. *Geochim. Cosmochim. Acta*, 42:833–846.

Luck J. M. and Allègre C. J. (1983). ^{187}Re-^{187}Os systematics in meteorites and cosmological consequences. *Nature*, 302:130–132.

Lumpkin G. R. and Ribbe P. H. (1983). Composition, order-disorder and lattice parameters of olivines: Relationships in silicate, germanate, beryllate, phosphate and borate olivine. *Amer. Mineral.*, 68:164–176.

Lux G. (1987). The behavior of noble gases in silicate liquids: Solution, diffusion, bubbles and surface effects, with application to natural samples. *Geochim. Cosmochim. Acta*, 51:1549–1560.

Lux H. Z. (1939). Acid and bases in fused salt bath: The determination of oxygen-ion concentration. *Z. Elektrochem.*, 45:303–309.

Mackrodt W. C. and Stewart R. F. (1979). Defect properties of ionic solids, II: Point defect energies based on modified electron gas potentials. *J. Phys.*, 12C:431–449.

Mah A. D. and Pancratz L. B. (1976). Thermodynamic properties of nickel and its inorganic compounds. *U.S. Bur. Mines Bull.*, 668:1–125.

Maier C. G. and Kelley K. K. (1932). An equation for the representation of high temperature heat content data. *J. Amer. Chem. Soc.*, 54:3243–3246.

Manacorda T. (1985). *Elementi di Meccanica Ondulatoria.* Istituto di Matematiche Applicate Ulisse Dini, Pisa.

Mantoura R. C. F., Dickson A., and Riley J. P. (1978). The complexation of metals with humic materials in natural waters. *Estuarine Coastal Mar. Sci.*, 6:387–408.

Maresch M. V., Mirwald P. W., and Abraham K. (1978). Nachweis einer Mischlüke in der Olivinreihe Forsterit (Mg_2SiO_4)-Tephroit (Mn_2SiO_4). *Fortschr. Miner.*, 56:89–90.

Marshall B. D. and De Paolo D. J. (1982). Precise age determination and petrogenetic studies using the K-Ca method. *Geochim. Cosmochim. Acta*, 46:2537–2545.

Marshall W. L. and Franck E. U. (1981). Ion product of water substance, 0–1000°C, 1–10,000 bars: New international formulation and its background. *J. Phys. Chem. Ref. Data*, 10:295–303.

Martin R. F. (1970). Cell parameters and infra-red absorption of synthetic high to low albites. *Contrib. Mineral. Petrol.*, 26:62–74.

Martin R. F. (1974). Controls of ordering and subsolidus phase relations in the alkali feldspars.

In *The feldspars,* W. S. MacKenzie and J. Zussman, eds. Manchester: Manchester University Press.

Mason R. A. (1982). Trace element distributions between the perthite phases of alkali feldspars from pegmatites. *Min. Mag.,* 45:101–106.

Masse D. P., Rosèn E., and Muan A. (1966). Activity composition relations in Co_2SiO_4-Fe_2SiO_4 solid solution at 1180°C. *J. Amer. Ceram. Soc.,* 49:328–329.

Masson C. R. (1965). An approach to the problem of ionic distribution in liquid silicates. *Proc. Roy. Soc. London,* A287:201–221.

Masson C. R. (1968). Ionic equilibrium in liquid silicates. *J. Amer. Ceram. Soc.,* 51:134–143.

Masson C. R. (1972). Thermodynamics and constitution of silicate slags. *Jour. Iron Steel Inst.,* 210:89–96.

Matsui Y. and Banno S. (1962). Intercrystalline exchange equilibrium in silicate solid solutions. *Proc. Japan Acad.,* 41:461–466.

Matsui Y. and Nishizawa O. (1974). Iron magnesium exchange equilibrium between coexisting synthetic olivine and orthopyroxene. *Amer. Jour. Sci.,* 267:945–968.

Matsui Y. and Syono Y. (1968). Unit cell dimension of some synthetic olivine group solid solutions. *Geochem. Jour.,* 2:51–59.

Matthews A. (1980). Influences of kinetics and mechanism in metamorphism: A study of albite crystallization. *Geochim. Cosmochim. Acta,* 44:387–402.

Mattioli G. S. and Bishop F. (1984). Experimental determination of chromium-aluminum mixing parameter in garnet. *Geochim. Cosmochim. Acta,* 48:1367–1371.

Mayer J. G. (1933). Dispersion and polarizability and the Van Der Waals potential in the alkali halides. *J. Chem. Phys.,* 1:270–279.

McCallister R. H. and Yund R. A. (1977). Coherent exsolution in Fe-free pyroxenes. *Amer. Mineral.,* 62:721–726.

McCallum I. S. and Charette M. P. (1978). Zr and Nb partition coefficients: Implications for the genesis of mare basalts, KREEP and sea floor basalts. *Geochim. Cosmochim. Acta,* 42:859–870.

McClure D. S. (1957). The distribution of transition metal cations in spinels. *J. Phys. Chem. Solids,* 3:311–317.

McConnell D. (1966). Propriétes physique des grenats:calcul de la dimension de la maille unite a partir de la composition chimique. *Geochim. Cosmochim. Acta,* 31:1479–1487.

McDougall I. and Harrison T. M. (1988). *Geochronology and Thermochronology by the $^{40}Ar/^{39}Ar$ Method.* Oxford Monographs on Geology and Geophysics, no. 9. New York: Oxford University Press.

McGregor I. D. (1974). The system MgO-Al_2O_3-SiO_2: Solubility of Al_2O_3 in enstatite for spinel and garnet peridotite composition. *Amer. Mineral.,* 59:110–119.

McIntire W. L. (1963). Trace element partition coefficients—a review of theory and applications to geology. *Geochim. Cosmochim. Acta,* 27:1209–1269.

McKay G. A. and Weill D. F. (1976). Petrogenesis of KREEP. *Proc. Seventh Lunar Sci. Conf.,* 2427–2447.

McMillan P. (1985). Vibrational spectroscopy in the mineral sciences. In *Reviews in Mineralogy,* vol. 14, P. H. Ribbe (series ed.), Mineralogical Society of America.

Meadowcroft T. R. and Richardson F. D. (1965). Structural and thermodynamic aspects of phosphate glasses. *Trans. Faraday Soc.,* 61:54–70.

Meagher E. P. (1975). The crystal structures of pyrope and grossularite at elevated temperatures. *Amer. Mineral.,* 60:218–228.

Meagher E. P. (1982). Silicate garnets. In *Reviews in Mineralogy,* vol. 5, P. H. Ribbe (series ed.), Mineralogical Society of America.

Medaris L. G. (1969). Partitioning of Fe^{2+} and Mg^{2+} between coexisting synthetic olivine and orthopyroxene. *Amer. Jour. Sci.,* 267:945–968.

Megaw H. D. (1974). The architecture of the feldspars. In *The Feldspars,* W. S. MacKenzie and J. Zussman, eds. Manchester: Manchester University Press.

Mel'nik Y. P. (1972). Thermodynamic parameters of compressed gases and metamorphic reactions involving water and carbon dioxide. *Geochem. Int.*, 9:419–426.

Mendeleev D. (1905). *Principles of Chemistry* London.

Mendeleev D. (1908). *Versuch einen chemisschen Aufflassung des Weltäthers* St. Petersburg.

Menzer G. (1926). Die kristallstructur von granat. *Zeit. Kristallog.*, 63:157–158.

Merkel G. A. and Blencoe J. G. (1982). Thermodynamic procedures for treating the monoclinic/triclinic inversions as a high-order phase transition in equations of state for binary analbite-sanidine feldspars. In *Advances in Physical Geochemistry*, vol. 2, S. K. Saxena (series ed.). New York: Springer-Verlag.

Merlino S. (1976). Framework silicates. *Izvj. Jugoslav. Centr. Krist.*, 11:19–37.

Merrihue C. and Turner G. (1966). Potassium-argon dating by activation with fast neutrons. *J. Geophys. Res.*, 71:2852–2857.

Millero F. J. (1972). The partial molal volumes of electrolytes in aqueous solutions. In *Water and Aqueous Solutions*, R. A. Horne (series ed.), New York: Wiley Interscience.

Millero F. J. (1977). Thermodynamic models for the state of metal ions in sea water. In *The Sea*, vol. 6, E. D. Goldberg et al., eds. New York: Wiley Interscience.

Misener D. J. (1974). Cationic diffusion in olivine to 1400°C and 35 kbar. In *Geochemical Transport and Kinetics*, A. W. Hoffmann, B. J. Giletti, H. S. Yoder Jr., and R. A. Yund, eds., Carnegie Institution of Washington.

Misra N. K. and Venkatasubramanian V. S. (1977). Strontium diffusion in feldspars—a laboratory study. *Geochim. Cosmochim. Acta*, 41:837–838.

Moore P. B. and Smith J. V. (1970). Crystal structure of β-Mg_2SiO_4: Crystal chemical and geophysical implications. *Phys. Earth Planet. Int.*, 3:166–177.

Moore R. K., White W. B., and Long T. V. (1971). Vibrational spectra of the common silicates, I: The garnets. *Amer. Mineral.*, 56:54–71.

Mori T. (1977). Geothermometry of spinel Iherzolites. *Contrib. Mineral. Petrol.*, 59:261–279.

Mori T. and Green D. H. (1978). Laboratory duplication of phase equilibria observed in natural garnet Iherzolites. *J. Geol.*, 86:83–97.

Morimoto N. and Tokonami M. (1969). Oriented exsolution of augite in pigeonite. *Amer. Mineral.*, 54:1101–1117.

Morimoto N., Appleman D. E., and Evans H. T. Jr. (1960). The crystal structures of clinoenstatite and pigeonite. *Zeit. Kristallogr.*, 67:587–598.

Morioka M. (1980). Cation diffusion in olivine, I: Cobalt and magnesium. *Geochim. Cosmochim. Acta*, 44:759–762.

Morioka M. (1981). Cation diffusion in olivine, II: Ni-Mg, Mn-Mg, Mg, and Ca. *Geochim. Cosmochim. Acta*, 45:1573–1580.

Morioka M. and Nagasawa H. (1991). Ionic diffusion in olivine. In *Advances in Physical Geochemistry*, vol. 8, J. Ganguly, ed. New York-Heidelberg-Berlin: Springer-Verlag.

Morlotti R. and Ottonello G. (1982). Solution of rare earth elements in silicate solid phases: Henry's law revisited in light of defect chemistry. *Phys. Chem. Minerals*, 8:87–97.

Morlotti R. and Ottonello G. (1984). The solution of trace amounts of Sm in fosteritic olivine: An experimental study by EMF galvanic cell measurements. *Geochim. Cosmochim. Acta*, 48:1173–1181.

Morris R. V. and Haskin L. A. (1974). EPR measurement of the effect of glass composition on the oxidation states of europium. *Geochim. Cosmochim. Acta*, 38:1435–1445.

Morse S. A. (1976). The lever rule with fractional crystallization and fusion. *Amer. Jour. Sci.*, 276:330–346.

Morse S. A. (1980). *Basalts and Phase Diagrams*. New York, Heidelberg, and Berlin: Springer-Verlag.

Mott N. F. and Littleton M. S. (1938). Conduction in polar crystals, I: Electrolitic conduction in solid salts. *Trans. Faraday Soc.*, 34:485–499.

Mueller R. F. (1960). Compositional characteristics and equilibrium relation in mineral assemblages of a metamorphosed iron formation. *Amer. Jour. Sci.*, 258:449–497.

Mueller R. F. (1961). Analysis of relations among Mg, Fe and Mn in certain metamorphic minerals. *Geochim. Cosmochim. Acta,* 25:267–296.

Mueller R. F. (1962). Energetics of certain silicate solid solutions. *Geochim. Cosmochim. Acta,* 26:581–598.

Mueller R. F. (1972). Stability of biotite: A discussion. *Amer. Mineral.,* 57:300–316.

Mukherjee K. (1965). Monovacancy formation energy and Debye temperature of close-packed metals. *Phil. Mag.,* 12:915–918.

Mukhopadhyay D. K. and Lindsley D. H. (1983). Phase relations in the join kischsteinite $(CaFeSiO_4)$-faialite (Fe_2SiO_4). *Amer. Mineral.,* 62:1089–1094.

Muller R. A. (1979). Radioisotope dating with accelerators. *Phys. Today,* 32:23–28.

Mulliken R. S. (1934). A new electroaffinity scale; together with data on valence states and on valence ionization potentials and electron affinities. *J. Chem. Phys.,* 2:782–793.

Mulliken R. S. (1935). Electronic structure of molecules, XI: Electroaffinity, molecular orbitals, and dipole moments. *J. Chem. Phys.,* 3:573–591.

Munk W. H. (1966). Abyssal reciper. *Deep Sea Res.,* 13:707–730.

Munoz J. L. (1984). F-OH and Cl-OH exchange in micas with applications to hydrothermal ore deposits. In *Reviews in Mineralogy.* vol. 13., P. H. Ribbe (series ed.), Mineralogical Society of America.

Munoz J. L. and Ludington S. D. (1974). Fluorine-hydroxil exchange in biotite. *Amer. Jour. Sci.,* 274:396–413.

Munoz J. L. and Ludington S. D. (1977). Fluorine-hydroxil exchange in synthetic muscovite and its application to muscovite-biotite assemblages. *Amer. Mineral.,* 62:304–308.

Myers W. D. and Swiatecki W. J. (1966). Nuclear masses and deformations. *Nucl. Phys.,* 81:1–16.

Mysen B. O. (1976a). Nickel partitioning between upper mantle crystals and partial melts as a function of pressure, temperature, and nickel concentration. *Carnegie Inst. Wash. Yb.,* 75:662–668.

Mysen B. O. (1976b). Rare earth partitioning between crystals and liquid in the upper mantle. *Carnegie Inst. Wash. Yb.,* 75:656–659.

Mysen B. O. (1977a). Partitioning of nickel between liquid, pargasite and garnet peridotite minerals and concentration limit of behavior according to Henry's law at high pressure and temperature. *Amer. Jour. Sci.,* 278:217–243.

Mysen B. O. (1977b). Experimental determination of cesium, samarium and thulium partitioning between hydrous liquid, garnet peridotite minerals and pargasite. *Carnegie Inst. Wash. Yb.,* 76:588–594.

Mysen B. O. (1978). Limits of solution of trace elements in minerals according to Henry's law: Review of experimental data. *Geochim. Cosmochim. Acta,* 42:871–886.

Mysen B. O., Virgo D., and Scarfe C. M. (1980). Relation between the anionic structure and viscosity of silicate melts: A Raman spectroscopic study. *Amer. Mineral.,* 65:690–710.

Nafziger R. H. and Muan A. (1967). Equilibrium phase compositions and thermodynamic properties of olivines and pyroxenes in the system $MgO-FeO-SiO_2$. *Amer. Mineral.,* 52:1364–1384.

Nakamura A. and Schmalzried H. (1983). On the nonstoichiometry and point defects in olivine. *Phys. Chem. Minerals,* 10:27–37.

Nakamura A. and Schmalzried H. (1984). On the Fe^{2+}-Mg^{2+} interdiffusion in olivine. *Phys. Chem.,* 88:140–145.

Nath P. and Douglas R. W. (1965). Cr^{3+}/Cr^{6+} equilibrium in binary alkali silicate glasses. *Phys. Chem. Glasses,* 6:197–202.

Naumov G. B., Ryzhenko B., and Khodakovsky I. L. (1971). *Handbook of Thermodynamic Data* Atomizdat, Moscow.

Navrotsky A. (1985). Crystal chemical constraints on the thermochemistry of minerals. In *Reviews in Mineralogy,* vol. 14, P. H. Ribbe (series ed.), Mineralogical Society of America.

Navrotsky A. (1994). *Physics and Chemistry of Earth Materials* Cambridge University Press.

Navrotsky A. and Akaogi M. (1984). a-b-c phase relations in Fe_2SiO_4-Mg_2SiO_4 and Co_2SiO_4-Mg_2SiO_4: Calculation from thermochemical data and geophysical applications. *J. Geophys. Res.*, 89:10135–10140.

Navrotsky A. and Loucks D. (1977). Calculations of subsolidus phase relations in carbonates and pyroxenes. *Phys. Chem. Minerals*, 1:109–127.

Navrotsky A., Geisinger K. L., McMillan P., and Gibbs G. V. (1985). The tetrahedral framework in glasses and melts: Inferences from molecular orbital calculations and implications for structure, thermodynamics and physical properties. *Phys. Chem. Minerals*, 11:284–298.

Navrotsky A., Peraudeau G., McMillan P., and Coutures J. P. (1982). A thermochemical study of glasses and crystals along the joins silica-calcium aluminate and silica-sodium aluminate. *Geochim. Cosmochim. Acta*, 46:2039–2047.

Nernst W. (1891). Distribution of a substance between two solvents and between solvent and vapor. *Z. Phis. Chem.*, 8:110–139.

Neumann H., Mead J., and Vitaliano C. J. (1954). Trace element variation during fractional crystallization as calculated from the distribution law. *Geochim. Cosmochim. Acta*, 6:90–99.

Newnham R. E. and Megaw H. D. (1960). The crystal structure of celsian (barium feldspar). *Acta Cryst.*, 13:303–312.

Newton R. C. (1987). Thermodynamic analysis of phase equilibria in simple mineral system. In *Reviews in Mineralogy.* vol. 17, P. H. Ribbe (series ed.), Mineralogical Society of America.

Newton R. C. and Haselton H. T. (1981). Thermodynamics of the garnet-plagioclase-Al_2SiO_5-quartz geobarometer. In *Thermodynamics of Minerals and Melts*, R. C. Newton, A. Navrotsky, and B. J. Wood, eds. New York: Springer-Verlag.

Newton R. C. and Perkins D. (1982). Thermodynamic calibration of geobarometers based on the assemblages garnet-plagioclase-orthopyroxene (clinopyroxene)-quartz. *Amer. Mineral.*, 67:203–222.

Newton R. C. and Wood B. J. (1980). Volume behavior of silicate solid solutions. *Amer. Mineral.*, 65:733–745.

Newton R. C., Charlu T. V., and Kleppa O. J. (1977). Thermochemistry of high pressure garnets and clinopyroxenes in the system CaO-MgO-Al_2O_3-SiO_2. *Geochim. Cosmochim. Acta*, 41:369–377.

Newton R. C., Charlu T. V., and Kleppa O. J. (1980). Thermochemistry of high structural state plagioclases. *Geochim. Cosmochim. Acta*, 44:933–941.

Newton R. C., Geiger C. A., Kleppa O. J., and Brousse C. (1986). Thermochemistry of binary and ternary garnet solid solutions. *I.M.A., Abstracts with Program*, 186.

Newton W. and McCready N. (1948). Thermodynamic properties of sodium silicates. *C. Phys. Coll. Chem.*, 52:1277–1283.

Nishizawa O. and Akimoto S. (1973). Partitioning of magnesium and iron between olivine and spinel, and between pyroxene and spinel. *Contrib. Mineral. Petrol.*, 41:217–240.

Nordlie B. E. (1971). The composition of the magmatic gas of Kilauea and its behavior in the near surface environment. *Amer. Jour. Sci.*, 271:417–473.

Novak G. A. and Colville A. (1975). A linear regression analysis of garnet chemistries versus cell parameters (abstract). *Geol. Soc. Amer. S. W. Section Meeting*, Los Angeles, 7:359.

Novak G. A. and Gibbs G. V. (1971). The crystal chemistry of silicate garnets. *Amer. Mineral.*, 56:791–825.

Nylen P. and Wigren N. (1971). *Stechiometria* CEDAM, Padova.

Oba T. (1980). Phase relations in the tremolite-pargasite join. *Contrib. Mineral. Petrol.*, 71:247–256.

Ohmoto H. (1972). Systematics of sulfur and carbon isotopes in hydrothermal ore deposits. *Econ. Geol.*, 67:551–578.

Ohmoto H. and Lasaga A. C. (1982). Kinetics of reactions between aqueous sulfates and sulfides in hydrothermal systems. *Geochim. Cosmochim. Acta*, 46:1727–1745.

Ohmoto H. and Rye R. O. (1979). Isotopes of sulfur and carbon. In *Geochemistry of Hydrothermal Ore Deposits,* 2d ed., H. L. Barnes, ed. New York: John Wiley.

Oka Y. and Matsumoto T. (1974). Study on the compositional dependence of the apparent partitioning coefficient of iron and magnesium between coexisting garnet and clinopyroxene solid solutions. *Contrib. Mineral. Petrol.,* 48:115–121.

Okamura F. P. S., Ghose S., and Ohashi H. (1974). Structure and crystal chemistry of calcium Tschermak's pyroxene $CaAlAlSiO_6$. *Amer. Mineral.,* 59:549–557.

O'Neill H.St. C. and Wood B. J. (1979). An experimental study of Fe-Mg partitioning between garnet and olivine and its calibration as a geothermometer. *Contrib. Mineral. Petrol.,* 70:59–70.

O'Neill M. J. and Fyans R. L. (1971). Design of differential scanning calorimeters and the performance of a new system. Norwalk, Connecticut, Perkin-Elmer Corp., 38 pp.

O'Nions R. K. and Smith D. G. W. (1973). Bonding in silicates: An assessment of bonding in orthopyroxene. *Geochim. Cosmochim. Acta,* 37:249–257.

Onken H. (1965). Verfeinerung der Kristallstruktur von Monticellite. *Tsch. Min. Petr. Mitt.,* 10:34–44.

Onuma N., Higuchi H., Wakita H., and Nagasawa H. (1968). Trace element partition between two pyroxenes and host volcanic rocks. *Earth Planet. Sci. Letters,* 5:47–51.

Orville P. M. (1972). Plagioclase cation exchange equilibria with aqueous chloride solution: Results at 70°C and 2000 bars in the presence of quartz. *Amer. Jour. Sci.,* 272:234–272.

Osborn E. F. (1942). The system $CaSiO_3$-diopside-anorthite. *Amer. Jour. Sci.,* 240:751–788.

Osborn E. F. and Tait D. B. (1952). The system diopside-forsterite-anorthite. *Amer. Jour. Sci., Bowen Volume,* 413–434.

Ottonello G. (1980). Rare earth abundance and distribution in some spinel peridotite xenoliths from Assab (Ethiopia). *Geochim. Cosmochim. Acta,* 44:1885–1901.

Ottonello G. (1983). Trace elements as monitors of magmatic processes (1) limits imposed by Henry's law problem and (2) Compositional effect of silicate liquid. In *The Significance of Trace Elements in Solving Petrogenetic Problems and Controversies,* S. Augusthitis, ed. Theophrastus Publications, Athens.

Ottonello G. (1986). Energetics of multiple oxides with spinel structure. *Phys. Chem. Minerals.,* 13:79–90.

Ottonello G. (1987). Energies and interactions in binary (Pbnm) orthosilicates: A Born parametrization. *Geochim. Cosmochim. Acta,* 51:3119–3135.

Ottonello G. (1992). Interactions and mixing properties in the (C2/c) clinopyroxene quadrilateral. *Contrib. Mineral. Petrol.,* 111:53–60.

Ottonello G. and Morlotti R. (1987). Thermodynamics of nickel-magnesium olivine solid solution. *J. Chem. Thermodyn.,* 19:809, 818.

Ottonello G. and Ranieri G. (1977). Effetti del controllo petrogenetico sulla distribuzione del Ba nel processo di anatessi crustale. *Rend. S.I.M.P.,* 33:741–753.

Ottonello G., Bokreta M. and Sciuto P. F. (1996). Parameterization of energy and interactions in garnets: End-member properties. *Amer. Mineral.,* 81:429–447.

Ottonello G., Bokreta M., and Sciuto P. F. (in preparation). Parameterization of energy and interactions in garnets: Mixing properties.

Ottonello G., Della Giusta A., and Molin G. M. (1989). Cation ordering in Ni-Mg olivines. *Amer. Mineral.,* 74:411–421.

Ottonello G., Piccardo G. B. and Ernst W. G. (1979). Petrogenesis of some Ligurian peridotites, II: Rare earth element chemistry. *Geochim. Cosmochim. Acta,* 43:1273–1284.

Ottonello G., Princivalle F., and Della Giusta A. (1990). Temperature, composition and fO_2 effects on intersite distribution of Mg and Fe^{2+} in olivines. *Phys. Chem. Minerals,* 17:301–312.

Ottonello G., Della Giusta A., Dal Negro A., and Baccarin F. (1992). A structure-energy model for C2/c pyroxenes in the system Na-Mg-Ca-Mn-Fe-Al-Cr-Ti-Si-O. In *Advances in Physical Geochemistry,* S. K. Saxena (series ed.), vol. 10. Berlin-Heidelberg-New York: Springer-Verlag.

Ottonello G. Piccardo G. B., Mazzucotelli A., and Cimmino F. (1978). Clinopyroxene-orthopyroxene major and REE partitioning in spinel peridotite xenoliths from Assab (Ethiopia). *Geochim. Cosmochim. Acta.*, 42:1817–1828.

Papike J. J. and Clark J. R. (1968). The crystal structure and caution distributions of glaucophane. *Amer. Mineral.*, 53:1156–1173.

Papike J. J. and Ross M. (1970). Gedrites: Crystal structures and intracrystalline cation distributions. *Amer. Mineral.*, 55:1945–1972.

Parson I. (1978). Alkali feldspars: Which solvus? *Phys. Chem. Minerals*, 2:199–213.

Patchett J. P. (1983). Importance of the Lu-Hf isotopic system in studies of planetary chronology and chemical evolution. *Geochim. Cosmochim. Acta*, 47:81–91.

Paul A. and Douglas R. W. (1965a). Ferrous-ferric equilibrium in binary alkali silicate glasses. *Phys. Chem. Glasses*, 6:207–211.

Paul A. and Douglas R. W. (1965b). Cerous-ceric equilibrium in binary alkali silicate glasses. *Phys. Chem. Glasses*, 6:212–215.

Pauli W. (1933). Die allgemeinen Prinzipien der Wellenmechanik. *Handbuch der Physik*, 24/1:83–227.

Pauling L. (1927a). The sizes of ions and the structure of ionic crystals. *Jour. Amer. Chem. Soc.*, 49:765–792.

Pauling L. (1927b). The theoretical prediction of the physical properties of many-electron atoms and ions. Mole refraction, diamagnetic susceptibility, and extension in space. *Proc. Roy. Soc.*, A114:181–189.

Pauling L. (1929). The principles determining the structure of complex ionic crystals. *Jour. Amer. Chem. Soc.*, 51:1010–1026.

Pauling L. (1932). Nature of the chemical bond. IV. The energy of single bonds and the relative electronegativity of atoms. *Jour. Amer. Chem. Soc.*, 54:3570–3582.

Pauling L. (1947). Atomic radii and interatomic distances in metals. *Jour. Amer. Chem. Soc.*, 69:542–553.

Pauling L. (1948). *The Nature of the Chemical Bond.* Cornell University Press, 2d ed., Ithaca, N.Y.

Pauling L. (1960). *The Nature of the Chemical Bond.* Cornell University Press, Ithaca, N.Y.

Pauling L. (1980). The nature of silicon-oxygen bonds. *Amer. Mineral.*, 65:321–323.

Pauling L. and Sherman J. (1932). Screening constants for many-electrons atoms: The calculation and interpretation of X-ray term values and the calculation of atomic scattering factors. *Zeit. Krist.*, 81:1–29.

Payne J. A. and Drever R. (1965). An investigation of the beta decay of rhenium to osmium with high temperature proportional counters. Ph.D. diss., University of Glasgow.

Peacor D. R. (1973). High temperature single-crystal study of the cristobalite inversion. *Zeit. Krist.*, 138:274–298.

Perchuk L. L. (1977). Thermodynamic control of metamorphic processes. In *Energetics of Geological Processes*, S. K. Saxena and S. Batacharji, eds. New York: Springer-Verlag.

Perchuk L. L. (1991). Derivation of a thermodynamically consistent set of geothermometers and geobarometers for metamorphic and magmatic rocks. In *Progress in Methamorphic and Magmatic Petrology*, L. L. Perchuk, ed. Cambridge University Press, London.

Perchuk L. L. and Aranovich L. Ya. (1979). Thermodynamics of variable composition: Andradite-grossularite and pistacite-clinozoisite solid solutions. *Phys. Chem. Minerals*, 5:1–14.

Perkins D. (1983). The stability of Mg-rich garnet in the system $CaO-MgO-Al_2O_3-SiO_2$ at 1000 to 1300°C and high pressure. *Amer. Mineral.*, 68:355–364.

Phillips J. C. (1970). Ionicity of the chemical bond in crystals. *Rev. Mod., Phys.*, 42(3).:317.

Phillips M. W. and P. H. Ribbe (1973). The structure of monoclinic potassium-rich feldspars. *Amer. Mineral.*, 58:263–270.

Philpotts A. R. (1979). Silicate immiscibility in tholeiitic basalts. *J. Petrol.*, 20:99–118.

Philpotts J. A. and Schnetzler C. C. (1970). Phenocryst-matrix partition coefficients for K, Rb,

Sr and Ba, with application to anorthosite and basalt genesis. *Geochim. Cosmochim. Acta,* 34:307–322.

Piccardo G. B. and Ottonello G. (1978). Partial melting effects on coexisting mineral compositions in upper mantle xenoliths from Assab (Ethiopia). *Rend. S.I.M.P.,* 34:499–526.

Pitzer K. S. (1973). Thermodynamics of electrolytes. I Theoretical basis and general equations. *J. Phys. Chem.,* 77:268–277.

Pitzer K. S. (1983). Dieletric constants of water at very high temperature and pressure. *Proc. Natl. Acad. Sci. U.S.A.,* 80:4575–4576.

Plummer L. N., Wigley T. M. L. and Parkhurst D. L. (1978). The kinetics of calcite dissolution in CO_2-water system at 5° to 60°C and 0.0 to 1.0 atm CO_2. *Amer. Jour. Sci.,* 278:179–216.

Pluschkell W. and Engell H. J. (1968). Ionen und Elektronenleitung in Magnesiumorthosilikat. *Ber. Dtsch. Keram. Ges.,* 45:388–394.

Poldervaart A. and Hess H. H. (1951). Pyroxenes in the crystallization of basaltic magmas. *J. Geol.,* 59:472–489.

Pople J. A. (1951). The structure of water and similar molecules. *Proc. Roy. Soc.,* A202:323–336.

Popp R. K., Gilbert M. C. and Craig J. R. (1976). Synthesis and X-ray properties of Fe-Mg orthoamphiboles. *Amer. Mineral.,* 61:1267–1279.

Potter R. W. and Clynne M. A. (1978). The solubility of noble gases He, Ne, Ar, Kr and Xe in water up to the critical point. *J. Soln. Chem.,* 7:837–844.

Pourbaix M. (1966). *Atlas of Electrochemical Equilibria in Aqueous Solutions* Pergamon Press, Oxford.

Powell M. and Powell R. (1977a). Plagioclase-alkali feldspar thermometry revisited. *Min. Mag.,* 41:253–256.

Powell R. and Powell M. (1977b). Geothermometry and oxygen barometry using iron-titanium oxides: A reappraisal. *Min. Mag.,* 41:257–263.

Powenceby M. I., Wall V. J. and O'Neill H. St. C. (1987). Fe-Mn partitioning between garnet and ilmenite: Experimental calibration and applications. *Contrib. Mineral. Petrol.,* 97:116–126.

Prausnitz J. M., Lichtenhaler R. N., and De Azvedo E. G. (1986). *Molecular Thermodynamics of Fluid Phase Equilibria.* New York: Prentice-Hall.

Presnall D. C. (1969). The geometrical analysis of partial fusion. *Amer. Jour. Sci.,* 267:1178–1194 Prewitt C. T., Papike J. J. and Ross M. (1970). Cummingtonite. A reversible, nonquenchable transition from $P2_1/m$ to C2/m symmetry. *Earth. Planet. Sci. Letters,* 8:448–450.

Prigogine I. (1955). *Introduction to Thermodynamics of Irreversible Processes.* 2d ed. New York: Interscience Publishers.

Provost A. and Bottinga Y. (1974). Rates of solidification of Apollo 11 basalts and hawaiian tholeiite. *Earth. Planet. Sci. Letters,* 15:325–337.

Putnis A. and McConnell J. D. C. (1980). *Principles of Mineral Behavior.* Blackwell Scientific Publications, Oxford.

Rae A. I. M. (1973). A theory for the interactions between closed shell systems. *Chem. Phys. Lett.,* 18:574–577.

Raheim A. (1975). Mineral zoning as a record of P, T history of Precambrian metamorphic rocks in W. Tasmania. *Lithos,* 8:221–236.

Raheim A. (1976). Petrology of eclogites and surrounding schists from the Lyell Highway–Collingwood River area, W. Tansmania. *Geol. Soc. Austr. Jour.,* 23:313–327.

Raheim A. and Green D. H. (1974). Experimental determination of the temperature and pressure dependence of the Fe-Mg partition coefficient for coexisting garnet and clinopyroxene. *Contrib. Mineral. Petrol.,* 48:179–203.

Raheim A. and Green D. H. (1975). P, T paths of natural eclogites during metamorphism, a record of subduction. *Lithos,* 8:317–328.

Rakai R. J. (1975). Crystal structure of spessartite and andradite at elevated temperatures. M.Sc. diss., University of British Columbia.

Ramberg H. (1952). Chemical bonds and the distribution of cations in silicates. *J. Geol.*, 60:331–355.

Ramberg H. and Devore G. (1951). The distribution of Fe^{2+} and Mg in coexisting olivines and pyroxenes. *J. Geol.*, 59:193–210.

Rammensee W. and Fraser D. G. (1982). Determination of activities in silicate melts by Knudsen cell mass spectrometry, I: The system $NaAlSi_3O_8$-$KAlSi_3O_8$. *Geochim. Cosmochim. Acta*, 46:2269–2278.

Rankama K. (1954). *Isotope Geology* Pergamon Press, London.

Ravizza G. and Turekian K. K. (1989). Application of the ^{187}Re-^{187}Os system to black shale geochronometry. *Geochim. Cosmochim. Acta*, 53:3257–3262.

Rayleigh J. W. S. (1896). Theoretical considerations respecting the separation of gases by diffusion and similar processes. *Philos. Mag.*, 5th ser., 42:493–498.

Reddy K. P. R., Oh S. M., Major L. D. and Cooper A. R. (1980). Oxygen diffusion in forsterite. *J. Geophs. Res.*, 85:322–326.

Redlich O. and Kister A. T. (1948). Thermodynamics of non-electrolyte solutions: x-y-t relations in a binary system. *Ind. Eng. Chem.*, 40:341–345.

Reed M. and Spycher N. (1984). Calculation of pH and mineral equilibria in hydrothermal waters with application to geothermometry and studies of boiling and dilution. *Geochim. Cosmochim. Acta*, 48:1479–1492.

Reuter J. H. and Perdue E. M. (1977). Importance of heavy metal-organic matter interactions in natural waters. *Geochim. Cosmochim. Acta*, 41:325–334.

Ribbe P. H. (1983a). The chemistry, structure and nomenclature of feldspars. In *Reviews in Mineralogy*, vol. 2 (2d ed.), P. H. Ribbe (series ed.), Mineralogical Society of America.

Ribbe P. H. (1983b). Aluminium-silicon order in feldspars: Domain textures and diffraction patterns. In *Reviews in Mineralogy*, vol. 2 (2d ed.), P. H. Ribbe (series ed.), Mineralogical Society of America.

Richardson F. D. (1956). Activities in ternary silicate melts. *Trans. Faraday Soc.*, 52:1312–1324.

Richardson S. W. (1968). Staurolite stability in a part of the system Fe-Al-Si-O-H. *J. Petrol.*, 9:467–488.

Richardson S. W., Gilbert M. C. and Bell P. M. (1969). Experimental determination of kyanite-andalusite and andalusite-sillimanite equilibria: The aluminum silicate triple point. *Amer. Jour. Sci.*, 267:259–272.

Richet P. and Bottinga Y. (1983). Verres, liquides, et transition vitreuse. *Bull. Mineral.*, 106:147–168.

Richet P. and Bottinga Y. (1985). Heat capacity of aluminium-free liquid silicates. *Geochim. Cosmochim. Acta*, 49:471–486.

Richet P. and Bottinga Y. (1986). Thermochemical properties of silicate glasses and liquids: A review. *Rev. Geophys. Space Phys.*, 24:1–25.

Rickard D. T. (1975). Kinetics and mechanism of pyrite formation at low temperatures. *Amer. Jour. Sci.*, 275:636–652.

Rimstidt J. D. and Barnes H. L. (1980). The kinetics of silica-water reactions. *Geochim. Cosmochim. Acta*, 44:1683–1699.

Ringwood A. E. (1969). Phase transformations in the mantle. *Earth Planet. Sci. Letters*, 5:401–412.

Ringwood A. E. and Major A. (1970). The system Mg_2SiO_4-Fe_2SiO_4 at high pressures and temperatures. *Phys. Earth Planet. Int.*, 3:89–108.

Rittner E. S., Hutner R. A. and Du Pré F. K. (1949). Concerning the work of polarization in ionic crystals of the NaCl type. II. Polarization around two adjacent charges in the rigid lattice. *J. Chem. Phys.*, 17:198–203.

Robie R. A., Hemingway B. S. and Fisher J. R. (1978). Thermodynamic properties of minerals and related substances at 298.15 K and 1 bar (10^5 pascal) pressure and at higher temperatures. *U.S.G.S. Bull.*, 1452, 456 pp.

Robie R. A., Bethke P. M., Toulmin M. S. and Edwards J. L. (1966). X-ray crystallographic data, densities and molar volumes of minerals. In *Handbook of Physical Constants,* S. Clark, ed. *Geol. Soc. Amer. Mem.,* 97:27–73.

Robie R. A., Zhao-Bin, Hemingway B. S. and Barton M. D. (1987). Heat capacity and thermodynamic properties of andradite garnets, $Ca_3Fe_2Si_3O_{12}$, between 10 and 1000 K and revised values for ΔG_f^o (298.15 K) of hedembergite and wollastonite. *Geochim. Cosmochim. Acta,* 51:2219–2236.

Robinson C. C. (1974). Multiple sites for Er^{3+} in alkali glasses. *Jour. Non-Cryst. Solids,* 15:1–10.

Robinson G. R. Jr. and Haas J. L. Jr. (1983). Heat capacity, relative enthalpy and calorimetric entropy of silicate minerals: An empirical method of prediction. *Amer. Mineral.,* 68:541–553.

Robinson K., Gibbs G. V., Ribbe P. H. and Hall M. R. (1973). Cation distribution in three hornblendes. *Amer. Jour. Sci.,* 273A:522–535.

Robinson P. (1982). The composition space of terrestrial pyroxenes. Internal and external limits. In *Reviews in Mineralogy,* vol. 7, P. H. Ribbe (series ed.), Mineralogical Society of America.

Robinson P., Spear F. S., Schumacher J. C., Laird J., Klein C., Evans B. W. and Doolan B. L. (1982). Phase relations of metamorphic amphibole, natural occurrence and theory. In *Reviews in Mineralogy,* vol. 9B, P. H. Ribbe. (series ed.), Mineralogical Society of America.

Roedder E. (1951). Low temperature liquid immiscibility in the system K_2O-FeO-Al_2O_3-SiO_2. *Amer. Mineral.,* 36:282–286.

Roedder E. (1956). The role of liquid immiscibility in igneous petrogenesis: A discussion. *J. Geol.,* 64:84–88.

Roedder E. and Wieblen P. W. (1971). Lunar petrology of silicate melt inclusions, Apollo 11 rocks. *Proc. Apollo 11 Lunar Sci. Conf.,* 1:801–837.

Romanek C. S., Grossman E. L. and Morse J. W. (1989). Carbon isotope fractionation in aragonite and calcite: Experimental study of temperature and kinetic effects. *Geol. Soc. Amer. Abstr. Prog.,* 21:A76.

Romanek C. S., Grossman E. L. and Morse J. W. (1992). Carbon isotopic fractionation in synthetic aragonite and calcite: Effects of temperature and precipitation rate. *Geochim. Cosmochim. Acta,* 56:419–430.

Roozeboom P. (1899). Erstarrungpunkte der Mischkrystalle Zweier Stoffe. *Zeitschr. Phys. Chem.,* 385:30–32.

Rosenbaum J. M. (1994). Stable isotope fractionation between carbon dioxide and calcite at 900°C. *Geochim. Cosmochim. Acta,* 58:3747–3753.

Rosenhauer M. and Eggler D. H. (1975). Solution of H_2O and CO_2 in Diopside melt. *Carnegie Inst. Wash. Yb.,* 74:474–479.

Rosholt J. N., Zartman R. E. and Nkomo I. T. (1973). Lead isotope systematics and uranium depletion in the Granite Mountains, Wyoming. *Geol. Soc. Amer. Bull.,* 84:989–1002.

Ross M. and Huebner J. S. (1975). A pyroxene geothermometer based on composition temperature relationship of naturally occurring orthopyroxene, pigeonite and augite. In *International Conference on Geothermometry and Geobarometry,* The Pennsylvania State University.

Ross M. and Ree F. H. (1980). Repulsive forces of simple molecules and mixtures at high density and temperature. *J. Chem. Phys.,* 73:6146–6152.

Rossmann G. R. (1982). Pyroxene spectroscopy. In *Reviews in Mineralogy,* vol. 7, P. H. Ribbe (series ed.), Mineralogical Society of America.

Rothbauer R. (1971). Untersuchung eines 2M-Muskovit mit Neutronenstrahlen. *N. Jahrb. Min. Monat.,* 4:143–154.

Roux J. (1971a). Fixation du rubidium et du cesium dans la nepheline et dans l'albite a 600°C dans les conditions hydrothermales. *C. R. Acad. Sci. Fr. Ser. D,* 222:1469–1472.

Roux J. (1971b). *Etude experimentale sur la distribution des elemets alcalines Cs, Rb et K entre l'albite, la néphéline et les solutions hdrothermales* These 3 cycle, Université de Paris Sud.

Roux J. (1974). Etude des solutions solides des néphélines (Na,K)AlSiO$_4$ et (Na,Rb)AlSiO$_4$. *Geochim. Cosmochim. Acta,* 38:1213–1224.

Rubey W. W. (1951). Geologic history of seawater. *Geol. Soc. Amer. Bull.,* 62:1111–1147.

Russell W. A., Papanastassiou D. A. and Tombrello T. A. (1978). Ca isotope fractionation on the earth and other solar system materials. *Geochim. Cosmochim. Acta,* 42:1075–1090.

Rye R. O. and Ohmoto H. (1974). Sulfur and carbon isotopes and ore genesis: A review. *Econ. Geol.,* 69:826–842.

Ryerson F. J. and Hess P. C. (1978). Implications of liquid distribution coefficients to mineral-liquid partitioning. *Geochim. Cosmochim. Acta,* 42:921–932.

Ryerson F. J. and Hess P. C. (1980). The role of P$_2$O$_5$ in silicate melts. *Geochim. Cosmochim. Acta,* 44:611–624.

Sack R. O. (1980). Some constraints on the thermodynamic mixing properties of Fe-Mg orthopyroxenes and olivines. *Contrib. Mineral. Petrol.,* 71:257–269.

Sahama Th. G. and Torgeson D. R. (1949). Some examples of applications of thermochemistry to petrology. *J. Geol.,* 57:255–262.

Sakai H. (1968). Isotopic properties of sulfur compounds in hydrothermal processes. *Geochem. J.,* 2:29–49.

Sakai H. and Tsutsumi M. (1978). D/H fractionation factors between serpentine and water at 1000° to 500°C and 2000 bar water pressure and the D/H ratios of natural serpentines. *Earth Planet. Sci. Letters,* 40:231–242.

Salje E. (1985). Thermodynamics of sodium feldspar I: Order parameter treatment and strain induced coupling effects. *Phys. Chem. Minerals,* 12:93–98.

Salje E. (1988). Structural phase transitions and specific heat anomalies. In *Physical Properties and Thermodynamic Behaviour of Minerals,* E. K. H. Salje, ed. Reidel Publishing Company.

Salje E., Kuscholke B., Wruck B., and Kroll H. (1985). Thermodynamics of sodium feldspar, II: Experimental results and numerical calculations. *Phys. Chem. Minerals,* 12:99–107.

Samsonov V. (1968). *Handbook of the Physicochemical Properties of the Elements.* New York-Washington: IFI/Plenum.

Sanderson R. T. (1960). *Chemical Periodicity.* New York: Reinhold.

Sanderson R. T. (1966). Bond energies. *J. Inorg. Nucl. Chem.,* 28:1553–1565.

Sanderson R. T. (1967). *Inorganic Chemistry.* New York: Reinhold.

Sartori F. (1976). The crystal structure of 1M lepidolite. *Tsch. Min. Petr. Mitt.,* 23:65–75.

Sarver J. V. and Hummel F. A. (1962). Solid solubility and eutectic temperature in the system Zn$_2$SiO$_4$-Mg$_2$SiO$_4$. *J. Amer. Ceram. Soc.,* 45:304–314.

Sasaki S., Fujino K., Takeuchi Y., and Sadanaga R. (1980). On the estimation of atomic charges by the X-ray method for some oxides and silicates. *Acta Cryst.,* A36:904–915.

Sasaki S. Y., Takeuchi Y., Fujino K., and Akimoto S. (1982). Electron density distribution of three orthopyroxenes, Mg$_2$Si$_2$O$_6$, Co$_2$Si$_2$O$_6$ and Fe$_2$Si$_2$O$_6$. *Zeits. Krist.,* 158:279–297.

Sato Y., Akaogi M., and Akimoto S. (1978). Hydrostatic compression of the synthetic garnets pyrope and almandine. *J. Geophys. Res.,* 83:335–338.

Saxena S. K. (1968). Distribution of iron and magnesium between coexisting garnet and clinopyroxene in rocks of varying metamorphic grade. *Amer. Mineral.,* 53:2018–2021.

Saxena S. K. (1969). Silicate solid solution and geothermometry. 3. Distribution of Fe and Mg between coexisting garnet and biotite. *Contrib. Mineral. Petrol.,* 22:259–267.

Saxena S. K. (1972). Retrival of thermodynamic data from a study of intercrystalline and intracrystalline ion-exchange equilibrium. *Amer. Mineral.,* 57:1782–1800.

Saxena S. K. (1973). *Thermodynamics of Rock-Forming Crystalline Solutions.* Berlin-Heidelberg-New York: Springer-Verlag.

Saxena S. K. (1977). A new electronegativity scale for geochemists. In *Energetics of Geological Processes,* S. K. Saxena and S. Bhattacharji, eds. New York-Heidelberg-Berlin: Springer-Verlag.

Saxena S. K. (1979). Garnet-clinopyroxene geothermometer. *Contrib. Mineral. Petrol.,* 70:229–235.

Saxena S. K. (1983). Exsolution and Fe^{2+}-Mg order-disorder in pyroxenes. In *Advances in Physical Geochemistry,* vol. 3, S. K. Saxena (series ed.). Berlin-Heidelberg-New York: Springer-Verlag.

Saxena S. K. (1989). *INSP and THERMO Computer Routines and Corresponding Data-base* Uppsala University, Sweden.

Saxena S. K. and Ghose S. (1971). Mg^{2+}-Fe^{2+} order disorder and the thermodynamics of the orthopyroxene-crystalline solution. *Amer. Mineral.,* 56:532–559.

Saxena S. K. and Ribbe P. H. (1972). Activity-composition relations in feldspars. *Contrib. Mineral. Petrol.,* 37:131–138.

Saxena S. K. and Chatterjee N. (1986). Thermochemical data on mineral phases I. The system $CaO-MgO-Al_2O_3-SiO_2$. *J. Petrol.,* 27:827–842.

Saxena S. K. and Fei Y. (1987). Fluids at crustal pressures and temperatures I. Pure species. *Contrib. Mineral. Petrol.,* 95:370–375.

Saxena S. K. and Fey Y. (1988). Fluid mixtures in the C-H-O system at high pressure and temperature. *Geochim. Cosmochim. Acta,* 52:505–512.

Saxena S. K., Sykes J., and Eriksson G. (1986). Phase equilibria in the pyroxene quadrilateral. *J. Petrol.,* 27:843–852.

Saxena S. K., Chatterjee N., Fei Y., and Shen G. (1993). *Thermodynamic Data On Oxides and Silicates.* New York: Springer-Verlag.

Schairer J. F. (1950). The alkali feldspar join in the system $NaAlSiO_4-KAlSiO_4-SiO_2$. *J. Geol.,* 58:512–517.

Schairer J. F. and Bowen N. L. (1938). The system leucite-diopside-silica. *Amer. Jour. Sci.,* Ser.5, 35:507–528.

Schairer J. F. and Bowen N. L. (1955). The system $K_2O-Al_2O_3-SiO_2$. *Amer. Jour. Sci.,* 253:681–747.

Schairer J. F. and Bowen N. L. (1956). The system $Na_2O-Al_2O_3-SiO_2$. *Amer. Jour. Sci.,* 254:129–195.

Schairer J. F. and Yoder H. S. (1960). The nature of residual liquids from crystallization, with data on the system nepheline-diopside-silica. *Amer. Jour. Sci.,* 258A:273–283.

Schilling G. J. and Winchester J. W. (1966). Rare earths in Hawaian basalts. *Science,* 153:867–869.

Schmalzried H. (1978). Reactivity and point defects of double oxides with emphasis on simple silicates. *Phys. Chem. Minerals,* 2:279–294.

Schott J., Berner R. A. and Sjoberg E. L. (1981). Mechanism of pyroxene and amphibole weathering, I. Experimental studies of iron-free minerals. *Geochim. Cosmochim. Acta,* 45:2123–2135.

Schreiber H. D. (1976). The experimental determination of redox states, properties, and distribution of Chromium in synthetic silicate phases and application to basalt petrogenesis. Ph.D. diss., University of Wisconsin.

Schwartz K. B. and Burns R. G. (1978). Mossbauer spetroscopy and crystal chemistry of natural Fe-Ti garnets. *Trans. Amer. Geophys. Union,* 59:395–396.

Schweitzer E. (1977). The reactions pigeonite = diopside + enstatite at 15 kbar. M.S. diss., State University of New York, Stony Brook.

Schwerdtfeger K. and Muan A. (1966). Activities in olivine and pyroxenoid solid solution of the system Fe-Mn-Si-O at 1150°C. *Trans. Met. Soc. AIME,* 236:201–211.

Sciuto P. F. and Ottonello G. (1995a). Water-rock interaction on Zabargad Island (Red Sea)., a case study: I). application of the concept of local equilibrium. *Geochim. Cosmochim. Acta,* 59:2187–2206.

Sciuto P. F. and Ottonello G. (1995b). Water-rock interaction on Zabargad Island (Red Sea)., a case study: II). from local equilibrium to irreversible exchanges. *Geochim. Cosmochim. Acta,* 59:2207–2214.

Seck H. A. (1971). Koexisterend Alkali-feldspate und Plagioclase im System $NaAlSi_3O_8$-$KAlSi_3O_8$-$CaAl_2Si_2O_8$-H_2O bei temperaturen von 650°C bis 900°C. *N. Jahrb. Min. Abh.,* 155:315–395.

Seck H. A. (1972). The influence of pressure on the alkali-feldspar solvus from peraluminous and persilicic materials. *Fortschr. Mineral.*, 49:31–49.

Seifert S. and O'Neill H.St. C. (1987). Experimental determination of activity-composition relations in Ni_2SiO_4-Mg_2SiO_4 and Co_2SiO_4-Mg_2SiO_4 olivine solid solutions at 1200 K and 0.1 MPa and 1573 K and 0.5 GPa. *Geochim. Cosmochim. Acta*, 51:97–104.

Seward T. M. (1981). Metal complex formation in aqueous solutions at elevated temperatures and pressures. *Phys. Chem. Earth*, 13:113–132.

Shannon R. D. (1976). Revised effective ionic radii and systematic studies of interatomic distances in halides and chalcogenides. *Acta Cryst.*, A32:751–767.

Shannon R. D. and Prewitt C. T. (1969). Effective ionic radii in oxides and fluorides. *Acta Cryst.*, B25:925–946.

Shaw D. M. (1970). Trace element fractionation during anatexis. *Geochim. Cosmochim. Acta*, 34:237–243.

Shaw D. M. (1978). Trace element behaviour during anatexis in the presence of a fluid phase. *Geochim. Cosmochim. Acta*, 42:933–943.

Shmulovich K. I., Shmonov V. M., and Zharikov V. A. (1982). The thermodynamics of supercritical fluid systems. In *Advances in Physical Geochemistry*. vol. 2, S. K. Saxena (series ed.). New York: Springer-Verlag.

Shock E. L. and Helgeson H. C. (1988). Calculations of the thermodynamic and transport properties of aqueous species at hight pressures and temperatures: Correlation algorithms for ionic species and equations of state predictions to 5Kb and 1000°C. *Geochim. Cosmochim. Acta*, 52:2009–2036.

Shock E. L. and Helgeson H. C. (1990). Calculations of the thermodynamic and transport properties of aqueous species at high pressures and temperatures: Standard partial molal properties of organic species. *Geochim. Cosmochim. Acta*, 54:915–946.

Shock E. L. and Koretsky C. M. (1995). Metal-organic complexes in geochemical processes: Estimation of standard partial molal thermodynamic properties of aqueous complexes between metal cations and monovalent organic acid ligands at high pressures and temperatures. *Geochim. Cosmochim. Acta*, 59:1497–1532.

Shock E. L., Helgeson H. C. and Sverjensky B. A. (1989). Calculation of thermodynamic and transport properties of aqueous species at high pressures and temperatures: Standard partial molal properties of inorganic neutral species. *Geochim. Cosmochim. Acta*, 53:2157–2183.

Simkin T. and Smith J. V. (1970). Minor element distribution in olivine. *J. Geol.*, 78:304–325.

Simons B. (1986). *Temperatur und Druckabhangigkeit der Fehlstellenkonzentration der Olivine und Magnesiowustite* Habilitationsschrift, Christian-Albrechts Universitat, Kiel.

Sipling P. J. and Yund R. A. (1976). Experimental determination of the coherent solvus for sanidine-high albite. *Amer. Mineral.*, 61:897–906.

Sjoberg E. L. (1976). A fundamental equation for calcite dissolution kinetics. *Geochim. Cosmochim. Acta*, 40:441–447.

Skinner B. J. (1956). Physical properties of end-members of the garnet group. *Amer. Mineral.*, 41:428–436.

Skinner B. J. (1966). Thermal expansion. In *Handbook of Physical Constants*, S. P. Clark Jr., ed. Memoir 97, The Geological Society of America.

Slavinskiy V. V. (1976). The clinopyroxene-garnet geothermometer. *Dokl. Akad. Nauk USSR*, 231:181–184.

Smith D. M. and Stocker R. L. (1975). Point defects and non-stoichiometry in forsterite. *Phys. Earth. Planet. Int.*, 10:183–192.

Smith J. V. (1983). Some chemical properties of feldspars. In *Reviews in Mineralogy*, vol. 2 (2d ed.), P. H. Ribbe (series ed.), Mineralogical Society of America.

Smith J. V. and Yoder H. S. Jr. (1956). Experimental and theoretical studies of the mica polymorphs. *Min. Mag.*, 31:209–235.

Smyth J. R. (1973). An orthopyroxene structure up to 850 °C. *Amer. Mineral.*, 58:636–648.

Smyth J. R. (1974). The high temperature crystal chemistry of clinohyperstene. *Amer. Mineral.*, 59:1069–1082.

Smyth J. R. and Bish D. L. (1988). *Crystal Structures and Cation Sites of the Rock-Forming Minerals* Allen and Unwin, Boston.

Smyth J. R. and Hazen R. M. (1973). The crystal structure of forsterite and hortonolite at several temperatures up to 900°C. *Amer. Mineral.*, 58:588–593.

Smyth J. R., Smith J. V., Artioli G., and Kvick A. (1987). Crystal structures of coesite at 15 K and 298 K from neutron and X-ray single crystal data: Test of bonding models. *J. Phys. Chem.*, 91:988–992.

Sockel H. G. and Hallwig D. (1977). Ermittlung kleiner Diffusionkoeffizienten mittels SIMS in oxydischen Verbindungen. *Mikrokim. Acta*, 7:95–107.

Soddy F. (1914). *The Chemistry of the Radio-Elements. Pt. II* Longmans & Green, London.

Speer J. A. (1984). Micas in igneous rocks. In *Reviews in Mineralogy*, vol. 13, P. H. Ribbe (series ed.), Mineralogical Society of America.

Speidel D. H. and Osborn E. F. (1967). Element distribution among coexisting phases in the system $MgO\text{-}FeO\text{-}Fe_2O_3\text{-}SiO_2$ as a function of temperature and oxygen fugacity. *Amer. Mineral.*, 52:1139–1152.

Spencer K. J. and Lindsley D. H. (1981). A solution model for coexisting iron-titanium oxides. *Amer. Mineral.*, 66:1189–1201.

Stebbins J. F., Carmichael I. S. E. and Moret L. K. (1984). Heat capacities and entropies of silicate liquids and glasses. *Contrib. Mineral. Petrol.*, 86:132–148.

Steele I. M. and Smith J. V. (1982). Ion probe analysis of plagioclase in three howardites and three eucrites. *Geochim. Cosmochim. Acta*, 42:959–971.

Steele I. M., Hutcheon I. D. and Smith J. V. (1980a). Ion microprobe analysis and petrogenetic interpretations of Li, Mg, Ti, K, Sr, Ba in lunar plagioclase. *Geochim. Cosmochim. Acta*, Suppl. 14, 571–590.

Steele I. M., Hutcheon I. D. and Smith J. V. (1980b). Ion microprobe analysis of plagioclase feldspar $(Ca_{1-x}Na_xAl_{2-x}Si_{2+x}O_8)$. for major, minor and trace elements. *VIII Int. Congr. X-ray Optics Microanalysis*, Pendell Pub. Co, Midland, Michigan.

Steele I. M., Smith J. V., Raedeke L. D. and McCallum I. S. (1981). Ion probe analysis of Stillwater plagioclase and comparison with lunar analyses. *Lunar Planet. Sci.*, 12:1034–1036.

Stevenson F. J. (1976). Stability constants of Cu^{2+}, Pb^{2+} and Cd^{2+} complexes with humic acids. *Soil. Sci. Soc. Amer. Jour.*, 40:665–672.

Stevenson F. J. (1982). *Humus Chemistry: Genesis, Composition, Reactions.* New York: Wiley Interscience.

Stevenson F. J. (1983). Trace metal-organic matter interaction in geologic environment. In *The Significance of Trace Elements in Solving Petrogenetic Problems and Controversies*, S. Augustitis, ed. Theophrastus Publications, Athens.

Stewart D. B. and Ribbe P. H. (1969). Structural explanation for variations in cell parameters of alkali feldspars with Al/Si ordering. *Amer. Jour. Sci.*, 267A:444–462.

Stewart R. F., Whitehead M. A. and Donnay D. (1980). The ionicity of the Si-O bond in quartz. *Amer. Mineral.*, 65:324–326.

Stockmayer W. H. (1941). Second virial coefficients of polar gases. *J. Chem. Phys.*, 9:398–402.

Stolper E. (1982). The speciation of water in silicate melts. *Geochim. Cosmochim. Acta*, 46:2609–2620.

Stolper E., Fine G., Johnson T., and Newmann S. (1987). Solubility of carbon dioxide in albitic melt. *Amer. Mineral.*, 72:1071–1085.

Stolper E., Walker D., Hager B. H. and Hays J. F. (1981). Melt segregation from partially molten source regions: The importance of melt density and source region size. *J. Geophys. Res.*, 86:6261–6271.

Stormer J. C. Jr. (1975). A practical two feldspar thermometer. *Amer. Mineral.*, 60:667–674.

Strip K. F. and Kirkwood J. G. (1951). Asymptotic expansion of the rotational partition function of the asymmetric top. *J. Chem. Phys.*, 19:1131–1133.

Stull D. R. and Prophet H. (1971). *JANAF Thermochemical Tables* Data series, Washington D.C., 37:1–1141.

Stumm W. and Morgan J. J. (1981). *Acquatic Chemistry.* New York: John Wiley.

Sueno S., Papike J. J. and Prewitt C. T. (1973). The high temperature crystal chemistry of tremolite. *Amer. Mineral.,* 58:649–664.

Sueno S., Papike J. J., Prewitt C. T. and Brown G. E. (1972). Crystal structure of high cummingtonite. *J. Geophys. Res.,* 77:5767–5777.

Suess H. (1965). Secular variations of the cosmic-ray produced carbon-14 in the atmosphere and their interpretation. *J. Geophys. Res.,* 70:5937–5952.

Sumino Y. (1979). The elastic constants of Mn_2SiO_4, Fe_2SiO_4 and Co_2SiO_4, and the elastic properties of olivine group minerals at high temperature. *J. Phys. Earth,* 27:209–238.

Sumino Y., Nishizawa O., Goto T., Ohno I., and Ozima I. (1977). Temperature variation of elastic constants of single crystal forsterite between 190 and 400°C. *J. Phys. Earth,* 28:273–280.

Sutin N., Rowley J. K. and Dodson W. (1961). Chloride complexes of iron (III) ions and the kinetics of chloride-catalyzed exchange reactions between iron (II) and iron (III) in light and heavy water. *J. Phys. Chem.,* 65:1248–1252.

Sutton S. R., Jones K. W., Gordon B., Rivers M. L., Bajt S., and Smith J. V. (1993). Reduced chromium in olivine grains from lunar basalt 15555: X-ray Absorption Near Edge Structure (XANES). *Geochim. Cosmochim. Acta,* 57:461–468.

Suzuki I. and Anderson O. L. (1983). Elasticity and thermal expansion of a natural garnet up to 1000 K. *J. Phys. Earth,* 31:125–138.

Suzuki I., Seya K., Takei H., and Sumino Y. (1981). Thermal expansion of fayalite, Fe_2SiO_4. *Phys. Chem. Minerals,* 7:60–63.

Suzuoki T. and Epstein S. (1976). Hydrogen isotope fractionation between OH-bearing minerals and water. *Geochim. Cosmochim. Acta,* 40:1229–1240.

Szilagyi M. (1971). Reduction of Fe^{3+} ion by humic acid preparations. *Soil Sci.,* 3:233–235.

Tanaka T. and Masuda A. (1982). La-Ce geochronometer: A new dating method. *Nature,* 300:515–518.

Tanger J. C. and Helgeson H. C. (1988). Calculation of the thermodynamic and transport properties of aqueous species at high pressures and temperatures: Revised equation of state for the standard partial molal properties of ions and electrolytes. *Amer. Jour. Sci.,* 288:19–98.

Tardy Y. (1979). Relationship among Gibbs energies of formation of compounds. *Amer. Jour. Sci.,* 279:217–224.

Tardy Y. and Garrels R. M. (1974). A method of estimating the Gibbs energies of formation of layer silicates. *Geochim. Cosmochim. Acta,* 38:1101–1116.

Tardy Y. and Garrels R. M. (1976). Prediction of Gibbs energies of formation: I-Relationships among Gibbs energies of formation of hydroxides, oxides and aqueous ions. *Geochim. Cosmochim. Acta,* 40:1015–1056.

Tardy Y. and Garrels R. M. (1977). Prediction of Gibbs energies of formation of compounds from the elements, II: Monovalent and divalent metal silicates. *Geochim. Cosmochim. Acta,* 41:87–92.

Tardy Y. and Gartner L. (1977). Relationships among Gibbs energies of formation of sulfates, nitrates, carbonates, oxides and aqueous ions. *Contrib. Mineral. Petrol.,* 63:89–102.

Tardy Y. and Viellard Ph. (1977). Relationships among Gibbs energies and enthalpies of formation of phosphates, oxides and aqueous ions. *Contrib. Mineral. Petrol.,* 63:75–88.

Taylor H. P. Jr. (1968). The oxygen isotope geochemistry of igneous rocks. *Contrib. Mineral. Petrol.,* 19:1–71.

Taylor H. P. Jr. and Epstein S. (1962). Relation between $^{18}O/^{16}O$ ratio in coexisting minerals of igneous and metamorphic rocks. Part I. Principles and experimental results. *Geol. Soc. Amer. Bull.,* 73:461–480.

Temkin M. (1945). Mixtures of fused salts as ionic solutions. *Acta Phys. Chim. URSS,* 20:411–420.

Temkin M. (1963). The kinetics of stationary reactions. *Akad. Nauk. SSSR Doklady,* 152:782–785.

Tequì C., Robie R. A., Hemingway B. S., Neuville D. R. and Richet P. (1991). Melting and thermodynamic properties of pyrope ($Mg_3Al_2Si_3O_{12}$). *Geochim. Cosmochim. Acta,* 55:1005–1010.

Tetley N. W. (1978). *Geochronology by the $^{40}Ar/^{39}Ar$ technique using HIFAR reactor* Ph.D. Dissertation, Australia National University, Canberra.

Thompson A. B. (1976). Mineral reactions in pelitic rocks, II: Calculations of some P-T-X (Fe-Mg). phase relations. *Amer. Jour. Sci.,* 276:425–444.

Thompson J. B. Jr. (1969). Chemical reactions in crystals. *Amer. Mineral.,* 54:341–375.

Thompson J. B. Jr. (1970). Chemical reactions in crystals: Corrections and clarification. *Amer. Mineral.,* 55:528–532.

Thompson J. B. Jr. and Hovis G. L. (1978). Triclinic feldspars: Angular relations and the representation of feldspar series. *Amer. Mineral.,* 63:981–990.

Thompson J. B. Jr. and Waldbaum D. R. (1969). Mixing properties of sanidine crystalline solutions. III Calculations base on two phase data. *Amer. Mineral.,* 54:811–838.

Thompson J. B. Jr., Waldbaum D. R. and Hovis G. L. (1974). Thermodynamic properties related to ordering in end member alkali feldspars. In *The Feldspars,* W. S. MacKenzie and J. Zussman, eds. Manchester: Manchester University Press.

Thompson P. and Grimes N. W. (1977). Madelung calculation for the spinel structure. *Phil. Mag.,* 36:501–505.

Thorstenson D. C. (1970). Equilibrium distribution of small organic molecules in natural waters. *Geochim. Cosmochim. Acta,* 34:745–770.

Tilton G. R. (1960). Volume diffusion as a mechanism for discordant lead ages. *J. Geophys. Res.,* 65:2933–2945.

Tokonami M., Horiuchi H., Nakano A., Akimoto S., and Morimoto N. (1979). The crystal structure of pyroxene type $MnSiO_3$. *Min. Jour.,* 9:424–426.

Toop G. W. and Samis C. S. (1962a). Some new ionic concepts of silicate slags. *Can. Met. Quart.,* 1:129–152.

Toop G. W. and Samis C. S. (1962b). Activities of ions in silicate melts. *Trans. Met. Soc. AIME,* 224:878–887.

Tosi M. P. (1964). Cohesion of ionic solids in the Born model. *Solid State Phys.,* 16:1–120.

Tossell J. A. (1980). Theoretical study of structures, stabilities and phase transition in some metal dihalide and dioxide polymorphs. *J. Geophys. Res.,* 85:6456–6460.

Tossell J. A. (1981). Structures and cohesive properties of hydroxides and fluorides calculated using the modified electron gas ionic model. *Phys. Chem. Minerals,* 7:15–19.

Tossell J. A. (1985). Ab initio SCF MO and modified electron gas studies of electron deficient anions and ion pairs in mineral structures. *Physica,* 131B:283–289.

Tossell J. A. (1993). A theoretical study of the molecular basis of the Al avoidance rule and of the spectral characteristics of Al-O-Al linkages. *Amer. Mineral.,* 78:911–920.

Tossell J. A. and Vaughan D. J. (1992). *Theoretical Geochemistry: Application of Quantum Mechanics in the Earth and Mineral Sciences.* New York-Oxford: Oxford University Press.

Tossell J. A., Vaughan D. J. and Johnson K. H. (1973). X-ray photoelectron, x-ray emission and UV spectra of SiO_2 calculated by the SCF-$X\alpha$. scattered wave method. *Chem. Phys. Lett.,* 20:329–334.

Tossell J. A., Vaughan D. J. and Johnson K. H. (1974). The electronic structure of rutile, wustite and hematite from molecular orbital calculations. *Amer. Mineral.,* 59:319–334.

Touret J. and Bottinga Y. (1979). Equation d'etat pour le CO_2; application aux inclusion carboniques. *Bull. Mineral.,* 102:577–583.

Treuil M. (1973). Critères petrologiques geochimiques et structuraux de la genèse et de la differenciation des magmas basaltiques: Exemple de l'Afar. Ph.D. diss., University of Orleans.

Treuil M. and Joron J. L. (1975). Utilisation des elements hydromagmatophiles pour la semplification de la modelisation quantitative des processes magmatiques. Exemples de l'Afar et de la dorsale medioatlantique. *Rend. SIMP,* 31:125–174.

Treuil M. and Varet (1973). Criteres volcanologiques, petrologiques, et geochimiques de la genèse et de da differenciation des magmas basaltiques: Exemple de l'Afar. *Bull. Soc. Geol. France,* 15:506–540.

Treuil M., Joron J. L., Jaffrezic H., Villemant B., and Calas G. (1979). Geochimie des elements hydromagmatophiles, coefficiens de partage mineraux/liquides et proprietes structurales de ces elements dans les liquides magmatiques. *Bull. Mineral.,* 102:402–409.

Turekian K. K. (1969). The oceans, streams, and atmosphere. In *Handbook of Geochemistry,* vol. 1, K. H. Wedepohl, ed. Berlin-Heidelberg-New York: Springer-Verlag.

Turner D. R., Whitfield M., and Dickson A. G. (1981). The equilibrium speciation of dissolved components in fresh water and seawater at 25°C and 1 atm pressure. *Geochim. Cosmochim. Acta,* 45:855–881.

Turner G. (1968). The distribution of potassium and argon in chondrites. In *Origin and Distribution of the Elements,* L. H. Ahrens, ed. Pergamon Press, London.

Turner G., Miller J. A. and Grasty R. L. (1966). The thermal history of the Bruderheim meteorite. *Earth Planet. Sci. Letters,* 1:155–157.

Turnock A. C., Lindsley D. M. and Graver J. E. (1973). Synthesis and unit cell parameters of Ca-Mg-Fe pyroxenes. *Amer. Mineral.,* 58:50–59.

Ubbelhode A. R. (1965). *Melting and Crystal Structure* Oxford University Press, London.

Ungaretti L. (1980). Recent developments in X-ray single crystal diffractometry applied to the crystal-chemical study of amphiboles. *15th Conf. Yugoslav. Centre Crystallog.,* Bar, Yugoslavia.

Ungaretti L., Mazzi F., Rossi G., and Dal Negro A. (1978). Crystal-chemical characterization of blue amphiboles. *Proceedings of The 1th I. M.A. General Meeting,* Novosibirsk.

Ungaretti L., Smith D. C. and Rossi G. (1981). Crystal-chemistry by X-ray structure refinement and electron microprobe analysis of a series of sodic calcic to alkali-amphiboles from the Nybo eclogite pod, Norway. *Bull. Mineral.,* 104:400–412.

Urey H. C. (1947). The thermodynamic properties of isotopic substances. *J. Chem. Soc. (London).* 1:562–581.

Urey H. C., Brickwedde F. G. and Murphy G. M. (1932). An isotope of hydrogen of mass 2 and its concentration (abstract). *Phys. Rev.,* 39:864.

Urusov V. S., Lapina I. V., Kabala Yu. K. and Kravchuk I. F. (1984). Isomorphism in the forsterite-tephrolite series. *Geokhimiya,* 7:1047–1055.

van der Waals J. D. (1890). Molekulartheorie eines Korpers, der aus zwei verschiedenen Stoffen besteht. *Zeit. Phys. Chem.,* 5:133–167.

Vaughan D. J., Burns R. G. and Burns V. M. (1971). Geochemistry and bonding of thiospinel minerals. *Geochim. Cosmochim. Acta.* 35:365–381.

Verschure R. H., Andriessen P. A. M., Boelrijk N. A., Hebeda E. H., Maijer C., Priem H. N. A. and Verdurmen E. A. (1980). On the thermal stability of Rb-Sr and K-Ar biotite systems. *Contrib. Mineral. Petrol.,* 74:245–252.

Verwey E. J. W. (1941). The charge distribution in the water molecule and calculation of the intermolecular forces. *Recl. Trav. Chim. Pays Bas Belg.,* 60:887–896.

Viellard P. (1982). Modele de calcul des energies de formation des mineraux bati sur la connaissance affinee des structures cristallines. *CNRS Memoirs,* 69, Université Louis Pasteur, Strasbourg.

Virgo D., Mysen B. O. and Kushiro I. (1980). Anionic constitution of 1-atmosphere silicate melts: Implications for the structure of igneous melts. *Science,* 208:1371–1373.

Viswanathan K., and Ghose S. (1965). The effect of Mg^{2+}-Fe^{2+} substitution on the cell dimension of cummingtonite. *Amer. Mineral.,* 30:1106–1112.

Volfinger M. (1970). Partage de Na et Li entre sanidine muscovite et solution hydrothermale a 600°C et 1000 bars. *C. R. Acad. Sci. Fr. Ser. D.*, 271:1345–1347.

Volfinger M. (1976). Effect de la temperature sur les distributions de Na, Rb et Cs entre la sanidine la muscovite la phlogopite et une solution hydrothermale sous une pression de 1 kbar. *Geochim. Cosmochim. Acta*, 40:267–282.

Waffe H. S. and Weill D. F. (1975). Electrical conductivity of magmatic liquid effects of temperatures, oxygen fugacity and composition. *Earth Planet. Sci. Letters*, 28:254–260.

Wagman D. D., Evans W. H., Parker W. B., Halow I., Bailey S. M. and Stumm R. H. (1981). Selected values of the chemical thermodynamic properties. *NBS Tech. Note*, 270-8:1–134.

Wagman D. D., Evans W. H., Parker W. B., Stumm R. H., Halow I., Bailey S. M., Churney K. L. and Nuttall R. L. (1982). The NBS tables of chemical thermodynamic properties. Selected values for inorganic and C1 and C2 organic substances in SI units. *J. Phys. Chem. Ref. Data*, 11, Suppl. 2.

Wagner R. C. (1952). *Thermodynamics of Alloys* Addison-Wesley, Reading, Mass.

Wainwright J. E. and Starkey J. (1971). A refinement of the structure of anorthite. *Z. Kristallogr.*, 133:75–84.

Wall V. J. and England R. N. (1979). Zn-Fe spinel-silicate-sulphide reactions as sensors of metamorphic intensive variables and process. *Geol. Soc. Amer. Abstr. Progr.*, 11:534.

Walther J. V. and Helgeson H. C. (1977). Calculation of the thermodynamic properties of aqueous silica and the solubility of quartz and its polymorphs at high pressures and temperatures. *Amer. Jour. Sci.*, 277:1315–1351.

Warner R. D. and Luth W. C. (1973). Two-phase data for the join monticellite ($CaMgSiO_4$)-forsterite (Mg_2SiO_4): Experimental results and numerical analyses. *Amer. Mineral.*, 58:998–1008.

Warren B. and Bragg W. L. (1928). XII. The structure of diopside $CaMg(SiO_3)_2$. *Z. Kristallogr.*, 69:168–193.

Warren B. and Modell D. I. (1930). 1. the structure of enstatite $MgSiO_3$. *Z. Kristallogr.*, 75:1–14.

Wasastjerna J. (1923). On the radii of ions. *Soc. Sci. Fenn. Comm. Phys. Math.*, 38:1–25.

Watson E. B. (1976). Two-liquid partition coefficients: Experimental data and geochemical implications. *Contrib. Mineral. Petrol.*, 56:119–134.

Watson E. B. (1977). Partitioning of manganese between forsterite and silicate liquid. *Geochim. Cosmochim. Acta*, 41:1363–1374.

Weill D. F. and McKay G. A. (1975). The partitioning of Mg, Fe, Sr, Ce, Sm, Eu and Yb in lunar igneous system and a possible origin of KREEP by equilibrium partial melting. *Proc. Sixth Lunar Sci. Conf.*, 1143–1158.

Wells P. R. A. (1977). Pyroxene thermometry in simple and complex systems. *Contrib. Mineral. Petrol.*, 62:129–139.

Wenner D. B. and Taylor H. P. Jr. (1971). Temperatures of serpentinization of ultramafic rocks based on $^{18}O/^{16}O$ fractionation between coexisting serpentine and magnetite. *Contrib. Mineral. Petrol.*, 32:165–185.

Wetherill G. W. (1956). Discordant uranium-lead ages. *Trans. Amer. Geophys. Union*, 37:320–326.

White B. S., Breakley M., and Montana A. (1989). Solubility of argon in silicate liquids at high pressures. *Amer. Mineral.*, 74:513–529.

Whiteway S. G., Smith I. B. and Masson C. R. (1970). Theory of molecular size distribution in multichain polymers. *Can. Jour. Chem.*, 48:33–45.

Whitfield M. (1975a). An improved specific interaction model for sea water at 25°C and 1 atmosphere total pressure. *Mar. Chem.*, 3:197–213.

Whitfield M. (1975b). The extension of chemical models for sea water to include trace components at 25°C and 1 atm. pressure. *Geochim. Cosmochim. Acta*, 39:1545–1557.

Whitney J. A. and Stormer J. R. (1977). The distribution of $NaAlSi_3O_8$ between coexisting microcline and plagioclase and its effect on geothermometric calculations. *Amer. Mineral.*, 62:687–691.

Whittaker E. J. W. and Muntus R. (1970). Ionic radii for use in geochemistry. *Geochim. Cosmochim. Acta*, 34:945–956.

Wilhem E., Battino R., and Wilcock R. J. (1977). Low pressure solubility of gases in liquid water. *Chem. Rev.*, 77:219–262.

Will T. M. and Powell R. (1992). Activity-composition relationships in multicomponent amphiboles: An application of Darken's quadratic formalism. *Amer. Mineral.*, 77:954–966.

Willaime C. and Gaudais M. (1977). Electron microscope study of plastic defects in experimentally deformed alkali feldspar. *Bull. Soc. Fr. Mineral. Cristallogr.*, 100:263–271.

Williams R. J. (1971). Reaction constants in the system Fe-MgO-SiO$_2$-O$_2$ at 1 atm between 900 and 1300°C: Experimental results. *Amer. Jour. Sci.*, 270:334–360.

Winchell A. N. (1933). *Elements of Optical Mineralogy.* New York: John Wiley.

Winter J. K., Okamura F. P. and Ghose S. (1979). A high temperature structural study of high albite, monalbite, and the analbite-monalbite phase transition. *Amer. Mineral.*, 64:409–423.

Witte H. H. and Wölfel E. (1955). Electron distribution in rock salt. *Zeit. Phys. Chem.*, 3:296–329.

Wohl K. (1946). Thermodynamic evaluation of binary and ternary liquid system. *Trans. Amer. Inst. Chem. Eng.*, 42:215–249.

Wohl K. (1953). Thermodynamic evaluation of binary and ternary liquid systems. *Chem. Eng. Progr.*, 49:218–219.

Wolery T. J. (1983). *EQ3NR. A Computer Program for Geochemical Aqueous Speciation-Solubility Calculations: User's Guide and Documentation* Lawrence Livermore Laboratory, Livermore, Cal., UCRL-53414.

Wolery T. J. (1986). *Some Forms of Transition State Theory, Including Non-Equilibrium Steady State Forms* Lawrence Livermore Laboratory, Livermore, Cal., UCRL-94221.

Wones D. R. (1967). A low pressure investigation of the stability of phlogopite. *Geochim. Cosmochim. Acta*, 31:2248–2253.

Wones D. R. (1972). Stability of biotite: A reply. *Amer. Mineral.*, 57:316–317.

Wones D. R. and Dodge F. C. W. (1977). The stability of phlogopite in the presence of quartz and diopside. In *Thermodynamics in Geology*, D. G. Fraser, ed. Dordrecht-Holland, Reidel.

Wones D. R. and Eugster H. P. (1965). Stability of biotite: Experiment theory and application. *Amer. Mineral.*, 50:1228–1272.

Wood B. J. (1974). Crystal field spectrum of Ni^{2+} in olivine. *Amer. Mineral.*, 59:244–248.

Wood B. J. (1976a). The reaction phlogopite + quartz = enstatite + sanidine + H$_2$O. In *Progress in Experimental Petrology*, G. M. Biggar, ed. Natural Environment. Res. Council:Publ. series D.

Wood B. J. (1976b). The partitioning of iron and magnesium between garnet and clinopyroxene. *Carnegie Inst. Wash. Yb.*, 75:571–574.

Wood B. J. (1976c). Samarium distribution between garnet and liquid at high pressure. *Carnegie Inst. Wash. Yb.*, 75:659–662.

Wood B. J. (1980). Crystal field electronic effects on the thermodynamic properties of Fe^{2+} minerals. In *Advances in Physical Geochemistry*, vol. 1, S. K. Saxena (series ed.), New York-Heidelberg-Berlin: Springer-Verlag.

Wood B. J. (1987). Thermodynamics of multicomponent systems containing several solid solutions. In *Reviews in Mineralogy.* vol. 17, P. H. Ribbe (series ed.), Mineralogical Society of America.

Wood B. J. (1988). Activity measurements and excess entropy-volume relationships for pyrope-grossular garnets. *J. Geol.*, 96:721–729.

Wood B. J. and Banno S. (1973). Garnet-orthopyroxene and orthopyroxene-clinopyroxene relationships in simple and complex systems. *Contrib. Mineral. Petrol.*, 42:109–124.

Wood B. J. and Fraser D. G. (1976). *Elementary Thermodynamics for Geologists* Oxford University Press.

Wood B. J. and Kleppa O. J. (1981). Thermochemistry of forsterite-fayalite olivine solutions. *Geochim. Cosmochim. Acta*, 45:529–534.

Wood B. J. and Kleppa O. J. (1984). Chromium-aluminum mixing in garnet: A thermochemical study. *Geochim. Cosmochim. Acta*, 48:1373–1375.

Wood B. J. and Strens R. G. J. (1972). Calculation of crystal field splittings in distorted coordination polyhedra: Spectra and thermodynamic properties of minerals. *Min. Mag.*, 38:909–917.

Wood B. J., Hackler R. T. and Dobson D. P. (1994). Experimental determination of Mn-Mg mixing properties in garnet, olivine and oxide. *Contrib. Mineral. Petrol.*, 115:438–448.

Wood J. A. (1979). *The Solar System* Prentice-Hall, Englewood Cliffs, N.J.

Woodland A. B. and O'Neill H. St. C. (1993). Synthesis and stability of garnet and phase relations with solutions. *Amer. Mineral.*, 78:1002–1015.

Wriedt H. A. and Darken L. S. (1965). Lattice defects and solubility of nitrogen in deformed ferritic steel. *Trans. Met. Soc. AIME*, 233:111–130.

Yeh H. W. (1980). D/H ratios and late-stage dehydration of shales during burial. *Geochim. Cosmochim. Acta*, 341–352.

Yoder H. S. and Eugster H. P. (1954). Phlogopite synthesis and stability range. *Geochim. Cosmochim. Acta*, 6:157–185.

York D. (1969). Least-squares fitting of a straight line with correlated errors. *Earth Planet. Sci. Letters*, 5:320–324.

Yoshimori A. and Kitano Y. (1956). Theory of the lattice vibration of graphite. *J. Phys. Soc. Japan*, 11:352–361.

Young J. A. and Koppel J. U. (1965). Phonon spectrum of graphite. *J. Chem. Phys.*, 42:357–364.

Yund R. A. (1974). Coherent exsolution in the alkali feldspars. In *Geochemical Transport and Kinetics*, Hofmann, Giletti, Yoder, and Yund, eds. New York: Carnegie Inst. Washington and Academic Press.

Yund R. A. (1983). Diffusion in feldspars. In *Reviews in Mineralogy*, vol. 2 (2d ed.), P. H. Ribbe (series ed.), Mineralogical Society of America.

Yund R. A. and Anderson T. F. (1974). Oxygen isotope exchange between potassium feldspar and KCl solution. In *Geochemical Transport and Kinetics*, Hofmann, Giletti, Yoder, and Yund, eds. New York: Carnegie Inst. Washington and Academic Press.

Yund R. A. and Tullis J. (1983). Subsolidus phase relations in the alkali feldspars with emphasis on coherent phases. In *Reviews in Mineralogy*, vol. 2 (2d ed.), P. H. Ribbe (series ed.), Mineralogical Society of America.

Zhang Z. and Saxena S. K. (1991). Thermodynamic properties of andradite and application to skarn with coexisting andradite and hedenbergite. *Contrib. Mineral. Petrol.*, 107:255–263.

Zheng Y. F. (1991). Calculation of oxygen isotope fractionation in metal oxides. *Geochim. Cosmochim. Acta*, 55:2299–2307.

Zheng Y. F. (1993a). Calculation of oxygen isotope fractionation in anhydrous silicate minerals. *Geochim. Cosmochim. Acta*, 57:1079–1091.

Zheng Y. F. (1993b). Calculation of oxygen isotope fractionation in hydroxyl-bearing silicates. *Earth Planet. Sci. Letters*, 120:247–263.

INDEX

Absolute
  reaction rates, in the transition state theory, 598
  thermodynamic properties, in the HKF model, 521
  velocity, of a reaction, 588
Acetic acid. *See* CH_3COOH
Acidity: Lux-Flood, of oxides in melts, 417–19, 674, 676
Acmite
  coulomb energy, 284
  dispersive energy, 284
  enthalpy of formation from the elements, 284
  heat capacity, 146
  lattice energy, 284
  molar volume, 273
  repulsive energy, 284
  structural class, 268–69
  structural properties, 273
Acoustic
  branches, 137
  vibrational modes, definition, 137
Actinolite
  Fe-Mg intracrystalline distribution, 315
  formula unit, 300
  mixing with cummingtonite, 315
  structural class, 300
Activation energy
  and reaction rate, 598
  of reactions, 599–600
Active surface area: definition, 592
Activity coefficient
  in a simple mixture, 163
  in reciprocal mixtures, 168
  in Redlich-Kister expansions, 169
  in subregular mixtures, 169
  in the Henry's law range, 116
  in the quasi-chemical model, 164
  in Wohl's model, 171

individual ionic, for aqueous solutes, 493, 497
mean ionic, for electrolytes in solution, 494, 497–98
of a gaseous component in mixture, 625
of a trace component in solution, 658, 664
of compositional terms in mixture, 162
Activity diagrams, 582
  availability of in literature, 587
  construction of, 585
Activity product: in hydrolitic equilibria, 579
Adiabatic flashing: effect on fluid's chemistry, 580
Adularia
  compressibility, 354
  thermal expansion, 354
Advection: definition, 608
Affinity to equilibrium
  and solubility product, 518
  in hydrolitic equilibria, 579
  overall, of a system, 120
Age equation
  $^{14}_{6}C$, 766
  K-Ar, 747
  of total Ar release, 751
Al-clinopyroxene
  bulk modulus, 278
  enthalpy of formation from the elements, 283
  heat capacity function, 283
  molar volume, 278
  standard state entropy, 283
  structural class, 269
  thermal expansion, 278
Al-orthopyroxene
  bulk modulus, 278
  enthalpy of formation from the elements, 283

Printed in the USA
CPSIA information can be obtained
at www.ICGtesting.com
JSHW052025301024
72691JS00005B/24